STAR CLUSTERS: BASIC GALACTIC BUILDING BLOCKS THROUGHOUT TIME AND SPACE

IAU SYMPOSIUM No. 266

COVER ILLUSTRATION: ORION AND THE TRAPEZIUM CLUSTER.

This composite *Spitzer* and *Hubble Space Telescope* image exemplifies the chaotic environment in the Orion Nebula, created by having caught the Galactic star-formation process in the act.
Swirls of green in *Hubble*'s ultraviolet and visible-light view reveal hydrogen and sulfur gas that have been heated and ionized by intense ultraviolet radiation from the massive stars in the Trapezium cluster, seen as the yellow stars in the centre of the image. Meanwhile, *Spitzer*'s infrared view exposes carbon-rich molecules (polycyclic aromatic hydrocarbons) in the cloud. These organic molecules have been illuminated by the Trapezium cluster's stars, and are shown in the composite as wisps of red and orange.
Stellar winds from clusters of newborn stars scattered throughout the cloud etched all of the well-defined ridges and cavities in Orion. The large cavity near the right of the image was most likely carved by winds from the Trapezium's stars.

This image is a false-colour composite where blue corresponds to light detected at wavelengths of 0.43, 0.50 and 0.53 μm and green to wavelengths of 0.6, 0.65, and 0.91 μm. Light emitted at 3.6 μm is orange and at 8.0 μm is red.

Image and paraphrased text courtesy of the *Spitzer* Science Center (NASA/JPL-Caltech/T. Megeath & M. Robberto).

CAMBRIDGE UNIVERSITY PRESS
The Edinburgh Building, Cambridge CB2 8RU, United Kingdom
32 Avenue of the Americas, New York, NY 10013-2473, USA
477 Williamstown Road, Port Melbourne, VIC 3207, Australia
Ruiz de Alarcón 13, 28014 Madrid, Spain
Dock house, The Waterfront, Cape Town 8001, South Africa

First published 2010

Printed in the United Kingdom at the University Press, Cambridge

Typeset in System $\LaTeX 2_\varepsilon$

A catalogue record for this book is available from the British Library

Library of Congress Cataloguing in Publication data

This journal issue has been printed on FSC-certified paper and cover board. FSC is an independent, non-governmental, not-for-profit organization established to promote the responsible management of the world's forests. Please see www.fsc.org for information

ISBN 9780521764995 hardback
ISSN 1743–9213

Table of Contents

Session VI. Cluster dynamics
Chair: Simon Goodwin

Session VII. Star clusters as laboratories for stellar evolution
Chairs: Léo Girardi and Licai Deng

Symposium summary

Part 2: Posters

Preface: Star clusters as basic galactic building blocks

It is widely accepted that stars do not form in isolation, but result from the fragmentation of molecular clouds, which in turn leads to star cluster formation. Over time, clusters dissolve or are destroyed by interactions with molecular clouds or tidal striping, and their members become part of the general field population. Star clusters are thus among the basic building blocks of galaxies.

In turn, star cluster populations, from young associations and open clusters to old globulars, are powerful tracers of the formation, assembly and evolutionary history of their parent galaxies. Although their importance (e.g., in mapping out the Milky Way) had been recognised for decades, major progress in this area has only become possible in recent years, both for Galactic and extragalactic cluster populations. This area has seen a major recent investment in time and effort, largely thanks to significant new resources in theory, simulations and observations, including breakthroughs in computational power, the advent of *Hubble Space Telescope*-driven science, deep and more precise data for large numbers of Galactic clusters, and an explosion of astrometric data (e.g., *Hipparcos*, UCAC2).

In the summer of 2009, several years had passed since a number of very successful IAU symposia were held on aspects related to the build-up of galaxies and their basic building blocks. IAU Symposium 254 on 'The Galactic Disk in Cosmological Context' (June 2008), IAU Symposium 241 on 'Stellar populations as basic building blocks of galaxies' (December 2006) and IAU Symposium 235 on 'Galaxy Evolution across the Hubble Time' (August 2006) addressed galaxy formation and evolution in a wider sense, while IAU Symposium 246 on 'Dynamical Evolution of Dense Stellar Systems' (September 2007) and IAU Symposium 207 on 'Extragalactic Star Clusters' (March 2001) focussed on star clusters in their own right.

While most of these past Symposia addressed individual aspects of galaxy formation and evolution, we felt the need for a new Symposium encompassing many of the individual aspects in a more general approach, and in particular focussing on star clusters (including the full range from very young associations and open clusters to young massive star clusters and old globulars), and star cluster populations in the wider context of their parent galaxies. Putting these results and new developments related to star clusters as individual entities into the broader context of galaxy evolution is the next logical step, which requires the combined efforts of theorists, observers and modellers working on a large variety of spatial scales, and spanning a very wide range of expertise. We are now reaching the stage that we are within reach of answering a number of fundamental questions that will have a significant impact on our understanding of numerous related issues as well, ranging from the formation, assembly and evolution of galaxies, to the details of the process of star formation itself. These two issues are the backbone of research in modern astrophysics. Thus, we proposed to focus on the role of star clusters, of any size and age, and their stellar populations in the overall context of galaxy evolution, across space (from local to high redshift) and time (from currently forming to fossil remnants).

Star clusters are the observational foundation for stellar astrophysics and evolution, provide essential tracers of galactic structure, and are unique stellar dynamical environments. Star formation, stellar structure and stellar evolution continue to benefit and improve tremendously from the study of these systems. Additionally, fundamental quantities such as the initial mass function can be successfully derived from modelling either the Hertzsprung–Russell diagrams or the integrated velocity structures (i.e., for massive clusters, the velocity dispersions, leading to dynamical mass estimates, and combined with integrated luminosities and independent age estimates) of, respectively, resolved and unresolved clusters and cluster populations. Star cluster studies thus span the fields of Galactic and extragalactic astrophysics, while heavily affecting our detailed understanding of the process of star formation in dense environments.

Globular clusters are regarded as fossil records of the earliest epochs of galaxy formation, including that of our own Milky Way. At the same time, the young massive star clusters currently being formed in the most violent starburst environments may be viable protoglobular clusters. Thus, star clusters form an ideal testbed for stellar evolution theories and are in fact among the best tools to study extreme stellar populations, such as X-ray binaries or 'blue stragglers', both of which are likely products of dynamical interactions within clusters. However, dynamical modelling of clusters and entire cluster systems, at any age, still poses a considerable challenge for both theory and computational requirements.

Recent advances in instrumentation are driving a renaissance in the study of Galactic clusters, while extragalactic cluster studies are significantly aided by the development of new instrumentation supporting ever wider fields of view. New wide-field imaging cameras on several 4–8m telescopes offer the unique opportunity to study entire cluster populations to very faint magnitudes in a single pointing, both in the optical and near-infrared (e.g., *VISTA*). Complementing these photometric capabilities are a new generation of multi-object spectrographs and multiplexed integral-field units on 8–10m telescopes. In addition, from *Chandra/XMM* and *GALEX* at short wavelengths to the *Spitzer Space Telescope* in the mid- and far-infrared, our observational window in which to study both the star clusters and their stellar populations are unsurpassed in terms of wavelength coverage and spatial resolution at the present time. With major efforts being expended on the planning for possible 30–42m 'extremely large telescopes', now is the opportune time to look forward to future progress in mapping a representative slice of the local Universe at the highest possible resolution and thus in the greatest detail ever achieved. Combining the emerging, unprecedented understanding of local stellar populations with observations of galaxies ('composite' stellar populations) at ever higher redshifts, we now have a fighting chance to constrain the evolution of the basic galactic building blocks throughout space and time.

With all of these exciting efforts ongoing at the present time, we believed that it was an opportune time to organise a Symposium on 'Star Clusters as Basic Galactic Building Bocks Throughout Time and Space' at the XXVII$^{\rm th}$ General Assembly of the IAU in the summer of 2009. Looking back at a very successful meeting, we are strengthened in our resolve to make this meeting happen and would like to thank all participants for a hugely rewarding and thought-provoking week in Rio de Janeiro.

Richard de Grijs and Jacques Lépine
November 2009

SCIENTIFIC ORGANISING COMMITTEE

Co-chairs:

Richard de Grijs	(China & UK)	Jacques Lépine	(Brazil)

Members:

Beatriz Barbuy	(Brazil)	Giovanni Carraro	(Italy & Chile)
Licai Deng	(China)	Michael Dopita	(Australia)
Yu Gao	(China)	Doug Geisler	(Chile)
Rosa González Delgado	(Spain)	John Lattanzio	(Australia)
Stephen L. W. McMillan	(USA)	André Moitinho	(Portugal)
Anas Osman	(Egypt)	Philippe Prugniel	(France)
Ata Sarajedini	(USA)	Alison Sills	(Canada)

Acknowledgements

The Symposium was sponsored and supported by IAU Divisions VII (Galactic System), VIII (Galaxies) and X (Radio Astronomy); and by IAU Commissions No. 28 (Galaxies), No. 33 (Structure and Dynamics of the Galactic System), No. 34 (Interstellar Matter), No. 37 (Star Clusters and Associations), No. 40 (Radio Astronomy) and No. 55 (Communicating Astronomy with the Public).

Financial support was provided by the International Astronomical Union in the form of travel grants for 21 participants, selected from the invited and contributing speakers, PhD students and scientists from less-favoured countries. The Local Organising Committee consisted of the Local Organising Committee of the IAU's XXVIIth General Assembly, to whom we are grateful for structural and logistical support.

The editors of these proceedings are grateful to Joana Oliveira and Jacco van Loon for their indispensable assistance with the technical aspects of putting this volume together.

Introductory review

Star clusters: Basic galactic building blocks throughout time and space

Star clusters: basic galactic building blocks
Proceedings IAU Symposium No. 266, 2009
R. de Grijs & J. R. D. Lépine, eds.

doi:10.1017/S1743921309990809

The nature and nurture of star clusters

Bruce G. Elmegreen

IBM Research Division, T. J. Watson Research Center, 1101 Kitchawan Road,
Yorktown Heights, NY 10598 USA
email: bge@us.ibm.com

Abstract. Star clusters have hierarchical patterns in space and time, suggesting formation processes in the densest regions of a turbulent interstellar medium. Clusters also have hierarchical substructure when they are young, which makes them all look like the inner mixed parts of a pervasive stellar hierarchy. Young field stars share this distribution, presumably because some of them came from dissolved clusters and others formed in a dispersed fashion in the same gas. The fraction of star formation that ends up in clusters is apparently not constant, but may increase with interstellar pressure. Hierarchical structure explains why stars form in clusters and why many of these clusters are self-bound. It also explains the cluster mass function. Halo globular clusters share many properties of disk clusters, including what appears to be an upper cluster cutoff mass. However, halo globulars are self-enriched and often connected with dwarf galaxy streams. The mass function of halo globulars could have initially been like the power-law mass function of disk clusters, but the halo globulars have lost their low-mass members. The reasons for this loss are not understood. It could have happened slowly over time as a result of cluster evaporation, or it could have happened early after cluster formation as a result of gas loss. The latter model explains best the observation that the globular cluster mass function has no radial gradient in galaxies.

Keywords. open clusters and associations: general, solar neighborhood, galaxies: star clusters, stars: formation

---

1. Introduction

Star formation is hierarchical in time and space, suggesting that turbulent processes partition the gas. Star formation is also clustered to varying degrees, suggesting the action of some dimensionless quantity such as the turbulent Mach number. Here, we review recent observations of clusters and associations with emphasis on hierarchical structure for the young regions, the resulting mass functions for open and globular clusters, and the likely dwarf galaxy origin for some halo globular clusters in the Milky Way.

2. Hierarchical structure in young clusters

2.1. *Spatial correlations*

Piskunov *et al.* (2006) identified several 'open cluster complexes' according to groupings of position, velocity, and age of clusters in the solar neighborhood (see also Lyngå 1982; Kharchenko *et al.* 2005). Examples of these complexes included one of intermediate age consisting of seven clusters in Perseus–Auriga, an older one with six clusters in the Hyades, and a younger one with 23 members that is essentially Gould's Belt. A typical complex spans a distance of several hundred parsecs. They are probably the same objects that were called 'star complexes' by Efremov (1995), who found them as groupings of Cepheid variables and red supergiants. Ivanov (2005) catalogued star complexes in M33 based on clusterings of blue stars, HII regions, and Wolf–Rayet stars.

De la Fuente Marcos & de la Fuente Marcos (2008) identified five open cluster complexes from cluster positions and velocities within 2.5 kpc of the Sun: at a Galactic longitude of $l = 12°$ and a distance of 1300 pc, there is the Scutum–Sagittarius complex; at $l = 75°$ and 1400 pc, the Cygnus complex; at $l = 132°$ and 2000 pc, the Cassiopeia–Perseus complex; at $l = 200°$ and 500 pc, the Orion complex, and at $l = 295°$ and 2000 pc, the Centaurus–Carina complex. They suggest that open cluster complexes are fragments from a common gas cloud. This is the usual explanation for star complexes too.

The gas clouds that form star complexes contain $\sim 10^7$ $M_\odot$ of HI or H_2, and are fragmented into giant molecular clouds (GMCs) and molecular cloud cores that form OB associations and OB subgroups or clusters, respectively (e.g., Elmegreen 2007). These enormous regions usually occur in galactic spiral arms with a separation of several kiloparsecs (McGee & Milton 1964; Elmegreen & Elmegreen 1987). Presumably, the largest clouds are formed by galactic-scale processes (i.e., their mass is the Jeans mass of the ambient interstellar medium; ISM), and then fragment into GMCs, GMC clumps, and eventually star-forming cores. Most GMCs in the Milky Way (Grabelski *et al.* 1987) and M33 (Engargiola *et al.* 2003) are found in these gas giants, which are primarily composed of HI. In M51, where the gas is more molecular overall because of the higher pressure, the gas giants are mostly molecular (Rand & Kulkarni 1990).

Not all star formation produces dense clusters. Elias *et al.* (2009) showed that clustering varies spatially in Gould's Belt, with a larger fraction of stars ending up in clusters in Orion than in the Sco–Cen region. This gradient in cluster fraction corresponds to a gradient in young cluster density, suggesting that the dense, high-pressure centers of star complexes produce stars more efficiently than low-pressure peripheral regions. Higher efficiency means that the final ratio of star-to-gas mass is higher, implying that the region is more likely to end up self-bound as a cluster after the gas leaves.

Hierarchical structure extends from star complexes to embedded clusters to individual young stars inside embedded clusters. Feitzinger *et al.* (1984, 1987) were among the first to recognize hierarchical or fractal structure in large-scale star-forming regions. An important point about hierarchical structure is that the average density of gas increases down the hierarchy, i.e., towards smaller fragment masses. The mass fraction represented by dense star-forming clumps increases along this sequence too. Bound regions require a high mass fraction for stars and so appear only at the bottom of the hierarchy. This is the primary reason why bound clusters are dense (much denser than the background tidal limit, which is all they would need for self-boundedness at zero velocity dispersion). With hierarchical structure, the average gas density is high in regions where the star-formation efficiency is high. This is the reason why most star formation begins in the form of a dense embedded cluster (Elmegreen 2008).

Scheepmaker *et al.* (2009) studied clusters in M51. Autocorrelation functions for these clusters in three age bins show that the youngest sample is well correlated: it is hierarchical with a fractal dimension of ~ 1.6. The autocorrelation means that clusters are distributed in cluster pairs and triplets, and these are inside clusters complexes and so on up to scales greater than 1 kpc. Clusters in the Antennae galaxy are also autocorrelated up to ~ 1 kpc (Zhang *et al.* 2001). Sánchez & Alfaro (2008) studied HII regions in a number of galaxies. They found that the positions of the brightest HII region have the smallest fractal dimensions, which means they are still hierarchical. For the galaxy NGC 6946, the high-brightness HII regions have a fractal dimension $D_c = 1.64$, those of medium brightness have $D_c = 1.82$, and those with low brightness have $D_c = 1.79$. Sánchez & Alfaro (2008) also found that D_c decreases slightly with galaxy brightness, which means that low-luminosity galaxies have more hierarchical structure in the positions of their H II regions. Presumably these low-luminosity galaxies have less shear to smooth out the hierarchical birth positions of the H II regions.

Elmegreen *et al.* (2006) studied the size distribution of star-forming regions in NGC 628 by box counting, which is a common technique for measuring fractal dimensions. They used a high-resolution image of this galaxy from the ACS instrument on *HST* and blurred it in successive stages. The number of regions at each blurring stage was counted with SExtractor. They found that the cumulative size distribution of star-forming regions is $n(R)\mathrm{d}R \sim R^{-2.5}\mathrm{d}R$ for size R, and they fit this to a projected fractal Brownian-motion density distribution having a 3D power spectrum with the Kolmogorov slope, -3.66. Elmegreen *et al.* (2003) measured azimuthal intensity profiles of optical light from whole galaxies and found that they also have power-law power spectra like turbulence. For example, the young stars and dust clouds seen in optical images of the flocculent galaxy NGC 5055 have the same scale-free distribution as the H I gas in the Large Magellanic Cloud (LMC). Both give a Kolmogorov power spectrum. Block *et al.* (2009) presented power spectra of *Spitzer Space Telescope* images of several galaxies, using the near-infrared passbands. They found that the passbands dominated by stars (channels 1 and 2) had power spectra from noise plus the point-spread function that comes from the unresolved stellar images, while the passbands dominated by dust and polycyclic aromatic hydrocarbon emission (channel 4) had Kolmogorov power spectra, clearly different in slope from the star-dominated power spectra. Images of the range of Fourier components that give the power-law power spectrum in channel 4 show the resolved hierarchical part of the galaxy.

2.2. *Time correlations*

Clusters are also correlated in time in the sense that the age difference increases with separation. Efremov & Elmegreen (1998) found this for ~ 600 clusters in the LMC, and de la Fuente Marcos & de la Fuente Marcos (2009a) found it for local Milky Way clusters. The slope of the correlation is about 0.5. This is interpreted as an indication that the duration of star formation in a complex increases as the square root of the size (Elmegreen & Efremov 1996). Such a relation between time and size also applies to self-gravitating clouds with a common pressure: self-gravity means that $GM/R \sim 5V^2$, and a common pressure P means that $0.1GM^2/R^4 \sim P$ (for mass M, radius R, and velocity dispersion V). Eliminating M gives $V^2/R = (0.4PG)^{0.2}$, which is constant. As a result, the crossing time is $R/V = R^{0.5}\,(0.4PG)^{-0.25}$. This is the same form as the observed relation between star-formation duration and region size. For typical $P = 10^6 k_{\mathrm{B}}$, $R/V = 2.3 \times 10^4 R^{0.5}$ in cgs units, or $R/V = 1.3R^{0.5}$ Myr for R in pc and V in km s^{-1}. Efremov & Elmegreen (1998) found, for an age range of 1–100 Myr in the LMC, that the time separation equals 30 Myr times the spatial separation in degrees to the power 0.33 (for the LMC, 1 degree = 960 pc). This is about the same relation as in the above simple derivation. For the Milky Way, de la Fuente Marcos & de la Fuente Marcos (2009a) found a time separation equal to 11.1 Myr times the spatial separation in parsecs to the power 0.16. Both the LMC and Milky Way exhibit good correlations, but the slopes are different. De la Fuente Marcos & de la Fuente Marcos (2009a) suggest that if cluster disruption is taken into account, the slopes are a bit steeper, between ~ 0.3 for the Milky Way and ~ 0.5 for the LMC. If clusters within 2 kpc of the Sun are considered, the local slope could be ~ 0.4.

2.3. *Cluster pairs*

The time–distance correlation among young clusters implies that there should be pairs of clusters born at about the same time and place. Indeed, such pairs are well observed. They were originally discovered in the LMC by Bhatia & Hatzidimitriou (1988) and in the Small Magellanic Cloud (SMC) by Hatzidimitriou & Bhatia (1990). Dieball *et al.* (2002) studied them again in the LMC. De la Fuente Marcos & de la Fuente Marcos (2009b) studied cluster pairs in the Milky Way and assessed whether they are interacting. For example, they found that NGC 3293 and NGC 3324 near η Carinae are weakly interacting; their age difference is 4.7 Myr. Similarly, NGC 659 and NGC 663 are weakly interacting and their age difference is 19.1 Myr. Cluster pairs are the closest members of the hierarchy of clusters found by autocorrelation studies, power spectra, fractal analysis, and box-counting techniques.

2.4. *Individual stars*

Individual young stars are correlated even if they are not in clusters. The two-point correlation function for stars in Taurus has a power-law form (Gomez *et al.* 1993; Larson 1995). Bastian *et al.* (2009) looked at the positions and ages of stars in the LMC. They included 2000 sources in each of several age intervals in the color–magnitude diagram. The two-point correlation function for these field stars has a slope that gradually changes from -0.5 at the youngest age to 0 at an age of 175 Myr. The coefficient in the two-point correlation function goes systematically to zero along this age sequence too. They also studied the Cartwright & Whitworth (2004) Q parameter, which is the ratio of the average minimum-spanning-tree length to the average correlation length. When a distribution of objects is fragmented into pieces, Q is less than 0.8 or so; when it is smooth or with a smooth gradient, Q is closer to 1. Bastian *et al.* (2009) found that Q also shows a gradual variation with stellar age, from ~ 0.55 to ~ 0.75 as age increases to 175 Myr. This is consistent with the trend in the two-point correlation function: older field stars are less clumped. Gieles *et al.* (2008) did the same analysis for SMC stars, finding the same age trends in the two-point correlation function and Q out to about 100 Myr.

Odekon (2006) determined a 'correlation dimension,' d_c, for stars seen by *HST* in dwarf galaxies. The correlation dimension comes from the equation $N(r) \propto r^{d_c}$, where $N(r)$ is the average number of stars in a region of radius r. She found that the brightest dwarf has the largest d_c and the faintest dwarf has the smallest d_c. This is consistent with the trend for HII-region clustering in galaxies found by Sánchez & Alfaro (2008).

2.5. *Internal cluster structure*

The hierarchy of young stellar structure often persists even inside currently forming clusters. Testi *et al.* (2000) found substructure in an embedded infrared cluster in Serpens, Smith *et al.* (2005) found it in the ρ Oph region, and Dahm & Simon (2005) found four subclusters with $\sim$ Myr age differences in NGC 2264. Azimuthal profiles of young clusters have non-Poissonian distributions as well (Gutermuth *et al.* 2005). A recent X-ray map of young stars in NGC 6334 (Feigelson *et al.* 2009) shows substantial substructure in the positions of soft and hard X-ray sources (which correspond to embedded populations with less than and greater than 10 magnitudes of visual extinction, respectively). The X-ray maps are nearly complete for stars more massive than the Sun. Sánchez & Alfaro (2009) show that for 16 Milky Way clusters, the stars in the younger, bigger clusters are more clumped. Evidently, as a cluster ages, its subclusters mix and the substructure smooths out.

Schmeja *et al.* (2008) studied clumpy structure in four embedded clusters. For IC 348, NGC 1333, and the ρ Oph region, Q is lower (more clumpy) for class 0/1 objects (young) than for class 2/3 (old). Also, among four subclumps in ρ Oph, Q is lower and it is more gaseous where class 0/1 objects dominate, and Q is also lower for class 0/1 alone than for class 2/3. For an even younger stage, Johnstone *et al.* (2000, 2001), Enoch *et al.* (2006), and Young *et al.* (2006) found that the millimeter-wave pre-stellar clumps in several regions are spatially correlated.

2.6. *Summmary: 'clusters' are the dense cores of a pervasive hierarchy of star formation*

Hierarchal structure in the ISM presumably comes from self-gravity and turbulence. This gas structure continues to substellar scales, as shown by high-resolution molecular observations. The densest regions, which are where individual stars form, are often clustered into the next-denser regions. Stars form in the densest regions, at first somewhat independently it seems, and then they move around, possibly interact, and ultimately mix together inside the next-lower-density region. That mixture is the 'cluster.' More and more subclusters mix over time until the cloud disrupts. The net efficiency of star formation (fractional star mass) is automatically high on small and dense scales because of the hierarchy of structures.

Clusters are the inner mixed parts of the hierarchy of young stellar structures. The hierarchy that is present on larger scales was also present on smaller scales before this mixing. If we consider a cluster's 'Nature' to be its internal properties, and its 'Nurture' to be its external properties, then Nature and Nurture for a star cluster are essentially the same thing. Internal and external depend on the extent to which gravitational mixing has occurred at the time of observation.

3. Unclustered star formation

Barbá *et al.* (2009) studied the giant star-forming region NGC 604 in M33 with *HST*/NICMOS to look for obscured young clusters. They found that it contains almost no clusters but that most star formation is distributed, in agreement with Hunter *et al.* (1996). Giant molecular clouds also show some star formation in a distributed form (Megeath *et al.* 2004; Jørgensen *et al.* 2006, 2007), although in the solar neighborhood, most star formation is clustered. Maíz–Apellániz (2001) generalized the discussion by considering three types of clustering, (i) compact clusters with weak unclustered halos typically measuring 50×50 pc^2, (ii) compact clusters with strong halos, measuring 100×100 pc^2, and (iii) purely hierarchical star formation with no clusters.

A possible reason for variations in young stellar clustering is explored by Elmegreen (2008). The basic point was outlined above, that hierarchical star formation automatically leads to high efficiencies and bound stellar clustering at high densities. The point now is to find the mass fraction of star formation that has a sufficiently high efficiency to produce a bound cluster after the gas leaves. This fraction can be calculated from the density probability density function (pdf) in a GMC. The density pdf has a log-normal form in isothermal turbulence and a power-law form at high density when self-gravity is important (Klessen 2001). An integral over density times the density pdf gives the mass. The integral illustrates the importance of the breadth of the density pdf: for a narrow pdf (i.e., with a low dispersion), the critical efficiency is reached at a high density and a low mass fraction, while for a wide pdf, the critical efficiency is reached at a low density and a high mass fraction. What matters is the slope of the pdf at the density where the efficiency becomes critical for producing a bound cluster. Thus the breadth of the pdf enters into this slope. Also entering is the centroid density. Generally, the pdf is in the high-density falling part at the critical value for high efficiency. Then a shift in the pdf towards higher density will also make the slope shallower there (for the lognormal case). Thus, either a broader pdf or a higher average density will lead to a higher fraction of the star-formation mass in the form of bound clusters. A broader pdf could arise from a higher turbulent Mach number. It follows that higher-pressure regions are more likely to form bound clusters.

4. Cluster mass functions

The mass function for bound clusters is approximately $n(M)\mathrm{d}M \propto M^{-2}\mathrm{d}M$ for a given age range (Elmegreen & Efremov 1997; Zhang & Fall 1999; de Grijs *et al.* 2003; de Grijs & Anders 2006). This function is also observed for various ages, although the lower limit to the power law depends on the age because of cluster-magnitude detection limits and cluster fading with age (Elmegreen & Efremov 1997; Hunter *et al.* 2003; de Grijs & Anders 2006).

The cluster mass function resembles the mass function of cluster-forming cloud clumps (Reid & Wilson 2005; Rathborne *et al.* 2006), and both presumably get their form from the hierarchical structure of interstellar matter. In a hierarchy, all of the mass is represented at all levels, so the total mass in each level (which has logarithmic intervals of mass), $Mn(\log M)\mathrm{d}\log M$, is the same as the total mass in any other level. It follows that $Mn(\log M)\mathrm{d}\log M = \mathrm{constant}\times\mathrm{d}\log M$, $n(\log M) \propto M^{-1}$, and $n(M) \propto M^{-2}$. Similarly, if subclouds are randomly selected from a hierarchical cloud, their distribution function will be M^{-2} too. Or, if we generate a fractal Brownian-motion cloud, the mass distribution of clumps will be close to M^{-2}, with a preferred value of the power spectrum given by the Kolmogorov-turbulence law (Elmegreen *et al.* 2006).

Recent observations suggest that the power-law mass function sometimes has an upper mass cutoff, making it a Schechter function (as originally applied to galaxies; Schechter 1976). These cutoffs have been observed in M51 (Gieles *et al.* 2006a,b) and several other galaxies (Waters *et al.* 2006; Larsen 2009).

Before we turn to halo globular clusters, the essential points of the preceding sections can be summarized as follows:

- Gas is hierarchical in space and time, presumably as a result of turbulence and self-gravity.
- Therefore star formation is hierarchical.
- Therefore the efficiency of star formation ($M_{\mathrm{stars}}/M_{\mathrm{total}}$) increases with average density.
- Therefore bound objects (which require high efficiency) form at high density. The result is a cluster.
- Finally, the mass function of all this structure is a power law with a slope of -2 for equal intervals of mass.
- The origin of the upper cutoff mass for clusters is unexplained.

5. Globular clusters

The mass functions of galactic clusters, reviewed above, resemble the upper parts of the mass functions of globular clusters, which appear to have lost their low-mass members. Jordán *et al.* (2007) fit power laws to the upper parts of the globular cluster mass functions in Virgo galaxies ($3 - 20 \times 10^5$ $\mathrm{M}_\odot$ clusters). They found that the upper-mass power law is steeper for lower-luminosity galaxies, and the width of the lognormal mass function is also smaller for lower-luminosity galaxies. This result implies that low-mass galaxies have a lower upper-mass cutoff for their clusters. Waters *et al.* (2006) measured the mass function of M87 globular clusters with *HST* and also found that the best fit was an evolved Schechter function with a cutoff mass of $M_{\mathrm{c}} = 10^{6.4}$ $\mathrm{M}_\odot$. The evolution models in Waters *et al.* assumed various mass dependencies for cluster evaporation, with the best fit having no mass dependence. For either this standard model of mass-loss rate, dM/dt = constant, or the Lamers *et al.* (2006) model, $dM/dt \propto M^{0.38}$, the low-mass globular clusters evaporate first, leaving only the high-mass clusters after a Hubble time.

The problem with evaporation models is that globular cluster mass functions are independent of galactocentric radius in galaxies (Kundu *et al.* 1999; Jordán *et al.* 2007), while the evaporation rate depends on the tidal density, which depends on galactocentric radius. Thus, the outer regions of galaxies should have more low-mass globular clusters remaining than the inner regions, but they do not. One proposed solution is that globular cluster orbits are highly radial (Fall & Zhang 2001), but this disagrees with the observed profile of globular cluster velocity dispersion versus radius (Vesperini *et al.* 2003).

McLaughlin & Fall (2008) separated the Milky Way globular cluster luminosity function into three groups according to the density at half-light radius. The peak mass was found to depend on the density as expected for dM/dt = constant evaporation of a Schechter function, and the mass functions are independent of position, as required. Chandar *et al.* (2007) found the same for M104. However, low-density globular clusters are of lower mass anyway because they all have about the same radius (e.g., McLaughlin 2000). Perhaps density-dependent evaporation is the reason they all have the same radius, but this has not been demonstrated independently.

Another option is that the globular cluster mass function was peaked from a young age (Vesperini 2000; Parmentier & Gilmore 2007) and has not evolved much since then. Evaporation tends to preserve a peak in the mass function once it forms (Vesperini 1998). Parmentier *et al.* (2008) showed that cluster disruption during initial gas removal can convert a power-law mass function into a lognormal mass function at a very early stage. This would satisfy the observations that show no radial gradients in the mass function. The problem with this model is that modern clusters do not lose their initial power-law mass functions so quickly. The difference could be that halo globular clusters formed with a lower efficiency ($< 25\%$) than disk clusters ($\sim 40\%$) because cluster disruption is more likely during gas loss for low efficiencies. We note that, in the hierarchical model of cluster formation, lower efficiency for a given cluster density (required to define an object as a cluster) follows from lower ISM pressures (Elmegreen 2008), and that is consistent with globular cluster formation in dwarf galaxies.

Halo globular clusters differ from disk open clusters in several ways. Globulars are more massive and this presumably allows self-enrichment of heavy elements from red-giant winds. Globular clusters are also older, and this means that evolutionary effects are more prominent, such as evaporation and core collapse. Globular clusters are lower in metallicity than disk clusters, and this means that stellar winds are weaker and any self-enrichment is more obvious. Globular clusters are also in the halo and many appear to come from dissolved companion dwarf galaxies. Others could come from the starburst phase of a merger when the disk and its clusters are scattering into the halo. Yet others could come from star formation in small galaxies before the large galaxies formed.

Harris (2009) used the *HST* to observe six giant-elliptical galaxies within 40 Mpc. He found 7800 globular clusters with half-light radii larger than 1.5 pc. The blue (low metallicity) and red (high metallicity) sequences found previously for many globular cluster systems are clear in these new data. The blue sequence turns slightly redder at high mass, suggesting self-enrichment for $M > 10^6$ $M_\odot$. Normal blue globulars have $[Fe/H] \sim -1.5$ dex, while in his survey, the most massive blue globulars have $[Fe/H] \sim -1$ dex. This redward trend toward higher mass among the blue globulars was also found in M87 by Peng (2009). Bailin & Harris (2009) suggest supernova ejecta are trapped in the more massive clusters during the gas-rich formation phase. Similar observations are found in Harris *et al.* (2006), Strader *et al.* (2006), and Mieske *et al.* (2006).

Bedin *et al.* (2004) found multiple main sequences and main-sequence turnoffs in ω Cen. The multiple main sequence implies there is a range of He abundances (D'Antona *et al.* 2002; Norris 2004), while the multiple subgiant branches mean there is a range of ages (Milone *et al.* 2008) or CNO abundances (Cassisi *et al.* 2008). Self-enrichment is likely. Ventura *et al.* (2009) found a split subgiant branch in NGC 1851. They modeled the second generation with a CNO abundance three times that of the first generation. They suggested that massive asymptotic giant branch stars in the first generation have ejecta with five times the CNO abundances of the stars themselves, and these ejecta were diluted by 50% with pristine gas to keep the He abundance low. CNO enrichment is usually so large in globular clusters that the progenitor stars have to outnumber the cluster stars. Either the former cluster was much more massive (D'Ercole *et al.* 2008) or the globular cluster was the core of a dwarf galaxy which collected ejecta from many other clusters (Bekki & Norris 2006).

Marcolini (2007) ran hydrodynamic models of star formation and metal production in a dwarf-spheroidal galaxy with an ω Cen-type cluster forming in the nucleus. They assumed star formation had a high rate for 1 Gyr and then a lower rate for another 0.6 Gyr. The α elements form from Type II supernovae (SNII) quickly and Fe forms from Type Ia SNe more slowly. The SNII push out and mix with the nuclear gas, producing a uniform and growing Fe abundance. This gas resettles to the center when the massive stars are gone. SNIa then become important, and they pollute the central region locally, leading to pockets of low α/Fe. The result is a wide range of [Fe/H] in globular cluster stars, with high α/Fe ratios at low [Fe/H], and an increasing dispersion towards low α/Fe at high [Fe/H] abundances, as observed.

Another possibility is that small clusters merge in the nucleus of a dwarf galaxy. Georgiev *et al.* (2009) found two globular clusters in the dwarf irregular galaxy UGCA and thought they should merge in 0.4 Gyr through dynamical friction and form a nuclear globular cluster.

The Sagittarius (Sgr) dwarf galaxy is a good example of a source for Milky Way globular clusters, having Terzan 7, Terzan 8, Arp 2, and M54 as part of its orbital debris. Carraro *et al.* (2007) suggested that the globular cluster Whiting 1 is also likely associated with the Sgr dwarf, which would make six in total. The age of Whiting 1 is 6.5 ± 0.7 Gyr an $[\mathrm{Fe/H}] = -0.4$ to -1.1 dex, which are consistent with the age–metallicity pattern in the Sgr dwarf. Whiting 1 also has an extended luminosity profile, presumably a tidal tail, and its position, distance, and velocity place it in the Sgr stream.

Casetti–Dinescu *et al.* (2009) considered the Virgo stellar stream and integrated its stellar orbits back for 5 Gyr. They found that the globular cluster NGC 2419 could be in this stream. Gao *et al.* (2007) found common streams for many Milky Way globular clusters, based on common energy, angular momentum, and orbital poles. Gao *et al.* suggest that 20% of globular clusters are in common streams.

Smith *et al.* (2009) noted that four globular clusters lie in a halo stellar kinematic overdensity. Their metallicities are $[\mathrm{Fe/H}] = -2.22$, -1.54, -1.58 and -1.65 dex; one of them, NGC 5466, is disrupting (Odenkirchen & Grebel 2004; Belokurov *et al.* 2006). Other related studies are by Dinescu *et al.* (1999), Palma *et al.* (2002), and Mackey & Gilmore (2004).

6. Summary

Stars form from gas in hierarchical patterns. Clusters are the inner mixed part of this star-formation distribution. In such a hierarchy, the star-formation efficiency is automatically high on small scales, and if it is high enough, the cluster remains bound after the gas leaves. The cluster mass function, $dN/dM \sim M^{-2}$, follows from the hierarchy too. There is a possible upper cutoff mass, M_c, making the mass function something like $M^{-2}e^{-M/M_c}$. Halo globular clusters may have a similar upper-mass cutoff and they all drop towards lower mass, without the power law. It is not yet understood if the missing low-mass clusters were removed by a long-term process of evaporation, or if they were destroyed early on by some other process, including gas loss after star formation. Globular clusters often show multiple star-formation events, which implies self-enrichment over a period of 0.1 to 1 Gyr. Perhaps their high mass and low metallicity is enough to produce this enrichment, which is not generally observed for low-mass clusters today. Or perhaps globular clusters were dwarf galaxy nuclei that collected independent clusters with a wide range of metallicities. Many Milky Way globular clusters orbit in the stellar debris streams from former dwarf galaxy companions.

References

Bailin, J. & Harris, W. E. 2009, *ApJ*, 695, 1082

Barbá, R. H., Maíz Apellániz. J., Pérez, E., Rubio, M., Bolatto, A., Fariña, C., Bosch, G., & Walborn, N. R. 2009, in: E. Pérez, R. de Grijs, & R. M. González Delgado, *Young massive star clusters: initial conditions and environments*, *ApSS*, 324, 309, Dordrecht: Springer

Bastian, N., Gieles, M., Ercolano, B., & Gutermuth, R. 2009, *MNRAS*, 392, 868

Bedin, L. R. 2004, *ApJ* (Letters), 605, L125

Bekki, K. & Norris, J. E. 2006, *ApJ* (Letters), 637, L109

Belokurov, V., Evans, N. W., Irwin, M. J., Hewett, P. C., & Wilkinson, M. I., 2006, *ApJ* (Letters), 637, L29

Bhatia, R. K. & Hatzidimitriou, D. 1988, *MNRAS*, 230, 215

Block, D. L., Puerari, I., Elmegreen, B. G., Elmegreen, D. M., Fazio, G. G., & Gehrz, R. D. 2009, *ApJ*, 694, 115

Carraro, G., Zinn, R., & Moni Bidin, C. 2007, *A&A*, 466, 181

Cartwright, A. & Whitworth, A. P., 2004, *MNRAS*, 348, 589

Casetti–Dinescu, D. I., Girard, T. M., Majewski, S. R., Vivas, A. K., Wilhelm, R., Carlin, J. L., Beers, T. C., & van Altena, W. F. 2009, *ApJ* (Letters), 701, L29

Cassisi, S., Salaris, M., Pietrinferni, A., Piotto, G., Milone, A. P., Bedin, L. R., & Anderson, J. 2008, *ApJ* (Letters), 672, L115

Chandar, R., Fall, S. M., & McLaughlin, D. E. 2007, *ApJ* (Letters), 668, L119

Dahm, S. E. & Simon, T. 2005, *AJ*, 129, 829

D'Antona, F., Caloi, V., Montalbán, J., Ventura, P., & Gratton, R. 2002, *A&A*, 395, 69

D'Ercole, A., Vesperini, E., D'Antona, F., McMillan, S. L. W., & Recchi, S. 2008, *MNRAS*, 391, 825

de Grijs, R., Anders, P., Bastian, N., Lynds, R., Lamers, H. J. G. L. M., & O'Neil Jr., E. J. 2003, *MNRAS*, 343, 1285

de Grijs, R. & Anders, P. 2006, *MNRAS*, 366, 295

de la Fuente Marcos, R. & de la Fuente Marcos, C. 2008, *ApJ*, 672, 342

de la Fuente Marcos, R. & de la Fuente Marcos, C. 2009a, *ApJ*, 700, 436

de la Fuente Marcos, R. & de la Fuente Marcos, C. 2009b, *A&A* (Letters), 500, L13

Dieball, A., Müller, H., & Grebel, E. K. 2002, *A&A*, 391, 547

Dinescu, D. I., Girard, T. M., & van Altena, W. F., 1999, *AJ*, 117, 1792

Efremov, Yu. N. 1995, *AJ*, 110, 2757

Efremov, Yu. N. & Elmegreen, B. G. 1998, *MNRAS*, 299, 588

Elias, F., Alfaro, E. J., & Cabrera–Caño, J. 2009, *MNRAS*, 397, 2

Elmegreen, B. G. 2007, *ApJ*, 668, 1064

Elmegreen, B. G. 2008, *ApJ*, 672, 1006

Elmegreen, B. G. & Elmegreen, D. M. 1987, *ApJ*, 320, 182

Elmegreen, B. G. & Efremov, Yu. N. 1996, *ApJ*, 466, 802

Elmegreen, B. G. & Efremov, Yu. N. 1997, *ApJ*, 480, 235
Elmegreen, B. G., Elmegreen, D. M., & Leitner, S. N. 2003, *ApJ*, 590, 271
Elmegreen, B. G., Elmegreen, D. M., Chandar, R., Whitmore, B., & Regan, M. 2006, *ApJ*, 644, 879
Engargiola, G., Plambeck, R. L., Rosolowsky, E., & Blitz, L. 2003, *ApJS*, 149, 343
Enoch, M. L., *et al.* 2006, *ApJ*, 638, 293
Fall, S. M. & Zhang, Q. 2001, *ApJ*, 561, 751
Feigelson, E. D., Martin, A. L., McNeill, C. J., Broos, P. S., & Garmire, G. P. 2009, *AJ*, 138, 227
Feitzinger, J. V. & Braunsfurth, E. 1984, *A&A*, 139, 104
Feitzinger, J. V. & Galinski, T. 1987, *A&A*, 179, 249
Gao, S., Jiang, B.-W., & Zhao, Y.-H. 2007, *ChJAA*, 7, 111
Georgiev, I. Y., Hilker, M., Puzia, T. H., Goudfrooij, P., & Baumgardt, H. 2009, *MNRAS*, 396, 1075
Gieles, M., Bastian, N., & Ercolano, E. 2008, *MNRAS* (Letters), 391, L93
Gieles, M., Larsen, S. S., Bastian, N., & Stein, I. T. 2006a, *A&A*, 450, 129
Gieles, M., Larsen, S. S., & Sheepmaker, R. A. 2006b, *A&A* (Letters), 446, L9
Gomez, M., Hartmann, L., Kenyon, S. J., & Hewett, R. 1993, *AJ*, 105, 1927
Grabelsky, D. A., Cohen, R. S., Bronfman, L., Thaddeus, P., & May, J. 1987, *ApJ*, 315, 122
Gutermuth, R. A., Megeath, S. T., Pipher, J. L., Williams, J. P., Allen, L. E., Myers, P. C., & Raines, S. N. 2005, *ApJ*, 632, 397
Harris, W. E. 2009, *ApJ*, 699, 254
Harris, W. E., Whitmore, B. C., Karakla, D., Okoń, W., Baum, W. A., Hanes, D. A., & Kavelaars, J. J. 2006, *ApJ*, 636, 90
Hatzidimitriou, D. & Bhatia, R. K. 1990, *A&A*, 230, 11
Hunter, D. A., Baum, W. A., O'Neil Jr., E. J., & Lynds, R. 1996, *ApJ*, 456, 174
Hunter, D. A., Elmegreen, B. G., Dupuy, T. J., & Mortonson, M. 2003, *AJ*, 126, 1836
Ivanov, G. R. 2005, *Publ. Astron. Soc. Rudjer Boskovic*, 5, 75
Johnstone, D., Wilson, C. D., Moriarty–Schieven, G., Joncas, G., Smith, G., Gregersen, E., & Fich, M. 2000, *ApJ*, 545, 327
Johnstone, D., Fich, M., Mitchell, G. F., & Moriarty–Schieven, G. 2001, *ApJ*, 559, 307
Jordán, A., McLaughlin, D. E., Côté, P., Ferrarese, L., Peng, E. W., Mei, S., Villegas, D., Merritt, D., Tonry, J. L., & West, M. J. 2007, *ApJS*, 171, 101
Jørgensen, J. K., *et al.* 2006, *ApJ*, 645, 1246
Jørgensen, J. K., Johnstone, D., Kirk, H., & Myers, P. C. 2007, *ApJ*, 656, 293
Kharchenko, N. V., Piskunov, A. E., Röser, S., Schilbach, E., & Scholz, R.-D. 2005, *A&A*, 438, 1163
Klessen, R. 2001, *ApJ*, 556, 837
Kundu, A., Whitmore, B. C., Sparks, W. B., Macchetto, F. D., Zepf, S. E., & Ashman, K. M. 1999, *ApJ*, 513, 733
Lamers, H. J. G. L. M. & Gieles, M. 2006, *A&A* (Letters), 455, L17
Larsen, S. S. 2009, *A&A*, 494, 539
Larson, R. B. 1995, *MNRAS*, 272, 213
Lynga, G. 1982, *A&A*, 109, 213
Mackey A. D. & Gilmore G. F., 2004, *MNRAS*, 355, 504
Maíz–Apellániz, J. 2001, *ApJ*, 563, 151
Marcolini, A., Sollima, A., D'Ercole, A., Gibson, B. K., & Ferraro, F. R. 2007, *MNRAS*, 382, 443
McGee, R. X. & Milton, J. A. 1964, *Austr. J. Phys.*, 17, 128
McLaughlin, D. E. 2000, *ApJ*, 539, 618
McLaughlin, D. E. & Fall, S. M. 2008, *ApJ*, 679, 1272
Megeath, S. T., Allen, L. E., Gutermuth, R. A., Pipher, J. L., Myers, P. C., Calvet, N., Hartmann, L., Muzerolle, J., & Fazio, G. G. 2004 *ApJS*, 154, 367
Mieske, S. 2006, *ApJ*, 653, 193
Milone, A. P. 2008, *ApJ*, 673, 241

Norris, J. E. 2004, *ApJ* (Letters), 612, L25
Odekon, M. C. 2006, *AJ*, 132, 1834
Odenkirchen, M., & Grebel. E. K., 2004, in: F. Prada, D. Martínez Delgado & T. J. Mahoney (eds.), *Satellites and Tidal Streams*, ASP Conf. Ser. Vol. 327, p. 284, San Francisco: ASP
Palma C., Majewski S. R., & Johnston K. V., 2002, *ApJ*, 564, 736
Parmentier, G. & Gilmore, G. F. 2007, *MNRAS*, 377, 352
Parmentier, G., Goodwin, S. P., Kroupa, P., & Baumgardt, H. 2008, *ApJ*, 678, 347
Peng, E. W., Jordán, A., Blakeslee, J. P., Mieske, S., Côté, P., Ferrarese, L., Harris, W. E., Madrid, J. P., & Meurer, G. R. 2009, *ApJ*, 703, 42
Piskunov, A. E., Kharchenko, N. V., Röser, S., Schilbach, E., & Scholz, R.-D. 2006, *A&A*, 445, 545
Rand, R. J. & Kulkarni, S. R. 1990, *ApJ* (Letters), 349, L43
Rathborne, J. M., Jackson, J. M., & Simon, R. 2006, *ApJ*, 641, 389
Reid, M. A. & Wilson, C. D. 2005, *ApJ*, 625, 891
Sánchez, N., & Alfaro, E. J. 2008, in: *Highlights of Spanish Astrophysics V*, in press (arXiv:0810.0478)
Sánchez, N. & Alfaro, E. J. 2009, *ApJ*, 696, 2086
Scheepmaker, R. A., Lamers, H. J. G. L. M., Anders, P., & Larsen, S. S. 2009, *A&A*, 494, 81
Schechter, P. 1976, *ApJ*, 203, 297
Schmeja, S., Kumar, M. S. N., & Ferreira, B. 2008, *MNRAS*, 389, 1209
Smith, M. D., Gredel, R., Khanzadyan, T., & Stanke, T. 2005, *MmSAI*, 76, 247
Smith, M. C., Evans, N. W., Belokurov, V., Hewett, P. C., Bramich, D. M., Gilmore, G., Irwin, M. J., Vidrih, S., & Zucker, D. B. 2009, *MNRAS*, 399, 1223
Strader, J., Brodie, J. P., Spitler, L., & Beasley, M. A. 2006, *AJ*, 132, 2333
Testi, L., Sargent, A. I., Olmi, L., & Onello, J. S. 2000, *ApJ* (Letters), 540, L53
Ventura, P., Caloi, V., D'Antona, F., Ferguson, J., Milone, A., & Piotto, G. 2009, *MNRAS*, 399, 934
Vesperini, E. 1998, *MNRAS*, 299, 1019
Vesperini, E. 2000, *MNRAS*, 318, 841
Vesperini, E., Zepf, S. E., Kundu, A., & Ashman, K. M. 2003, *ApJ*, 593, 760
Waters, C. Z., Zepf, S. E., Lauer, T. R., Baltz, E. A., & Silk, J. 2006, *ApJ*, 650, 885
Young, K. E., *et al.* 2006, *ApJ*, 644, 326
Zhang, Q. & Fall, S. M. 1999, *ApJ* (Letters), 527, L81
Zhang, Q., Fall, S. M., & Whitmore, B. C. 2001, *ApJ*, 561, 727

Session I

Physics and modes of star cluster formation

Star clusters: basic galactic building blocks
Proceedings IAU Symposium No. 266, 2009
R. de Grijs & J. R. D. Lépine, eds.

doi:10.1017/S1743921309990810

Embedded and open star clusters: basic questions

Hans Zinnecker

Astrophysikalisches Institut Potsdam, An der Sternwarte 16, D-14482 Potsdam, Germany
email: hzinnecker@aip.de

Abstract. This paper provides an introduction to IAU Symposium 266 on star clusters as basic building blocks in space and time. We define clusters as bound systems and discriminate them from general stellar clusterings or groups and unbound associations. We give a few examples of young, embedded, compact clusters which may evolve into looser, open clusters after dynamical relaxation due to mass loss and secular relaxation processes. We ask how and where star clusters form (in normal and interacting galaxies) and provide statistics of open clusters in terms of cluster masses and ages in the solar neighborhood (where observational data are most complete). Finally, we list a number of basic questions for current and future star cluster research and discuss the prospects for cluster studies with the next generation infrared and submillimeter telescopes (*Herschel; JWST, E–ELT; ALMA, NOEMA*).

Keywords. galaxies: star clusters, stars: formation, HII regions, infrared: stars, X-rays: stars

1. Introduction

As the title of this Symposium suggests, star clusters are basic building blocks of galaxies throughout space and time. Ideally, the first question to ask in this context would be how to define a star cluster. I prefer a strict definition for clusters to be gravitationally bound systems, and suggest to discriminate clusters from stellar clusterings or groups and unbound associations. Other immediate questions would be: Where, across the face of galaxies, do star clusters form (mostly in spiral arms?) and do they form continuously or at certain periods in time when galaxies interact (e.g., during close gravitational encounters)? And how did globular clusters form?

As the first speaker at this symposium, I consider it to be my task to whet your appetite for the subject and to raise some of the basic questions in this topical field of research. In an appendix at the end of this paper, I list 12 questions: too many to discuss in this short review, but some of them will be addressed.

The outline of my contribution is as follows. We start by showing and discussing a representative example (NGC 7538), where an evolutionary sequence from dense embedded near-infrared (near-IR) clusters to an optically visible cluster and HII region can be seen in a single, large-scale near-IR image. Then, we briefly describe the emergence of optical open clusters from obscured compact protoclusters. Next, we summarize some of the open cluster statistics, and discuss present-day and initial mass functions of open clusters as well as the distribution of their lifetimes. This then leads to the question as to which fraction of the local field stars has originated from disintegrated open clusters over the course of Galactic history. We also mention the problem of isolated field OB stars and whether they can all be explained as cluster-born runaway stars. We conclude with an outlook at what we can expect from future new IR telescopes such as the *JWST* and

Figure 1. True-color (JHK) near-IR image of NGC 7538 (see text for details). (Credit: J. Rayner, M. McCaughrean, H. Zinnecker.)

E–ELT and submillimeter (submm) imaging arrays such as *ALMA* and *NOEMA* (the upgraded *Plateau de Bure Interferometer, PdBI*).

2. The NGC 7538 HII region: an instructive example

In Figure 1, we show the HII region NGC 7538, part of the Cas OB2 complex at a distance of ~ 2.7 kpc. This is one of the first large-scale true-color composite infrared-array JHK images (McCaughrean *et al.* 1991; Zinnecker *et al.* 1993), and actually made it to the front cover of the August 1991 edition of *Sky and Telescope*. The image was made by combining three separate mosaics of the region, one each at J (1.2μm), H (1.65 μm), and K' (2.1μm). By having these three images represent blue, green, and red respectively, the colors in the resulting composite give some idea of what this region might look like if our eyes were sensitive at near-IR rather than optical wavelengths. In this kind of image, the colors impart real wavelength-dependent information about the various sources. For instance, red objects are cool or embedded in dust, while blue objects are relatively hot.

The image in Figure 1 covers about 12×12 arcmin^2 (i.e., $\sim 9.5 \times 9.5$ pc^2 at 2.7 kpc), at a scale of 0.77 arcsec pixel^{-1}. Each of the J, H, and K' images is a mosaic of 25 overlapping subimages taken, using a NICMOS3 HgCdTe 256×256-pixel array camera on the University of Hawaii 88-inch telescope on Mauna Kea in October 1990. The effective integration time is 2 minutes per pixel per filter, yielding a point-source detection limit of $K' \sim 17$ mag (or absolute K-band magnitude of ~ 5, corresponding to young stellar objects with masses as low as 0.3 $M_\odot$ at an age of 1 Myr). The image is more or less centered on the OB stars that illuminate the optical HII region NGC 7538, while the famous IR sources (Werner et al. 1979) are to the south-east. IRS 1, 2, and 3 are in the very bright red core, and IRS 9 is at the apex of the complex reflection nebula. Several new, fainter reflection nebulae are seen, with quite intriguing morphologies. The diffuse HII-region emission seen in purple and pink is probably a combination of thermal Bremsstrahlung and reflected light, both strongest in J, along with some ionized hydrogen-line emission in the K' filter (Brackett γ at 2.166 μm). There are several newly discovered stellar groups in this image, most notably between the HII region and the compact core hosting IRS 1–3, and the very red small clusterings to the south, some seemingly aligned in a filamentary substructure. Finally, there are also several dark lanes seen in projection against bright nebulosity, most markedly in the IRS 9 reflection nebulosity.

The overall morphology is of three major condensations, which may form a sequence in age: the diffuse HII region (northwest, oldest), the compact IR core (center), and the less compact IR reflection nebula (southeast, youngest). It is also possible that while star formation has propagated from the HII region to the IRS 1–3 core, the complex to the southeast has evolved independently from the same molecular cloud. The dense cluster between the optical HII region and IRS 1–3 also seems to be evidence for sequential star formation (Elmegreen & Lada 1977), with the cluster forming from the shock front from the HII region compressing the surrounding molecular material. The cluster is only revealed now, as the OB stars act to clear away the obscuring gas and dust. A similar case is the NGC 3603 cluster and HII region clearing the view to the NGC 3603–IRS 9 subgroup (see the near- and mid-IR *VLT* images of Nürnberger 2003); see also the beautiful *Spitzer* images of the Rosette and RCW 49 clusters. Other examples include the M17 cluster (Hoffmeister *et al.* 2008; see their *JHK VLT* image and analysis) and the Sh 2–212 HII region triggering secondary massive star formation at its border (Deharveng *et al.* 2008).

Whatever story is revealed by subsequent detailled analysis of these and other images, there is the striking impression that we are seeing different evolutionary stages in a large star-forming-region complex with considerable substructure. Some of the substructure may ultimately merge with the final main cluster, giving rise to an age spread in the cluster stellar population.

While we have emphasized infrared imaging to detect and study young, embedded star clusters, X-ray imaging can also penetrate dust extinction in molecular clouds and reveal embedded populations: the optical depth at 2μm is about the same as at 2 keV! A splendid example is the *Chandra* image of the W3 star-forming complex (Feigelson & Townsley 2008), revealing diverse stellar populations, some of them dense clusters and some more like loose associations.

In conclusion, it is apparent that near-IR and X-ray imaging are capable of revealing a wealth of information, giving us a much clearer view deep inside embedded, dust-obscured stellar nurseries.

3. From embedded to open clusters

The example of the NGC 7538 complex above was chosen to illustrate the emergence of an optically visible open cluster and HII region as the end stage of an energetic, internal dynamical evolution. In fact, the birth of optically visible star clusters is reminscent of chicken hatching from their eggs: internal action breaks up the cocoon-like shell. For star clusters, the internal action is due to massive stars, which energize the dense parent cloud with ionizing photons, which in turn generate the gas pressure of an HII region (with a temperature of about 10 000 K). The pressure gradient drives the parent gas and dust away from the cluster center and creates a bubble, inside of which the cluster finally becomes transparent to optical radiation (e.g., Hα). The expansion of the dense gas and dust is assisted by the powerful stellar winds and outflows from the massive OB stars. Note that supernova remnants do not yet play a role at this stage. They only set in after some 5–10 Myr have elapsed, but they may be a powerful agent to trigger additional (coeval and coherent) star formation in the wider vicinity of the primary cluster or group, a fine example being the Upper Sco subgroup of the Sco OB2 association (Preibisch & Zinnecker 2007).

Many embedded star clusters do not survive the expansion of their HII region, because too much gas and hence binding mass is lost, as first discussed by von Hoerner (1968). In their influential review, Lada & Lada (2003) argue that only $\sim$ 10% of all embedded clusters survive. The rest is immediately released as a field-star population. However, I think the jury is still out as to the true 'infant mortality' by cluster mass fraction. It could be 50 instead of 90% (still high), in line with the *Spitzer* data for the Orion A and B molecular clouds, where small groups and clusterings of stars make up about half the total number of young stars, while the rest is found in several big clusters, including the Trapezium (T. Megeath 2009, priv. comm.). Theoretically speaking, the infant-mortality rate depends on the cluster mass and the star-formation efficiency, as described in detail by Parmentier (these proceedings). Those clusters that survive as bound entities may nevertheless lose a fraction of their stars in the process and end up as smaller and less dense clusters. It seems possible that a compact and fairly massive cluster like NGC 3603 may end up as a much looser and lower-mass system, depending on slight changes in the star-formation efficiency and the precise timing of the onset of massive-star-ionization feedback. These speculations can be tested with cluster-formation simulations, where massive stars form last (probably by competitve accretion near the center). We refer to the calcuations by Kroupa *et al.* (2001), who suggest that the Orion Trapezium cluster may ultimately evolve into a cluster like the Pleiades.

4. Open cluster statistics: masses and ages

How do we go about determining the integrated masses of open clusters and from there the present-day and initial mass function (IMF) of open clusters?

There a several methods to infer the masses of star clusters, including direct star counts (lower limits only) or indirect scaling with the stellar IMF (extrapolating the bright end of the IMF to the full IMF, assuming the lower-mass section is known, i.e., assuming a Salpeter or Kroupa-type IMF down to a given lower-mass limit). Dynamical mass determinations are also possible. For example, one can, in principle, use the cluster size and stellar velocity dispersion (hard to obtain in practical terms) to derive virial masses based on the assumption that the cluster is in virial equilibrium (not true for the youngest clusters). This method is frequently used to obtain the masses of molecular clouds in which the star clusters form. Another dynamical method, less well-known, uses tidal masses. At a given location in the Galaxy, the tidal field is known (it can be expressed in terms of Oort constants; cf. King 1962) and hence an observed cluster tidal radius immediately provides the mass enclosed inside this radius. Tidal masses depend on the tidal radius cubed, and the mass accuracy thus depends sensitively on how well the tidal radius can be deduced from King-profile fits of the cluster stellar-surface-density distribution from a limited number of bright member stars. A further caveat is the implicit assumption that the cluster population fills its tidal radius. This is likely true for old clusters but unlikely for very young clusters, which are more compact, with sizes about one tenth the tidal radii (1 pc versus 10 pc, roughly speaking). The tidal and the virial method of determining cluster masses are independent of the total cluster luminosity, the latter being the only means to infer extragalactic cluster masses (e.g., Zhang & Fall 1999; Larsen 2002; de Grijs 2004) using an age-dependent cluster luminosity–mass relation.

Piskunov *et al.* (2008) were courageous enough to derive cluster mass functions in the solar neighborhood from a compilation of tidal cluster masses in the range from a few tens to a few hundred thousand solar masses. The data are based on clusters selected from the ASCC–2.5 catalogue, which is complete to apparent integrated cluster V-band magnitude of 8, with 440 clusters (and compact associations) above this completeness limit. The completeness area in the solar neigborhood corresponds to an effective radius of 1 kpc around the Sun. The brightest clusters can be seen far beyond this completeness radius. The local cluster mass function can be fitted with a power-law slope (with an index of -2), in agreement with data for extragalactic clusters (derived by converting cluster luminosity into cluster mass functions). The local present-day cluster mass function shows a broad maximum between cluster masses from 30 to 300 $M_{\odot}$.

The sample of open clusters can be binned into several age bins, with as youngest bin that containing clusters of 4–8 Myr. This allowed Piskunov *et al.* (2008) to derive the IMF of open clusters. The result is a two-segment power law, with a slope of $-5/3$ (i.e., different from -2) for the high-mass section above 2000 $M_{\odot}$, while the lower-mass part is characterized by a slope of $-4/5$ down to 50 $M_{\odot}$. The interpretation of these results, especially the fact that the initial cluster mass function seems to be different from (flatter than) the overall cluster mass function averaged over all ages, remains to be understood in terms of cluster dynamical evolution. Note that Lada & Lada (2003) suggested a power-law slope of -2 for their sample of embedded clusters between 50 and 1000 $M_{\odot}$ within 500 pc from the Sun.

Let us end this paragraph with a brief comment about the age determination and distribution of open clusters. The half lifetime for clusters, i.e., the typical open cluster decay timescale, is about 300 Myr, again based on the analysis of Piskunov *et al.* (2006). This is somewhat longer than the original estimate of Wielen (1971). It implies some 30 generations of open star clusters since the formation of the Galactic disk. The surface density of clusters of all ages in the solar neighborhood is about 100 kpc^{-2}. The birthrate of young clusters (age bin 4–8 Myr) is about 0.4 kpc^{-2} Myr^{-1}. Age determination of open clusters is done based on the main-sequence turnoff method. Young clusters can be age-dated using pre-main-sequence tracks and the corresponding K-band luminosity function, which shows a feature at the main-sequence turn-on (cf. Zinnecker *et al.* 1993).

5. Open clusters as the source of Galactic field stars

For a long time, the prevailing view has been that about 10% of all Galactic field stars in the solar neighborhood originated from disintegrating open clusters (Wielen 1971). Recent re-analysis has shown that this fraction is likely as high as 50% (Piskunov *et al.* 2008; Zinnecker *et al.* 2009). The reason for this is that the average mass of young clusters in a statistically complete, magnitude-limited sample is much higher than thought before. Piskunov *et al.* (2008) estimated the average cluster mass of present-day open clusters of all ages at 700 $M_{\odot}$, while the average initial cluster mass (for clusters after the infant-mortality phase of dynamical protocluster gas-mass loss) is 4500 $M_{\odot}$. This shows how much mass open clusters lose during the later stages due to internal and external processes (due to stellar winds and secular stellar dynamical evaporation versus tidal disruption caused by passing molecular clouds; see Gieles, these proceedings).

A related issue is the origin of field OB stars. There appear to be a number of Galactic O stars which do not occur in embedded or open clusters. The question arises to what extent these field O stars are all runaway stars, either ejected from massive binaries after one star exploded as a supernova or ejected from dense clusters due to internal dynamical slingshot interactions involving transfer of potential energy from multiple systems into kinetic energy several times the escape speed from the parent cluster. A recent study by Schilbach & Röser (2008) concluded that most O-type runaway stars with good proper-motion determinations and radial velocities can be traced back directly to some young parent cluster in a Galactic potential. A similar study for B-type runaway stars is not yet available. The existence of randomly located field OB stars is important for widely distributed effective supernova energy input into the interstellar medium as a source of turbulence.

6. Future prospects with next-generation telescopes

Let us try to cast an eye at the future of young star cluster research, both ground-based and space-borne. The *ALMA* submm array in Chile will be operational in 2012 and from that point will give us detailed high-angular-resolution images of protocluster clouds, at various evolutionary stages. It will resolve cold, dense protostellar clumps destined to become stars. A glimpse can already be gleaned from the results of current millimeter interferometers such as *CARMA*, *SMA*, and *PdBI* (the latter will be upgraded from 6 to 12 telescopes to become the *Northern Extended Millimeter Array, NOEMA*, in the near future). The *Herschel* far-IR satellite observatory with its PACS and SPIRE imaging cameras, now in operation, will soon provide us with clump-mass distributions and fragmentation patterns in protoclusters and offer us clues as to the origin of the stellar IMF in clusters (Ph. André 2009, priv. comm.).

Further down the line, *JWST*—because of its extreme sensitivity in the thermal infrared—will likely spatially resolve not only the luminous, but also the faint energy sources in nearby protoclusters and protostellar groups, shedding light on massive-star-formation processes, such as in the Kleinmann–Low nebula (IRC 2) behind the Orion Nebula Cluster. It is still unclear what is going on in this nearest but most heavily obscured site of massive star formation. *JWST* (both with NIRCAM and MIRI) may also be instrumental to address the question of initial mass segregation in protoclusters, as it can map the location of massive protostellar objects with respect to their lower-masss siblings at very early, very deeply embedded stages before dynamical evolution can change the birth configuration. Recent studies of young infrared star clusters suggested there is no evidence for mass segregation (Ascenso *et al.* 2008), contrary to expectation based on the model of competitive accretion (Bonnell *et al.* 2007). The ultimate facility to study the dense centers of massive star clusters will be the *E–ELT* (first light expected around 2018). Adaptive-optics observations with a 42 m diameter telescope in the near-IR K band (2μm) will go a long way to penetrate even 100 magnitudes or more of visual extinction and resolve even very compact clusters, in both imaging and velocity space, in distant Galactic star-forming regions, including the massive Sgr B2 millimeter continuum cores near the Galactic Centre at 8 kpc where analogs of the Arches or Quintuplet massive clusters may be forming now (J. Bally 2009, priv. comm.). The science case for *E–ELT* near/thermal-IR and integral-field spectroscopy is further described in the Proceedings of IAU Symposium 232 (Zinnecker 2006).

Appendix: 12 questions

- Definition of a 'cluster' (bound versus unbound, clusterings or groups);
- Origin of the cluster mode of star formation versus distributed mode;
- Statistics of embedded clusters in the solar neighborhood (1–2 kpc);
- 'Infant mortality' (dynamical evolution during early gas-mass loss);
- Open clusters as survivors (NGC 3603 precursor of the Pleiades?);

- Determination of open cluster masses, ages, and lifetimes (methods);
- Statistics of open clusters in the solar neighborhood (masses, ages);
- What fraction of field stars originated in open clusters? binaries?
- OB clusters versus OB associations: an evolutionary sequence or not?
- Do all massive stars form in clusters? what about runaway OB stars?
- Progenitors of massive Galactic starburst (Wolf–Rayet/red-supergiant) clusters;
- Origin of Population II globular clusters (minimum metallicity $Z = 0.01\,Z_{\odot}$?).

Acknowledgements

I thank Richard de Grijs for the invitation to present this introductory contribution, Charles Lada and Anatoly Piskunov for insightful discussions, and the German DAAD for providing travel funds to participate in the IAU General Assembly in Rio de Janeiro and this particular IAU Symposium.

References

Ascenso, J., Alves, J., & Lago, M. T. V. T. 2009, *A&A*, 495, 147
Bonnell, I. A., Larson, R. B., & Zinnecker, H. 2007, in: B. Reipurth, D. Jewitt & K. Keil (eds.), *Protostars and Planets V* (Tucson: University of Arizona Press), p. 147
de Grijs, R. 2004, in: H. J. G. L. M. Lamers, L. J. Smith & A. Nota (eds.), *The Formation and Evolution of Massive Young Star Clusters*, ASP Conf. Ser. (San Francisco: ASP), Vol. 322, p. 29
Deharveng, L., Lefloch, B., Kurtz, S., Nadeau, D., Pomarès, M., Caplan, J., & Zavagno, A. 2008, *A&A*, 482, 585
Elmegreen, B. G. & Lada, C. J. 1977, *ApJ*, 214, 725
Feigelson, E. D. & Townsley, L. K. 2008, *ApJ*, 673, 354
Hoffmeister, V. H., Chini, R., Scheyda, C. M., Schulze, D., Watermann, R., Nürnberger, D., & Vogt, N. 2008, *ApJ*, 686, 310
King, I. 1962, *AJ*, 67, 471
Kroupa, P., Aarseth, S., & Hurley, J. 2001, *MNRAS*, 321, 699
Lada, C. J. & Lada, E. A. 2003, *ARA&A*, 41, 57
Larsen, S. S. 2002, *AJ*, 124, 1393
McCaughrean, M. J., Rayner, J. T., & Zinnecker, H. 1991, *MemSAIt*, 62, 715
Nürnberger, D. 2003, *A&A*, 404, 255
Piskunov, A. E., Kharchenko, N. V., Röser, S., Schilbach, E., & Scholz, R.-D. 2006, *A&A*, 445, 545
Piskunov, A. E., Kharchenko, N. V., Schilbach, E., Röser, S., Scholz, R.-D., & Zinnecker, H. 2008, *A&A*, 487, 557
Preibisch, T. & Zinnecker, H. 2007, in: B. G. Elmegreen & J. Palous (eds.), *Triggered Star Formation in a Turbulent ISM*, IAU Symp. 237, p. 270
Schilbach, E. & Röser, S. 2008, *A&A*, 489, 105
von Hoerner, S. 1968, in: Y. Terzian (ed.), *Interstellar Ionized Hydrogen* (New York: Benjamin), p. 101
Werner, M. W., Becklin, E. E., Gatley, I., Matthews, K., Neugebauer, G., & Wynn–Williams, C. G. 1979, *MNRAS*, 188, 463
Wielen, R. 1971, *A&A*, 13, 309
Zhang, Q. & Fall, S. M. 1999, *ApJ* (Letters), 527, L81
Zinnecker, H. 2006, in: P. Whitelock, M. Dennefeld & B. Leibundgut (eds.), *The Scientific Requirements for Extremely Large Telescopes*, IAU Symp. 232, p. 324
Zinnecker, H. 2009, in: J. Andersen, J. Bland-Hawthorn & B. Nordstroem (eds.) *The Galaxy Disk in Cosmological Context*, IAU Symp. 254, p. 221
Zinnecker, H., McCaughrean, M. J., & Wilking, B. A. 1993, in: E. H. Levy & J. I. Lunine (eds.), *Protostars and Planets III* (Tucson: University of Arizona Press), p. 429

Star clusters: basic galactic building blocks
Proceedings IAU Symposium No. 266, 2009
R. de Grijs & J. R. D. Lépine, eds.

doi:10.1017/S1743921309990822

The origin of mass segregation in NGC 3603

Xiaoying Pang, Eva K. Grebel and Martin Altmann[1]

Astronomisches Rechen-Institut, Zentrum für Astronomie der Universität Heidelberg, Mönchhofstrasse 12–14, 69120 Heidelberg, Germany
email: xiaoying@ari.uni-heidelberg.de

Abstract. NGC 3603 is one of the most massive, compact young star clusters in the Milky Way. The cluster has an age of only about 1 Myr and is embedded in a giant molecular cloud with ongoing star formation. We have analyzed deep imaging data obtained with the Wide Field and Planetary Camera 2 aboard the *Hubble Space Telescope*. We have obtained two epochs separated by 10 years, from which we derived proper motions which we used to determine cluster membership. After the removal of field stars, the resulting color–magnitude diagram shows a main sequence in addition to another clear sequence of pre-main-sequence stars. The cluster shows pronounced mass segregation and appears to have a very short crossing timescale. Our photometric, astrometric and kinematic data help us to evaluate the dissolution timescale of NGC 3603 and whether the mass segregation is likely to be primordial or evolutionary.

Keywords. galaxies: star clusters, Galaxy: kinematics and dynamics

1. Introduction

The NGC 3603 giant molecular cloud in the Carina spiral arm contains one of the most massive, young compact clusters in the Milky Way. This is the HD 97950 star cluster, which is often referred to by the NGC number of the surrounding, extended structure. The cluster has an age of approximately 1 Myr. This makes it an interesting target to study the open question of the impact of primordial versus dynamical mass segregation, assuming that dynamical effects would not yet have had enough time to significantly alter the young cluster's structure. Another interesting question is that of the duration of star formation in star clusters. It is commonly assumed that such objects do not show any appreciable age spread, but that all stars formed in one single, brief burst. Very young clusters may permit us to investigate the duration of such a burst of intense star formation.

2. Observations and data reduction

Deep imaging data of NGC 3603 were obtained repeatedly with the Wide Field and Planetary Camera 2 (WFPC2) aboard the *Hubble Space Telescope*. One set of observations was obtained in 1997 using the F547M, F675W and F814W filters. The other set was observed in May 2007 using the F555W, F675W and F814W filters. In both cases, the cluster center is located on the high-resolution PC chip. The 10 years of epoch difference provide us with the opportunity of obtaining the cluster's proper motion.

Both sets of data were reduced using HSTphot (Dolphin 2000), a program developed for crowded-field stellar photometry of WFPC2 data. The incompleteness in the central regions of the cluster turned out to be fairly high due to high stellar densities and a concentration of very luminous stars in the center. We identified 445 common stars on the PC chip in the exposures of the two epochs. Here, we only concentrate on the PC chip ($r \sim 20''$) for which there is ~ 90 percent overlap between two epochs. Also, with $\sim 0.05''$ pixel^{-1}, the PC chip has twice the resolution of the three other chips.

3. Results

3.1. *Proper motion*

By comparing the position of common stars between the two epochs, we derived relative proper motions for these stars. These proper motions help us to select cluster members quite effectively, since the cluster stars have a smaller proper motion than the largely much closer field stars. The resulting relative proper-motion distribution of member stars on the PC chip is shown in Figure 1. The uncertainty in the proper motion is on average 0.05 pixels, corresponding to a velocity of 9 km s^{-1}.

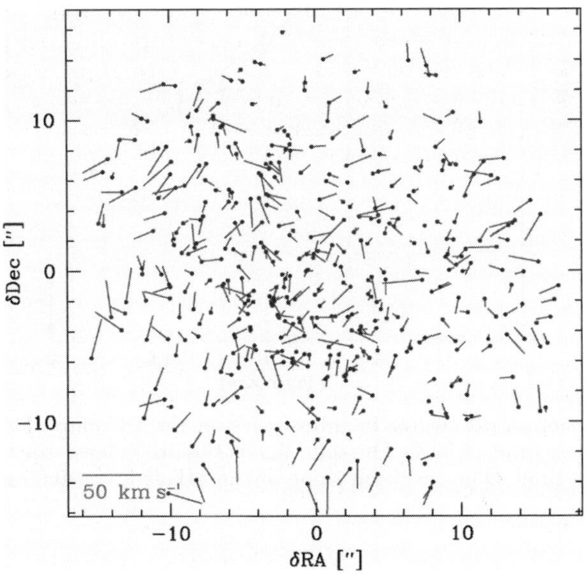

Figure 1. Map of measured proper motions of member stars on the PC chip. Each filled dot represents the position of a cluster member star. The lines point in the direction of the proper motion.

The color–magnitude diagram (CMD) of the proper-motion-selected cluster member stars shows a well-populated main sequence (MS) and pre-main sequence (PMS). From isochrone fitting we find an age of $\sim$ 1 Myr for the MS and PMS stars. The wide distribution of PMS stars suggests a possible age spread of up to 3 Myr.

3.2. *Mass function*

The mass function (MF) of NGC 3603 was derived only for member stars on the PC chip. We count stars in absolute V-magnitude bins spaced such that we cover mass bins with a logarithmic bin size of 0.2 mag to measure the MF of NGC 3603. Using the same method as applied by Grebel & Chu (2000), we find the absolute magnitude corresponding to the mass bins using isochrones of solar metallicity and an age of 1 Myr. For the MS stars we chose an isochrone from Lejeune & Schaerer (2001), while for the PMS stars we used Siess *et al.* (2000). The center of NGC 3603 is severely affected by crowding and by additional incompleteness introduced by the saturation of luminous stars. The incompleteness was quantified by magnitude- and position-dependent artificial-star experiments and corrected for.

Fitting the corrected number counts for MS and PMS stars detected on the PC chip within the mass range of 1–100 $M_\odot$ (Figure 2) results in a MF slope of $\Gamma = -0.95 \pm 0.21$. Our result agrees well with the earlier studies of Sung *et al.* (2004) and Stolte *et al.* (2006). Even though the low-mass end of our MF is unreliable due to low completeness (lower than 50%), it follows the slope of the more massve stars very well.

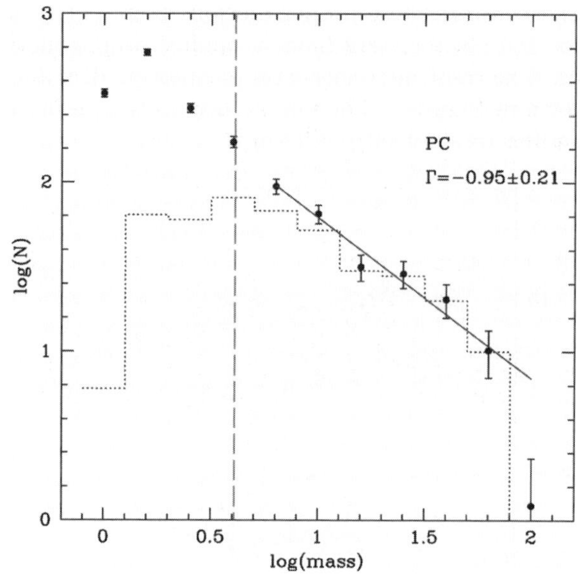

Figure 2. Mass function of all cluster member stars on the PC chip. The vertical dashed line indicates a completeness limit of 50%. The solid line is the linear least-squares fit to points above the 50% completeness limit. The resulting slope of the MF is $\Gamma = -0.95 \pm 0.21$.

3.3. *Dynamical status of NGC 3603*

As already observed in earlier studies (Nürnberger *et al.* 2002; Grebel 2004; Stolte *et al.* 2006; Harayama *et al.* 2008), NGC 3603 shows significant mass segregation. Mass segregation is also observed in the young star clusters ONC (Orion Nebula Cluster; Hillenbrand & Hartmann 1998), NGC 2244 and NGC 6530 (Chen *et al.* 2007). In Figure 3, we show the radial variation of the NGC 3603 MF as measured in different annuli around the cluster center. The MF slope increases as we move outwards, revealing the presence of more massive stars in the center compared to the outskirts of the cluster. However, the origin of this mass segregation is still under debate. Bonnell & Davies (1998) ran simulations of an ONC-like cluster to investigate the timescale of mass segregation. It turned out that the timescale needed for dynamical mass segregation is longer than the ONC's age, probably $\sim 3t_{\rm cross}$. Massive stars need to have originated within the innermost 10% of the stellar distribution if they are to form a Trapezium-like object within $3 - 5\,t_{\rm cross}$. Other work by Moeckel & Bonnell (2000) ran simulations with initial conditions of 'no initial segregation,' 'moderate initial segregation,' 'heavy initial segregation' and 'only massive stars segregated.' By comparing the simulation results with the observations for the ONC and NGC 6231, the simulations with 'only massive stars segregated' appeared to be consistent with the available observational data.

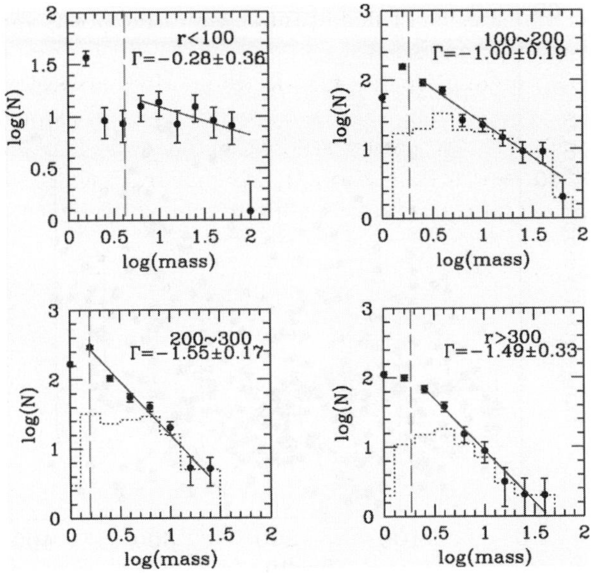

Figure 3. Radial variation of the MF of all member stars on the PC chip measured in different annuli. The dashed line is the 50% completeness limit. The solid lines are weighted linear least-squares fits to the data points above the 50% completeness limit.

To investigate the origin of mass segregation, we calculate the mean tangential velocity for the same annuli as in Figure 3. Figure 4 shows the distribution of tangential velocities of all member stars on the PC chip. There is a slight indication for increasing velocities beyond a radius of 280 pixels, but considering the size of the error bars this may not be significant. The velocity dispersion converted from proper motions is $\sigma = 3.44$ km s^{-1}. If we assume an isotropic distribution, the intrinsic velocity dispersion of NGC 3603 is $\sigma_{\rm int} = 4.87$ km s^{-1}, which is larger than that of the ONC (2.5 km s^{-1}; Bonnell & Davies 1998). Van den Bergh (1979) derived an even larger heliocentric radial velocity for NGC 3603 of about 6 km s^{-1}. Harayama *et al.* (2008) found a core radius of ~ 0.14 pc. Combining $\sigma_{\rm int}$ with this estimate, the cluster crossing timescale is 3×10^4 yr. In other words, the age of NGC 3603 is almost 30 $t_{\rm cross}$. According to the simulations of Bonnell & Davies (1998), this is long enough to produce significant mass segregation through dynamical evolution.

The relaxation timescale from Binney & Tremaine (1987) can differ for very massive stars to low-mass stars. For NGC 3603, which contains stellar masses from subsolar to about 100 M$_\odot$, the relaxation timescale will differ dramatically for different stellar masses. Stars of 1 M$_\odot$ will have a relaxation timescale on the order of 10^6 yr, but for 100 M$_\odot$ stars it will be on the order of 10^4 yr. Hence, massive stars would already have undergone dynamical evolution while less massive stars are still little affected. Allison *et al.* (2009) also found that the dense cores of clusters can mass segregate within $\sim 0.1 - 0.2$ Myr, which for NGC 3603 corresponds to about 10–20 crossing times. Moreover, these authors concluded that this time is sufficient to mass segregate the most massive stars and that a cluster can dynamically segregate stars above 2–4 M$_\odot$. Consequently, this mass-segregation time already amounts to ~ 10 crossing times of the NGC 3603 core, so that this cluster is actually dynamically old. Given these estimates, it is therefore reasonable to assume that massive stars already sank to the center of NGC

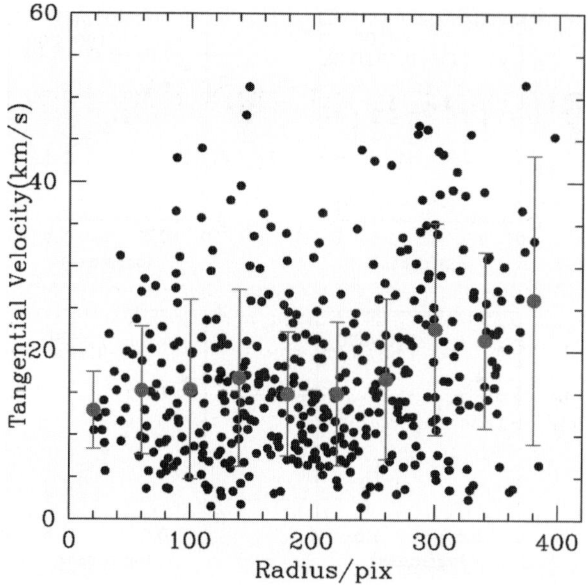

Figure 4. Tangential velocity distribution of all member stars on the PC chip. The small black dots indicate stars. The larger dots with error bars show the mean tangential velocity in each annulus. The radius of each annulus is 40 pixels.

3603 as a result of dynamical evolution. For low-mass stars, on the other hand, it is still difficult to distinguish primordial and evolutionary effects.

References

Allison, R. J., Goodwin, S. P., Parker, R. J., de Grijs, R., Portegies Zwart, S. F., & Kouwenhoven, M. B. N. 2009, *ApJ* (Letters), 700, 99
Binney, J. & Tremaine, S. 1987, *Galactic Dynamics*, Princeton: Princeton University Press
Bonnell, I. A. & Davies, M. B. 1998, *MNRAS*, 295, 691
Chen, L., de Grijs, R., & Zhao, J. L. 2007, *AJ*, 134, 1368
Dolphin, A. E. 2000, *PASP*, 112, 1383
Grebel, E. K. 2004, in: H. J. G. L. M. Lamers, L. J. Smith & A. Nota (eds.), *The Formation and Evolution of Massive Young Star Clusters*, ASP Conf. Ser. vol. 322, p. 101, San Francisco: Astron. Soc. Pac.
Grebel, E. K. & Chu, Y.-H. 2000, *AJ*, 119, 787
Harayama, Y., Eisenhauer, F., & Martins, F. 2008, *ApJ*, 675, 1319
Hillenbrand, L. A. & Hartmann, L. E. 1998, *ApJ*, 492, 540
Lejeune, T. & Schaerer, D. 2001, *A&A*, 366, 538
Moeckel, N. & Bonnell, I. A. 2009, *MNRAS*, 400, 657
Nürnberger, D. E. A. & Petr–Gotzens, M. G. 2002, *A&A*, 382, 537
Siess, L., Dufour, E., & Forestini, M. 2000, *A&A*, 358, 593
Sung, H. & Bessell, M. 2004, *ApJ*, 127, 1014
Stolte, A., Brandner, W., Brandl, B., Zinnecker, H., & Grebel, E. K. 2004, *ApJ*, 128, 765
Stolte, A., Brandner, W., Brandl, B., & Zinnecker, H. 2006, *ApJ*, 132, 253
van den Bergh, S. 1979, *A&A*, 63, 275

Star clusters: basic galactic building blocks
Proceedings IAU Symposium No. 266, 2009
R. de Grijs & J. R. D. Lépine, eds.

doi:10.1017/S1743921309990834

Physics and modes of star cluster formation

Matthew R. Bate

School of Physics, University of Exeter, Stocker Road, Exeter EX4 4QL, UK
email: mbate@astro.ex.ac.uk

Abstract. I review the progress made in understanding the physics and modes of star cluster formation through the use of direct self-gravitating hydrodynamical simulations, including those that have recently been performed incorporating radiative transfer and magnetic fields.

Keywords. gravitation, hydrodynamics, MHD, radiative transfer, stellar dynamics, methods: numerical, stars: formation, stars: low-mass, brown dwarfs, stars: luminosity function, mass function

---

1. Introduction

Understanding star formation is a difficult problem. On the theoretical side, there is a huge range of scales involved (roughly 10 orders of magnitude in spatial scale and 12 orders of magnitude in temporal scale) and a large number of physical processes (gravity, supersonic turbulent fluid dynamics, magnetic fields, radiative transfer, chemistry). On the observational side, many stages of star formation (particularly massive star formation) are obscured and the entire process of star cluster formation takes millions of years, meaning that we need to try to piece together a complete theory using snapshots of the process taken from different regions.

Over the past decade, it has become possible to begin simulating the formation of clusters of stars directly using self-gravitating hydrodynamical calculations. This has opened up the possibility of trying to understand the origin of the statistical properties of stellar systems by conducting 'numerical experiments' in which the initial conditions and/or physical processes included are varied and the effects of these changes on the outputs of the star formation process (i.e., the statistical properties of the stellar systems) are measured.

In this contribution, I review the progress that has been made in simulating star cluster formation directly and what we have learnt to date of how the star-formation process depends on various physical processes and the initial conditions in molecular clouds.

2. Hydrodynamical simulations

The starting point for all direct simulations of star formation is self-gravitating compressible fluid dynamics. This is the minimum amount of physics required even to begin modelling the collapse and fragmentation of a molecular cloud to form a stellar cluster. However, even in this limited case, modelling the formation of a stellar cluster is beyond current computational resources because—as the collapse to form the first star ensues—the time steps required to evolve the calculation decrease by $\sim 10-12$ orders of magnitude. Thus, the calculation would grind to a halt as the first star forms. To get past this point, an approximation must be made. The standard procedure now used in both particle-based (Bate *et al.* 1995) and grid-based numerical methods (Krumholz *et al.* 2004) is to replace collapsing protostars by sink particles. Here, the dense gas within a specified radius of the centre of the protostar, the accretion radius, is combined into a single point mass with the same total mass and momentum as the gas it replaces. Gas that subsequently falls within this radius is accreted by the sink particle. This method allows the very short time steps that would be required to evolve the gas deeper inside the protostar to be avoided.

Klessen *et al.* (1998) used this method to follow the formation of a cluster of protostars from a cloud of molecular gas. Their calculation began with a static cloud in a periodic box with Gaussian density perturbations and produced 55 protostars following a roughly lognormal mass function. Subsequent calculations showed that the peak of the mass function was located near the mean Jeans mass of the initial clouds (Klessen & Burkert 2001a,b). Klessen (2001) also investigated clouds with driven turbulence and found that the resulting stellar mass function was broader and flatter if the turbulence was driven on small scales than for large-scale driving. Offner *et al.* (2008) showed that there is not much difference between decaying turbulence and large-scale driving. Bonnell *et al.* (2003) performed a calculation of the collapse of an isolated 1000 $M_\odot$ molecular cloud containing decaying turbulence. This calculation pointed to the formation of large clusters occurring hierarchically, with the large stellar cluster containing more than 400 stars being formed through the merging of many subclusters.

A limitation of these initial models is that they only resolved the collapsing protostars to scales of several hundreds of AU (i.e., the sink particles had accretion radii of hundreds of AU). Thus, they did not capture the opacity limit for fragmentation that occurs when collapsing gas becomes optically thick to its own radiation, and so did not capture all of the expected fragmentation (leading to incomplete mass functions), or resolve brown dwarfs, multiple systems and discs.

Bate *et al.* (2002a,b, 2003) performed the first cluster-formation calculation that resolved the opacity limit for fragmentation, thus capturing all fragmentation and resolving even the lowest-mass brown dwarfs. The calculation also resolved binaries with separations as close as 1 AU and discs with radii down to $\approx$ 10 AU. The calculation demonstrated that star formation in clusters could be a higher chaotic and dynamical environment, with discs being truncated by dynamical encounters, fragmenting to form multiple systems, and stars and brown dwarfs being ejected from unstable multiple systems and escaping the cluster's gravitational potential. The calculation produced a modest 50 stars and brown dwarfs, enough for a crude comparison with observations. Within the uncertainties, the properties of the stars and brown dwarfs, such as the initial mass function (IMF), stellar multiplicity and the fraction of close binaries, were found to be in reasonable agreement with observations. As with the earlier, more poorly resolved, simulations, the stellar masses were found to originate through a process of competitive accretion, with the characteristic stellar mass being roughly the Jeans mass. All objects began with low masses and accreted from the cloud, typically to the mean thermal Jeans mass, but with a few stars in the centres of subclusters reaching higher masses and forming the high-mass end of the IMF. For those objects with final masses much lower than the typical Jeans mass (i.e., brown dwarfs), it was found that dynamical interactions and ejections from multiple systems were the key to their low masses (Bate *et al.* 2002a). They began with low masses (as did those objects that ended up with stellar masses), but their accretion was terminated when they were involved in dynamical interactions in small groups, which increased their velocities (typically to a few km s^{-1}) and ejected them from the dense cores in which they began forming. Thus, they were unable to accrete to the typical Jeans mass and ended up with substellar masses. Approximately three quarters of the brown dwarfs were found to originate from the fragmentation of massive circumstellar discs, while the remainder formed in dense filaments, fell into existing multiple systems and were ejected before they had accreted much mass.

This first calculation of star cluster formation that resolved the opacity limit for fragmentation was followed by three similar calculations that investigated the dependence of stellar properties on the initial conditions in the molecular clouds and variations in the opacity limit for fragmentation. Bate & Bonnell (2005) performed an identical calculation to that of Bate *et al.* (2003), but for a cloud with a smaller radius and nine times higher density. Thus, the mean thermal Jeans mass in the cloud was one third of that in the original calculation. The calculation produced a median stellar mass a factor of three lower than the original calculation, exactly matching the change in the mean Jeans mass. This confirmed the results of the earlier, more poorly resolved calculations, which indicated that the typical stellar mass was similar to the mean thermal Jeans mass. More recently, Jappsen *et al.* (2005) and Bonnell *et al.* (2006), using calculations with nonisothermal equations of state at very low molecular densities, showed that the transition from atomic-line to dust cooling could set an appropriate Jeans mass, which in turn produced the characteristic mass of the IMF.

Bate (2005) performed a third opacity-limited calculation, identical to that of Bate *et al.* (2003), except that the opacity limit for fragmentation was moved to a lower density by a factor of nine (i.e., the transition from an isothermal collapse to the gas becoming optically thick to its own radiation was assumed to occur earlier in the collapse). This increased the minimum mass by a factor of three to ≈ 9 Jupiter masses. Such a change may occur, for example, in lower-metallicity gas which cools less effectively. Bate found that apart from increasing the minimum mass of a brown dwarf, this change to the equation of state produced no significant change in the rest of the IMF or the other stellar properties.

Finally, Bate (2009c) performed a fourth opacity-limited calculation, identical to that of Bate *et al.* (2003), except that the power spectrum of the initial turbulent velocity field was changed from $P(k) \propto k^{-4}$ (chosen to match the Larson (1981) scaling relations for velocity dispersion with length scale observed in molecular clouds) to $P(k) \propto k^{-6}$, which has more power on larger scales. The structures produced in the cloud during the evolution were very different to those found in the original calculation with large shocks and little small-scale structure. However, despite this difference, the stellar properties obtained were indistinguishable from those of the original calculation. In particular, the IMFs were almost identical.

Together, these four opacity-limited calculations show that the IMFs produced by hydrodynamical simulations using barotropic equations of state depend primarily on the mean thermal Jeans mass of the progenitor clouds and not on the exact value of the opacity limit for fragmentation or on the nature of the turbulent motions in the molecular gas. This is because, as long as there is sufficient structure in the gas and the equation of state allows fragmentation to produce many objects that interact dynamically, the processes of competitive accretion (Bonnell *et al.* 1997) and dynamical interactions and ejections (Bate *et al.* 2003) do not depend significantly on the initial structure and turbulent motions in the molecular cloud. Other properties, such as stellar multiplicity, were found not to vary significantly between any of the calculations for the same reason.

The calculations discussed above either did not resolve the opacity limit for fragmentation, and thus did not resolve low-mass objects or many multiple systems, or produced only a few dozen stars and brown dwarfs. Thus, making detailed comparisons with the observed properties of stars and brown dwarfs was not possible.

Recently, Bate (2009a) performed a hydrodynamical simulation of star cluster formation that resolved the opacity limit for fragmentation *and* produced well over 1000 stars and brown dwarfs. The calculation was identical to that of Bate *et al.* (2003), except that the cloud was an order of magnitude more massive (500 $M_\odot$). With so many objects, the statistical properties of low-mass stars and brown dwarfs are well determined and detailed comparison with observed stellar properties is possible. Bate found that many stellar properties were in good agreement with observations. For example, stellar multiplicity was found to be a strongly increasing function of primary mass with the multiplicity of very-low-mass objects ($0.03 - 0.10$ $M_\odot$), M dwarfs, and solar-type stars increasing from ≈ 20 to 60%, as observed. The trends for low-mass binaries to have smaller separations and equal-mass components were also reproduced, and even the distribution of relative orientations of the orbital planes of triple systems were found to be in agreement with the observed distribution. The two main areas of disagreement with observed stellar properties were that the calculation produced a much higher ratio of brown dwarfs to stars than observed and there was a deficit of unequal-mass solar-type binaries.

However, the fact that a calculation only including gravity and fluid dynamics (without the more complicated physics of radiative transfer and magnetic fields) can reproduce many of the observed statistical properties of stellar systems implies that the origin of these properties is primarily due to dissipative gravitational dynamics and not significantly altered by additional processes. In many ways, this is surprising because, for example, magnetic fields are thought to be highly effective at transporting angular momentum and radiative feedback is likely to inhibit fragmentation near existing protostars.

3. Radiation hydrodynamical simulations

Thus, the question moves on to what the role of additional physical processes is in the star-formation process. Bate (2009b) recently repeated the two cluster formation calculations of Bate *et al.* (2003) and Bate & Bonnell (2005), this time including a realistic equation of state and radiative transfer in the flux-limited diffusion approximation. The calculations modelled gas to within 0.5 AU of each protostar and the energy released down to such scales, but did not include the radiative feedback coming from the stars themselves or discs within 0.5 AU. Despite this, the inclusion of radiative feedback from the forming protostars back into the cloud had a huge effect on the fragmentation. Massive circumstellar discs, in particular, were found to be much hotter and more resistant to fragmentation than in the barotropic calculations. The result was that in both calculations, the numbers of objects formed were reduced by a factor of ≈ 4 from the original calculations. This, in turn, led to fewer dynamical interactions between objects and fewer ejections. Since the ejected objects tend to end up as low-mass stars and brown dwarfs, the ratio of brown dwarfs to stars was reduced by the inclusion of radiative feedback, bringing the IMFs produced by the calculations into good agreement with observations.

In addition, as mentioned in the previous section, using a barotropic equation of state, the stellar IMF was found to scale with the mean Jeans mass of the clouds. However, with radiative transfer, the IMFs produced from the two clouds were found to be indistinguishable despite their differing mean Jeans masses. Thus, Bate (2009b) proposed that radiative feedback from protostars self-regulates the star-formation process and erases, or severely weakens, the dependence of the IMF on the mean thermal Jeans mass of the progenitor cloud. This may help to explain the observed invariance of the IMF.

Following these calculations, Offner *et al.* (2009) also performed radiation hydrodynamical calculations of the collapse of molecular clouds to form low-mass stars. Their calculations were performed using adaptive-mesh refinement while Bate used smoothed-particle hydrodynamics. Unlike Bate, Offner *et al.* included radiative feedback from the stellar objects themselves and their accretion luminosity, but they used much larger accretion radii of ≈ 130 AU. Despite their differences, both studies found that radiative feedback dramatically reduced the number of objects formed relative to calculations performed using a barotropic equation of state and, therefore, that even in the case of low-mass star formation, it is crucial to include the effects of radiative feedback.

4. Magnetohydrodynamical simulations

Magnetic fields have long been recognised as a potentially important physical process in star formation. However, while their role in some aspects of star formation is clearly crucial (e.g., protostellar jets), their role in star cluster formation is less clear. Recent ideal magnetohydrodynamical (MHD) calculations have confirmed that magnetic fields can play a crucial role in protostellar-disc formation (Price & Bate 2007; Hennebelle & Fromang 2008) and the fragmentation of isolated molecular-cloud cores to form binary and multiple systems (Price & Bate 2007; Hennebelle & Teyssier 2008). However, their role in cluster formation has only recently begun to be investigated numerically.

Price & Bate (2008) recently repeated the cluster-formation calculation of Bate *et al.* (2003), this time including magnetic fields of varying strengths using the ideal MHD approximation. While even at low strengths, the magnetic field was found to significantly influence the star-formation process, for magnetic fields where the ratio of gas to magnetic pressure was less than unity (i.e., plasma $\beta < 1$) the results were substantially different to the hydrodynamic case. Anisotropic turbulent motions and column-density striations aligned with the magnetic-field lines were found, both of which have recently been observed in the Taurus molecular cloud (Goldsmith *et al.* 2008). In addition, large-scale magnetically supported voids were produced. The additional large-scale support provided by the magnetic field strongly suppressed collapse in the clouds, leading to up to a 75% reduction in the amount of mass converted into stars over the course of the calculations. This led to a more quiescent mode of star formation and Price & Bate found that the relative formation efficiency of brown dwarfs was lower in the strongly magnetized runs because of a reduction in the importance of protostellar ejections.

5. Radiation magnetohydrodynamical simulations

Most recently, Price & Bate (2009) combined their treatments of magnetic fields and radiation hydrodynamics to perform the first radiation magnetohydrodynamical calculations of star cluster formation. Again, the calculations were of small 50 $M_\odot$ molecular clouds as modelled by Bate *et al.* (2003), Bate (2009b) and Price & Bate (2008).

The effects of radiative feedback and magnetic fields, found separately in the earlier studies, were combined in these calculations. The main effect of radiative feedback was to inhibit fragmentation on small scales, while the main effect of magnetic fields was to provide support to the low-density gas on large scales, decreasing the star-formation rate. With strong magnetic fields and radiative feedback, the net result was an inefficient star-formation process with a star-formation rate of $\approx$ 10% per free-fall time. This is much less than the rates found without magnetic fields (typically $\approx$ 50% per free-fall time) and approaches the observed rate of $\sim 3-6$% (Evans *et al.* 2009).

6. The future

In summary, calculations of star cluster formation that only take into account gravity and compressible fluid dynamics can reproduce many of the observed properties of stellar systems, including many binary properties. They also produce stellar mass distributions in qualitative agreement with the observed IMF. Such properties do not appear to depend sensitively on the properties of the molecular-cloud turbulence as long as the turbulence is decaying or driven on large scales (as seems to be observed; Brunt *et al.* 2009). This is due to the nature of competitive accretion and dynamical interactions between protostars, which determine the spectrum of stellar masses in such calculations. These are local processes that have little memory of the large-scale initial conditions.

However, it is clear that both radiative transfer and magnetic fields need to be incorporated into future calculations of star cluster formation. Radiative feedback, even from low-mass protostars, is crucial to obtain quantitative agreement with the observed IMF and may help to explain why star formation does not depend sensitively on other initial conditions in molecular clouds (such as their density and temperature). Dynamically important magnetic fields seem to be required to explain the low rate of star formation and many of the structures observed in molecular clouds.

The challenges for the future are clear: to incorporate the effects of radiative transfer and magnetic fields into larger calculations that are capable of making precise predictions of stellar properties. Ideally, the methods for including radiative transfer and magnetic fields should also be improved. For radiative transfer, this means going beyond the flux-limited diffusion approximation, while for magnetic fields the nonideal effects of ambipolar diffusion, Hall conduction and resistivity should be included.

References

Bate, M. R. 2005, *MNRAS*, 363, 363
Bate, M. R. 2009a, *MNRAS*, 392, 590
Bate, M. R. 2009b, *MNRAS*, 392, 1363
Bate, M. R. 2009c, *MNRAS*, 397, 232
Bate, M. R. & Bonnell, I. A. 2005, *MNRAS*, 356, 1201
Bate, M. R., Bonnell, I. A., & Bromm, V. 2002a, *MNRAS* (Letters), 332, L65
Bate, M. R., Bonnell, I. A., & Bromm, V. 2002b, *MNRAS*, 336, 705
Bate, M. R., Bonnell, I. A., & Bromm, V. 2003, *MNRAS*, 339, 577
Bonnell, I. A., Bate, M. R., Clarke, C. J., & Pringle, J. E. 1997, *MNRAS*, 285, 201
Bonnell, I. A., Bate, M. R., & Vine, S. G. 2003, *MNRAS*, 343, 413
Bonnell, I. A., Clarke, C. J., & Bate, M. R. 2006, *MNRAS*, 368, 1296
Brunt, C. M., Heyer, M. H., & Mac Low, M.-M. 2009, *MNRAS*, 504, 883
Evans, N. J., *et al.* 2009, *ApJS*, 181, 321
Goldsmith, P. F., Heyer, M., Narayanan, G., Snell, R., Li, D., & Brunt, C. 2008, *ApJ*, 680, 428
Hennebelle, P. & Fromang, S. 2008, *A&A*, 477, 9
Hennebelle, P. & Teyssier, R. 2008, *A&A*, 477, 25
Jappsen, A.-K., Klessen, R. S., Larson, R. B., Li, Y., & Mac Low, M.-M. 2005, *A&A*, 435, 611
Klessen, R. S. 2001, *ApJ*, 550, 77
Klessen, R. S. & Burkert, A. 2000, *ApJS*, 128, 287
Klessen, R. S. & Burkert, A. 2001b, *ApJ*, 549, 386
Klessen, R. S., Burkert, A., & Bate, M. R. 1998, *ApJ* (Letters), 501, L205
Krumholz, M. R., McKee, C. F., & Klein, R.I. 2004, *ApJ*, 611, 399
Larson, R. B. 1981, *MNRAS*, 194, 809
Offner, S. S. R., Klein, R. I., & McKee, C. F. 2008, *ApJ*, 686, 1174
Offner, S. S. R., Klein, R. I., McKee, C. F., & Krumholz, M. R. 2009, *ApJ*, 703, 131
Price, D. J. & Bate, M. R. 2007, *MNRAS*, 377, 77
Price, D. J. & Bate, M. R. 2008, *MNRAS*, 385, 1820
Price, D. J. & Bate, M. R. 2009, *MNRAS*, 398, 33

Star clusters: basic galactic building blocks
Proceedings IAU Symposium No. 266, 2009
R. de Grijs & J. R. D. Lépine, eds.

doi:10.1017/S1743921309990846

The *VLT*–FLAMES Tarantula Survey

C. J. Evans,[1] N. Bastian, Y. Beletsky, I. Brott, M. Cantiello, J. S. Clark, P. A. Crowther, A. de Koter, S. E. de Mink, P. L. Dufton, P. Dunstall, M. Gieles, G. Gräfener, V. Hénault–Brunet, A. Herrero, I. D. Howarth, N. Langer, D. J. Lennon, J. Maíz Apellániz, N. Markova, F. Najarro, J. Puls, H. Sana, S. Simón–Díaz, S. J. Smartt, V. E. Stroud, W. D. Taylor, C. Trundle, J. Th. van Loon, J. S. Vink and N. R. Walborn

[1]UK ATC, Royal Observatory Edinburgh, Blackford Hill, Edinburgh, EH9 3HJ, UK
email: chris.evans@stfc.ac.uk

Abstract. The Tarantula Survey is an ambitious ESO Large Programme that has obtained multi-epoch spectroscopy of over 1000 massive stars in the 30 Doradus region in the Large Magellanic Cloud. Here, we introduce the scientific motivations of the survey and give an overview of the observational sample. Ultimately, quantitative analysis of every star, paying particular attention to the effects of rotational mixing and binarity, will be used to address fundamental questions in both stellar and cluster evolution.

Keywords. stars: early-type, stars: fundamental parameters, binaries: spectroscopic, open clusters and associations: individual (30 Doradus)

---

1. Introduction

The Tarantula Nebula (30 Doradus, NGC 2070) in the Large Magellanic Cloud (LMC) is the brightest and most massive H II region in the Local Group. It is a beautiful and very intricate region, far removed from a 'simple single stellar population.' Indeed, Walborn & Blades (1997) identified at least five distinct populations (cf. Walborn 2009):

(*a*) The central 'Carina Phase' concentration, rich in early O-type stars and including the dense cluster R136;

(*b*) A younger, likely triggered, 'Orion Phase' to the north and west of R136;

(*c*) A 'Sco OB1 Phase' of early-type supergiants throughout the central field;

(*d*) An older 'h and χ Persei Phase' in Hodge 301, containing cooler, more evolved supergiants, to the northwest of the centre; and

(*e*) A separate 'Sco OB1 Phase' surrounding the luminous blue variable R143.

With its rich stellar populations, 30 Dor is the ideal laboratory in which to investigate a number of important outstanding questions regarding the physics, evolution, binary fraction and chemical enrichment of the most massive stars. Building on the successes of the *VLT*–FLAMES Survey of Massive Stars (Evans *et al.* 2005), here we introduce a new multi-epoch spectral survey of over 1000 massive stars in the 30 Dor region.

In the broader context, 30 Dor is at the northern end of a large column of molecular gas which extends south for over 2000 pc (Cohen *et al.* 1988; Fukui *et al.* 2008). N-body models examining the recent edge-on motion of the LMC through the halo of the Milky Way suggest significant star formation in the eastern part of the LMC, as manifested by 30 Dor, due to ram pressure (Mastropietro *et al.* 2009). With the reservoir of gas to the south, the region seems destined to become an even more spectacular star-formation complex over the next few million years.

2. Multiplicity in massive stars

The effects of binarity/multiplicity on the formation and subsequent evolution of high-mass stars is a vibrant area of research. Indeed, one of the key ingredients missing from current theories of both star formation and cluster evolution is a robust binary fraction of massive stars and the distribution of the mass ratios in these systems. Some motivation in this direction was provided by Zinnecker & Yorke (2007): "The future of spectroscopic massive binary research lies in the near-infrared and in multi-epoch radial velocity surveys of embedded massive stars." These words were primarily concerned with the earliest stages of star formation, but they coincide with growing interest in multi-epoch spectroscopic studies in open clusters, aimed at identification and characterisation of their binary populations (Table 1). The most pertinent of these is the study of 50 early-type stars in 30 Dor by Bosch *et al.* (2009). From *Gemini* spectroscopy at seven epochs they found a binary fraction of $\geqslant$50%, noting that the data were not inconsistent with it being 100%. Recent multi-epoch adaptive-optics-corrected SINFONI observations found (tentative) evidence for a short-period companion in only one of the six central Wolf–Rayet (WR) stars at the core of R136 (Schnurr *et al.* 2009), but it is clear that there is a very rich binary population in 30 Dor.

Table 1. Selected multi-epoch spectroscopic surveys in open clusters.

Cluster	Binary fraction	Reference
IC 1805	$\geqslant$0.20	De Becker *et al.* (2006)
NGC 6231	$\geqslant$0.63	Sana *et al.* (2008)
NGC 6611	$\geqslant$0.44	Sana *et al.* (2009)
NGC 2244	$\geqslant$0.17	Mahy *et al.* (2009)
30 Dor	$\geqslant$0.50	Bosch *et al.* (2009)

One of the serendipitous aspects of the FLAMES Survey of Massive Stars was the large number of spectroscopic binaries discovered (Table 2). The time sampling of the service-mode observations did a reasonable (but not thorough) job of binary detection, with lower limits to the binary fraction of ~30% in three of the target clusters. Adopting the same methods as Sana *et al.* (2009), we have calculated the detection probabilities for short-, intermediate- and long-period binaries for each cluster field. The aggregated detection probabilities (for systems with periods of two days to ten years) are given in the final column of Table 2. The similarity in detection probabilities suggests that the lower fraction found in NGC 330 is genuinely different to the others. While it is unfair to compare the NGC 330 observations with the rich, younger cluster fields of NGC 346 and N11, NGC 2004 is its LMC cousin. This difference in the binary fraction is intriguing and the subject of ongoing work.

Table 2. Spectroscopic binaries from Evans *et al.* (2006).

Cluster	Galaxy	# O+Early B	# Binary	Binary fraction	Detection prob. [2d–10yr]
NGC 346	SMC	103	27	$\geqslant$ 26%	0.66
NGC 330	SMC	104	4	$\geqslant$ 4%	0.71
NGC 2004	LMC	105	24	$\geqslant$ 23%	0.64
N11	LMC	120	43	$\geqslant$ 36%	0.64

The relationship of the binary fraction with density and the spatial extent of a cluster is still unclear (e.g., Mahy *et al.* 2009), while the binary fraction in OB associations is often similar to clusters, but with fewer short-period systems (Zinnecker & Yorke 2007).

3. The Tarantula Survey

The new survey comprises 160 hr of *VLT*–FLAMES spectroscopy in the 30 Dor region (PI Evans). Most of the observations (142 hr) have now been completed, with the remainder scheduled for the coming semester.

One of the prime drivers for this survey was the issue of binarity, shaping the multi-epoch observational strategy. It is clear that identification of binaries, and the mass ratios in those systems, is an important empirical result for N-body models of star and cluster formation/evolution. Moreover, in many clusters, e.g., NGC 6231 (Sana *et al.* 2008), the majority of O-type stars are members of a binary system. Thus, to gain a true understanding of the upper Hertzsprung–Russell diagram (HRD), the effects of binarity need to be fully included in theoretical models of stellar evolution (e.g., de Mink *et al.* 2009). Although we have focussed on this aspect for this Symposium, the genesis of the survey arose from a much broader range of other scientific motivations, including:

- The role of stellar rotation in the chemical enrichment and evolution of massive stars. Hunter *et al.* (2008) revealed new challenges for theory in B-type stars; we seek to investigate these effects in the more dominant, massive O-type stars.
- Determination of the rotational velocity distribution in 30 Dor. Are there sufficient high-mass, rapidly rotating stars to provide a channel for long-duration γ-ray bursts (cf. Yoon *et al.* 2006)?
- Armed with precise radial velocities and identification of binaries, do we see kinematic evidence of mass segregation and/or infant mortality in and around R136?
- A more holistic objective of a near-complete census of the closest 'proto-starburst,' with applications in the context of population synthesis methods and interpretation of spectra of unresolved massive stars clusters at Mpc distances.

4. GIRAFFE observations

The primary dataset comprises spectroscopy of 1000 stars using the GIRAFFE spectrograph, which is fed by 132 MEDUSA fibres available for science (or sky) observations across a 25′ field (Pasquini *et al.* 2002). Targets were selected from unpublished imaging with the Wide-Field Imager (WFI) on the *ESO/MPG 2.2m* telescope, and from Brian Skiff's reworking of the Selman *et al.* (1999) photometric catalogue in the central 90″. To obtain a representative sample of the upper part of the HRD, including evolved luminous stars, no colour cut was applied to potential targets but a faint cutoff ($V < 17$ mag) was enforced to ensure sufficient signal-to-noise ratio for each star.

Nine MEDUSA configurations were observed, each of which was observed at three wavelength settings (see Table 3). This yields full coverage of the classical blue–optical region used for spectroscopic classification and analysis, combined with higher resolution spectroscopy of the Hα region to enable determination of the stellar-wind intensity. From inspection of initial reductions, the minimum signal-to-noise ratio in the stacked spectra for the faintest stars is ~50, i.e., the spectra are suitable for quantitative analysis as well as radial velocity monitoring.

Table 3. Summary of FLAMES–GIRAFFE observations.

GIRAFFE setting	λ coverage (Å)	R	Exposures
LR02	3980–4535	6,500	6×(2×1815s)
LR03	4505–5050	7,500	3×(2×1815s)
HR15N	6470–6790	17,000	2×(2×2265s)

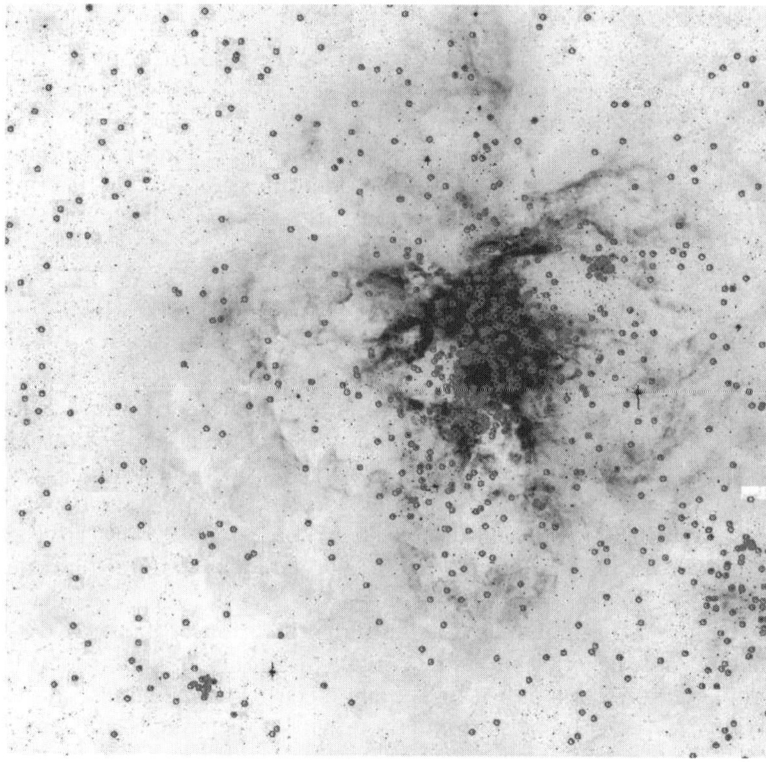

Figure 1. $14' \times 14'$ V-band WFI image showing the FLAMES–GIRAFFE targets in and around 30 Dor (north to the top, east to the left).

The distribution of the majority of the MEDUSA targets is shown in Figure 1. The survey samples the full extent of 30 Dor and outwards into the 'field' population and other nearby OB associations to fully exploit the FLAMES field of view and spare fibres, thus bolstering the observational sample.

4.1. *Preliminary classification*

Work is now progressing in earnest with the final science reductions of the GIRAFFE spectra. To characterise the spectral content of the survey, we extracted the reduced spectra from one pair of LR02 observations for each MEDUSA configuration. In advance of the full reductions, it was not possible to classify ~150 stars from just one observation (although most are likely B-type stars) but from visual inspection of the spectra the sample contains:

- In excess of 300 O-type stars and ~20 WR/'slash' stars. This is a hugely significant improvement in terms of sampling the upper HRD, e.g., compared to the analysis of 28 O-type stars in the LMC by Mokiem *et al.* (2007). Each star will be studied for binary companions, and then analysed to obtain physical and stellar-wind parameters, including the first large-scale study of nitrogen enrichment in O-type stars.
- Over 400 B-type spectra, which will be used to establish the baseline chemical abundances in 30 Dor and, with such a large sample in one field, will be used to revisit the role of rotationally induced mixing on surface nitrogen enrichment (cf. Hunter *et al.* 2008).
- ~150 cooler stars with spectral types of A, F and later. Some will be foreground objects to be discarded, but the majority will be evolved, luminous stars which will

be used to investigate the short lifetimes of these evolutionary phases via population synthesis models.

4.2. *Binary detection probabilities*

Using the methods from Sana *et al.* (2009), we have calculated the detection probabilities for binaries from the actual time sampling of the nine observed MEDUSA configurations. In these calculations, we assume a $\Delta v_{\rm r}$ threshold of 20 km s^{-1}, requiring a radial velocity precision of $\sim$5 km s^{-1} (which should be achievable at the resolving power of the new spectroscopy for all but the fastest-rotating stars). The detection probabilities, as a function of orbital period, for the first MEDUSA configuration are shown by the black line in Figure 2. We are relatively complete up to periods of a few 10s of days, with a steep decline beyond 100 days. The inclusion of one additional epoch in the coming observing season significantly helps with the detection of both intermediate- and long-period binaries (red/grey line). By quantifying our detection probabilities using such simulations, we will be able to put firm limits on the observed binary fraction.

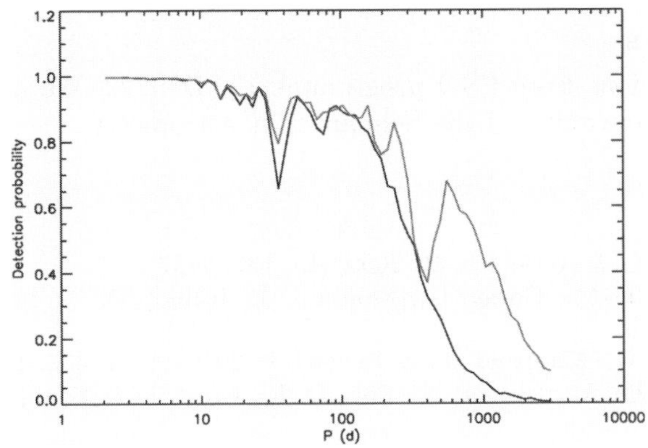

Figure 2. Detection probability of binary companions for one of the MEDUSA configurations. The black line shows the detection results for the five epochs already executed; the red (grey) line illustrates the increased probability obtained from a sixth epoch, scheduled for observation in the coming semester, i.e., separated by approximately one year from the other epochs.

5. Supplementary data

5.1. *ARGUS and UVES observations of R136*

R136 is too dense for effective use of the MEDUSA fibres, so a 15″ exclusion radius around the core was employed in the fibre allocations. To investigate the dynamics and binarity of stars in and around R136, as part of the Large Programme we have observed five pointings with the ARGUS integral field unit (IFU) which delivers a 12″ × 7″ field of view. Each pointing has been observed with the LR02 GIRAFFE setting (which delivers a resolving power of $\sim$10,000 in IFU mode) at five epochs.

In parallel to the ARGUS observations, we used the fibre feed to the red arm of UVES to observe 25 stars that were not included in the MEDUSA configurations. The λ5200 standard setup was used, delivering spectral coverage of $\sim$ λλ4200–6200 at R = 47,000.

5.2. *VLT-SINFONI K-band spectroscopy*

The majority of the known WR and extreme O-type emission-line stars in 30 Dor are in the central regions. Near-infrared IFU observations with SINFONI (12 hr; PI Gräfener) will be used to obtain K-band spectroscopy of the central arcminute around R136. The stellar-wind lines in the K band, principally from Brackett γ and He II, are more sensitive than the optical lines at low mass-loss rates, enabling a more precise determination of the physical parameters of the most extreme stars.

5.3. *Faulkes photometric follow-up*

In the longer term, the spectroscopy in the central ~10 arcminutes will be supplemented with multiband photometric monitoring with the *Faulkes Telescope South*, as part of their schools education programme. *Faulkes* has a $4'.7 \times 4'.7$ field of view, so the main body of 30 Dor will be mapped with several pointings, delivering multi-epoch photometry that will, for example, assist with the analysis of identified binary systems.

6. Summary

We have an exceptional and unique data resource available to us to investigate the massive-star population in 30 Dor, now the hard work begins!

Acknowledgements

Based on observations from ESO programme 182.D–0222. We are indebted to Brian Skiff for his careful reworking of the Selman *et al.* astrometry.

References

Bosch, G., Terlevich, E., & Terlevich, R. 2009, *AJ*, 137, 3437
Cohen, R. S., Dame, T. M., Garay, G., Montani, J., Rubio, M., & Thaddeus, P. 1988, *ApJ* (Letters), 331, L95
De Becker, M., Rauw, G., Manfroid, J., & Eenens, P. 2006, *A&A*, 456, 1121
de Mink, S. E., Cantiello, M., Langer, N., Pols, O. R., Brott, I., & Yoon, S.-C. 2009, *A&A*, 497, 243
Evans, C. J., *et al.* 2005, *A&A*, 437, 467
Evans, C. J., Lennon, D. J., Smartt, S. J., & Trundle, C. 2006, *A&A*, 456, 623
Fukui, Y., Kawamura, A., Minamidani, T., Mizuno, Y., Kanai, Y., Mizuno, N., Onishi, T., Yonekura, Y., Mizuno, A., Ogawa, H., & Rubio, M. 2008, *ApJS*, 178, 56
García, B. & Mermilliod, J. C. 2001, *A&A*, 368, 122
Hunter, I., Brott, I., Lennon, D. J., Langer, N., Dufton, P. L., Trundle, C., Smartt, S. J., de Koter, A., Evans, C. J., & Ryans, R. S. I. 2008, *ApJ* (Letters), 676, L29
Mahy, L., Nazé, Y., Rauw, G., Gosset, E., De Becker, M., Sana, H., & Eenens, P. 2009, *A&A*, 502, 937
Mastropietro, C., Burkert, A., & Moore, B. 2009, *MNRAS*, 399, 2004
Mokiem, M. R., *et al.* 2007, *A&A*, 465, 1003
Pasquini, L., *et al.* 2002, *Msngr*, 110, 1
Sana, H., Gosset, E., Nazé, Y., Rauw, G., & Linder, N. 2008, *MNRAS*, 386, 447
Sana, H., Gosset, E., & Evans, C. J. 2009, *MNRAS*, in press (arXiv:0909.0486)
Schnurr, O., Chené, A.-N., Casoli, J., Moffat, A. F. J., & St-Louis, N. 2009, *MNRAS*, 397, 2049
Selman, F., Melnick, J., Bosch, G., & Terlevich, R. 1999, *A&AS*, 341, 98
Walborn, N. R. & Blades, J. C. 1997, *ApJS*, 112, 457
Walborn, N. R. 2009, in: Leitherer *et al.* (eds.), *Hot & Cool: Bridging Gaps in Massive Star Evolution*, ASP Conf. Ser., in press
Yoon, S.-C., Langer, N., & Norman, C. 2006, *A&A*, 460, 199
Zinnecker, H. & Yorke, H. W. 2007, *ARA&A*, 45, 481

Star clusters: basic galactic building blocks
Proceedings IAU Symposium No. 266, 2009
R. de Grijs & J. R. D. Lépine, eds.

doi:10.1017/S1743921309990858

Stellar feedback and triggered star formation

Jan Palouš,[1] Richard Wünsch,[1] James E. Dale,[1] Vojtěch Sidorin[1] and Anthony Whitworth[2]

[1]Astronomical Institute, Academy of Sciences of the Czech Republic,
Boční 1401, CZ-14061, Prague 4, Czech Republic
email: palous@ig.cas.cz

[2]School of Physics and Astronomy, Cardiff University, Queens Buildings,
The Parade, Cardiff CF24 3AA, UK
email: Anthony.Whitworth@astro.cf.ac.uk

Abstract. Young, massive stars influence their ambient medium through winds and radiation. The outcome of this feedback depends on the number of massive stars in a star cluster and on the density of the ambient medium. This contribution is based on a comparison of observations to the results of numerical simulations. We discuss the gravitational fragmentation of feedback-driven shells expanding from young stellar clusters. The thin-shell approximation is compared to 3D hydrodynamical simulations with smoothed-particle hydrodynamics and adaptive-mesh refinement codes. We explore the influence of external pressure and propose a thick-shell dispersion relation, where the pressure of the external medium is included. The mass spectrum of the shell fragments is constructed and we speculate about the origin of the deficit of low-mass objects.

Keywords. stars: formation, ISM: structure, galaxies: star clusters

1. Introduction

In the Milky Way (McClure–Griffiths *et al.* 2002; Ehlerová & Palouš 2005) and in nearby galaxies within 15 Mpc (Walter *et al.* 2008), HI observations reveal a plethora of structures, some of which are shells with sizes of 10–1000 pc and expansion velocities of 5–40 km s^{-1}. Using the *Spitzer* GLIMPSE survey, Churchwell *et al.* (2006, 2007) found many smaller (0.1–10 pc) and younger structures (< 1 Myr), frequently surrounding young stellar clusters. In some cases (e.g., RCW 79 Deharveng *et al.* 2007), the shells are fragmented and some of the fragments harbor young stellar clusters visible in the near-infrared (near-IR).

In a search for HI counterparts of the IR structures revealed by the GLIMPSE survey, Sidorin (2008) identified several regions with shapes showing striking similarities. The high-angular-resolution HI data collected with the *Arecibo* telescope as part of the I-GALFA project have shown an even better correspondence (see Figure 1). The identification of IR structures in HI observations provides radial velocities, which may be used to estimate distances, sizes and also thicknesses of the shell-like structures. For N107, we derive a distance of 4.7 kpc, a size of 15.5 pc and a thickness of 3.1 pc. Several young stellar objects are connected to this bubble.

We conclude that there is a growing collection of expanding shells seen in HI, and at optical and IR wavelengths, with evidence of fragmentation and triggered formation of new stellar clusters along their edges.

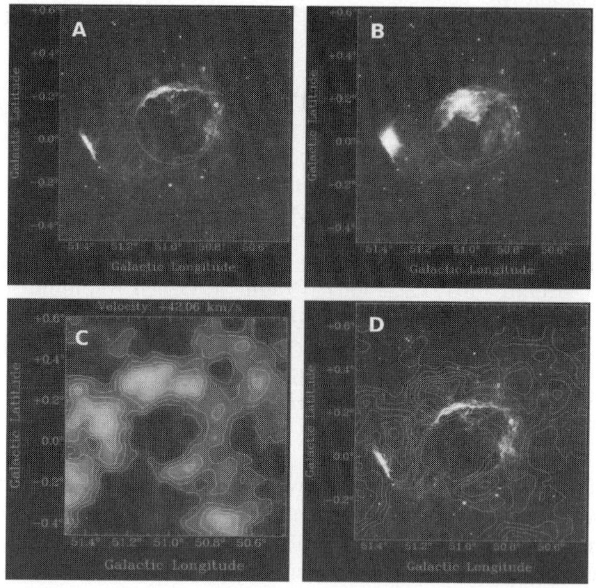

Figure 1. Bubble N107 (green circle) of Churchwell *et al.* (2006) as seen in HI. *(a) Spitzer*/IRAC, 8μm. (b) *Spitzer*/MIPS, 24μm. (c) *Arecibo*, I-GALFA, HI line (21 cm), radial velocity 42 km s^{-1}: similar to the radial velocities of the HII regions associated with the bubble. (d) Background: *Spitzer*/IRAC, 8 μm; contours: *Arecibo*, I-GALFA, HI.

2. SPH and AMR simulations: pressure confinement

The investigation of gravitational fragmentation of expanding shells using analytical and numerical analysis by Dale *et al.* (2009a) reveals the constraints of the thin-shell approximation as developed by Vishniac (1983), Elmegreen (1994), Whitworth *et al.* (1994) and Wünsch & Palouš (2001). With two different numerical approaches, Eulerian stationary adaptive-mesh refinement (AMR) and Lagrangian smoothed particle hydrodynamics (SPH), several 3D numerical experiments have been performed. The setup of these experiments assumes momentum-driven shells expanding into a very-low-density environment with low or high thermal pressure. We do not study the influence of ram pressure. Its effect can differ from that of thermal pressure and it will introduce other instabilities, which we try to avoid here.

The results of the AMR and SPH codes and the analytical prediction of the linear thin-shell analysis are compared. We observe the expanding-shell fragmentation process and arrive at the following conclusions (Dale *et al.* 2009a):

- the AMR and SPH results agree extremely well;
- the thermal pressure of the external environment strongly affects the fragmentation process;
- expansion into a low-pressure environment forms high-mass fragments only;
- confinement of the shell with high external pressure leads to the formation of fragments with a mass spectrum predicted by the thin-shell analysis.

3. Thick-shell dispersion relation

In an attempt to disentangle the role of external pressure, we introduce a shell in hydrostatic equilibrium between internal pressure and its own gravity, surrounded by an environment imparting external pressure. Such a thick shell, with a given surface density,

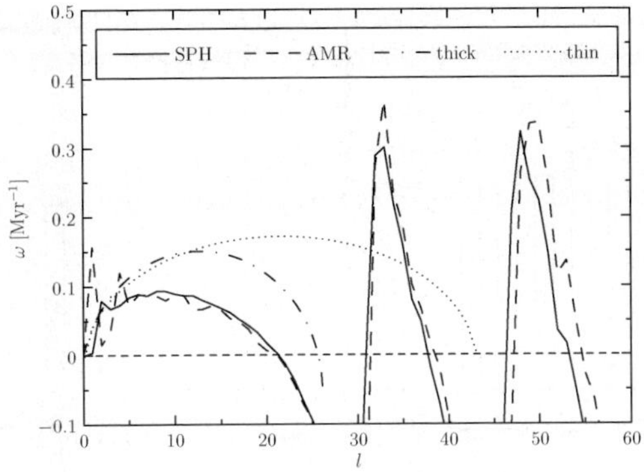

Figure 2. Dispersion relation of a shell expanding into a low-pressure environment. The thick- and thin-shell dispersion relations are compared to dispersion relations derived from AMR and SPH numerical experiments.

thickness and internal sound speed is subjected to perturbations and we derive the thick-shell dispersion relation. The growth rate of the fragments is not only a function of the shell's expansion velocity, its radius, surface density and internal sound speed, like in the thin-shell case. In the thick-shell dispersion relation, the external pressure is also included. The detailed derivation of this relation is given in Wünsch *et al.* (in prep.).

In Figure 2 we show a comparison of the numerical results from our AMR and SPH codes for thin-shell and thick-shell dispersion relations for a shell expanding into a low-pressure environment. The thick-shell dispersion relation shows much better agreement with the results of the numerical simulations. It also predicts a deficit of low-mass fragments.

The thick-shell dispersion relation leads to the following conclusions:

- for low external pressure, low-mass fragments are underabundant;
- there is a critical value of external pressure, which is a function of the shell's surface density, when the shell-fragmentation process is in agreement with the thin-shell prediction;
- for a given shell surface density, at the critical value of the external pressure, the thickness of the shell is greatest, compared to other values of the external pressure;
- for high external pressure, low-mass fragments are overabundant.

4. Mass spectrum of fragments

Palouš (2007) used linear thin-shell analysis to derive a mass spectrum following a Salpeter-like power law at the high-mass end and a turnover at low masses. In our numerical experiments with the AMR and SPH codes, we also analyze the mass spectrum of the fragments. We follow the fragmentation process using the potential-based clump-finding algorithm of Smith *et al.* (2008).

The different stages of shell evolution are represented in Figure 3. In the early stages, the mass spectrum shows a high-mass end decline similar to the Salpeter slope. However, later on the mass spectrum is top heavy, with a strong deficit of low-mass fragments. The mass evolution of individual fragments in the numerical simulations shows how the early fragments accrete mass, leaving increasingly lower-density environments for those fragments that form later. A more detailed analysis of the mass-accretion process onto the fragments forming from an expanding shell is the subject of a forthcoming article (Dale *et al.*, in prep.).

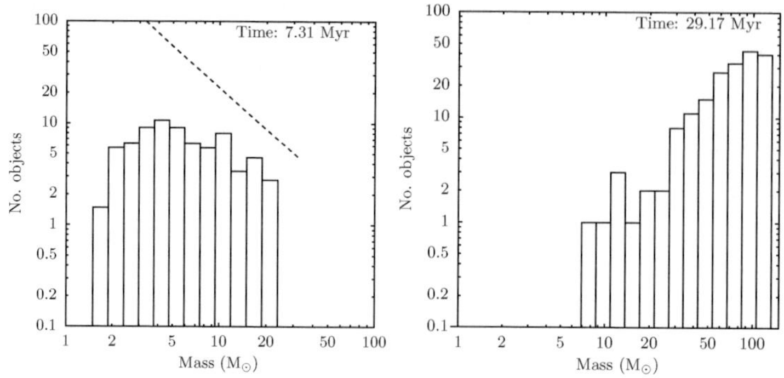

Figure 3. Time evolution of the fragment mass spectrum. *(left)* Early stage. The dashed line represents the Salpeter-like slope with a spectral index of −2.35. *(right)* Late stage.

The preliminary results of our clump-evolution analysis show that:

• the main fraction of the shell mass is accreted by the first objects, while the late-comers are starved since not much gas is left by the objects that formed earlier;

• shells expanding into a low-density, high-pressure environment develop top-heavy fragment mass functions;

• for mass replenishment from a dense shell environment (e.g., if the shell expands into a dense molecular cloud), more low-mass objects may be formed.

5. Conclusions

We conclude from our analysis of the fragmentation of expanding shells that:

(*a*) the pressure in the shell environment plays an important role and the thin-shell analysis is applicable only when the external pressure has a certain predicted value;

(*b*) linear analysis on the basis of the dispersion relations for thin or thick shells describes the early stages of the fragmentation process only. Subsequently, nonlinear processes dominate mass accretion;

(*c*) the accretion process makes the fragment mass spectrum top heavy. Low-mass objects may form through accumulation of new mass by the shell from its environment.

Acknowledgements

JED acknowledges support by the EU FP6 'Constellation' project (MRTN-CT-2006-035890). RW, VS and JP acknowledge support from the Institutional Research Plan AV0Z10030501 of the Academy of Sciences of the Czech Republic and project LC06014 Centre for Theoretical Astrophysics of the Ministry of Education, Youth and Sports of the Czech Republic.

References

Churchwell, E., *et al.* 2006, *ApJ*, 649, 759
Dale, J. E., Wünsch, R., Whitworth, A., & Palouš, J. 2009, *MNRAS*, 398, 1537
Deharveng, L., Zavagno, A., Lefloch, B., Caplan, J., & Pomarès 2007, *Proc. IAU Symp.* No. 237, p. 212
Ehlerová, S & Palouš, J. 2005, *A&A*, 437, 101
Elmegreen, B. G. 1994, *ApJ*, 427, 384
McClure–Griffiths, N. M., Dickey, J. M., Gaensler, B. M., & Green, A. J. 2002, *ApJ*, 578, 176
Palouš, J. 2007, *Proc. IAU Symp.* No. 237, p. 114

Sidorin, V. 2008, *IR, optical and X-ray counterparts of H*I *shells in the Milky Way*, MSc Thesis, Charles University, Prague

Smith, R., Clark, P., & Bonnell, I. 2008, *MNRAS*, 391, 1091

Vishniac, E. T. 1983, *ApJ*, 274, 152

Walter, F., Brinks, E., de Blok, W. J. G., Bigiel, F., Kennicutt Jr., R. C., Thornley, M. D., & Leroy, A. 2008, *AJ*, 136, 2563

Whitworth, A. P., Bhattal, A. S., Chapman, S. J., Disney, M. J., & Turner, J. A. 1994, *MNRAS*, 268, 291

Wünsch, R. & Palouš, 2001, *A&A*, 374, 746

Session II

Massive star clusters: formation, evolution, feedback and destruction

Star clusters: basic galactic building blocks
Proceedings IAU Symposium No. 266, 2009
R. de Grijs & J. R. D. Lépine, eds.

doi:10.1017/S174392130999086X

Young massive star clusters: achievements and challenges

Richard de Grijs

Kavli Institute for Astronomy and Astrophysics, Peking University, Beijing 100871, China
email: grijs@kiaa.pku.edu.cn

Department of Physics & Astronomy, The University of Sheffield, Sheffield S3 7RH, UK
email: R.deGrijs@sheffield.ac.uk

National Astronomical Observatories, Chinese Academy of Sciences, Beijing 100012, China

Abstract. In spite of significant recent and ongoing research efforts, most of the early evolution and long-term fate of young massive star clusters remain clouded in uncertainties. Here, I discuss our understanding of the initial conditions of star cluster formation and the importance of initial substructure for the subsequent dynamical-evolution and mass-segregation timescales. I also assess our current understanding of the (initial) binary fraction in star clusters and the shape of the stellar initial mass function at the low-mass end in the low-metallicity environment of the Large Magellanic Cloud. Finally, I question the validity of our assumptions leading to dynamical cluster mass estimates. I conclude that it seems imperative that observers, modellers and theorists combine efforts and exchange ideas and data freely for the field to make a major leap forward.

Keywords. stellar dynamics, methods: N-body simulations, binaries: general, stars: luminosity function, mass function, open clusters and associations: general, galaxies: star clusters, Magellanic Clouds

1. Introduction

It is now widely accepted that stars do not form in isolation, at least for stellar masses above $\sim$0.5 $M_\odot$. In fact, 70–90% of stars may form in a clustered mode (cf. Lada & Lada 2003). Star formation results from the fragmentation of molecular clouds, which in turn preferentially leads to star cluster formation. Over time, clusters dissolve or are destroyed by interactions with molecular clouds or tidal stripping by the gravitational field of their host galaxy.

A significant amount of recent research has focused on whether at least some of the young massive star clusters (YMCs) associated with the most violent starburst events in the local Universe may evolve into counterparts of the ubiquitous globular clusters (GCs) observed in almost all local galaxies. The main motivation for these studies was essentially twofold. First, for any compact, bound cluster to survive for a cosmologically significant length of time, its stellar initial mass function (IMF) must have a sufficient number of low-mass stars (acting as dynamical 'glue') to keep it together for so long. This places strong constraints on the IMF shape of any surviving YMCs and GCs. Second, as presumed hallmarks of the most violent galaxy-wide starburst episodes, YMCs trace the star-formation (and, to some extent, the assembly) histories of their host galaxies. As such, they serve as the proverbial light houses in the dark.

Rather than regurgitating the well-known issues affecting our detailed understanding of YMC-to-GC evolution (for reviews see, e.g., de Grijs & Parmentier 2007; de Grijs 2010), in this contribution I will focus on recent results as regards the shape of the low-mass IMF in cluster environments (Section 2), the effects of the initial conditions on cluster dynamics (Section 3) and the binary contributions in star clusters as a function of age (Section 4). In Section 5, I will conclude this contribution by highlighting some of the remaining uncertainties hampering a more detailed understanding of the underlying physics driving star cluster evolution and its observational interpretation.

2. The low-mass initial mass function

The shape of the stellar IMF in the solar neighbourhood for stellar masses > 1 M$_\odot$ has essentially remained unchallenged since Salpeter's (1955) seminal study. Nevertheless, the origin of its apparent universality is still hotly debated (e.g., Bonnell *et al.* 2007; Goodwin & Kouwenhoven 2009). Better constraining the physical origin of the IMF will have a major impact on, e.g., our understanding of the conditions prevailing in a wide range of starburst events, and the formation of the first stars and clusters in the early Universe. However, for the latter we would need to follow the full radiative cooling processes from primordial gas and the subsequently formed metallic elements in full detail!

At low masses, most current models agree that the solar-neighbourhood IMF flattens. This can be modelled by either multiple power-law or lognormal mass distributions (cf. Kroupa 2001; Chabrier 2003). While the former provides a mathematically convenient and observationally useful scaling law, the latter is supported by realistic numerical simulations in an attempt to understand the underlying physics (Hennebelle & Chabrier 2008). These simulations take into account dynamical depletion of the lowest-mass stars and replace the idea of a single Jeans mass for all newly formed stars in a given molecular cloud by a distribution of local Jeans masses which are representative of the lognormal density distribution of the turbulent, fragmenting gas. As statistically significant samples of roughly coeval stars, rich young star clusters play a major role in constraining the low-mass IMF. Open questions remaining in this field relate to, among others, the initial structure of newly formed clusters and whether the ubiquitous mass segregation observed in clusters of any age is dynamical or perhaps primordial (i.e., related to the process of star formation).

Note, however, that Goodwin & Kouwenhoven (2009) recently argued that a universal IMF does not necessarily provide unambiguous information about the star-formation activity from which the individual stars originated. One needs to take into account the initial binary fraction and mass-ratio distribution as well as the core-mass function and star-formation efficiency. They showed convincingly that very different (binary) mass-ratio distributions can produce very similar IMFs from very similar *core*-mass functions, while the resulting IMFs are also (to first order) insensitive to the binary fraction.

Ignoring the Goodwin & Kouwenhoven (2009) suggestions for the moment, preliminary clues as to the shape of the low-mass IMF (down to $\sim$0.15–0.30 M$_\odot$) in the low-metallicity ($Z \sim 0.4$ Z$_\odot$) environment of young ($\sim$4–45 Myr) Large Magellanic Cloud (LMC) clusters have recently been uncovered on the basis deep *Hubble Space Telescope* imaging observations (e.g., Da Rio *et al.* 2009; Liu *et al.* 2009a,b). Da Rio *et al.* (2009) reach a lower-mass limit of $\simeq 0.43$ M$_\odot$ in constructing their IMF of the LH 95 stellar association, while Liu *et al.* (2009a,b) push their method as low as 0.15–0.30 M$_\odot$ in the populous, young ($\sim$20–45 Myr-old) clusters NGC 1805 and NGC 1818. These analyses are particularly challenging in view of the stellar population's evolutionary state, requiring pre-main-sequence modelling to derive robust stellar mass estimates, which is notoriously difficult.

While these results must therefore be taken with a degree of caution, they appear to imply that the IMFs of these young clusters are essentially the same as that in the solar

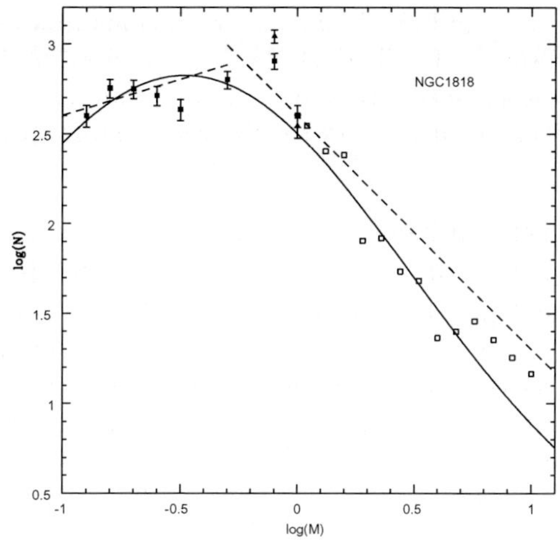

Figure 1. Complete mass function of NGC 1818. The dashed lines show the standard Kroupa (2001) IMF while the solid line represents the best-fitting lognormal distribution. See Liu *et al.* (2009a) for more details.

neighbourhood, although the characteristic stellar masses are somewhat higher. Figure 1 shows the complete mass function of NGC 1818 (Liu *et al.* 2009a) down to the 50% completeness limit. One would ideally want to probe younger star-forming regions to reach firmer conclusions, but these are inevitably obscured by large amounts of dust, hence requiring deep and often wide-field infrared (IR), (sub)millimetre, radio and X-ray surveys (as well as pointed observations) that are now coming online (e.g., the *Spitzer Space Telescope's* GLIMPSE survey or the *UKIRT* IR deep-sky survey, UKIDSS; e.g., Benjamin *et al.* 2003; Lucas *et al.* 2008) and which probe the low-mass stellar mass distribution in particular (see, e.g., Rathborne *et al.* 2009).

3. Early dynamical evolution

Simulations of star cluster evolution almost always assume that the stars are initially smoothly distributed and in dynamical equilibrium. However, both observations and the theory of star formation tell us that this is not how clusters form. We recently investigated the effects of substructure and initial clumpiness on the early evolution of clusters (Allison *et al.* 2009; see also references therein). Comparisons with observations will allow us to constrain how much initial substructure can be present. The most massive stars in young star clusters are almost always observed to be mass segregated (e.g., Hillenbrand & Hartmann 1998; de Grijs *et al.* 2002a,b,c; Gouliermis *et al.* 2004). A crucial question triggered by this observation relates to the physical origin of this characteristic mass distribution. Do massive stars form in the centres of clusters, or do they migrate there over time due to gravitational interactions with other cluster members? In smooth, relaxed clusters it has been argued that the most massive stars must form in the cores (e.g., Bonnell & Davies 1998; and references therein), which is therefore often referred to as primordial mass segregation (but see Ascenso *et al.* 2009). But does substructure perhaps play an important role?

Both observational evidence (e.g., Larson 1995; Testi *et al.* 2000; Elmegreen 2000; Lada & Lada 2003; Gutermuth *et al.* 2005; Allen *et al.* 2007) and theorical considerations suggest that young star clusters tend to form with a significant amount of substructure.

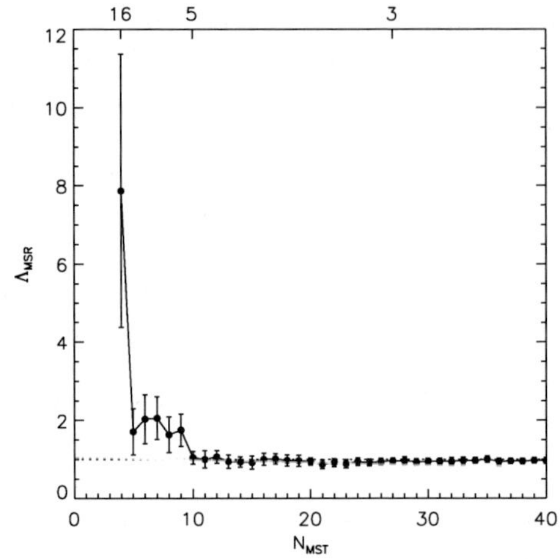

Figure 2. Mass-segregation ratio, $\Lambda_{\rm MST}$, for the Orion Nebula Cluster. The dashed line indicates the absence of mass segregation. See Allison *et al.* (2009) for more details.

Their progenitor molecular clouds are observed to have significant levels of substructure in both density and kinematics (e.g., Carpenter & Hodapp 2008), which is likely induced by the supersonic turbulence thought to dominate molecular cloud structure (e.g., Mac Low & Klessen 2004; Ballesteros–Paredes *et al.* 2007). Observations also support a scenario in which young clusters lose their substructure on timescales of < 2 Myr (e.g., Cartwright & Whitworth 2004; Schmeja *et al.* 2008). Simulations suggest that the only way in which this could happen is if clusters are born dynamically cool (Goodwin *et al.* 2004; Allison *et al.* 2009). On the basis of these arguments, Allison *et al.* (2009) recently performed an ensemble of *N*-body simulations aimed at exploring the earliest phases of cluster evolution. They find that cool, substructured clusters appear to mass segregate dynamically for stellar masses down to a few $M_\odot$ on timescales of a few Myr. This is reminiscent of the observational status of the Orion Nebula Cluster (e.g., Bonnell & Davies 1998; Allison *et al.* 2009; Moeckel & Bonnell 2009).

Allison *et al.* (2009) modelled an initially highly substructured cluster (using multiple realisations to assess the numerical uncertainties), characterised by a ratio of the kinetic to potential energy of 0.3 (where 0.5 is virialised) and a fractal dimension of 1.6 (where 3 corresponds to a spherically symmetric distribution), containing 1000 stars drawn from a Kroupa (2001) IMF spanning the mass range from 0.08 to 50 $M_\odot$. Given the cluster's nonsphericity, we used a novel approach to determine the degree of mass segregation, quantified by using the concept of the 'minimum spanning tree' (MST). A sample's MST corresponds to the path connecting all points in the sample with the shortest possible path length, and which contains no closed loops (see, e.g., Prim 1957). We defined a 'mass-segregation ratio',

$$\Lambda_{\rm MSR} = \frac{\langle l_{\rm norm} \rangle}{l_{\rm massive}} \pm \frac{\sigma_{\rm norm}}{l_{\rm massive}}, \tag{3.1}$$

where $\langle l_{\rm norm} \rangle$ is the average length of the MST of sets of $N_{\rm MST}$ random stars and $l_{\rm massive}$ is the length of the MST of the $N_{\rm MST}$ most massive stars. The dispersion associated with the average length of the random MSTs is roughly Gaussian and can therefore be quantified by the standard deviation, $\sigma_{\rm norm}$. Using this definition, if $\Lambda_{\rm MST} > 1$, a given sample of cluster stars is mass segregated.

Figure 2 shows an application of this method to the 900 stars in the Orion Nebula Cluster for which masses are available (Hillenbrand 1997). The method clearly identifies the mass segregation of the central Trapezium system at $N_{\rm MST} = 4$, $\Lambda_{\rm MST} = 8.0 \pm 3.5$, but it also shows that there appears to be a secondary level of mass segregation involving the nine most massive stars > 5 M$_\odot$ (cf. Hillenbrand & Hartmann 1998; see Allison *et al.* 2009 for additional details).

More work is required to systematically address the most likely initial conditions for cluster formation leading to the observed configurations. In particular, we have thus far not included binary stars, although they are expected to have an effect on the MST lengths. If a binary is resolved, it is very likely that the two components will be linked as a node, as will higher-order systems. Allison *et al.* (2009) suggest that this raises the possibility that MSTs could be very useful in locating binary and multiple systems by looking for short links within the MST (but see Cartwright & Whitworth 2005). In addition, many cluster observations suffer from significant incompleteness, particularly near the most massive stars where low-mass stars cannot be detected even if they are present. This poses a significant problem as it is impossible to know if there are many low-mass stars in the 'central' regions (in whatever way defined in the presence of substructure).

4. The contribution of binaries to star cluster evolution

More often than not, simulations of star clusters neglect the presence of binary stars. Observations of local star-forming regions lead us to suspect that all, or nearly all, stars form in binary or triple systems (Goodwin & Kroupa 2005; Duchêne *et al.* 2007; Goodwin *et al.* 2007). Such systems significantly affect the dynamical evolution of the cluster, yet the initial binary fractions in dense star clusters are poorly known. Almost all studies of binarity have been limited to nearby solar-metallicity populations (see Duchêne 1999 and Duchêne *et al.* 2007 for reviews). However, it might be expected that metallicity (e.g., through its effects on cooling and hence on the opacity limit for fragmentation) will play a role in the fragmentation of cores to produce binary systems (Bate 2005; Goodwin *et al.* 2007).

The binary fractions in more distant, massive clusters have not yet been studied thoroughly, because of observational limitations. Note, however, that statistical colour-magnitude analysis based on artificial-star tests offers a promising alternative (e.g., Romani & Weinberg 1991; Rubenstein & Bailyn 1997; Bellazzini *et al.* 2002; Cool & Bolton 2002; Zhao & Bailyn 2005; Davis *et al.* 2008). In addition, all clusters thus far studied in this way are old stellar systems, in which dynamical evolution is expected to have altered the initial binary population significantly. Efforts have begun to address this issue for the much more distant young populous clusters in the LMC (e.g., Elson *et al.* 1998). Hu *et al.* (2009) estimate that the binary fraction in NGC 1818 in the mass range between 1.3 and 1.6 M$_\odot$ is $\sim$0.35 for systems with an approximately flat mass-ratio distribution, q, for $q > 0.4$. This is consistent with a *total* binary fraction of F stars of 0.6 to unity.

In Figure 3 (Hu *et al.* 2009), I show the cumulative distribution function (CDF) with pseudo-colour (defined as the colour change along the main sequence in the colour-magnitude diagram) as observed for the entire magnitude range of interest in NGC 1818 (solid line) and for a stellar population with photometric errors but no binaries (dotted line). This is clearly a very bad fit to the data. The figure also shows the best-fitting (r.m.s.) binary fraction ($f_{\rm b}$), assuming a uniform mass-ratio distribution, of $f_{\rm b} = 0.62 \pm 0.05(1\sigma)$ (dashed line). Note that the fit is poor at larger pseudo-colours. This is always the case and is due to the presence of unresolved higher-order systems.

The combination of the results presented in this and the previous sections triggers a number of questions. Do high binary fractions affect mass segregation at early times or the relaxation of substructure? Do they leave observational signatures? NGC 1818 is

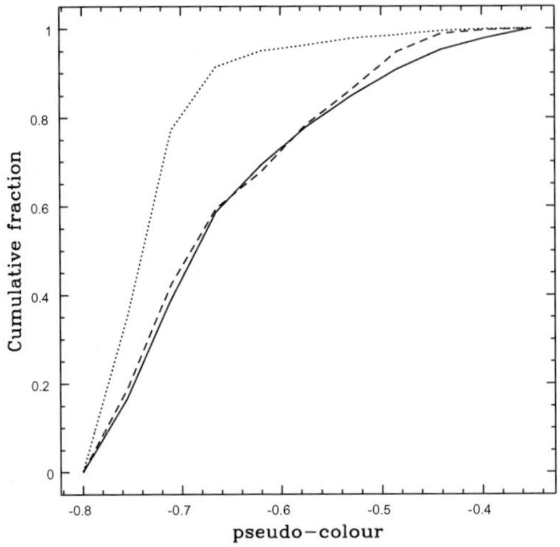

Figure 3. Observed cumulative distribution function with pseudo-colour in NGC 1818 (solid line; full stellar sample) compared with an artificial stellar population with zero binary fraction (f_b, dotted line), and the best-fitting (r.m.s.) uniform mass-ratio distribution of $f_b = 0.62 \pm 0.05$ (dashed line). See Hu *et al.* (2009) for more details.

several crossing times old, so that the binary population should have been modified by dynamical interactions. In particular, soft (i.e., wide) binaries are expected to have been destroyed by this age. Therefore, the high binary fraction found for F stars suggests that these binaries are relatively 'hard' and able to survive dynamical encounters.

5. Uncertainties abound

For any cluster, but particularly for the lower-mass end of the cluster mass function, it seems clear that the effect of binaries, mass segregation and the dynamical alteration of mass functions by two-body relaxation are important factors that cannot be ignored.

We recently explored the usefulness of the diagnostic age versus mass-to-light-ratio diagram in the context of YMCs and Galactic open clusters (e.g., Smith & Gallagher 2001; Mengel *et al.* 2002; McCrady *et al.* 2003, 2005; Larsen *et al.* 2004; Bastian *et al.* 2006; Goodwin & Bastian 2006; Moll *et al.* 2008; see de Grijs & Parmentier 2007 for a review). This diagram is often used in the field of extragalactic young to intermediate-age massive star clusters to constrain the shape of their stellar IMF, as well as their stability and the likelihood of their longevity. Based on high-resolution spectroscopy to obtain the objects' integrated velocity dispersions, σ, and on high-spatial-resolution imaging to obtain accurate projected half-light radii, $r_{\rm hl}$, most authors then apply Spitzer's (1987) equation,

$$M_{\rm dyn} = \eta \frac{r_{\rm hl}\sigma^2}{G}, \tag{5.1}$$

to obtain the dynamical cluster masses, $M_{\rm dyn}$ (G is the gravitational constant and $\eta \approx 9.75$ is a dimensionless parameter which is usually assumed to be constant; but see Fleck *et al.* 2006; Kouwenhoven & de Grijs 2008).

Despite a number of simplifying assumptions (see, e.g., de Grijs & Parmentier 2007 for a review; Moll *et al.* 2008), one can get at least an initial assessment as to whether a given cluster may be (i) significantly out of virial equilibrium, in particular 'super-virial', (ii) significantly over- or underabundant in low-mass stars, or (iii) populated by a significant fraction of binary and higher-order multiple systems. This has led to suggestions that, in the absence of significant external perturbations, young massive clusters (YMCs) located in the vicinity of the simple stellar population models and aged $\geqslant 10^8$ yr may survive for a Hubble time and eventually become old GC-like objects (e.g., Larsen *et al.* 2004; Bastian *et al.* 2006; de Grijs & Parmentier 2007).

Using the massive young Galactic cluster Westerlund 1 as a key example, we cautioned that stochasticity in the IMF introduces significant additional uncertainties (de Grijs *et al.* 2008). For Galactic open clusters, the effect of binaries within clusters may well account for most of the displacement of the observed cluster positions to below the model curves in de Grijs *et al.* (2008). Kouwenhoven & de Grijs (2008) pointed out that if the velocity dispersion of binary systems were similar to the velocity dispersion of the cluster as a whole, the *observationally measured* velocity dispersion would overestimate the mass of a cluster. Based on a comparison with Kouwenhoven & de Grijs (2008), in de Grijs *et al.* (2008) we concluded that the vast majority of our sample of open clusters are indeed expected to be binary dominated.

We also note that the cluster masses may well have been overestimated by factors of a few through the universal use of Eq. (5.1). In particular, for highly mass-segregated clusters containing significant binary fractions, a range of stellar IMF representations, and for combinations of characteristic relaxation timescales and cluster half-mass radii, the adoption of a single scaling factor $\eta \approx 9.75$ introduces systematic offsets. To correct for these, we would need to adopt smaller values of η (e.g., Fleck *et al.* 2006; Kouwenhoven & de Grijs 2008), and this would thus lead to dynamical mass overestimates if $\eta = 9.75$ were assumed.

We have now reached a stage in star cluster studies in which it is imperative to combine observational, theoretical and numerical modelling efforts to make the next major leap in our physical understanding. The level of detail required to make significant progress necessitates the combined forces of modellers in the fields of stellar population synthesis, *N*-body simulations and smooth-particle hydrodynamics, as well as observers with a good grasp of the intricacies of curent-generation data problems and reduction issues. It looks, therefore, that the conventional small research teams of the past may soon need to expand and become more inclusive to make significant and exciting headway.

Acknowledgements

I thank the Scientific Organising Committee of IAU Symposium 266 for their vote of support to deliver this review talk. I would also like to highlight that most of the results presented in this contribution are based on PhD thesis work by Richard Allison (University of Sheffield), Qiang Liu and Yi Hu (National Astronomical Observatories, Chinese Academy of Sciences). I acknowledge funding to attend this meeting from the Royal Society, the International Astronomical Union and the Learned Societies Fund of the University of Sheffield.

References

Allen, L., Megeath, S. T., Gutermuth, R., Myers, P. C., Wolk, S., Adams, F. C., Muzerolle, J., Young, E., & Pipher, J. L. 2007, in: B. Reipurth, D. Jewitt & K. Keil (eds.), *Protostars and planets V* (Tucson: University of Arizona Press), p. 361

Allison, R. J., Goodwin, S. P., Parker, R. J., de Grijs, R., Portegies Zwart, S. F., & Kouwenhoven, M. B. N. 2009, *ApJ* (Letters), 700, L99

Ascenso, J., Alves, J., & Lago, M. T. V. T. 2009, *A&A*, 495, 147

Ballesteros–Paredes, J., Klessen, R. S., Mac Low, M.-M., & Vázquez-Semadeni, E. 2007, in: B. Reipurth, D. Jewitt & K. Keil (eds.), *Protostars and planets V* (Tucson: University of Arizona Press), p. 63

Bate, M. R. 2005, *MNRAS*, 363, 363
Bastian, N., Saglia, R. P., Goudfrooij, P., Kissler-Patig, M., Maraston, C., Schweizer, F., & Zoccali, M. 2006, A&A, 448, 881
Bellazzini, M., Fusi Pecci, F., Messineo, M., Monaco, L., & Rood, R. T. 2002, *AJ*, 123, 1509
Benjamin, R. A. *et al.* 2003, *PASP*, 115, 953
Bonnell, I. A. & Davies, M. B. 1998, *MNRAS*, 295, 691
Bonell, I. A., Larson, R. B., & Zinnecker, H. 2007, in: B. Reipurth, D. Jewitt & K. Keil (eds.), *Protostars and planets V* (Tucson: University of Arizona Press), p. 149
Carpenter, J. M. & Hodapp, K. W. 2008, in: B. Reipurth (ed.), *Handbook of star forming regions, Vol. I: The northern sky* (San Francisco: ASP), p. 899
Cartwright, A. & Whitworth, A. P. 2004, *MNRAS*, 348, 589
Chabrier, G. 2003, *PASP*, 115, 763
Cool, A. M. & Bolton, A. S. 2002, in M. M. Shara (ed.), *Stellar collisions, mergers and their consequences* (San Francisco: ASP), p. 163
Da Rio, N., Gouliermis, D. A., & Henning, T. 2009, *ApJ*, 696, 528
Davis, D. S., Richer, H. B., Anderson, J., Brewer, J., Hurley, J., Kalirai, J. S., Rich, R. M., & Stetson, P. B. 2008, *AJ*, 135, 2155
de Grijs, R. 2010, *Phil. Trans R. Soc. A*, in press (arXiv:0911.0778)
de Grijs, R. & Parmentier, G. 2007 *ChJA&A*, 7, 155
de Grijs, R., Johnson, R. A., Gilmore, G. F., & Frayn, C. M. 2002a, *MNRAS*, 331, 228
de Grijs, R., Gilmore, G. F., Johnson, R. A., & Mackey, A. D. 2002b, *MNRAS*, 331, 245
de Grijs, R., Gilmore, G. F., Mackey, A. D., Wilkinson, M. I., Beaulieu, S. F., Johnson, R. A., & Santiago, B. X. 2002c, *MNRAS*, 337, 597
de Grijs, R., Goodwin, S. P., Kouwenhoven, M. B. N., & Kroupa, P. 2008, A&A, 492, 685
Duchêne, G. 1999, *A&A*, 341, 547
Duchêne, G., Delgado-Donate, E., Haisch Jr, K. E., Loinard, L., & Rodríguez, L. F. 2007, in: B. Reipurth, D. Jewitt & K. Keil (eds.), *Protostars and planets V* (Tucson: University of Arizona Press), p. 379
Elmegreen, B. G. 2000, *ApJ*, 530, 277
Elson, R. A. W., Sigurdsson, S., Davies, M., Hurley, J., & Gilmore, G. 1998, *MNRAS*, 300, 857
Fleck, J.-J., Boily, C. M., Lançon, A., & Deiters, S. 2006, *MNRAS*, 369, 1392
Goodwin, S. P. & Kouwenhoven, M. B. N. 2009, *MNRAS* (Letters), 397, L36
Goodwin, S. P. & Kroupa, P. 2005, *A&A*, 439, 565
Goodwin, S. P., Whitworth, A. P., & Ward-Thompson, D. 2004, *A&A*, 414, 633
Goodwin, S. P., Kroupa, P., Goodman, A., & Burkert, A. 2007, in: B. Reipurth, D. Jewitt & K. Keil (eds.), *Protostars and planets V* (Tucson: University of Arizona Press), p. 133
Gouliermis, D., Keller, S. C., Kontizas, M., Kontizas, E., & Bellas-Velidis, I. 2004, *A&A*, 416, 137
Gutermuth, R. A., Megeath, S. T., Pipher, J. L., Williams, J. P., Allen, L. E., Myers, P. C., & Raines, S. N. 2005, *ApJ*, 632, 397
Hennebelle, P. & Chabrier, G. 2008, *ApJ*, 684, 395
Hillenbrand, L. A. 1997, *AJ*, 113, 1733
Hillenbrand, L. A. & Hartmann, L. W. 1998, *ApJ*, 492, 540
Hu, Y., Deng, L., de Grijs, R., Goodwin, S. P., & Liu, Q. 2009, *ApJ*, submitted (arXiv:0801.2814).
Kouwenhoven, M. B. N. & de Grijs, R. 2008, A&A, 480, 103
Kroupa, P. 2001, *MNRAS*, 322, 231
Lada, C. J. & Lada, E. A. 2003, *ARA&A*, 41, 57
Larsen, S. S., Brodie, J. P., & Hunter, D. A. 2004, *AJ*, 128, 2295
Larson, R. B. 1995, *MNRAS*, 272, 213
Liu, Q., de Grijs, R., Deng, L. C., Hu, Y., Baraffe, I., & Beaulieu, S. F. 2009a, *MNRAS*, 396, 1665
Liu, Q., de Grijs, R., Deng, L. C., Hu, Y., & Beaulieu, S. F. 2009b, *A&A*, 503, 469
Lucas, P. W. *et al.* 2008, *MNRAS*, 391, 136
Mac Low, M.-M. & Klessen, R. S. 2004, *Rev. Modern Phys.*, 76, 125
McCrady, N., Gilbert, A. M., & Graham, J. R. 2003, *ApJ*, 596, 240

McCrady, N., Graham, J. R., & Vacca, W. D. 2005, *ApJ*, 621, 278
Mengel, S., Lehnert, M. D., Thatte, N., & Genzel, R. 2002, *A&A*, 383, 137
Moeckel, N. & Bonnell, I. A. 2009, *MNRAS*, 396, 1864
Moll, S. L., Mengel, S., de Grijs, R., Smith, L. J., & Crowther, P. A. 2008, *MNRAS*, 382, 1877
Prim, R. C. 1957, *Bell Systems Tech. J.*, 36, 1389
Rathborne, J. M., Lada, C. J., Muench, A. A., Alves, J. F., Kainulainen, J., & Lombardi, M. 2009, *ApJ*, 699, 742
Romani, R. W. & Weinberg, M. D. 1991, *ApJ*, 372, 487
Rubenstein, E. P. & Bailyn, C. D. 1997, *ApJ*, 474, 701
Salpeter, E. E. 1955, *ApJ*, 121, 161
Schmeja, S., Kumar, M. S. N., & Ferreira, B. 2008, *MNRAS*, 389, 1209
Smith, L. J. & Gallagher III, J. S. 2001, *MNRAS*, 326, 1027
Spitzer Jr., L. 1987, Dynamical Evolution of Globular Clusters (Princeton: Princeton University Press)
Testi, L., Sargent, A. I., Olmi, L., & Onello, J. S. 2000, *ApJ* (Letters), 540, L53
Zhao, B. & Bailyn, C. D. 2005, *AJ*, 129, 1934

Star clusters: basic galactic building blocks
Proceedings IAU Symposium No. 266, 2009
R. de Grijs & J. R. D. Lépine, eds.

doi:10.1017/S1743921309990871

Nuclear star clusters

Torsten Böker

European Space Agency, Keplerlaan 1, 200AG Noordwijk, the Netherlands
email: tboeker@rssd.esa.int

Abstract. The centers of most galaxies in the local Universe are occupied by compact, barely resolved sources. Based on their structural properties, position in the Fundamental Plane, and integrated spectra, these sources clearly have a stellar origin. They are therefore called 'nuclear star clusters' (NCs) or 'stellar nuclei'. NCs are found in galaxies of all Hubble types, suggesting that their formation is intricately linked to galaxy evolution. Here, I review some recent studies of NCs, describe ideas for their formation and subsequent growth, and touch on their possible evolutionary connection with both supermassive black holes and globular clusters.

Keywords. galaxies: nuclei, galaxies: star clusters

1. Introduction

The nuclei of galaxies are bound to provide 'special' physical conditions because they are located at the bottom of the potential well of their host galaxies. This unique location manifests itself in various distinctive phenomena such as supermassive black holes (SMBHs), active galactic nuclei (AGN), central starbursts, or extreme stellar densities. The evolution of galactic nuclei is closely linked to that of their host galaxies, as inferred from a number of global-to-nucleus scaling relations discovered in the last decade.

Recently, observational and theoretical interest has been refocussed onto the compact and massive star clusters found in the nuclei of galaxies of all Hubble types. These 'nuclear star clusters' (NCs) are intriguing objects that are linked to a number of research areas: (i) they are a promising environment for the formation of massive black holes because of their extreme stellar densities, (ii) they may also constitute the progenitors of at least some halo globular clusters through 'NC capture' following the tidal disruption of a satellite galaxy, and (iii) their formation process is influenced by (and important for) the central potential, which in turn governs the secular evolution of their host galaxies.

I will briefly summarize what has been learned about NCs over the last few years, describe some proposed formation mechanisms of NCs, and discuss the new paradigm of 'central massive objects', which links NCs with SMBHs in galactic nuclei. Finally, I briefly mention a scenario in which NCs may be the progenitors of (some) globular clusters.

2. Properties of nuclear star clusters

Extragalactic star clusters are compact sources, and in general their study requires high spatial resolution, afforded only by the *Hubble Space Telescope* or large ground-based telescopes using adaptive optics. Over the last decade, a number of studies—based on both imaging and spectroscopic observations—have contributed to the following picture of NCs:

(*a*) NCs are common: the fraction of galaxies with an unambiguous NC detection is 75% in late-type (Scd–Sm) spirals (Böker *et al.* 2002), 50% in earlier-type (Sa–Sc) spirals (Carollo *et al.* 1997), and 70% in spheroidal (E and S0) galaxies (Côté *et al.* 2006). These

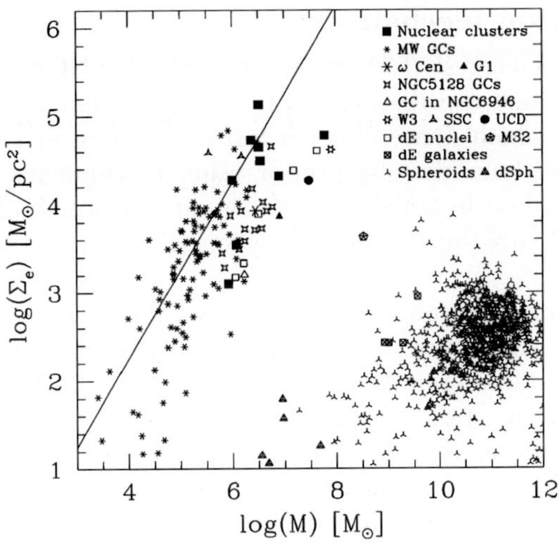

Figure 1. Mean projected mass density of various stellar systems inside the effective radius, r_e, plotted against total mass. This is similar to a face-on view of the Fundamental Plane. NCs occupy the high end, a region populated by other types of massive stellar clusters, and are well separated from elliptical galaxies and spiral bulges. The solid line represents a constant cluster size, i.e., $r_e = 3$ pc (from Walcher *et al.* 2005).

numbers are likely lower limits, although for different reasons. In the latest-type disks, it is sometimes not trivial to locate the galaxy center unambiguously so that no particular source can be identified with it. In contrast, many early-type galaxies have very steep surface brightness profiles (SBPs) that make it difficult to detect even luminous clusters against this bright background.

(*b*) NCs are much more luminous than 'normal' globular clusters (GCs). With typical absolute I-band magnitudes between -14 and -10 (Böker *et al.* 2002; Côté *et al.* 2006), they are roughly 40 times more luminous than the average Milky Way globular cluster (Harris 1996).

(*c*) However, NCs are as compact as Milky Way GCs. Their half-light radii are typically $2 - 5$ pc, independent of galaxy type (Geha *et al.* 2002; Böker *et al.* 2004; Côté *et al.* 2006).

(*d*) Despite their compactness, NCs are very massive: their typical dynamical mass are 10^6 – 10^7 $M_\odot$ (Walcher *et al.* 2005), i.e., at the extreme high end of the globular cluster mass function.

(*e*) Their mass density clearly separates NCs from compact galaxy bulges. This is demonstrated in Figure 1, which compares the mass and mass density of NCs to those of other spheroidal stellar systems. The clear gap between bulges/ellipticals on the one hand, and NCs on the other makes a direct evolutionary connection between the two classes of objects unlikely.

(*f*) The star-formation history of NCs is complex, as evidenced by the fact that most NCs have stellar populations composed of multiple generations of stars (Walcher *et al.* 2005; Rossa *et al.* 2006). While all NCs show evidence for an underlying old ($\gtrsim 1$ Gyr) population of stars, most also have a young generation with ages below 100 Myr. This is strong evidence that NCs experience frequent and repetitive star-formation episodes (Walcher *et al.* 2005).

(*g*) NCs obey similar scaling relations with host-galaxy properties as SMBHs. This has triggered a very active research area, but its implications are still to be understood fully, as discussed in more detail in Section 4.

3. Possible formation mechanisms

There are a number of suggested formation scenarios for NCs, and so far few have been ruled out. In principle, one can distinguish between two main categories: (i) migratory formation scenarios in which dense clusters form elsewhere in the galaxy and then fall into the center through dynamical friction or other mechanisms (Andersen *et al.* 2008; Cappuzzo–Dolcetta & Miocchi 2008), and (ii) in situ cluster buildup through (possibly episodic) gas infall and subsequent star formation within a few parsecs from the galaxy center.

The processes that funnel gas onto NCs in nearby galaxies have recently been studied in some detail, enabled by significant improvements in the sensitivity and spatial resolution of millimeter-wave interferometers (e.g., Schinnerer *et al.* 2006, 2007). In general, bar-shaped asymmetries in the disk potential can lead to a prolonged influx of molecular gas into the central few pc, thus providing the reservoir for an intense burst of star formation and leading to the rejuvenation of an existing NC. The starburst is, however, self-regulating in the sense that mechanical feedback from stellar winds and/or supernova explosions can expel the remaining gas and even temporarily change the gas-flow pattern (Schinnerer *et al.* 2008). This scenario naturally leads to episodic star formation, thus explaining the presence of multiple stellar populations in NCs.

Less clear, however, are the reasons as to why gas accumulates in the nucleus of a shallow disk galaxy *in the absence* of a prominent central mass concentration, i.e., how the 'seed clusters' form initially. A few studies have attempted to provide an explanation for this puzzle. For example, Milosavlević (2004) suggests the magneto-rotational instability in a differentially rotating gas disk as a viable means to transport gas towards the nucleus and support (semi-)continuous star formation there.

More recently, Emsellem & van de Ven (2008) pointed out that the tidal field becomes compressive in shallow density profiles, causing gas to collapse onto the nucleus of a disk galaxy. If correct, then NC formation is indeed expected to be a natural consequence of galaxy formation, which would go a long way towards explaining at least some of the observed scaling relations between NCs and their host galaxies.

The question of when a particular NC (i.e., its 'seed' cluster) has formed is equivalent to asking how old its oldest stars are. This question is extremely difficult to answer in all galaxy types, albeit for different reasons. In late-type spirals, for example, the NC nearly always contains a young stellar population which dominates the spectrum and thus makes detection of an underlying older population challenging, not to mention its accurate age determination.

Early-type galaxies, on the other hand, have much steeper surface brightness profiles and, therefore, a low contrast between the NC and the galaxy body. This makes spectroscopic studies of NCs in ellipticals and S0s exceedingly difficult. The few published studies have focused on the NCs of dE,N galaxies and showed that even these can have stellar populations that are significantly younger than the rest of the host galaxy (Butler & Martínez–Delgado 2005; Chilingarian *et al.* 2007; Koleva *et al.* 2009). In general, stellar population fits as well as the rather high dynamical mass-to-light ratios of NCs indicate that they contain a significant population of evolved (at least 1 Gyr old) stars, i.e., they have been in place for a long time (Walcher *et al.* 2005).

4. Central black holes and nuclear star clusters

A number of recent studies (Rossa *et al.* 2006; Wehner & Harris 2006; Ferrarese *et al.* 2006; Balcells *et al.* 2007; Graham & Driver 2007) have demonstrated that NCs follow similar scaling relations with their host galaxies as SMBHs and typically extend these

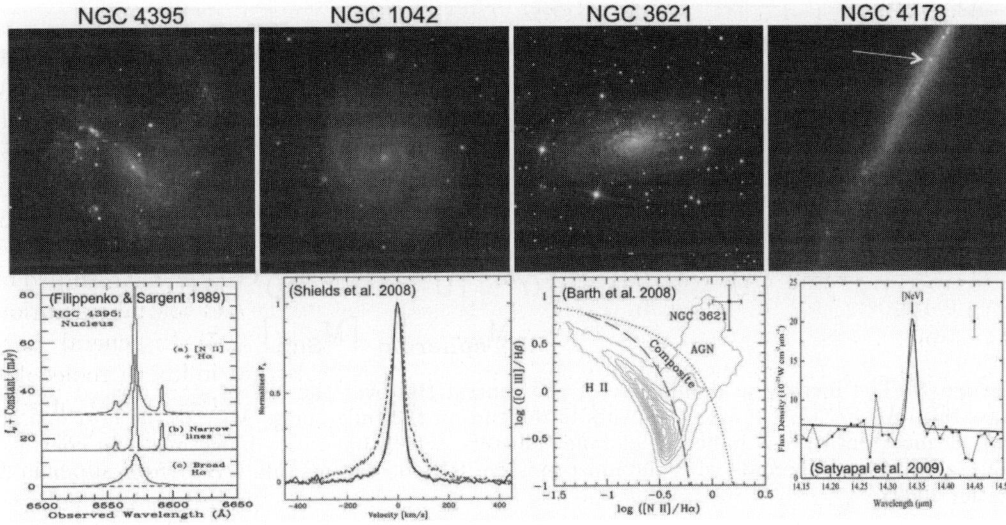

Figure 2. Four bulgeless disk galaxies with evidence for an active galactic nucleus (AGN; references in the lower panels). The case of NGC 4395 has long been thought unique, but detailed observations of other late-type disks have shown that such low-luminosity AGN are easily missed in optical surveys. Nevertheless, the AGN fraction in bulgeless disks appears to be lower than in galaxies with more massive bulges.

relations to lower SMBH masses. This has triggered speculation about a common formation mechanism of NCs and SMBHs, which is governed mostly by the mass of the host galaxy's spheroid. The idea put forward is that NCs and SMBHs are two incarnations of a 'central massive object' (CMO). Galaxies above a certain mass threshold ($\approx 10^{10}$ $M_{\odot}$) form predominantly SMBHs while lower-mass galaxies form NCs.

On the other hand, it is well established that many galaxies contain *both* a NC and a SMBH (Seth *et al.* 2008). This is true even at the extreme late end of the Hubble sequence, i.e., in galaxies that have no bulge component at all. A famous example known for a long time is the 'mini Seyfert' NGC 4395 (Filippenko & Sargent 1989), but a number of similar cases have been found recently, as demonstrated in Figure 2. These active galactic nuclei (AGN) are often missed in spectra taken with relatively wide apertures because the AGN signatures are faint compared to those of their surroundings, especially in the presence of (circum)nuclear star formation.

However, this does not imply that *all* NCs harbor a SMBH. On the contrary, a recent survey of high-ionization [NeV] emission in late-type galaxies (Satyapal *et al.* 2009) indicates that AGN in bulgeless disks are indeed rare: only about 5% of spirals with Hubble type Sd–Sm show [NeV] emission. Interestingly, all these low-luminosity 'mini AGN' are found in galaxies that also host an NC, possibly suggesting that the presence of an NC is necessary (but not sufficient) for the formation of a SMBH.

Why, then, do some NCs contain a SMBH, but not all? What is the mass ratio between NCs and SMBHs and how is it regulated? Important observational constraints come from Graham & Spitler (2009), who identified all galaxies with reliable measurements of both NC mass and SMBH mass. They conclude that the ratio $M_{\rm BH}/(M_{\rm BH}+M_{\rm NC})$ is a function of bulge mass, as illustrated in Figure 3. Massive bulges only host SMBHs, but no NCs. At the other end of the spectrum, in 'pure' disk galaxies, the mass of the SMBH (if it exists at all) is negligible with respect to the NC mass. In between, there is a transition region in which galaxies host both NCs and SMBHs with comparable masses.

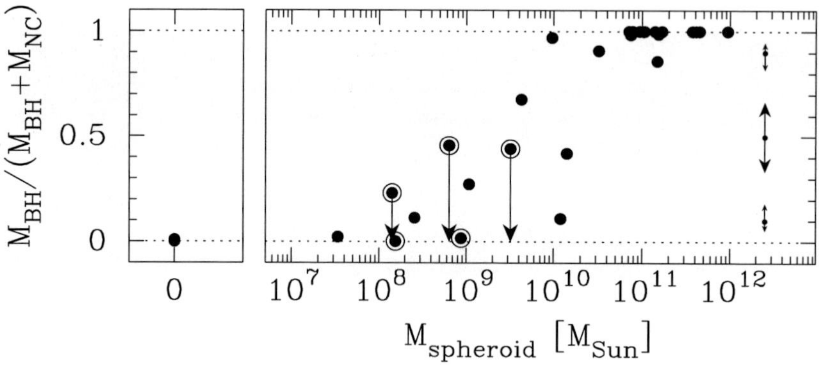

Figure 3. The increasing dominance of the central BH over the NC of stars, traced by the mass ratio $M_{\rm BH}/(M_{\rm BH} + M_{\rm NC})$ appears to depend on the bulge mass $M_{\rm sph}$ of the host galaxy. The leftmost data point indicates globular clusters, which have zero bulge mass. In contrast, the highest-mass spheroids with the most massive BHs do not contain a NC (from Graham & Spitler 2009).

A theoretical explanation as to why this may be so has been offered by Nayakshin *et al.* (2009). They speculate that 'competitive feedback' between the SMBH and the gas inflow feeding star formation during the NC's buildup determines which of the two components can grow more efficiently. The outcome of this 'race' between the SMBH (which grows on the Salpeter timescale) and the NC (which should grow on the dynamical timescale regulating gas inflow) is decided by the bulge mass, i.e., the stellar velocity dispersion, σ. Below a value of $\sigma \approx 150$ km s^{-1}, the BH cannot grow efficiently, while above this value it grows fast enough that its radiative feedback hinders NC growth. The presence of a SMBH also has important consequences for the dynamical evolution of an NC, since it prevents core collapse and might even disrupt the NC (Merritt 2009).

One important question in this context, which has been neglected so far, is why some galaxies apparently contain *neither* NC nor SMBH. In other words, how can a galaxy avoid having a CMO? Progress along these lines will require a better understanding of the formation and survival of 'pure' disk galaxies, a problem that is still challenging for current models of structure formation.

5. Nuclear clusters as precursors of globular clusters

As mentioned in Section 3, most NCs likely formed a long time ago. In fact, some theories of structure formation suggest that already the first protogalaxies underwent rapid nucleation (Cen 2001) and formed a dense star cluster in their center. If these protogalaxies are gas rich, the NC will most likely experience multiple bursts of star formation similar to the present-day NCs in late-type disks. This process continues until the protogalaxy is destroyed in a merger. Because of its compactness and high stellar density, the NC will survive the merger and from that moment on will passively age in the halo of the merger product.

That this process indeed occurs is best demonstrated by the case of M 54. This Milky Way globular cluster is believed to be the nucleus of the Sagittarius dwarf galaxy (Layden & Sarajedini 2000) which is currently being 'swallowed' by the Milky Way. Another plausible example is ω Cen which has long been thought to be the remnant nucleus of an accreted dwarf galaxy because of it extreme mass and multiple stellar populations.

Assuming that globular clusters spent some part of their history at the bottom of a galaxy's potential well might naturally explain the multiple stellar populations observed in a number of globulars in the Galaxy (Böker 2008). It is also consistent with the roughly constant specific frequency of globular clusters and the observed universal mass fraction of globular cluster systems in the local Universe (McLaughlin 1999).

References

Andersen, D. R., Walcher, C. J., Böker, T., Ho, L. C., van der Marel, R. P., Rix, H.-W., & Shields, J. C. 2008, *ApJ*, 688, 990
Balcells, M., Graham, A. W., & Peletier, R. F. 2007, *ApJ*, 665, 1104
Barth, A., Strigari, L. E., Bentz, M. C., Greene, J. E., & Ho, L. C. 2009, *ApJ*, 690, 1031
Böker, T., Laine, S., van der Marel, R. P., Sarzi, M., Rix, H.-W., Ho, L. C., & Shields, J. C. 2002, *AJ, 123*, 1389
Böker, T., Sarzi, M., McLaughlin, D. E., van der Marel, R. P., Rix, H.-W., Ho, L. C., & Shields, J. C. 2004, *AJ, 127*, 105
Böker, T. 2008, *ApJ* (Letters), 672, L111
Butler, D. J. & Martínez–Delgado, D. 2005, *AJ, 129*, 2217
Cappuzzo–Dolcetta, R., & Miocchi, P. 2008, *ApJ*, 681, 1136
Carollo, C. M., Stiavelli, M., de Zeeuw, P. T., & Mack, J. 1997, *AJ, 114*, 2366
Cen, R. 2001, *ApJ*, 560, 592
Chilingarian, I. V., Sil'Chenko, O. K., Afanasiev, V. L., & Prugniel, Ph. 2007, *AstL*, 33, 292
Côté, P., et al. 2006, *ApJS*, 165, 57
Emsellem, E. & van de Ven, G. 2008, *ApJ*, 674, 653
Ferrarese, L. *et al.* 2006, *ApJ* (Letters), 644, L21
Filippenko, A. V. & Sargent, W. L. W. 1989, *ApJ* (Letters), 342, L11
Geha, M., Guhathakurta, P., & van der Marel, R. P. 2002, *AJ, 124*, 3073
Graham, A. W. & Driver, S. P. 2007, *ApJ*, 655, 77
Graham, A. W. & Spitler, L. R. 2009, *MNRAS*, 397, 2148
Harris, W. E. 1996, *AJ, 112*, 1487
Koleva, M., de Rijcke, S., Prugniel, Ph., Zeilinger, W. W., & Michielsen, D. 2009, *MNRAS*, 396, 2133
Layden, A. C. & Sarajedini, A. 2000, *AJ, 119*, 1760
McLaughlin, D. E. 1999, *AJ, 117*, 2398
Merritt, D. 2009, *ApJ*, 694, 959
Milosavlević, M. 2004, *ApJ* (Letters), 605, L13
Nayakshin, S., Wilkinson, M. I., & King, A. 2009, *MNRAS*, 398, 54
Rossa, J., van der Marel, R. P., Böker, T., Gerssen, J., Ho, L. C., Rix, H.-W., Shields, J. C., & Walcher, C.-J. 2006, *AJ, 132*, 1074
Satyapal, S., Böker, T., Mcalpine, W., Ghozzi, M., Abel, N. P., & Heckman, T. 2009, *ApJ*, 704, 439
Shields, J. C., Walcher, C. J., Böker, T., Ho, L. C., Rix, H.-W., & van der Marel, R. P. 2008, *ApJ*, 682, 104
Schinnerer, E., Böker, T., Emsellem, E., & Lisenfeld, U. 2006, *ApJ*, 649, 181
Schinnerer, E., Böker, T., Emsellem, E., & Downes, D. 2007, *A&A* (Letters), 462, L27
Schinnerer, E., Böker, T., Meier, D. S., & Calzetti, D. 2008, *ApJ* (Letters), 684, L21
Seth, A., Agüeros, M., Lee, D., & Basu–Zych, A. 2008, *ApJ*, 678, 116
Walcher, C. J., van der Marel, R. P., McLaughlin, D., Rix, H.-W., Böker, T., Häring, N., Ho, L. C., Sarzi, M., & Shields, J. C. 2005, *ApJ*, 618, 237
Walcher, C. J., Böker, T., Charlot, S., Ho, L. C., Rix, H.-W., Rossa, J., Shields, J. C., & van der Marel, R. P. 2006, *ApJ*, 649, 692
Wehner, E. H. & Harris, W. E. 2006, *ApJ* (Letters), 644, L17

Star clusters: basic galactic building blocks
Proceedings IAU Symposium No. 266, 2009
R. de Grijs & J. R. D. Lépine, eds.

doi:10.1017/S1743921309990883

Colliding galaxies and the rapid assembly of clouds into proto-super star clusters

Guillermo Tenorio–Tagle,[1] Enrique Pérez,[2] Casiana Muñoz–Tuñón,[3] Sergiy Silich[1] and Jan Palouš[4]

[1]Instituto Nacional de Astrofísica Optica y Electrónica, AP 51, 72000 Puebla, Mexico
email: gtt@inaoep.mx

[2]Instituto de Astrofísica de Andalucía (CSIC), Camino bajo de Huetor 50, E-18080 Granada, Spain

[3]Instituto de Astrofísica de Canarias, E-38200 La Laguna, Tenerife, Spain

[4]Astronomical Institute, Academy of Sciences of the Czech Republic, Boční II 1401, 141 31 Prague, Czech Republic

Abstract. We present a short summary of several 2D hydrodynamic calculations that suggest that upon the collision of galaxies two physical mechanisms lead to the formation of proto-super star clusters. These are condensation, induced by radiative cooling, and implosion caused by the shocked intercloud medium. Even in the absence of gravity, these lead to storage and compression of the dense cloud component into massive and compact gravitationally unstable condensations. The resulting entities exhibit enhanced surface densities that are several hundred times higher than their initial values. These are here postulated as the cradles of very efficient and rapid star-formation episodes, able to withstand the negative feedback effects associated with star formation, while leading to the formation of massive and compact super star clusters.

Keywords. galaxies: star clusters, galaxies: interactions

1. Introduction

Hubble Space Telescope observations (Ho 1997; Melo *et al.* 2005; Gilbert & Graham 2007; Walcher *et al.* 2005) led to the discovery of massive, young and compact collections of stars, today called super star clusters, the most spectacular modality of star formation. They are found in a large variety of galaxies, from blue compact dwarfs to colliding spirals. Their stellar masses range from a few $\times 10^5$ to several $\times 10^6$ $M_\odot$, all within a small volume of just a few pc across. Their discovery led immediately to a large number of questions. Perhaps the most important among these is (Tenorio–Tagle *et al.* 2003) how to collect all of the interstellar matter that will be turned into stars into such a small volume? Giant molecular gas complexes, with masses from several $\times 10^5$ to a few $\times 10^6$ $M_\odot$ soon became the favorite suspect sites where super star clusters could form. However, it is well known that molecular clouds far exceed the dimensions of super star clusters and so the question remains unanswered. How can so many stars form unaffected by the negative feedback produced by their neighboring stars? The question points particularly to massive stars. All stars with masses greater than 20 $M_\odot$ evolve rapidly while dumping an enormous amount of ionizing radiation into interstellar space, capable of heating the collapsing cloud, and thus of inhibiting star formation. Furthermore, all stars more massive than 8 $M_\odot$ exhibit strong stellar winds and end up exploding as supernovae. The worry is that such a large and continuous mechanical energy deposition may rapidly disrupt the collapsing cloud and inhibit further star formation. Thus, the central question today is how to avoid the negative feedback produced by massive stars or, alternatively, how do super star clusters form?

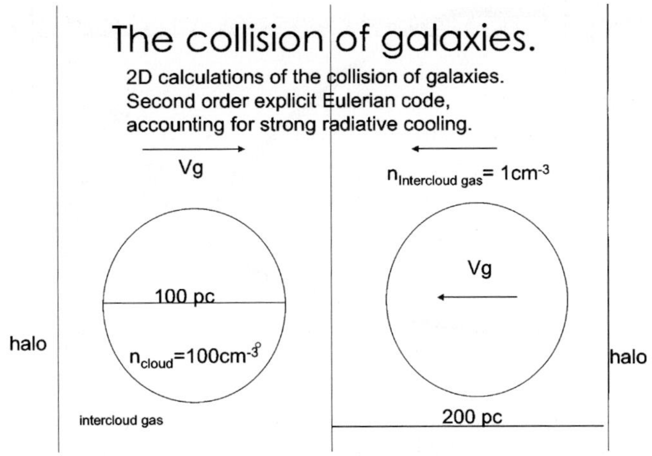

Figure 1. Input model. Two galactic disks approaching each other at high speed (V_g), with their molecular clouds set to undergo a head-on collision. The dimensions and assumed geometry are indicated in the figure.

2. The model

Based on the empirical evidence from the Antennae system (which now presents more than 1000 young super star clusters) and other colliding and merging galaxies (Whitmore *et al.* 1993; Gilbert & Graham 2007), we have started a series of numerical simulations that consider direct collisions of molecular clouds and their surrounding intercloud medium. The calculations assume a collection of molecular clouds immersed in the disks of their host galaxies. The galaxies approach each other and interact at high speed, which leads to the appearance of two strong shocks that process both the intercloud medium and the molecular clouds. Our first results showed that dense clouds can cool rapidly through radiation, long before they are fully processed by their respective shock. On the other hand, if the intercloud gas is initially of low density, say, less than 1 atom cm^{-3}, then its cooling time will be much longer than the collision time ($\sim 2 \times 10^5$ yr). In this case, the shocked intercloud medium would remain hot and highly pressurized, and thus it would be able to implode both the as yet unprocessed cloud and the condensations or fragments that result from the direct cloud–cloud collision. The final result is a pile up of all initial cloud material into massive, compact condensations.

This all happens in a time that is much shorter than the gravitational collapse time of the original clouds. The timescale is also much shorter than the time required for the most massive stars to become supernovae and sufficiently short to allow production of only a small fraction of the total ionizing radiation expected from massive clusters. Star formation should then be able to proceed unaffected by negative feedback. This would lead to a very efficient star-formation episode.

However, this would only happen if a large-scale disturbance of the galaxies' interstellar medium can rapidly aggregate all of the required matter. Or, as phrased by Larson (1987), "If the gas fueling a starburst is turned into stars on a short timescale, then the gas must also be assembled into the starburst region in a similar timescale."

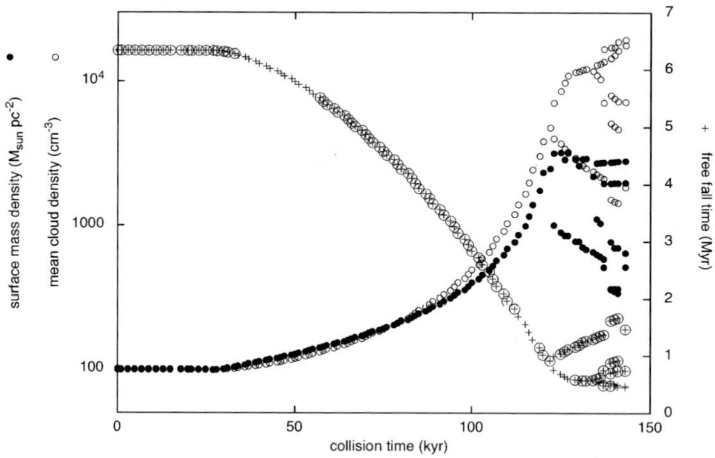

Figure 2. Rapid assembly of the cloud component into super star cluster building blocks. The left axis indicates surface mass density ($M_\odot$ pc^{-2}; filled circles) and the cloud's mean particle density (cm^{-3}; open circles). The right axis marks the corresponding free-fall time (τ_{ff} in Myr; encircled plus signs indicate Jeans-unstable clouds), all shown as a function of time. A few very small clouds have not been plotted for reasons of clarity.

3. First results

Figure 1 shows the initial condition for one of our 2D calculations. In this case, the clouds with a diameter of 100 pc and a density of 100 cm^{-3} (and thus a mass of 1.2×10^6 $M_\odot$ each), are located in the plane of their respective galaxy. They are surrounded by an intercloud medium with a density of 1 cm^{-3} that extends to 100 pc on either side of the galactic planes and all the way to the edge of the computational grid in the direction parallel to the plane of the interaction. The initial setup is such that the clouds will suffer a head-on collision, with an intensity defined only by the velocity of approach of the two galaxies, here 1000 km s^{-1}. The calculations were performed using a 2D second-order explicit Eulerian code, which accounts for strong radiative cooling (Tenorio–Tagle & Muñoz–Tuñón 1997, 1998). The code does not account for gravitational terms and so the resulting flows are only due to radiation hydrodynamics.

Figure 2 plots the run of density (cm^{-3}), surface density ($M_\odot$ pc^{-2}) and free-fall time ($\tau_{ff} = (G\rho)^{-1/2} = 2\, n_3^{-1/2}$ Myr; where n_3 is the number density in units of 10^3 cm^{-3}) as a function of the evolution time (in kyr) for all clouds ($n \geqslant 100$ cm^{-3}) recognized within the computational grid. The figure also indicates clouds that become Jeans unstable to gravitational collapse whenever their τ_{ff} is shorter than their sound-crossing time ($\tau_s = 11\, r\, T^{-1/2}$ Myr, where r is the fragment size in pc and T their mean temperature, in degrees K).

The figure shows how the resulting condensations end up with densities enhanced by a factor of between 10 and several hundreds from their original value. The same happens to their surface density, which is enhanced by factors of a few to several tens from the initial value and, correspondingly, the collapse time is drastically reduced from the initial value (6.2 Myr) to a few $\times 10^5$ yr. The latter implies that during the collision time (2×10^5 yr), the hydrodynamical events act much faster than gravity, leaving all of the dense gas in place within dense and compact condensations ready to collapse, fragment and form stars within a few $\times 10^5$ yr.

4. Conclusions

Given the structure of the interstellar medium and all of its different phases, one of the main questions regarding super star clusters is how to store the large amount of interstellar matter that is turned into stars into such compact volumes.

We have shown that the strong compression caused by the collision of galaxies and, in particular, direct cloud collisions, condensation and fragmentation induced by strong radiative cooling and implosion driven by the hot, shocked intercloud gas, lead to storage and compression of the dense cloud component into massive and compact gravitationally unstable condensations.

The time required to do so is about 10^5 yr, much shorter than the initial collapse time of the individual clouds ($\tau_{\rm ff} \sim 6.25 \times 10^6$ yr). This implies that a large-scale disturbance of a galaxy's interstellar medium may collect and compress the matter required to build super star clusters much faster than gravity alone. In our case, the large-scale disturbance is provided by a galaxy collision. The time required to assemble the cloud component into super star cluster building blocks is also similar to the final free-fall time of the resulting condensations (a few $\times 10^5$ yr). Having all matter in place on such short timescales has profound implications for feedback. This is because the assembly time is much shorter than the time when the first supernova (originating from the most massive star expected to form in the cluster) occurs. On this short timescale, only a small fraction of the total ultraviolet radiation expected from a massive cluster is produced. Star formation under such conditions can then proceed freely, without being affected by negative star-formation feedback, leading instead to a very efficient star-formation episode and possibly to the formation of a super star cluster.

Several more calculations including a structured interstellar medium in the disks and halos of the colliding galaxies, as well as grazing clouds, will be the subject of a forthcoming article.

Acknowledgements

This study was supported by CONACYT México under research grants 60333 and 82912, by the Spanish Ministry of Science and Innovation under grant AYA2007-64712, the ESTALLIDOS collaboration (grant AYA2007-67965-C03-01) and Consolider–Ingenio 2010 program grant CSD2006-00070: First Science with the GTC. The authors acknowledge support from the Institutional Research Plan AV0Z10030501 of the Academy of Sciences of the Czech Republic and from the project LC06014 Center of Theoretical Astrophysics of the Ministry of Education, Youth and Sports of the Czech Republic as well as the CONACYT Mexico and Czech Academy of Sciences research grant 2009–2010.

References

Gilbert, A. & Graham, J. 2007, *ApJ*, 668, 168

Ho, L. C. 1997, *Rev. Mex. A&A Conf. Ser.*, 6, 5

Larson, R. B. 1987 in: T. X. Thuan, T. Montmerle & J. T. T. Van: *Starbursts and galaxy evolution*, p. 467, Gif-sur-Yvette: Editions Frontières

Melo, V. P., Muñoz–Tuñón, C., Maíz–Apellániz, J., & Tenorio–Tagle, G. 2005, *ApJ*, 619, 270

Tenorio–Tagle, G. & Muñoz–Tuñón, C. 1997, *ApJ*, 478, 134

Tenorio–Tagle, G. & Muñoz–Tuñón, C. 1998, *MNRAS*, 293, 299

Tenorio–Tagle, G., Palouš, J., Silich, S., Medina–Tanco, G. A., & Muñoz–Tuñón, C. 2003, *A&A*, 411, 397

Walcher, C. J., van der Marel, R. P., McLaughlin, D., Rix, H.-W., Böker, T., Häring, N., Ho, L. C., Sarzi, M., & Shields, J. C. 2005, *ApJ*, 618, 237

Whitmore, B. C., Schweizer, F., Leitherer, C., Borne, K., & Robert, C. 1993, *AJ*, 106, 1354

Star clusters: basic galactic building blocks
Proceedings IAU Symposium No. 266, 2009
R. de Grijs & J. R. D. Lépine, eds.

doi:10.1017/S1743921309990895

Star cluster disruption

Mark Gieles

European Southern Observatory, Casilla 19001, Santiago, Chile
email: mgieles@eso.org

Abstract. Star clusters are often used as tracers of major star-formation events in external galaxies as they can be studied out to much greater distances than individual stars. It is vital to understand their evolution if they are used to derive, for example, the star-formation history of their host galaxy. More specifically, we want to know how cluster lifetimes depend on their environment and on structural properties such as mass and radius. This review presents a theoretical overview of the early evolution of star clusters and the consequent long-term survival chances. It is suggested that clusters forming with initial densities of $\gtrsim 10^4\ \mathrm{M_\odot\,pc^{-3}}$ survive the gas expulsion, or 'infant mortality,' phase. At $\sim$10 Myr, they are bound and have densities of $\sim 10^{3\pm1}\ \mathrm{M_\odot\,pc^{-3}}$. After this time, they are stable against expansion through stellar evolution, encounters with giant molecular clouds and will most likely survive for another Hubble time if they are located in a moderate tidal field. Clusters with lower initial densities ($\lesssim 100\ \mathrm{M_\odot\,pc^{-3}}$) will disperse into the field within a few 10s of Myrs. Some discussion is given on how extragalactic star cluster populations, and especially their age distributions, can be used to gain insight into disruption.

Keywords. galaxies: star clusters, open clusters and associations: general

1. Introduction

In the Milky Way, the majority of star formation occurs in embedded clusters (Lada & Lada 2003). At the same time, only a small fraction of stars in the disc resides in (open) clusters, indicating that most of the clusters/associations are not long-lived and disruption on timescales on the order of the ages of open clusters is important.

Historically, studies of the lifetimes of star clusters focussed on the open clusters in the Milky Way. At the end of the 1950s, the scarcity of old ($\gtrsim$ Gyr) open clusters was noted nearly simultaneously in three studies: van den Bergh (1957), von Hoerner (1958) and Oort (1958). It was already speculated then by these authors that this was due to their finite lifetimes. The first quantitative explanation for this observation came from Spitzer (1958), who showed that interactions with giant molecular clouds (GMCs) are a plausible cause for the disruption of open clusters. In his model, a cluster with a density of 1 $\mathrm{M_\odot\,pc^{-3}}$ survives the periodic heating due to passing GMCs for about 200 Myr. Wielen (1971) constructed an age distribution of open clusters within 1 kpc of the Sun and derived a median lifetime of 200 Myr. Because of this remarkable agreement with the theoretical predictions of Spitzer, GMC encounters have since been held as the dominant disruption mechanism for open clusters. Another supporting argument for this idea came from the fact that the old open clusters are strongly concentrated towards the Galactic anticentre, where the density of GMCs is low, and their survival chances high (van den Bergh & McClure 1980 and also Figure 1).

Later, Elson & Fall (1985) and Hodge (1987) constructed age distributions for the Large and Small Magellanic Cloud (LMC and SMC) clusters, respectively. They compared both to Wielen's result and showed that the age distributions in the Magellanic Clouds extend

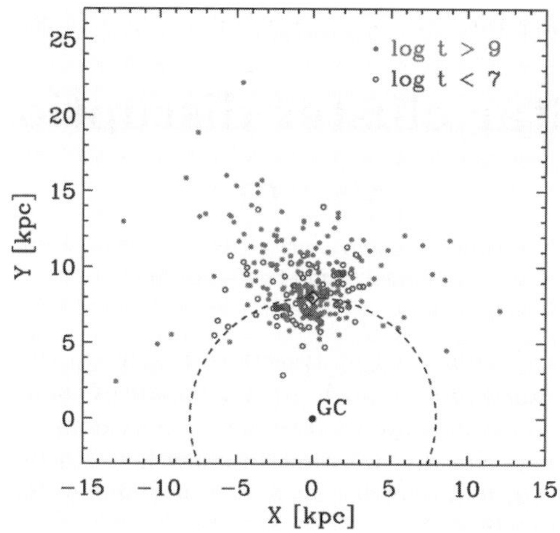

Figure 1. Distribution of young and old open clusters in the Galactic plane, based on the catalogue of Dias *et al.* (2002). The old open clusters ($\geqslant 1$ Gyr) are found preferentially towards the Galactic anticentre.

to older ages, and consequently concluded that these clusters survive longer than their Milky Way counterparts.

This review attempts to summarise work done on the disruption of star clusters since the seminal studies mentioned above. Three time spans are considered in which different destruction mechanisms are important:

- First few Myr: expulsion of residual gas from the star-formation process;
- First few 100 Myr: mass loss due to stellar evolution;
- First Gyr: interactions with GMCs and two-body relaxation in a tidal field.

These points will be discussed from a theoretical and observational point of view in Sections 2 and 3, respectively.

2. Theoretical considerations

2.1. *The first few Myr: gas expulsion and 'infant mortality' of clusters*

A lot of attention has recently been paid to the disruption of star clusters just after their formation, due to the removal of residual gas from the star-formation process by feedback from hot stars. This was already considered by Tutukov (1978) and Lada & Lada (2003) argue it could be the explanation for the strong drop at a few Myr in the age distribution of Galactic embedded and open clusters. The arguments as to why this is destructive are as follows: if the embedded stars are in virial equilibrium with their surroundings, then the stellar velocities are too high once the gas is removed. The cluster consequently expands or even completely dissolves. If we define the total mass of the system as $M_{\rm tot} = M_{\rm g} + M_*$, where M_* is the stellar mass and $M_{\rm gas}$ the gas mass, virial equilibrium before gas expulsion can be written as

$$v_{\rm rms}^2 = \frac{GM_{\rm tot}}{2r_{\rm v}}. \tag{2.1}$$

Here, G is the gravitational constant, $v_{\rm rms}$ the rms velocity of the stars and $r_{\rm v}$ the virial radius of the (embedded) cluster. The latter relates to the half-mass radius as $r_{\rm v} \approx 1.25 r_{\rm h}$, slightly dependent on the density profile. Since $v_{\rm rms}^2$ is three times the one-dimensional velocity dispersion squared, $\sigma_{\rm 1d}^2$, and $r_{\rm h}$ is 4/3 times the half-mass radius in projection,

$r_{\rm eff}$, Equation (2.1) can also be written in observable quantities as

$$\sigma_{\rm 1d}^2 = \frac{GM_{\rm tot}}{\eta r_{\rm eff}}, \tag{2.2}$$

with $\eta \approx 10$. In most models it is assumed that the gas-expulsion timescale, $t_{\rm exp}$, is much shorter than the crossing time of the stars, $t_{\rm cr}$. The latter is defined as

$$t_{\rm cr} = 2\sqrt{2}\left(\frac{r_{\rm v}^3}{GM}\right)^{1/2} \tag{2.3}$$

$$\approx 0.65\,{\rm Myr}\left(\frac{\rho_{\rm h}}{10^3\,{\rm M_\odot pc^{-3}}}\right)^{-1/2}. \tag{2.4}$$

For $t_{\rm cr} \gg t_{\rm exp}$ the positions of the stars are static during the change of the potential and the cluster's response can be calculated using the *impulsive approximation*.

The velocities of the stars right after gas removal are still defined by the total mass of the embedded cluster (Equation 2.1) and hence the cluster will expand to find a new equilibrium. Let's take $M_* = \epsilon M_{\rm tot}$, such that ϵ is the star-formation efficiency (SFE). The final radius can then be expressed in terms of ϵ and the initial radius, $r_{\rm v}(0)$, as (e.g., Hills 1980)

$$\frac{r_{\rm v}}{r_{\rm v}(0)} = \frac{\epsilon}{2\epsilon - 1}. \tag{2.5}$$

From this, it can be seen that for $\epsilon \leqslant 0.5$ the cluster dissolves. Since this value is high for a high SFE, gas expulsion is a plausible cause for the disruption of a large fraction of embedded clusters. A value of $\epsilon = 0.5$, however, is (for several reasons) an upper limit for complete destruction of the cluster:

- *Escape of unbound stars:* For $\epsilon \leqslant 0.5$, the system is globally unbound since the total energy (kinetic plus potential) is positive. However, most of the positive energy is carried away by escaping stars, which have velocities much higher than $v_{\rm rms}$. Almost all dynamical models find that for $\epsilon \gtrsim 0.33$ a fraction of the stars will remain bound (see Figure 2).
- *Subvirial initial conditions:* This was first discussed by Lada *et al.* (1984) who considered a collapsing cloud in which the stars are dynamically 'cold' while the gas is removed (see also Proszkow *et al.* 2009). If the stellar velocities are only a fraction, f, of the virial velocities then Equation (2.5) becomes $r_{\rm v}/r_{\rm v}(0) = \epsilon/(2\epsilon - f^2)$, i.e., the condition for complete disruption becomes $\epsilon = f^2/2$, which for $f = 0.5$ is $\epsilon = 0.125$ (Goodwin 2008), much lower than the analytical estimate of $\epsilon = 0.5$ for initially virialised, embedded clusters. The combination of ϵ and the initial virial state, or the *effective* SFE (Goodwin & Bastian 2008), thus determines the survival chance.
- *Clumpy initial conditions:* Fellhauer *et al.* (2009) showed that if stars form in a clumpy fashion, and if these clumps merge quickly, then the survival chance of the merger remnant is enhanced.
- *Slow gas removal:* If the gas-expulsion timescale is longer than $t_{\rm cr}$, the cluster will expand adiabatically, which can result in a bound cluster for ϵ as low as 0.1 (e.g., Geyer & Burkert 2001; Baumgardt & Kroupa 2007 and Figure 2).

Therefore, the importance of the gas-removal phase in disrupting star clusters depends on many parameters: the initial density of the embedded cluster, the gas-removal timescale, the initial virial state of the embedded cluster, the star-formation efficiency and the initial configuration of the stars (clumpy or smooth density profile). This has led to several different predictions. For example, most models assume that gas removal is instantaneous. If ϵ is the same for all clusters then the disruption rate is independent

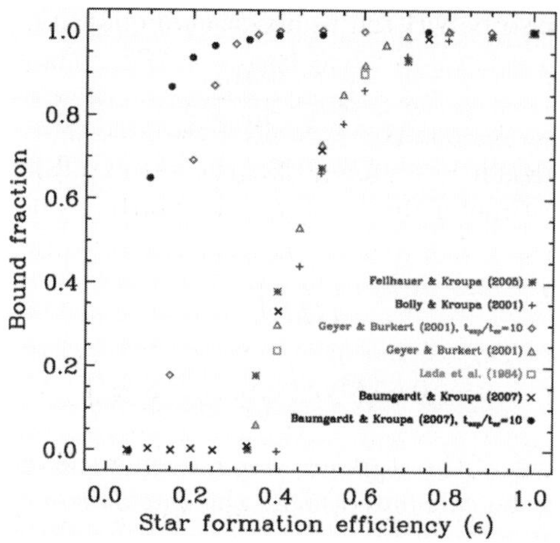

Figure 2. Fraction of stellar mass that remains bound after gas expulsion for different SFEs, found by several authors using N-body simulations. (Data taken from Baumgardt & Kroupa 2007.)

of cluster mass. Baumgardt *et al.* (2008) advocate, however, that gas expulsion is less efficient in disrupting massive clusters because of their higher binding energy per unit mass. This makes the general prediction that the exposed cluster mass function should be shallower than the embedded cluster mass function (see Parmentier *et al.* 2008).

Since the embedded phase is brief ($\sim$1 Myr), most of these parameters are poorly constrained from observations and the interpretation of the models is therefore sensitive to the assumptions that are made with regard to the initial conditions. In Section 3 possible observational signatures of this early disruption will be discussed.

2.2. *The first few 100 Myr: mass loss due to stellar evolution*

Clusters that survive the gas-expulsion phase continue to lose mass through the evolution of the member stars. For a Kroupa-type initial mass function (IMF) between 0.1 and 100 $M_\odot$, the cluster mass reduces roughly by 10/20/30% in the first 10/100/500 Myr. Since these stellar evolution timescales are much longer than typical values of t_{cr}, the cluster will, in most cases, react by adiabatic expansion. This expansion is analogous to the increase of the orbital separation of a binary when one of the members loses mass. It is far less sensitive to the fractional mass loss than in the impulsive approximation, since

$$\frac{r_v(t)}{r_v(0)} = \frac{M(0)}{M(t)} \tag{2.6}$$

$$= \frac{1}{\mu}. \tag{2.7}$$

In the last step, μ is used for the fraction of mass that remains and this can be compared to ϵ in the instantaneous-mass-loss case (Equation 2.5). For $\mu = 0.5$, the cluster will expand by a factor of two and not lose mass, whereas for impulsive mass loss the cluster nearly dissolves, as we saw in Section 2.1.

The situation is in reality somewhat more complicated than this simple prediction. The expansion is more severe and can even result in complete disruption if the cluster is mass segregated before the bulk of the stellar evolution takes place (Vesperini *et al.* 2009). If the cluster is not primordially mass segregated, it will still segregate while the stars evolve (Applegate 1986) and this can eventually lead to enhanced expansion at later times. Finally, the fate of the cluster also depends on the Jacobi (or tidal) radius (r_J).

The combined effect of mass loss through stellar evolution and dynamical evolution in a tidal field was considered by Fukushige & Heggie (1995). They show that when clusters expand to a radius of $\sim 0.5\, r_{\rm J}$, they lose equilibrium and most of their stars overflow $r_{\rm J}$ in a few crossing times. As was the case with gas expulsion, a lot more observational constraints on the initial conditions of star clusters, such as the degree of primordial mass segregation and the initial radius with respect to the tidal radius, are needed before we can make general statements about the effect of stellar evolution on the evolution of star clusters.

2.3. *Interaction with giant molecular clouds*

An external disruptive factor, already considered by Spitzer (1958), is represented by cluster encounters with GMCs. Since GMCs are typically more massive than clusters, the cluster is more affected by an encounter than the cloud (Theuns 1991). The cluster lifetime due to periodic heating of passing clouds is inversely proportional to the volume density of molecular gas, $\rho_{\rm gas}$, and proportional to the density of the cluster, a typical result for disruption by external tidal 'shocks' (e.g., Ostriker *et al.* 1972)

$$t_{\rm dis}^{\rm GMC} \approx 1\,{\rm Gyr} \left(\frac{\rho_{\rm gas}^{\odot}}{\rho_{\rm gas}}\right) \left(\frac{\rho_{\rm h}}{10\,{\rm M_{\odot}\,pc^{-3}}}\right). \tag{2.8}$$

Here, $\rho_{\rm gas}^{\odot} \approx 0.03\ {\rm M_{\odot}\ pc^{-3}}$ is the molecular-gas density in the solar neighbourhood and the constant is taken from Gieles *et al.* (2006) and is equal, within a factor of two, to the original result of Spitzer (1958). The dependence of $t_{\rm dis}$ on $\rho_{\rm gas}$ means that lifetimes of star clusters also scale roughly inversely with the observable surface density of molecular gas, $\Sigma_{\rm gas}$, enabling us to make order-of-magnitude predictions for the lifetimes of clusters in other galaxies. In spiral galaxies, disruption by GMC encounters is especially important in the early stages of evolution, since clusters form in the thin gaseous disc where $\rho_{\rm gas}$ is high. Older clusters are typically more associated with the thick disc, where $\rho_{\rm gas}$ is low and GMC encounters are less frequent. Since young ($\lesssim$1 Gyr) clusters in spiral galaxies have a near-constant radius (e.g., Larsen 2004), Gieles *et al.* (2006) argue that the dissolution time due to GMC encounters is longer for more massive clusters, due to their higher density. It is not clear whether this lack of a mass–radius relation is a universal property imprinted by formation, or the result of an evolutionary effect. This will be addressed in Section 3.

2.4. *Two-body relaxation in the tidal field of the host galaxy*

On somewhat longer timescales, stars are lost from clusters through two-body relaxation in which stars get accelerated to the escape velocity, $v_{\rm esc}$ (e.g., Ambartsumian 1938; Spitzer 1940). The relevant timescale of escape is, to first order, the half-mass relaxation time (Spitzer 1987),

$$t_{\rm rh} = 0.138 \frac{N^{1/2} r_{\rm h}^{3/2}}{\sqrt{\bar{m} G} \ln \Lambda}, \tag{2.9}$$

where N is the number of stars, $\bar{m}$ the mean stellar mass and $\ln\Lambda$ the Coulomb logarithm: $\ln\Lambda \approx \ln 0.11N$ (Giersz & Heggie 1994). The relaxation time is approximately the time stars need to establish a Maxwellian velocity distribution. A fraction $\xi_{\rm e}$ of the stars in the tail of the distribution have velocities larger than $v_{\rm esc}$ and consequently escape. Assuming that this high-velocity tail is refilled every $t_{\rm rh}$, the dissolution timescale is $t_{\rm dis} = t_{\rm rh}/\xi_{\rm e}$. For isolated clusters, $v_{\rm esc} = 2\, v_{\rm rms}$. For a Maxwellian velocity distribution a fraction $\xi_{\rm e} = 0.0074$ has $v > 2\, v_{\rm rms}$ and then $t_{\rm dis} = 137\, t_{\rm rh}$. For tidally limited clusters $\xi_{\rm e}$ is higher since $v_{\rm esc}$ is lower. For a typical cluster density profile $\xi_{\rm e} \approx 0.033$, implying

$t_{\rm dis} \approx 30\,t_{\rm rh}$ (Spitzer 1987). The escape fraction $\xi_{\rm e}$ is often taken as constant (e.g., Gnedin & Ostriker 1997). However, $\xi_{\rm e}$ depends on $r_{\rm h}$ (through $v_{\rm rms}$) and on the strength of the tidal field, or $r_{\rm J}$ (through $v_{\rm esc}$). Effectively, $\xi_{\rm e}$ depends on the ratio $r_{\rm h}/r_{\rm J}$ (e.g., Spitzer & Chevalier 1973; Wielen 1988). Gieles & Baumgardt (2008) show that $\xi_{\rm e} \propto (r_{\rm h}/r_{\rm J})^{3/2}$ for $r_{\rm h}/r_{\rm J} \gtrsim 0.05$ (the tidal regime). Combined with Equation (2.9), we then find for clusters on circular orbits in the tidal regime that $t_{\rm dis} \propto N/\omega$, apart from the slowly varying Coulomb logarithm. Here, $\omega \equiv V_{\rm G}/R_{\rm G}$ is the angular frequency in the galaxy and $R_{\rm G}$ and $V_{\rm G}$ are the galactocentric distance and the velocity around the galaxy centre, respectively. So, for a flat rotation curve, we find $t_{\rm dis} \propto R_{\rm G}$ for a cluster of a given mass (e.g., Chernoff & Weinberg 1990; Vesperini & Heggie 1997). This linear dependence of $t_{\rm dis}$ on $R_{\rm G}$ makes it difficult to explain the universality of the globular cluster mass function from dynamical evolution of a power-law initial cluster mass function (e.g., Vesperini *et al.* 2003), but this will not be discussed here. Note that ω can also be written as $\omega \propto \rho_{\rm J}^{1/2}$, where $\rho_{\rm J}$ is the density within $r_{\rm J}$, which is how Lee & Ostriker (1987) present it.

Fukushige & Heggie (2000) show that the linear scaling of $t_{\rm dis}$ with N is slightly affected by the finite time it takes stars to find the Lagrangian 'exit' points. Baumgardt (2001) showed that $t_{\rm cr}$ is of importance and finds, for equal-mass clusters, that $t_{\rm dis} \propto t_{\rm rh}^{3/4} t_{\rm cr}^{1/4}$. Baumgardt & Makino (2003) find that this scaling also holds for models of clusters with a mass spectrum, stellar evolution and for different types of orbits in a logarithmic potential. Their result for $t_{\rm dis}$ can be summarised as

$$t_{\rm dis} \approx 2\,{\rm Myr}\left(\frac{N}{\ln\Lambda}\right)^{3/4}\frac{R_{\rm G}}{\rm kpc}\left(\frac{V_{\rm G}}{220\,{\rm km\,s^{-1}}}\right)^{-1}(1-\varepsilon), \tag{2.10}$$

where ε is the eccentricity of the orbit. If the Coulomb logarithm is taken into account, the scaling is something like $t_{\rm dis} \propto N^{0.65}$ (Lamers *et al.* 2005).

3. Observational signatures of cluster disruption

3.1. *The velocity dispersion of young clusters*

Ho & Filippenko (1996) were the first to demonstrate that one can empirically determine the dynamical mass, $M_{\rm dyn}$, of star clusters beyond the Local Group using their integrated light. They used high-resolution spectroscopy to measure $\sigma_{\rm 1d}$ from the width of the spectral lines and high-angular-resolution (*HST*) imaging to measure $r_{\rm eff}$. The dynamical mass then follows from the virial relation (Equation 2.2). They did this for the ~10 Myr-old super star cluster 'A' in the dwarf starburst galaxy NGC 1569 and found that its $M_{\rm dyn}$ is comparable to the masses of Milky Way globular clusters, concluding that cluster 'A' is a young globular cluster. Since then, this analysis has been applied to several dozen more clusters with different ages and masses and in different galaxies (e.g., Larsen *et al.* 2004; Bastian *et al.* 2006; McCrady & Graham 2007; Mengel *et al.* 2008).

3.2. *Supervirial and dispersing?*

The empirically derived $M_{\rm dyn}$ relies on the assumption that the cluster is in virial equilibrium. One way of verifying the validity of this assumption is by comparing $M_{\rm dyn}$ with the photometric mass, $M_{\rm phot}$. The latter can be derived from the cluster's total luminosity and age and comparison to an SSP model. Bastian *et al.* (2006) provide a compilation of $M_{\rm phot}$ and $M_{\rm dyn}$ values of 19 clusters. They find that for the somewhat older clusters ($\gtrsim 50-100$ Myr), there is good agreement between $M_{\rm phot}$ and $M_{\rm dyn}$. For many of the young (~10 Myr) star clusters $M_{\rm dyn} \gg M_{\rm phot}$. Goodwin & Bastian (2006) suggest

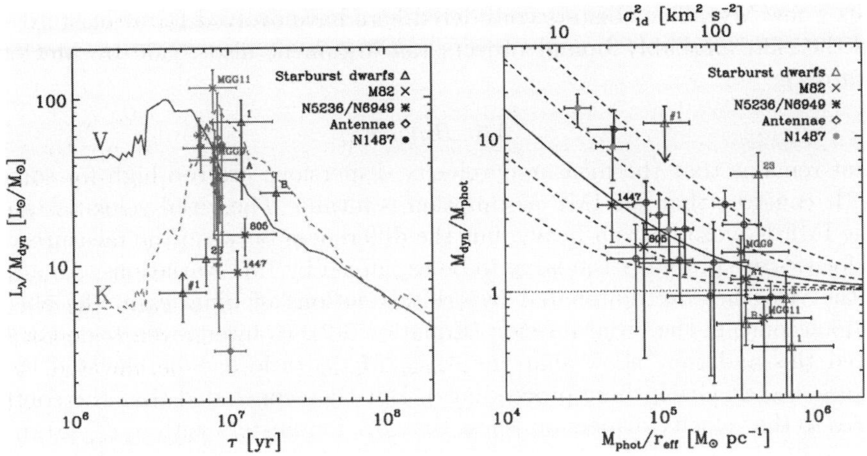

Figure 3. *(left)* Light-to-dynamical-mass ratio for 24 clusters from the literature. The red and blue points refer to studies done in the optical and near-infrared, respectively. Different symbols correspond to different galaxy types. The lines show the photometric evolution of an SSP model in two filters. *(right)* Dynamical mass over photometric mass for the same clusters, shown as a function of $M_{\rm phot}/r_{\rm eff}$, which is a proxy of $\sigma_{\rm 1d}$ (Equation 2.2). The full line is a prediction of the effect of binaries on $M_{\rm dyn}$ (Equation 3.1), with 1σ and 2σ variations due to stochastic fluctuations shown as dashed lines. (Figure from Gieles *et al.* 2009.)

that this is a signature of gas expulsion and the 'infant mortality' scenario discussed in Section 2.1, i.e., they conclude that these clusters are supervirial.

In the left panel of Figure 3, an updated version of figure 5 in Bastian *et al.* (2006) is given. It shows the light-to-dynamical-mass ratio of 24 young star clusters in different galaxies. The data are compiled from the literature (McCrady *et al.* 2003 ; Bastian *et al.* 2006; Smith *et al.* 2006; McCrady & Graham 2007; Moll *et al.* 2007; Östlin *et al.* 2007; Larsen *et al.* 2008; Mengel *et al.* 2008) and will be discussed in more detail in Gieles *et al.* (2009) and Portegies Zwart *et al.* (2010). The observations taken in the visible are shown in blue and those taken in the near-infrared are shown in red. The photometric evolution in the V and K bands from the Bruzual & Charlot (2003) SSP models using a Chabrier IMF is shown as a full (blue) line and a dashed (red) line, respectively. All clusters have an age close to 10 Myr. This is because of observational constraints: younger clusters are more obscured and at 10 Myr the red-supergiant phase sets in, making the clusters brighter (especially in the near-infrared; see Figure 3) and therefore easier to study.

It is important to realise what initial conditions are required for the gas-expulsion scenario to work. The models discussed in Section 2.1 show that after instantaneous gas removal it takes roughly 10 initial crossing times to either completely disrupt or expand to a new virial equilibrium. In the models of Goodwin & Bastian (2006) this corresponds to $\sim$25 Myr since their clusters have $t_{\rm cr} = 2.5$ Myr in the embedded phase. Only with an initial $t_{\rm cr}$ of a few Myr is it possible to 'catch' an unbound, expanding cluster at 10 Myr. This corresponds to an initial density (stars and gas) of $\sim$100 M$_\odot$ pc^{-3} (Equation 2.4). Since the clusters considered here have densities of $\sim$10^3 M$_\odot$ pc^{-3} (see Figure 4), gas expulsion cannot be responsible for the high velocity dispersions. The densities in the embedded phase were at least a factor $1/\epsilon$ higher because of the presence of gas, but probably even more since they have also expanded since then. A similar conclusion can be drawn from the measured velocities, which are on the order of 20 km s^{-1}. With such velocities, an unbound cluster dissolves into the field in a few Myr, incompatible with their radii being only a few pc. We conclude that the 'infant mortality' phase of dense clusters only lasts a few Myr. The clusters considered here have evolved for at least $10 - 100\,t_{\rm cr}$ and are therefore probably bound objects (an argument also made by McCrady and collaborators).

3.3. *Binaries*

The point remains that the measured velocity dispersions are too high for some of the clusters. It could be that the IMF assumption is invalid. The virial velocities are higher when the IMF is more bottom-heavy, but the differences between the measured and the virial velocities are probably too large to be explained by IMF variations. An alternative is that the velocities are dominated by orbital motions of binaries in the cluster, not taken into account in the virial relation (Equation 2.2). Kouwenhoven & de Grijs (2008) considered this and they show that the $M_{\rm dyn}/M_{\rm phot}$ ratio can be elevated by several factors (see also de Grijs, these proceedings). But they concluded that the contribution of binaries to the velocity dispersion is less than 5% for clusters with $\sigma_{\rm 1d} \gtrsim 10\,{\rm km\,s^{-1}}$. So binaries cannot be responsible for the observed velocities according to their model. But Kouwenhoven & de Grijs (2008) did not include the mass-dependent stellar mass-to-light ratio in their models and also did not take into account stars more massive than 20 $M_\odot$. Therefore, the effect will be stronger for the clusters considered in Figure 3, since the light is dominated by massive stars ($\sim$15 $M_\odot$) for which (primordial) multiplicity is high and short periods and mass ratios close to unity are common.

Gieles *et al.* (2009) derive a simple analytic expression for the additional velocity dispersion due to orbital motions of stars in binaries, $\sigma_{\rm bin}$, relative to the dynamical dispersion of a virialised cluster, $\sigma_{\rm 1d}$,

$$\frac{\sigma_{\rm bin}^2}{\sigma_{\rm 1d}^2} \approx \left(\frac{f}{0.25}\right)\left(\frac{q}{0.6}\right)^{\frac{3}{2}}\left(\frac{m_1}{15{\rm M}_\odot}\right)^{\frac{2}{3}}\left(\frac{10^3\ {\rm days}}{P}\right)^{\frac{2}{3}}\left(\frac{M/r_{\rm eff}}{10^5{\rm M}_\odot{\rm pc}^{-1}}\right)^{-1}. \tag{3.1}$$

Here, f is the binary fraction, q the mass ratio, m_1 the mass of the primary star (the contribution of the secondary star is ignored) and P the orbital period. The normalisation constants are based on properties of massive binaries in the Milky Way and appropriate for an age of 10 Myr. Since the cluster light is dominated by the most massive stars, the measured velocity dispersion is affected by binaries according to this relation. The prediction for $M_{\rm dyn}$ including the velocities of binaries ($M_{\rm dyn} \propto \sigma_{\rm 1d}^2 + \sigma_{\rm bin}^2$) is shown in the right panel of Figure 3 (full line). The dashed lines show the 1σ and 2σ spread in the model due to stochastics. The discrepancy between $M_{\rm dyn}$ and $M_{\rm phot}$ can largely be explained by virialised clusters with a modest binary population. Assuming that all clusters host a similar binary population, the clusters with a low dynamical dispersion, for which we use $M_{\rm phot}/r_{\rm eff}$ as a proxy (Equation 2.2), are affected most. This could be the reason why the data show a decreasing $M_{\rm dyn}/M_{\rm phot}$ ratio for an increasing $M_{\rm phot}/r_{\rm eff}$ ratio.

This suggests that these clusters are survivors of the gas-expulsion phase and are at the beginning of the stellar evolution phase. In the next Hubble time, these clusters will adiabatically expand by a factor of $\sim$2 due to mass loss through stellar evolution if they are not mass segregated (Section 2.2) and it remains to conclude that these clusters are good young globular cluster candidates if the tidal field they are in is not too strong.

3.4. *The mass–radius relation of star clusters*

Several studies have pointed out that there is a very weak dependence of cluster radius on mass/luminosity (e.g., Zepf *et al.* 1999; Larsen 2004; Bastian *et al.* 2005). For the clusters in Figure 4, this seems not to be the case. Although these are only 24 clusters, and it is far from a tight relation, these objects seem more consistent with a constant

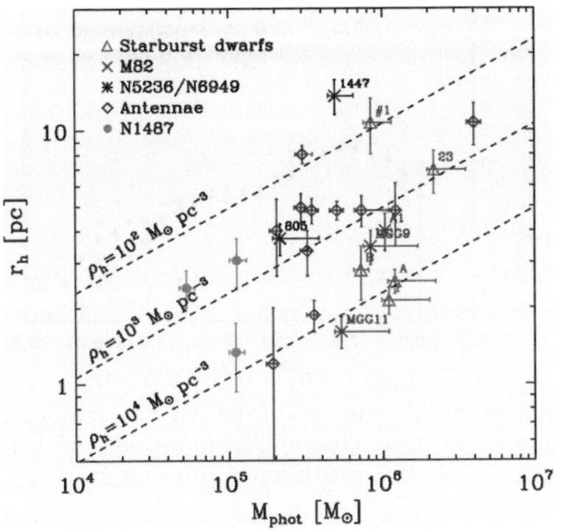

Figure 4. Mass–radius relation for the same 24 clusters shown in Figure 3, using the same colour and symbol coding and $r_{\rm h} = (4/3)r_{\rm eff}$. Lines of constant half-mass density, $\rho_{\rm h} \equiv 3M/(8\pi r_{\rm h}^3)$, are overplotted. The cluster densities imply $0.2 \lesssim t_{\rm cr}/{\rm Myr} \lesssim 2$ (Equation 2.4).

density of $\rho_{\rm h} \approx 10^{3\pm1}$ M$_\odot$, or $r_{\rm h} \approx (M/10^4\,{\rm M}_\odot)^{1/3}$, within a factor of two. The difference between this figure and the studies mentioned above is that the clusters considered here are confined to a narrow (young) age range. The mass–radius relation is of importance for disruption since $t_{\rm dis}^{\rm GMC} \propto \rho_{\rm h}$ (Section 2.3). So, for a constant cluster density, GMCs will destroy all clusters equally fast. If the mass–radius relation evolves with time, for example to a near-constant radius as found by several authors, then $t_{\rm dis}^{\rm GMC}$ will become mass dependent. In any case, the densities of these clusters imply that $t_{\rm dis}^{\rm GMC}$ is much longer than a Hubble time, even when $\rho_{\rm gas}$ is 10 times higher than in the solar neighbourhood (Equation 2.8), so GMC encounters will not play an important role in the disruption of these clusters at this stage.

In a solar-neighbourhood-type tidal field all clusters in Figure 4 have $r_{\rm h}/r_{\rm J} \approx 0.03$ within a factor of two. This implies that these clusters are probably also very stable against expansion due to mass loss through stellar evolution, since in Section 2.2 it was argued that rapid dissolution due to mass loss through stellar evolution only occurs when $r_{\rm h}/r_{\rm J}$ increases to ∼0.5. A similar idea was recently put forward by Pfalzner (2009), who discovered two evolutionary sequences in young Galactic star clusters. A dense group that starts with a density of $\sim 10^5$ M$_\odot$ pc^{-3} at a few Myr containing the Arches cluster, NGC 3603 and Trumpler 14, to which they refer as 'starburst clusters.' They seem to expand along a sequence of constant mass, where at 10 − 20 Myr there are the red-supergiant clusters (RSGC01 and RSGC02) with a density of $\sim 10^3$ M$_\odot$ pc^{-3}. The second sequence also starts at a few Myr, but at much lower densities (~ 10 M$_\odot$ pc^{-3}) and expands along a $M \propto 1/r_{\rm h}$ sequence to densities comparable to the field-star density. These are referred to as 'leaky clusters.' The clusters in Figure 4 could be compared to the red-supergiant clusters in the Milky Way, i.e., the end stage of the starburst sequence.

4. Discussion

Until the 1990s, the open clusters in the Milky Way and the populous clusters in the Magellanic Clouds were the targets of studies of disruption. *HST* has enabled us to establish the properties of larger populations of clusters containing more massive clusters in quiescent spiral galaxies (e.g., Larsen & Richtler 2000), interacting galaxies

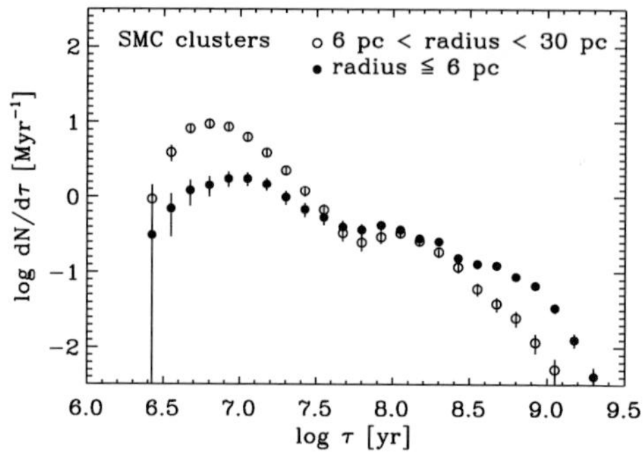

Figure 5. Age distribution of SMC clusters based on the catalogue of Chiosi *et al.* (2006). The sample is split into small and large clusters/associations, with the boundary at 6 pc. The histograms are made using a 0.5 dex bin width with different starting values (boxcar averaging).

(e.g., Bastian *et al.* 2005; Whitmore *et al.* 1999) and merger remnants (e.g., Miller *et al.* 1997).

The primary tool used for studies of cluster disruption is the age distribution. Different groups give varying weights to the various aspects described in Section 2 in the interpretation of the results. There are two flavours of empirically based cluster disruption models: one is based on *externally driven cluster disruption*. In this model, the dissolution time depends on the mass of the cluster and its environment. It was based on the age and mass distributions of luminosity-limited cluster samples in different galaxies (Boutloukos & Lamers 2003). The variations in dissolution timescales are explained by differences in tidal-field strength (Lamers *et al.* 2005) and GMC density (Gieles *et al.* 2006). The second model assumes that *internally driven cluster disruption* is most important. In this model, roughly 80–90% of the number of clusters is destroyed each age dex resulting in a (mass-limited) age distribution that declines roughly as τ^{-1}. In this model, it is assumed that the infant mortality of clusters proceeds for a few 100 Myr and that the disruption rate is equally fast for all masses. This model is based on the cluster population of the Antennae galaxies (Fall *et al.* 2005; Whitmore *et al.* 2007) and it was argued by these authors that their model is universal and should be able to describe the age distributions of cluster populations in other galaxies as well.

The universal cluster dissolution scenario has led to some controversy. Chandar *et al.* (2006) show that the age distribution of SMC clusters declines as $\tau^{-0.85}$, almost the same as in the Antennae. Gieles *et al.* (2007) used the same data set and show that the age distribution of massive clusters ($\gtrsim 10^{3.5}$ $M_\odot$) is flat in the first few 100 Myr. They reproduce the $\tau^{-0.85}$ result for the full sample and conclude that the decline is caused by detection incompleteness, confirmed by de Grijs & Goodwin (2008) using similar arguments but an independent data set. Boutloukos & Lamers (2003) show that for a constant formation history and no disruption, the age distribution of a luminosity-limited sample declines as $\tau^{-\zeta(\alpha-1)}$, where $-\alpha$ is the index of the cluster initial mass function. The index ζ describes the cluster fading with age: $F_\lambda(\tau) \propto \tau^{-\zeta}$, with F_λ being the flux at wavelength λ of a cluster with constant mass. For the $U(V)$ band, $\zeta \approx 1.0(0.7)$, resulting in $\tau^{-\eta}$, with $0.7 \lesssim \eta \lesssim 1.0$ due to fading alone when $\alpha = 2$. If the cluster mass function is a Schechter function (Gieles 2009; Larsen 2009), the age distribution of a luminosity-limited sample not affected by disruption is as steep as $0.9 \lesssim \eta \lesssim 1.4$ (see also Konstantopoulos, these proceedings). Both models rely on the assumption that the cluster-formation history has been constant in the age range considered. Bastian *et al.*

(2009) recently pointed out that this is invalid for interacting galaxies like the Antennae galaxies and that a recent burst can be confused by mass-independent cluster disruption.

Another issue, especially important for distant cluster populations, is what we actually define as a star cluster. Figure 5 shows the age distribution of clusters and associations in the central region of the SMC (Chiosi *et al.* 2006). The sample is split in two subsamples of ~200 clusters, based on their size. The age distribution of large clusters falls off much more quickly than that of the compact group. The median radii of both samples are 4.5 pc and 9 pc, respectively. At a distance of 20 Mpc, it will be very hard to distinguish between these two groups. A speculative suggestion is that more of the short-lived 'associations' or 'leaky clusters' are included in cluster samples in distant galaxies, which would not be considered genuine star clusters if they could be resolved.

Acknowledgement

I thank Brad Whitmore and Iraklis Konstantopoulos for an interesting discussion over lunch at Copacabana beach during this conference and the organisers of IAUS 266 for an interesting meeting!

References

Applegate, J. H. 1986, *ApJ*, 301, 132
Ambartsumian, V. A. 1938, *Sci. Mem. Leningrade State Univ. #22, Ser. Math. Sci.*, 4, 19
Bastian, N., Gieles, M., Lamers, H. J. G. L. M., Scheepmaker, R. A., & de Grijs, R. 2005 *A&A*, 431, 905
Bastian, N., Saglia, R. P., Goudfrooij, P., Kissler–Patig, M., Maraston, C., Schweizer, F., & Zoccali, M. 2006, *A&A*, 448, 881
Bastian, N., Trancho, G., Konstantopoulos, I. S., & Miller, B. W. 2009, *ApJ*, 701, 607
Baumgardt, H. 2001, *MNRAS*, 325, 1323
Baumgardt, H. & Kroupa, P. 2007, *MNRAS*, 380, 1589
Baumgardt, H., Kroupa, P., & Parmentier, G. 2008, *MNRAS*, 384, 1231
Baumgardt, H. & Makino, J. 2003, *MNRAS*, 340, 227
Boutloukos, S. G. & Lamers, H. J. G. L. M. 2003, *MNRAS*, 338, 717
Bruzual, G. & Charlot, S. 2003, *MNRAS*, 344, 1000
Chandar, R., Fall, S. M., & Whitmore, B. C. 2006, *ApJ* (Letters), 650, L111
Chernoff, D. F. & Weinberg, M. D. 1990, *ApJ*, 351, 121
Chiosi, E., Vallenari, A., Held, E. V., Rizzi, L., & Moretti, A. 2006, *A&A*, 452, 179
de Grijs, R. & Goodwin, S. P. 2008, *MNRAS*, 383, 1000
Dias, W. S., Alessi, B. S., Moitinho, A., & Lépine, J. R. D. 2002, *A&A*, 389, 871
Elson, R. A. W. & Fall, S. M. 1985, *ApJ*, 299, 211
Fall, S. M., Chandar, R., & Whitmore, B. C. 2005, *ApJ* (Letters), 631, L133
Fellhauer, M., Wilkinson, M. I., & Kroupa, P. 2009, *MNRAS*, 397, 954
Fukushige, T. & Heggie, D. C. 1995, *MNRAS*, 276, 206
Fukushige, T. & Heggie, D. C. 2000, *MNRAS*, 318, 753
Geyer, M. P. & Burkert, A. 2001, *MNRAS*, 323, 988
Gieles, M. & Baumgardt, H. 2008, *MNRAS*, 389, L28
Gieles, M., Lamers, H. J. G. L. M., & Portegies Zwart, S.F. 2007, *ApJ*, 668, 268
Gieles, M., Portegies Zwart, S. F., Baumgardt, H., Athanassoula, E., Lamers, H. J. G. L. M., Sipior, M., & Leenaarts, J. 2006, *MNRAS*, 371, 793
Gieles, M. 2009, *MNRAS*, 394, 2113
Gieles, M., Sana, H., & Portegies Zwart, S. F. 2009, *MNRAS*, in press (arXiv:0911.1557)
Giersz, M. & Heggie, D. C. 1994, *MNRAS*, 268, 257
Gnedin, O. Y. & Ostriker, J. P. 1997, *ApJ*, 474, 223
Goodwin, S. P 2008, in: E. Pérez, R. de Grijs, & R. M. González Delgado (eds.), *Young massive star clusters – Initial conditions and environments*, *ApSS*, 324, 259

Goodwin, S. P. & Bastian, N. 2006, *MNRAS*, 373, 752

Hills, J. G. 1980, *ApJ*, 235, 986

Ho, L. C. & Filippenko, A. V. 1996, *ApJ* (Letters), 466, L83

Hodge, P. 1987, *PASP*, 99, 724

Kouwenhoven, M. B. N., & de Grijs, R. 2008, *A&A*, 480, 103

Lada, C. J. & Lada, E. A. 2003, *ARA&A*, 41, 57

Lada, C. J., Margulis, M., & Dearborn, D. 1984, *ApJ*, 285, 141

Lamers, H. J. G. L. M., Gieles, M., & Portegies Zwart, S. F. 2005, *A&A*, 429, 173

Larsen, S. S. 2004, *A&A*, 416, 537

Larsen, S. S. 2009, *A&A*, 494, 539

Larsen, S. S., Origlia, L., Brodie, J., & Gallagher III, J. S. 2008, *MNRAS*, 383, 263

Larsen, S. S., Brodie, J. P., & Hunter, D. A. 2004, *AJ*, 128, 2295

Larsen, S. S., & Richtler, T. 2000 *A&A*, 354, 836

Lee, H. M., & Ostriker, J. P. 1987, *ApJ*, 322, 123

McCrady, N., Gilbert, A. M., & Graham, J. R. 2003, *ApJ*, 596, 240

McCrady, N. & Graham, J. R. 2007, *ApJ*, 663, 844

Mengel, S., Lehnert, M. D., Thatte, N. A., Vacca, W. D., Whitmore, B., & Chandar, R. 2008, *A&A*, 489, 1091

Miller, B. W., Whitmore, B. C., Schweizer, F., Fall, S. M. 1997, *AJ*, 114, 2381

Moll, S. L., Mengel, S., de Grijs, R., Smith, L. J., & Crowther, P. A. 2007, *MNRAS*, 382, 1877

Östlin, G., Cumming, R. J., & Bergvall, N. 2007, *A&A*, 461, 471

Oort, J. H. 1958, in: D. J. K. O'Connell (ed.), *Proc. Ricerche Astronomiche*, 5, 63

Ostriker, J. P., Spitzer Jr., L. J., & Chevalier, R. A. 1972, *ApJ* (Letters), 176, L51

Parmentier, G., Goodwin, S. P., Kroupa, P., & Baumgardt, H. 2008, *ApJ*, 678, 347

Pfalzner, S. 2009, *A&A* (Letters), 498, L37

Portegies Zwart, S. F., McMillan, S. L. W., & Gieles, M. 2010, *ARA&A*, in prep.

Proszkow, E.-M., Adams, F. C., Hartmann, L. W., & Tobin, J. J. 2009, *ApJ*, 697, 1020

Smith, L. J., Westmoquette, M. S., Gallagher III, J. S., O'Connell, R. W., Rosario, D. J., & de Grijs, R. 2006, *MNRAS*, 370, 513

Spitzer Jr., L. J. 1940, *MNRAS*, 100, 396

Spitzer Jr., L. J. 1958, *ApJ*, 127, 17

Spitzer Jr., L. 1987, *Dynamical evolution of globular clusters*, Princeton: Princeton University Press

Spitzer Jr., L. J. & Chevalier, R. A. 1973, *ApJ*, 183, 565

Theuns, T. 1991, *Mem. Soc. Astron. Ital.*, 62, 909

Tutukov, A. V. 1978, *A&A*, 70, 57

van den Bergh, S. 1957, *ApJ*, 125, 445

van den Bergh, S., & McClure, R. D. 1980, *A&A*, 88, 360

Vesperini, E. & Heggie, D. C. 1997, *MNRAS*, 289, 898

Vesperini, E., Zepf, S. E., Kundu, A., & Ashman, K. M. 2003, *ApJ*, 593, 760

Vesperini, E., McMillan, S. L. W., & Portegies Zwart, S. 2009, *ApJ*, 698, 615

von Hoerner, S. 1958, *Zeitschr. Astrophys.*, 44, 221

Whitmore, B. C., Chandar, R., & Fall, S. M. 2007, *AJ*, 133, 1067

Whitmore, B. C., Zhang, Q., Leitherer, C., Fall, S. M., Schweizer, F., & Miller, B. W. 1999, *AJ*, 118, 1551

Wielen, R. 1971, *A&A*, 13, 309

Wielen, R. 1988, in: J. E. Grindlay & A. G. D. Philp (eds.), *Proc. IAU Symp.* No. 126, p. 393

Zepf, S. E., Ashman, K. M., English, J., Freeman, K. C., & Sharples, R. M. 1999, *AJ*, 118, 752

Star clusters: basic galactic building blocks
Proceedings IAU Symposium No. 266, 2009
R. de Grijs & J. R. D. Lépine, eds.

doi:10.1017/S1743921309990901

On the temporal evolution of the stellar mass function of Galactic clusters

Guido De Marchi,[1] Francesco Paresce[2] and Simon Portegies Zwart[3]

[1]ESA, Space Science Department, Keplerlaan 1, 2200 AG Noordwijk, the Netherlands
email: gdemarchi@rssd.esa.int

[2]INAF – IASF, Via Gobetti 101, 40129 Bologna, Italy

[3]Sterrewacht Leiden, Leiden University, Postbus 9513, 2300 RA Leiden, the Netherlands

Abstract. We show that we can obtain a good fit to the present-day stellar-mass functions of a large sample of young and old Galactic clusters with a tapered Salpeter power-law distribution function with an exponential truncation of the form $dN/dm \propto m^{\alpha}\,[1-\exp(-m/m_c)^{\beta}]$. The average value of the power-law index α is ~ -2.2, very close to the Salpeter value of -2.3, while the characteristic mass, m_c, is in the range 0.1–0.6 $M_{\odot}$ and does not seem to vary in any systematic way with the present cluster parameters such as metal abundance, total cluster mass or central concentration. However, the characteristic mass shows a remarkable correlation with the dynamical age of the cluster, namely $m_c/M_{\odot} \simeq 0.15 + 0.5 \times t_{\rm dyn}^{3/4}$, where $t_{\rm dyn}$ is the dynamical time, taken as the ratio of cluster age and dissolution time. The small scatter around this correlation is likely due to uncertainties on the estimated value of $t_{\rm dyn}$. We attribute the observed trend to the onset of mass segregation through two-body relaxation in a tidal environment, causing preferential loss of low-mass stars from the cluster and hence a drift of the characteristic mass towards higher values. If dynamical evolution is indeed at the origin of the observed trend, it seems plausible that globular clusters, now with $m_c \simeq 0.35\,M_{\odot}$, were born with a stellar mass function very similar to that measured today in the youngest Galactic clusters and with a value of m_c around 0.15 $M_{\odot}$. This is consistent with the absence of a turn-over in the mass function of the Galactic bulge down to the observational limit at $\sim 0.2\,M_{\odot}$ and argues for the universality of the initial mass function of Population I and II stars.

Keywords. globular clusters: general, open clusters and associations: general, stars: luminosity function, mass function

---

1. Introduction

As stated in the scientific rationale of this Symposium, there is general consensus that stars do not form in isolation but rather from the fragmentation of molecular clouds, leading to star cluster formation (e.g., Elmegreen, these proceedings). This makes stellar clusters ideal places to study the properties of star formation and its end result, the stellar initial mass function (IMF). However, it is also equally well established today that the vast majority of stars, in our Galaxy and elsewhere, are not in clusters but in the field, since clusters disrupt over time (see, e.g., Gieles, these proceedings). Therefore, any attempt to set constraints on the star-formation mechanisms from the analysis of the present-day stellar mass function (MF) of but the youngest clusters cannot ignore the consequences of their dynamical evolution.

This issue becomes particularly important if we want to compare the results of star formation in physically different environments that are noncoeval. The obvious example is addressing differences in the way low-mass stars ($< 1\,M_{\odot}$) formed in globular clusters (GCs), at $z \simeq 5$, and in young clusters in the local Universe. While in both cases the raw data show a broad plateau in the mass distribution, suggesting a characteristic mass on the order a few tenths of $M_\odot$ (e.g., Elmegreen *et al.* 2008), until the effects of dynamical evolution are properly understood no conclusion can be drawn as to the uniformity of the star-formation process across time, i.e., on the universality of the IMF.

On the other hand, even though the two-body relaxation process that governs the cluster's dynamical evolution and leads to preferential loss of low-mass stars from the cluster is today rather well understood (e.g., Spitzer 1987), it is in general not possible to roll back the effects of dynamics and derive the IMF from the present-day MF of a cluster, since we cannot trace back the trajectories of stars that have escaped from the cluster. It is, however, possible and statistically meaningful to study the evolution of the stellar MF on a global scale by looking at the differences between clusters at different evolutionary stages. This requires a homogeneous sample of high-quality observations of Galactic (open and globular) clusters, treated in a uniform way and with reliable errors. Early in this decade, this type of homogeneous study became possible for GCs (Paresce & De Marchi 2000), mostly thanks to the *Hubble Space Telescope (HST)*. Meanwhile, high-quality data have become available for young clusters (YCs) as well, mostly from wide-field ground-based surveys, thereby making this study possible on a global scale. We present here the preliminary results of this work.

2. The tapered power law

Since our goal is to detect and quantify changes in the shape of the MF in different environments, in particular in GCs and YCs, we need a functional form for the distribution of stellar masses that is flexible enough to adapt to the variety of observed MFs, yet simple enough to be described by a small number of parameters over a wide mass range. Kroupa (2002) proposed a segmented, multipart power law for this purpose, but we cannot adopt it here because this approach fixes the mass points between which the slope is fitted, thereby making it impractical or even impossible to detect variations in the characteristic mass which, as we will see, are the signature of dynamical evolution.

Our choice falls instead on a tapered power-law (TPL) distribution of the type

$$f(m) = \frac{dN}{dm} \propto m^{\alpha} \left[1 - e^{(-m/m_c)^{\beta}}\right], \tag{2.1}$$

where m_c is the characteristic mass, α the index of the power-law portion for high masses and β the tapering exponent that defines the shape of the MF below m_c. The TPL combines in one expression the ubiquitous power-law shape above $1\,M_\odot$ (e.g., Salpeter 1955) with the plateau and drop observed near the hydrogen-burning limit (e.g., Chabrier 2003). As we already showed (De Marchi *et al.* 2005), the TPL fits remarkably well the MFs of both YCs and GCs over the entire stellar mass range. We stress here that it is irrelevant whether the IMF resulting from star formation takes on the shape of a TPL, as long as it suitably represents the observed mass distribution.

3. The data sample

The data used in this work come from a number of different sources. As regards GCs, we consider the entire *HST* sample of Paresce & De Marchi (2000), which consists of twelve relatively dense clusters, and add to it four low-concentration GCs studied by our team in recent years (see Table 1). As De Marchi *et al.* (2007) showed, a strong trend exists between a cluster's concentration and the shape of its present MF, so it is important to cover both dense and loose objects. As for the YCs, the data cover

Table 1. The sample of YCs, dense and loose GCs used in this study.

Name	α	β	m_c	$\log t$	t_{dyn}	Ref	Name	α	β	m_c	$\log t$	t_{dyn}	Ref
ρ Oph	−1.8	1.8	0.15	5.7	0.001	I	Orion	−2.2	2.4	0.20	6.0	0.002	G,M
Taurus	−2.3	2.1	0.33	6.0	0.25	H	IC 348	−2.0	3.0	0.14	6.3	0.004	K
σ Ori	−2.0	2.6	0.18	6.5	0.006	D	λ Ori	−1.8	2.4	0.10	6.7	0.01	A
Chamaeleon 1	−2.1	2.0	0.22	6.7	0.01	H	IC 2391	−2.1	2.6	0.16	7.7	0.06	B
Blanco 1	−1.7	2.0	0.20	8.1	0.09	J	Pleiades	−2.2	2.3	0.27	8.1	0.06	J
M 35	−1.8	2.4	0.33	8.2	0.08	B	Coma Ber	−1.3	1.7	0.16	8.6	0.48	G
Praesepe	−2.0	3.5	0.26	8.8	0.18	E	Hyades	−2.1	2.8	0.45	8.8	0.37	C
NGC 104	−2.3	2.4	0.35	10.1	0.15	L	NGC 5139	−2.3	2.7	0.35	10.1	0.23	L
NGC 5272	−2.3	2.9	0.35	10.1	0.15	L	NGC 6121	−2.3	3.1	0.37	10.1	0.89	L
NGC 6254	−2.3	2.4	0.35	10.1	0.52	L	NGC 6341	−2.3	2.3	0.33	10.1	0.50	L
NGC 6397	−2.3	2.7	0.35	10.1	0.68	L	NGC 6656	−2.3	2.7	0.33	10.1	0.31	L
NGC 6752	−2.3	2.9	0.37	10.1	0.40	L	NGC 6809	−2.3	2.5	0.33	10.1	0.61	L
NGC 7078	−2.3	2.7	0.28	10.1	0.15	L	NGC 7099	−2.3	2.2	0.29	10.1	0.52	L
NGC 2298	−2.3	3.5	0.6	10.1	0.72	F	NGC 6218	−2.3	3.5	0.6	10.1	0.60	F
NGC 6712	−2.3	3.7	0.8	10.1	0.77	F	NGC 6838	−2.3	3.2	0.55	10.1	0.78	F

Notes: Units for the characteristic mass m_c are $M_\odot$, the age $\log t$ is in yr, while t_{dyn} is the ratio of the cluster age and the time to dissolution. An age of 12.7 Gyr is assumed for all GCs. For bibliographic references (Ref) see the corresponding letters in the References.

high-quality observations of about a dozen star-forming regions and associations of various ages from the most recent literature, as indicated in Table 1.

To prevent biases in the result, the data should refer to the global MF (GMF), i.e., that of the cluster as a whole. For GCs, where complete cluster coverage is not always possible, we use information on mass stratification and mass segregation to derive the GMF from the local MF (see, e.g., De Marchi *et al.* 2006). Alternatively, we use the MF measured near the half-light radius, since it has been shown to reflect quite reliably the properties of the GMF (De Marchi *et al.* 2000). For YCs, we specifically selected those objects for which the coverage is as complete as possible.

The values of the TPL parameters (α, β, m_c) that best fit the data are given in Table 1. Since the GC data do not constrain the mass range $> 0.8\,M_\odot$, we imposed $\alpha \equiv 2.3$, in line with the average value for YCs. The typical uncertainties in α and β are, respectively, 0.1 and 0.2, while that associated with m_c is $0.05\,M_\odot$.

4. Evolution of the mass function

Despite the widely different physical properties of the clusters in our sample, e.g., in terms of chemical composition, total mass, stellar density and age, the parameters α and β span a relatively narrow range of values, with no significant differences between YCs and GCs: for the former, $\alpha = -2.1 \pm 0.2$ and $\beta = 2.4 \pm 0.4$, while for the latter, $\alpha \equiv -2.3$ and $\beta = 2.6 \pm 0.3$. The characteristic mass shows a much wider spread. The average value of m_c for dense globular clusters is $0.34 \pm 0.04\,M_\odot$ and for loose globulars it grows to $\sim 0.64 \pm 0.10\,M_\odot$, although the statistics are limited in this latter case. The variation in m_c is even more pronounced for YCs, ranging from 0.15 to $0.6\,M_\odot$. In principle, this variation could reflect differences in the star-formation process, possibly induced by differences in environmental conditions. However, there does not seem to be a correlation between m_c and the clusters' physical properties such as total mass, concentration or metallicity. For example, the dense GCs in our sample span almost two decades in metallicity and total mass, but have the same m_c, while all YCs have practically the same metallicity but a wide range of m_c values. Therefore, the environment does not seem to play a role in the shape of the MF of the clusters in our sample.

Another possibility to consider to explain the observed spread in m_c is the role of dynamical evolution, which could alter the shape of the MF over time because of the preferential loss of low-mass stars caused by two-body relaxation. It is thus worth investigating whether our data show any correlation or trend between the cluster's characteristic mass and dynamical state. For this reason, we list in Table 1 the dynamical age $t_{\rm dyn}$ of each cluster, defined as $t/t_{\rm dis}$ or the ratio of the cluster age and the time to dissolution $t_{\rm dis}$, i.e., the time at which the cluster has lost 95% of its original mass. Using N-body models, Baumgardt & Makino (2003) showed that $t_{\rm dis} \propto t_{\rm rh}^x t_{\rm cr}^{1-x}$ where $t_{\rm rh}$ is the half-mass relaxation time, $t_{\rm cr}$ the crossing time and x depends on the initial concentration of the cluster. Gieles *et al.* (2005), using the same N-body models, found that $t_{\rm dis}$ can be expressed as a function of the total initial mass of the cluster $M_{\rm ini}$, as $t_{\rm dis} \propto t_0 M_{\rm ini}^{0.62}$ (see also Gieles, these proceedings), where t_0 is a constant that depends on the properties of the galaxy and on the cluster orbit. The values of $t_{\rm dyn}$ in Table 1 come from Baumgardt *et al.* (2008) for the specific GCs in our sample. For YCs, we have estimated their initial mass and the corresponding value of $t_{\rm dis}$ following Baumgardt & Makino (2003; in particular their figures 2 and 3) from the current mass given by Piskunov *et al.* (2008). For clusters younger than 100 Myr we have assumed $t_{\rm dis} \simeq 500$ Myr but, as we will see later, our conclusions would not change if $t_{\rm dis}$ were larger or smaller by a factor of two (hereafter we will assume an uncertainty of 1/3 on $t_{\rm dis}$). The case of Taurus, however, is different: Ballesteros Paredes *et al.* (2009) conducted a gravitational analysis of its orbit and concluded that the Taurus molecular cloud must be suffering significant tidal disruption, in spite of its young age (1 Myr). We conservatively assigned to it $t_{\rm dyn} = 1/4$, although the actual value could be larger.

The run of $m_{\rm c}$ as a function of $t_{\rm dyn}$ is shown in Figure 1, where all clusters are labeled individually, except for the GCs (see caption for details). The figure reveals a rather remarkable trend of increasing $m_{\rm c}$ with dynamical age over the entire range covered by the data. The dashed line is a purely empirical eyeball fit to the data of the relationship

$$m_{\rm c}/{\rm M}_\odot \simeq 0.15 + 0.5 \times t_{\rm dyn}^{3/4}, \qquad (4.1)$$

although an index of 1/2 would still provide an acceptable fit. The residual scatter seen in Figure 1 is dominated by two effects: the uncertainty in $t_{\rm dyn}$ and incomplete coverage of the clusters. As an example of the latter, we show in Figure 1 also the data point corresponding to only the central regions of IC 348 (from the MF of Muench *et al.* 2003). In spite of the young age of the cluster, the MF in its core shows strong signs of mass segregation, with $m_{\rm c} = 0.33\,{\rm M}_\odot$ as opposed to $m_{\rm c} = 0.14\,{\rm M}_\odot$ for the cluster as a whole. This can be due to primordial mass segregation and/or to very fast dynamical evolution (see, e.g., Allison *et al.* 2009), but in either case this data point should be excluded from the sample so as not to skew the results. In spite of our attempts to select only clusters with complete radial coverage, however, this information might not be completely available or fully correct in the literature, and this is likely to account for some of the residual scatter that we observe.

5. Discussion and conclusions

At least qualitatively, the trend seen in Figure 1 is consistent with the onset of mass segregation through two-body relaxation in a tidal environment, causing the preferential loss of low-mass stars from the cluster and hence a drift of the characteristic mass towards higher values. Vesperini & Heggie (1997) showed (using N-body models) that stellar evaporation, integrated over the cluster's orbit and further enhanced by the presence of the Galactic tidal field, causes a flattening of the MF, i.e., a selective depletion at the low-mass end. Their models do not reveal a drift of characteristic mass towards higher values, but their IMF in all cases is a pure power law. A drift of $m_{\rm c}$ is visible in the

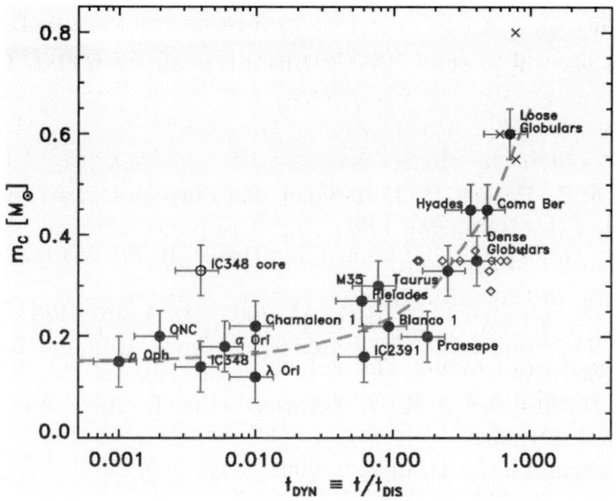

Figure 1. Run of the characteristic mass, m_c, as a function of dynamical age. Dense and loose GCs are indicated, respectively, with diamonds and crosses, with a filled circle providing the average value for each class. The core of IC 348 is indicated as an open circle to distinguish it from the whole cluster. The dashed line is an eyeball fit to the data.

N-body models of Portegies Zwart *et al.* (2001), but it proceeds very slowly and only appears when the study of the MF is limited to the regions inside the half-mass radius. On the other hand, De Marchi *et al.* (2007) showed that dense and loose GCs have today systematically different GMFs, consistent with a much stronger selective loss of low-mass stars in the latter. The fact that m_c is systematically higher in loose GCs proves that its value must drift as a result of the cluster's dynamical evolution, even though the analysis of the N-body model results has so far failed to detect it.

An important consequence of the trend shown in Figure 1, if dynamical evolution is indeed at its origin, is that it seems then perfectly plausible that all GCs, having now $m_c \simeq 0.35\,\mathrm{M}_\odot$ if they are dense or $\sim 0.65\,\mathrm{M}_\odot$ if they are loose, were born with a stellar MF like that measured today in the youngest Galactic clusters, with $m_c \simeq 0.15\,\mathrm{M}_\odot$.

This conclusion is fully consistent with the absence of a turnover in the MF of the Galactic bulge down to the observational limit at $\sim 0.2\,\mathrm{M}_\odot$ (Zoccali *et al.* 2000). Regardless as to whether the bulge is primarily the result of fast formation at early epochs (e.g., Ballero *et al.* 2007) or a collection of the stars lost from disrupted clusters throughout the life of the Galaxy (e.g., Gnedin & Ostriker 1997), its very location at the bottom of the Galaxy's potential well makes it very hard for low-mass stars to escape. Therefore, its present MF should not have been altered by dynamical evolution, unlike the case of the comparably old GCs, and should still reflect the properties of the IMF. Forthcoming *HST* observations of the Galactic bulge (Brown *et al.* 2009) will provide insights into the MF of the bulge below $0.15\,\mathrm{M}_\odot$, where we expect a turnover.

If the picture sketched above is correct and dynamical evolution is the main source of MF variations in Galactic clusters, it would argue quite strongly in favour of the universality of the IMF of Population I and II stars. Previous claims in this sense (e.g., Gilmore 2001; Elmegreen *et al.* 2008) were based on the observation that the stellar MFs in different environments have *similar* m_c values, within factors of a few. Here, we show for the first time that the value of m_c may in fact be exactly *the same*, suggesting that environmental conditions do not play a significant role in the process of star formation, or at least regarding its end result. Compelling theories have already been presented as to why this should be the case (Bate 2009; Goodwin & Kouwenhoven 2009).

Acknowledgements

SPZ acknowledges the support of NWO through grant # 639.073.803.

References

Allison, R., Goodwin, S. P., Parker, R. J., de Grijs, R., Portegies Zwart, S. F., & Kouwenhoven, M. B. N. 2009, *ApJ* (Letters), 700, L99

Ballesteros–Paredes, J., Gómez, G. C., Loinard, L., Torres, R. M., & Pichardo, B. 2009, *MNRAS* (Letters), 395, L81

Ballero, S., Matteucci, F., Origlia, L., & Rich, M. 2007, *A&A*, 467, 123

Barrado y Navascués, D., Stauffer, J. R., Bouvier, J., Jayawardhana, R., & Cuillandre, J.-C. 2004, *ApJ*, 610, 1064 (A)

Barrado y Bavascués, D., Stauffer, J. R., & Jayawardhana, R. 2004, *ApJ*, 614, 386 (B)

Bate, M. 2009, *MNRAS*, 397, 232

Baumgardt, H., De Marchi. G., & Kroupa, P. 2008, *ApJ*, 685, 247

Baumgardt, H. & Makino, J. 2003, *MNRAS*, 340, 227

Bouvier, J., *et al.* 2008, *A&A*, 481, 661 (C)

Brown, T., *et al.* 2009, *AJ*, 137, 3172

Caballero, J. 2008, *A&A*, 478, 667 (D)

Chabrier, G. 2003, *PASP*, 115, 763

Chappelle, R., Pinfield, D. J., Steele, I. A., Dobbie, P. D., & Magazzù, A. 2005, *MNRAS*, 361, 1323 (E)

De Marchi, G., Paresce, F., & Portegies Zwart, S. 2005, in: E. Corbelli, F. Palla, & H. Zinnecker (eds.), *The Initial Mass Function 50 years later* (Dordrecht: Springer), p. 77

De Marchi, G., Paresce, F., & Pulone, L. 2000, *ApJ*, 530, 342

De Marchi, G., Paresce, F., & Pulone, L. 2007, *ApJ* (Letters), 656, L65 (F)

De Marchi, G., Pulone, L., & Paresce, F. 2006, A&A, 449, 161

Elmegreen, B., Klessen, R., & Wilson, C. 2008, *ApJ*, 681, 365

Gieles, M., Bastian, N., Lamers, H. J. G. L. M., & Mout, J. N. 2005, *A&A*, 441, 949

Gilmore, G. 2001, in: L. Tacconi & D. Lutz (eds.), *Starburst Galaxies* (Heidelberg: Springer), p. 34

Gnedin, O. & Ostriker, J. 1997, *ApJ*, 474, 223

Goodwin, S. & Kouwenhoven, M. 2009, *MNRAS* (Letters), 397, L36

Kraus, A. & Hillenbrand, L. 2007, *AJ*, 134, 2340 (G)

Kroupa, P. 2002, in: E. Grebel & W. Brandner (eds.), *Modes of Star Formation and the Origin of Field Populations*, ASP Conf. Ser. 285, (San Francisco: ASP), p. 86

Luhman, K. 2007, *ApJS*, 173, 104 (H)

Luhman, K. & Rieke, G. 1999, *ApJ*, 525, 440 (I)

Moraux, E., Bouvier, J., Stauffer, J. R., Barrado y Navascués, D., & Cuillandre, J.-C. 2007, *A&A*, 471, 499 (J)

Muench, A., Lada, E. A., Lada, C. J., Elston, R. J., Alves, J. F., Horrobin, M., Huard, T. H., Levine, J. L., Raines, S. N., & Román–Zúñiga, C. 2003, *AJ*, 125, 2029 (K)

Paresce, F. & De Marchi, G. 2000, *ApJ*, 534, 870 (L)

Piskunov, A., Schilbach, E., Kharchenko, N. V., Röser, S., & Scholz, R.-D. 2008, *A&A*, 477, 165

Portegies Zwart, S., McMillan, S. L. W., Hut, P., & Makino, J. 2001, *MNRAS*, 321, 199

Salpeter, E. 1955, *ApJ*, 121, 161

Slesnick, C., Hillenbrand, L. A., & Carpenter, J. M. 2004, *ApJ*, 610, 1045 (M)

Spitzer, L. 1987, *Dynamical evolution of globular clusters*, (Princeton: Princeton University Press)

Vesperini, E. & Heggie, D. 1997, *MNRAS*, 289, 898

Zoccali, M., Cassisi, S., Frogel, J. A.; Gould, A., Ortolani, S., Renzini, A., Rich, R. M., & Stephens, A. W. 2000, ApJ, 530, 418

Star clusters: basic galactic building blocks
Proceedings IAU Symposium No. 266, 2009
R. de Grijs & J. R. D. Lépine, eds.

doi:10.1017/S1743921309990913

Early dynamical evolution of star cluster systems

Geneviève Parmentier

Institute of Astrophysics & Geophysics, Allée du 6 Août 17, B-4000 Liège, Belgium
Argelander Institut für Astronomie, Auf dem Hügel 71, D-53121 Bonn, Germany
email: gparm@astro.uni-bonn.de

Abstract. Violent relaxation, the protocluster dynamical response to the expulsion of its residual star-forming gas, is a short albeit crucial episode in the evolution of star clusters and star cluster systems. Because it is heavily driven by cluster-formation and environmental conditions, it is a potentially highly rewarding phase in terms of probing star formation and galaxy evolution. In this contribution, I review how cluster-formation and environmental conditions affect the shape of the young cluster mass function and the relation between the present star-formation rate of galaxies and the mass of their young, most massive cluster.

Keywords. stellar dynamics, stars: formation, galaxies: star clusters

1. Introduction

Compact, gas-embedded clusters are observed to be the formation sites of a significant fraction of stars in the local Universe. Following star formation, feedback processes—ranging from protostellar outflows, photoionization-driven gas overpressure and stellar winds to, eventually, supernovae—expel the gas that has not yet been converted into stars. That is, the efficiency with which a cluster parent core turns its gas into stars is always less than 100%. Not only does gas expulsion terminate star formation, it also leaves the newly-formed cluster out of equilibrium, thereby forcing it into violent relaxation. In an attempt to regain a state of equilibrium, the cluster expands, loses a fraction of its stars or even undergoes complete disruption. These processes are referred to as cluster *infant weight loss* and *infant mortality*, respectively.

The dynamical response of clusters to gas expulsion implies that the cluster age distribution, the cluster mass function, the ratio of the mass in clusters and that in stars, as well as the distribution function of cluster half-mass radii all offer promising diagnostic tools of star cluster formation conditions. Baumgardt & Kroupa (2007) have built and made available to the community an N-body model grid describing the temporal evolution of cluster mass loss and spatial expansion for a wide range of local star-formation efficiencies (SFEs; i.e., the mass fraction of dense molecular gas that cluster-forming cores turn into stars), gas-expulsion timescales and external tidal-field impacts. To date, this is the most comprehensive grid of model clusters through violent relaxation. Combined with the increasing quality of observational data sets (including of the much needed cluster half-mass radii: see, e.g., Scheepmaker *et al.* 2007), the time is now ripe to decipher observed properties of young star clusters to probe their formation and environmental conditions.

The outline of this contribution is as follows. Following a brief reminder of the main parameters involved in violent relaxation in Section 2, I highlight how formation and environmental conditions affect the shape of the cluster mass function in Section 3. I also present some preliminary results of ongoing work regarding the relation between the luminosity of the brightest young cluster and the present star-formation rate (SFR) of spiral galaxies, the SFR–$M_V^{\rm brightest}$ relation (Section 4).

2. Gas expulsion: main parameters

In what follows, the mass of a cluster is defined as the stellar mass enclosed within its tidal radius. We assume that the impact of the external tidal field is solely due to the smooth gravitational potential of a Milky Way-type galaxy, i.e., at this stage we neglect the potential impact on the early evolution of a star cluster of its natal giant molecular cloud. We adopt the Jacobi radius as an estimate of the cluster's tidal radius (see Baumgardt *et al.* 2009; their equation 1).

In the course of violent relaxation, the mass $m_{\rm cl}$ of a star cluster evolves rapidly and must therefore be defined as an instantaneous quantity. It obeys

$$m_{\rm cl} = F_{\rm b}(t/\tau_{\rm cross}, \varepsilon, \tau_{\rm GExp}/\tau_{\rm cross}, r_{\rm h}/r_{\rm t}) \times \varepsilon \times m_{\rm c}, \tag{2.1}$$

where $m_{\rm c}$ is the mass of the progenitor molecular core, ε the local SFE† and $F_{\rm b}$ the stellar mass fraction of the initially gas-embedded cluster still contained within the cluster tidal radius at time t. When using the term '(gas-)embedded cluster mass,' we refer to the stellar mass $m_{\rm ecl} = \varepsilon \times m_{\rm c}$. The local SFE is assumed to be uniform among all cores of a given simulation (i.e., no core-mass dependent SFE). The bound fraction, $F_{\rm b}$, depends on the time $t/\tau_{\rm cross}$ elapsed since gas expulsion and on the gas-expulsion timescale, $\tau_{\rm GExp}/\tau_{\rm cross}$ (both expressed in units of the molecular-core crossing time), on the local SFE, ε, and on the impact of the external tidal field. The latter is quantified as the ratio of the half-mass radius, $r_{\rm h}$, and the tidal radius, $r_{\rm t}$, of the gas-embedded cluster (see below for details). The older the cluster and/or the lower the SFE and/or the quicker gas expulsion and/or the stronger the host galaxy's tidal field, the lower $F_{\rm b}$. Building on the motion of a supershell propagating through the cluster-forming core and collecting the residual star-forming gas, Parmentier *et al.* (2008) derived an expression for the gas-expulsion timescale $\tau_{\rm GExp}/\tau_{\rm cross}$ (their eq. 6; see also figure 3 in Parmentier & Fritze 2009). Because a cluster expands following gas expulsion, the mass fraction of stars it loses depends on how compact the gas-embedded cluster is compared to its tidal radius. The impact of the external tidal field is therefore accounted for by the ratio $r_{\rm h}/r_{\rm t}$. As will be shown in Section 3, the cluster-forming core mass–radius relation is therefore of paramount importance for the outcome of violent relaxation, since it influences both the depth of the core potential well, and hence the gas-expulsion timescale, and the sensitivity to the external tidal field of the exposed cluster (see also Parmentier 2009).

3. Young cluster mass functions

In this section, I show how specific combinations of the core mass–radius relation, local SFE and external tidal field lead to cluster mass functions markedly different from the core mass function and, therefore, to mass-dependent cluster infant mortality. Cluster-forming cores are spatially resolved in the nearby regions of our Galaxy only, and the slope of the core mass–radius relation remains therefore poorly determined [see Parmentier *et al.* (submitted) for a discussion]. I consider two extreme cases in turn, i.e., constant core radii and constant core surface densities.

† We explicitly assume that stars have had sufficient time to reach virial equilibrium with the gas potential (i.e., gas expulsion occurs after a few crossing times) and, hence, that the local SFE closely matches the effective SFE (see Goodwin 2009 for a discussion).

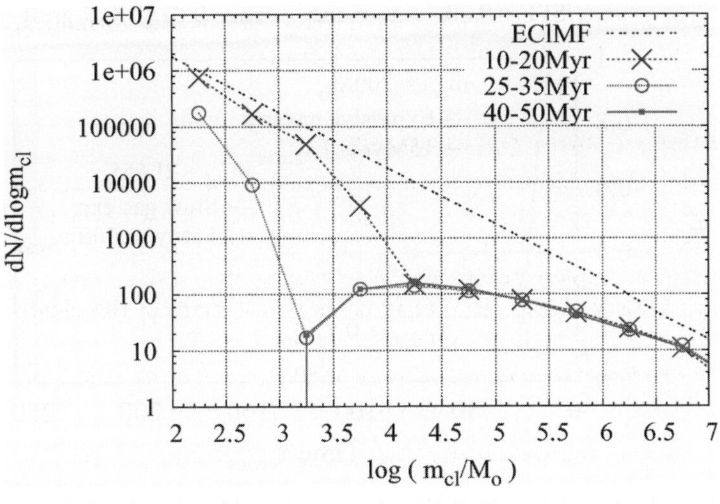

Figure 1. Time evolution of the cluster mass function for constant cluster-forming core radii (r_c = 1 pc) and the local SFE is 20%. The symbol-free dotted (black) line is the power-law embedded-cluster mass function ('ECIMF'). The lines with crosses, open circles and filled squares (blue, pink and red, respectively) show the cluster mass functions integrated over the age ranges quoted in the legend.

Constant cluster-forming core radii and mass-varying gas expulsion timescale. Under the assumption of constant radii (r_c = 1 pc), a sequence of increasing core mass equates to one of increasing densities and deepening potential wells. As a result, high-mass cores undergo slower gas expulsion and retain a larger fraction of bound stars than their low-mass counterparts. Under the assumption of a weak tidal field, Parmentier *et al.* (2008) showed that the shape of the post-violent-relaxation cluster mass function is highly sensitive to the local SFE (see their figures 2 and 3). While a local SFE $\varepsilon \simeq 0.4$ is conducive to a cluster mass function mirroring the core mass function (mass-independent infant mortality and weight loss), a local SFE $\varepsilon \simeq 0.2$ results in the survival of the (initially) highest-mass embedded clusters only. For an exposed cluster to survive such a low local SFE, gas must be expelled adiabatically, a condition met by high-mass cores and their deep potential wells. Only those produce bound clusters at the end of violent relaxation, rendering cluster infant mortality heavily mass dependent. As a result, a power-law core-mass function gives rise to a bell-shaped cluster mass function.

An aspect which Parmentier *et al.* (2008) did not investigate is the time evolution of the power-law mass function of embedded clusters in the bell-shaped cluster mass function. The topic of evolving secularly (i.e., when violent relaxation effects have faded away) a power-law cluster mass function in a bell-shape is extensively covered in the literature. Starting from a power-law at an age of ~50–100 Myr, internal two-body relaxation and external tidal stripping preferentially removes low-mass clusters, thereby carving a turnover in the mass function. As time goes by, this turnover moves towards higher cluster masses and smaller mass-function amplitudes. However, the evolution of relevance here is violent relaxation instead of secular evolution and, as we shall see, the cluster mass function evolves differently.

This is the perfect example to illustrate that the evolutionary stage of a cluster is not determined by its age in an absolute sense. As for the gas-expulsion timescale, what matters is the age expressed in units of core crossing times (see Equation 2.1). This implies that, keeping all other parameters the same, clusters born in cores of higher

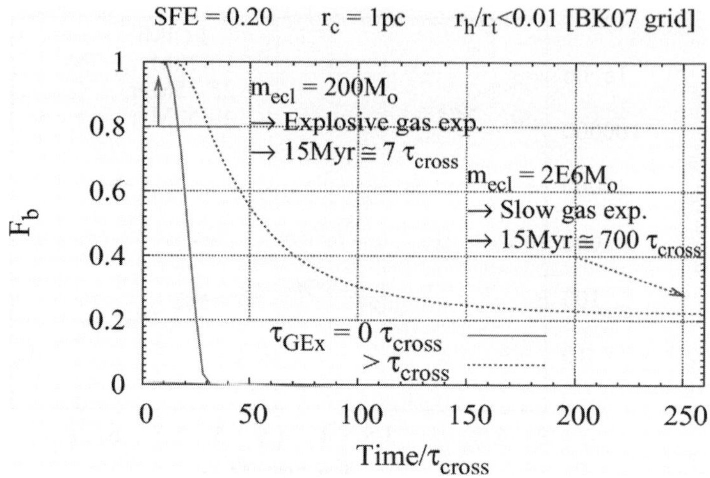

Figure 2. Time evolution of the bound fraction of stars for an SFE of 20% and two distinct regimes of cluster gas expulsion, i.e., explosive ($\tau_{\rm GExp} \ll \tau_{\rm cross}$) and adiabatic ($\tau_{\rm GExp} = 3\,\tau_{\rm cross}$). Under the assumption of constant core radii, high-mass clusters have a higher density, which results in slower gas expulsion, shorter crossing time and—once exposed—faster evolution than their low-mass counterparts.

densities evolve—once exposed—at a faster rate than those formed in low-density cores. Figure 1 shows the embedded-cluster mass function (i.e., the core-mass function shifted by -0.7 in $\log m$ to account for the constant 20% local SFE) and snapshots of the cluster mass function at ages of $\simeq$ 15, 30 and 45 Myr. The early cluster mass function (age $\simeq$ 15 Myr; dotted line with $\times$ signs) shows prominent substructure, with cluster masses not yet significantly affected by infant weight loss in the low- and high-mass regimes ($m_{\rm cl} < 3000\,{\rm M}_\odot$ and $m_{\rm cl} > 2 \times 10^6\,{\rm M}_\odot$, respectively). They originate from how $F_{\rm b}$ evolves with time and as a function of gas-expulsion timescale. In Figure 2, the solid and dotted lines show the $F_{\rm b}$ time evolutions for explosive ($\tau_{\rm GExp} \ll \tau_{\rm cross}$) and adiabatic (here $\tau_{\rm GExp} = 3\,\tau_{\rm cross}$) gas expulsion, respectively. A massive embedded cluster (say, $m_{\rm ecl} = 2 \times 10^6\,{\rm M}_\odot$) expels its residual gas on an adiabatic timescale and hence retains a bound core of stars in spite of the low SFE of 20%. Also, by virtue of the high density of its natal core, a young age of 15 Myr equates to several hundreds of core crossing times. Accordingly, cluster violent relaxation is over by this young age. This is why most of the bell shape, which arises from clusters initially more massive than $m_{\rm ecl} \sim 10^6\,{\rm M}_\odot$, is already carved by an age of 15 Myr.

By contrast, low-mass clusters (e.g., $m_{\rm ecl} = 200\,{\rm M}_\odot$) have low core densities, long core crossing times and the same age of 15 Myr equates to a few $\tau_{\rm cross}$ only, a stage at which the instantaneous bound fraction $F_{\rm b}$ is still close to unity. Consequently, the low-mass regime of the cluster mass function at an age of 15 Myr sticks to the embedded-cluster mass function. As clusters age to 50 Myr, the cluster mass function in the high-mass regime does not evolve anymore and the low-mass regime of the mass function disappears as low-mass objects experience infant weight loss until their eventual dissolution.

Constant cluster-forming core surface densities and mass-varying tidal-field impact.

Following Parmentier *et al.* (2008), molecular cores of constant surface density have an (almost) constant gas-expulsion timescale. Yet, the infant-mortality and infant-weight-loss rates of their embedded clusters are markedly mass dependent due to the tidal field impact. We insist that this effect takes place even if all cores are located within the same

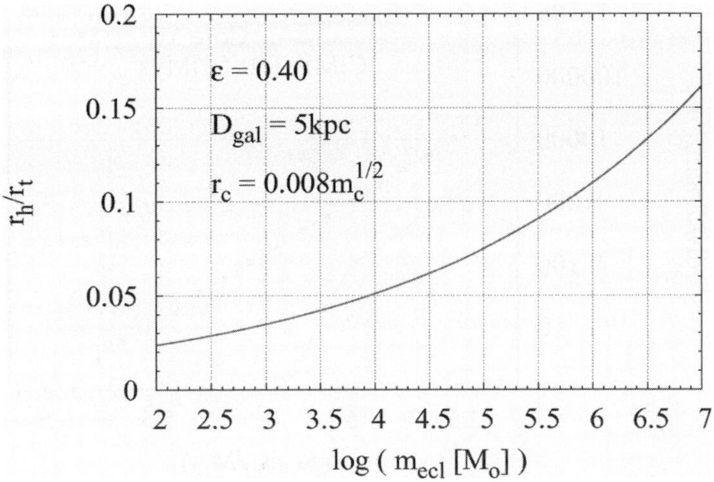

Figure 3. Dependence of the ratio of the half-mass radius, $r_{\rm h}$, and the tidal radius, $r_{\rm t}$ of the embedded cluster on its stellar mass, $m_{\rm ecl}$, for constant-surface-density cores. Because of their large spatial extent, high-mass clusters exhibit a greater sensitivity to the external tidal field and present a greater likelihood of disruption following gas expulsion.

limited region of a galaxy within which the external tidal field does not vary markedly. The ratio of the initial half-mass and tidal radii of these embedded clusters scales as $r_{\rm h}/r_{\rm t} \propto m_{\rm c}^{1/2}/m_{\rm c}^{1/3} \propto m_{\rm c}^{1/6}$ and, therefore, high-mass clusters have a greater sensitivity to the external tidal field since the volume of space in which they can expand freely is smaller than for low-mass clusters compared to the initial cluster size. This is illustrated in Figure 3, where the core mass–radius relation is $r_{\rm c} = 0.008\,(m_{\rm c}/1\,{\rm M}_\odot)^{1/2}$ pc†, the local SFE is $\varepsilon = 0.4$ and the galactocentric distance is $D_{\rm gal} = 5$ kpc in a Milky Way-like potential. As a result, a power-law core-mass function evolves in a post-violent-relaxation cluster mass function depleted in high-mass objects (see Figure 4).

What does it all mean?

The key point to retain from this section is that the shape of the cluster mass function, the mass–radius relation of cluster-forming cores, their local SFE and the external tidal field are intimately linked issues. That most observed mass functions of young clusters are power laws with a spectral index of −2 [see Parmentier & Gilmore (2007), and references therein; but see also Anders *et al.* (2007) for a possible turnover in the cluster mass function of the Antennae merger, NGC 4038/39] therefore tells us something about cluster formation conditions which are common to different types of cluster environments. The high sensitivity to the external tidal field of high-mass cores of constant surface density may be a hint that, contrary to theoretical expectations, the index β of the mass–radius relation $r_{\rm c} \propto m_{\rm c}^{\beta}$ of cluster-forming cores is shallower than $\beta = 1/2$. More simulations are required to settle this point conclusively. In particular, greater attention must be paid to cluster radius-related quantities, e.g., the distribution function of young cluster half-mass radii and the reason why clusters lack a clear mass versus half-mass radius correlation (Larsen 2004). We will investigate these aspects in a forthcoming paper.

† Core mass–radius relations adopted in this contribution are taken from Parmentier & Fritze (2009). An update on these relations will be presented in Parmentier *et al.* (submitted).

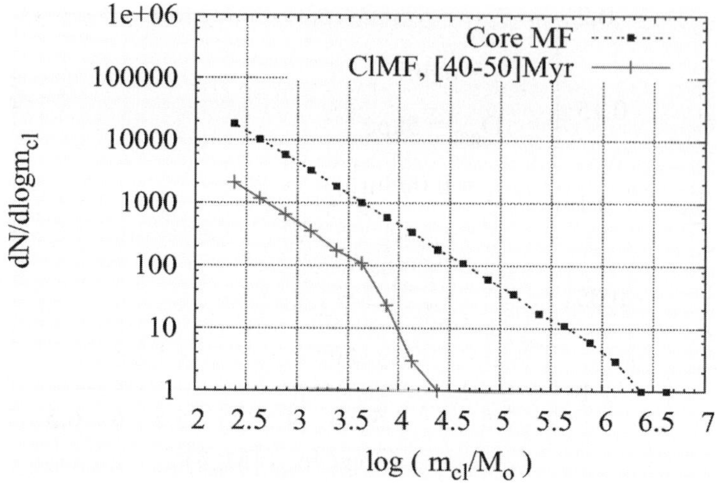

Figure 4. Constant-surface-density cores: when more massive objects are more vulnerable. The dotted (black) line is the core-mass function and the solid (red) line is the cluster mass function at an age of $\simeq 50$ Myr. As can be understood from Figure 3, cluster infant mortality acts predominantly in the high-mass regime. Model parameters are the same as in Figure 3.

4. Brightest young star clusters in galaxies

Star clusters are chronometers of major star-formation events in their host galaxies (Kroupa 2002; Kotulla *et al.* 2008). When the SFR of a galaxy increases, not only does the galaxy form a larger number of gas-embedded star clusters, the cluster mass function is also sampled up to a higher cluster mass, partly as a result of the size-of-sample effect. Time spans in the history of galaxies characterized by vigorous star formation therefore show up prominently in cluster age histograms for two reasons: (i) the number of clusters is larger and (ii) these more numerous clusters also include more massive ones, which experience longer dissolution timescales and are therefore more likely to be observed.

Observationally, this effect is best known as a correlation between the present SFR of galaxies and the absolute visual magnitude, $M_V^{\rm brightest}$, of their brightest cluster (Larsen 2002; Weidner *et al.* 2004; Bastian 2008). Building on synthetic cluster populations, Bastian (2008) shows that *if* galaxies form all their stars in gas-embedded clusters, a universal mass fraction of 8% of clusters survive the transition from their gas-embedded stage to being exposed.

That the ratio of the mass in clusters and stars depends sensitively on the mean local SFE (Parmentier & Fritze 2009) prompted Parmentier *et al.* (submitted) to investigate which cluster-formation conditions are the main drivers of the SFR–$M_V^{\rm brightest}$ relation. The absolute magnitude of a cluster depends on its age-dependent mass-to-light ratio, on the stellar mass of its parent gas-embedded cluster and on the amount of infant weight loss the latter has experienced. Section 3 demonstrates that cluster-mass-related quantities are heavily shaped by cluster-formation conditions. We have therefore divided our study of how the SFR–$M_V^{\rm brightest}$ relation responds to input parameter variations into two main parts. First, we consider the limited age range of 10 Myr over which we explicitly assume the constancy of the integrated cluster mass-to-light ratio, i.e., we investigate how the *mass of the most massive cluster* responds to cluster-formation conditions, specifically the core-mass function, core mass–radius relation, local SFE and strength of an external tidal field (Parmentier *et al.*, submitted). In a second forthcoming paper, we will extend the cluster age range to 100 Myr and assess how mass-to-light ratio variations, and hence

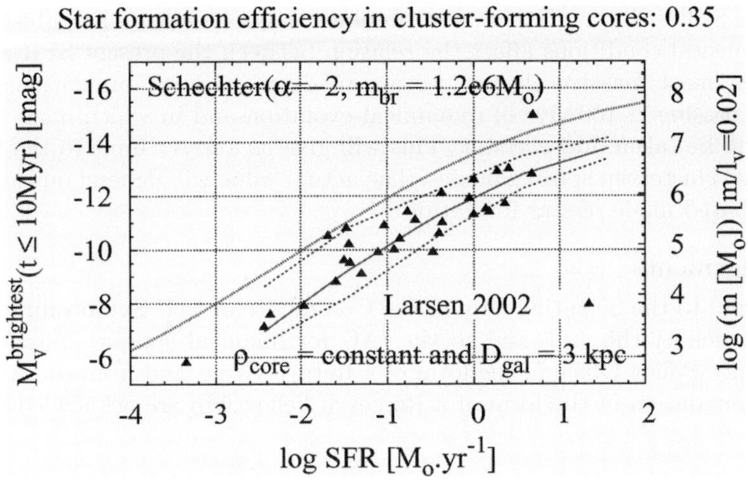

Figure 5. Observed relation between the absolute visual magnitude of the brightest cluster of spiral galaxies and their present SFR (filled triangles). The dotted (pink) line is the relation between the SFR and the mass of the most massive core for the adopted Schechter core-mass function ($\alpha = -2$, $m_{\rm br} = 1.2 \times 10^6$ M$_\odot$). Solid (red) and dashed (blue) lines represent an iso-SFE model and its 1σ limits. We caution that, at the exploratory stage of this model, the evolutionary time span of the model (10 Myr) is narrower than that of the observations ($\gtrsim$ 100 Myr), thereby preventing us from drawing conclusions about the best local SFE estimate (see text for details).

the age of the observed brightest cluster, contribute to shape the SFR–$M_V^{\rm brightest}$ relation. As for the first part of the study, our findings are as follows. The *vertical location* of the SFR–$M_V^{\rm brightest}$ relation depends on the external tidal field and local SFE. Stronger tidal fields and/or smaller SFE are conducive to smaller mass/brightness for the most massive/brightest clusters. As for the *shape* of the SFR–$M_V^{\rm brightest}$ relation, it is dictated mostly by the core-mass function (see Figure 5) and the core mass–radius relation. For constant-surface-density cores, the SFR–$M_V^{\rm brightest}$ relation becomes shallower when the SFR increases because of the greater sensitivity to the external tidal field of high-mass embedded clusters and hence the larger amount of infant weight loss they experience. By contrast, for constant core radii, it becomes steeper towards higher SFR as, in this case, higher-mass cores experience slower gas expulsion and thus retain a larger bound fraction of stars.

In the case illustrated in Figure 5, cluster-forming cores have a constant volume density ($r_{\rm c} = 0.026\,(m_{\rm c}/1\,{\rm M}_\odot)^{1/3}$ pc), the local SFE is 35% and the synthetic cluster population is located at a galactocentric distance of 3 kpc in a Milky Way-like potential. The core-mass function is a Schechter function with a spectral index of -2 and its break mass, $m_{\rm br}$, has been adjusted so that, following star formation and infant weight loss, the break mass of the cluster mass function is about $\simeq 2 \times 10^5$ M$_\odot$, as observed for spiral galaxies (Larsen 2009). The solid and dotted lines are the model and its 1σ limits, respectively. The filled triangles are the data points from Larsen (2002) for spiral galaxies. Figure 5 suggests that an iso-SFE model provides a good fit to the ensemble of the data, provided that other conditions—core mass–radius relation (its slope *and* normalization), external tidal field—do not vary significantly from one spiral galaxy to another. We emphasize once again that the simulation time span (10 Myr) is shorter than the age range of clusters shown in this plot ($\gtrsim$ 100 Myr). Since the model vertical location is also age dependent through both the amount of infant weight loss and the mass-to-light ratio, the limited time span of our simulations prevents us from drawing firm conclusions about the best local SFE estimate. Having now obtained a firm handle on how cluster formation and environmental conditions affect the relation between the present SFR and the mass of the young, most massive cluster, we are ready to perform simulations on a larger scale, encompassing $\simeq 100$ Myr of dynamical evolution and in which mass-to-light ratio variations will be taken into account. This will give us a *lower* limit to the local SFE in gas-embedded clusters in spiral galaxies. The actual value will depend on the importance of the distributed mode of star formation.

Acknowledgements

I am grateful to the Scientific Organising Committee of IAU Symposium 266 for their invitation to deliver this talk and to the IAU for financial support. Support from the Belgian Science Policy Office in the form of a Return Grant and from the Alexander von Humboldt Foundation in the form of a Research Fellowship are acknowledged.

References

Anders, P., Bissantz, N., Boysen, L., de Grijs, R., Fritze–v. Alvensleben, U. 2007, *MNRAS*, 377, 91
Bastian, N. 2008, *MNRAS*, 390, 759
Baumgardt, H. & Kroupa P. 2007, *MNRAS*, 380, 1589
Baumgardt, H., Parmentier, G., Gieles, M., & Vesperini E., 2009, *MNRAS*, in press (arXiv:0909.5696)
Goodwin, S. P. 2009, in: E. Pérez, R. de Grijs & R. M. González Delgado, *Young massive stars clusters – Initial conditions and environments*, *ApSS*, 324, 259
Kotulla, R., Fritze, U., & Anders, P. 2008, *MNRAS*, 387, 1149
Kroupa, P. 2002, *MNRAS*, 330, 707
Larsen, S. S. 2002, *AJ*, 124, 1393
Larsen, S. S. 2004, *A&A*, 416, 537
Larsen, S. S. 2009, *A&A*, 494, 539
Parmentier, G. & Gilmore, G. 2007, *MNRAS*, 377, 352
Parmentier, G., Goodwin, S., Kroupa, P., & Baumgardt, H. 2008, *ApJ*, 678, 347
Parmentier, G. & Fritze, U. 2009, *ApJ*, 690, 112
Parmentier, G. 2009, *Rev. Mod. Astron.*, 21, in press (arXiv:0901.3140)
Parmentier, G., Kroupa, P., & Baumgardt, H., *ApJ*, submitted
Scheepmaker, R. A., Haas, M. R., Gieles, M., Bastian, N., Larsen, S. S., & Lamers, H. J. G. L. M. 2007, *A&A*, 469, 925
Weidner, C., Kroupa, P., & Larsen, S. S. 2004, *MNRAS*, 350, 1503

Session III

Star cluster systems

Star clusters: basic galactic building blocks
Proceedings IAU Symposium No. 266, 2009
R. de Grijs & J. R. D. Lépine, eds.

doi:10.1017/S1743921309990925

The distance and space distribution of bulge globular clusters

Sergio Ortolani,[1] Beatriz Barbuy[2] and Eduardo Bica[3]

[1]Dipartimento di Astronomia, University of Padova,
Vicolo dell'Osservatorio 2, 35122, Padova, Italy
email: sergio.ortolani@unipd.it

[2]Universidade de São Paulo, IAG, Rua do Matao 1226, Cidade Universitaria,
São Paulo 05508–900, Brazil
email: barbuy@astro.iag.usp.br

[3]Universidade Federal do Rio Grande do Sul, Departamento de Astronomia CP 15051, RS,
Porto Alegre 91501–970, Brazil
email: bica@if.ufrgs.br

Abstract. We present the distribution of Galactic bulge globular clusters and a method based on simultaneous detection of field and cluster horizontal branches to derive the cluster distances. This method has the advantage of being independent of both reddening and the reddening law, $R_V = A_V/E(B-V)$. The vast majority of clusters projected in the direction of the Galactic bulge are located on the near side of the Galactic Center. Deviations from the reddening law do not seem to be responsible for this peculiarity. We need to introduce a peculiar, steep dependence of the absolute horizontal-branch magnitude with metallicity in the metal-rich regime if we want to reproduce a symmetrical distribution. Instead, if the observed distribution is correct, we expect a rather large number of bulge globular clusters are still to be discovered.

Keywords. Galaxy: bulge, globular clusters: general, Galaxy: structure

---

1. Introduction

The Galactic globular clusters are strongly concentrated towards the Galactic Center. In a $20°\times20°$ field, there are about 74 known globular clusters, of which 60 are located closer than 4 kpc (in projection) from the Galactic Center. They represent almost half of the total number of globular clusters known in the entire Galaxy. A recent review describes the properties of the globular clusters located in a $5°\times5°$ field (Barbuy *et al.* 1998), showing that the vast majority, possibly all, belong to the family of bulge clusters. Now, more than 10 years since their discovery, we may have uncertain distances because of the high reddening and the relatively poorly known absolute magnitude of the horizontal branch (HB), which is currently used as standard candle for most of the clusters. Here, we present a detailed overview of a larger field ($6°\times6°$) containing 23 clusters, and investigate their distance distribution, apply checks using different methods, and discuss possible biases and the consequences of the most common assumptions.

2. Overview

The distances to the clusters in the present study are based on the HB magnitude–metallicity relation, $M_V(\mathrm{HB}) = 0.16[\mathrm{Fe/H}] + 0.98$. The HB visual magnitudes and reddening values are taken from individual published studies. The results show that all but two clusters are located on the near side of the Galactic Center (assuming a conservative distance to the Galactic Center of 8 kpc), and at an average distance of about 6.3 kpc from the Sun (see Figure 1).

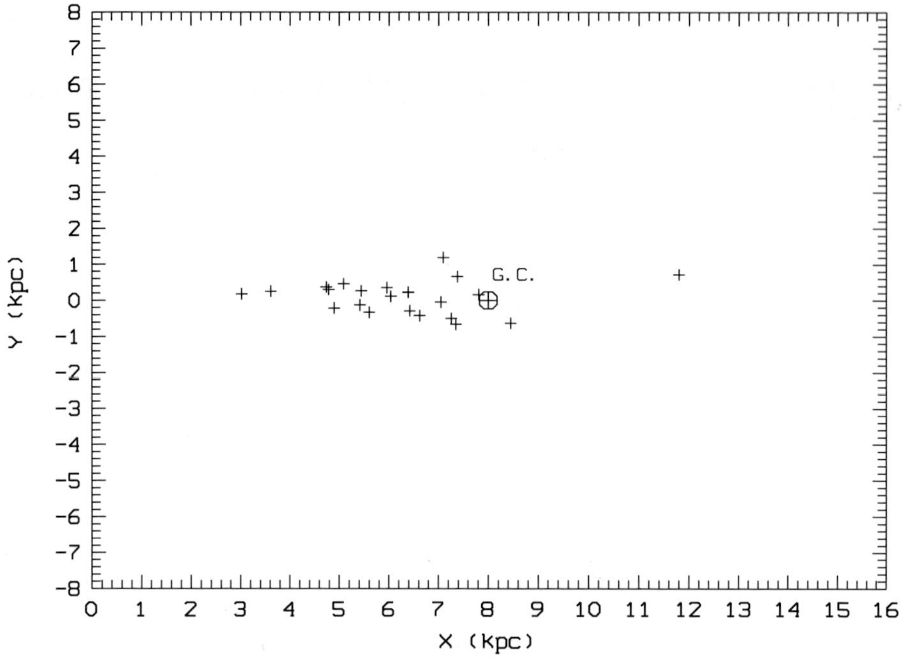

Figure 1. Distribution of the globular clusters in a $6° \times 6°$ field projected onto the Galactic plane.

Considering that there is no trend of distance with reddening (Barbuy *et al.* 1998), we checked the relation between visual magnitude and reddening to see if the slope of the reddening law might be significantly different from the standard $R_V = A_V/E(B-V) = 3.1$. As expected from the long line of sight through the Galactic disk, the average value of 3.1 seems a reasonable interpolation of the data. If we exclude that this choice of reddening law is responsible for the asymmetrical distribution of the globular clusters around the Galactic Center, we need to check the assumed relation for the absolute magnitude of the HB. A small change in the coefficient or in the zeropoint (0.1–0.2 mag) does not change the results significantly. For an average distance of 8 kpc, one must impose a zeropoint differing by about 0.6 mag from the assumed value, which is not compatible with the distribution of the halo clusters. Another alternative could be a higher luminosity dependence with metallicity, combined with a lower zeropoint, to match the zeropoint for the lower-metallicity halo clusters. A coefficient as high as about 0.5 would be required, still incompatible with the much lower coefficients derived from halo clusters and field stars. It is clear that an independent check of the distances is needed. For this purpose, we adopted a different approach, based on the simultaneous detection of stellar field and cluster HBs in the same frame, i.e., affected by the same reddening. We assume that the field HB (appearing as a red clump) corresponds to the bulge field HB and that the barycentric position of the bulge field coincides with the Galactic Center.

We determined the luminosity differences between the HBs of the clusters and the field for all clusters where the field HB was clearly defined in the same frame. In some cases, we used proper-motion separation of the two components (notably for NGC 6553 and NGC 6528). This method is independent of both the reddening and the reddening law adopted. The results from a subsample of 12 clusters indicate that the cluster HBs are systematically brighter than the field by an average of 0.31 mag in the V band. Only three clusters have a fainter HB than the corresponding field stars. On average, the magnitude difference decreases with our original distance measurements discussed above, indicating that the measured luminosity difference is the result of a real distance effect and not something peculiar.

If we interpret this result as purely an effect of distance differences between the baricentric position of the field population and the cluster positions, we again obtain a similar result to that already derived using the standard calibration and the individual reddening for each cluster. However, in doing so we ignore differential metallicity effects on the HB luminosity. From the literature, we derive that the sample clusters have an average metallicity of about [Fe/H] $= -0.9$. The bulge field has a metallicity around solar, or somewhat higher (Zoccali *et al.* 2008). It is easy to see that if we want to match the average cluster distance modulus with that of the field, we need a coefficient for the HB luminosity dependence on metallicity of about 0.4 between about [Fe/H] $= -1.3$ and solar metallicity. This is marginally compatible with the halo clusters, and only if we assume a sudden change in slope around a metallicity between [Fe/H] $= -1.3$ and -1.0. A change of slope in this interval was proposed by McNamara (1999). An alternative is that the currently known sample of globular clusters is biased due to a peculiar combination of distance and reddening and that we are missing most of the clusters on the other side of the Galactic Center in a relatively narrow cone subtending at least 10–20 degrees.

3. Implications

Studies of the distances of Galactic bulge clusters show an anomalous distribution, with a concentration on the near side of the Galactic Center. This asymmetric distribution is apparently confirmed by the systematically brighter HBs of the clusters compared to the bulge field HBs projected in the same direction. To match their distances, we need a rather different calibration equation for the absolute magnitude of the HB of the bulge clusters from that representing the halo clusters. An alternative explanation is that we are missing most of the low-latitude clusters projected in the direction of the Galactic bulge due to reddening, distance issues and crowding effects. Future near-infrared surveys may reveal these clusters, provided that crowding is not a major problem.

References

Barbuy, B., Bica, E., & Ortolani, S. 1998, *A&A*, 333, 117

McNamara, D. H. 1999, *PASP*, 111, 489

Zoccali, E., Hill, V., Lecureur, A., Barbuy, B., Renzini, A., Minniti, D., Gomez, A., & Ortolani, S. 2008, *A&A*, 486, 177

Star clusters: basic galactic building blocks
Proceedings IAU Symposium No. 266, 2009
R. de Grijs & J. R. D. Lépine, eds.

doi:10.1017/S1743921309990937

The range of validity of cluster masses and ages derived from broad-band photometry

J. Maíz Apellániz

Instituto de Astrofísica de Andalucía–CSIC, Granada, Spain
email: jmaiz@iaa.es

Abstract. I analyze the stochastic effects introduced by sampling the stellar initial mass function (SIMF) in the derivation of the individual masses and the cluster mass function (CMF) from broad-band visible/near-infrared unresolved photometry. The classical method of using unweighted *UBV* photometry to simultaneously establish ages and extinctions of stellar clusters is found to be unreliable for clusters older than ≈ 30 Myr, even for relatively large cluster masses. On the other hand, augmenting the filter set to include longer-wavelength filters and using weights for each filter increases the range of masses and ages that can be measured accurately with unresolved photometry. Nevertheless, a relatively large range of masses and ages is found to be dominated by SIMF sampling effects that render the observed masses useless, even when using *UBVRIJHK* photometry.

Keywords. methods: analytical, methods: numerical, methods: statistical, open clusters and associations: general, globular clusters: general, galaxies: star clusters

1. Description

This work is the third in a series in which we are analyzing the possible biases present in mass functions (Maíz Apellániz & Úbeda 2005; Maíz Apellániz 2008). I have used a combination of analytical approximations and Monte Carlo simulations to study the effect of stochastically sampling the stellar IMF (SIMF; assumed to be of Kroupa type) in the determination of masses and ages of unresolved stellar custers from broad-band photometry. For this purpose, I have used the new version (3.1) of CHORIZOS (Maíz Apellániz 2004), which incorporates an evolutionary synthesis module. For a given mass and age, I generated a minimum of 10 000 realizations of the SIMF using solar-metallicity (a) Geneva isochrones for massive stars and (b) Padova isochrones with new asymptotic giant branch (AGB) treatment for low and intermediate masses. In each case, I measured the mass (and age, if appropriate) assuming that the cluster has a well-sampled SIMF. From the individual realizations I derived (a) the observed cluster mass and age distributions for clusters of a (fixed) real mass and age and (b) the observed cluster mass functions (CMFs) for a real, truncated-power-law CMF with a slope of $\gamma = -2.0$. Three different cases in order of complexity were considered: (i) single-filter observations of clusters with known age and extinction, (ii) multifilter observations of clusters with unknown age and known extinction and (iii) multifilter observations of clusters with unknown age and extinction. In the first case, masses were computed directly by converting filter-convolved luminosities to masses, while in the second and third cases a χ^2 minimization code (CHORIZOS) was used to derive the masses and ages. The reader is referred to Maíz Apellániz (2009) for further details.

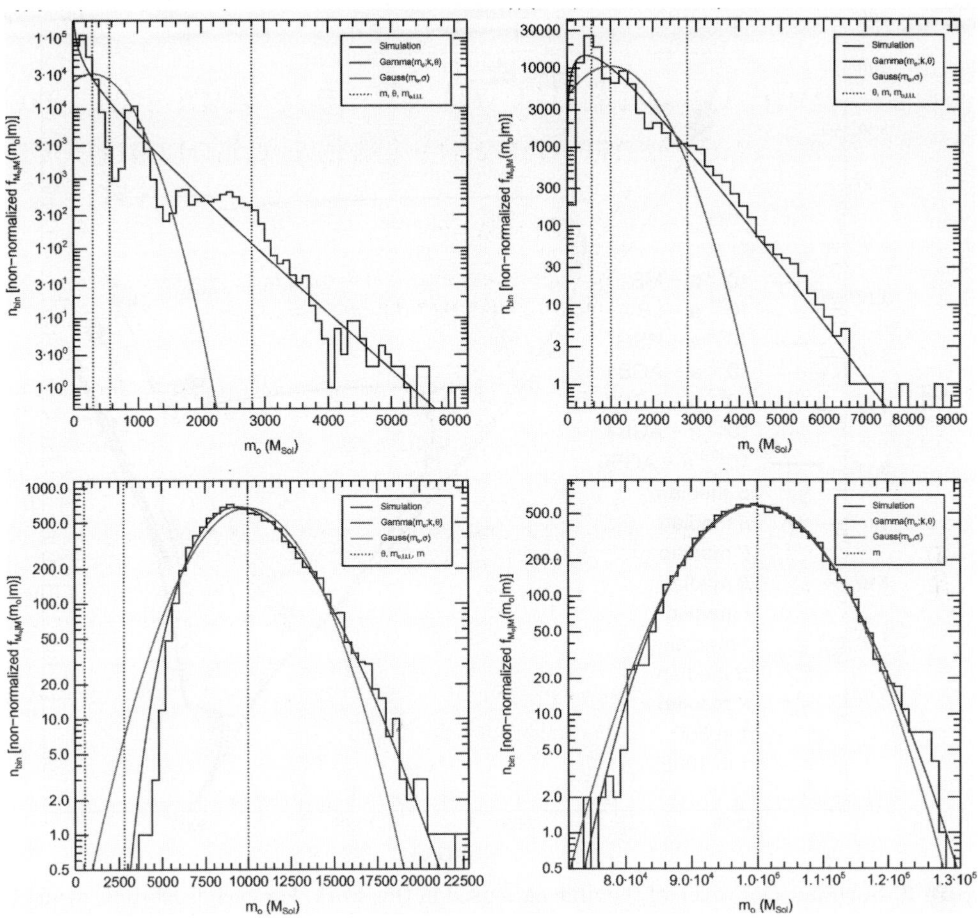

Figure 1. Results of the Monte Carlo simulations of $f_{M_{\rm o}|M}(m_{\rm o}|m)$ and Gaussian and Gamma fits for 10 Myr-old clusters observed in V with four real masses: 300 M$_\odot$ *(top left)*, 1000 M$_\odot$ *(top right)*, 10 000 M$_\odot$ *(bottom left)* and 100 000 M$_\odot$ *(bottom right)*. The Poisson mass θ for this filter and age is 566 M$_\odot$. For large real masses, both fits provide reasonable approximations, but for small values of m the Gamma approximation is significantly better. The vertical dotted lines show the values of the real mass, m, and (when visible) of θ and $m_{\rm o,LLL}$ (the observed cluster mass obtained when one mistakes the brightest possible star in the isochrone for a cluster).

2. Case 1: single filter, known age and extinction

• For a given cluster mass (m), the observed cluster mass ($m_{\rm o}$) distribution can be better approximated by a Gamma function than by a Gaussian distribution (Figure 1):

$$f_{M_{\rm o}|M}(m_{\rm o}|m) = A m_{\rm o}^{m/\theta - 1} e^{-m_{\rm o}/\theta} \theta^{-m/\theta}. \tag{2.1}$$

• The Poisson mass, θ, is a measurement of the stochasticity and depends (strongly) on the age and the filter used to derive $m_{\rm o}$ (Table 2).

$$\overline{m_{\rm o}} = m \quad ; \quad \sigma_{m_{\rm o}} = \sqrt{\theta m}. \tag{2.2}$$

• The large differences in θ are a consequence of the different fractions of stars located above the median mass for the luminosity in each filter, for a given isochrone (Figure 2).

• The observed CMF, $f_{M_{\rm o}}(m_{\rm o})$, differs from the real (truncated power-law) CMF, $f_M(m)$, in two aspects (Figure 3): (i) For large masses, $f_{M_{\rm o}}(m_{\rm o})$ is similar to $f_M(m)$ but,

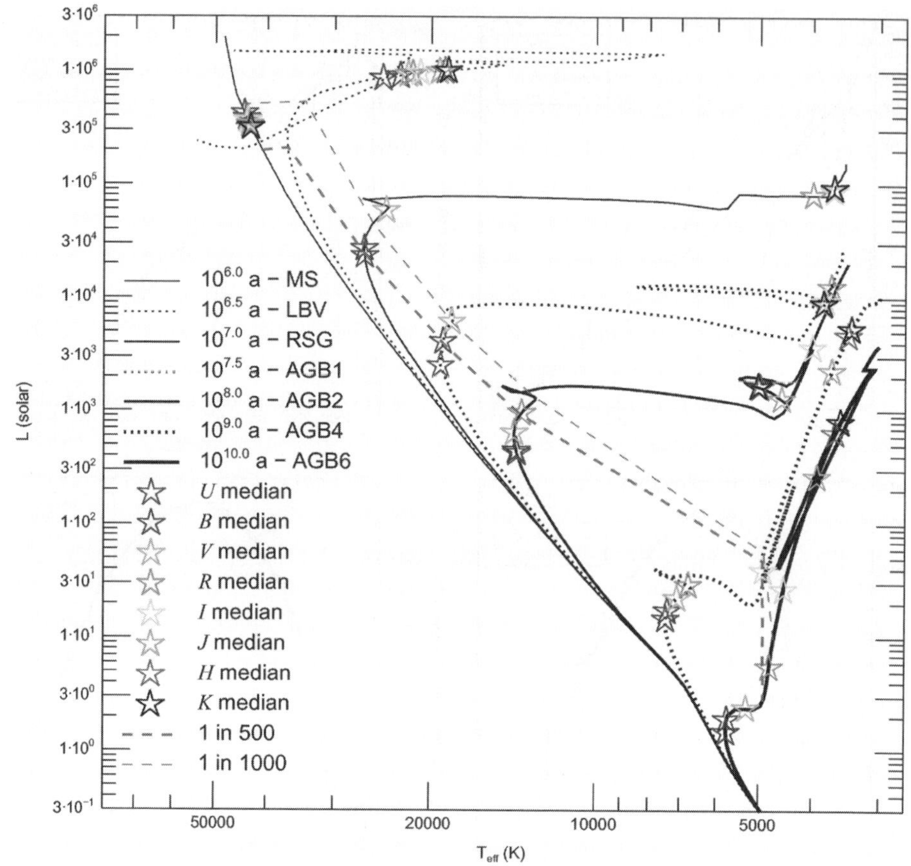

Figure 2. Isochrones for seven of the nine ages used in this work. For each isochrone, symbols of different colors are used to indicate the median mass for the luminosity in the eight filters U, B, V, R, I, J, H and K, assuming a Kroupa SIMF between 0.1 and 120 $M_\odot$. The two dashed lines join the points in each isochrone beyond which 0.2 and 0.1%, respectively, of the remaining stars in a well-sampled Kroupa SIMF are located. The two remaining isochrones are not included for the sake of clarity.

as the cluster mass decreases, at one point a 'hump' or overdensity appears. (ii) To the left of the hump, $f_{M_o}(m_o)$ asymptotically approaches a power law with a slope greater than -1.0, with some clusters having m_o values lower than the real cutoff in m.

• Using an analytical approximation, it can be shown that if one is willing to tolerate a systematic error in the CMF slope γ of up to $\Delta\gamma$, one can only analyze observed masses

$$m_o > 0.5\gamma(\gamma + 1)\theta/\Delta\gamma, \tag{2.3}$$

e.g., for $\gamma = -2.0$ and $\Delta\gamma = 0.1$, $m_o > 10\,\theta$.

3. Case 2: multiple filters, unknown age, and known extinction

I used three execution types:

1. *UBVRIJHK* photometry with single-band θ-dependent weights for each filter;
2. *UBVRIJHK* photometry with constant weights;
3. *UBV* photometry with single-band θ-dependent weights for each filter.

Table 1. Values of θ (in $M_\odot$) for different ages and filters for clusters observed with a single filter and known age and extinction. The labels below each age indicate the type of the brightest possible star. For a given filter, θ starts at a low value in the main-sequence (MS) phase, grows rapidly in the luminous-blue-variable (LBV) phase, decreases during the red-supergiant (RSG) and early AGB phases and then experiences a moderate growth when the cluster becomes old. For a given age, θ increases as a function of wavelength with the only exception of the MS phase, where the effect is the opposite (but rather weak).

Filter	1 Myr MS	3.16 Myr LBV	10 Myr RSG	31.6 Myr AGB1	100 Myr AGB2	316 Myr AGB3	1 Gyr AGB4	3.16 Gyr AGB5	10 Gyr AGB6
U	235	945	160	149	53	33	23	26	42
B	200	1314	192	305	51	32	27	52	92
V	190	1555	566	370	89	38	51	106	163
R	186	1722	1034	443	151	72	109	173	242
I	178	1949	1585	616	273	463	476	477	581
J	157	2741	2240	1170	708	3598	2930	2702	2445
H	146	3081	2476	1640	1281	5184	4828	4499	4123
K	140	3333	2542	1771	1527	5717	5688	5512	5058

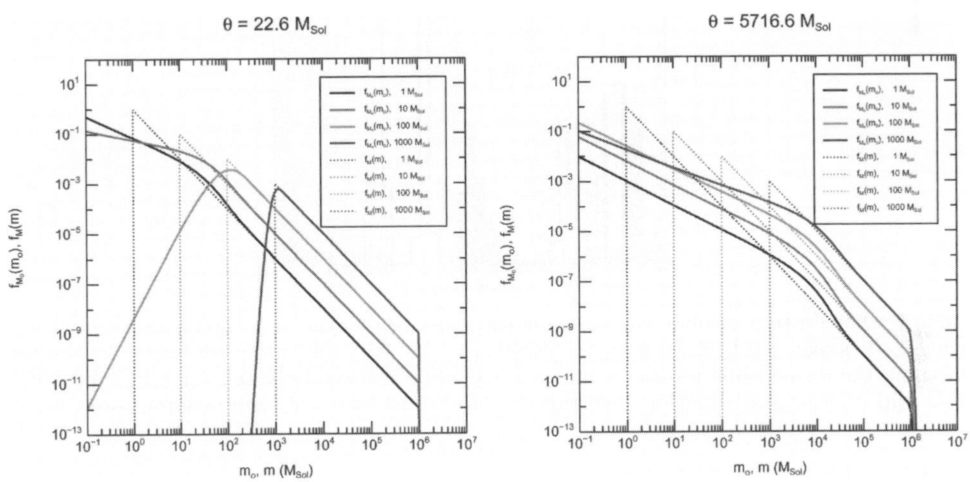

Figure 3. Observed (continuous lines) and real (dotted lines) CMFs for the minimum and maximum values of θ in Table 2. The left panel corresponds to 1 Gyr-old clusters observed in U and the right panel to 316 Myr-old clusters observed in K. Each panel shows four cases of the lower mass cutoff for $f_M(m)$ from 1 to 1000 $M_\odot$. As θ increases, the hump becomes more pronounced and moves towards the right.

The observed age distributions for 10^4 $M_\odot$ clusters are shown in Figure 4. The main results are:

- The ages calculated with $UBVRIJHK$ photometry and constant weights (i.e., the standard method) are highly uncertain, even for 10^4 $M_\odot$ clusters.
- Using weights based on the stochasticity of each filter significantly narrows the observed age distribution for 10^4 $M_\odot$ clusters, thus allowing to obtain ages with relatively small uncertainties.
- UBV and $UBVRIJHK$ photometry provide relatively similar results, with the former having the advantage for old clusters and the latter for their intermediate-age counterparts.

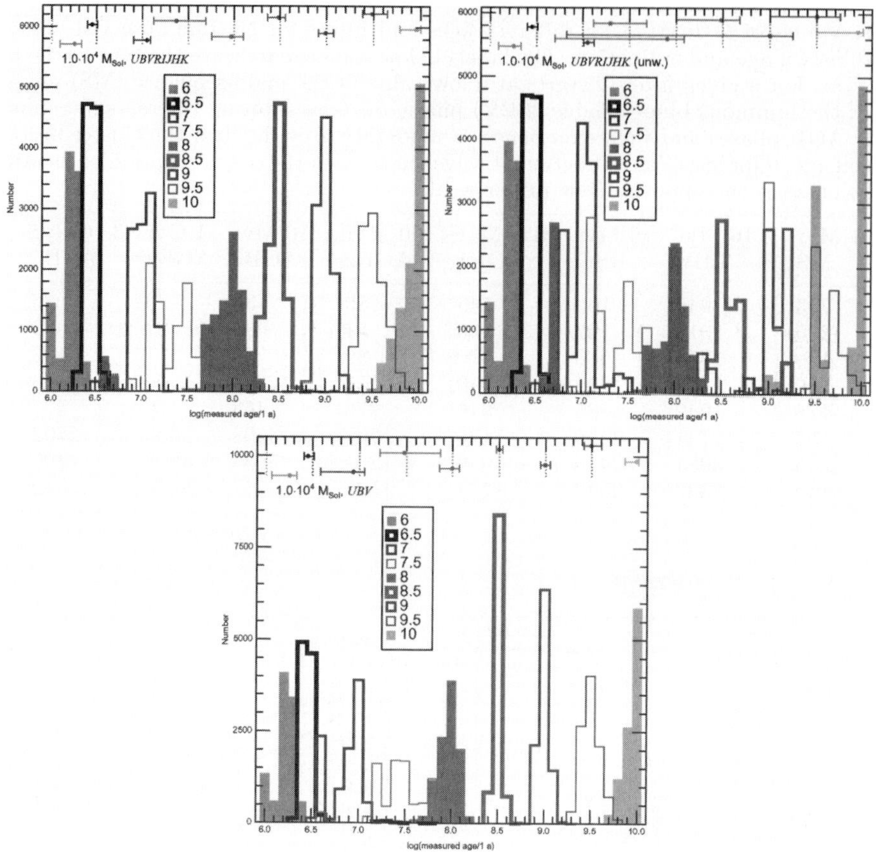

Figure 4. Distribution of observed ages for the nine input ages and three execution types *(1: upper left, 2: upper right, 3: bottom)* for clusters of 10^4 $M_\odot$ and unknown ages. At the top of each panel, the dashed line indicates the real age while the symbols and error bars provide the median and inferior and superior uncertainties (1σ equivalents) of each distribution.

- The AGB1 stage (31.6 Myr) has significantly greater uncertainties than the other evolutionary phases.

4. Case 3: multiple filters, unknown age and extinction

I used two execution types:

1. *UBVRIJHK* photometry with single-band θ-dependent weights for each filter;
2. *UBV* photometry with single-band θ-dependent weights for each filter.

The observed age distributions for 10^4 and 10^5 $M_\odot$ clusters are shown in Figure 5. The main results are:

- As opposed to the cases where extinction is known, the addition of *RIJHK* photometry provides a significant improvement in the observed ages.
- *UBV* photometry alone is not sufficient to accurately determine the ages of unresolved stellar clusters, even for massive ones, if extinction is unknown.
- For some ages (1–10 Myr, 100 Myr, 10 Gyr), *UBVRIJHK* photometry provides relatively accurate observed ages for $m = 10^4$ $M_\odot$. For other ages, higher masses are needed.

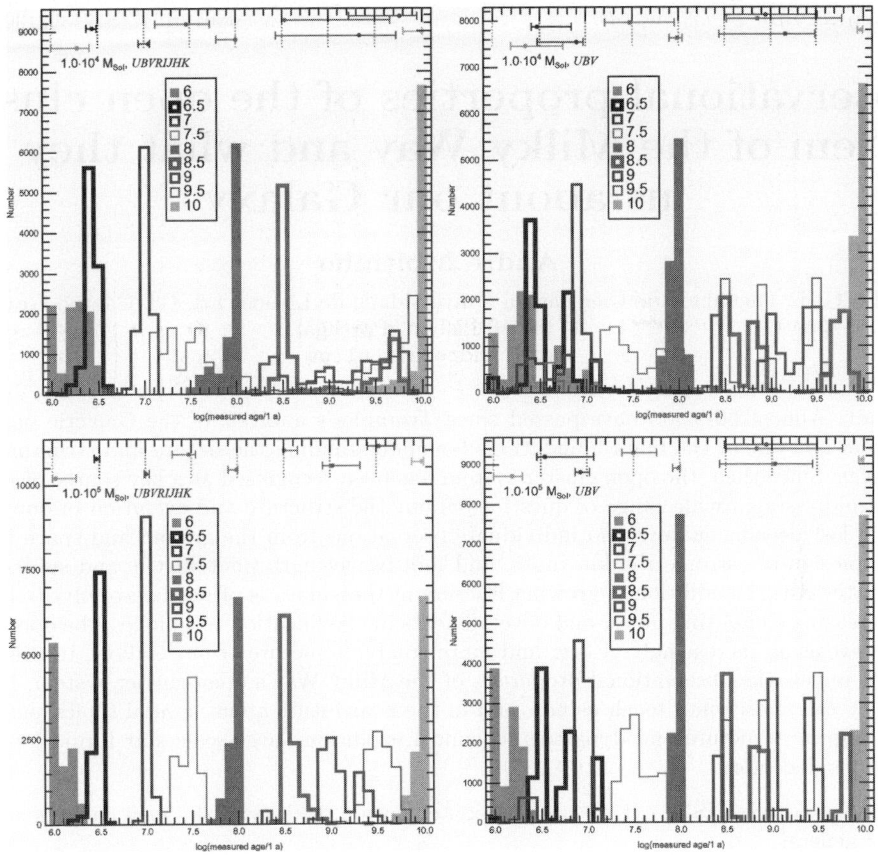

Figure 5. Distribution of observed ages for the nine input ages and two execution types *(1: left plots, 2: right plots)* for clusters of 10^4 *(top)* and 10^5 $M_\odot$ *(bottom)* and unknown ages and extinctions. At the top of each panel, the dashed line indicates the real age, while the symbols and error bars provide the median and inferior and superior uncertainties (1σ equivalents) of each distribution.

5. Conclusions

- SIMF sampling effects can introduce large biases in the determination of masses and ages from unresolved photometry.
- One should use weights based on the relative stochasticity of each filter when using χ^2-minimization techniques to calculate cluster properties from unresolved photometry.
- The addition of a wide photometric baseline (U to K) significantly reduces uncertainties when extinction is determined from the data.
- For clusters younger than 100 Myr, a critical (post-RSG) stage exists around 30 Myr where ages are especially difficult to determine.

References

Maíz Apellániz, J. 2004, *PASP*, 116, 859
Maíz Apellániz, J. 2008, *ApJ*, 677, 1278
Maíz Apellániz, J. 2009, *ApJ*, 699, 1938
Maíz Apellániz, J. & Úbeda, L. 2005, *ApJ*, 629, 873

Star clusters: basic galactic building blocks
Proceedings IAU Symposium No. 266, 2009
R. de Grijs & J. R. D. Lépine, eds.

doi:10.1017/S1743921309990949

Observational properties of the open cluster system of the Milky Way and what they tell us about our Galaxy

André Moitinho

SIM/FCUL, Faculdade de Ciências da Universidade de Lisboa, Ed. C8, Campo Grande, 1749-016 Lisbon, Portugal
email: andre@sim.ul.pt

Abstract. Almost 80 years have passed since Trumpler's analysis of the Galactic open cluster system laid one of the main foundations for understanding the nature and structure of the Milky Way. Since then, the open cluster system has been recognised as a key source of information for addressing a wide range of questions about the structure and evolution of our Galaxy. Over the last decade, surveys and individual observations from the ground and space have led to an explosion of astrometric, kinematic and multiwavelength photometric and spectroscopic open cluster data. In addition, a growing fraction of these data is often time-resolved. Together with increasing computing power and developments in classification techniques, the open cluster system reveals an increasingly clearer and more complete picture of our Galaxy. In this contribution, I review the observational properties of the Milky Way's open cluster system. I discuss what they can and cannot teach us now and in the near future about several topics such as the Galaxy's spiral structure and dynamics, chemical evolution, large-scale star formation, stellar populations and more.

Keywords. catalogs, ISM: extinction, Galaxy: evolution, Galaxy: disk, open clusters and associations: general

1. Preamble

In 1930, Robert Trumpler published his seminal paper in which he proves the existence of the interstellar medium (ISM). In that paper, Trumpler showed how distances to open clusters (OCs) derived from the apparent magnitudes of cluster stars of known spectral types are systematically greater than those derived from their apparent sizes, and how the effect increases with distance. He correctly attributed this phenomenon to the presence of an intervening medium. This was the first paper using OCs as tracers of Galactic structure.

2. Overview

Since the work of Trumpler, the open cluster system has been used to characterise, study and analyse a variety of aspects related to the structure, composition, dynamics, formation and evolution of the Milky Way. Not being exhaustive, these aspects include:

- Spatial scales: scale height of disk(s), height of the Sun above the Galactic plane, distance to the Galactic Centre.
- Spiral structure: tracing spiral arms.
- Kinematics and dynamics: rotation of spiral pattern, Galactic rotation curve (velocity of the 'local standard of rest', Oort constants), orbits of OCs and their relation to cluster survival/disruption.

• Galactic formation and evolution: ages of the oldest clusters, abundance pattern, inflow rate.

• Distribution (absorption) and kind (extinction law) of interstellar matter.

The potential of using OCs for studying the Milky Way is well known and sentences like "open clusters are valuable objects for Galactic studies" appear in the introduction of almost every OC paper. Leaving aside the astrophysical reasons, the main advantage of using OCs is the precision with which one can derive their reddening values, distances, ages, velocities and metallicities, compared to individual stars. This arises not only from improved statistics, as OCs are ensembles of stars, but also from the distribution of cluster stars in photometric diagrams. In fact, the colour–magnitude and two-colour diagrams (CMDs and TCDs, respectively) of star clusters can exhibit a rich structure that provides bench marks to constrain those parameters beyond a simple statistical average. An example is the 'elbow' displayed in optical CMDs for early A-type stars, which is very useful for constraining distances.

There are also some disadvantages, or at least difficulties and systematics in studying the Galaxy on the basis of OCs. Confusion against crowded backgrounds in the Galactic plane and dissolution of OCs likely produce selection effects that conspire against the identification of faint and/or poorly populated clusters. These effects will produce a selection against old clusters, which are fainter and potentially poorer (due to dynamical evolution) than their younger counterparts, and against the low-mass end of the OC population. The extent of such a selection is far from well known, but it must be kept in mind when deriving conclusions from the observed global properties of the OC system.

Using OCs for studying the Galaxy relies on a huge amount of painstaking data gathering. Each cluster is just a point in parameter space, but requires measuring many stars. Traditionally, stars were measured one by one using photoelectric photometers and spectrographs. Modern CCDs, infrared arrays and multi-object spectrographs (MOSs) have the ability of observing hundreds or thousands of stars in one shot, but reducing the data has proved to be very time consuming. In the last few years, increased computing power and development of automated or semi-automated reduction procedures have alleviated the reduction burden. But it took several years for the community to master software like DAOPHOT (Stetson *et al.* 1990). I guess that many of us have known what it is to reduce images one by one, selecting by hand the adequate PSF stars, testing the PSF model and redoing it if necessary. In the mid 1990s, the workstation I had at work would take several hours just to run ALLSTAR on a 2048 × 2048-pixel frame. Now, this can be done in a few minutes, or less, and most of us have used experience to devise automatic ways to select the proper PSF stars, so that a whole observing run can be reduced fairly quickly.

The actual job of studying the Galaxy requires a catalogue of OC parameters (reddening values, distances, ages, velocities, metallicities, diameters). Thousands of studies of individual OCs exist in the literature, often with nonconcordant results which must be critically inspected and collated. Despite the care put into the process, the variety of techniques used by different authors to observe, reduce and derive parameters, lead to inhomogeneous compilations that can limit their usefulness. This problem is well known and has motivated several attempts of deriving homogeneous parameter sets, usually by adopting some well-defined analysis procedure, reference curves (e.g., the zero-age main sequence, ZAMS, and isochrones) and measurements. Still, the problem persists for most objects.

It is therefore clear that assembling the observational basis for studying the Galaxy is an incremental process. It is built up slowly by adding the results of individual cluster studies, each one contributing one probing point. It is therefore natural that qualitative leaps in our understanding of the Galaxy happen in intervals of many years. Roughly, such jumps take about 20 years.

Very schematically, in the 1980s there was the Lund catalogue (Lyngå 1987), which contained around 1200 clusters of which a few hundred had some of their parameters determined. The classic review on OCs and Galactic structure is that of Janes & Adler (1982) and its update (Janes *et al.* 1988). The 1990s brought us the on-line WEBDA OC database (formerly BDA; Mermilliod 1992) which not only included a list of clusters and their parameters (essentially the Lund catalogue), but also the actual measurements of their stars (photometry, astrometry, spectral types, etc.). From the observational point of view, CCDs and infrared arrays became of common use, *Hipparcos* provided a wealth of astrometric and kinematic data and settled the distance scale through accurate parallaxes of the Hyades and Pleiades. Although the new data did not bring in new significant results regarding the spatial structure of the Milky Way, the number of abundance measurements and discovery of more old OCs shifted the focus on the origin and chemical evolution of the Galaxy. The canonical review on the subject is that of Friel (1995) and its updates. The 2000s brought the Dias *et al.* (2002) catalogue of OCs, which updates the Lund catalogue. The latest version includes 1787 clusters and has eliminated a number of entries shown to be nonexistent. The 2MASS survey (Cutrie *et al.* 2003) has triggered searches of new clusters (optically detectable and embedded, e.g., Bica *et al.* 2003b; Dutra *et al.* 2003; Froebrich *et al.* 2007) but very few of them have actually been studied. So far, this decade has not yet provided a fresh review on our subject, although some refined recalculations taking advantage of currently available data have been published (Bonatto *et al.* 2006; Piskunov *et al.* 2006). However, the rapidly growing number of spectroscopic observations allowed by MOSs, availability of reduction pipelines and ongoing large-scale photometric surveys are expected to soon take the understanding of our Galaxy to the next step.

3. The sample

As discussed above, knowledge of OCs is compiled in catalogues. The most complete catalogue of optically revealed OCs is that of Dias *et al.* (2002)†. Its latest version (2.10) lists 1784 objects of which 969 (54%) have simultaneous reddening, distance and age determinations. The number of clusters with distance, ages, radial velocities and proper motions is much smaller (430, or 24%) and only 178 (10%) have their metallicities determined. Lists of infrared (IR) OCs and candidates have been compiled by Bica *et al.* (2003a,b), Dutra *et al.* (2003) and Froebrich *et al.* (2007), based on searches using 2MASS data products. These objects, detected in the IR, add about 700 clusters to the optically identified family. However, very few have been studied so far and many might not be real OCs. The main references for the IR cluster lists are Bica *et al.* (2003a,b) and Dutra *et al.* (2003). The systematic IR search of Froebrich *et al.* (2007) has produced a list of around 1000 candidates, which they estimate suffers from 50% false-detection contamination. Once considering the number of already identified IR clusters, the Froebrich *et al.* (2007) list still contributes with the discovery of about 150 OCs (Glushkova *et al.* 2009).

The distribution of optical and IR clusters is shown in Figure 1. It is noticeable how the IR clusters are mostly located close to the $b = 0°$ plane, while the optical clusters define a thicker disk. This is expected, since most obscuring material is defining what is known as the extreme thin disk. The number distribution of clusters is not uniform along

† http://www.astro.iag.usp.br/~wilton/

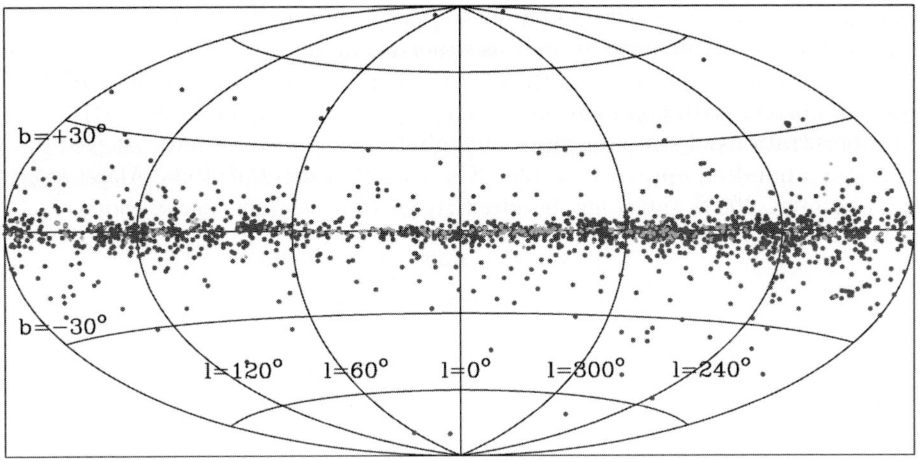

Figure 1. Aitoff distribution of known open clusters and candidates on the Galactic plane. Black: optically visible clusters. Grey: infrared clusters.

the Galactic longitude range. Moreover, clusters in different age groups exhibit different distributions (e.g., figure 3 of Bonatto *et al.* 2006).

At least two obvious biases modulate the distribution of *identified clusters*. The first is introduced by the extinction produced by dense interstellar material. Figure 2 clearly shows how the OCs delineate the wavy dust disposition. That extinction affects the detectability of sources is well known, but the kind of patterns it introduces in the distribution of OCs is often overlooked in studies of the Galaxy using OCs (and other objects). The other bias is produced by crowded backgrounds, against which an OC might contribute little to the density of field stars and can thus remain unidentified. This effect also includes nearby sparse OCs that occupy large areas on the sky.

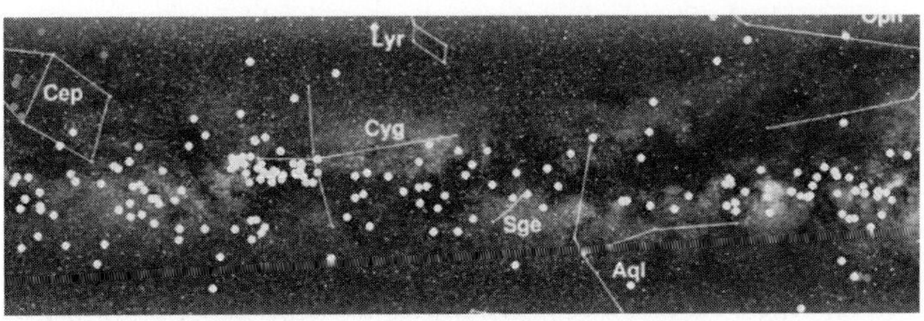

Figure 2. Optically visible OCs (circles) plotted over an optical image of the first quadrant of the Milky Way (credit A. Mellinger).

But there are also biases in the sample of *studied clusters*. On the one hand, interest has focussed on studying specific regions of the Galaxy (e.g., Cygnus, Carina and Puppis–Vela), thus artificially increasing the number of studied clusters in those regions with respect to others. On the other hand, many clusters have been observed for reasons other than studying the Galaxy. This was especially true in the past, when focus was mostly on studying stellar evolution. The tendency has been to study bright and/or well-populated clusters, which would provide much clearer Hertzsprung–Russell diagrams. Given the aforementioned difficulties in data reduction, there was little motivation to study poorer (potentially older) and fainter (potentially older/more distant/more reddened).

Although these biases exist in principle, all-sky proper-motion and photometric surveys can help to overcome them. The *Hipparcos*/Tycho-2 (ESA 1997; Høg *et al.* 2000) and UCAC2 (Zacharias *et al.* 2004) astrometric (and photometric) catalogues, the 2MASS *JHK* photometry and images, as well as other available data have enabled identification of OCs beyond simple visual recognition. Searches for concentrations of stars at a common distance or with common proper motion, cluster sequences in CMDs or simply spatial concentrations of stars obeying some photometric criteria have allowed discovery of more than a hundred optically visible OCs (e.g., Platais *et al.* 1998; Alessi *et al.* 2003; Kharchenko *et al.* 2005) and a few hundred IR clusters, as mentioned above.

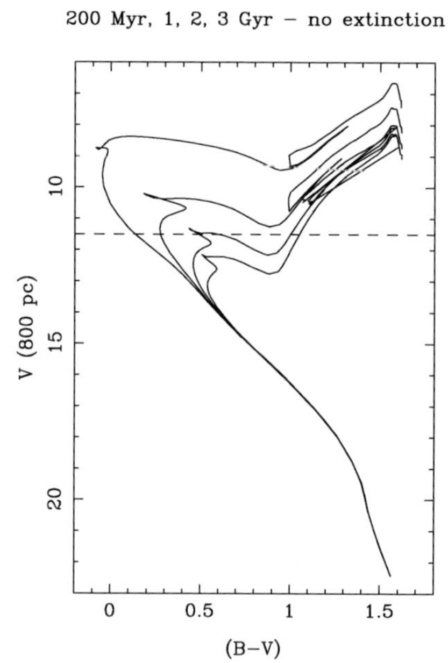

Figure 3. 0.2, 1, 2 and 3 Gyr isochrones (Girardi *et al.* 2000) at a distance of 800 pc from the Sun. No extinction is applied. The dashed line indicates the 11.5 mag completeness limit of the stellar data catalogue used in Piskunov *et al.* (2006).

It has been argued that the sample of known OCs is complete within 850 pc (Piskunov *et al.* 2006; 1 kpc is also quoted). The argument is based on the fact that the number of clusters kpc^{-1} starts to decrease beyond 850 pc. It seems, however, that even within 850 pc the sample of OCs might be missing a significant number of poorly populated intermediate-age and old objects. Figure 3 shows a series of isochrones, from moderately young to moderately old, placed at 800 pc from the Sun, with no extinction applied. It is readily seen that at 1 Gyr and older, only the evolved portion of the cluster sequence is above the dashed line, which marks the completeness limit of the astrometric and photometric catalogue used in Piskunov *et al.* (2006). Apart from the red clump, the other post-main-sequence regions of the CMD will be barely populated. The red clump will only be noticeable for fairly populated clusters. For the not-so-rich clusters, the number of clump stars might not be enough to produce a visible cluster signature either in proper-motion vector-point diagrams or in CMDs. The ability to identify a cluster based on kinematics will also be hampered along directions dominated by the solar motion. Furthermore, in reality, clusters will be affected by extinction, which has the effect of making the sequences fainter, thus pushing them even further below the completeness line. For the 200 Myr example, the red clump is very sparsely populated and the presence of 1 mag of absorption will leave the out the main sequence, thus making it very hard to detect (unless it is visually obvious on a sky chart).

As for the total number of OCs in the Milky Way, estimates depend on many hypotheses, such as how well the number of clusters per volume in the solar neighbourhood represents its position relative to the Galactic Centre, the density-decay law and cluster completeness. In any case, estimates indicate numbers on the order of 10^5 OCs (Bonatto *et al.* 2006; Piskunov *et al.* 2006).

4. Spatial scales

We now proceed to see what the statistical properties of the OC distribution tell us about the spatial scales of the Galactic disk and our position within it. The increased samples of OCs analysed in recent studies essentially recover the classic results from the 1980s, although in a clearer way. In particular, it is found that the disk scale height (SH) defined by the young clusters is smaller than that derived from older clusters, with the young and intermediate-age OCs (< 200 Myr versus < 1 Gyr) having SHs of 48 pc and ~ 150 pc, respectively. For older clusters, the vertical distribution is essentially uniform, so that no SH can be derived. Also, the distribution of OCs younger than ~ 1 Gyr clearly shows how the disk outside the solar circle is $1.4 - 2.0\times$ wider than inside this radius (Bonatto *et al.* 2006). Assuming that OCs are symmetrically distributed above and below the plane, the zero point of the height distribution gives a measure of the distance of the Sun from the Galactic plane. The current values determined from OCs range from about 15 to 22 pc (Bonatto *et al.* 2006; Piskunov *et al.* 2006; Joshi 2007).

These seem to be well-established properties of the OC system that persist as the sample grows. Interpretation is, however, not so straightforward and unambiguous. One of the first things that call our attention is that the SH defined by thin-disk stars (~ 200 pc) is much larger that that defined by the young OCs, although it is more or less compatible with that of intermediate-age OCs. This could be seen as a disk evolutionary effect, by which the disk would thicken with age due to some heating mechanism such as collisions with external systems or by different conditions in the ancient disk (Janes & Phelps 1994). However, it could also be an effect of the selective destruction of OCs at lower heights. Indeed, the range of heights of the different OC age groups are very similar. What is different is the proportion of clusters at small to large heights (see, e.g., figure 7 of Bonatto *et al.* 2006). This might as well be simply reflecting the destruction of clusters that tend to be close to the plane, which would trim the distribution peak, leading to a flatter distribution and hence a larger SH. This scenario seems consistent with the apparent flatness of the old OC distribution, which is not seen in the smaller sample of Janes & Phelps (1994), who find a SH of ~ 375 pc. Verification requires a kinematic analysis.

From the perspective of OCs as progenitors of the field population, the link between the OC and field-star SHs remains to be firmly established. Kroupa (2002) has shown how (kinematically hot) stars escaping star clusters, mostly while emerging from their parental molecular clouds, can thicken Galactic disks and even produce thick disks. This mechanism could provide the link. Another interesting consequence is that if this mechanism were the sole avenue for producing a thick disk, then we would not expect to find thick-disk OCs. Finding these clusters would mean that other processes (such as capture from an accreted galaxy) would (also) operate.

As for the the thicker SH outside the solar circle, the result is derived from a global comparison of the SH of outer versus inner clusters. Its precise meaning is not straightforward and is most likely a combination of effects such as the flaring of the disk and the amplitude of the warp. Isolating both effects requires larger samples of OCs to analyse the SH at smaller Galactic longitude intervals.

OCs have also been used for estimating the distance to the Galactic Centre. From a kinematic analysis of 301 OCs within 3 kpc from the Sun, Shen & Zhu (2007) derive $R_{\rm GC} \sim 7.95$ kpc, which is consistent with the the value the same authors derive from OB stars (8.25 kpc). Although based on a simple Oort model, these determinations agree with the well-known results of Genzel *et al.* (2000; 7.8–8.2 kpc) from analysis of radial velocities of more than 100 stars around the black hole Sgr A*, and of Eisenhauer *et al.* (2003), who found $R_{\rm GC} = 7.94 \pm 0.42$ kpc based on astrometric and radial-velocity observations of star S2 around Sgr A*. These results support the long-known indications that the standard IAU value, $R_{\rm GC} \equiv 8.5$ kpc, should be revised and lowered by ~ 500 pc. A synthesis of the determinations of $R_{\rm GC}$ in the past decades is shown in figure 2 of Shen & Zhu (2007).

Another aspect concerning the spatial scales of the Galactic disk is the extent of the warp. By looking along low-extinction windows in the third Galactic quadrant (3GQ), Vázquez *et al.* (2008) have been able use OCs to trace the disk out to Galactocentric distances greater than 16 kpc. In their figure 6, one can see how the southern warp reaches distances greater than 1 kpc below the formal Galactic plane. Finally, Carraro *et al.* (2009) reported OCs at large Galactocentric distances, with VdB–Hagen 4 reaching 20 kpc. This value is significantly larger than the $\sim$ 14 kpc truncation limit usually adopted and sets a lower limit to the extent of the stellar disk as probed by OCs.

5. Spiral structure

Young OCs have long been used for tracing spiral structure (e.g., Becker & Fenkart 1970; Moffat *et al.* 1979). Since star formation in the Milky Way occurs mainly in the spiral arms, it is expected that the younger clusters, which have not yet departed significantly from their birth sites, will trace the local spiral structure. Compared to gaseous spiral indicators such as H I, H II and CO, which are observed at radio wavelengths and for which we rely on rotation curves to derive their distances, OC distances are much better determined. Furthermore, it is well known that the velocity field in spiral arms does not correspond exactly to that implied by the general rotation curve. Indeed, spiral arms appear as glitches in empirical rotation curves. Thus, one arrives at the situation of tracing spiral structure using velocity–distance relations that do not apply well to the regions where they are used. Despite this source of uncertainty, radio observations roughly reveal the general spiral structure of the Milky Way, with the advantage of being insensitive to obscuration and thus being able to reach most of the Galaxy. OC distances are more precise, but are mainly obtained from observations in the optical domain and cannot probe beyond a few kpc due to interstellar obscuration (except for a few absorption windows at certain Galactic longitudes).

When mentioning young OCs as spiral tracers, the question arises as to what is meant by young. Dias & Lépine (2005), in their figure 1, show that OCs up to $\sim$ 12 Myr remain in their parent arms. At $\sim$ 20 Myr, they are already drifting away from the arms, joining the general disk population, although spiral traces are still identifiable. This figure is illustrative of spiral structure as traced by OCs and also shown in many other studies. In particular, the Carina–Sagittarius, local (or Orion) and Perseus arms are prominent in the solar vicinity up to distances of $\sim 4 - 5$ kpc.

Expanding on this classical view, the recent papers of Carraro *et al.* (2005), Moitinho *et al.* (2006) and Vázquez *et al.* (2008) have revealed the spiral structure in the 3GQ based on OCs, CO clouds and early-type field stars. These studies support a four-armed pattern and have shown that the Outer (Norma–Cygnus) arm crosses the whole 3GQ, reaching distances beyond 10 kpc from the Sun and that the local (Orion) arm is more than a simple spur, reaching the Outer arm. As for Perseus, it is not traced by young OCs in this region, being apparently disrupted by the crossing of the local arm, but is then again traced by OCs in the 4GQ. An aspect of these analyses that is worthwhile emphasising is the treatment of early-type field stars. At sufficiently large distances, early-type stars belonging to spiral arms can be considered as spatially confined along their lines of sight. These stars will thus produce cluster-like sequences in photometric diagrams that allow determining average reddening values and distances using the same techniques as used for analysing clusters. Interestingly, these sequences had previously been identified and baptised as 'blue plumes' (because of their plume-like appearance on the blue side of CMDs) and were interpreted as the signature of a dwarf galaxy cannibalised by the Milky Way (the Canis Major dwarf galaxy; Martínez–Delgado *et al.* 2005). However, these blue plumes have been shown not only to trace the spiral structure in the 3GQ, but also across the 2GQ (Carraro *et al.* 2008), where due to superposition with evolved OCs they have often been interpreted as blue stragglers.

6. Kinematics and dynamics

The 1990s and 2000s witnessed an explosion of astrometric data. The main players being the *Hipparcos* and Tycho-2 catalogues (ESA 1997; Høg *et al.* 2000) and the UCAC2 compilation (Zacharias *et al.* 2004). The new UCAC3 catalogue is being released during this IAU General Assembly.

These data have led to many studies of OCs. From the individual cluster point of view, they have allowed to determine cluster membership probabilities, cluster proper motions, internal dynamics and to firmly set the Hyades and Pleiades luminosity scales, which are crucial for calibrating the cosmological distance ladder and the ages provided by stellar evolution models. These astrometric data have also been widely used for studying star-forming regions (which are out of the scope of this review) containing OCs. But regarding large-scale Galactic structure, they have not triggered great qualitative progress. The main reason for this is the very limited distance range, of a few hundred pc, that they can probe.

OCs have often been used for studying the local behaviour of the Galactic rotation curve. However, a proper rotation curve covering a sizable range of the Galaxy is still lacking. Work in progress in this area was recently reported by Frinchaboy & Majewski (2008). The outcome of most kinematic research has been to improve the determination of parameters such as the local standard of rest rotation speed, Oort's constants and the velocity of the Sun. An example of such an analysis is summarised in table 2 of Piskunov *et al.* (2006). These improvements do not only come from the available high-quality astrometric data, but also from the quantity of data, which allows to resolve local structures such as star-formation complexes and stellar streams and eliminate the systematics they introduce (e.g., Piskunov *et al.* 2006).

In addition to these more classical calculations, OCs have also been used recently to determine the rotation speed of the spiral pattern. Dias & Lépine (2005) confirmed that spiral arms rotate as rigid bodies, and determined the pattern speed by direct integration of cluster orbits and by comparing the positions of clusters in different age groups, leading to an adopted value of 24 km s^{-1} kpc^{-1}. As a an additional result, they find that the co-rotation radius is $(1.06 \pm 0.08)\times$ the solar Galactocentric distance. It is also interesting to note that Bobylev *et al.* (2007) find a different speed for the Orion arm compared to Carina–Sagittarius and Perseus. Another basic parameter is the epicyclic frequency (the residual circular velocity after subtracting the Galactic rotation curve) which Lépine *et al.* (2008) find to be 42 ± 4 km s^{-1} kpc^{-1} at the solar radius.

7. Formation and evolution

The ages of the oldest OCs place a lower limit on the age of the Galactic disk. Presently, the oldest-known clusters are (still) NGC 6791 and Be 17 which have ages of around 10 Gyr (Carraro *et al.* 1999; Salaris *et al.* 2004), which is younger than the halo, although they almost fall in the range of the youngest globular clusters.

Few old open clusters have been identified in the inner Galaxy. This is usually attributed to the harsher conditions that would destroy OCs at small Galactocentric radii. Curiously, the orbit of NGC 6791, which is one of the oldest known OCs and is presently outside the solar circle, can reach a distance of only 4.6 kpc from the Galactic Centre.

The most discussed aspects of Galactic evolution scenarios are those related to chemical-abundance gradients. Most observational studies in the last two decades find a smoothly decreasing metallicity–Galactic radius gradient. Recent studies determine a gradient of ~ -0.06 dex kpc^{-1} (Friel *et al.* 2002; Chen *et al.* 2003). However, controversy has been generated by the work of Twarog *et al.* (1997), who reexamined the metallicity scales of published observations and found the radial metallicity distribution to be better described by a step function with the transition occurring at around 10 kpc from the Galactic Centre. The inner region would be characterised by a shallow gradient (with an average [Fe/H] = 0.0 dex), followed by a -0.3 dex step and then an almost flat metallicity distribution. On the theoretical side, models usually yield simple gradients, although the recent paper of Magrini *et al.* (2009) is able to predict a sudden change at around 7 kpc, followed by an almost constant plateau starting at around 11 kpc. From a dynamical point of view, the simulations of Corder & Twarog (2001) show how orbital diffusion can smooth the discontinuity (assuming it is real) and how using younger clusters enhances the detection probability of the step. Another perspective is offered by Mishurov *et al.* (2002) and Lépine *et al.* (2003) who show how the co-rotation resonance can break a smooth abundance gradient.

Regarding a possible age–metallicity relation, most studies fail to find such a dependence over a cluster age range of almost 10 Gyr, although there is marginal evidence from clusters at Galactocentric radii larger that 10 kpc (see Friel 1995; Friel *et al.* 2002). The most natural explanation remains that chemical enrichment must have essentially taken place early in the Galaxy's evolution.

For a more in-depth view of OCs as probes of the origin and chemical evolution of the Milky Way the reader is referred to the reviews of Janes & Phelps (1994), Friel (1995) and Friel *et al.* (2002).

8. Interstellar extinction

We now briefly return to the point from which this review started: extinction. Joshi (2005) studied the interstellar-absorption pattern within $\sim 4 - 5$ kpc using catalogued reddening values towards 722 OCs. He found that absorption approximately follows a sinusoidal curve with Galactic longitude, with the highest values occurring at $l \sim 30$–$50°$ and the lowest values around $l \sim 220$–$250°$. Extinction is generally low in the 3GQ, increasing only ~ 0.07 mag kpc^{-1} between $160° < l < 280°$, in agreement with the analysis by Vázquez *et al.* (2008). Except for the opaque barrier at $l \sim 265°$ due to the Vela molecular ridge, the low-absorption window continues in the 4GQ towards Carina. In general, for the remaining longitude range, absorption is higher and happens mostly between 1 and 2 kpc from the Sun.

Joshi (2005) also determined that the dust plane is tilted by 0.6° with respect to the $b = 0°$ plane. The maximum and minimum height of the dust plane occur at ~ 54 and 234°, respectively, coincident with the maximum and minimum of the Galactic warp.

Although these and other studies reveal the distribution and amount of interstellar matter, they do not give any idea of the nature of the absorbing material. This is due to most OC reddening and absorption values having been derived assuming a standard interstellar extinction law. A study of the variations of the interstellar law would certainly be interesting.

9. Prospects

As insistently advocated in this contribution, an important goal for the future is to have an unbiased, or at least well-understood, sample of OCs. This will require mitigating extinction artifacts, which demands analysing the mostly unstudied infrared clusters. On the way, we will learn much about the kind, quantity and distribution of interstellar material. Success will, however, depend on the development of IR photometric analysis techniques capable of yielding precise cluster parameters.

Another well-known problem to be addressed is that of the homogeneity of OC-parameter compilations. Homogeneity of the data itself is also certainly an issue, but of second order when compared to the analysis. However, a word of caution should be issued: homogeneous methods do not necessarily produce homogeneous catalogues. As an example, *Hipparcos* parallaxes are great for determining the distance to close clusters such as the Pleiades. For a cluster at 300 pc, parallaxes will still be available but will not do a very good job. In this case, a good ZAMS fit would produce a distance on a scale in better agreement with the Pleiades distance from parallaxes.

As for the future, one cannot emphasise enough how important *Gaia* will be to Galactic astronomy. *Gaia* will deliver excellent distance estimates and kinematics for single stars, but it will not end the use of OCs for studying the Galaxy. Clusters are still unique when we need ages, and *Gaia* will help improving them. Also, *Gaia* is not an IR mission and will therefore not be able to penetrate dense regions.

Many large-area surveys are either ongoing or planned for the near future using ground-based facilities: UKIDSS, SEGUE, IPHAS, VPHAS+, VVV, VHS, LSST, PAN-STARRS and others. These surveys will produce a wealth of data from the optical to the near-IR, including in selected bands such as Hα. In several cases, observations will even be time-resolved.

The next few years promise to offer a profusion of intense emotions to OC researchers.

Acknowledgements

Although I appear as the sole author of this contribution, it has also been the product of years of stimulating discussions with several colleagues. I am particularly grateful to Wilton Dias, Giovanni Carraro, Rubén Vázquez and Jacques Lépine. This work was partially supported under a bilateral FCT/CAPES agreement.

References

Alessi, B. S., Moitinho, A., Dias, W. S., & Lépine, J. R. D. 2003, *A&A*, 410, 565

Becker, W. & Fenkart, R. B. 1970, in: W. Becker & G. I. Kontopoulos (eds.), *The Spiral Structure of our Galaxy*, IAU Symp. (Dordrecht: Reidel) vol. 38, p. 205

Bica, E., Dutra, C. M., & Barbuy, B. 2003a, *A&A*, 397, 177

Bica, E., Dutra, C. M., Soares, J., & Barbuy, B. 2003b, *A&A*, 404, 223

Bobylev, V. V., Bajkova, A. T., & Lebedeva, S. V. 2007, As. Lett., 33, 720

Bonatto, C., Kerber, L. O., Bica, E., & Santiago, B. X. 2006, *A&A*, 446, 121

Carraro, G., Girardi, L., & Chiosi, C. 1999, *MNRAS*, 309, 430

Carraro, G., Vázquez, R. A., Costa, E., Perren, G., & Moitinho, A. 2009, *ApJ*, submitted
Carraro, G., Vázquez, R. A., & Moitinho, A. 2008, *A&A*, 482, 777
Carraro, G., Vázquez, R. A., Moitinho, A., & Baume, G. 2005, *ApJ* (Letters), 630, L153
Chen, L., Hou, J. L., & Wang, J. J. 2003, *AJ*, 125, 1397
Corder, S. & Twarog, B. A. 2001, *AJ*, 122, 895
Cutri, R. M., *et al.* 2003, explanatory Supplement to the 2MASS All Sky Data Release
Dias, W. S., Alessi, B. S., Moitinho, A., & Lépine, J. R. D. 2002, *A&A*, 389, 871
Dias, W. S. & Lépine, J. R. D. 2005, *ApJ*, 629, 825
Dutra, C. M., Bica, E., Soares, J., & Barbuy, B. 2003, *A&A*, 400, 533
Eisenhauer, F., Schödel, R., Genzel, R., Ott, T., Tecza, M., Abuter, R., Eckart, A., & Alexander, T. 2003, *ApJ* (Letters), 597, L121
ESA 1997, The *Hipparcos* and Tycho Catalogues (European Space Agency)
Friel, E. D. 1995, *ARA&A*, 33, 381
Friel, E. D., Janes, K. A., Tavarez, M., Scott, J., Katsanis, R., Lotz, J., Hong, L., & Miller, N. 2002, *AJ*, 124, 2693
Frinchaboy, P. M. & Majewski, S. R. 2008, *AJ*, 136, 118
Froebrich, D., Scholz, A., & Raftery, C. L. 2007, *MNRAS*, 374, 399
Genzel, R., Pichon, C., Eckart, A., Gerhard, O. E., & Ott, T. 2000, *MNRAS*, 317, 348
Girardi, L., Bressan, A., Bertelli, G., & Chiosi, C. 2000, *A&AS*, 141, 371
Glushkova, E., Koposov, S., Zolotukhin, I., Beletsky, Y., Vlasov, A., & Leonova, S. 2009, *Ast. Lett.*, in press (arXiv:0910.1330)
Høg, E., Fabricius, C., Makarov, V. V., Urban, S., Corbin, T., Wycoff, G., Bastian, U., Schwekendiek, P., & Wicenec, A. 2000, *A&A* (Letters), 355, L27
Janes, K. A. & Adler, D. 1982, *ApJS*, 49, 425
Janes, K. A. & Phelps, R. L. 1994, *AJ*, 108, 1773
Janes, K. A., Tilley, C., & Lyngå, G. 1988, *AJ*, 95, 771
Joshi, Y. C. 2005, *MNRAS*, 362, 1259
Joshi, Y. C. 2007, *MNRAS*, 378, 768
Kharchenko, N. V., Piskunov, A. E., Röser, S., Schilbach, E., & Scholz, R. 2005, *A&A*, 440, 403
Kroupa, P. 2002, *MNRAS*, 330, 707
Lépine, J. R. D., Acharova, I. A., & Mishurov, Y. N. 2003, *ApJ*, 589, 210
Lépine, J. R. D., Dias, W. S., & Mishurov, Y. 2008, *MNRAS*, 386, 2081
Lyngå, G. 1987, Computer-based Catalogue of Open Cluster Data, 5th ed., Tech. rep., Observatoire de Strasbourg, Centre de Données Stellaires, Strasbourg
Magrini, L., Sestito, P., Randich, S., & Galli, D. 2009, *A&A*, 494, 95
Martínez–Delgado, D., Butler, D. J., Rix, H.-W., *et al.* 2005, *ApJ*, 633, 205
Mermilliod, J. C. 1992, *Bull. Inform. CDS*, 40, 115
Mishurov, Y. N., Lépine, J. R. D., & Acharova, I. A. 2002, *ApJ* (Letters), 571, L113
Moffat, A. F. J., Jackson, P. D., & Fitzgerald, M. P. 1979, *A&AS*, 38, 197
Moitinho, A., Vázquez, R. A., Carraro, G., Baume, G., Giorgi, E. E., & Lyra, W. 2006, *MNRAS* (Letters), 368, L77
Piskunov, A. E., Kharchenko, N. V., Röser, S., Schilbach, E., & Scholz, R. 2006, *A&A*, 445, 545
Platais, I., Kozhurina–Platais, V., & van Leeuwen, F. 1998, *AJ*, 116, 2423
Salaris, M., Weiss, A., & Percival, S. M. 2004, *A&A*, 414, 163
Shen, M. & Zhu, Z. 2007, *ChJA&A*, 7, 120
Stetson, P. B., Davis, L. E., & Crabtree, D. R. 1990, in: G. H. Jacoby (ed.), *CCDs in astronomy*, ASP Conf. Ser. (San Francisco: ASP), vol 8, p. 289
Trumpler, R. J. 1930, *PASP*, 42, 214
Twarog, B. A., Ashman, K. M., & Anthony–Twarog, B. J. 1997, *AJ*, 114, 2256
Vázquez, R. A., May, J., Carraro, G., Bronfman, L., Moitinho, A., & Baume, G. 2008, *ApJ*, 672, 930
Zacharias, N., Urban, S. E., Zacharias, M. I., Wycoff, G. L., Hall, D. M., Monet, D. G., & Rafferty, T. J. 2004, *AJ*, 127, 3043

Star clusters: basic galactic building blocks
Proceedings IAU Symposium No. 266, 2009
R. de Grijs & J. R. D. Lépine, eds.

doi:10.1017/S1743921309990950

A new era for the globular cluster system in M31

Myung Gyoon Lee,[1] Sang Chul Kim,[2] Ho Seong Hwang,[3] Hong Soo Park,[1] Doug Geisler,[4] Ata Sarajedini[5] and William E. Harris[6]

[1]Department of Physics and Astronomy, Seoul National University, Seoul 151-742, Korea
email: [mglee, hspark]@astro.snu.ac.kr

[2]Korea Astronomy and Space Science Institute, Daejeon, Korea
email: sckim@kasi.re.kr

[3]CEA Saclay, Gif-sur-Yvette, France
email: hoseong.hwang@cea.fr

[4]Department of Physics, University of Concepción, Concepción, Chile
email: dgeisler@astro-udec.cl

[5]Department of Astronomy, University of Florida, Gainesville, FL, USA
email: ata@astro.ufl.edu

[6]Department of Physics and Astronomy, McMaster University, Hamilton, ON, Canada
email: harris@physics.mcmaster.ca

Abstract. The globular cluster system in M31 is an ideal laboratory for studying the formation and evolution of M31 as well as the globular clusters themselves. There have been numerous surveys and studies of the globular clusters in M31. However, only recently has the entire body of M31 been searched for globular clusters using wide-field CCD images by our group. A new era for the M31 globular cluster system has begun with the advent of wide-field CCD surveys of M31. We have discovered more than 100 new globular clusters in M31. Our catalog currently includes more than 500 globular clusters confirmed either based on spectra or *HST* images, many more than in the Milky Way. We present the structure, kinematics and chemical abundance of the M31 globular cluster system based on this large sample, and the implications for the formation and evolution of M31.

Keywords. galaxies: spiral, Local Group, galaxies: stellar content, galaxies: evolution, galaxies: star clusters

1. Introduction

M31, the great spiral galaxy in Andromeda, has long been considered a twin of our Galaxy. Since Hubble's seminal work in 1932 (Hubble 1932), numerous surveys of globular clusters in M31 have appeared (see Galleti *et al.* 2006; and references therein). Some examples of extensive studies of M31 globular clusters based on these surveys include Barmby *et al.* (2000), Perrett *et al.* (2002) and Puzia *et al.* (2005). Yet, these surveys were based on photographs or small-field CCD images, and many globular clusters in M31 remain to be discovered.

We recently finished a wide-field survey covering a $3^\circ \times 3^\circ$ field centered on M31, using CCD images obtained at the *KPNO 0.9m* and spectra taken with the Hydra multi-fiber spectrograph at the *WIYN 3.5m* telescope. We found 113 new globular clusters, 258 probable globulars, and confirmed the nature of 383 globular clusters known from previous studies (Kim *et al.* 2007). At present, the total number of known globular clusters in M31 is greater than 500. Combining our results with those from the literature, we produced a master catalog of ∼500 globular clusters with velocity data, and ∼400 globulars with spectroscopically measured metallicities (Lee *et al.* 2008; Lee *et al.*, in prep.). This is the first survey of the entire M31 system based on CCD imaging, opening up a new era for the study of M31 globular clusters. Our survey was followed by several new surveys (e.g., Huxor *et al.* 2008; Caldwell *et al.* 2009; Alves–Brito *et al.* 2009).

2. Kinematics and metallicity of the M31 globular cluster system

Figure 1 shows a map of the globular clusters in M31, and rotation curves for the entire sample, the disk sample (located closer than 1 kpc from the major axis) and the halo sample (located farther than 5 kpc from the major axis). The disk sample follows the galaxy's rotation very well. Surprisingly, the halo sample shows weak but measurable rotation, while it has a much larger velocity dispersion.

The existence of a halo in M31 has long been controversial. From analysis of the kinematic data we showed that there is indeed a dynamically hot halo in M31 that is rotating but primarily pressure supported. We derived a dynamical mass of $(5.5 \pm 0.4) \times 10^{11}$ $M_{\odot}$ out to a radius of 55 kpc, and $(1.9 \pm 0.1) \times 10^{12}$ $M_{\odot}$ out to 100 kpc, using the projector and tracer mass estimators, respectively.

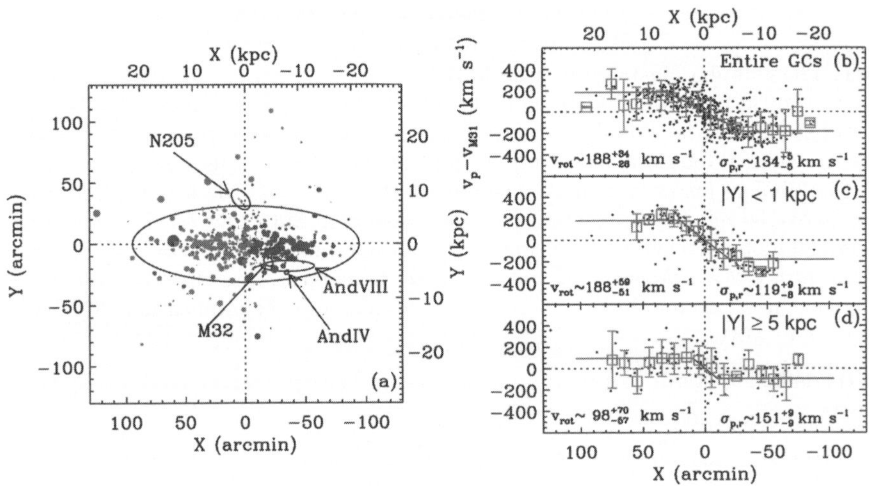

Figure 1. *(a)* Map of the globular clusters in M31. The symbol size is proportional to the velocity deviation. *(b)* Relative velocity with respect to the M31 center versus X, representing distance along M31's major axis for the entire globular cluster system. *(c)* Same as *(b)* but for the globular clusters located closer than 1 kpc from the major axis. Y represents the distance from the major axis. *(d)* Same as *(b)* but for the globular clusters in the halo of M31 located farther than 5 kpc from the major axis. Squares represent the mean values, and the solid lines represent the reference rotation curve (Lee *et al.* 2008).

In addition, we have identified 50 'friendless' globular clusters that are distinct from neighbors in space and kinematics, as shown in Figure 2. These probably have a different origin from the other globular clusters.

Figure 3 displays spatial distributions and rotation curves for the metal-rich and metal-poor globular clusters in M31. It is found that not only the metal-rich but also the metal-poor globular clusters show strong rotation in the inner region, and that the metal-rich globular clusters in the central region exhibit a large velocity dispersion, indicating that they belong to the galactic bulge. The rotation curve for the metal-poor globular clusters

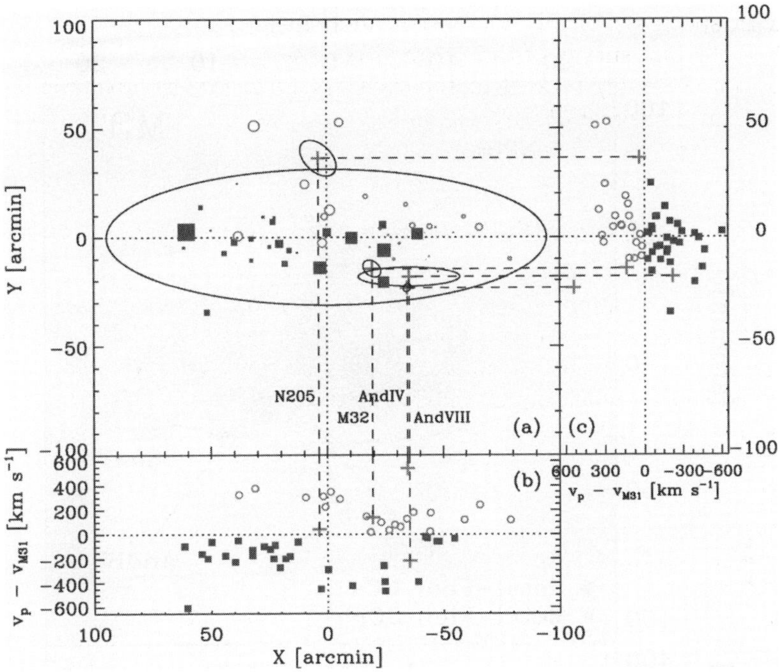

Figure 2. *(a)* Spatial distribution of friendless globular clusters in M31. The symbol size is proportional to the velocity deviation. Radial velocities versus projected radii along the *(b)* major and *(c)* minor axis. Filled squares and open circles represent the approaching and receding 'friendless' globular clusters, respectively. The large ellipse represents the optical extent of M31, i.e., R_{25} (Lee *et al.* 2008).

shows an asymmetric feature at $X < -30'$, with a maximum velocity much greater than the expected value from the opposite side.

We derived metallicities and ages for the globular clusters from the line indices measured in the Hydra spectra, using both the Brodie & Huchra (1990) method and the Lick-index grid method following Puzia *et al.* (2005). Figure 4 displays Lick line indices for hydrogen Balmer lines (Hβ, Hγ and Hδ) versus [MgFe]$'$ and $\langle$Fe$\rangle$ versus Mg_2, compared with the simple stellar population models for various values of [Z/H] and age given by Thomas *et al.* (2004). Balmer lines are sensitive to age and [MgFe]$'$ is sensitive to metallicity. We also plotted the data for M31 and Galactic globular clusters from the literature (see for details, Lee *et al.*, in prep.). Most of the M31 globular clusters are very old, although some are much younger. Old globular clusters show a wide range in metallicity. Most have $[\alpha/\mathrm{Fe}] \approx 0.2$ dex.

Figure 5 displays the metallicity distribution of the entire sample, the bulge, thin-disk, thick-disk and halo samples. The metallicity distribution of the entire M31 globular cluster system can be fit by two or three Gaussian components. The M31 globular cluster system shows an excess of intermediate-metallicity globular clusters with $-1.5 < [\mathrm{Fe/H}] < -1.0$ dex, compared with the Galactic globular clusters. Most of the bulge sample are metal-rich, while most of the halo sample are metal-poor. The intermediate-metallicity component is distinguishable only in the thin- and thick-disk samples, and it has no counterpart in our Galaxy.

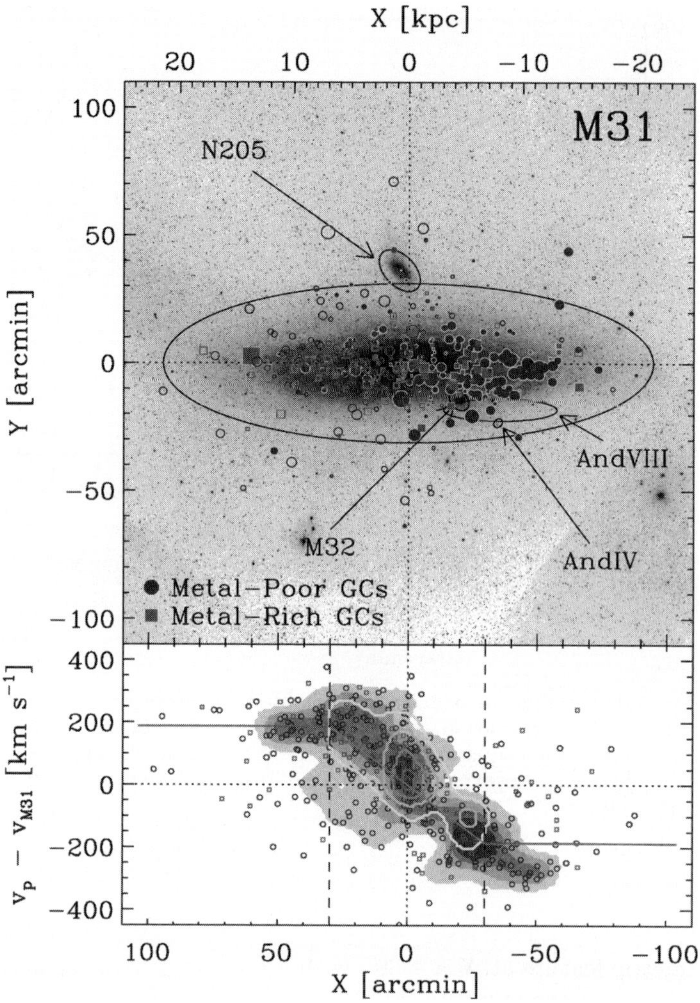

Figure 3. *(top)* Spatial distribution of the metal-rich (squares) and metal-poor (circles) globular clusters in M31. Open and closed symbols represent globular clusters receding and approaching with respect to M31, respectively. Their size represents relative velocity. *(bottom)* Radial velocity with respect to the center of M31 versus projected distance along the major axis for the metal-poor (pseudo-color map) and metal-rich globular clusters (contours).

3. Formation and evolution of M31

Our findings for the structure, kinematics and metallicity of the M31 globular cluster system provide very useful clues to understanding the formation and evolution of M31. The existence of a rotating, metal-rich central bulge, the radial metallicity gradient and the extended, smooth, metal-poor halo in M31 can be explained by the rapid dissipative collapse of a protogalactic cloud. On the other hand, the extended bulge, substructure and friendless globular clusters are evidence of accretion and merging of small-scale filaments. This is also consistent with the recent finding that the distribution of red giants in the outer parts of M31 is remarkably extended with complicated structures, and may even be connected to M33 (McConnachie *et al.* 2009).

The structure of the metal-poor halo and extended bulge in M31 is similar to the double-halo structure recently found in our Galaxy, consisting of an outer and inner halo

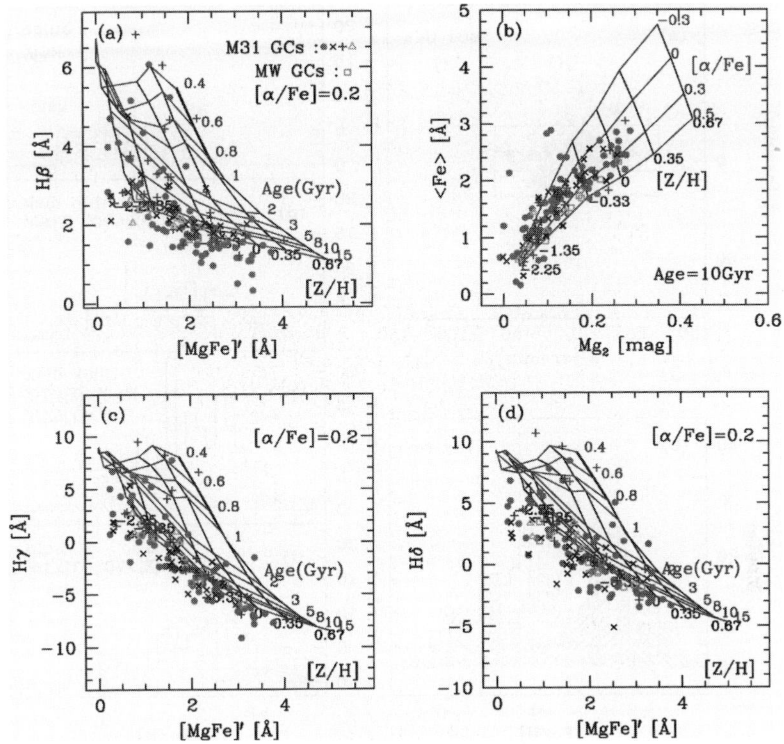

Figure 4. Lick line indices for hydrogen Balmer lines (Hβ, Hγ and Hδ) versus [MgFe]′ *(a, c* and *d)*, and ⟨Fe⟩ versus Mg_2 *(b)* for M31 globular clusters measured in this study (circles) and from the literature (crosses: Puzia *et al.* 2005; pluses: Beasley *et al.* 2004; triangles: Trager *et al.* 1998). Open squares represent Galactic globular clusters. Grids represent the simple stellar population models for various values of [Z/H] and age (Gyr) given by Thomas *et al.* (2004).

(Carollo *et al.* 2007). This double-halo structure was also found from the spectroscopic study of red giants in M31 (Koch *et al.* 2008). However, the uniqueness of the M31 halo globular system is that it is rotating faster than the Galactic halo, although it is mainly pressure supported. The presence of a metal-poor halo in M31 rotating faster than that of our Galaxy is a mystery. It indicates that the galaxy probably suffered a major merger in the past.

Therefore, it is concluded that the formation of M31 is likely linked to three processes, including (i) a rapid monolithic collapse in the early phase, (ii) a major merger in the middle phase and (iii) sporadic accretion and minor mergers throughout its entire evolution.

Acknowledgements

This work is supported in part by grant R01-2007-000-20336-0 from the Basic Research Program of the Korea Science and Engineering Foundation.

References

Alves-Brito, A., Forbes, D. A., Mendel, J. T., Hau, G. K. T., & Murphy, M. T. 2009, *MNRAS* (Letters), 395, L34

Barmby, P., Huchra, J. P., Brodie, J. P., Forbes, D. A., Schroder, L. L., & Grillmair, C. J. 2000, *AJ*, 119, 727

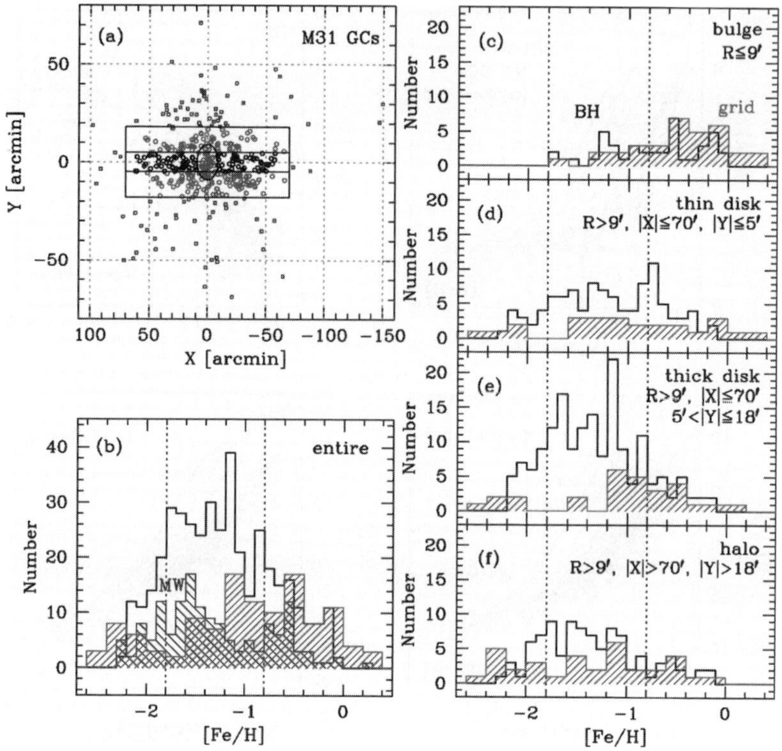

Figure 5. Metallicity distribution of M31 globular clusters. *(a)* Map of the globular clusters. Lines represent the boundaries for each sample of the bulge, thin disk, thick disk and halo in M31. *(b), (c), (d), (e) and (f)* represent the metallicity distribution of the entire sample, the bulge, thin-disk, thick-disk and halo samples, respectively. Empty and hatched histograms represent data derived using the Brodie & Huchra (1990) method and the Lick-index grid method, respectively. In *(b)*, Galactic globular clusters are shown as the hatched histogram labeled 'MW.'

Brodie, J. P. & Huchra, J. P. 1990, *ApJ*, 362, 503

Beasly, Brodie, J. P., Strader, J., Forbes, D. A., Proctor, R. N., Barmby, P., & Huchra, J. P. 2004, *AJ*, 128, 1623

Caldwell, N., Harding, P., Morrison, H., Rose, J. A., Schiavon, R., & Kriessler, J. 2009, *AJ*, 137, 94

Carollo, D., *et al.* 2007, *Nature*, 450, 1020

Galleti, S., Federici, L., Bellazzini, M., Buzzoni, A., & Fusi Pecci, F. 2006, *A&A*, 456, 985

Hubble, E. 1932, *ApJ*, 76, 44

Huxor, A. P., Tanvir, N. R., Ferguson, A. M. N., Irwin, M. J., Ibata, R., Bridges, T., & Lewis, G. F. 2008, *MNRAS*, 385, 1989

Kim, S. C., Lee, M. G., Geisler, D., Sarajedini, A., Park, H. S., Hwang, H. S., Harris, W. E., Seguel, J. C., & von Hippel, T. 2007, *AJ*, 134, 706

Koch, A., *et al.* 2008, *ApJ*, 689, 958

Lee, M. G., Hwang, H. S., Kim, S. C., Park, H. S., Geisler, D., Sarajedini, A., & Harris, W. E. 2008, *ApJ*, 674, 886

McConnachie, A. W., *et al.* 2009, *Nature*, 461, 66

Perrett, K. M., Bridges, T. J., Hanes, D. A., Irwin, M. J., Brodie, J. P., Carter, D., Huchra, J. P., & Watson, F. G. 2002, *AJ*, 123, 2490

Puzia, T., Perrett, K. M., & Bridge, T. J. 2005, *A&A*, 434, 909

Thomas, D., Maraston, C., & Korn, A. 2004, *MNRAS* (Letters), 351, L19

Trager, S. C., Worthey, G., Faber, S. M., Burstein, D., & Gonzalez, J. J. 1998, *ApJS*, 116, 1

Star clusters: basic galactic building blocks
Proceedings IAU Symposium No. 266, 2009
R. de Grijs & J. R. D. Lépine, eds.

doi:10.1017/S1743921309990962

Proper-motion studies of Milky Way starburst clusters – a new definition of starburst cluster templates

Andrea Stolte[1] and Wolfgang Brandner[2]

[1]I. Physikalisches Institut, University of Cologne, Zülpicher Str. 77, 50937 Köln, Germany
email: astolte@ph1.uni-koeln.de

[2]Max-Planck-Institut für Astronomie, Königstuhl 17, 69117 Heidelberg, Germany
email: brandner@mpia.de

Abstract. Starburst clusters in the Milky Way have the advantage that individual stars down to subsolar masses can be resolved. Thus far, field contamination along the line of sight towards the Galactic Centre and spiral arms was the limiting factor in deriving an unbiased census of the stellar population in Milky Way starbursts and, hence, the spatial extent and initial mass function in starburst clusters. As the next generation of telescopes with higher sensitivity and spatial resolution are being developed, these resolved clusters become increasingly important as templates for young, massive extragalactic systems, which will be resolved at the high-mass end of the stellar mass function. With the aim to obtain a uniform characterisation of starburst cluster properties in the Milky Way, we have initiated a proper-motion membership survey. This technique became feasible for clusters out to distances of 8 kpc with diffraction-limited imaging using adaptive optics from the ground and with *Hubble Space Telescope* from space.

Keywords. techniques: high angular resolution, astrometry, stars: formation, stars: pre–main-sequence, stars: luminosity function, mass function, open clusters and associations: general, galaxies: star clusters

1. Introduction

Starburst clusters, here defined as young, massive clusters with a total stellar mass in excess of 10^4 $M_\odot$, form one of the major building blocks of galaxies. In extragalactic systems, such as the merging Antennae galaxies or galaxies with rich cluster populations such as M51, several hundreds of young, massive clusters with ages $<$ 100 Myr are observed (e.g., Whitmore & Schweizer 1995; Lee *et al.* 2005). At the low star-formation rate in the Milky Way today, starburst clusters are rare. Yet, as they contain the densest stellar environments outside the Galactic Centre, their stellar populations can only be resolved in the Galaxy in their entirety, and outside the immediate cluster cores in the Magellanic Clouds. Despite our proximity to the nearest starburst clusters at distances of 4 to 8 kpc from the Sun, several cluster characteristics are not well quantified. Among these are the full cluster extent, the amount of mass segregation, the core radii, and, most importantly, the stellar initial mass function (IMF). On theoretical grounds, one might expect that starburst cluster environments lead to an overproduction of high-mass stars as a consequence of increased temperatures, densities, accretion rates and possibly merging pre-stellar cores in the dense cluster centres. In practice, however, it is difficult to distinguish the effects of a flattened stellar *initial* mass function (MF) from the flattening induced by dynamical mass segregation. Hence, while present-day MFs in the Arches and NGC 3603 cluster cores were found to be flattened with respect

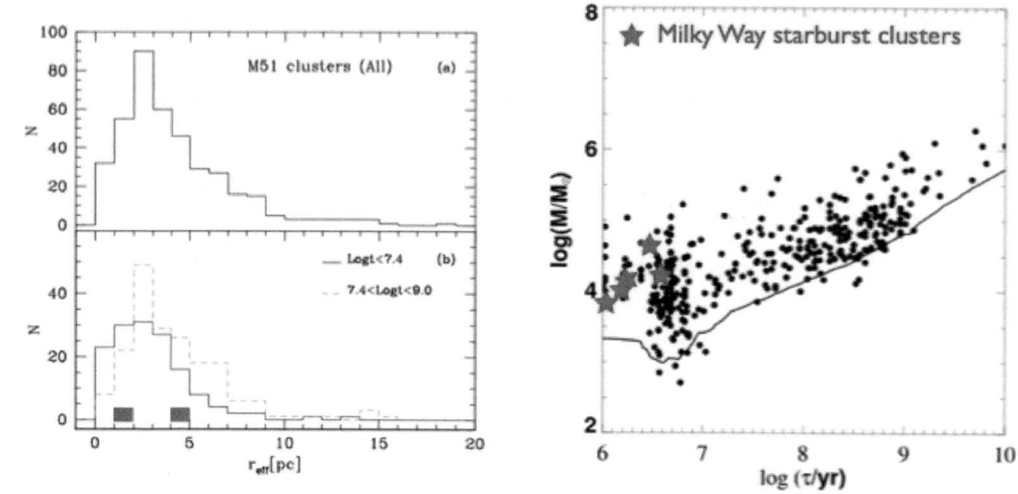

Figure 1. Comparison of Milky Way starburst cluster characteristics with young star clusters in the merging galaxy M51 (the Whirlpool galaxy). The four clusters in the survey span similar sizes as the young clusters in M51, and cover, in particular, the low-mass end of the cluster population, which will be detected with the next generation of high-resolution telescopes such as the *E–ELT* and the *TMT*. (Figures adapted from Lee *et al.* 2005.)

to a Salpeter IMF (Figer *et al.* 1999; Stolte *et al.* 2005, 2006; Harayama *et al.* 2008), conclusions on the IMF are difficult to deduce in the presence of incomplete cluster populations, unknown radial extent and rapid mass segregation (see Kim *et al.* 2006 for a detailed discussion). Recently, Espinoza *et al.* (2009) showed that incomplete samples caused by variable extinction across the field can mimic a flattening in the IMF. With different assumptions than previous studies, these authors derive a normal present-day MF for the Arches cluster. The deviating results emphasize the need to obtain complete samples and understand the possible biases in each particular data set. At the same time, a homogeneous treatment of the data would help the relative comparison between different clusters.

The fact that even the closest starburst cluster populations are not well constrained has important implications for our understanding of extragalactic star clusters. The light of unresolved extragalactic clusters is dominated by the most massive O-type stars, while the total cluster mass and number density are determined by low- and intermediate-mass stars. The interpretation of the observed integrated light profile, therefore, depends crucially on assumptions regarding the true cluster extent and the stellar IMF.

2. The Milky Way starburst cluster survey

Local clusters can serve as calibrators for extragalactic cluster systems, if their properties are well understood. A complete census of cluster members in each local starburst is crucial to quantify mass segregation, dynamical cluster dispersal, the IMF and the survival timescale before clusters are dispersed into the field. With the aim to map each cluster population as completely as possible, we initiated a proper-motion membership survey of four Milky Way starburst clusters (the Arches, Quintuplet, NGC 3603 and Westerlund 1). Extended regions around each cluster are monitored out to at least the tidal radius (1–2 pc), such that ejected stars and low-mass stars transported to the cluster outskirts can be detected.

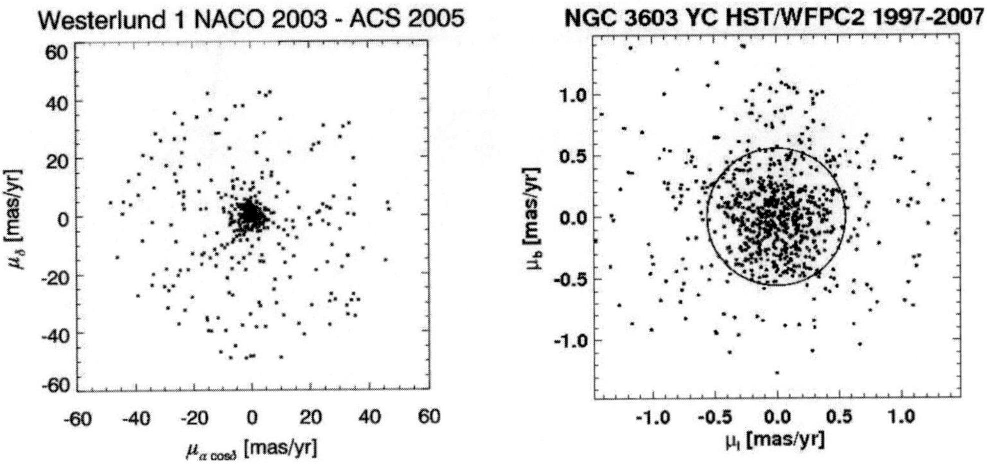

Figure 2. Proper-motion diagrams for Westerlund 1 *(left)* and NGC 3603 *(right)*. The colour–magnitude diagrams in Figure 3 were derived based on a 1σ membership-selection criterion. Field stars occupying a larger range of Galactic orbits and hence a large spread in velocities are efficiently removed from the cluster samples.

Proper-motion membership selections are used very efficiently in nearby star-forming regions (e.g., Kharchenko *et al.* 2009; and references therein). Recently, we were able to apply this technique for the first time to the Arches starburst cluster at a distance of 8 kpc, near the center of the Galaxy (Stolte *et al.* 2008). The clear distinction between field and cluster stars allows sampling the cluster stars among the dominant field population out to large radii, and hence enables determination of the cluster extent and the distribution of low-mass members in particular.

The goals of the survey are to obtain proper-motion membership out to the tidal radius in each cluster, determine the present cluster extent, and derive the unbiased stellar IMF. Furthermore, the internal cluster dynamics will constrain the tidal disruption and dynamical evolution of starburst clusters. In particular, the combination with radial velocities reveals possible deviations from virial equilibrium, which are crucial for measuring dynamical cluster masses. This is particularly important for extragalactic clusters, where the velocity dispersion and/or the mass-to-light ratio are employed to derive the total cluster mass and survival timescale, and to constrain the existence of intermediate-mass black holes at the centres of starburst clusters (e.g., Gürkan *et al.* 2004; Portegies Zwart *et al.* 2004; and references therein).

3. The method

With diffraction-limited imaging using adaptive optics from the ground (near-infrared) or *HST*/WFPC2 from space (optical), a spatial resolution of 50–80 milliarcseconds (mas) is achieved with pixel scales of 27–40 mas. In deep astrometric observations, a positional uncertainty of $\frac{1}{20}$ of a pixel is now routinely obtained, or a relative accuracy of 1–2 mas (rms). With a time baseline of only two years, this implies that cluster and field stars can be separated at the level of 30 km s^{-1}, sufficient to distinguish the concentrated cluster population from the diverse orbits of stars along the line of sight (Figure 2).

The survey sample contains four of the best-studied and most massive young starburst clusters in the Milky Way. With the Arches and Quintuplet, two clusters represent the Galactic Centre's star-forming environment, which is most representative for the central

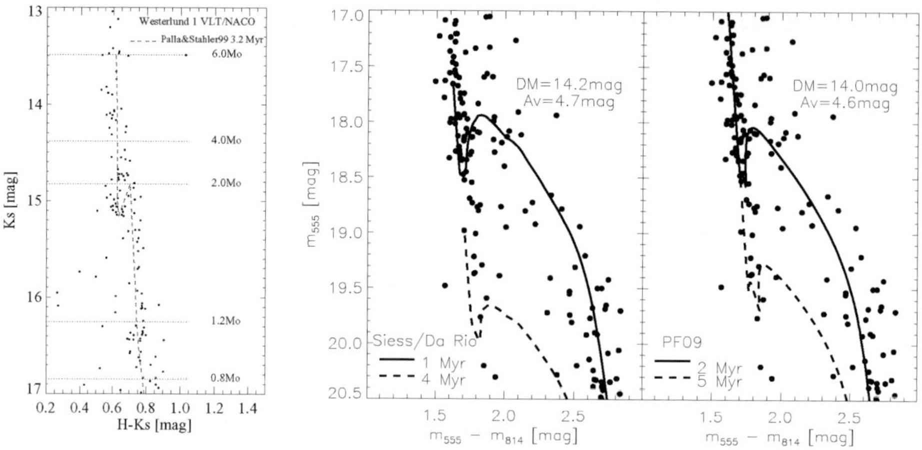

Figure 3. Colour–magnitude diagrams created from the membership selection shown in Figure 2. The membership selection in Westerlund 1 *(left)* allowed a significantly more accurate determination of the cluster distance and age. In NGC 3603 YC (Young Cluster), the sample selected from *HST*/WFPC2 imaging based on a 10-year baseline displays at least two distinctive hydrogen-burning turn-on points, indicating multiple star-formation epochs. The photometric precision and membership selection provide new constraints on pre-main-sequence evolution models. While the Siess models *(left)* do not reproduce the tip of the youngest hydrogen-burning turn-on point, the newest generation of simulations by the Pisa group (*right*; Tognelli *et al.*, in prep.) trace this phase extremely well.

regions of galaxies such as M31 (Andromeda) or M51 (the Whirlpool). NGC 3603 YC (Young Cluster) at the heart of the giant H II region NGC 3603 and the very massive cluster Westerlund 1 (Clark *et al.* 2005; Brandner *et al.* 2008) evolve in the more quiescent environment of the spiral arms. With ages of 2 to 5 Myr, these clusters span the age range where planetary-system formation takes place. Compared to the cluster sample in the star-forming merger M51 (Figure 1), the four starburst clusters in the survey display comparable cluster sizes and occupy the low-mass range of young extragalactic clusters. Hence, despite the small number of clusters available in the Milky Way, a significant range of ages and environments is covered to study cluster evolution.

For each of the four starburst clusters in the sample, we obtained first-epoch *VLT*/NAOS–CONICA (NACO) observations, which were combined with previous observations of each cluster core. In addition, several fields around the immediate core regions were observed to reach large clustercentric distances. For the extended cluster areas, a second-epoch observation will be obtained in the next few years. The astrometric precision can be optimised by deriving the optical distortion correction for NACO, which is currently being done by N. Kudryavtseva (in prep.). For NGC 3603 we obtained two epochs of *HST*/WFPC2 observations separated by a 10-year baseline. From these observations, the internal velocity dispersion could be measured for the first time (Rochau *et al.*, submitted).

4. Preliminary results

Constraining cluster membership implies that the global cluster properties can be derived with higher accuracy. In Brandner *et al.* (2008), we revised the age of Westerlund 1 from 4 ± 1 to 3.3 ± 0.2 Myr, and the cluster distance from 3.9 ± 0.7 to 3.5 ± 0.2 kpc. In the central starburst cluster in the giant H II region NGC 3603, we were able to resolve multiple star-forming events for the first time from the membership selection in the pre-main-seqeunce/main-sequence (PMS/MS) transition region (Figure 3). The main-sequence turn-on point stands out clearly, so that we are now in the position to constrain stellar evolution models. In addition, stars in transition from the PMS to the MS pass through the short-lived phase of the radiative-convective gap. Only starburst clusters have sufficient members to detect several sources in this important phase of stellar evolution and constrain the physics during the hydrogen-burning equilibration phase.

5. Summary and survey outlook

The proper-motion survey of Milky Way starburst clusters leads to accurate cluster properties such as ages, distances and cluster extent. The photometric accuracy obtained in the colour–magnitude diagrams with member selection allowed us to constrain PMS evolution models and select the model most representative of the observed PMS colours. With the second epoch of observations in the extended cluster areas, the cluster radii will be measured and a complete sample of cluster stars with individual extinction estimates will be derived, from which the cluster IMF can be deduced.

The dynamical state of a starburst cluster has severe implications for the derivation of masses and survival timescales of young, massive extragalactic clusters, where radial-velocity measurements and assumptions on the mass-to-light ratio have to be employed. To use local resolved starbursts as templates for extragalactic clusters, it is particularly critical that the astrometric precision allows us to observe the internal velocity dispersion in each cluster and its variation with radius. In comparison with cluster evolution models (Baumgardt & Kroupa 2008), deviations from virial equilibrium can be deduced. Quantifying the dynamical state of each cluster is also crucial to understand the effects of different environments, such as the Galactic Centre or spiral arms, on cluster evolution. Recently, Pfalzner (2009) suggested on the basis of the mass density versus radius relation in young star clusters that starburst clusters evaporate slowly with time, while low-mass clusters become rapidly unbound during the gas-expulsion phase. With a uniform sample of proper-motion-selected cluster members, the mass density and half-mass radius are more accurately determined and selection biases in the more distant starburst cluster sample compard to the sample of nearby, lower-mass star-forming clusters can be quantified.

In summary, the Milky Way starburst cluster survey will provide revised distances, ages and radial extents for the four starburst clusters the Arches, Quintuplet, NGC 3603 and Westerlund 1. The dynamical state and the stellar IMF in each cluster will be derived out to the tidal radii. These results can be applied directly when the integrated-light profiles of extragalactic young, massive clusters are interpreted. Galactic starburst cluster templates become increasingly important in view of the next generation of telescopes such as the *E–ELT* or *TMT*, with which the high-mass end ($M > 20\ M_\odot$) of the stellar population in young, massive clusters in all nearby galaxies out to distances of several Mpc will be resolved.

References

Baumgardt, H., & Kroupa P. 2007, *MNRAS*, 380, 1589

Brandner, W., Clark, J. S., Stolte, A., Waters, R., Negueruela, I., & Goodwin, S. P. 2008, *A&A*, 478, 137

Clark, J. S., Negueruela, I., Crowther, P. A., & Goodwin, S. P. 2005, *A&A*, 434, 949

Espinoza, P., Selman, F., & Melnick, J. 2008, *A&A*, 501, 563

Figer D. F., Kim, S. S., Morris, M., Serabyn E., Rich, R. M., & McLean, I. S. 1999, *ApJ*, 525, 750

Gürkan, M. A., Freitag, M., & Rasio, F. A. 2004, *ApJ*, 604, 632

Kharchenko, N. V., Piskunov, A. E., Röser, S., Schilbach, E., Scholz, R.-D., & Zinnecker, H. 2009, *A&A*, 504, 681

Kim, S. S., Figer, D. F., Kudritzki, R. P., & Najarro, F. 2006, *ApJ* (Letters), 653, 113

Harayama, Y., Eisenhauer, F., & Martins, F. 2008, *ApJ*, 675, 1319

Lee M. G., Chandar, R., & Whitmore, B. C. 2005, *AJ*, 310, 2128

Pfalzner, S. 2009, *A&A*, 498, L37

Portegies Zwart, S. F., Baumgardt, H., Hut, P., Makino, J., & McMillan, S. L. W. 2004, *Nature*, 428, 724

Stolte, A., Brandner, W., Grebel, E. K., Lenzen, R., & Lagrange, A.-M. 2005, *ApJ* (Letters), 628, L113

Stolte, A., Brandner, W., Brandl, B., & Zinnecker, H. 2006, *AJ*, 132, 253

Stolte, A., Ghez, A. M., Morris, M., Lu, J. R., Brandner, W., & Matthews, K. 2008, *ApJ*, 675, 1278

Whitmore, B. C. & Schweizer, F. 1995, *AJ*, 109, 960

Session IV

Globular cluster chemical evolution

Star clusters: basic galactic building blocks
Proceedings IAU Symposium No. 266, 2009
R. de Grijs & J. R. D. Lépine, eds.

doi:10.1017/S1743921309990974

Globular cluster abundances: the imprint of first-generation massive stars

Corinne Charbonnel

Geneva Observatory, University of Geneva, 51 Chemin des Maillettes, 1290 Versoix, Switzerland
email: Corinne.Charbonnel@unige.ch

Laboratoire d'Astrophysique de Toulouse–Tarbes, Université de Toulouse, CNRS, France

Abstract. Galactic globular cluster (GC) stars exhibit abundance patterns that are not shared by their field counterparts, namely the well-documented O–Na, C–N and Mg–Al anticorrelations. Recent observations provide compelling evidence that these abundance anomalies were already present in the intracluster gas from which the presently observed stars formed. The current explanation is that the gas was polluted very early in the history of the GC by material processed through H burning at high temperatures and then lost by stars more massive than the long-lived stars we still observe today. However the 'polluters' have not yet been unambiguously identified. Most studies have focused on asymptotic giant brach stars, but rotating massive stars present an interesting alternative. Here, we critically analyse the pros and cons of both potential stellar polluters. We discuss the constraints that the observational data provide on stellar nucleosynthesis and hydrodynamics, as well as on the formation and early evolution of very massive star clusters.

Keywords. stars: abundances, globular clusters: general

1. Chemical dissection of globular clusters

The classical paradigm describing globular clusters (GC) as fairly simple, large aggregates of coeval stars formed from homogeneous chemical material has been seriously challenged recently. There is indeed well-established evidence that GCs actually host multiple stellar populations. Much of this evidence comes from an increasing number of photometric observations in Galactic GCs but also in Magellanic Cloud clusters. Multiple populations manifest themselves in a split of different evolutionary sequences as observed in the cluster colour–magnitude diagrams.

The credit of this discovery, however, certainly goes to stellar spectroscopists, who have been collecting an incredible amount of data over the last three decades, revealing the striking chemical properties of GCs. At the beginning, only the brightest, most evolved stars were 'reachable'. A major step forward was achieved with the advent of 8–10m-class telescopes, which enabled access to abundance analysis in GC turnoff stars.

We refer to Gratton *et al.* (2004), Sneden (2005), and Decressin (2007) for extensive reviews of observed abundance trends in GCs. We briefly recall here the main points. (i) Individual GCs appear to be fairly homogeneous as far as Fe (Carretta *et al.* 2009a) and the iron-peak elements (Ni, Cu) are concerned (with the notable exception of ω Cen). (ii) They exhibit very low scatter and the same trends as field stars for the neutron-capture elements (both s- and r-process elements; Ba, La, Eu) and the α elements (Si, Ca). (iii) They exhibit, however, specific patterns and large star-to-star abundance variations in the lighter elements from C to Al, which are not shared by their field counterparts. Let us focus on these light-element 'anomalies'.

After the pioneering work of Osborn (1971), several studies showed that GCs are very heterogeneous in C and N, and that anticorrelated variations in molecular-band strengths of CN and CH exist all the way down to the main-sequence turnoff in all GCs where data are available. Star-to-star variations of the strengths of the Na and Al lines at the tip of the red-giant branch (RGB) were discovered at the end of the 1970s (Cohen 1978; Peterson 1980; Norris *et al.* 1981). An anti-correlation between O and Na abundances (see Pilachowski 1989; Drake *et al.* 1992; Sneden *et al.* 1991 for the earliest studies) exists among evolved turnoff and RGB stars in all GCs studied to date (Carretta *et al.* 2009b). Whenever all necessary data are available, N is found to be anticorrelated with O, which itself positively correlates with C, and the sum [(C+N+O)/Fe] appears to be constant within the observational errors (Decressin *et al.* 2009a; and references therein), with the possible exception of NGC 1851 (Yong *et al.* 2009). In addition, variations in Al and Mg, again correlated with each other, have been discovered (e.g., Shetrone 1996; Ivans *et al.* 1999; Ramirez & Cohen 2002; Carretta *et al.* 2009b). The sum Mg+Al is also found to be constant in several clusters (Shetrone 2006). Last but not least, observations in NGC 6752, M 13, and M 71 show that a large spread in the ratio ^{24}Mg:^{25}Mg:^{26}Mg exists, combined with a positive correlation between ^{26}Mg and Al as well as an anticorrelation between ^{24}Mg and Al, but no correlation between ^{25}Mg and Al (Yong *et al.* 2003, 2005, 2006). Additionally, the Li abundance was found to be anticorrelated with that of Na in turnoff stars in a couple of GCs (Pasquini *et al.* 2005; Bonifacio *et al.* 2007; Lind *et al.* 2009).

As an illustration, we show in Figure 1 the abundance patterns in NGC 6752, which is (to date) one of the best-studied GCs in terms of its chemical properties. The largest O values are similar to those of field-halo stars of the same metallicity as NGC 6752 ([Fe/H] $= -1.5$ dex). We can thus assume that this is the original O abundance of the intracluster gas. This is also true for the corresponding values of Na, Al, and Mg. We can consider then that all abundances corresponding to the highest O abundances (as indicated by open circles in Figure 1) are the initial values and that abundances at the other extreme (indicated by open squares) correspond to the most extremely processed material. For the initial abundances of C and N, we have to adopt [C/N] and [N/Fe] $= 0$, in view of the corresponding abundances of field-halo stars (Spite *et al.* 2005). We note that the corresponding initial abundances for Mg isotopes, based on observations of NGC 6752 do not correspond exactly to the scarce available observations of field-halo stars of the same metallicity. Pasquini *et al.* (2005) reported observations of Li in nine turnoff stars of NGC 6752 with UVES and found that Li is correlated with O and anticorrelated with Na. Stars with 'original' O (high) and Na (low) have a Li content similar to field-halo stars in the Spite plateau, i.e., $A(\mathrm{Li}) = \log(\mathrm{Li/H}) + 12 \sim 2.3$ dex. Stars with 'extreme' abundances (lowest O and highest Na) are not devoid of Li, but have a significant amount of $A(\mathrm{Li}) \sim 2$ dex.

Very importantly, high-resolution spectrographs on very large telescopes allow us to determine the abundances of the relevant elements in faint subgiant and turnoff stars in several GCs. The same chemical patterns are found independently of the evolutionary status of the stars, which is crucial, as discussed in Section 2. All these features are considered anomalous because they are not seen in field stars of similar metallicity (e.g., Gratton *et al.* 2000). However they could be considered 'universal' as far as GCs are concerned, since they have been found in all GCs where they have been looked for so far. Consequently, at a conference in Santa Barbara in January 2009, Piotto proposed a new definition for GCs that we should certainly adopt in future textbooks: "GCs are large aggregates of stars showing the O–Na anticorrelation".

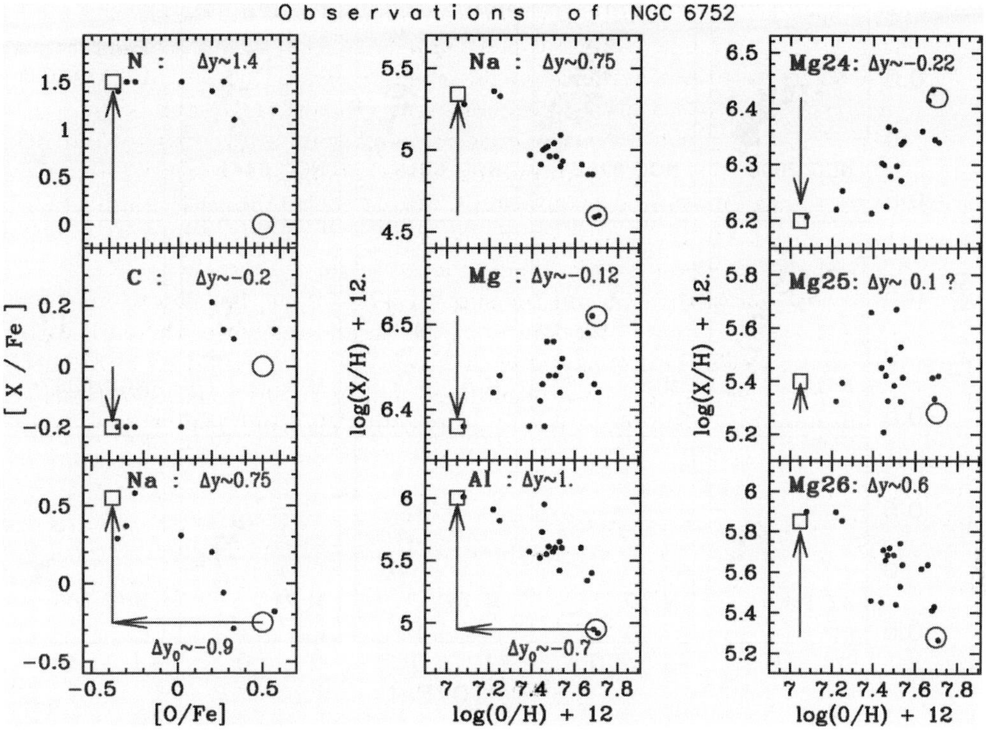

Figure 1. Abundances of C, N, Na, Mg, Al and Mg isotopes in NGC 6752 stars as a function of the corresponding oxygen abundance. Observations (filled symbols) in the *left column* (plotted as a function of [O/Fe]) are from Carretta *et al.* (2005) for turnoff stars, and in the other two columns (plotted as a function of O/H) from Yong *et al.* (2003) for bright giant stars. In all figures, the estimated initial abundances of the gas from which the cluster was formed are indicated by an open circle (i.e., it is assumed that they have the same composition as field-halo stars of the same metallicity—[Fe/H] = −1.5 dex—with, e.g., [O/Fe] = 0.5, [C/Fe] = 0, etc.). The most extreme abundances observed at present in NGC 6752 are indicated by open squares. Arrows indicate the magnitude and direction of the abundance spread, which is also given by Δy [in dex, where y = log(abundance)] in each panel. Δy_0 in the bottom panels indicates the corresponding variation in the oxygen abundance, larger in the case of Carretta *et al.* (2005; a factor of $\sim$ 8) than in Yong *et al.* (2003; a factor of 5). In the former case, the authors have tried to enhance the chance of including extreme cases by selecting stars with likely strong and weak CN bands using the Strömgren c_1 index. (Figure from Prantzos *et al.* 2007: PCI07.)

2. The self-enrichment scenario and the potential polluters

The observed abundance patterns are the signatures of the simultaneous operation of the CNO, NeNa and MgAl cycles of hydrogen burning (Denissenkov & Denissenkova 1990; Langer *et al.* 1993; Langer & Hoffman 1995). However, although the CNO cycle is activated for temperatures on the order of 20×10^6 K, the NeNa and MgAl chains require temperatures around 35×10^6 K and 50×10^6 K, respectively. Destruction of ^{24}Mg by proton capture needs still higher temperatures, around 70×10^6 K (e.g., Arnould *et al.* 1999). Prantzos *et al.* (2007; hereafter PCI07) actually showed that the observed patterns of all elements and isotopes in a typical GC like NGC 6752 require H burning in a narrow range of temperatures around 75×10^6 K. This is much higher than the internal temperature of the low-mass turnoff stars that we currently observe (see Figure 3). As a

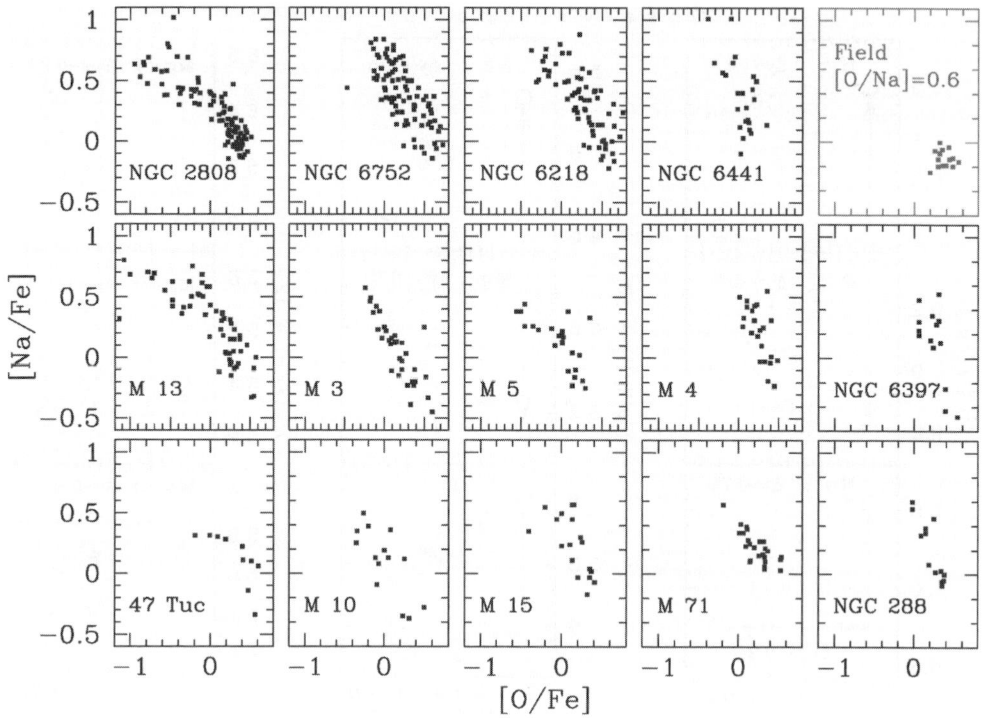

Figure 2. O–Na anticorrelation observed in several Galactic GCs and in field stars (for references, see Decressin 2007).

consequence, the abundance variations cannot be produced in situ, but certainly reflect the initial composition of the protostars. It is thus clear now that a large fraction of low-mass GC stars were formed from material processed through H burning at high temperatures and then lost by more massive and faster evolving stars, and partially mixed with some original gas.

In other words, compelling evidence leads us to believe that at least two generations of stars evolved in all Galactic GCs during their infancy. The first corresponds to the 'bulk' of normal stars born with the pristine composition of the protocluster gas. These objects are those with the highest O and Mg and the lowest Na and Al abundances also found in their field contemporaries. The second generation consists of stars formed from material polluted to various degrees by the ejecta of more massive and short-lived stars, and which present lower O and Mg and higher Na and Al abundances than their first-generation counterparts. Various aspects of this 'self-enrichment scenario' are developed in Prantzos & Charbonnel (2006, hereafter PC06) and Decressin *et al.* (2007a,b).

Because we have very strong nucleosynthesis constraints, we can get clues as to the potential polluters by looking at the temperature at which hydrogen burning occurs in various types of stars. Figure 3 shows the maximum temperature reached in the H-burning regions for stars of various initial masses at different stages of their evolution. The horizontal dotted lines delimit the temperature range of H burning within the progenitors of the chemical anomalies, constrained by the combination of the observational data for Na, Al and the Mg isotopes (see PCI07 for more details). One sees that there are actually two potential polluters: massive asymptotic giant branch (AGB) stars and massive stars.

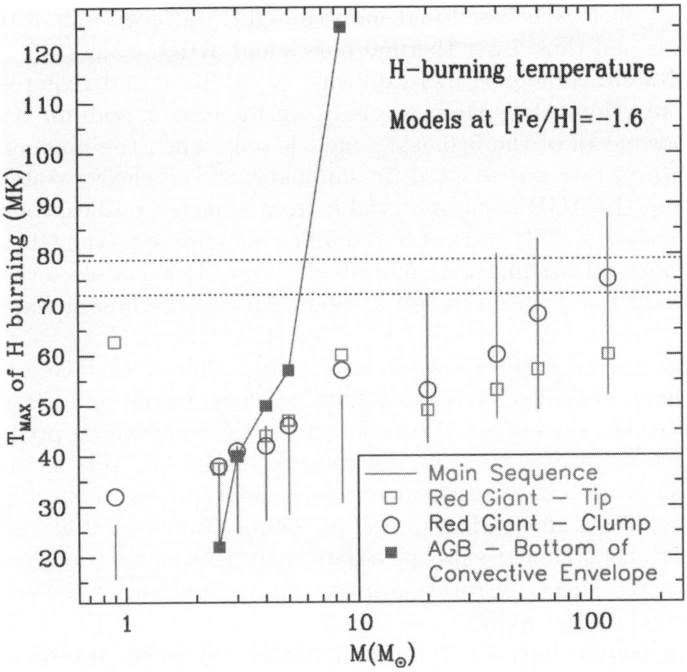

Figure 3. H-burning temperatures at various phases of a star's life as a function of initial stellar mass. The correspondence between symbols and evolutionary phases is displayed in the bottom right part of the figure. Vertical straight lines (for central H burning) start at the zero-age main sequence and end at central H exhaustion. For the other phases we show only the maximum temperature reached in the H-burning zones of interest. The two horizontal dotted lines enclose the temperature range reproducing the extreme abundances that are observed in NGC 6752. (Figure from PCI07.)

3. Were AGB stars the polluting agents?

Massive AGB stars that undergo efficient hot-bottom burning (HBB) during the thermal-pulse phase (TP–AGB) have been proposed as possible GC polluters in this so-called self-enrichment scenario (Ventura *et al.* 2001). The AGB hypothesis has been discussed extensively in the literature, first on a qualitative basis, and more recently with the help of custom-made standard (i.e., nonrotating) stellar models (see references in Decressin *et al.* 2009a). These studies have pointed out several difficulties in generating the observed chemical patterns in theoretical TP–AGB models. The main problem is due to the competition between the third dredge up (3DUP) which contaminates the AGB envelope with the helium-burning ashes produced in the thermal pulse and HBB which modifies the envelope abundances through the CNO cycle and the NeNa and MgAl chains. It is thus very difficult to obtain simultaneous O depletion and Na enrichment in the TP–AGB envelope, while keeping the C+N+O sum constant as required by the observations (for more details and references, see Charbonnel 2007).

To date, only massive (i.e., 5–6.5 $M_\odot$) AGB models manage to achieve simultaneously an encouraging agreement with the observed O depletion and Na enrichment (Ventura & D'Antona 2008a). These models include the 'full spectrum of turbulence' (FST) formulation for convection that strongly affects the O depletion once the stars enter the TP–AGB (Ventura & D'Antona 2005a). Compared to the classical mixing-length theory treatment, FST leads indeed to higher temperatures at the base of the convective envelope (resulting in more advanced nucleosynthesis) and induces higher surface luminosities, resulting in stronger mass loss and thus fewer thermal pulses and 3DUP events.

The level of Na enrichment is more difficult to estimate and requires fine tuning of the NeNa-cycle reaction rates. More precisely, an increase in sodium is achieved in the most oxygen-poor ejecta of the 5–6.5 $M_\odot$ models only when the maximum allowed values for the $^{22}Ne(p,\gamma)$ rate are adopted. In summary, and as clearly stated by Ventura & D'Antona (2008a), the AGB scenario is viable from a nucleosynthesis point of view, provided that only massive AGB stars of 5 to 6.5 $M_\odot$ contribute to the GC self-enrichment, and under the physical assumptions described above. This causes severe difficulties in terms of mass budget and initial mass function (IMF) of the first stellar generation (see PC06).

The previously quoted studies have focused only on physical uncertainties related to the TP–AGB phase, and their predictions have not been tested in different astrophysical contexts. In particular, the impact of rotation on the nucleosynthesis predictions for AGB stars had never been investigated in the context of the GC self-enrichment scenario, although rotation is often invoked to understand a wide variety of observations (see, e.g., Maeder & Meynet 2000, 2006; Chiappini *et al.* 2008). Decressin *et al.* (2009a) addressed this question for the first time. Using up-to-date treatment for rotation-induced processes, they investigated the effects of rotation up to the completion of the second dredge up (2DUP) and arrival on the AGB.

During central helium burning, rotational mixing efficiently transports primary ^{12}C and ^{16}O outside the convective core in the H-burning region, where these elements are processed by the CNO cycle, resulting in important production of primary ^{14}N. In the massive (initial mass greater or equal to $\sim 4\ M_\odot$) rotating models, after central He exhaustion the convective envelope penetrates into the layers affected by rotation-induced mixing. In contrast to standard models, the 2DUP then produces significant surface enrichment in total C+N+O that cannot be erased by hot-bottom (hydrogen) burning during the subsequent TP–AGB evolution. Third dredge-up episodes can only further increase the total C+N+O mass fraction, as they bring to the surface the products of He burning. As a consequence, if rotating massive AGB stars were responsible for the abundance patterns observed in GCs, one would expect large C+N+O differences between (O-rich and Na-poor) first-generation stars and (O-poor and Na-rich) second-generation stars, as shown in Figure 4.

This behaviour is in sharp contrast with what is observed in low- and intermediate-metallicity GCs. To date, indeed, no significant star-to-star variation of the total C+N+O has been detected, except in NGC 1851 (Yong *et al.* 2009). We note, however, that the stars observed by Yong and collaborators in NGC 1851 are very bright objects. In view of its position in the colour–magnitude diagram, the only star with C+N+O exceeding the typical error bar could actually be an AGB star. In that case, the high value of the total CNO may not have been inherited at birth by the star, but may be due to nuclear processes within the star itself. Its possible nature as an AGB star would also explain why this star is the only object in the sample to exhibit some enhancement in s-process elements. CNO measurements in NGC 1851 of unevolved stars (as done in most of the other clusters studied so far) are thus crucial to settle the issue.

Our (Decressin *et al.* 2009a) study leads us to the conclusion that massive, rotating AGB stars can be discarded as potential polluters in the self-enrichment scenario in GCs, unless the crowded environment prevented intermediate-mass stars from rotating. This latest hypothesis is highly improbable in view of observations finding higher stellar rotation velocities in clusters than in the field (see Decressin *et al.* 2009a; and references therein).

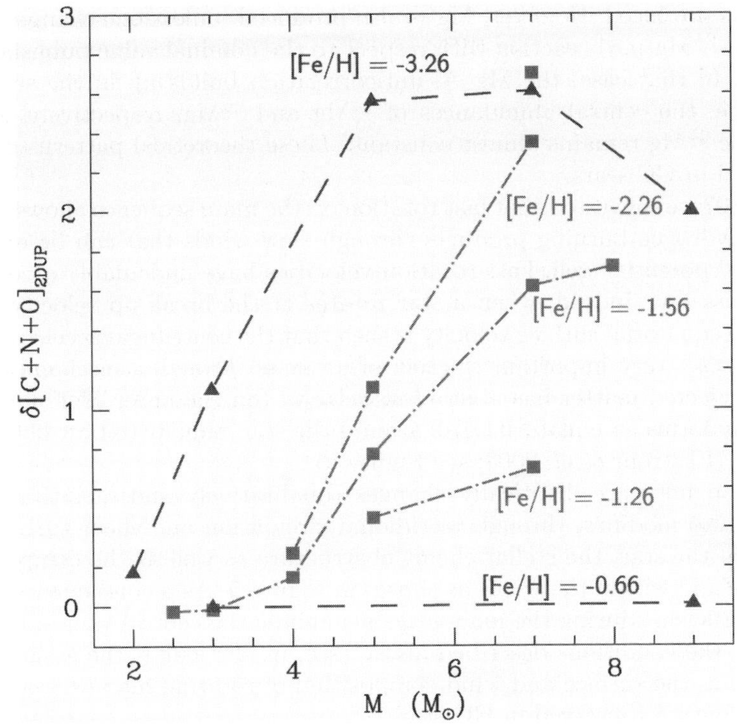

Figure 4. Surface C+N+O increase at the end of the 2DUP for rotating stars with various initial masses and metallicities. Squares and triangles indicate models computed with, respectively, STAREVOL and the Geneva code.

PC06 discuss other shortcomings of the AGB scenario, the main one being related to the peculiar IMF it requires. In addition, they underline the fact that the AGB scenario gives no satisfactory answer as to the role of stars more and less massive than the presumed polluters.

4. The alternative: the wind of fast-rotating massive stars (WFRMS) scenario

The idea that the winds of massive stars (hereafter WMS) may be at the origin of helium anomalies in GC stars was suggested by Norris (2004) and Maeder & Meynet (2006) to explain the blue main sequence of the cluster ω Cen: the high helium content of the stars on that sequence could originate from WMS which produce a large helium/metal ratio. The 'wind of fast-rotating massive stars' (WFRMS) scenario was then proposed by PC06 and further developed by Decressin *et al.* (2007a,b; hereafter DMCPE07 and DCM07) in the more general context of the GC abundance anomalies described in Section 1.

As shown in Figure 3, the central temperatures of main-sequence massive stars (i.e., with an initial mass higher than $\sim$ 25 $M_\odot$) reach the high values required to produce the observed abundance anomalies. In DMCPE07 we computed models of rotating massive stars with the Geneva code, using different prescriptions for the nuclear reaction rates. In the core of main-sequence massive stars, the temperature is such that the CNO cycle is in equilibrium. There, the C–N and O–Na anticorrelations are easily generated, although the extent of Na production depends on the adopted reaction rates for the NeNa chain. In addition, Al is produced. However, Mg is also produced, unless one changes significantly the rate of the ^{24}Mg(p,γ) reaction with respect to the nominal value published by Illiadis *et al.* (2001). In that case, the Mg–Al anticorrelation builds up in the stellar core. At the same time, the central abundances of ^{24}Mg and ^{26}Mg, respectively, decrease and increase, while ^{25}Mg remains almost constant. These theoretical patterns correspond to those observed in GC stars.

In DMCPE07 we suggested that fast rotation on the main sequence allows massive stars to eject the hydrogen-burning products through slow winds that can be easily retained within the GC potential well. Fast rotation velocities have undeniably a very important impact on mass loss. Indeed, when a star rotates at the break-up velocity (by this, we mean that the equatorial surface velocity is such that the centrifugal acceleration balances gravity), it loses a very important fraction of its mass through a mechanical wind. This mechanically ejected matter has a very low velocity (on the order of 20 km s^{-1} or less) and very likely forms an equatorial disk around the star, similar to that observed around Be-type stars (Ekström *et al.* 2008; see Figure 5).

Fast rotation not only drastically changes (qualitatively and quantitatively) stellar mass loss. It also modifies, through meridional circulation and shear turbulence that it induces within the star, the stellar chemical structure as well as the composition at the stellar surface and within the wind, as shown in Figure 5. As a consequence, the surface-abundance variations during the main sequence mimic the central values, but with some delay. Within the conditions described above (see, in particular, the assumption on the ^{24}Mg(p,γ) rate), the surface and wind composition of rotating massive stars present the H-burning signatures observed in GC stars.

The main reason as to why massive stars have been discarded in the past is related to the fact that they are expected to also produce heavier elements. Of course, within our framework, one has to avoid further contamination by the products of He burning and by the yields in heavy elements of Type II supernovae. Regarding the former, DMCPE07 note that once a star leaves the main sequence its rotation velocity suddenly decreases. Consequently, the wind velocities strongly increase and the corresponding ejected matter is probably lost from the GC. On the other hand, if black holes form through core collapse, they could swallow most of the pre-supernova yields (i.e., most of the heavy elements formed after the main-sequence phase). More probably, supernova ejection resulting from an initialy fast-rotating star, if it occurs, may favor ejection along jets aligned along the rotational axis. Of course, these ideas are highly speculative and need to be checked through detailed hydrodynamical simulations.

5. Dynamical issues

The schematic evolution of a GC is sketched in Figure 6. Based on the determination of the composition of giant stars in NGC 2808 by Carretta *et al.* (2006), PC06 determined that around 70% of stars exhibit abundance anomalies in this specific cluster. Decressin *et al.* (2007a) find similar results for NGC 6752 in their analysis of the data of Carretta *et al.* (2007): around 85% of the cluster stars (in a sample of 120 stars) show abundance anomalies. Therefore, most stars still evolving in globular clusters seem to be second-generation stars.

How can we produce such a high fraction of chemically peculiar stars? The main problem is that assuming a Salpeter (1955) IMF for the polluters, the accumulated mass of the slow winds ejected by the fast-rotating massive stars would only provide 10% of the total number of low-mass stars. To match the observations thus requires either (a) a flat IMF with a slope of 0.55 instead of the canonical Salpeter slope of 1.35 or (b) that 95% of the first-generation stars have escaped the cluster (Decressin *et al.* 2007a).

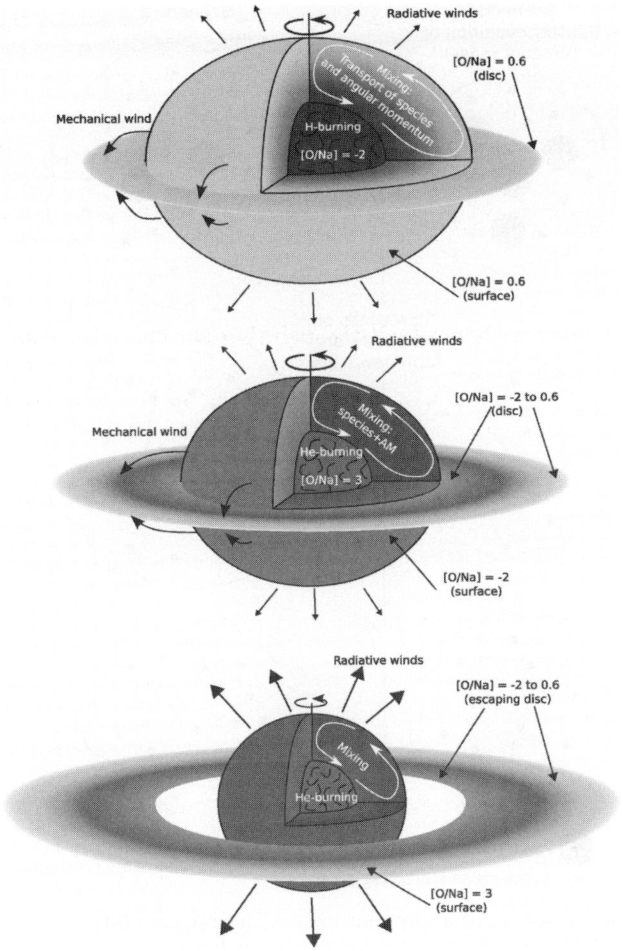

Figure 5. Schematic view of the evolution of fast-rotating massive stars. The colours reflect the chemical composition of the various stellar regions and of the disk: green corresponds to the initial chemical composition, blue and red are, respectively, for material loaded in H- and He-burning products. The typical [O/Na] value in the various stellar regions is indicated (it is ~ 0.6 dex in the first generation protostars and ~ -2 dex in second-generation stars). *(top)* During the main sequence, a slow outflowing equatorial disk forms and dominates matter ejection with respect to radiative winds. *(middle)* At the beginning of central He burning, the composition of the disk material spans the range in [O/Na] observed today in low-mass cluster stars. The star has already lost an important fraction of its initial mass. *(bottom)* Due to heavy mass loss, the star moves away from critical velocity and does not supply its disk anymore. A radiatively driven fast wind takes over before the products of He burning reach the stellar surface.

The proportion of anomalous to normal star observed today depends on at least two quantities: (i) the number of polluter stars and (ii) the dynamical history of the cluster which may lose during its lifetime first- and second-generation stars in different proportions. Decressin *et al.* (2008, 2009a) estimated these proportions based on dynamical models for GCs. When internal dynamical evolution and dissolution due to tidal forces are accounted for, starting from an initial fraction of anomalous stars of 10% produces a present-day fraction of about 25%, still too small compared with the observed 70–85%. If gas expulsion by supernovae is accounted for, a much higher fraction is expected to be produced, in better agreement with the observational requirements (Decressin *et al.* 2010).

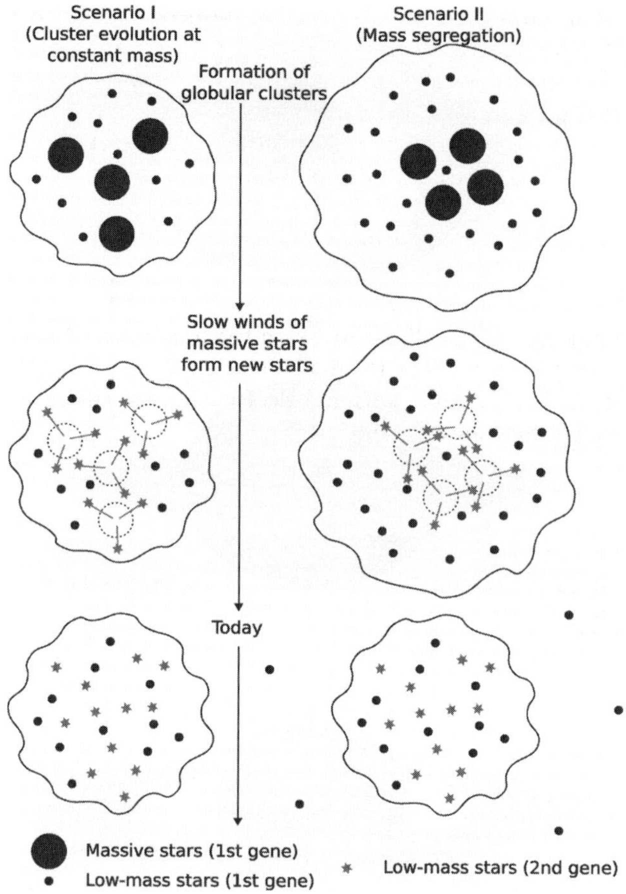

Figure 6. Schematic evolution of a globular cluster: a first generation of stars is formed from a giant molecular cloud. Massive first-generation stars evolve and give birth to a second generation of low-mass stars (dashed symbols in middle panels). Subsequently, the cluster evolves and today (lower panels) a mixture of first- and second-generation low-mass stars is present. In the right panel, the cluster is initially mass segregated and massive stars (and hence second-generation stars) are concentrated towards the cluster centre.

6. Conclusions

A significant fraction of stars in globular clusters (about 70–85%) exhibit peculiar chemical patterns with strong abundance variations in light elements, along with constant abundances in heavy elements. C–N, O–Na and Mg–Al anticorrelations, observed both in turnoff stars and in bright giants in GCs, call for a self-enrichment component of the intracluster gas. These features are not seen in field stars and appear to be intrinsic properties of GCs, related to cluster formation itself. They involve CNO cycles and NeNa and MgAl chains operating at relatively high temperatures.

Today, there is not a unique, neither a well-defined, nor even a fully consistent scenario for the self-enrichement of GCs. The most commonly invoked scenario assumes that the H-processed ejecta of massive AGB stars of a first generation form a second stellar generation. This idea, however, suffers from several drawbacks from the nucleosynthetic point of view, but also as far as mass budget and IMF of the polluters are concerned.

On the other hand, these abundance anomalies can be created in the H-burning core of a first generation of fast-rotating massive stars and the corresponding elements are conveyed to the stellar surface thanks to rotation-induced mixing. If the rotation of the stars is fast enough, this matter is ejected at low velocity through a mechanical wind at the equator. It then pollutes the interstellar medium from which a second generation of chemically anomalous stars can be formed. It is important to note that the 'winds of fast-rotating massive stars' scenario to explain the GC chemical properties is only one aspect, among many others, of the evolution of fast-rotating massive stars. For example, that kind of star might also be at the origin of the carbon-rich ultra-metal-poor stars or of the high helium abundance in a subpopulation of ω Cen.

Acknowledgements

I would like to thank my main collaborators on the 'Globular Cluster project': Thibaut Decressin, Georges Meynet and Nikos Prantzos. I acknowledge financial support from the IAU, from the French 'Programme National de Physique Stellaire' (CNRS/INSU) and from the Swiss National Science Foundation.

References

Arnould, M., Goriely, S., & Jorissen, A. 1999, *A&A*, 347, 572

Bonifacio, P., Pasquini, L., Molaro, P., Carretta, E., François, P., Gratton, R. G., James, G., Sbordone, L., Spite, F., & Zoccali, M. 2007, *A&A*, 470, 153

Carretta E., Gratton, R. G., Lucatello, S., Bragaglia A., & Bonifacio, P. 2005, *A&A*, 441, 131

Carretta, E., Bragaglia, A., Gratton, R. G., Leone, F., Recio–Blanco, A., & Lucatello, S. 2006, *A&A*, 450, 523

Carretta, E., Bragaglia, A., Gratton, R. G., Lucatello, S., & Momany, Y. 2007, *A&A*, 464, 927

Carretta, E., Bragaglia, A., Gratton, R., D'Orazi, V., & Lucatello, S. 2009a, *A&A*, in press

Carretta, E., Bragaglia, A., Gratton, R., & Lucatello, S. 2009b, *A&A*, 505, 139

Charbonnel, C. 2007, in: F. Kerschbaum, C. Charbonnel & R. F. Wing, *Why Galaxies Care About AGB Stars: Their Importance as Actors and Probes*, ASP Conf. Ser., Vol. 378, p. 416, San Francisco: ASP

Chiappini, C., Ekström, S., Meynet, G., Hirschi, R., Maeder, A., & Charbonnel, C. 2008, *A&A* (Letters), 479, L9

Cohen, J. G. 1978, *ApJ*, 223, 487

Decressin, T., 2007, PhD Thesis, Geneva University

Decressin, T., Baumgardt, H., & Kroupa, P. 2008, *A&A*, 492, 101

Decressin, T., Baumgardt, H., Kroupa, P., Meynet, G., & Charbonnel, C. 2009a, in: E. E. Mamajek, D. R. Soderblom, & R. F. G. Wyse, *The Ages of Stars*, Proc. IAU Symp. No. 258, p. 265

Decressin, T., Charbonnel, C., & Meynet, G. 2007a, *A&A*, 475, 859 (DCM07)

Decressin, T., Meynet, G., Charbonnel, C., Prantzos, N., & Ekström, S. 2007b, *A&A*, 464, 1029 (DMCPE07)

Decressin, T., Charbonnel, C., Siess, L., Palacios, A., Meynet, G., & Georgy, C. 2009a, *A&A*, 505, 727

Decressin, T., Baumgardt, Kroupa, P., Meynet, G., & Charbonnel, C. 2009b, in: E. E. Mamajek, D. R. Soderblom, & R. F. G. Wyse, *The Ages of Stars*, Proc. IAU Symp. No. 258, p. 265

Decressin, T., Baumgardt, Charbonnel, C., & Kroupa, P., in prep.

Denissenkov, P. A. & Denissenkova S. N. 1990, *SvA Lett.*, 16, 275

Drake, J. J., Smith, V. V., & Suntzeff, N. B. 1992, *ApJ* (Letters), 395, L95

Ekström, S., Meynet, G., Maeder, A., & Barblan, F. 2008, *A&A*, 478, 467

Gratton, R., Sneden, C., Carretta, E., & Bragaglia, A. 2000, *A&A*, 354, 169

Gratton, R., Sneden, C., & & Carretta, E. 2004, *ARA&A*, 42, 385

Illiadis, C., D'Auria, J. M., Starrfield, S., Thompson, W.J., & Wiescher, M. 2001, *ApJS*, 134, 151

Ivans, I. I., Sneden, C., Kraft, R. P., Suntzeff, N. B., Smith, V. V., Langer, G. E., & Fulbright, J. P. 1999, *AJ*, 118, 1273

Langer, G. E., Hoffman, R., & Sneden, C. 1993, *PASP*, 105, 301
Langer, G. E. & Hoffman, R., 1995 *PASP*, 107, 1177
Lind, K., Primas, F., Charbonnel, C., Grindahl, F., & Asplund, M. 2009, *A&A*, 503, 545
Maeder, A. & Meynet, G. 2000, *ARA&A*, 38, 143
Maeder, A. & Meynet, G. 2006, *A&A* (Letters), 448, L37
Norris, J. E., Cottrell, P. L., Freeman, K. C., & Da Costa, G. S. 1981, *ApJ*, 244, 205
Norris, J. E. 2004, *ApJ* (Letters), 612, L25
Osborn, W. 1971, *Observatory*, 91, 223
Pasquini, L., Bonifacio, P., Molaro, P, Francois, P, Spite, F., Gratton, R. G., Carretta, E., & Wolff, B. 2005, *A&A*, 441, 549
Peterson, R. C. 1980, *ApJ* (Letters), 237, L87
Pilachowski, C. A. 1989, in: G. Cayrel de Strobel, M. Spite & T. L. Evans (eds.), *The abundance spread within globular clusters*, Proc. IAU JCM5, p. 1
Prantzos, N. & Charbonnel, C. 2006, *A&A*, 458, 135 (PC06)
Prantzos, N., Charbonnel, C., & Iliadis, C. 2007, *A&A*, 470, 179 (PCI07)
Ramirez, S. V. & Cohen, J. G. 2002, *AJ*, 123, 3277
]Salpeter1955 Salpeter, E. E. 1955, *ApJ*, 121, 161
Shetrone, M. D. 1996, *AJ*, 112, 1517
Sneden, C. 2005, in: V. Hill, P. François & F. Primas (eds.), *From lithium to uranium: Element tracers of early cosmic evolution*, Proc. IAU Symp. No. 228, p. 337
Sneden, C., Gratton, R. G., & Crocker, D. A. 1991, *A&A*, 246, 354
Spite, M., Cayrel, R., & Plez, B. 2005, *A&A*, 430, 655
Ventura, P., D'Antona, F., Mazzitelli, I., & Gratton, R. 2001, *ApJ* (Letters), 550, L65
Ventura, P. & D'Antona, F. 2005a, *A&A*, 431, 279
Ventura, P. & D'Antona, F. 2008a, *A&A*, 479, 805
Yong, D., Grundahl, F., Lambert, D. L., Nissen, P. E., & Shetrone, M. D. 2003, *A&A*, 402, 984
Yong, D., Grundahl, F., Nissen, P. E., Jensen, H. R., & Lambert, D. L. 2005, *A&A*, 438, 875
Yong, D., Aoki, W., & Lambert, D. L. 2006, *ApJ*, 638, 1018
Yong, D., Grundahl, F., D'Antona, F., Karakas, A. I., Lattanzio, J. C., & Norris, J. E. 2009, *ApJ* (Letters), 695, L62

Star clusters: basic galactic building blocks
Proceedings IAU Symposium No. 266, 2009
R. de Grijs & J. R. D. Lépine, eds.

doi:10.1017/S1743921309990986

The lithium history of NGC 6397

Francesca Primas,[1] Karin Lind,[1] Corinne Charbonnel,[2] Frank Grundahl[3] and Martin Asplund[4]

[1]European Southern Observatory, Karl-Schwarzschild-Str. 2, D-85748, Garching, Germany
email: [fprimas, klind]@eso.org

[2]Geneva Observatory, 51 chemin des Mailettes, 1290 Sauverny, Switzerland
email: Corinne.Charbonnel@obs.unige.ch

[3]Department of Physics & Astronomy, Aarhus University, Ny Munkegade, 8000 Århus C, Denmark
email: fgj@phys.au.dk

[4]Max-Planck-institut für Astrophysik, Karl-Schwarzschild-Str. 1, D-85748 Garching, Germany
email: asplund@MPA-Garching.MPG.de

Abstract. The primordial lithium abundance inferred from *WMAP* and standard Big Bang nucleosysnthesis is approximately three times higher than the plateau value measured in old metal-poor Population II stars, suggesting that these stars have undergone atmospheric Li depletion. To constrain the physics responsible for such depletion, we conducted a homogeneous analysis of a large sample of stars in the metal-poor globular cluster NGC 6397, covering all evolutionary phases from below the main-sequence turnoff to high up the red-giant branch (RGB). The dwarf, turnoff, and early subgiant stars form a thin abundance plateau, with a sharpe edge in the middle of the subgiant branch, where Li dilution caused by the inward extension of the convective envelope starts (the beginning of the so-called first dredge up). A second steep abundance drop is seen at the RGB bump, again highlighting the need for the onset of nonstandard mixing in this evolutionary phase. Moreover, by also measuring the sodium abundances of the targets, we have gained insight into the degree of pollution by early cluster self-enrichement, and may separate highly polluted, Li-poor and Na-rich stars from stars formed from pristine material. Our observational findings strictly limit both the extent of lithium surface depletion, which in turn constrains the efficiency of mixing below the outer convection zone, and the resulting spread in lithium abundance in metal-poor turn-off stars.

Keywords. stars: abundances, stars: atmospheres, stars: evolution, globular clusters: individual (NGC 6397)

1. Introduction

From the time of its discovery in the stellar atmospheres of metal-poor Population II stars (Spite & Spite 1982), lithium has been considered a key diagnostic to test and improve our knowledge of the primordial Universe, stellar interiors and evolution, and spallation physics (especially so if lithium abundances are combined with beryllium and boron). Lithium-7 is one of the four primordial isotopes which were formed in observable quantities by nuclear reactions during the first minutes after the Big Bang (e.g., Olive *et al.* 2000; and references therein). Together with deuterium, helium-3, and helium-4, knowledge of its primordial abundance provides an important observational constraint on the baryon-to-photon ratio $\eta \propto \Omega_b h^2$, which is the main free parameter of standard Big Bang nucleosynthesis (SBBN).

Despite numerous observations of lithium in stars, we still lack a consistent picture and the primordial abundance remains under debate. Prior to the discovery of a notably flat and constant Li abundance—$A(\mathrm{Li}) = 12 + \log(N(\mathrm{Li})/N(\mathrm{H})) \approx 2.2$—among Galactic warm halo dwarf stars, spanning a wide range of effective temperatures and metallicities (the so-called Spite plateau), it was generally believed that the primordial Li abundance was in the range $A(\mathrm{Li}) = 3.0 - 3.3$, which is the value measured in meteorites and also the maximum value detected in Population I stars, both in the field and in open clusters (cf. Boesgaard & Steigman 1985). If this were the case, the close-to-constant but lower value of lithium along the Spite plateau would have required that all of the oldest stars in our Galaxy had suffered uniform depletion by about a factor of 10. If, instead, the plateau value represented the amount of Li produced during the Big Bang, the Galaxy should have been enriched in its Li content by a factor of at least 10 since its birth.

During the last decade, observations of cosmic-microwave-background (CMB) anisotropies, most particularly through the *Wilkinson Microwave Anisotropy Probe (WMAP)* experiment (Bennet *et al.* 2003; Spergel *et al.* 2003), have allowed unprecedented precision as regards the determination of the baryon density $\Omega_\mathrm{b}h^2$ and seem to reveal that the truth about Li lies between the two extreme solutions discussed above. With the most recent value of $\Omega_\mathrm{b}h^2$ derived from the five-year release of *WMAP* data ($\Omega_\mathrm{b}h^2 = 0.02273 \pm 0.00062$; Dunkley *et al.* 2009), updated SBBN predictions provide a precise determination of the primordial abundances of the light elements D, ^{3}He, ^{4}He, and ^{7}Li that can be compared with observations in low-metallicity environments. For ^{7}Li, the CMB-derived primordial abundance is clearly higher ($A(\mathrm{Li}) = 2.72 \pm 0.06$) than the current determinations in low-metallicity halo stars. This result seems to be very robust with respect to the nuclear uncertainties on the SBBN reactions. An alternative explanation is that the Li abundance seen on the surface of halo stars is not the pristine value, but that these stars have instead undergone surface-lithium depletion at some point during their evolution.

That lithium and other elements should be prone to atomic diffusion and thus be depleted from stellar atmospheres by gravitational settling was already predicted by Michaud *et al.* (1984). However, without the existence of macroscopic mixing acting below the outer convection zone, a thin and close-to-flat Spite plateau cannot be reproduced by stellar-structure models (Richard *et al.* 2005). As of today, the physical process behind the mixing has yet to be identified. So, the question is, can we—through detailed observations and analysis of lithium in Population II stars—place rigid constraints on the nature and efficiency of this nonstandard mixing?

2. Our project: definition and goals

To constrain the physics responsible for surface depletion, one needs to accurately establish the behaviour of surface Li abundance as a function of stellar effective temperature and evolutionary phase. Globular clusters are ideal environments, since relative stellar parameters can be determined precisely. NGC 6397 is one of the most extensively observed globular clusters and its lithium content has been documented down to the magnitude of the cluster turnoff point in several studies (Pasquini *et al.* 1996; Castilho *et al.* 2000; Thévenin *et al.* 2001; Bonifacio *et al.* 2002; Korn *et al.* 2007), although with poor number statistics at faint magnitudes.

Therefore, we decided to target this cluster by observing a large sample of stars and conduct a homogeneous analysis, fully covering all evolutionary phases from below the main-sequence turnoff (around $V = 17$ mag) to high up the RGB (around $V = 11.5$ mag). In total, 349 stars were observed, 160 of which sample the subgiant and the red-giant branch (80 each), and the rest (∼190) sample the turnoff point. All our targets were selected from Strömgren *uvby* photometry. The sample was observed with FLAMES, the high-resolution multiplex facility at the *VLT/UT2*, using the UVES+GIRAFFE/MEDUSA combined mode. For approximately one third of the sample we also covered the Na doublet at 8183Å and 8194Å, which is a fundamental ingredient to properly disentangle between in situ depletion and pre-enrichment effects (by testing the existence of a Li–Na anticorrelation, if any). More details about the observations and the reduction of the data set can be found in Lind *et al.* (2009b).

3. Stellar parameters

We derived effective temperatures from *uvby* photometry, gravities from the common relation between luminosity and mass (obtained from the cluster isochrone), and metallicity from measuring the equivalent widths of several FeI and FeII lines. The most critical parameter for lithium-abundance determination is the effective temperature: ±100 K in effective temperature corresponds to approximately ±0.07 dex in lithium abundance.

The colours were translated to effective temperatures using the MARCS synthetic-colour calibration of Önehag *et al.* (2009), as well as the infrared-flux-method calibrations of Ramírez & Meléndez (2005) and Alonso *et al.* (1996, 1999). In the turnoff region, our targets show a spread of 0.015 mag (1σ) in $v - y$ colour around the cluster sequence. With a sensitivity of $v - y$ of 37 K per 0.01 mag, this corresponds to a spread in effective temperature of 56 K. The more temperature-sensitive $b - y$ colour has a typical spread of 0.01 mag, which corresponds to 95 K. Moreover, the target colours could also be affected by differential reddening, and we circumvented the problem by constructing colour–magnitude fiducial sequences and interpolated the visual magnitude of each target onto the sequence. This is done to reduce artificial spreads in effective temperature caused by reddening and observational uncertainty, which would propagate into a significant uncertainty in lithium abundance.

We have compared our temperatures with the values determined in the recent study by González Hernández *et al.* (2009) for a group of subgiants and main-sequence stars in NGC 6397. These authors use profile fitting of the Hα wings to infer effective temperatures, with 3D hydrodynamical model atmospheres. The Hα-based values have a systematic offset towards higher temperatures by approximately 80 K, but the relative agreement is good, considering the fully independent methods adopted. Their temperatures show a scatter around the cluster sequence of 105 K for the subgiants and 85 K for the dwarfs, which may be a reasonable estimate of the true spread.

4. Abundance analysis

We carried out the abundance analysis using 1D, LTE (local thermodynamic equilibrium), plane-parallel and spherical MARCS model atmospheres (Gustafsson *et al.* 2008). Spectral synthesis was employed for the lighter elements (up to Al), while equivalent-width measurements were used for all other elements (e.g., Fe).

Lithium abundances were derived from minimisation of the χ^2 distance between synthesis and observations for a total of 305 stars (including a few upper limits). Sodium abundances were derived using the same technique, but for a subset of 117 stars (again, including upper limits). Both Li and Na abundances were corrected for non-LTE effects. For Li, we calculated the non-LTE corrections according to the recent study by Lind *et al.* (2009a); for Na we applied non-LTE corrections following the work of Mashonkina *et al.* (2000).

For completeness, it is important to mention that we derived many more abundances of α, iron-peak and heavy elements. These will be presented in a forthcoming paper (Lind *et al.*, in prep.) which will tackle the chemical evolution of this globular cluster.

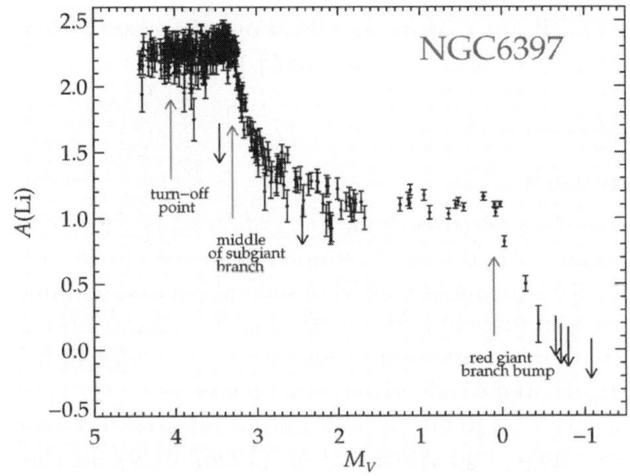

Figure 1. Non-LTE Li abundances versus absolute visual magnitude for all our NGC 6397 targets. Labels identify some of our main findings as described in the text.

5. Results and implications

The dwarf, turnoff, and early subgiant stars in our sample form a thin abundance plateau, with a sharp edge in the middle of the subgiant branch, where the Li dilution caused by the first dredge up sets in. A second steep abundance drop is seen at the RGB bump (cf. Figure 1), which likely is due to the onset of thermohaline mixing (Charbonnel & Zahn 2007).

A considerably large spread in Na abundance ($\sim$1 dex) confirms the findings of Carretta *et al.* (2005), that NGC 6397 has suffered from intracluster pollution in its infancy. The most lithium-poor objects show extreme Na enhancement (cf. Figure 2), and we deduce that these objects have suffered heavy pollution from a previous generation of stars. This is a very important result on its own, since it rules out other explanations for the Li deficiency (such as, e.g., abnormal rotational velocity or binarity; see, e.g., Ryan *et al.* 2001). Second, detecting lithium in highly polluted, sodium-rich stars points to the need for mixing with pristine matter, since stellar ejecta presumably are Li-free (e.g., Prantzos & Charbonnel 2006).

The Li–Na anticorrelation (cf. Figure 2) relies on a handful of the most polluted objects to be significant, i.e., it appears that Li abundances are essentially unaffected by relatively large degrees of pollution. No significant difference in the behaviour of mean lithium abundance with evolutionary phase is found if polluted stars (assumed to be those with a $A(\mathrm{Na}) \geqslant 3.9$) are separated from nonpolluted objects. Moreover, the spread in abundance among nonpolluted objects with similar stellar parameters ($\sigma = 0.09$ dex) is consistent with observational errors, and does not allow for significant star-to-star differences.

Figure 3 illustrates how the mean lithium abundance varies with visual magnitude in the cluster and also compares our findings to other, similar studies of NGC 6397. A feature seen in all three of the most recent studies is that subgiants appear more lithium-rich than less evolved stars, as already suggested by Charbonnel & Primas (2005) for field stars. Our observations reveal a small upturn in Li abundance, located just prior to the

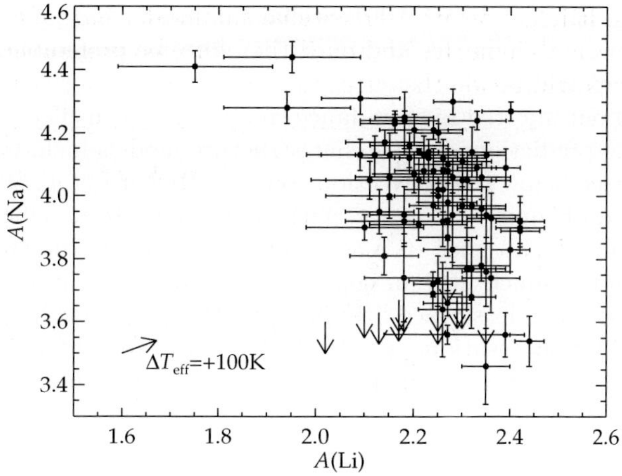

Figure 2. Non-LTE Li and Na abundances for the subsample of 100 dwarfs and subgiants with $M_V > 3.3$ mag. The arrows mark Na-abundance upper limits. If the effective temperature of a target is overestimated by 100 K, its corresponding location in this diagram will shift in the direction of the tilted arrow (or vice versa).

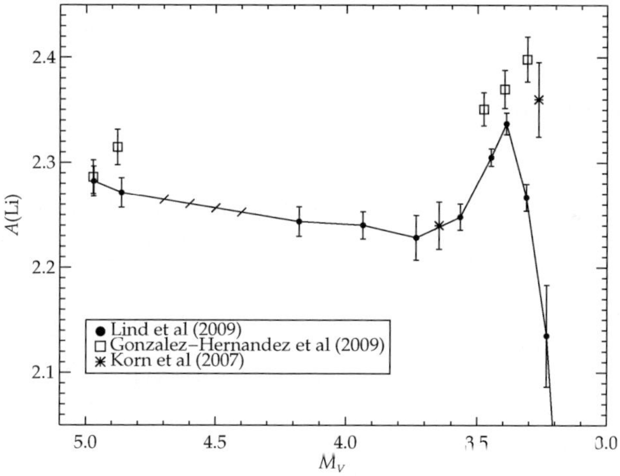

Figure 3. Group-averaged lithium abundance versus absolute visual magnitude for our targets and two other studies of NGC 6397, as indicated by the different symbols. The higher lithium abundances inferred by González Hernández *et al.* can be traced to their warmer effective-temperature scale (80 K).

stage where Li is strongly diluted, which may indicate that an overabundance is formed above Li-free layers in the stellar interior.

Since the effective-temperature determination has large influence on the derived lithium abundances, and as of yet no true consensus has been reached regarding the reliability of different photometric and spectroscopic methods for determining this fundamental stellar parameter for metal-poor turnoff stars (especially not within 100 K accuracy), trends of lithium with effective temperature are likely to contain a certain bias. Whereas Figure 3 points to a general agreement between different studies, a similar plot with effective temperature as reference axis would show significant discrepancies, as was also pointed out by González Hernández *et al.* (2009). Stellar luminosity has, of course, no influcence on the inferred lithium abundances and may, therefore, be preferable to use as reference for, e.g., model comparisons.

When comparing our inferred Li abundance trends with both effective temperature and stellar luminosity to predictions from stellar-structure models including atomic diffusion and ad hoc turbulence below the convection zone (cf. Richard *et al.* 2005), we find strong support for the need of some turbulence, with strict limits to its efficiency (cf. figure 10 in Lind *et al.* 2009b). In absence of macroscopic turbulence, the turnoff and subgiant stars would deplete too much lithium compared to more unevolved stars, giving rise to large ($\sim$0.6 dex) variations in surface-lithium abundance with evolutionary phase, which are not reproduced by observations. On the other hand, too efficient mixing erases any difference between the groups and cannot explain that subgiants appear more lithium-rich than less evolved stars.

References

Alonso, A., Arribas, S., & Martínez–Roger, C. 1996, *A&A*, 313, 873

Alonso, A., Arribas, S., & Martínez–Roger, C. 1999, *A&As*, 140, 261

Boesgaard, A. M. & Steigman G. 1985, *ARA&A*, 23, 319

Bennett, C. L., *et al.* *ApJS*, 148, 1

Bonifacio, P., *et al.* 2002, *A&A*, 390, 91

Carretta, E., Gratton, R. G., Lucatello, S., Bragaglia, A., & Bonifacio, P. 2005, *A&A*, 433, 597

Castilho, B. V., Pasquini, L., Allen, D. M., Barbuy, B., & Molaro, P. 2000, *A&A*, 361, 92

Charbonnel, C. & Primas, F. 2005, *A&A*, 442, 961

Charbonnel, C. & Zahn, J.-P. 2007, *A&A*, 467, L15

Cyburt, R. H., Fields, B. D., & Olive, K. A. 2008, *JCAPP*, 11, 12

Dunkley, J., *et al.* 2009, *ApJS*, 180, 306

González Hernández, J. I., Bonifacio, P., Caffau, E., Steffen, M., Ludwig, H.-G., Behara, N. T., Sbordone, L., Cayrel, R., & Zaggia, S. 2009, *A&A*, 505, 13

Gustafsson, B., Edvardsson, B., Eriksson, K., Jørgensen, U. G., Nordlund, Å, & Plez, B. 2008, *A&A*, 486, 951

Korn, A. J., Grundahl, F., Richard, O., Mashonkina, L., Barklem, P. S., Collet, R., Gustafsson, B., & Piskunov, N. 2007, *ApJ*, 671, 402

Lind, K., Asplund, M., & Barklem, P. S. 2009a, *A&A*, 503, 541

Lind, K., Primas, F., Charbonnel, C., Grundahl, F., & Asplund, M. 2009b, *A&A*, 503, 545

Mashonkina, L. I., Shimanskiĭ, V. V., & Sakhibullin, N. A. 2000, *AstRep*, 44, 790

Michaud, G., Fontaine, G., & Beaudet, G. 1984, *ApJ*, 282, 206

Olive, K. A., Steigman, G., & Walker, T. P. 2000, *Phys. Rev.*, 333, 389

Önehag, A., Gustafsson, B., Eriksson, K., & Edvardsson, B. 2009, *A&A*, 498, 527

Pasquini, L. & Molaro, P. 1996, *A&A*, 307, 761

Prantzos, N. & Charbonnel, C. 2006, *A&A*, 458, 135

Ramírez, I. & Meléndez, J. 2005, *ApJ*, 626, 465

Richard, O., Michaud, G., & Richer, J. 2005, *ApJ*, 619, 538

Ryan, S. G., Kajino, T., Beers, T. C., Suzuki, T. K., Romano, D., Matteucci, F., & Rosolankova, K. 2001, *ApJ*, 549, 55

Spergel, D. N., Verde, L., Peiris, H. V. 2003, *ApJS* 148, 175

Spite, M. & Spite, F. 1982, *Nature*, 297, 483

Thévenin, F., Charbonnel, C., de Freitas Pacheco, J. A., Idiart, T. P., Jasniewicz, G., de Laverny, P., & Plez, B. 2001, *A&A*, 373, 905

Star clusters: basic galactic building blocks
Proceedings IAU Symposium No. 266, 2009
R. de Grijs & J. R. D. Lépine, eds.

doi:10.1017/S1743921309990998

Globular cluster abundances and what they can tell us about galaxy formation

Judith G. Cohen

Palomar Observatory, Mail Stop 249–17, California Institute of Technology, Pasadena, CA 91125, USA
email: jlc@astro.caltech.edu

Abstract. We review the properties of globular clusters (GCs) which make them useful for studying the Galactic halo, Galactic chemical evolution and the early stages of the formation of the Milky Way. We review the evidence that GCs have a chemical inventory similar to those of halo field stars. We discuss the abundance ratios for dSph galaxies and show that it is possible to have formed at least part of the Galactic-halo field stellar population by dissolving GCs and/or accreting dSph galaxies, but only if this occurred at an early stage in the formation of the Galaxy. We review the constraints on halo-formation timescales deduced from the low magnesium isotopic ratios in metal-poor halo field dwarfs, which indicate that asymptotic giant-branch (AGB) stars did not have time to contribute significantly, while M71 contains two populations, one without and also one with a substantial AGB contribution. We review the limited evidence for GCs with a second population showing additional contributions from Type II supernovae, currently confined to ω Cen, M54 and M22, all of which may have been the nuclei or central regions of accreted galaxies. We check our own data for additional similar GCs and find preliminary indications that NGC 2419, a massive GC far in the outer Galactic halo, may also belong to this group.

Keywords. stars: abundances, Galaxy: formation, Galaxy: halo, globular clusters: general, galaxies: dwarf, Local Group

1. Introduction

Globular clusters (GCs) have unique attributes which make them laboratories for studying stellar evolution. Such stellar systems have well-determined distances and ages from their color–magnitude diagrams (CMDs), accompanied by detailed kinematic information based on large radial-velocity surveys. They show no evidence for the presence of dark matter, having mass-to-light ratios expected for old stellar systems. Their chemical inventories can be studied at high spectral resolution and with large samples of stars in a given cluster. Proper motions are available in some cases, leading to orbit determinations. Tidal tails are seen in a few cases, constraining mass models for the halo. Since Galactic GCs are old, with typical ages of $\sim$ 12 Gyr, and are, at least to first order, formed of stars with identical chemical compositions, they can provide unique insights into the formation of the Milky Way galaxy.

Some of the key issues involving GCs that are of great current interest are to what extent there are multiple populations in GCs and to what extent they are chemically homogenous. Also, whether the Milky Way halo could have been made at least in part by dissolving GCs and accreting dSph satellite galaxies, whether the GCs and the halo have the same metallicity distribution function (MDF), whether they have the same chemical inventory and what we can learn from them about early supernovae (SNe), star formation and formation of the Galactic halo.

2. Globular cluster abundance ratios versus halo field stars and dSph satellites

It has been clear for several years now that the abundance ratios [X/Fe] for Galactic halo field stars show trends with [Fe/H]. Broadly speaking, the ratios increase to a super-solar level as [Fe/H] decreases for the α elements, while they decline for Cr I and Mn I, with minimal scatter about the mean relations, implying that the halo is well mixed with many SNe contributing to the chemical inventory of each star. A major theme of work over the past few years has been to extend these relations to ever more metal-poor halo field stars and explore the subtle differences that may exist in chemical inventory between stars in the inner and outer parts of the Galactic halo, subjects not part of the current discussion. However, over the metallicity range characteristic of the Galactic GCs, it is now well established that the GCs, ignoring those associated with the dissolving Sgr dSph galaxy, follow the trends established by the halo field stars, see e.g. the review of Gratton *et al.* (2004). Detailed chemical-evolution models for the Galaxy, such as those of Matteucci (2009) or Kobayashi *et al.* (2006), can reproduce most of these trends. One long-standing issue, resolved in Schoerck *et al.* (2009) using the MDF for

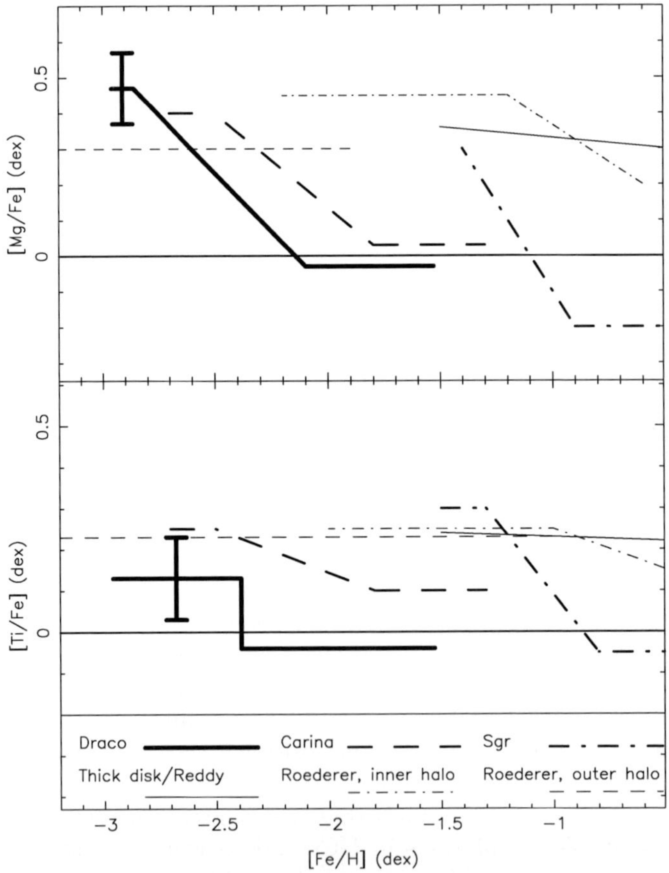

Figure 1. [Mg/Fe] *(top)* and [Ti/Fe] *(bottom)* in three dSph satelllites (Draco: Cohen & Huang 2010; Sgr: Monaco *et al.* 2005; Sbordone *et al.* 2007; and Carina: Koch *et al.* 2008) as a function of [Fe/H] compared to the trends for Galactic field halo stars from Roederer (2008) and thick-disk stars from Reddy *et al.* (2006) as well as for Galactic GCs. See figure 21 of Cohen & Huang (2009) for details.

Galactic halo field stars inferred from the Hamburg/ESO Survey with corrections for selection efficiency is that it is probably not statistically significant that there is no GC with [Fe/H] < -2.4 dex given the very low fraction of extremely metal-poor stars in the halo.

Another issue is that of the abundance ratios and MDFs for the satellite dSph galaxies of the Milky Way compared to the Milky Way halo. Helmi *et al.* (2006) summarized the position of the very large DART survey at the *VLT*, claiming that the dSph galaxies lacked the expected fraction of extremely metal-poor (EMP) stars, which are found in the Galactic halo, and that this and other related evidence from the behavior of abundance ratios of the α elements precluded the possibility of making the Galactic halo in whole or part from dissolved dSph galaxies. This controversial paper has been refuted, and now even the authors admit that it is flawed. It is clear from a large number of recent studies, including that of Cohen & Huang (2009) for the classical dSph galaxy Draco and of Kirby *et al.* (2008) for the recently discovered ultrafaint satellite galaxies, that the local dSph galaxies have an appropriate (low) fraction of EMP stars, and that at the lowest Fe metallicities, $[\alpha/\text{Fe}]$ in such low-luminosity satellite galaxies is quite close to the values for similar-metallicy stars in the Galactic halo. At later times, the star-formation history of the dSph galaxies diverged strongly from that of the halo and strong differences in chemical inventory and abundance ratios arose between them. Thus, the bulk of the accretion must have been early, as is expected from cold dark-matter hierarchical galaxy-formation models and simulations.

3. Elements known to vary within GCs and why

The anticorrelation between O and Na is ubiquitous among GCs. The range is very large, sometimes a factor of 10 in abundance of each, and is seen at all luminosities of the member stars. There is also an anticorrelation between Mg and Al, sometimes with a very large range in Al (a rather rare element, so burning only a small amount of Mg into Al will produce a big change in the Al abundance). GC stars show large star-to-star variations of C and N, almost always with C and N anticorrelated, see e.g. Cohen *et al.* (2005). None of this is seen among samples of halo field stars. All of this is tied to horizontal-branch characteristics, Galactic orbits, almost certainly to He abundances, and is evidence for at least two generations of star formation in GCs, as is discussed in a long series of papers by E. Carretta and collaborators, the most recent of which is Carretta *et al.* (2010).

Star-to-star variations of elements between Ca and Ni within GCs is discussed in Section 5. Moving on to the heavy neutron-capture elements beyond the Fe peak, M15 is the one GC known to show a star-to-star variation among them, first detected by Sneden *et al.* (1997) (see also Otsuki *et al.* 2006). I (unpublished) have verified that the variation in abundance among the heavy neutron-capture elements within M15 is about a factor of five, and the production mechanism for these heavy elements is the r-process, rather than the s-process. Results for a sample of 12 red giant-branch (RGB) stars in M15 are shown in Figure 2. The low minima in y indicate that the heavy elements in M15 originate in the r-process. The range in Δr (the x coordinate of the figure) of the minima for the curve for each M15 star indicates the range in the fractional r-process contribution to the chemical inventory of the M15 stars. These are very rare elements, $\epsilon(\text{Eu})/\epsilon(\text{Fe}) \sim 10^{-11}$ in M15, so small star-to-star variations within a very metal-poor GC should not be surprising.

4. Constraints from isotopic abundances

Isotopic abundance ratios can convey still more detailed information than atomic abundance ratios as production of a particular isotope requires specific nuclear reactions. The ratios among the three stable isotopes of Mg, all of which can be formed in massive stars (Woosley & Weaver 1995) are of particular interest as they provide a rough chronometer. ^{24}Mg, the most common isotope, is formed as a primary isotope from H, while ^{25,26}Mg are formed as secondary isotopes. The heaviest Mg isotopes are also produced in intermediate-mass AGB stars (Karakas & Lattanzio 2003), so the isotopic ratios ^{25,26}Mg/^{24}Mg increase with the onset of AGB stars. Therefore, Mg isotopic ratios in halo stars can be used to constrain the rise of the AGB stars in our Galaxy and—if the AGB contribution is not present—the minimum timescale for halo formation.

Measurement of the Mg isotopic ratios requires detection of weak contributions from the rarer heavier Mg isotopes in the wings of much stronger ^{24}MgH lines, hence exquisite spectra of very high spectral resolution (~100 000) and very high signal-to-noise ratio, as well as careful attention to details of the line list used for the spectral synthesis and to the analysis procedure.

This idea is applied in two papers written with my former postdoc, Jorge Meléndez, who did most of the work, Meléndez & Cohen (2007, 2009). We see no sign of an AGB contribution among our sample of metal-poor Galactic halo dwarfs. In the GC M71, at [Fe/H] −0.7 dex, we see two populations, as shown in Figure 3. The first has weak CN, normal O, Na, Mg and Al, and a low ratio of ^{26}Mg/Mg (~ 4%), consistent with models of Galactic chemical evolution with no contribution from AGB stars. The second population has enhanced Na and Al, accompanied by lower O and by higher

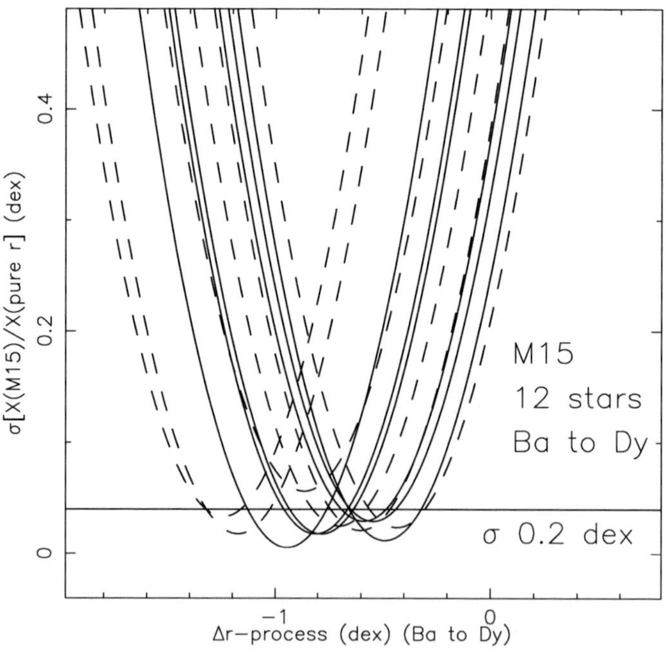

Figure 2. Deviations from a scaled r-process are shown for 12 stars in the GC M15. The y value for each star is the rms of the difference between [X/Fe] for the M15 star and that assuming a pure r-process distribution scaled by [Fe/H] for M15 with a fixed offset, Δr, over the elements between Ba and Dy for which abundances have been derived in the star, typically six elements.

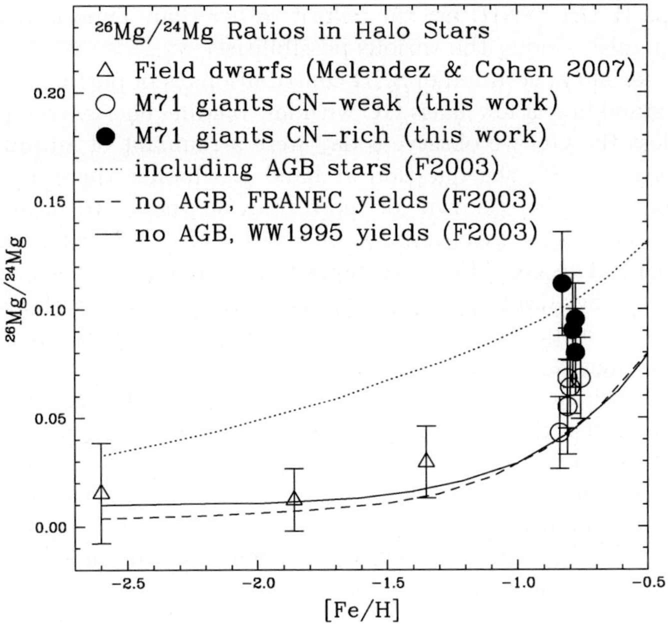

Figure 3. Our ^{26}Mg/^{24}Mg ratios in both field dwarfs (triangles; Meléndez & Cohen 2007) and M71 giants (Meléndez & Cohen 2009) as a function of [Fe/H]. Chemical-evolution models by Fenner *et al.* (2003) including (dotted line) and excluding (solid and dashed lines) AGB stars are shown. (Figure from Meléndez & Cohen 2009.)

^{26}Mg/Mg ($\sim$ 8%), consistent with models which incorporate ejecta from AGB stars via normal stellar winds. We therefore infer that the timescale for formation of the first generation of stars we see today in this GC must be sufficiently short to avoid a contribution from AGB stars, i.e., less than $\sim$ 0.3 Gyr.

5. Variation of Type II supernova contributions

Spectroscopic evidence for multiple populations in Galactic GCs has been known for a long time based on studies of the O–Na anticorrelation and on spreads in the CNO abundances among stars within a particular GC as reviewed by C. Charbonnel at this meeting. These require at least two generations of GC stars, with the second polluted by ejecta from either AGB stars or massive rapidly rotating stars.

The existence of multiple generations of stars within GCs has very recently been detected using photometry as well. The sample of GCs with CMDs based on ACS images taken with *HST* has grown rapidly, most recently through the ACS Survey of Galactic Globular Clusters (an *HST* Treasury project) (Sarajedini *et al.* 2007). This extremely accurate photometry has enabled detection of subtle features in the GC CMDs that were previously lost in the noise, including multiple main sequences, a double subgiant branch in NGC 1851 (Milone *et al.* 2008) and other such phenomena.

They are usually explained by small differences in age, He content (which, although difficult to detect directly, must occur in association with the O/Na anticorrelation and the CNO variations) or metal content. When invoking the latter, one must be careful to distinguish between variations among those elements such as CNO, which can be produced by ejecta from intermediate- or low-mass AGB stars, and variations in the heavier elements such as Ca or Fe, which are only produced in SNe. Presumably at such early times, Type II SNe (SNII) are the culprit rather than Type Ia SNe. Great care is required to distinguish among the various possibilities.

The issue of a second generation of SNII contributions is of particular interest, since it is hard to understand how a low-mass GC with low binding energy could retain energetic SNII ejecta, unless the GC we observe today were a remnant of an initally much more massive stellar system. So, the question at hand is whether there are any other such cases, in addition to ω Cen (known for more than 30 years), to have a wide intrinsic range in [Ca/H], [Fe/H], etc., extending over a range of $\sim$ 1.3 dex with multiple peaks (Norris *et al.* 1996). The GC M54 also shows this, but it is believed to be part of the central region of the Sgr dSph galaxy, currently being accreted by the Milky Way. Very recently M22, under suspicion for many years, was confirmed to have such by Marino *et al.* (2010), who suggest a range in [Fe/H] of perhaps $\sim$ 0.15 dex.

Are there more such cases? Is this the tip of the iceberg? Do other GCs all have more modest variations in [Fe/H]? A literature search for recent detailed abundance analyses of samples of GC stars, almost always luminous RGB stars, indicates that σ[Fe/H] is quite small in these clusters compared to the obvious cases described above; they show dispersions in Fe abundance of less than 0.05 dex (10%), and sometimes as low as 5%. A careful check of a large sample of GCs, particularly of the most massive objects, is desirable as some of the most massive Galactic GCs are not well studied.

As a preliminary step towards this goal, I have checked the spectra from *Keck*/LRIS slitmasks I took from 1998 to 2003 for a study of CH, CN and NH bands with P. Stetson and M. Briley. These cover the H and K lines of Ca II. I measured the strength of absorption in the 3933 Å line. Results are shown for four GCs in Figure 4. I have also located a substantial number of *Keck*/Deimos slitmasks of Galactic GCs which cover the 8500 Å Ca-triplet region, two of which were analyzed prior to this IAU Symposium, shown in Figure 5. These spectra were taken for other purposes and are not optimized for the present goal, but still yield upper limits to σ[Fe/H] $\sim$ 0.03 dex in the best cases.

Ignoring M71, which has substantial field-star contamination due to its low Galactic latitude, the most interesting result is that for NGC 2419, a massive GC far out in the halo at a distance of 84 kpc. The data in hand appear to show a substantial range of

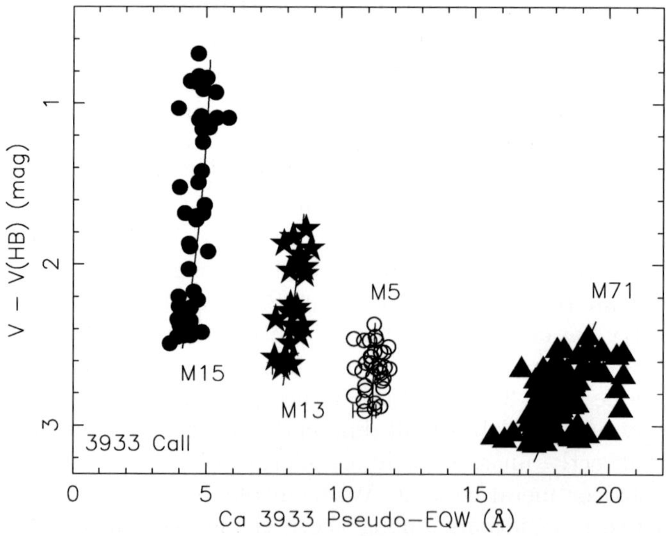

Figure 4. Equivalent width of the 3933 Å line of Ca for samples in four GCs as a function of luminosity below the horizontal branch.

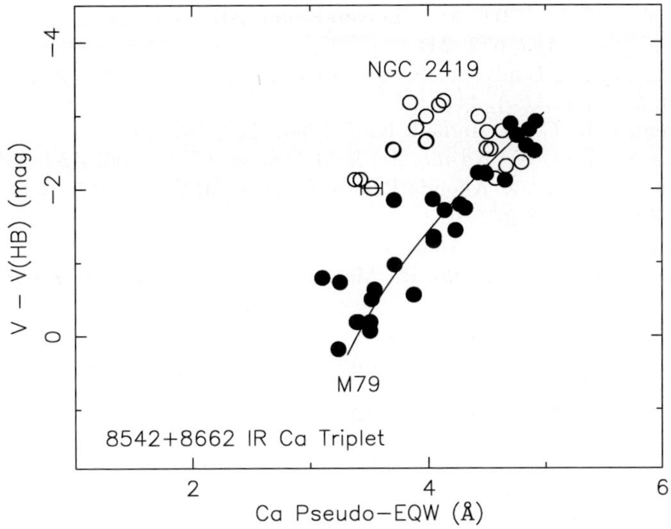

Figure 5. Sum of the equivalent width of the 8542 and the 8662 Å lines from the Ca triplet for samples of upper-RGB stars in M79 and NGC 2419, as a function of luminosity.

[Ca/H], which is probably not due to field-star contamination. Efforts are now underway both to establish the limiting accuracy that can be reached by such techniques and to verify this interesting result, as—if valid—NGC 2419 must represent the remnant of a former Milky Way satellite.

Acknowledgements

The work presented here was carried out in part with my former postdoctoral fellows Jorge Meléndez and Wenjin Huang.

References

D'Antona, F. D., Stetson, P. B., Ventura, P., Milone, A. P., Piotto, G., & Caloi, V. 2010, *MNRAS* (Letters), 399, L151

Carretta, E., Bragaglia, A., Gratton, R., & Lucatello, S. 2009, *A&A*, 505, 139

Cohen, J. G., Briley, M. M., & Stetson, P. B. 2005, *AJ*, 130, 1177

Cohen, J. G. & Huang, W. 2009, *ApJ*, 701, 1053

Fenner, Y., Gibson, B. K., Lee, H.-C., Karakas, A. I., Lattanzio, J. C., Chieffi, A., Limongi, M., & Yong, D. 2003, *PASA*, 20, 340

Gratton, R., Sneden, C., & Carretta, E. 2004, *ARA&A*, 42, 385

Helmi, A., *et al.*, 2006, *ApJ* (Letters), 651, L121

Karakas, A. I. & Lattanzio, J. C. 2003, *PASA*, 20, 279

Kirby, E. N., Simon, J. D., Geha, M., Guhathakurta, P., & Frebel, A. 2008, *ApJ* (Letters), 685, L43

Kobayashi, C., Umeda, H., Nomoto, K., Tominaga, N., & Ohkubo, T. 2006, *ApJ* (Letters), 653, 1145

Koch, A., Grebel, E. K., Gilmore, G. F., Wyse, R. F. G., Kleyna, J. T., Harbeck, D. R., Wilkinson, M. I., & Evans, N. W. *AJ*, 135, 1580

Merino, A. F., Milone, A. P., Piotto, G., Villanova, S., Bedin, L. R., Bellini, A., & Renzini, A. 2010, *A&A*, in press

Matteucci, F. 2009, in Grebel, E. & Moore, B. (eds), *The Origin of the Galaxy and the Local Group*, in press

Meléndez, J. & Cohen, J. G. 2007, *ApJ* (Letters), 659, L25

Meléndez, J. & Cohen, J. G. 2009, *ApJ* (Letters), 699, 2017

Milone, A. P., *et al.* 2008, *ApJ*, 673, 241

Monaco, L., Bellazzini, M., Bonifacio, P., Ferraro, F. R., Marconi, G., Pancino, E., Sbordone, L., & Zaggia, S. 2005, *A&A*, 441, 141

Norris, J. E., Freeman, K. C., & Mighell, K. J. 1996, *ApJ*, 462, 241

Otsuki, K., Honda, S., Aoki, W., Kajino, T., & Matthews, G. J. 2006, *ApJ* (Letters), 641, L117

Reddy, B. E., Lambert, D. L., & Allende Prieto, C. 2006, *MNRAS*, 367, 1329

Roederer, I. U. 2008, *AJ*, 137, 272

Sarajedini, A., *et al.* 2007, *AJ*, 133, 1658

Sbordone, L., Bonifacio, P., Buonanno, R., Marconi, G., Monaco, L., & Zaggia, S. 2007, *A&A*, 465, 815

Schoerck, T., *et al.* 2009, *A&A*, 507, 817

Sneden, C., Kraft, R. P., Shetrone, M. D., Smith, G. H., Langer, G. E., & Prosser, C. F. 1997, *AJ*, 114, 1964

Woosley, S. E. & Weaver, T. A. 1995, *ApJ*, 101, 181

Star clusters: basic galactic building blocks
Proceedings IAU Symposium No. 266, 2009
R. de Grijs & J. R. D. Lépine, eds.

doi:10.1017/S1743921309991001

Chemical abundances in tidally disrupted globular clusters

D. Yong,[1] J. Meléndez,[2] K. Cunha,[3] A. I. Karakas,[1] J. E. Norris[1] and V. V. Smith[3]

[1]Australian National University, Mount Stromlo Observatory, Australia
[2]Centro de Astrofísica da Universidade do Porto, Portugal
[3]National Optical Astronomy Observatories, USA

Abstract. We present abundance measurements in the tidally disrupted globular cluster NGC 6712. In this cluster, there are large star-to-star variations of the light elements C, N, O, F and Na. While such abundance variations are seen in every well-studied globular cluster, they are not found in field stars and indicate that clusters like NGC 6712 cannot provide many field stars and/or field stars do not form in environments with chemical-enrichment histories like those of NGC 6712. Preliminary analysis of NGC 5466, another tidally disrupted cluster, suggests little (if any) abundance variation for O and Na and the abundance ratios [X/Fe] are comparable to field stars at the same metallicity. Therefore, globular clusters like NGC 5466 may have been Galactic building blocks.

Keywords. Galaxy: abundances, globular clusters: individual (NGC 5466, NGC 6712), stars: abundances

---

Globular clusters are the oldest Galactic objects for which reliable ages can be obtained (e.g., Gratton *et al.* 2003) and have long been regarded to be the first bound systems to have formed in the protogalactic era (Peebles & Dicke 1968). As such, they have witnessed the subsequent formation and evolution of our Galaxy. However, were globular clusters innocent bystanders to the chaotic hierarchical assembly of our Galaxy, or were they active, albeit possibly minor, participants? After all, the vast majority of stars are believed to form in clusters (e.g., Lada & Lada 2003).

Several globular clusters are currently experiencing tidal disruption. SDSS images have revealed large tidal tails near Pal 5 (Odenkirchen *et al.* 2001) and NGC 5466 (Belokurov *et al.* 2006). Additional evidence for tidal stripping comes from flat mass functions which indicate significant depletions of low-mass stars, presumably stripped by the Galactic tidal field. [At this meeting, Balbinot *et al.* (2009) presented another globular cluster, NGC 6642, whose mass function shows evidence for tidal disruption.]

Chemical-abundance analyses provide a powerful constraint upon the fraction of field stars that may come from globular clusters and/or the types of globular clusters that may populate the field. Every well-studied Galactic globular cluster, including Pal 5 (Smith *et al.* 2002), shows large star-to-star abundance variations for the light elements from C to Al (e.g., Gratton *et al.* 2004; Yong *et al.* 2008b; Meléndez & Cohen 2009). However, such abundance variations have not yet been found in field stars.

NGC 6712 is the only cluster whose mass function decreases with decreasing mass (de Marchi *et al.* 2000). With an orbit penetrating deep into the bulge (Dinescu *et al.* 1999), tidal forces have stripped a large fraction of low-mass stars. Using the high-resolution infrared spectrograph Phoenix on the *Gemini South* telescope, we measured the abundances of C, N, O, F, Na and Fe (Yong *et al.* 2008a).

Within our sample of seven bright giants (all have radial velocities and Fe abundances consistent with cluster membership), we detect large star-to-star abundance variations for C, N, O, F and Na (see Figure 1). Such abundance patterns are not found in field stars and therefore clusters like NGC 6712 and Palomar 5 cannot provide many field stars and/or field stars do not form in environments with chemical-enrichment histories like those of NGC 6712 and Palomar 5.

NGC 6712 is only the second cluster in which F has been observed in more than two stars and both clusters show F abundance variations which may be produced in asymptotic giant branch stars with masses $M > 5$ M$_\odot$ (Smith *et al.* 2005; Karakas *et al.* 2008). Indeed, within the limited samples, the fluorine abundances in all globular cluster stars are considerably lower than in field and bulge stars at the same metallicity (see Figure 2), which highlights additional chemical differences between the field and cluster environment.

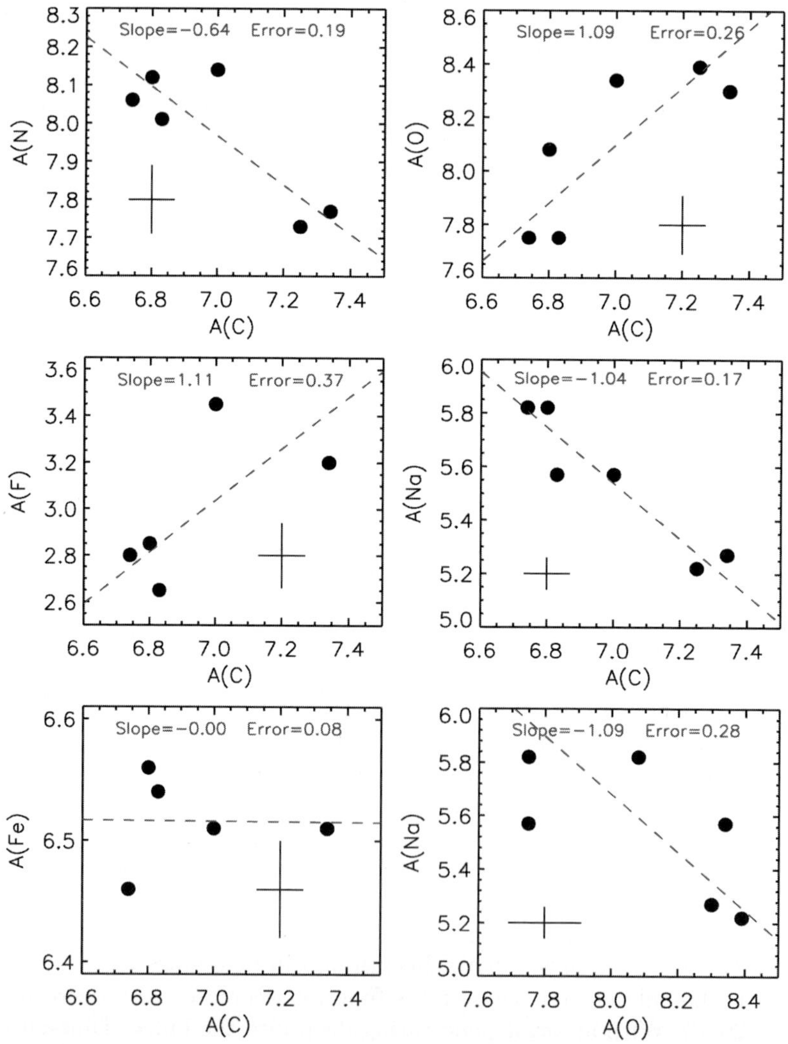

Figure 1. Elemental abundances A(X) versus A(C) as well as A(Na) versus A(O) *(bottom right panel)*. A representative error bar is shown. The dashed line is the linear least-squares fit to the data (slope and associated error are included).

The chemical abundances measured in this cluster offer the intriguing prospect of allowing us to estimate the initial mass of this cluster, and, therefore, the fraction of mass lost through tidal stripping. Carretta (2006) found a correlation between the amplitude of the abundance variations (actually, the interquartile range of, e.g., [O/Fe], [Na/Fe] and/or [O/Na] and the absolute magnitude). Based on the interquartile range of [O/Na], we tentatively confirm the calculations by Takahashi & Portegies Zwart (2000), which indicate that NGC 6712 was once one of the most massive clusters to have formed in our Galaxy.

Of great interest would be the chemical analysis of additional globular clusters that show evidence for tidal disruption to search for a cluster in which there are no light-element abundance variations. Equally important would be to expand the search for

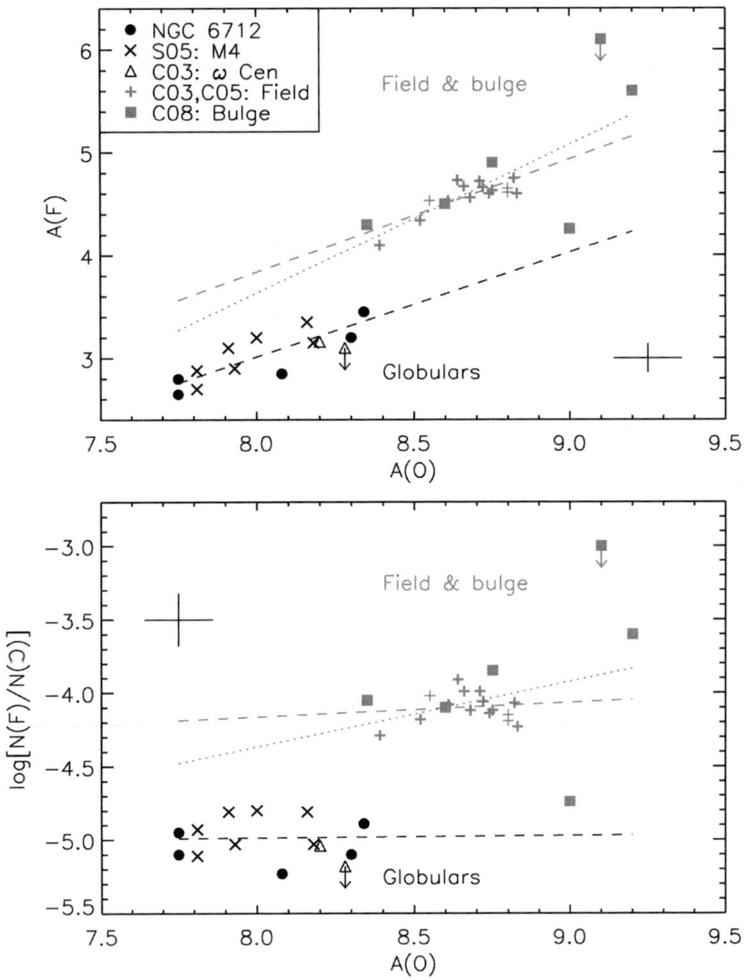

Figure 2. A(F) versus A(O) (top) and log[N(F)/N(O)] versus A(O) (bottom). NGC 6712 (black circles), M4 (black crosses: Smith *et al.* 2005), ω Cen (black triangles: Cunha *et al.* 2003), bulge stars (red squares: Cunha *et al.* 2008) and field stars (red plus signs: Cunha *et al.* 2003; Cunha & Smith 2005) are shown. The dashed lines are the linear least-squares fits to the field and bulge, and globular cluster data, respectively (excluding upper limits). The dotted line is the fit to the field and bulge data excluding the upper limits and the bulge star with A(O) = 9.0.

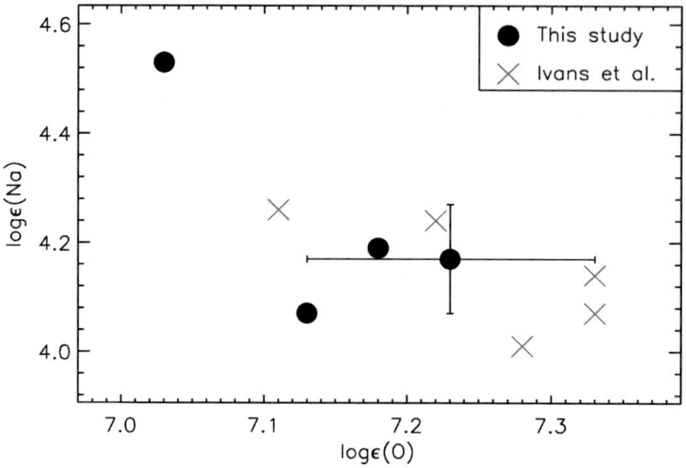

Figure 3. Preliminary abundances of Na versus O in NGC 5466.

field stars which exhibit the globular cluster light-element abundance patterns. These endeavors will enhance our understanding of the formation and evolution of our Galaxy.

To this end, we present preliminary results for NGC 5466, a tidally disrupted globular cluster. Our sample of nine stars includes five stars analysed by I. Ivans (priv. comm.). Excluding one O-rich, Na-poor star, the remaining stars have O and Na abundance measurements that may be regarded as constant within the measurement uncertainties (see Figure 3). All stars lie within the range exhibited by field stars at the cluster's metallicity. Therefore, clusters like NGC 5466 may have contributed to building the halo.

References

Belokurov, V., Evans, N. W., Irwin, M. J., Hewett, P. C., & Wilkinson, M. I. 2006, *ApJ* (Letters), 637, L29

Carretta, E. 2006, *AJ*, 131, 1766

Cunha, K. & Smith, V. V. 2005, *ApJ*, 626, 425

Cunha, K., Smith, V. V., & Gibson, B. K. 2008, *ApJ* (Letters), 679, L17

Cunha, K., Smith, V. V., Lambert, D. L., & Hinkle, K. H. 2003, *AJ*, 126, 1305

de Marchi, G., Leibundgut, B., Paresce, F., & Pulone, L. 1999, *A&A* (Letters), 343, L9

Dinescu, D. I., Girard, T. M., & van Altena, W. F. 1999, *AJ*, 117, 1792

Gratton, R. G., Bragaglia, A., Carretta, E., Clementini, G., Desidera, S., Grundahl, F., & Lucatello, S. 2003, *A&A*, 408, 529

Gratton, R., Sneden, C., & Carretta, E. 2004, *ARA&A*, 42, 385

Karakas, A. I., Lee, H. Y., Lugaro, M., Görres, J., & Wiescher, M. 2008, *ApJ*, 676, 1254

Lada, C. J. & Lada, E. A. 2003, *ARA&A*, 41, 57

Meléndez, J. & Cohen, J. G. 2009, *ApJ*, 699, 2017

Odenkirchen, M., *et al.*, 2001, *ApJ* (Letters), 548, L165

Peebles, P. J. E. & Dicke, R. H. 1968, *ApJ*, 154, 891

Smith, G. H., Sneden, C., & Kraft, R. P. 2002, *AJ*, 123, 1502

Smith, V. V., Cunha, K., Ivans, I. I., Lattanzio, J. C., Campbell, S., & Hinkle, K. H. 2005, *ApJ*, 633, 392

Takahashi, K. & Portegies Zwart, S. F. 2000, *ApJ*, 535, 759

Yong, D., Grundahl, F., Johnson, J. A., & Asplund, M. 2008, *ApJ*, 684, 1159

Yong, D., Meléndez, J., Cunha, K., Karakas, A. I., Norris, J. E., & Smith, V. V. 2008, *ApJ*, 689, 1020

Star clusters: basic galactic building blocks
Proceedings IAU Symposium No. 266, 2009
R. de Grijs & J. R. D. Lépine, eds.

doi:10.1017/S1743921309991013

Globular cluster abundances: the role of asymptotic giant branch stars

Amanda I. Karakas

Mount Stromlo Observatory, Australian National University, Weston Creek ACT 2611, Australia
email: akarakas@mso.anu.edu.au

Abstract. One of the more popular theories to account for the abundance anomalies in globular cluster stars is the 'self-pollution scenario,' where the polluters were a previous generation of intermediate-mass asymptotic giant branch (AGB) stars. This idea has proved attractive because: (i) the hot-bottom burning experienced by these objects qualitatively provides an ideal proton-capture environment to produce helium and convert C and O to N, Ne to Na and Mg to Al, and (ii) the slow winds from these stars allow their retention by the cluster's gravitational potential. New stellar yields from low-metallicity AGB models are presented and compared to abundances derived in globular clusters. We also discuss external pollution and inhomogeneous-pollution models that use AGB stars as polluters. Current models of AGB stars cannot match all observational features of globular cluster stars. However, stellar modelling uncertainties are considerable and suggest AGB stars should not be ruled out just yet.

Keywords. stars: abundances, stars: AGB and post-AGB, globular clusters: general

1. Introduction

Star-to-star abundance variations of the light elements Li, C, N, O, F, Na, Mg and Al have been observed in every well-studied Galactic globular cluster (GC) to date (Kraft 1994; Gratton *et al.* 2004 and references therein), but are not found in field stars of the same metallicity (Gratton *et al.* 2000). Hence, these abundance patterns are somehow the result of the cluster environment. The variations of the elements follow a similar pattern: C–N, O–Na and Mg–Al are anticorrelated (Shetrone 1996; Kraft *et al.* 1997; Cannon *et al.* 1998; Gratton *et al.* 2001; Cohen *et al.* 2005; Cohen & Meléndez 2005). Furthermore, Smith *et al.* (2005) noted that stars in M4 showed evidence for a Na–F anticorrelation (an F spread was also found in NGC 6712; Yong *et al.* 2008a), while Pasquini *et al.* (2005) and Bonifacio *et al.* (2007) found a Na–Li anticorrelation in NGC 6752 and 47 Tucanae, respectively. The Mg–Al anticorrelation is of particular interest, because from data for NGC 6752 (Yong *et al.* 2005), it is apparent that there is a larger spread in Al relative to Fe, on the order of ~ 1 dex or more, but only a modest spread of ~ 0.3 dex in Mg relative to Fe. A further important test for any model is determining how well it can reproduce the Mg isotopic ratios observed in GC stars (e.g., in NGC 6752; Yong *et al.* 2003), which show a positive correlation between ^{26}Mg and Al, and an anticorrelation between the dominant ^{24}Mg and Al.

The abundances of iron-peak and neutron-capture elements do not show the same star-to-star scatter nor do they vary *with* the light elements (Gratton *et al.* 2004, James *et al.* 2004; Yong *et al.* 2006, 2008b). There are exceptions noted in the literature, including the metal-rich cluster 47 Tuc (Wylie *et al.* 2006) and NGC 1851 (Yong *et al.* 2009). The other important exception is the massive cluster ω Centauri, whose age and metallicity spread, along with a rise in s-process element abundances with increasing [Fe/H], suggests that it evolved very differently from other GCs and may have an extragalactic origin (Smith *et al.* 2000).

For a typical globular cluster, the key points are that oxygen has been destroyed in some stars by up to 1 dex or more, the CNO abundances are constant within uncertainties, and the Na and Al abundances are enriched relative to field stars of the same metallicity. The Mg isotopes in the 'polluted' stars are also nonsolar, with ^{26}Mg enriched at the expense of ^{24}Mg, with no correlation between ^{25}Mg and Al. There is no evidence for large-scale variation of the neutron-capture elements.

The star-to-star abundance variations in C, N, O and Na have been observed in stars at or near the main-sequence turnoff (e.g., Gratton *et al.* 2001; Ramírez & Cohen 2003), supporting the hypothesis that the GC stars did not produce their abundances in situ. Instead, a previous generation of stars polluted the atmospheres of stars we observe today or provided part of the material from which those stars formed (Cottrell & Da Costa 1981). Because [Fe/H] is roughly constant in stars in a given GC, it has been assumed that the polluters were intermediate-mass AGB stars with initial masses between ~4 and 8 $M_{\odot}$ rather than supernovae, which produce Fe. The hot-bottom burning (HBB) experienced by these stars provides the hydrogen-burning environment (at least qualitatively) that can alter the abundances of the light elements via the CNO cycle, and the NeNa and MgAl hydrogen-burning chains. Another consequence is the production of a significant quantity of helium via the second dredge up as well as by HBB. Note that the slow winds from rapidly rotating massive stars have also been suggested as providing the hydrogen-burning environment. This hypothesis is not discussed here, but we refer to Charbonnel (these proceedings) for further details.

The mass lost via the slow winds of AGB stars could, in principle, have been retained by the cluster from which new stars may have been born (Thoul *et al.* 2002). Detailed AGB models have so far mostly failed to match the observed abundance trends (e.g., Denissenkov & Herwig 2003; Fenner *et al.* 2004; Karakas *et al.* 2006b). However, major uncertainties affect the model predictions and could leave room for an AGB solution (e.g., Ventura & D'Antona 2005; Renzini 2008).

It is the AGB solution that will be discussed further in light of new stellar models of intermediate-mass AGB stars.

2. Yields from asymptotic giant branch stars

In AGB stars with initial masses $\gtrsim 4\ M_{\odot}$, depending on metallicity, Z, the base of the convective envelope becomes hot enough to sustain proton-capture nucleosynthesis (HBB). HBB can change the surface composition because the entire envelope is exposed to the hot-burning region a few thousand times per interpulse period. One important reason that intermediate-mass AGB stars could have polluted GCs is that these stars have relatively short lifetimes ($\tau \lesssim 100$ Myr), compared to lower-mass AGB stars which have lifetimes of a few gigayears. Along with HBB, the repeated action of the third dredge up (TDU) will also change the surface composition. Following a thermal pulse, the convective envelope may move inward (in mass) to regions previously mixed by He-shell instabilities. This inward movement of the convective envelope is known as the TDU and is responsible for enriching the surface in ^{12}C and other He-burning products, as well as heavy elements produced by the *slow*-neutron-capture process.

Stellar nucleosynthesis predictions from low-metallicity, intermediate-mass AGB stars are available in the literature (e.g., Herwig 2004; Karakas & Lattanzio 2007; Ventura & D'Antona 2009). Here, we compare GC abundance trends using new AGB yields that are an update to the stellar yields presented in Karakas & Lattanzio (2007). These new

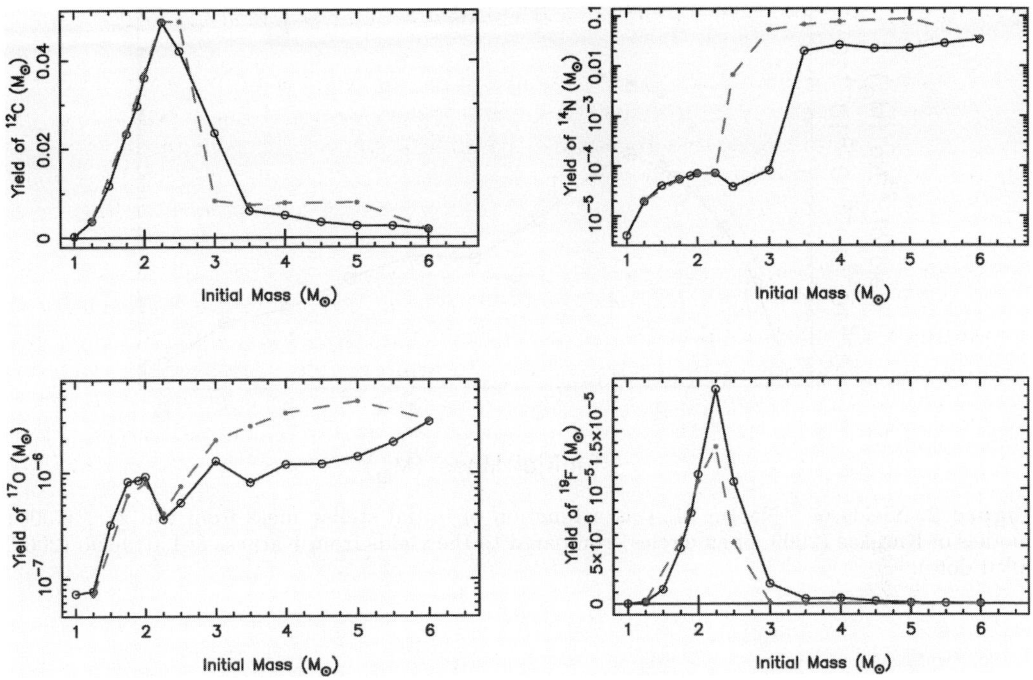

Figure 1. Yields of ^{12}C, ^{14}N, ^{17}O and ^{19}F (in $M_{\odot}$) as a function of initial stellar mass from the $Z = 0.0001$ models of Karakas (2009, open circles) compared to the yields from Karakas & Lattanzio (2007, filled dots).

yields are discussed in detail in Karakas (2009); in summary, these models use updated reaction rates and provide yields from scaled-solar initial abundances for the $Z = 0.008$ and 0.004 models.

In Figure 1 we show the yields of ^{12}C, ^{14}N, ^{17}O and ^{19}F as a function of initial stellar mass from the $Z = 0.0001$ models of Karakas (2009) compared to the yields from Karakas & Lattanzio (2007). The new yields of ^{14}N from the intermediate-mass AGB models have been reduced, while the yields of ^{12}C are reasonably consistent with the previous calculations. Even with the reductions in ^{14}N, the total CNO expelled from these AGB models, regardless of mass, is increased compared to the initial amount by at least a factor of a few. If we take the 5 $M_{\odot}, Z = 0.0001$ model as a representative example, the total CNO mass fraction at the surface increases by 2.5 dex during the AGB phase. This increase is driven by TDU mixing episodes, that mix primary ^{12}C from the He-burning shell to the envelope. Once there, HBB converts the ^{12}C to ^{14}N. It should be obvious that one of the important observational constraints found in GC stars has already been violated. Namely, if these clusters were polluted by a previous generation of $Z = 0.0001$ AGB stars as modelled here, the CNO content of the most 'polluted' present-day stars (i.e., O depleted and Na enriched) should be higher than the stars with a 'normal' halo-like composition.

In Figure 2 we show the yields of ^{23}Na from the $Z = 0.0001$ models of Karakas (2009) compared to the yields from Karakas & Lattanzio (2007). Here we can see a huge reduction in the sodium yields, by a few orders of magnitude for the most massive models. There are two main reasons for this. First, the NeNa and MgAl chain proton-capture rates have been updated, and in particular the ^{23}Na(p, γ)^{24}Mg and ^{23}Na(p, α)^{20}Ne rates from Hale *et al.* (2004) are faster than the rates we used previously (El Eid & Champagne

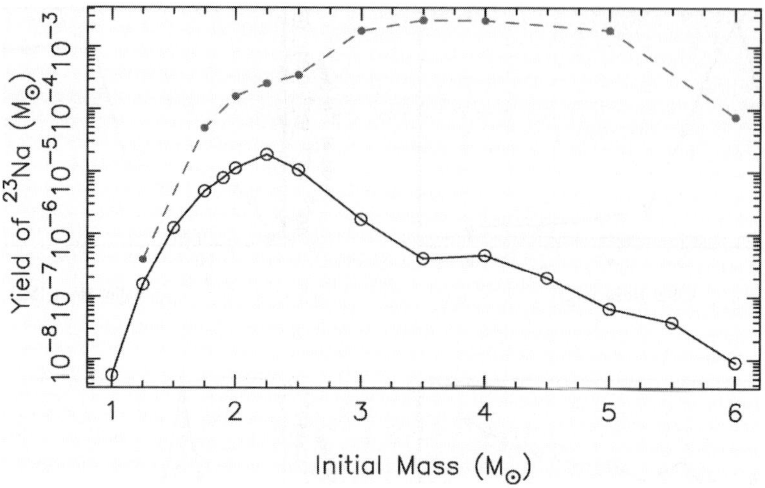

Figure 2. Yields of ^{23}Na (in $M_{\odot}$) as a function of initial stellar mass from the $Z = 0.0001$ models of Karakas (2009, open circles) compared to the yields from Karakas & Lattanzio (2007, filled dots).

1994). Second, we have computed new intermediate-mass $Z = 0.0001$ models with increased mass loss. These new models experience fewer thermal pulses and produce lower yields. It should be pointed out that the rates alone account for at least an order of magnitude decrease in sodium, and this can be seen at 2.5 $M_{\odot}$ where we are using the same stellar model as in Karakas & Lattanzio (2007).

The self-consistent chemical evolution model of NGC 6752 described in Fenner *et al.* (2004) used AGB yields computed using an initial composition determined by a burst of Population III supernovae. In this model, the general trend of the O–Na anticorrelation could be reproduced, however the AGB models produced too much Na, and none of the models destroyed enough O, except in the most massive 6.5 $M_{\odot}$ case. The Mg–Al anticorrelation was not reproduced at all, because Mg and Al were produced together in the AGB models. Furthermore, both neutron-rich Mg isotopes, ^{25}Mg and ^{26}Mg, are synthesized in the He-burning shells of AGB stars (e.g., Karakas *et al.* 2006a). For this reason, a correlation between these isotopes and Al was found, contrary to observations. The updated models of Karakas (2009) would not change these findings, because while less sodium was synthesized, the same trends between Mg and Al were observed. In summary, these AGB models cannot explain the abundance patterns observed in GC stars. First, the TDU mixes material from the He-burning shell to the envelope, resulting in increases in CNO, and Mg and Al. The TDU is also responsible for the large increases in Na: dredge up mixes (mostly primary) ^{22}Ne to the envelope and this is converted by proton capture to ^{23}Na through HBB. Second, the amount of O destruction required (1 dex or more) is not found, except in the most massive AGB models.

Intermediate-mass AGB stars are subject to serious modelling uncertainties, in particular as regards the convection model used, the treatment of convective boundaries, and the mass-loss rate used during the AGB phase. Nuclear-reaction-rate uncertainties also strongly impact some isotopes (e.g., ^{23}Na and ^{27}Al; see also Izzard et al 2007). In the next section, we examine other AGB model calculations in the literature. Can these do any better at reproducing the GC abundance patterns?

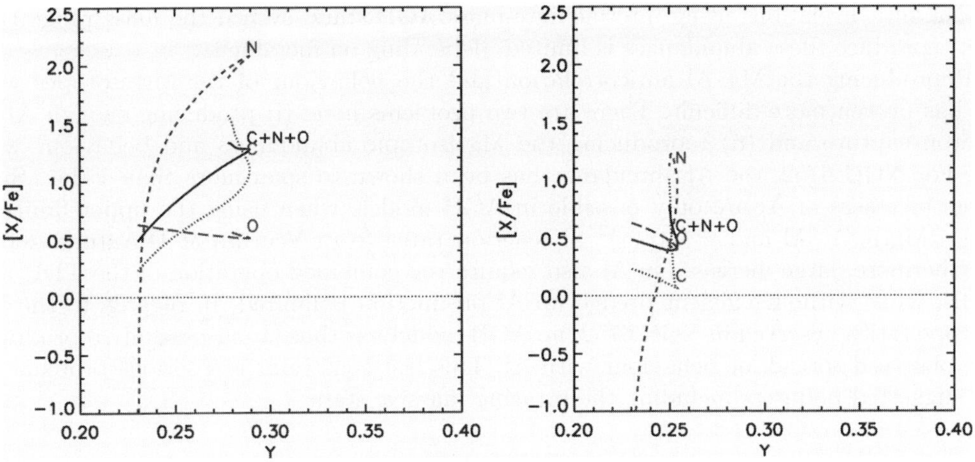

Figure 3. Temporal evolution of the CNO abundance ([CNO/Fe]) in the cluster gas as a function of helium, Y, assuming a standard Salpeter initial mass function. (*left*) Results using yields from Fenner *et al.* (2004) and (*right*) yields from Ventura *et al.* (2002). We use the standard spectroscopic notation $[X/Fe] = \log(X/Fe) - \log(X/Fe)_{\odot}$. (From Karakas *et al.* 2006b.)

3. Other AGB models in the literature

Here we consider the intermediate-mass AGB models published by Ventura & D'Antona (Ventura & D'Antona 2009; and references therein). These stellar interior models are distinct from most other models because they use the 'full spectrum of turbulence' (FST) convection model instead of the mixing-length theory (MLT) of convection. In Ventura & D'Antona (2005), the authors outlined the main differences between AGB models computed with FST versus the MLT. FST results in more efficient envelope convection, hence higher temperatures at the base of the convective envelope and consequently higher luminosities. Furthermore, the TDU is not predicted to be efficient. For this reason, little He-shell material is mixed into the envelope and this allows CNO to stay approximately constant, except at the lowest masses (≈ 3 $M_{\odot}$). Indeed, both C and O are destroyed in their most massive AGB models, contrary to the models of Karakas & Lattanzio (2007) which produce both of these elements at the lowest metallicities ($Z = 0.0001$) as a result of efficient dredge up. However the lack of dredge up in the Ventura & D'Antona models limits the amount of Na and Al that can be produced, so these authors have included a small amount of convective overshoot (e.g., Ventura & D'Antona 2008).

In Figure 3 we show the results from Karakas *et al.* (2006b) of a chemical evolution model that uses the stellar yields from Ventura *et al.* (2002) in the right panel, compared to the yields from Fenner *et al.* (2004) in the left panel. In each panel, we show the evolution of helium on the x axis versus C, N and O on the y axis. The combined CNO evolution is also shown. The main point of this figure is to show that the yields from Ventura *et al.* (2002) keep the CNO abundance approximately constant, to within 0.4 dex. In contrast, using the yields from Fenner *et al.* (2004), the CNO abundance increases by ~ 0.8 dex, well outside of observational uncertainties.

Owing to a higher efficiency of convection, the models of Ventura & D'Antona are able to destroy more oxygen and, because there is less He-burning material, produce much less sodium (indeed, producing too little sodium has been a problem for these authors). However, overall it has been shown to be possible to reproduce the C–N and

O–Na anticorrelations using intermediate-mass AGB stars, even if the mass range that can reproduce these abundances is limited, depending on metallicity.

Reproducing the Mg–Al anticorrelation and the behaviour of the Mg isotopes with Al has proven more difficult. There are two problems here: (i) producing enough Al by proton capture and (ii) reproducing the Mg isotopic abundances and behaviour with Al. For NGC 6752, the Al abundance has been shown to span more than 1 dex. Such large increases of Al are only possible in AGB models when using the upper limits of the ^{26}Mg(p,γ)^{27}Al and ^{26}Al(p,γ)^{27}Si reaction rates (e.g., Ventura & D'Antona 2008). Furthermore, large increases in Al also require the combined operation of the TDU and HBB, while without efficient dredge up Al production is limited. In regards to the Mg isotope ratios observed in NGC 6752, no AGB model has thus far successfully reproduced the observed spread, or behaviour with Al. This last constraint is a serious problem for *all* suggested polluters including the rotating massive stars.

4. External pollution and inhomogeneous-pollution models

Bekki & Norris (2006) and Bekki *et al.* (2007) explored the idea that the abundance properties of the Galactic GCs were formed by 'external pollution.' In this framework, GCs are assumed to form in low-mass dwarf galaxies embedded in dark-matter subhaloes at very high redshifts. Field AGB stars of the dwarf galaxies then chemically influence the early GC-formation processes. GCs within a dwarf galaxy therefore can be formed from the mixture of (i) gas expelled from the field AGB stars formed earlier in the dwarf and (ii) the interstellar gas falling into the central region of the dwarf. The idea of external pollution is most attractive because it naturally provides an explanation for the large fraction of polluted stars without recourse to an implausible initial mass function. For example, for the self-pollution hypothesis to explain the large enhancements of helium ($Y \lesssim 0.4$ in mass fraction) postulated in massive clusters such as ω Cen and NGC 2808, an unusual initial mass function had been proposed by D'Antona & Caloi (2004). Karakas *et al.* (2006) explored this idea further and found that a factor of 10 more intermediate-mass AGB stars is required to obtain $Y \gtrsim 0.36$, depending on the fraction of primordial gas that is allowed to mix with the AGB ejecta.

Recently, Marcolini *et al.* (2009) explored the idea of an *inhomogeneous-pollution* chemical evolution model. This model depends on the existence of (i) a peculiar pre-enrichment phase in the GC's parent galaxy associated with very-low-metallicity Type II supernovae (SNe) and (ii) localized inhomogeneous enrichment from a single Type Ia SN and intermediate-mass AGB field stars. GC formation is assumed to take place within this chemically inhomogeneous region. The first low-mass stars to form are those with a 'polluted' composition (i.e., O depleted and Na enhanced), while 'normal' stars (i.e., O rich and Na depleted) are formed in a second stage when self-pollution from supernovae occurs and the inhomogeneous pollution from the previous phase is dispersed. This model naturally accounts for the low [O/Fe] values of the most polluted stars, this is because such low [O/Fe] values arise in the inhomogeneous region polluted by the initial SN Ia, which produces much more Fe compared to O. Note that the model is also able to maintain Fe approximately constant and to account for the C–N, O–Na and Mg–Al anticorrelations. One serious drawback that the model suffers is that the AGB yields (taken from Karakas & Lattanzio 2007) still need to be adjusted to fit the data. The adjustments required include decreases in the yields of C and large increases in Al. From the discussion of the Ventura & D'Antona models it is clear that such large adjustments are within the scope allowed by stellar modelling uncertainties.

5. Summary

It has been postulated that low-metallicity, intermediate-mass AGB stars polluted the Galactic GCs while they were forming. This idea has proven attractive because these stars release slow winds that can be retained by the shallow potential well of GCs, they do not produce Fe and the HBB experienced by these stars provides the ideal hydrogen-burning environment (at least qualitatively) that can alter the abundances of the light elements.

We have explored this idea using the latest AGB models available in the literature. We have compared the models to the observed abundance trends and conclude that AGB models cannot, in general, account for all observed chemical features of GC stars. The latest models from Karakas (2009) suffer from the same problem as discussed by Fenner *et al.* (2004), that is, material from the He-burning region increases the total CNO content of the AGB ejecta. Also, these models cannot destroy enough oxygen to satisfy the observational requirements. The AGB models from Ventura & D'Antona provide a better match to most of the observed abundance trends, but still struggle to account for the large increases in Al and cannot reproduce the Mg isotopic abundances observed in NGC 6752.

However, major uncertainties affect the model predictions and could leave room for an AGB solution. Renzini (2008) argued convincingly that AGB stars are still the best candidates as polluters. He suggested major modifications to the AGB models including rapid ejection of the envelope following the second dredge up, which would allow little time for HBB and the TDU. Can these proposed modifications be tested against observations of the most luminous, oxygen-rich AGB stars in the Large and Small Magellanic Clouds (LMC and SMC, respectively)? The mean metallicity of the SMC is approximately the same as that of the stars in 47 Tuc ([Fe/H] ~ -0.7). If a generation of intermediate-mass AGB stars polluted 47 Tuc, then these polluters would have *looked* like the most massive, luminous AGB stars in the SMC. Observations of these stars suggest they are rich in Li and s-process elements (García–Hernández *et al.* 2009; Smith & Lambert 1989). This evidence indicates that these luminous SMC giants have experienced HBB and the TDU. Given that neutron-capture elements are not correlated with light elements in most GCs, can these observations be used to rule out AGB stars as polluters?

References

Bekki, K. & Norris, J. E. 2006, *ApJ* (Letters), 637, L109

Bekki, K., Campbell, S. W., Lattanzio, J. C., & Norris, J. E. 2007, *MNRAS*, 377, 335

Bonifacio, P., Pasquini, L., Molaro, P., Carretta, E., François, P., Gratton, R. G., James, G., Sbordone, L., Spite, F., & Zoccali, M. 2007, *A&A*, 470, 153

Cannon, R. D., Croke, B. F. W., Bell, R. A., Hesser, J. E., & Stathakis, R. A. 1998, *MNRAS*, 298, 601

Cohen, J. G., Briley, M. M., & Stetson, P. B. 2005, *AJ*, 130, 1177

Cohen, J. G. & Meléndez, J. 2005, *AJ*, 129, 303

Cottrell, P. L. & Da Costa, G. S. 1981, *ApJ* (Letters), 245, L79

D'Antona, F. & Caloi, V. 2004, *ApJ*, 611, 871

Denissenkov, P. A. & Herwig, F. 2003, *ApJ* (Letters), 590, L99

El Eid M. F. & Champagne A. E. 1995, *ApJ*, 451, 298

Fenner, Y., Campbell, S., Karakas, A. I., Lattanzio, J. C., & Gibson, B. K. 2004, *MNRAS*, 353, 789

García–Hernández, D. A., Manchado, A., Lambert, D. L., Plez, B., García–Lario, P., D'Antona, F. Lugaro, M., Karakas, A. I., & van Raai, M. 2009, *ApJ* (Letters), 705, L31

Gratton, R. G., Sneden, C., & Carretta, E. 2004, *ARA&A*, 42, 385

Gratton, R. G., *et al.* 2001, *A&A*, 369, 87
Gratton, R. G., Sneden, C., Carretta, E., & Bragaglia, A. 2000, *A&A*, 354, 169
Hale, S. E., Champagne, A. E., Iliadis, C., Hansper, V. Y., Powell, D. C., & Blackmon, J. C. 2004, Phys. Rev. C, 70, 045802
Herwig, F. 2004, *ApJ*, 605, 425
Izzard, R. G., Lugaro, M., Karakas, A. I., Iliadis, C., & van Raai, M. 2007, *A&A*, 466, 641
James, G. *et al.* 2004, *A&A*, 414, 1071
Karakas, A. I., Lugaro, M., Wiescher, M., Goerres, J., & Ugalde, C., 2006, *ApJ*, 643, 471
Karakas, A. I., Fenner, Y., Sills, A., Campbell, S. W., & Lattanzio, J. C. 2006, *ApJ*, 652, 1240
Karakas, A. I. & Lattanzio, J. C. 2007, *PASA*, 24, 103
Karakas, A. I. 2009, *MNRAS*, submitted
Kraft, R. P. 1994, *PASP*, 106, 553
Kraft, R. P., Sneden, C., Smith, G. H., Shetrone, M. D., Langer, G. E., & Pilachowski, C. A. 1997, *AJ*, 113, 279
Marcolini, A., Gibson, B. K., Karakas, A. I., & Sánchez-Blázquez, P. 2009, *MNRAS*, 395, 719
Pasquini, L., Bonifacio, P., Molaro, P., François, P., Spite, F., Gratton, R. G., Carretta, E., & Wolff, B. 2005, *A&A*, 441, 549
Renzini, A. 2008, *MNRAS*, 391, 354
Ramírez, S. V. & Cohen, J. G. 2003, *AJ*, 125, 224
Shetrone, M. D. 1996, *AJ* 112, 1517
Smith, V. V. & Lambert, D. L. 1989, *ApJ* (Letters), 345, L75
Smith, V. V., Suntzeff, N. B., Cunha, K., Gallino, R., Busso, M., Lambert, D. L., & Straniero, O. 2000, *AJ*, 119, 1239
Smith, V. V., Cunha, K., Ivans, I. I., Lattanzio, J. C., Campbell, S., & Hinkle, K. H. 2005, *ApJ*, 633, 392
Thoul, A., Jorissen, A., Goriely, S., Jehin, E., Magain, P., Noels, A., & Parmentier, G. 2002, *A&A*, 383, 491
Ventura, P. & D'Antona, F. 2005, *A&A*, 431, 279
Ventura, P. & D'Antona, F. 2008, *A&A*, 479, 805
Ventura, P. & D'Antona, F. 2009, *A&A*, 499, 835
Ventura, P., D'Antona, F., & Mazzitelli, I. 2001, *A&A*, 393, 215
Wylie, E. C., Cottrell, P. L., Sneden, C. A., & Lattanzio, J. C. 2006, *ApJ*, 649, 248
Yong, D., Grundahl, F., Lambert, D. L., Nissen, P. E., & Shetrone, M. D. 2003, *A&A*, 402, 985
Yong, D., Grundahl, F., Nissen, P. E., Jensen, H. R., & Lambert, D. L. 2005, *A&A*, 438, 875
Yong, D., Aoki, W., Lambert, D. L., & Paulson, D. B. 2006, *ApJ*, 639, 918
Yong, D., Meléndez, J., Cunha, K., Karakas, A. I., Norris, J. E., & Smith, V. V. 2008, *ApJ*, 689, 1020
Yong, D., Karakas, A. I., Lambert, D. L., Chieffi, A., & Limongi, M. 2008, *ApJ*, 689, 1031
Yong, D. Grundahl, F., D'Antona, F., Karakas, A. I., Lattanzio, J. C., & Norris, J. E. 2009, *ApJ* (Letters), 695, L62

Star clusters: basic galactic building blocks
Proceedings IAU Symposium No. 266, 2009
R. de Grijs & J. R. D. Lépine, eds.

doi:10.1017/S1743921309991025

Massive binaries and the enrichment of the interstellar medium in globular clusters

S. E. de Mink,[1] O. R. Pols,[1] N. Langer[2,1] and R. G. Izzard[3]

[1]Astronomical Institute Utrecht, Postbus 80000, 3508 TA Utrecht, The Netherlands

[2]Argelander-Institut für Astronomie, Auf dem Hügel 71, 53121 Bonn, Germany

[3]Université Libre de Bruxelles, Boulevard du Triomphe, B-1050 Brussels, Belgium
email: S.E.deMink@uu.nl

Abstract. Abundance anomalies observed in globular cluster stars indicate pollution with material processed by hydrogen burning. Two main sources have been suggested: asymptotic giant branch (AGB) stars and massive stars rotating near the break-up limit (spin stars). We discuss the idea that massive binaries may provide an interesting alternative source of processed material. We discuss observational evidence for mass shedding from interacting binaries. In contrast to the fast, radiatively driven winds of massive stars, this material is typically ejected with low velocity. We expect that it remains inside the potential well of a globular cluster and becomes available for the formation or pollution of a second generation of stars. We estimate that the amount of processed low-velocity material that can be ejected by massive binaries is larger than the contribution of the two previously suggested sources combined.

Keywords. globular clusters, binaries: close, stars: abundances, ISM: jets and outflows

1. Introduction

For a long time, star clusters have been considered idealized, single-age, chemically homogeneous stellar populations. However, it has recently become clear that some clusters show multiple main sequences and subgiant branches, and extended horizontal branches (e.g., Piotto *et al.* 2007). The presence of these features in some intermediate-age clusters may be explained by the effects of stellar rotation on the location of stars in the colour–magnitude diagram (Bastian & de Mink 2009). However, this provides no explanation for the features seen in old globular clusters, which seems to imply the existence of multiple populations within one cluster.

In addition, large star-to-star abundance variations are found for light elements such as C, N, O, Na and Al, while the composition of heavier elements (Fe-group and α elements) seems to be constant. Field stars with the same metallicity do not exhibit these abundance patterns (for a review see Gratton *et al.* 2004). These chemical variations have been interpreted as originating from the presence of both a 'normal' stellar population, exhibiting abundances similar to field stars of the same metallicity, and a second population of stars formed from material processed by hydrogen burning via the CNO cycle and the NeNa and MgAl chains (e.g., Prantzos *et al.* 2007). According to Carretta *et al.* (2009), 50–70% of the stars in gloular clusters belong to the second population.

Two sources of processed ejecta have been proposed: the slow winds of *massive AGB stars*, which enrich their convective envelopes with H-burning products (Ventura *et al.* 2001; Dantona *et al.* 2002; Denissenkov & Herwig 2003) and rapidly rotating massive stars (which we refer to as *spin stars*), which are believed to expel processed material centrifugally when they reach break-up rotation (Prantzos & Charbonnel 2006; Decressin

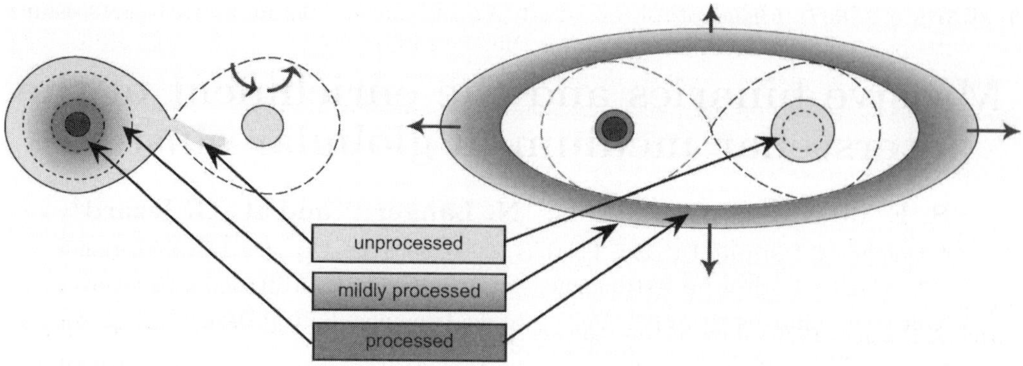

Figure 1. Cartoon representation of the proposed scenario. *(left)* A binary system at the onset of mass transfer. The deepest layers in the donor star have been processed by proton-capture reactions. The accreting star spins up as it accretes mass and angular momentum until it approaches the break-up limit. *(right)* The same system after the donor star has been stripped of its envelope. The companion star accreted just a fraction of the transferred mass, mainly unprocessed material originating from the outermost layers of the donor star. Material orginating from deeper layers in the donor star is shed into a circumbinary disk.

et al. 2007a). In this scenario a first generation of stars is formed from pristine material. Their low-velocity ejecta are trapped inside the potential well of the cluster and provide the material for the formation of a second generation of stars.†

Although both proposed sources are promising, matching the observed abundance patterns and providing enough ejecta for the formation of a second generation which outnumbers the first have proven to be two major challenges. Here, we propose *massive binaries* as a candidate for the internal pollution of globular clusters.

2. Binaries as sources of enrichment

Interacting binaries can shed large amounts of material processed by hydrogen burning into their surroundings. A clear example is the well-studied system RY Scuti. It is undergoing rapid mass transfer from a 7 $M_\odot$ supergiant to its 30 $M_\odot$ companion. Mass is lost from the system via the outer Lagrangian points into a circumbinary disk and a wider double-toroidal nebula. The nebula shows signatures of CNO processing: it is enriched in helium and nitrogen and depleted in oxygen and carbon (Smith *et al.* 2002; Grundstrom *et al.* 2007). In contrast to the high-velocity radiatively driven winds of massive stars, these ejecta have low velocities. Smith *et al.* (2001) measure expansion velocities ranging from 30 to 70 km s^{-1} in the nebula of RY Scuti, which are smaller than the present-day escape velocity of massive globular clusters. Furthermore, the nebula shows evidence for clumps (Smith *et al.* 2002) and dust in the outer parts (Gehrz *et al.* 2001), which may serve as seeds for the formation of a second generation of low-mass protostars.

Evidence for severe mass loss from interacting binaries comes from a wide variety of observed interacting and post-interaction systems: it appears to be a common phenomenon for many interacting binaries. Various authors have inferred highly nonconservative evolution for Algols, systems which are currently undergoing stable mass transfer (e.g., Refsdal *et al.* 1974; de Greve & Linnell 1994; Figueiredo *et al.* 1994; van Rensbergen *et al.* 2006;

† Glebbeek *et al.* (2009) suggested that a chain of multiple stellar collisions in the dense centre of a star cluster may also enrich the interstellar medium with processed material.

de Mink *et al.* 2007). Most notable are short-period binaries containing a compact object, e.g., cataclysmic variables, X-ray binaries, binary radio pulsars and double white-dwarf systems. Their formation requires a phase of severe mass and angular-momentum loss through ejection of a common envelope. Direct evidence for this type of evolution comes from planetary nebulae with close binary nuclei, which appear to have recently emerged from the common-envelope phase (for a review see Iben & Livio 1993).

Theoretical considerations support the idea that most interacting binaries shed large amounts of mass. Three-dimensional hydrodynamical simulations of the mass-transfer stream and accretion disk of the interacting binary β Lyrae predict that 50% of the transferred mass is lost (Bisikalo *et al.* 2000; Nazarenko & Glazunova 2006). In addition, Ulrich & Burger (1976) showed that the accreting star is driven out of thermal equilibrium and expands. This can lead to contact and strong mass and angular-momentum loss from the system (Flannery & Ulrich 1977). Furthermore, Packet (1981) noted that the accreting star reaches break-up rotation after gaining only a few percent of its own mass. Rapid rotation is found for many accreting stars in Algols (Barai *et al.* 2004) and this mechanism has been proposed to explain the formation of Be X-ray binaries (e.g., Pols *et al.* 1991). In principle, tides can counteract the effect of spin-up through mass transfer in close binaries. Petrovic *et al.* (2005) computed detailed binary evolution models, taking into account these effects. They find that massive binaries with initial periods as short as 3–6 days lose 70–80% of the transferred mass, on average. For wider and more massive systems, they expect even less conservative mass transfer, such that nearly the entire envelope of the primary is returned to the interstellar medium.

3. Composition of the ejecta

In de Mink *et al.* (2009b) we present the computations of the evolution of a 20 $M_\odot$ star in a close binary, taking into account the effects of nonconservative mass and angular-momentum transfer, rotation and tidal interaction. We find that this system sheds about 10 $M_\odot$ of material, nearly the entire envelope of the primary star. The ejecta are enriched in He, N, Na and Al, and depleted in C and O, similar to the abundance patterns observed in gobular cluster stars. However, Mg is not significantly depleted in the ejecta of this model. For a more detailed discussion, we refer to de Mink *et al.* (2009b).

4. Mass budget

One of the main challenges for the two previously proposed sources of pollution, massive AGB and spin stars, is to provide the large amount of ejecta needed to create a second population which is larger than the first. The population of low-mass stars (0.1–0.8 $M_\odot$), which can still be observed today, represents 38% of the stellar mass initially present in the cluster, assuming a standard Kroupa (2001) initial mass function (IMF) between 0.1 and 120 $M_\odot$(see Figure 2). The ejecta of AGB stars with initial masses between 4 and 9 $M_\odot$ represent up to 8.9% of the initial stellar mass (assuming an initial–final-mass relation as given by Ciotti *et al.* 1991). For spin stars, this fraction is 3.4% if one assumes that every massive star is single and born with a rotational velocity high enough to reach break-up rotation (using the models of Decressin *et al.* 2007a). These ejecta are not sufficient to create a second generation which is equally numerous as the first, even when we assume that the second generation consists only of low-mass stars and that star formation is very efficient (see Figure 2).

Two rather extreme solutions have been proposed.

(1) The IMF was highly anomalous, favouring the formation of the polluting stars with respect to the long-lived low-mass stars that we observe today. Even though we have no direct constraints on the IMF of globular clusters, Kroupa (2002) finds that the IMF is remarkably uniform in stellar populations with very different properties.
(2) Clusters were initially at least 10–20 times more massive and they have preferentially lost low-mass stars from the first generation as a result of dynamical evolution and tidal stripping (Decressin *et al.* 2007b, 2008; D'Ercole *et al.* 2008). In this section we investigate to what extent the ejecta of massive binaries can alleviate this conundrum.

Even though the current fraction of detected binaries in globular clusters is not high (e.g., Davies *et al.* 2008), this is not necessarily the case for the high-mass stars originally present in the cluster. Sana *et al.* (2008) and Mason *et al.* (2009) find a minimum binary fraction of 60–75% for O stars associated with clusters or OB associations. In globular clusters, these fractions may even be higher. In this environment, close binaries can be created and tightened during and after the star-formation process, for example by the dissipative interaction with gas (e.g., Bonnell & Bate 2005) and by three-body interactions such as the Kozai mechanism in combination with tidal friction (Fabrycky *et al.* 2007). With massive stars preferentially residing in the dense core of the cluster, where the dynamical encounters are most frequent, it is not unreasonable to assume that the large majority of massive stars interact through mass exchange.

Let us assume that every massive star is a member of an interacting binary. In Section 2 we argued that nearly the entire envelope of the donor is returned to the interstellar medium. For simplicity, we neglect the contribution of the secondary star after it has been spun up by mass transfer or during a possible phase of reverse mass transfer, and we assume that entire envelope of the primary becomes available for star formation. We assume helium core masses as in Prantzos & Charbonnel (2006) for stars more massive than 10 $M_{\odot}$. Under these assumptions, the slow ejecta of massive binaries represent 13%

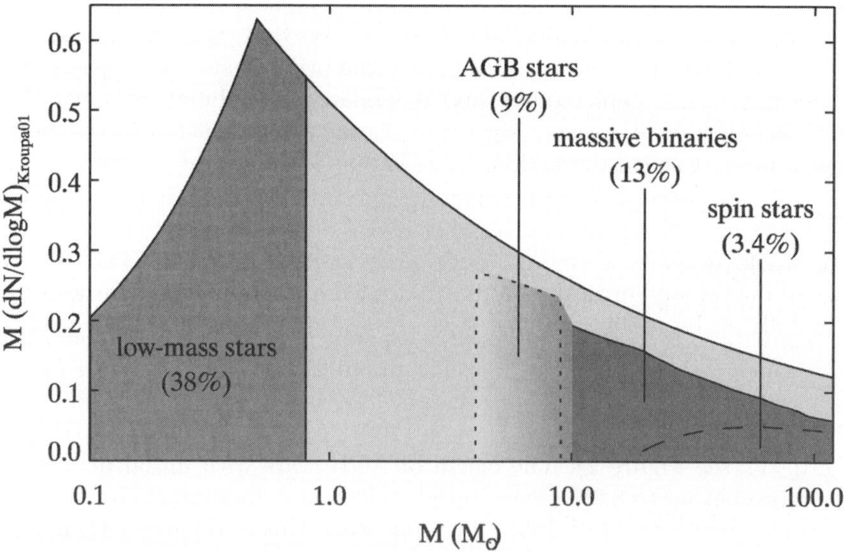

Figure 2. Mass weighted Kroupa (2001) IMF as a function of stellar mass. The surface areas indicate the mass contained in the first generation of long-lived, low-mass stars (dark blue), the ejecta of AGB stars (dotted line), spin stars, i.e., rapidly rotating massive stars (dashed line) and massive (red) and intermediate-mass (orange) binaries. Percentages indicate the fraction of mass relative to the total mass contained in stars of the first generation. See Section 4 for details. (Figure adapted from de Mink *et al.* 2009b.)

of the mass originally present in stars, i.e., more than the ejecta of AGB and spin stars combined.

Measurements of lithium suggest that the ejecta of the first generation are diluted with pristine gas (Pasquini *et al.* 2005). Together with an equal amount of pristine gas, the ejecta of binaries with donors more massive than 10 $M_\odot$ would represent 26% of the initial cluster mass (compared to 38% contained in the first generation of low-mass stars). The adopted lower-mass limit for our binary scenario is rather arbitrary. If we take into account the potential contribution of intermediate-mass stars (4–10 $M_\odot$), the ejecta would be sufficient to form a second population of chemically peculiar stars that outnumbers the first generation of normal stars. The assumptions in this scenario can be relaxed if the evaporation of stars from the cluster preferentially affects the first stellar generation, as suggested by Decressin *et al.* (2008).

5. Conclusions

We discussed the potential of massive binaries as the source of enrichment in globular clusters. The majority of massive stars are expected to be members of interacting binary systems. These return most of the envelope of their primary star to the interstellar medium during nonconservative mass transfer. We show that the amount of polluted material ejected by binaries may be larger than that of the two previously suggested sources, massive AGB stars and the slow winds of rapidly rotating massive stars. After dilution with pristine material, as lithium observations suggest, binaries could return enough material for the formation of a chemically enriched second generation that is equally numerous as the first generation of low-mass stars, without the need to assume a highly anomalous IMF, external polution of the cluster or significant loss of stars from the nonenriched first generation.

In addition to providing a new source of slowly ejected enriched material, binary interactions also affect the previously proposed scenarios. Binary mass transfer naturally produces a large number of rapidly rotating massive stars which may enrich their surroundings even further. Binary interaction will also affect the yields of intermediate-mass stars. Premature ejection of the envelope in 4–9 $M_\odot$ stars will result in ejecta with less pronounced anticorrelations as suggested in the AGB scenario. On the other hand, we expect that binary-induced mass loss may also prevent the dredge up of helium-burning products.

For a detailed comparison of the chemical predictions of this scenario, binary models for a range of masses and orbital periods are needed and population synthesis models are essential to fully evaluate the mass budget of the different sources. Some peculiar features, such as the apparant presence of distinct, chemically homogeneous subpopulations in ω Cen and NGC 2808 (e.g., Renzini 2008) deserve further attention.

References

Barai, P., *et al.* 2004, *ApJ*, 608, 989

Bastian, N. & de Mink, S. E. 2009, *MNRAS* (Letters), 398, L11

Bisikalo, D. V., Harmanec, P., Boyarchuk, A. A., Kuznetsov, O. A., & Hadrava, P. 2000, *A&A*, 353, 1009

Bonnell, I. A. & Bate, M. R. 2005, *MNRAS*, 362, 915

Carretta, E., Bragaglia, A., Gratton, R. G., *et al.* 2009, *A&A*, 505, 117

Ciotti, L., D'Ercole, A., Pellegrini, S., & Renzini, A. 1991, *ApJ*, 376, 380

D'Antona, F., Caloi, V., Montalbán, J., Ventura, P., & Gratton, R. 2002, *A&A*, 395, 69

Davis, D. S., Richer, H. B., Anderson, J., Brewer, J., Hurley, J., Kalirai, J. S., Rich, R. M., & Stetson, P. B. 2008, *AJ*, 135, 2155
de Greve, J. P., & Linnell, A. P. 1994, *A&A*, 291, 786
de Mink, S. E., Pols, O. R., & Hilditch, R. W. 2007, *A&A*, 467, 1181
de Mink, S. E., Pols, O. R., Langer, N., & Izzard, R. 2009a, *A&A*, in press
de Mink, S. E., Cantiello, M., Langer, N., Pols, O. R., Brott, I., & Yoon, S.-C. 2009b, *A&A*, 497, 243
Decressin, T., Baumgardt, H., & Kroupa, P. 2008, *A&A*, 492, 101
Decressin, T., Charbonnel, C., & Meynet, G. 2007a, *A&A*, 475, 859
Decressin, T., Meynet, G., Charbonnel, C., Prantzos, N., & Ekström, S. 2007b, *A&A*, 464, 1029
Denissenkov, P. A. & Herwig, F. 2003, *ApJ* (Letters), 590, L99
D'Ercole, A., Vesperini, E., D'Antona, F., McMillan, S. L. W., & Recchi, S. 2008, *MNRAS*, 391, 825
Fabrycky, D. & Tremaine, S. 2007, *ApJ*, 669, 1298
Figueiredo, J., de Greve, J. P., & Hilditch, R. W. 1994, *A&A*, 283, 144
Flannery, B. P. & Ulrich, R. K. 1977, *ApJ*, 212, 533
Gehrz, R. D., Smith, N., Jones, B., Puetter, R., & Yahil, A. 2001, *ApJ*, 559, 395
Glebbeek, E., Gaburov, E., de Mink, S. E., Pols, O. R., & Portegies Zwart, S. F. 2009, *A&A*, 497, 255
Gratton, R., Sneden, C., & Carretta, E. 2004, *ARA&A*, 42, 385
Grundstrom, E. D., Gies, D. R., Hillwig, T. C., McSwain, M. V., Smith, N., Gehrz, R. D., Stahl, O., & Kaufer, A. 2007, *ApJ*, 667, 505
Iben, I. J. & Livio, M. 1993, *PASP*, 105, 1373
Kroupa, P. 2001, *MNRAS*, 322, 231
Kroupa, P. 2002, *Science*, 295, 82
Mason, B. D., Hartkopf, W. I., Gies, D. R., Henry, T. J., & Helsel, J. W. 2009, *AJ*, 137, 3358
Nazarenko, V. V. & Glazunova, L. V. 2006, *Ast. Rep.*, 50, 380
Packet, W. 1981, *A&A*, 102, 17
Pasquini, L., Bonifacio, P., Molaro, P., Francois, P., Spite, F., Gratton, R. G., Carretta, E., & Wolff, B. 2005, *A&A*, 441, 549
Petrovic, J., Langer, N., & van der Hucht, K. A. 2005, *A&A*, 435, 1013
Piotto, G., Bedin, L. R., Anderson, J., King, I. R., Cassisi, S., Milone, A. P., Villanova, S., Pietrinferni, A., & Renzini, A. 2007, *ApJ* (Letters), 661, L53
Pols, O. R., Cote, J., Waters, L. B. F. M., & Heise, J. 1991, *A&A*, 241, 419
Prantzos, N. & Charbonnel, C. 2006, *A&A*, 458, 135
Prantzos, N., Charbonnel, C., & Iliadis, C. 2007, *A&A*, 470, 179
Refsdal, S., Roth, M. L., & Weigert, A. 1974, *A&A*, 36, 113
Renzini, A. 2008, *MNRAS*, 391, 354
Sana, H., Gosset, E., Nazé, Y., Rauw, G., & Linder, N. 2008, *MNRAS*, 386, 447
Smith, N., Gehrz, R. D., & Goss, W. M. 2001, *AJ*, 122, 2700
Smith, N., Gehrz, R. D., Stahl, O., Balick, B., & Kaufer, A. 2002, *ApJ*, 578, 464
Ulrich, R. K. & Burger, H. L. 1976, *ApJ*, 206, 509
van Rensbergen, W., de Loore, C., & Jansen, K. 2006, *A&A*, 446, 1071
Ventura, P., D'Antona, F., Mazzitelli, I., & Gratton, R. 2001, *ApJ* (Letters), 550, L65

Session V

Multiwavelength observations

Star clusters: basic galactic building blocks
Proceedings IAU Symposium No. 266, 2009
R. de Grijs & J. R. D. Lépine, eds.

doi:10.1017/S1743921309991037

Insights into star cluster formation from $\lambda \gtrsim 1\mu$m

Rémy Indebetouw,[1,2] Rosie Chen, Crystal Brogan, Barbara Whitney, Thomas Robitaille, Leslie Looney and the SAGE team

[1]Astronomy Department, University of Virginia, P.O. Box 400325, Charlottesville, VA 22904–4325, USA
email: remy@virginia.edu

[2]National Radio Astronomy Observatory, Charlottesville, VA, USA

Abstract. We would like to know how molecular clouds turn into stellar clusters, and with what efficiency massive stars form in those clusters, since massive stars are the main agents responsible for evolution of the interstellar medium of galaxies, and their subsequent star-formation history. The imprint of 'precluster' molecular cloud conditions can be observed, but only in the least evolved, most embedded clusters, necessarily at wavelengths that can penetrate more than 10 visual magnitudes of extinction. Mid-infrared photometric imaging, most recently and extensively from *Spitzer*, can be used to select young stellar objects in clustered star-formation environments in our Galaxy and nearby galaxies. Relatively sophisticated methods have been developed, but the fundamental principle remains the selection of sources that have excess infrared emission from circumstellar dust. By fitting radiative-transfer models to a source's spectral-energy distribution between ~ 1 and $\sim 100\mu$m, we constrain the circumstellar dust *distribution* and evolutionary state. We can explore many things with this protostellar distribution in mass/luminosity and time/evolutionary state. For example we do not see strong evidence for primordial mass segregation in initial studies. We find evidence of primordial hierarchical substructure, greater clustering at the youngest stages, and even imprints of the pre-stellar Jeans scale. We see correlation of the youngest sources with dense molecular clumps and constrain the timescales for chemical processing and dispersal of those clumps. We have only begun to mine the wealth of existing *Spitzer*, emerging *Herschel* and soon *ALMA* data.

Keywords. stars: formation, galaxies: star clusters, infrared: stars, submillimeter

1. Introduction and motivation

The youngest star clusters still being formed are a critical laboratory to understand physics ranging from the formation of individual massive stars through the star-formation history of galaxies. Consider first galaxy scales. The evolution of a galaxy depends on the efficiency with which it converts interstellar material into stars and how effectively those stars process the clouds and either inhibit or enhance subsequent star formation. One of the most commonly used empirical tools in extragalactic astronomy is the Schmidt–Kennnicutt 'law', a proportionality between the surface density of dense gas and the spatially resolved star-formation rate (Kennicutt 1998). Recent work has improved the calibration of the star-formation-rate determination, and it is now relatively well accepted that either optical recombination-line emission (e.g., Hα) or ultraviolet continuum, plus mid-infrared continuum (e.g., 24μm) can accurately represent, respectively, exposed and embedded star formation, and an appropriate linear combination is a good measure of the total rate (Calzetti *et al.* 2007; Kennicutt *et al.* 2009). Understanding the physics which leads to the Schmidt–Kennicutt correlation depends more critically on the abcissa, i.e., whether one should measure the surface density of molecular gas, of atomic gas, or the total. Recent work by the THINGS collaboration (Walter *et al.* 2008) has clarified this, showing that in denser parts of galaxies, such as the inner parts of spirals, interstellar gas is primarily in the molecular phase, and that star formation follows the surface density of that star-forming molecular gas tightly. In the outer parts of spirals, and lower-surface-brightness galaxies, the interstellar medium phase balance shifts towards the atomic and the neutral-hydrogen column density must be taken into account (Bigiel *et al.* 2008; Leroy *et al.* 2008). A simple theoretical model including the phase state of interstellar gas and the formation of stars from molecular material can reproduce the observations fairly well (Krumholz *et al.* 2009).

A fundamental question is at what size scale the Schmidt–Kennicutt correlation breaks down—clearly, at size scales relevant to individual stars, the average star-formation rate and gas surface density no longer make sense. However, we would like to know if relatively universal behaviour persists on the scale of a star cluster: given a molecular cloud of a particular surface density or mass, is the formation of a cluster from that cloud relatively well determined, or does it depend on detailed physical conditions such as the clumpiness, heating by nearby stars, penetration of dissociating or even ionizing radiation, turbulent mixing by nearby stars, supernovae, supershells or kiloparsec-scale instabilities? Detailed study of the distribution of new star clusters relative to dense molecular material, on *regional* size scales $\lesssim 100$ pc should show these effects and resolve the physics that we expect to break the Schmidt–Kennicutt correlation. In particular, the displacements of protoclusters from CO clumps, and the relative location of emission from denser gas tracers such as HCN, will reveal what physics is important and what molecular tracers most closely probe that physics.

It is possible that a $\sim 10^6$ $M_\odot$ molecular cloud forms a star cluster in a relatively universal way and that the fundamental physics determining star formation on galaxy scales is then merely the physics of forming a molecular cloud. This too has been an active area of recent research, with the prevalent theories being the formation of molecular material in the overdense regions of colliding or converging atomic flows: see Heitsch *et al.* (2009) for a recent study. These flows are subject to instabilities which fragment and drive turbulent motions in the molecular cloud formation zone, i.e., turbulence and substructure which, if stars and star clusters then form promptly ($\lesssim 10^6$ yr, rapidity for which there is significant evidence), will be preserved in the spatial distribution of nascent protostars.

Next, we consider smaller scales, and the formation of one massive ($\gtrsim 10$ $M_\odot$) star. It is apparent that nearly all massive stars form in clusters (de Wit *et al.* 2005). What is less clear is whether that is *necessary*. Simple arguments based on accretion rate and Kelvin–Helmholz time show that a massive protostar will have a central object close to the main sequence (having a high enough effective temperature to emit significant ultraviolet radiation) by the time ~ 15 $M_\odot$ have accreted, even if the accretion rate is still very high and the star is destined to be much more massive. Simple models of the formation of solar-mass stars in isolation break down when required to treat such extreme conditions in the protstellar core, and solutions require a combination of high pressure (likely nonthermal) in the core, accretion through a disk and may even require interactions with other protstars. In the clustered/cooperative-accretion picture, for example, gas drawn into the gravitational potential of the entire cluster is what enables the required very high accretion rates onto the most massive members (Bonnell & Bate 2006). A different theory posits that a turbulent core separates dynamically from the cluster-forming cloud before it accretes significantly onto its central protostar (Tan *et al.* 2006). These scenarios predict different interclump motions in the precluster cloud and different distributions (clustering versus fractal substructure) for the distribution of massive and less massive protostars. There are also questions regarding the relative ages of formation of the massive and less massive members, which can be addressed by observing the spatial and temporal distribution of protostars in protoclusters.

For all of these problems we need to study protoclusters $\lesssim 10^6$ yr old, because dynamical interactions between protostars will very rapidly erase the signatures we wish to detect (Allison *et al.* 2009; Fellhauer *et al.* 2009; Moeckel & Bonnell 2009).

2. Infrared study of Galactic protoclusters

Mid-infrared photometry is particularly powerful to identify young stellar objects. Even at 1–2 μm, somewhat more evolved but still protostellar objects (those denoted Class III for solar-mass protostars) will not show much excess above their Rayleigh–Jeans photospheric spectrum, but it has been long known that selection based on 3–5 μm excess results in a much more complete protostellar census (Strom *et al.* 1989).

In Galactic star-formation regions and protoclusters, certain parts of color–color space have been identified as mostly populated by young stellar objects (YSOs; Allen *et al.* 2004; Megeath *et al.* 2005). Such color selection of candidate YSOs can be made more sophisticated by explicit, simultaneous consideration of different color planes (Gutermuth *et al.* 2009) or by considering and fitting the spectral-energy distribution (SED) of each source over as wide a wavelength range as possible. Our group has generally used the latter, developing a set of 10 000 YSO radiative-transfer models and rapid code to find the best-fitting models for a set of data (Robitaille *et al.* 2006, 2007). The methods are not widely discrepant—we explore the differences in detail in the Eagle Nebula in Indebetouw *et al.* (2007)—because the SED is mathematically interchangeable with the position of the source in a set of color–color plots spanning all fluxes.

The Eagle Nebula study (Indebetouw *et al.* 2007) illustrates the power of this kind of analysis in a relatively modest stellar cluster. We clearly re-identify the young cluster NGC 6611 as containing mostly Class II or T Tauri-type protostars. In protostellar terms, that is young, but not extremely young. We discover a less evolved cluster to the north east of NGC 6611, still cospatial with the molecular cloud. NGC 6611 has already largely dispersed its end of that cloud, leaving only fragments—the famous pillars of creation—behind. We also show quantitatively how the younger sources are more clustered than older ones, a result now seen in many *Spitzer* studies of Galactic regions (e.g., Gutermuth *et al.* 2009). We also see that the spatial separation of protostellar objects is consistent with the fragmentation scales for the warm molecular material in the neighborhood of NGC 6611. Finally, we find a massive YSO to the north west that has not been discussed extensively in the literature.

3. Infrared study of extragalactic protoclusters

Moving to extragalactic studies, we developed a set of criteria to select protostars in the color–magnitude spaces formed by 2MASS and SAGE *Spitzer* IRAC and MIPS point-source photometry of the Large Magellanic Cloud (LMC; Meixner *et al.* 2006; Whitney *et al.* 2008). The cuts were designed to reject the majority of evolved stars found in the LMC by Blum *et al.* (2006), but accept the majority of our grid of 10 000 radiative-transfer models of dusty protostars (Robitaille *et al.* 2006, 2007). Figure 1 shows a figure from Whitney *et al.* (2008) of one color–magnitude diagram and the location of various types of sources. Gruendl & Chu (2009) used a somewhat different color cut, designed to more conservatively reject evolved stars, at the expense of also rejecting some more evolved (Class/Type II–III) young stellar objects. In Indebetouw *et al.* (2008) we also explored different color–magnitude cuts, this time accounting for somewhat extended sources such as ultracompact and compact H^+ regions, which can also be considered 'young stellar objects'. whitney *et al.* (2008) found over 1000 massive YSO candidates in the LMC, 50 times more than previously known. The correlation with interstellar gas is striking and quantitatively significant (Figure 2). Yang *et al.* (2007) used a similar YSO candidate list generated by Gruendl & Chu (2009) to assess the global stability of the LMC and found significant numbers of YSO candidates in regions considered stable from a Toomre-type stability analysis of the molecular and atomic gas. The distribution of protostellar candidates can be much better explained if one includes the gravitational potential of the older stellar population in the analysis, i.e., one must consider where stars already are in predicting where more stars will form in a galaxy.

We can also perform detailed regional studies of star formation in the Magellanic system. In N44, Chen *et al.* (2009) found a concentration of massive YSOs in the central molecular cloud, which may be triggered by a neighboring bubble H^+ region. A cloud of similar mass to the south also contains massive protostars, but with a sparser spatial distribution. Differences in evolutionary state between the YSOs in the two different molecular clouds may correlate with those different formation mechanisms. The N159 region is more striking, also containing two highly active molecular clouds and a third very quiescent one of equal or greater mass. The protostellar distribution in space and evolutionary state in N159 suggests that one of the active clouds may have been triggered, while the other probably formed stars spontaneously (Chen *et al.* 2009, submitted). We are further exploring the physical properties of dense (HCN, HCO^+-emitting) gas in the quiescent cloud to determine why it has so little massive star formation (Chen *et al.*, in prep.)

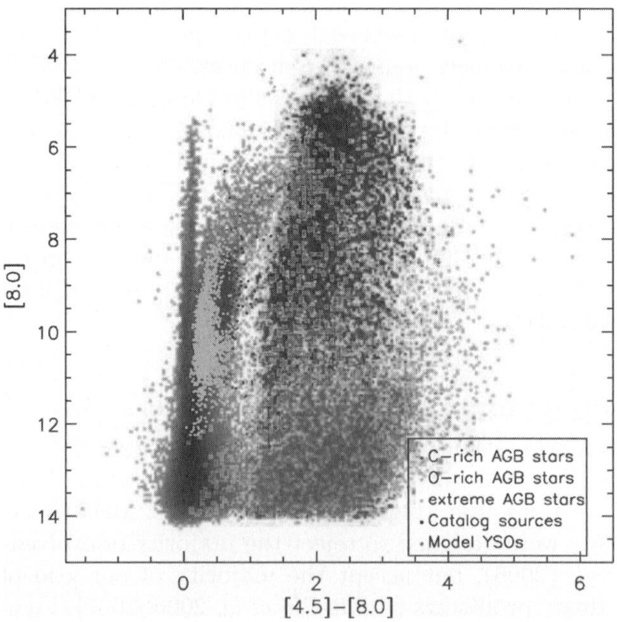

Figure 1. Color–magnitude diagram of sources in the LMC, showing how a cut at about [4.5]–[8.0] > 1 mag, somewhat increasing in color with decreasing magnitude, would select YSO candidates and exclude most evolved stars.

The molecular ridge south of 30 Doradus in the LMC contains nearly one third of the molecular gas of the entire galaxy (Mizuno *et al.* 2001), but has very little apparent activity in Hα or near-infrared tracers. The region is also rather discrepant when plotted on the Schmidt–Kennicutt correlation: given the surface density of molecular gas, the star-formation rate calculated from optical tracers is at least an order of magnitude below the trend. We explored the region using *Spitzer* data, and found that it does contain a number of modest mid-infrared star-forming sites (Indebetouw *et al.* 2008). Analysis of these sites, both with individual protostar models and modeling them as groups of stars or protoclusters, reveals that they are *poor* clusters of total mass $\lesssim 10^5$ M$_\odot$, too small to fully sample the stellar initial mass function up to massive OB stars. Thus, although there is a reasonable total mass of solar-mass star formation in the ridge (and enough to make it consistent within uncertainties with the Schmidt–Kennicutt correlation), the activity that is taking place is low-mass and poor. Thus, given the steep mass–luminosity and even steeper mass–ionizing flux relations for stars, the *integrated* Hα and infrared emission does not fully reflect the total star formation in the cloud. Why this giant cloud is only forming lower-mass stars, similar to Taurus in the Galaxy, remains a mystery.

4. Beginning to observe the gas as well

With millimeter receivers on the *Australia Telescope Compact Array* (Wong & Melatos 2002), we recently started to be able to observe dense gas tracers such as HCN and HCO$^+$ in the Magellanic system, at parsec-scale spatial resolution, comparable to the infrared studies. For example, in N44, we find that the least-evolved massive YSOs are most closely spatially associated with dense molecular clumps. Even within 10^6 years, the slightly more evolved (Type II) massive YSOs have begun processing their molecular material by heating and changing the HCN-to-HCO$^+$ ratio, and dispersing those clumps, leaving a weaker spatial association (Chen *et al.* 2010). Observations

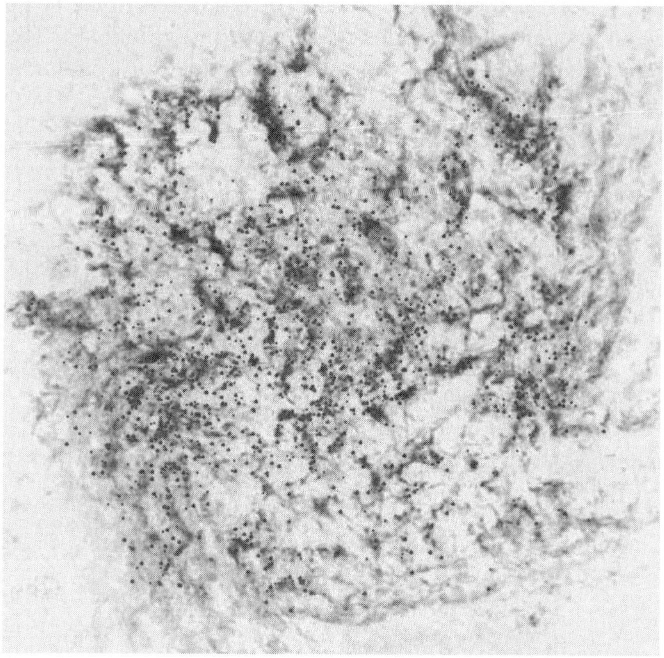

Figure 2. HI in the LMC with positions of YSO candidates.

of N159 in the LMC (Ott *et al.*, in prep.; Chen *et al.*, in prep.) promise to be even more interesting.

5. A sobering lesson from detailed Galactic studies

The largest caveat in the analysis discussed, especially using *Spitzer* data, which at 24μm has a spatial resolution of more than a parsec in the Magellanic system, is the small-scale structure of massive YSOs. We see repeatedly in submillimeter and infrared studies of Galactic regions such as NGC 6334 (Hunter *et al.* 2006; Brogan *et al.* 2009), that a parsec-scale region, which would be identified as a point source extragalactically, *nearly always* contains multiple massive members, *nearly always* in rather different apparent evolutionary states. In the very young region NGC 6334I(N), there are four massive sources within 10 000 AU. Only one has an H^+ region, two have prolific submillimeter line emission (hot-core chemistry) and one shows very little line emission, perhaps not yet warm enough to drive intense chemistry (Brogan *et al.* 2009). Clearly, modeling this as a single massive protostar has limitations. We have investigated this carefully in some Magellanic regions for which we have higher-resolution near-infrared imaging, even with adaptive optics, to better identify the multiplicity. Because of the steep mass–luminosity relation for stars, we find that the physical parameters (evolutionary state) determined from fitting two massive members of a small group or cluster do not differ very much from those determined by fitting the unresolved SED. More careful investigation with high-resolution observations is still required, but at least for the small number we have been able to look at, the main change in conclusions is that splitting the source into N sources for the same total luminosity, results in the expected $\simeq N^{2/3}$ increase in total mass for the group.

Clearly, we need to continue these studies of protoclusters with high-resolution infrared observations from adaptive-optics-enabled telescopes like *Gemini* and the *VLT*, and then with *JWST*. *ALMA* will soon permit direct observation in nearby galaxies of the molecular gas on the subparsec scales relevant to individual star formation and submillimeter continuum detection of the very most embedded protostars and protoclusters. Together, we expect to make great progress with these data in unraveling the complex and (optically) invisible process of cluster formation.

References

Allen, L. E., *et al.* 2004, *ApJS*, 154, 363
Allison, R. J., Goodwin, S. P., Parker, R. J., de Grijs, R., Portegies Zwart, S. F., & Kouwenhoven, M. B. N. 2009, *ApJ* (Letters), 700, L99
Blum, R. D., *et al.* 2006, *AJ*, 132, 2034
Bigiel, F., Leroy, A., Walter, F., Brinks, E., de Blok, W. J. G., Madore, B., & Thornley, M. D. 2008, *AJ*, 136, 2846
Bonnell, I. A. & Bate, M. R. 2006, *MNRAS*, 370, 488
Brogan, C. L., Hunter, T. R., Cyganowski, C. J., Indebetouw, R., Beuther, H., Menten, K. M., & Thorwirth, S. 2009, *ApJ*, 707, 1
Calzetti, D., *et al.* 2007, *ApJ*, 666, 870
Chen, C.-H. R., Chu, Y.-H., Gruendl, R. A., Gordon, K. D., & Heitsch, F. 2009, *ApJ*, 695, 511
Fellhauer, M., Wilkinson, M. I., & Kroupa, P. 2009, *MNRAS*, 397, 954
Gruendl, R. A. & Chu, Y.-H. 2009, *ApJS*, 184, 172
Gutermuth, R. A., Megeath, S. T., Myers, P. C., Allen, L. E., Pipher, J. L., & Fazio, G. G. 2009, *ApJS*, 184, 18
Heitsch, F., Stone, J. M., & Hartmann, L. W. 2009, *ApJ*, 695, 248

Hunter, T. R., Brogan, C. L., Megeath, S. T., Menten, K. M., Beuther, H., & Thorwirth, S. 2006, *ApJ*, 649, 888
Indebetouw, R., Robitaille, T. P., Whitney, B. A., Churchwell, E., Babler, B., Meade, M., Watson, C., & Wolfire, M. 2007, *ApJ*, 666, 321
Indebetouw, R., *et al.* 2008, *AJ*, 136, 1442
Kennicutt Jr., R. C. 1998, *ApJ*, 498, 541
Kennicutt, R. C., Hao, C.-N., Calzetti, D., Moustakas, J., Dale, D. A., Bendo, G., Engelbracht, C. W., Johnson, B. D., & Lee, J. C. 2009, *ApJ*, 703, 1672
Krumholz, M. R., McKee, C. F., & Tumlinson, J. 2009, *ApJ*, 699, 850
Leroy, A. K., Walter, F., Brinks, E., Bigiel, F., de Blok, W. J. G., Madore, B., & Thornley, M. D. 2008, *AJ*, 136, 2782
Megeath, S. T., Hartmann, L., Luhman, K. L., & Fazio, G. G. 2005, *ApJ* (Letters), 634, L113
Meixner, M., *et al.* 2006, *AJ*, 132, 2268
Mizuno, N., Rubio, M., Mizuno, A., Yamaguchi, R., Onishi, T., & Fukui, Y. 2001, *PASJ* (Letters), 53, L45
Moeckel, N. & Bonnell, I. A. 2009, *MNRAS*, 1370
Tan, J. C., Krumholz, M. R., & McKee, C. F. 2006, *ApJ* (Letters), 641, L121
Robitaille, T. P., Whitney, B. A., Indebetouw, R., Wood, K., & Denzmore, P. 2006, *ApJS*, 167, 256
Robitaille, T. P., Whitney, B. A., Indebetouw, R., & Wood, K. 2007, *ApJS*, 169, 328
Strom, K. M., Strom, S. E., Edwards, S., Cabrit, S., & Skrutskie, M. F. 1989, *AJ*, 97, 1451
Walter, F., Brinks, E., de Blok, W. J. G., Bigiel, F., Kennicutt, R. C., Thornley, M. D., & Leroy, A. 2008, *AJ*, 136, 2563
Whitney, B. A., *et al.* 2008, *AJ*, 136, 18
de Wit, W. J., Testi, L., Palla, F., & Zinnecker, H. 2005, *A&A*, 437, 247
Wong, T. & Melatos, A. 2002, *PASA*, 19, 475
Yang, C.-C., Gruendl, R. A., Chu, Y.-H., Mac Low, M.-M., & Fukui, Y. 2007, *ApJ*, 671, 374

Star clusters: basic galactic building blocks
Proceedings IAU Symposium No. 266, 2009
R. de Grijs & J. R. D. Lépine, eds.

doi:10.1017/S1743921309991049

An optical/near-infrared survey of GCs in early-type galaxies

A. L. Chies–Santos,[1] S. S. Larsen,[1] E. M. Wehner,[1] H. Kuntschner,[2] J. Strader,[3] J. P. Brodie[4] and J. F. C. Santos Jr.[5]

[1]Sterrenkundig Instituut, University of Utrecht, The Netherlands, [2]ST–ECF/ESO, Germany, [3]Harvard CfA, USA, [4]UCO/Lick Observatory, USA, [5]UFMG, Brazil
email: a.l.chies@uu.nl

Abstract. Optical/near-infrared observations of 14 globular cluster (GC) systems in early-type galaxies are presented. We investigate the recent claims (Yoon *et al.* 2006) of colour bimodality in GC systems being an artefact of the nonlinear colour–metallicity transformation driven by horizontal-branch morphology. Taking advantage of the fact that the combination of optical and near-infrared colours can, in principle, break the age/metallicity degeneracy we also analyse age distributions in these systems.

Keywords. galaxies: star clusters, galaxies: elliptical and lenticular, cD, galaxies: star clusters

1. Introduction

The ubiquity of globular clusters (GCs) around all major galaxies and the fact that young star clusters are being observed in many mergers and starbursts we observe today suggests that GC formation traces important star-formation events in the histories of their host galaxies (Brodie & Strader 2006).

GC systems generally exhibit a colour distribution that is bimodal (sometimes multimodal) in optical colours. Spectroscopic studies (e.g., Strader *et al.* 2005; Cenarro *et al.* 2007) show that this bimodality is due to old subpopulations with ages greater than 10 Gyr that differ in metallicity. Bimodal metallicity distributions may indicate distinct formation mechanisms, with age differences ~ 2 Gyr still allowed within the uncertainties on current age estimates. This metallicity bi/multimodality seems to hold for the majority of galaxies, except for the low-mass galaxies which appear to have only a metal-poor GC subpopulation. Recently, Yoon *et al.* (2006) challenged this metallicity-bimodality interpretation saying that it is an artefact of horizontal-branch morphologies. These authors argue that nonlinear colour–metallicity relations may transform a unimodal metallicity distribution into a bimodal optical colour distribution. This issue has been investigated in more detail by Cantiello & Blakeslee (2007), who concluded that combinations of optical and near-infrared (NIR) colours are much less prone to this effect. Metallicity bimodality was thus contested and it is, therefore, important to investigate this topic in more detail. On the other hand, if metallicity bi/multimodality is genuine, there could be multiple episodes/mechanisms of GC formation in a galaxy with a possible age difference. It is thus equally important to study the age distributions of these systems.

It is still an open issue in the literature whether some early-type galaxies known to contain old stellar populations from integrated light studies can host a high percentage of intermediate-age (2–3 Gyr) GCs. The two strongest cases are NGC 4365 (Puzia *et al.* 2002; Larsen *et al.* 2005) and NGC 5846 (Hempel *et al.* 2003). Comparison of two-colour plots of the GC systems with simple stellar population (SSP) models (e.g., Bruzual &

Charlot 2003; Maraston 2005) suggested the presence of intermediate-age GCs in these galaxies.

With the aim of investigating ages and metallicity distributions in GC systems of early-type galaxies, we examine 14 E/S0 systems, with $(m - M) < 32$ mag and $M_B < -19$ mag using optical and NIR imaging. We obtained deep ($\sim$ 3.4 hr) K-band photometry with the LIRIS instrument on the *William Herschel Telescope* at La Palma, which has a field of view of 4.2×4.2 arcmin^2. This was combined with optical imaging extracted from the *Hubble Space Telescope* public archive in the g (F475W) and z (F850LP) bands. GC candidates were selected automatically in galaxy-subtracted images through standard routines. The following cuts were applied, defining the final sample: $g < 23$ mag, $0.5 < (g - z) < 2$ mag, effective radii $1 < R_{\rm eff} < 15$ pc and magnitude error $K_{\rm err} < 0.5$ mag. With these restrictions, typical magnitude errors for the optical g and z bands were $\sim$ 0.05 mag. For the NIR band, the magnitude errors were measured as $\sim$ 0.06 mag for a $K = 19$ mag GC and $\sim$ 0.2 mag for a $K = 20$ mag GC. Artificial-star tests, however, show that the K-band magnitude errors can be underestimated by a factor of 2. In what follows, it is shown what knowledge was gained from these data regarding the age and colour (metallicity) distributions of the GC systems studied here.

2. Ages

The combination of optical/NIR imaging is an alternative to spectroscopy in age dating GCs. The technique relies on comparing two-colour diagrams with the predictions of SSP models. In Figure 1 two-colour diagrams $(g - k)$ versus $(g - z)$ for the GCs that satisfy the criteria mentioned above for the 14 galaxies are shown. In the left panel, a model grid from Maraston (2005) is overplotted with model sequences of constant age as solid lines: 2, 3, 4, 5, 6, 8, 11 and 14 Gyr. If one examines the case of NGC 4365, the galaxy that showed evidence of having intermediate-age clusters, one concludes that this galaxy is no different from any other in the sample. In the right panel, the same data as the left panel are shown, but now overplotted with Padova SSPs with Marigo *et al.* (2008) isochrones and a new treatment of the thermally pulsing asymptotic giant branch (TP–AGB) phase with the same age range as for the first panel. These latter models for old ages ($\sim 11, 14$ Gyr) are a close match to the data, while Maraston (2005) models show an offset that would imply intermediate ages ($\sim 2, 3$ Gyr). Charlot & Bruzual (in prep.) preliminary models yield basically the same conclusion as the Maraston (2005) models. the $(g - k)$ SSPs are $\sim$ $0.2 - 0.3$ mag too blue. A similar offset is also the case for all older SSP models. The only indication of intermediate ages is the presence of half a dozen GCs with $\sim (g - k) \sim 3.7$ mag in NGC 4406. The data for the NGC 4382 and NGC 4473 GC systems should be taken with caution as they were observed in nonphotometric conditions. The data spread along $(g - k)$ is attributed to observational uncertainties or is intrinsic to the systems. For M87, the GCs present a standard deviation of ± 0.75 mag along $(g - k)$ from the best-fit line to $(g - k)$ and $(g - z)$ and the observational errors reach a maximum of ± 0.4 mag (based on Artificial-star tests). One might think that $[\alpha/{\rm Fe}]$ ratios could perhaps account for part of this spread, but this of course would only explain the scatter if the ratios varied among clusters. Stochastic effects in the integrated colours are investigated following Santos & Frogel (1997). It is found that the fluctuations in the number of bright stars can account for only $\sim$ 0.1 mag of the spread in $(g - k)$ for a 14 Gyr-old SSP and a $\sim$ 0.2 mag for a 2 Gyr one, assuming that all clusters have 10^6 stars. We conclude that age dating GC systems as old ($\sim 11, 14$ Gyr) or of intermediate age ($\sim 2, 3$ Gyr) depends on the choice of the SSP model. Padova SSPs with Marigo *et al.* (2008) isochrones for old ages are in good agreement with the data. Moreover, there is no evidence for significant differences in the GC age distributions among the galaxies studied here. NGC 4406 could be an exception.

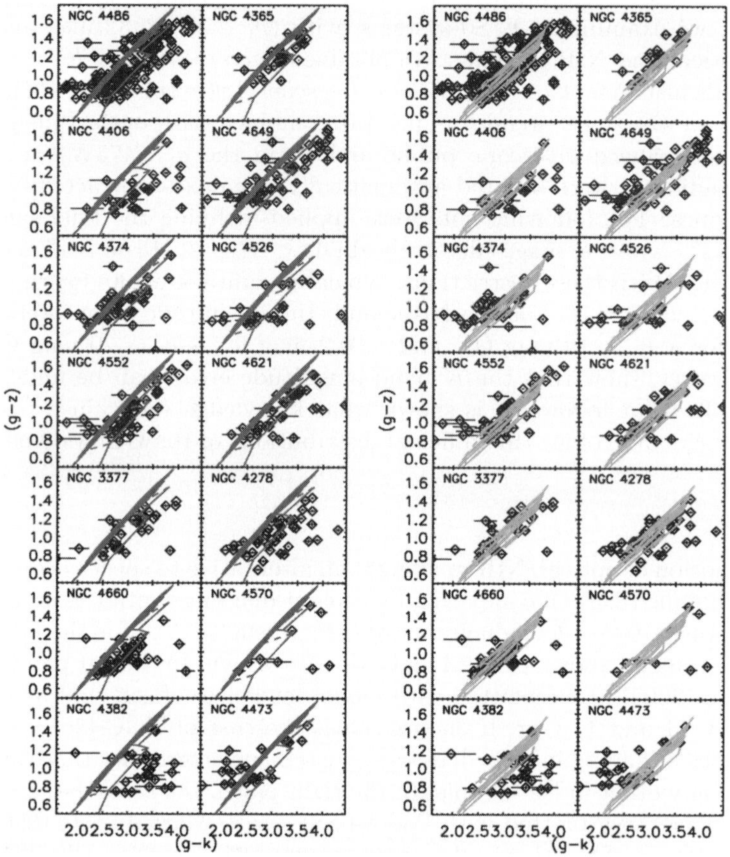

Figure 1. Two-colour diagrams for the GC systems of the 14 galaxies: $(g-k)$ versus $(g-z)$. *(left)* A model grid from Maraston (2005) is overplotted, with model sequences of constant age as solid lines: 2, 3, 4, 5, 6, 8, 11 and 14 Gyr. *(right)* A model grid of Padova SSPs with Marigo et al. (2008) isochrones is now overplotted with the same ages as in the left panel.

3. Bimodality

The combination of optical and NIR colours is a better indicator of the true underlying metallicity distribution than optical colours alone (Cantiello & Blakeslee 2007). At redder wavelengths, the stellar population will include decreasing contributions from horizontal-branch stars. If the metallicity distribution of a GC system is genuinely bimodal, this (as seen through optical colours) should be maintained as one moves to redder wavelengths. We thus test the alternative Yoon *et al.* (2006) scenario for colour bimodality.

In Figure 2. the $(g-z)$, $(g-k)$ and $(z-k)$ colour distributions for the GC systems of the 14 E/S0s shown in Figure 1. are plotted in the first, second and third panels, respectively. Note that the bimodality becomes less evident in $(g-k)$ compared to $(g-z)$ and even less pronounced in $(z-k)$. This is the colour that should have the smallest contribution from horizontal-branch stars. It is evident from this figure that the galaxies that have more than one peak in $(g-z)$ appear more strongly single peaked in $(z-k)$, while the multiple peaks in the $(g-k)$ distributions still appear more clearly. Note how this effect is pronounced in NGC 4486 (M 87) the most cluster-rich galaxy in the sample. This result fits very nicely into the Yoon *et al.* (2006) picture where the horizontal-branch morphology would induce a nonlinear behaviour in the colour–metallicity relation, and this would drive a unimodal metallicity distribution into a bimodal (optical) colour distribution. If in the reddest of the available colours, $(z-k)$, the best metallicity indicator available here, the distribution is unimodal, it seems straightforward to conclude that horizontal-branch morphology could be causing this. Taking into account that $(z-k)$ is the most sensitive colour among the ones available in this study to uncertainties in the K band, we tested whether this observational scatter could blur the bimodality. We carried out a linear transformation of the $(g-z)$ colour distribution to $(g-k)$ and $(z-k)$ that preserved bimodality in the absence of noise. We then simulated blue- and red-peaked Gaussian distributions, adding noise proportional to more realistic errors than the ones estimated by PHOT (calculated based on Artificial-star tests). From the results of these simulations it was found by running KMM tests that it is very likely to have a bimodal distribution, appearing more and more unimodal in the transition from the bluest $(g-z)$ to the reddest $(z-k)$ colours. Whether the absence of horizontal-branch stars in $(z-k)$ can indeed be responsible for the weakening of bimodality in this colour as opposed to $(g-z)$ (and $(g-k)$) and supports the scenario proposed by Yoon *et al.* (2006) cannot be concluded with the current data set, as observational uncertainties in the K band can also be the mechanism behind this.

3.1. *Horizontal-branch morphology*

In Figure 3, two-colour plots, $(g-k)$ versus $(z-k)$, for GCs of M87 and NGC 4649 (indicated by different symbols) are shown. Note the wavy feature the data presents around $(g-k) \sim 3.2$ and $(z-k) \sim 2$ mag. For the sake of clarity, in the left panel we show the data points alone with their corresponding error bars. A SPoT–Teramo 14 Gyr SSP with a realistic treatment of horizontal-branch morphology (Raimondo *et al.* 2005) is overplotted in the middle panel and a Padova 14 Gyr SSP with Marigo *et al.* (2008) isochrones is overplotted on the right. This wavy feature is also present in the SPoT–Teramo SSP model but does not fit well with the data, especially for the redder part. Other SSP models (Charlot & Bruzual, in prep., Maraston 2005 and Padova) that do not take into account horizontal-branch morphology do not show this behaviour, as can be seen in the right panel for the case of Padova. By comparing Figure 3 to figure 1 of Yoon *et al.* (2006), one concludes that $(z-k)$ is indeed a good indicator for [Fe/H] and that $(g-k)$ still contains some contribution of horizontal-branch stars. Higher signal-to-noise ratio, deeper data, sampling more GCs would be necessary to understand whether the nonlinear behaviour in the colour–metallicity relation induced by horizontal-branch morphology, as attested by Figure 3. would indeed drive a unimodal metallicity distribution into a bimodal (optical) colour distribution, as proposed by Yoon *et al.* (2006).

4. Conclusions

A sample of GC systems in E/S0s was observed in the K band with *WHT*/LIRIS and combined with archival *HST*/ACS imaging in the g and z bands. Based on these observations, we intended to study the overall ages and metallicity distributions of GC systems. We find that age dating GCs as old or intermediate age depends on the choice of the model. Padova SSPs for old ages ($\sim 12, 14$ Gyr) with Marigo *et al.* (2008) isochrones with a new treatment for the TP–AGB phase fit the data well. Maraston (2005) and

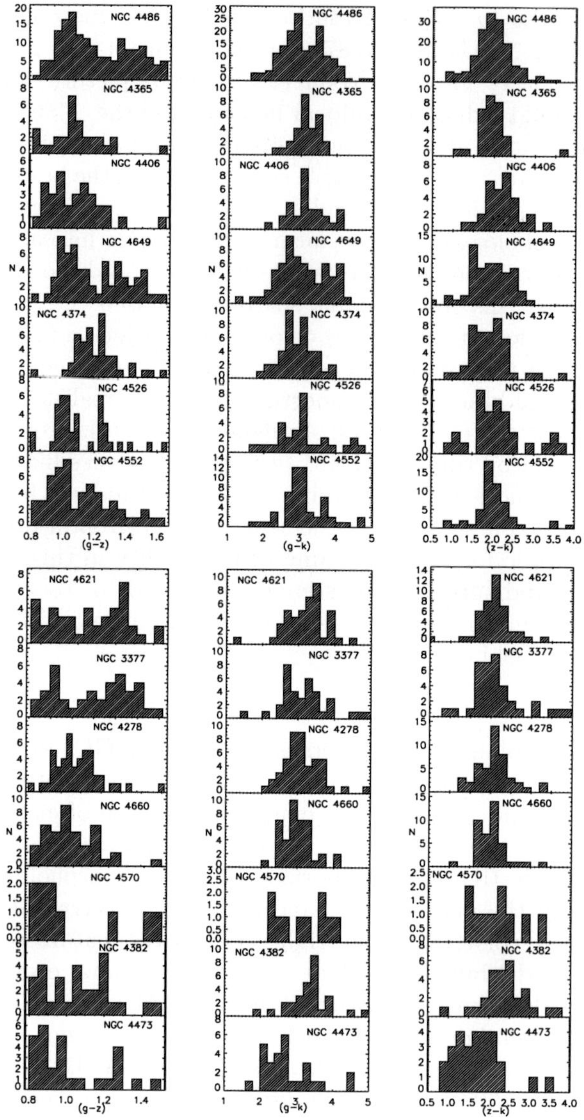

Figure 2. $(g-z)$, $(g-k)$ and $(z-k)$ colour distributions for the GC systems of the 14 E/S0s shown in Figure 1. Note that the bimodality becomes less evident in $(g-k)$ compared to $(g-z)$ and even less pronounced in $(z-k)$. The latter is the colour that should have the smallest contribution from horizontal-branch stars.

Charlot & Bruzual (in prep.) models yield intermediate ages ($\sim 2-3$ Gyr) for the majority of the GCs. The NGC 4365 GC system is not found to be an exception. In general, all GC systems in the two-colour diagrams look very much alike. NGC 4406 is found to be the strongest candidate for having intermediate ages.

As far as metallicity distributions are concerned, the bimodality becomes less evident in $(g-k)$ compared to $(g-z)$ and even less pronounced in $(z-k)$. The latter colour is the one that should have the smallest contribution from horizontal-branch stars. The disappearance of bimodality in this colour, while evident in the optical $(g-z)$ colour, could be attributed to the Yoon *et al.* (2006) effect, although the observational uncertainties could also account for it. The relation $(g-k)$ versus $(z-k)$ shows a wiggle around $(g-k) \sim 3.2$

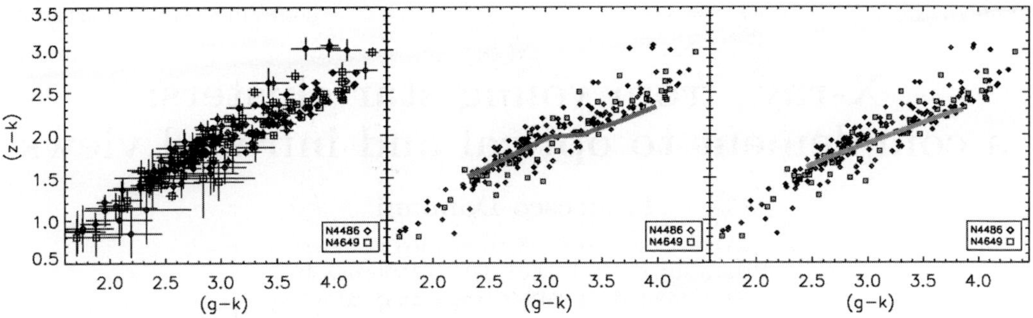

Figure 3. Two-colour plots, $(g-k)$ versus $(z-k)$, for GCs of M87 and NGC 4649, indicated by different symbols. The error bars are only presented in the left panel, for clarity. Note the wavy feature the data present around $(g-k) \sim 3.2$ and $(z-k) \sim 2$ mag. The SPoT–Teramo 14 Gyr SSP with a realistic treatment of horizontal-branch morphology (Raimondo *et al.* 2005) is overplotted in the middle panel and the Padova 14 Gyr SSP with Marigo *et al.* (2008) isochrones is overplotted in the right panel.

and $(z-k) \sim 2$ mag, which is present in the 14 Gyr SPoT–Teramo SSP models, which include a proper treatment of horizontal-branch morphology. It is concluded that redder colours, such as $(z-k)$, are ideal metallicity indicators.

This study calls attention to the importance of choice of SSP models. How well a set of models fits the data seems to be linked to which stellar evolutionary stages are incorporated in it and how this is done. As far as deriving overall ages of GCs with the optical/NIR imaging technique, it becomes clear that different kinds of TP–AGB treatments yield different results. SSPs including recent TP–AGB models are consistent with old GC ages. The addition (or not) of a realistic treatment of the horizontal-branch morphology can also make a big difference. Whether or not this evolutionary phase is responsible for optical colour bimodality cannot be answered with the data here presented.

Acknowledgements

We would like to thank G. Bruzual, S. Charlot and M. Cantiello for kindly providing access to their models prior to publication.

References

Brodie, J. P. & Strader, J. 2006, *ARA&A*, 44, 193
Bruzual, G. & Charlot, S. 2003, *MNRAS*, 344, 1000
Cantiello, M. & Blakeslee, J. P. 2007, *ApJ*, 669, 982
Cenarro, A. J., Beasley, M. A., Strader, J., Brodie, J. P., & Forbes, D. A. 2007, *AJ*, 134, 391
Emsellem, E., *et al.* 2007, *MNRAS*, 379, 401
Hempel, M., Hilker, M., Kissler–Patig, M., Puzia, T. H., Minniti, D., & Goudfrooij, P. 2003, *A&A*, 405, 487
Larsen, S. S., Brodie, J. P., & Strader, J. 2005, *A&A*, 443, 413
Maraston, C. 2005, *MNRAS*, 362, 799
Marigo, P., Girardi, L., Bressan, A., Groenewegen, M. A. T., Silva, L., & Granato, G. L. 2008, *A&A*, 482, 883
Puzia, T. H., Zepf, S. E., Kissler–Patig, M., Hilker, M., Minniti, D., & Goudfrooij, P. 2002, *A&A*, 391, 453
Raimondo, G., Brocato, E., Cantiello, M., & Capaccioli, M. 2005, *AJ*, 130, 2625
Santos Jr., J. F. C. & Frogel, J. F. 1997, *ApJ*, 479, 764
Strader, J., Brodie, J. P., Cenarro, A. J.,Beasley, M. A., & Forbes, D. A. 2005, *AJ*, 130, 1315
Yoon, S. J., Yi, S. K., & Lee, Y. W. 2006, *Science*, 311, 1129

Star clusters: basic galactic building blocks
Proceedings IAU Symposium No. 266, 2009
R. de Grijs & J. R. D. Lépine, eds.

doi:10.1017/S1743921309991050

X-rays from young star clusters: a complement to optical and infrared views

Francesco Damiani

INAF – Osservatorio Astronomico di Palermo,
Piazza del Parlamento 1, 90134 Palermo, Italy
email: damiani@astropa.inaf.it

Abstract. In the last few years, X-ray observational studies of young star clusters have advanced significantly, mainly thanks to the great capabilities of current X-ray observatories such as *Chandra* and *XMM/Newton*. In addition to enabling a detailed study of coronae of individual bright stars, high-spatial-resolution X-ray observations of many young clusters and star-forming regions, even massive and distant ones, have led to the detection of large populations of X-ray-bright members, often down to subsolar masses, and despite strong absorption. The peculiar ability of X-ray emission to select young, low-mass cluster stars against a crowded Galactic-plane field-star background has permitted better studies of global cluster properties, with respect to optical/infrared studies alone, including of cluster initial mass functions (across wide mass ranges), star-formation histories (with indication of age spreads—or even sequences—in many clusters) and morphologies (various degrees of symmetry and dynamical relaxation), sometimes with evidence of mass segregation. Also, the complementary availability of X-ray and optical/infrared data has enabled to place constraints on lifetimes and depletion mechanisms of pre-main-sequence circumstellar disks.

Keywords. stars: pre–main-sequence, open clusters and associations: general, X-rays: stars

1. Introduction: X-ray emission of young stars

X-ray observations have become an important tool to study stellar clusters, especially the youngest ones, including star-formation regions (SFRs). This is related to the properties of X-ray emission of young stars, which have been the subject of extensive reviews elsewhere (e.g., Feigelson & Montmerle 1999; Güdel 2004; Feigelson *et al.* 2007; Güdel & Nazé 2009). Very briefly, X-ray emission from solar-type stars was found to decrease strongly with stellar age, being highest at the earliest stellar evolutionary stages, during the pre-main-sequence (PMS) phase. This result was already found very clearly from the earliest studies of open clusters (Pleiades, Hyades) and SFRs (Taurus–Auriga, Chamaeleon) done with the *Einstein* observatory (see, e.g., Damiani *et al.* 1995; and references therein). More recent, deeper X-ray data have yielded slightly different (lower) mean X-ray luminosities, L_X, for some star clusters or groups, but the decrease with age is clearly still there. Therefore, very young (PMS) stars stand out clearly in X-ray images with respect to older field stars, which are $\sim$ 1000 times fainter. One decade after the *Einstein* observations, X-ray data obtained with the *ROSAT* satellite also permitted to discover genuine new clusters, such as those around σ Ori (Walter *et al.* 1997) or η Cha (Mamajek *et al.* 1999).

2. *XMM–Newton* and *Chandra* observations of young clusters

With respect to *Einstein* and *ROSAT*, the much better sensitivity, spatial and spectral resolution of the newest generation of X-ray space observatories with good imaging capabilities, *XMM–Newton* and *Chandra*†, both operating since late 1999–early 2000, have permitted significant advances in X-ray studies of young clusters. Both the number of SFRs studied, and the amount of detailed X-ray data which have become available for each SFR, have grown considerably.

To better appreciate the amount of new X-ray observational data now available for such clusters, and the many projects making use of them, often very recent or still ongoing, Tables 1 and 2 report lists of all young clusters observed to date by *XMM–Newton* and *Chandra*, respectively. In these lists, observations of small PMS star groups (e.g., near Herbig Ae/Be stars) or observations of individual PMS stars are not included. A number of the listed papers, moreover, are not studies of the properties of the entire population of a cluster, but only of some selected members, and therefore they will not be discussed further.

Clearly, the *Chandra* cluster list is nearly twice as long as the *XMM–Newton* list. It also includes a larger percentage of giant SFRs, often at large distances (say, > 1.5 kpc), hosting rich and crowded stellar populations: in these cases, the very high spatial resolution offered by *Chandra* images (PSF $\sim 0.5''$ FWHM on-axis) is really needed to spatially resolve the clusters' stellar X-ray sources, while *XMM–Newton* images (PSF $\sim 4 - 5''$ FWHM) of such clusters are often affected by source confusion, with many of the fainter sources being missed, and are more useful for X-ray studies of bright, massive cluster stars. As an example, NGC 6231 ($d \sim 1.3$ kpc) was observed using both the *XMM–Newton* EPIC (170 ks exposure; ~ 800 detected sources in a $\sim 30' \times 30'$ field of view) and *Chandra* ACIS instruments (120 ks exposure; ~ 1500 sources in a $\sim 17' \times 17'$ field). Despite the shorter exposure time and smaller field of view, *Chandra* has detected almost twice as many sources as *XMM–Newton*.

2.1. *X-rays versus other selection methods*

It may be useful to give some rough numbers on the relative efficiency in selecting various types of young stars, by using different methods, as it emerges from the literature listed in Tables 1 and 2:

- X-rays: up to 90% success rate (especially for rich clusters; effective for Class I–III stars).
- Infrared (IR)/Hα emission: widely variable 5–70% success rate (but useful for objects belonging to Classes I–II only, since it misses Class III stars).
- Proper motion: in principle, a better membership indicator, but available only for relatively few stars in not many clusters.
- Lithium (plus radial velocity) test: 'definitive' test, but observing with sufficiently high spectral resolution over 10 000 stars in a typical $20' \times 20'$ Galactic-plane cluster field down to, e.g., $V = 20$ mag takes a very long time.

An example of a high-mass SFR, studied in detail at many wavelengths including X-rays, is the cluster NGC 6530, in the HII region M8 (Lagoon Nebula). This is a typical 'blister' SFR, where the massive star cluster NGC 6530 excavates a cavity in the molecular cloud. For this cluster, sequential star formation was suggested by Lada *et al.* (1976). The *Chandra* ACIS images of this region show a complex structure (Figure 1), with the bulk of the cluster members found in a central, older, almost spherical cluster (Damiani *et al.* 2004), and younger subclusters of stars lying both to the west (around the well-known O7.5 star Herschel 36, embedded in the Hourglass Nebula) and to the south-east (near the massive young stellar object M8–East IR). About 1500 X-ray point sources are

† Other X-ray observatories have observed PMS stars between the *ROSAT* and *XMM–Newton/Chandra* eras, such as *ASCA* and *BeppoSAX*; their results are, however, more relevant for studies of coronal physics, not for cluster studies, and will not be discussed here.

Table 1. *XMM–Newton* observations of young clusters

Taurus–Auriga	Many papers, e.g., Favata *et al.* (2003); XEST project: Güdel *et al.* (2007); new members in Scelsi *et al.* (2008)
ρ Oph	Ozawa *et al.* (2005); Giardino *et al.* (2007)
Lupus	Gondoin (2006)
Upper Sco–Cen	Argiroffi *et al.* (2006)
Cha I	Stelzer *et al.* (2004); Robrade & Schmitt (2007)
TW Hya	Stelzer & Schmitt (2004); Argiroffi *et al.* (2005)
CrA	Hamaguchi *et al.* (2005); Forbrich *et al.* (2006)
NGC1333	Preibisch (2003)
IC348	Preibisch & Zinnecker (2004)
IC2391	Marino *et al.* (2005)
IC2602	–
NGC2547	Jeffries *et al.* (2006)
Serpens	Preibisch (2003)
25 Ori	–
Barnard 30/35	–
L1251	Simon (2006)
Barnard 335	–
OMC 2/3	–
σ Ori	Franciosini *et al.* (2006)
λ Ori	–
NGC2024	–
NGC2023	–
NGC2071	Skinner *et al.* (2007)
NGC2264	Simon & Dahm (2005)
IC5146	Skinner *et al.* (2009)
NGC6383	Rauw *et al.* (2003)
γ^2 Vel	Jeffries *et al.* (2009)
NGC2362	–
NGC6530	Rauw *et al.* (2002)
NGC6231	Sana *et al.* (2006, 2007)
NGC6604	De Becker *et al.* (2005)
M17	–
Cyg OB2	Linder *et al.* (2009)
χ Per (NGC884)	–
Tr16/14/Car OB1	Albacete Colombo *et al.* (2003); Antokhin *et al.* (2008)
Westerlund 1	Muno *et al.* (2007)

detected, down to about 0.3 $M_{\odot}$. Deep follow-up optical photometry (Prisinzano *et al.* 2005) obtained with WIFI (at the *ESO 2.2m* telescope) has shown a well-defined X-ray-selected cluster sequence down to $V \sim 21$ mag, which has been used to compute ages, masses and the initial mass function (IMF; see next section). A *VLT*/Giraffe spectroscopic study of radial velocity/lithium (Prisinzano *et al.* 2007) yielded an X-ray selection completeness of $\geqslant$90% down to $V \sim 18$ mag.

3. Recent results

The many new observational studies listed in Tables 1 and 2 have provided important information on the stellar populations in those clusters. In most instances, the bulk of the cluster members (often many hundreds) were selected individually for the first time, with little or no confusion, and reliable optical/IR identifications with very few (< 1%) ambiguous cases. This permitted to address a number of questions, which I outline here.

3.1. *Cluster morphologies*

Detection of a sizable population of a cluster permits to study its morphology, which may offer clues as to the cluster's dynamical state: quasi-spherical clusters are likely relaxed dynamically, while those with an irregular distribution of stars are dynamically less evolved. Very few young clusters are actually spherical, and those with the most regular morphologies still show an asymmetry of some sort. In this category should be placed, e.g., NGC 2362, NGC 6231, NGC 6611, W3 Main, NGC 2244 and the Orion Nebula Cluster (ONC), to a limited extent.

Table 2. *Chandra* observations of young clusters

Taurus–Auriga	Many papers on individual stars
η Cha	Evans *et al.* (2003, 2004)
ε Cha	Feigelson *et al.* (2003); Testa *et al.* (2008)
Cha I	Feigelson & Lawson (2004)
CrA + Coronet cluster	Hamaguchi *et al.* (2005); Forbrich *et al.* (2007)
ρ Oph	Imanishi *et al.* (2001); Gagné *et al.* (2004)
TW Hya association	Gizis *et al.* (2004)
NGC1333	Getman *et al.* (2002)
IC348	Preibisch & Zinnecker (2001, 2002)
L1251B	Simon (2009)
L1448N-A	Tsujimoto *et al.* (2005)
L1415	—
L1527	—
LkHα 101	Osten & Wolk (2009)
Orion Nebula Cluster	Garmire *et al.* (2000); Flaccomio *et al.* (2003a,b); Feigelson *et al.* (2003); COUP project: Getman *et al.* (2005), and entire *ApJS* vol. 160
Orion flanking fields	Ramirez *et al.* (2004); Rebull *et al.* (2006)
M78/NGC2068	Grosso *et al.* (2004)
NGC2071	Skinner *et al.* (2009)
NGC2024	Skinner *et al.* (2003)
OMC 2-3	Tsuboi *et al.* (2001); Tsujimoto *et al.* (2002)
λ Ori	—
σ Ori	Skinner *et al.* (2008)
L1630 (HH 24–26)	Simon *et al.* (2004)
Mon R2	Kohno *et al.* (2002); Nakajima *et al.* (2003)
NGC2244 (Rosette)	Townsley *et al.* (2003); Wang *et al.* (2008, 2009)
NGC2264 North/South	Ramirez *et al.* (2004); Rebull *et al.* (2006); Flaccomio *et al.* (2006)
NGC2362	Damiani *et al.* (2006); Delgado *et al.* (2006)
NGC7129	Stelzer & Scholz (2009)
IC1396–N	Getman *et al.* (2007)
CG12	Getman *et al.* (2008)
h Per	Currie *et al.* (2009)
DB2001CL–123	–
NGC281	–
Cep B	Getman *et al.* (2006)
M8/NGC6530	Damiani *et al.* (2004)
M20/NGC6514	Rho *et al.* (2004)
M16/NGC6611	Linsky *et al.* (2007); Guarcello *et al.* (2007)
M17/NGC6618	Townsley *et al.* (2003); Broos *et al.* (2007)
NGC6231	Damiani *et al.* (2009)
NGC6334	Ezoe *et al.* (2006); Feigelson *et al.* (2009)
NGC6357	Wang *et al.* (2007)
RCW 38	Wolk *et al.* (2006)
W3	Hofner *et al.* (2002); Feigelson & Townsley (2008)
W51	Koo *et al.* (2005)
Cyg OB2	Albacete Colombo *et al.* (2007)
S106	Giardino *et al.* (2004)
Berkeley 87	–
Serpens	Giardino *et al.* (2007); Winston *et al.* (2007)
RCW 36	–
W49A	Tsujimoto *et al.* (2006)
NGC7538	–
IC1805	–
Sh2–187	–
NGC1893	Caramazza *et al.* (2008)
DR21	–
W75N	–
GGD 27	Pravdo *et al.* (2009)
W3 (Main/North/OH)	Feigelson & Townsley (2008)
RCW 108/NGC6193	Skinner *et al.* (2005); Wolk *et al.* (2008)
Trumpler 14	Townsley (2006)
Trumpler 16	Evans *et al.* (2004); Albacete Colombo *et al.* (2008)
Carina complex	–
NGC3576	Townsley (2006)
NGC3603	Moffat *et al.* (2002); Poteet *et al.* (2004)
Westerlund 1	Skinner *et al.* (2006); Clark *et al.* (2008)
RCW 49 (Westerlund 2)	Tsujimoto *et al.* (2007); Nazé *et al.* (2008)
Sgr B2	Takagi *et al.* (2002)
Arches	Yusef–Zadeh *et al.* (2002); Wang *et al.* (2006)
RSG2	–

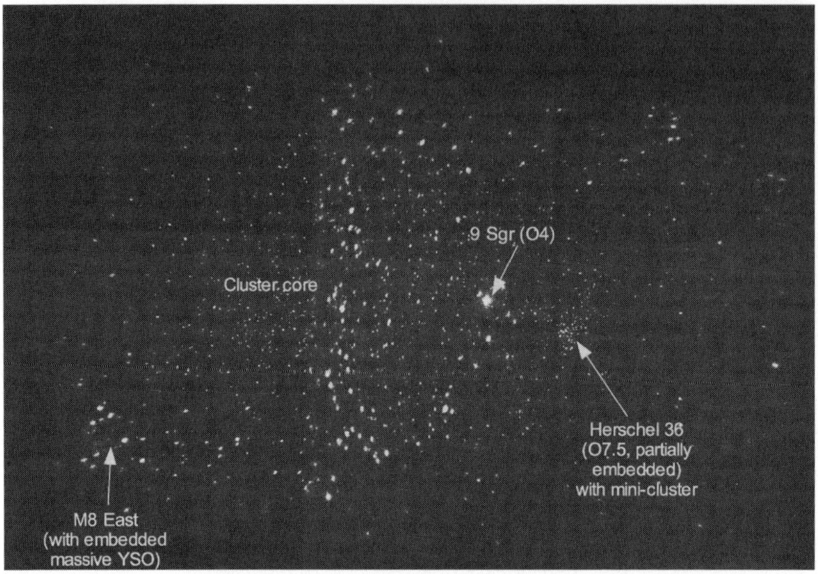

Figure 1. NGC 6530 and M8 in X-rays. *Chandra* ACIS mosaic ($\sim 20' \times 30'$).

Many other clusters instead show either an elongated, irregular shape, or even multiple star-forming subclusters, which can be taken as indicative that stars in each subcluster have not traveled very far since their formation, and are thus still near their birth places. Examples include NGC 6530/M8, NGC 6618/M17, Cyg OB2, NGC 6334, NGC 1893, W3 (OH) and Cep B.

3.2. *Cluster subpopulations*

The X-ray discovery of individual cluster members across a wide range of masses and extinctions (and, by inference, PMS evolutionary stage) permits comparison of different subpopulations in some clusters.

Low-mass versus massive stars: The low-mass stellar density peaks usually around OB stars. This is true in both near-spherical and clumpy clusters, where (often) each clump has one (or more) massive star(s) at its center. However, sometimes O stars are found off stellar-density peaks: examples are η Car in Trumpler 16, 9 Sgr in NGC 6530, HD 46223 in NGC 2244, the O star in W3 North (Albacete Colombo *et al.* 2008; Damiani *et al.* 2004; Wang *et al.* 2008, 2009; Feigelson & Townsley 2008). The reason of these 'anomalies' is still unknown: they are perhaps caused by dynamical evolution (runaway stars), or are signs of a different mode of massive-star formation.

Obscured/embedded versus unobscured stars: Cluster subpopulations characterized by large soft-X-ray absorption, indicative of heavy obscuration ($n_{\rm H} \geqslant 10^{22}\,{\rm cm}^{-2}$) are probably embedded, very young populations. They are often spatially distinct from more exposed ($n_{\rm H} \leqslant 10^{21}\,{\rm cm}^{-2}$) subpopulations in the same cluster: this was found in, e.g., NGC 6334, Cep B, NGC 6618 and NGC 6357 and is probably an indication of different episodes of star formation in the same region (Feigelson *et al.* 2009; Getman *et al.* 2006; Broos *et al.* 2007; Wang *et al.* 2007).

The nature of the obscured X-ray-selected cluster members is not easily assessed on the basis of available photometry in the (near-) IR (these stars are usually undetected in optical bands). A typical 2MASS color–magnitude diagram (CMD) of such X-ray-selected stars shows their strong and nonuniform extinction, and that the identification of a cluster sequence would be extremely hard on the basis of the IR data alone: for such a cluster population, X-ray membership information is evidently crucial. Moreover, the newly selected members from X-rays are not confined to the low-mass part of the cluster population: from de-reddening the positions in the IR CMD one also finds new *massive* candidate cluster members, e.g., in NGC 6618 (Broos *et al.* 2007) or NGC 6193 (Wolk *et al.* 2008), which look like very red objects (and were overlooked in previous studies) because of substantial extinction. In similar cases, unambiguous determinations of stellar masses/ages are impossible based on photometry alone, or at best highly uncertain.

3.3. *Age spread/sequences*

The study of X-ray-selected populations has also permitted to find differences in star-formation histories between clusters (based on isochronal stellar ages when feasible, or using other age indicators):

• Formation during *an extended period of time* (5–10 Myr): e.g., NGC 6530 (Damiani *et al.* 2004; Prisinzano *et al.* 2005), NGC 6231 (Damiani *et al.* 2009);

• Nearly *simultaneous* formation (within $\sim$ 1 Myr): NGC 2362 (Damiani *et al.* 2006; in agreement with the previous study by Moitinho *et al.* 2001);

• Formation along *mass sequences*: e.g., NGC 6231 (low-mass stars $\sim$ 10 Myr ago, massive stars $\sim$ 2 Myr ago; Damiani *et al.* 2009), W3 Main (low-mass stars $\sim$ 1 Myr ago, massive stars still in ultracompact H II region, $\leqslant$ 0.1 Myr ago; Feigelson & Townsley 2008).

These studies complement, therefore, similar studies of cluster stellar ages in nearer, better-studied clusters such as Taurus–Auriga and Orion (of lower masses, however), with comparable results.

3.4. *Triggering*

In a few cases, the characteristics of the X-ray-detected population in a SFR have been found compatible with those expected for triggered star formation by one or more massive stars. One piece of evidence in favor of this hypothesis is star formation through spatial sequences, with spatial subgroups having different average ages, as derived from either location in the CMD, varying morphology (relaxed versus knotty), differing amount of absorption or other youth indicators (disks, ultracompact H II regions, etc.).

Examples include NGC 6530/M8, where both isochronal ages and the evolutionary stages of the objects suggest a time sequence from the older cluster core, to the younger Hourglass Nebula, to the youngest M8 East IR protostar (Damiani *et al.* 2004; supporting earlier suggestions by Lada *et al.* 1976). Another example is the Cep B/Cep OB3b/Sh 155 complex (Getman *et al.* 2006, 2009), where the suggested sequence runs from northwest (outside the H II region Sh 155, older stars) to southeast (inside Sh 155, younger embedded cluster), on the basis of both the degree of X-ray absorption and disk frequency (both proxies for the stellar evolutionary stage). The two youth indicators agree with one another, and their spatial location also agrees qualitatively with the comet-head shape of Sh 155, and the position of the massive ionizing O7 star HD 217086, which the head of the Sh 155 nebula points to.

Finally, Getman *et al.* (2007) have found evidence of triggering also in the IC 1396N globule. Here, a spatial age sequence is found as a string of embedded Class I/0 objects (inside the globule), Class II (midway), and X-ray-selected Class III stars (outside the globule towards the exciting star HD 206267). This agrees with a radiation-driven implosion (RDI) model (e.g., Lefloch & Lazareff 1994) for triggered star formation in bright-rimmed clouds. On the other hand, the same authors (Getman *et al.* 2008) find that the simple RDI model is not compatible with their *Chandra* X-ray observation of another small star-forming globule, CG 12, where many distinct episodes of triggering might have taken place instead.

3.5. *Initial mass function*

The IMF has been studied intensively for many clusters, benefitting from the relatively easy and unbiased selection of large populations of members using X-ray data. Despite having been studied for decades, important questions still remain: Is the IMF universal? If not, how does it depend on ambient conditions? Are there clusters with top-heavy IMFs?

Different X-ray-based methods have been devised to address these questions: (a) Find optical counterparts to X-ray sources (especially low-mass; massive stars have often better-characterized membership), derive their masses from the position in the CMD and PMS evolutionary models (when feasible) and make a mass histogram. (b) Alternatively, compare the hard-band (less sensitive to absorption) X-ray luminosity function with that of well-studied, template clusters (e.g., the ONC), test shape similarity and normalize. Of course, these procedures are not free of uncertainties, and a few warnings are in order to obtain reliable results:

1. An X-ray IMF determination must be corrected for X-ray (in)completeness, especially in lowest-mass bins (e.g., Prisinzano *et al.* 2005 for NGC 6530).

2. If many cluster stars have IR excesses, photometrically derived masses are likely inaccurate, and one must resort to spectroscopy (Prisinzano *et al.* 2007; again for NGC 6530).

3. If many optically/IR unidentified X-ray sources remain (which still show clear spatial clustering, and are thus mostly members) one can try to include them by deriving stellar masses from X-ray luminosities, L_X, using the 'saturation' relation for low-mass stars $L_X \sim 10^{-3.5} L_{bol}$ (e.g., Randich *et al.* 2000; Flaccomio *et al.* 2003) and the L_{bol}–mass relation appropriate to the cluster age/reddening (e.g., Damiani *et al.* 2009: NGC 6231).

4. Optical glare from OB stars may prevent finding faint, low-mass stars as counterparts to detected X-ray sources: it is much less so in X-rays, since $L_X/L_{bol} \sim 10^{-3.5}$ for low mass stars, but $L_X/L_{bol} \sim 10^{-7}$ for OB stars (Pallavicini *et al.* 1981). Therefore, the brightness contrast is reduced from $\sim 10^6$ in the optical to only $10^2 - 10^3$ at X-rays, which are thus an especially good tool to pinpoint low-mass stars near bright and massive ones in a dense cluster (e.g., NGC 2244: Wang *et al.* 2009).

Results of studies of cluster IMFs include, for many clusters, the *first* IMF determination, often ranging down to (sub)solar-mass stars: up to 2 orders of magnitude in mass may be covered (e.g., NGC 6231: Damiani *et al.* 2009). A comparison between the IMFs of many clusters, derived using X-ray data, shows that the upper part of the IMF is generally found to be consistent with a Salpeter IMF. A turnoff is found at some mass, generally subsolar (but NGC 2362 is a possible exception; Damiani *et al.* 2006). However, the turnoff mass appears to be slightly different from cluster to cluster.

A comparison with a previous optical/Hα IMF determination for NGC 6231 (Sung *et al.* 1998) shows that this latter misses weak-line T Tauri stars entirely (90% of the low-mass stars in this case), yielding an apparent deficit of low-mass stars in this cluster. The rich, low-mass population of NGC 6231 is instead clearly detected in the X-ray data (Damiani *et al.* 2009).

X-ray luminosity functions (XLFs) have also been computed for many high-mass clusters and are another useful tool to examine possible IMF differences from cluster to cluster. Analogously to IMFs, XLFs of different massive young clusters are generally similar (see Wang *et al.* 2008 for a comparison among NGC 6357, NGC 2244, Cep B,

Table 3. Circumstellar-disk frequencies

Cluster	Age (Myr)	Diagnostics	Mass range	Frequency	Notes
ONC	1	near-IR	$M > 1$ M$_\odot$	20.6%	(1)
Cyg OB2	2	near-IR	$M > 1$ M$_\odot$	4.4%	very low
NGC2264	3	Hα		42%	
NGC1893	3–4	*Spitzer* SED		67%	
NGC2362	5	Hα (IR)		5% (12%)	
NGC6530	2–3	near-IR	$M > 0.5$ M$_\odot$	20%	
NGC6611	1	near-IR	$M > 0.5$ M$_\odot$	20%	
Tr 16	3	near-IR	$M > 0.5 - 1$ M$_\odot$	15%	
NGC6231	5–7	near-IR	$M > 1$ M$_\odot$	4%	very low

(1): The frequency becomes higher using more indicators/a larger mass range.

ONC and M17 XLFs). When scaling the observed L_X distributions to that of the ONC, no low-mass-star deficit is found. There appears to be no top-heavy IMF in any of the clusters studied by this method.

3.6. *Mass segregation*

X-ray data also permit to address the issue of mass segregation in high-mass clusters. The question is well posed only for those clusters with approximate spherical shape, where a meaningful center exists. Here, the X-ray-traced low-mass population can be compared to the (usually optically selected) massive stars.

Mass segregation has been found in a few X-ray-studied clusters (e.g., NGC 6231: Damiani *et al.* 2009; NGC 2362: Damiani *et al.* 2006), while it has not been found in others (e.g., NGC 2244: Wang *et al.* 2008, 2009). Whether this difference is due to dynamical evolution, or it is of primordial origin, remains unknown.

3.7. *Disk frequency versus age and environment*

Since different diagnostics of the presence and type of PMS stellar disks (inner disk from near-IR, outer disk from far-IR), or of classical T Tauri star status (accretion through Hα, optical/ultraviolet veiling) are found in the literature, it is not always possible to use a uniform disk indicator for a comparison of disk frequencies among many clusters. Using X-ray-selected diskless members, and various disk/accretion indicators, Table 3 compares the circumstellar-disk frequency (in a selected mass range) for several clusters.

The low disk frequencies found in Cyg OB2 (Albacete Colombo *et al.* 2007) and perhaps NGC 6231 (Damiani *et al.* 2009) are possibly caused by fast disk evolution under the effect of the radiation field of OB stars (photoevaporation). Another clue to the same effect has been found in NGC 6611 by Guarcello *et al.* (2007): disks are significantly less frequent in regions in which the OB stars' integrated ultraviolet flux is more intense, within the same cluster. This suggests strongly that age is not the only relevant parameter for disk evolution.

3.8. *Diffuse X-ray emission*

Genuinely diffuse X-ray emission from hot gas has been found in X-ray images of several high-mass clusters (Orion Nebula: Güdel *et al.* 2008; NGC 2024: Ezoe *et al.* 2006; M17: Townsley *et al.* 2003; Rosette: Wang *et al.* 2009; Westerlund 1: Muno *et al.* 2006; Carina Nebula: Ezoe *et al.* 2009; RCW38: Wolk *et al.* 2002, 2006; NGC 6334: Ezoe *et al.* 2006; 30 Dor in the LMC: Townsley *et al.* 2006), in excess of the extrapolated contribution of weak, unresolved point sources, often in a different location and with a different X-ray spectrum than the point sources.

The favored picture for the origin of diffuse X-rays envisages that winds from massive stars collide and coalesce. Then, a 'champagne flow' of hot gas flows out of the opening in the blister molecular cloud. Thus, a molecular cloud is *no longer a cold place* once massive stars form.

The X-ray spectrum of such diffuse X-ray emission is often soft (in Orion, M17, Rosette, 30 Dor); sometimes, a hard spectral component is also present (NGC 2024, RCW38, Westerlund 1), and a complex spatial/spectral structure was found in the giant molecular complexes of Carina and NGC 6334.

3.9. *The largest young clusters in our Galaxy in X-rays*

Most stars in the Galaxy form in large clusters, containing $> 10\,000$ stars, rather than in 'smallish' SFRs like Taurus–Auriga, with $\sim 100 - 200$ members. The study of stellar formation (and early evolution) in such giant clusters is therefore very important to understand the condition at birth of the 'average' stars in the Galaxy. However, the largest clusters in the Galaxy lie mostly at large distances, and behind considerable Galactic-plane absorption, which make their rich (and crowded) low-mass population hard to study. Three of such Galactic clusters have been studied in X-rays with *Chandra*:

- *Westerlund 1* (mass $\sim 10^5$ $M_\odot$, age $\sim 4 - 5$ Myr, distance ~ 5 kpc): two exposures of $18 + 42$ ks have detected 46 O and Wolf-Rayet (WR) stars, diffuse X-ray emission and 45 candidate PMS stars, out of an inferred population of $>$32 000 low-mass PMS stars (Clark *et al.* 2008).
- *Westerlund 2 (RCW49)* (age ~ 2 Myr, uncertain distance in the range 2–8 kpc (!)): hosts 12 known early-O stars, 2 WR stars, including WR20a ($82 + 83$ $M_\odot$). One exposure of 40 ks has detected more than 350 new candidate low-mass (PMS) members and also 30 candidate OB members. Possible triggered star formation has been suggested in the southeastern region (Tsujimoto *et al.* 2007; Nazé *et al.* 2008).
- *NGC 3603* (age $\sim 0.3 - 1$ Myr, distance ~ 7 kpc): analogous to starburst clusters in, e.g., M82. One exposure of 50 ks has detected 348 X-ray sources, of which 40 are OB and WR stars, and most of the others are candidate PMS members (Moffat *et al.* 2002).

In these clusters, current X-ray data have only detected a minor fraction of the (estimated) low-mass cluster population, because of the insufficient depth of the observations, and deeper observations capable of detecting the bulk of the cluster stars and pushing to their limits the capabilities of current X-ray instruments would be highly desirable.

4. Large projects with focus on star-formation regions

A number of recent or ongoing projects of X-ray studies of SFRs exist, based on X-ray observations, either very deep exposures or covering thoroughly a given region.

- COUP: '*Chandra* Orion Ultradeep Project' (PI E. Feigelson): a 850 ks ($\sim$ 10 days) nearly uninterrupted (spacecraft-operations permitting) deep pointing of the Orion Nebula Cluster, presented by Getman *et al.* (2005); all of *ApJS* vol. 160 was devoted to the project. In total, 1616 X-ray sources were found (hundreds of them heavily obscured), as well as 75 new cluster members.
- XEST: 'X-ray Extended Survey of Taurus' (PI M. Güdel): a survey with *XMM–Newton*, using more than 20 medium-depth (50 ks) pointings covering most of Taurus–Auriga (Güdel *et al.* 2007; and *A&A* vol. 468); 169 PMS stellar systems were observed, of which 136 were detected in X-rays.
- DROXO: 'Deep Rho Ophiuchi X-ray Observation' (PI S. Sciortino): a 500 ks deep *XMM–Newton* pointing of ρ Oph; 111 sources down to $\log L_X$ [erg s^{-1}] $= 28.3$ (Giardino *et al.* 2007; Pillitteri *et al.*, in prep.).
- *NGC 1893* (PI G. Micela): a 450 ks deep *Chandra* observation of the massive SFR NGC 1893, at $d \sim 3$ kpc in the anticenter direction, aimed at studying with sufficient

sensitivity stellar formation and early evolution in the outer part of the Galaxy (Caramazza *et al.* 2008, in prep.).

• SOXS '*Spitzer* Orion–A *XMM–Newton* Survey' (PI S. Wolk): 7 (+2) 50 ks *XMM–Newton* pointings in southern Orion, complemented with new *Spitzer* data. Data analysis is in progress, with detection of ~ 1700 Class I–II young stellar objects, and ~ 1100 X-ray sources (Wolk *et al.* 2009; Pillitteri *et al.*, this volume).

• *The Carina Nebula survey* (PI L. Townsley): a wide-angle *Chandra* survey of the giant Carina SFR, with 25 pointings covering among others the clusters Trumpler 16 (with η Car), Trumpler 14 and Collinder 228, and overall > 70 O and WR stars.

It is also worth mentioning in this context the recently approved 1 Ms mosaic of *Chandra* observations of the Cygnus OB2 giant SFR (PI J. Drake), although observations have not yet started. This will undoubtedly provide better insights into this very massive and rich cluster, where more than 1000 X-ray sources have already been discovered based on a shallower 100 ks observation (Albacete Colombo *et al.* 2007; Wright and Drake, this volume).

Globally, these large projects will permit more detailed studies of star formation occurring in very different environments, and at various evolutionary stages. A wealth of new results are likely to come once they are completed.

Acknowledgement

I acknowledge support from ASI–INAF contract I/088/06/0 (Italy).

References

Albacete Colombo, J. F., Méndez, M., & Morrell, N. I. 2003, *MNRAS*, 346, 704
Albacete Colombo, J. F., Flaccomio, E., Micela, G., Sciortino, S., & Damiani, F. 2007, *A&A*, 464, 211
Albacete Colombo, J. F., Damiani, F., Micela, G., Sciortino, S., & Harnden Jr., F. R. 2008, *A&A*, 490, 1055
Antokhin, I. I., Rauw, G., Vreux, J.-M., van der Hucht, K. A., & Brown, J. C. 2008, *A&A*, 477, 593
Argiroffi, C., Maggio, A., Peres, G., Stelzer, B., & Neuhäuser, R. 2005, *A&A*, 439, 1149
Argiroffi, C., Favata, F., Flaccomio, E., Maggio, A., Micela, G., Peres, G., & Sciortino, S. 2006, *A&A*, 459, 199
Broos, P. S., Feigelson, E. D., Townsley, L. K., Getman, K. V., Wang, J., Garmire, G. P., Jiang, Z., & Tsuboi, Y. 2007, *ApJS*, 169, 353
Caramazza, M., Micela, G., Prisinzano, L., Rebull, L., Sciortino, S., & Stauffer, J. R. 2008, *A&A*, 488, 211
Clark, J. S., Muno, M. P., Negueruela, I., Dougherty, S. M., Crowther, P. A., Goodwin, S. P., & de Grijs, R. 2008, *A&A*, 477, 147
Currie, T., Evans, N. R., Spitzbart, B. D., Irwin, J., Wolk, S. J., Hernandez, J., Kenyon, S. J., & Pasachoff, J. M. 2009, *AJ*, 137, 3210
Damiani, F., Micela, G., Sciortino, S., & Harnden Jr., F. R. 1995, *ApJ*, 446, 331
Damiani, F., Flaccomio, E., Micela, G., Sciortino, S., Harnden Jr., F. R., & Murray, S. S. 2004, *ApJ*, 608, 781
Damiani, F., Micela, G., Sciortino, S., Huélamo, N., Moitinho, A., Harnden Jr., F. R., & Murray, S. S. 2006, *A&A*, 460, 133
Damiani, F., Micela, G., Sciortino, S., & Harnden Jr., F. R. 2009, *AIP Conf. Ser.*, 1094, 916
De Becker, M., Rauw, G., Blomme, R., Pittard, J. M., Stevens, I. R., & Runacres, M. C. 2005, *A&A*, 437, 1029
Delgado, A. J., González–Martín, O., Alfaro, E. J., & Yun, J. 2006, *ApJ*, 646, 269
Evans, N. R., Seward, F. D.; Krauss, M. I., Isobe, T., Nichols, J., Schlegel, E. M., & Wolk, S. J. 2003, *ApJ*, 589, 509

Evans, N. R., Schlegel, E. M., Waldron, W. L., Seward, F. D., Krauss, M. I., Nichols, J., & Wolk, S. J. 2004, *ApJ*, 612, 1065

Ezoe, Y., Kokubun, M., Makishima, K., Sekimoto, Y., & Matsuzaki, K. 2006, *ApJ*, 638, 860

Ezoe, Y., Kokubun, M., Makishima, K., Sekimoto, Y., & Matsuzaki, K. 2006, *ApJ* (Letters), 649, L123

Ezoe, Y., Hamaguchi, K., Gruendl, R. A., Chu, Y.-H., Petre, R., & Corcoran, M. F. 2009, *PASJ*, 61, S123

Favata, F., Giardino, G., Micela, G., Sciortino, S., & Damiani, F. 2003, *A&A*, 403, 187

Feigelson, E. D. & Montmerle, T. 1999, *ARA&A*, 37, 363

Feigelson, E. D., Gaffney III, J. A., Garmire, G., Hillenbrand, L. A., & Townsley, L. 2003, *ApJ*, 584, 911

Feigelson, E. D., Lawson, W. A., & Garmire, G. P. 2003, *ApJ*, 599, 1207

Feigelson, E. D. & Lawson, W. A. 2004, *ApJ*, 614, 267

Feigelson, E., Townsley, L., Güdel, M., & Stassun, K. 2007, in: Reipurth, B., Jewitt, D., & Keil, K., *Protostars and Planets V*, p. 313 (Tucson: University of Arizona Press)

Feigelson, E. D. & Townsley, L. K. 2008, *ApJ*, 673, 354

Feigelson, E. D., Martin, A. L., McNeill, C. J., Broos, P. S., & Garmire, G. P. 2009, *AJ*, 138, 227

Flaccomio, E., Damiani, F., Micela, G., Sciortino, S., Harnden Jr., F. R., Murray, S. S., & Wolk, S. J. 2003, *ApJ*, 582, 382

Flaccomio, E., Damiani, F., Micela, G., Sciortino, S., Harnden Jr., F. R., Murray, S. S., & Wolk, S. J. 2003, *ApJ*, 582, 398

Flaccomio, E., Micela, G., & Sciortino, S. 2006, *A&A*, 455, 903

Forbrich, J., Preibisch, T., & Menten, K. M. 2006, *A&A*, 446, 155

Forbrich, J. & Preibisch, T. 2007, *A&A*, 475, 959

Franciosini, E., Pallavicini, R., & Sanz–Forcada, J. 2006, *A&A*, 446, 501

Gagné, M., Skinner, S. L., & Daniel, K. J. 2004, *ApJ*, 613, 393

Garmire, G., Feigelson, E. D., Broos, P., Hillenbrand, L. A., Pravdo, S. H., Townsley, L., & Tsuboi, Y. 2000, *AJ*, 120, 1426

Getman, K. V., Feigelson, E. D., Townsley, L., Bally, J., Lada, C. J., & Reipurth, B. 2002, *ApJ*, 575, 354

Getman, K. V., *et al.* 2005, *ApJS*, 160, 319

Getman, K. V., Feigelson, E. D., Townsley, L., Broos, P., Garmire, G., & Tsujimoto, M. 2006, *ApJS*, 163, 306

Getman, K. V., Feigelson, E. D., Garmire, G., Broos, P., & Wang, J. 2007, *ApJ*, 654, 316

Getman, K. V., Feigelson, E. D., Lawson, W. A., Broos, P. S., & Garmire, G. P. 2008, *ApJ*, 673, 331

Getman, K. V., Feigelson, E. D., Luhman, K. L., Sicilia–Aguilar, A., Wang, J., & Garmire, G. P. 2009, *ApJ*, 699, 1454

Giardino, G., Favata, F., & Micela, G. 2004, *A&A*, 424, 965

Giardino, G., Favata, F., Pillitteri, I., Flaccomio, E., Micela, G., & Sciortino, S. 2007, *A&A*, 475, 891

Giardino, G., Favata, F., Micela, G., Sciortino, S., & Winston, E. 2007, *A&A*, 463, 275

Gizis, J. E. & Bharat, R. 2004, *ApJ* (Letters), 608, L113

Gondoin, P. 2006, *A&A*, 454, 595

Grosso, N., Montmerle, T., Feigelson, E. D., & Forbes, T. G. 2004, *A&A*, 419, 653

Guarcello, M. G., Prisinzano, L., Micela, G., Damiani, F., Peres, G., & Sciortino, S. 2007, *A&A*, 462, 245

Güdel, M. 2004, *A&A Rev.*, 12, 71

Güdel, M., *et al.* 2007, *A&A*, 468, 353

Güdel, M., Briggs, K. R., Montmerle, T., Audard, M., Rebull, L., & Skinner, S. L. 2008, *Science*, 319, 309

Güdel, M. & Nazé, Y. 2009, *A&A Rev.*, 17, 309

Hamaguchi, K., Corcoran, M. F., Petre, R., White, N. E., Stelzer, B., Nedachi, K., Kobayashi, N., & Tokunaga, A. T. 2005, *ApJ*, 623, 291

Hofner, P., Delgado, H., Whitney, B., Churchwell, E., & Linz, H. 2002, *ApJ* (Letters), 579, L95
Imanishi, K., Koyama, K., & Tsuboi, Y. 2001, *ApJ*, 557, 747
Jeffries, R. D., Evans, P. A., Pye, J. P., & Briggs, K. R. 2006, *MNRAS*, 367, 781
Jeffries, R. D., Naylor, T., Walter, F. M., Pozzo, M. P., & Devey, C. R. 2009, *MNRAS*, 393, 538
Kastner, J. H., Huenemoerder, D. P., Schulz, N. S., Canizares, C. R., & Weintraub, D. A. 2002, *ApJ*, 567, 434
Kohno, M., Koyama, K., & Hamaguchi, K. 2002, *ApJ*, 567, 423
Koo, B.-C., Lee, J.-J., Seward, F. D., & Moon, D.-S. 2005, *ApJ*, 633, 946
Lada, C. J., Gottlieb, C. A., Gottlieb, E. W., & Gull, T. R. 1976, *ApJ*, 203, 159
Lefloch, B. & Lazareff, B. 1994, *A&A*, 289, 559
Linder, N., Rauw, G., Manfroid, J., Damerdji, Y., De Becker, M., Eenens, P., Royer, P., & Vreux, J.-M. 2009, *A&A*, 495, 231
Linsky, J. L., Gagné, M., Mytyk, A., McCaughrean, M., & Andersen, M. 2007, *ApJ*, 654, 347
Mamajek, E. E., Lawson, W. A., & Feigelson, E. D. 1999, *ApJ* (Letters), 516, L77
Marino, A., Micela, G., Peres, G., Pillitteri, I., & Sciortino, S. 2005, *A&A*, 430, 287
Moffat, A. F. J., *et al.* 2002, *ApJ*, 573, 191
Moitinho, A., Alves, J., Huélamo, N., & Lada, C. J. 2001, *ApJ* (Letters), 563, L73
Muno, M. P., Law, C., Clark, J. S., Dougherty, S. M., de Grijs, R., Portegies Zwart, S., & Yusef–Zadeh, F. 2006, *ApJ*, 650, 203
Muno, M. P., Gaensler, B. M., Clark, J. S., de Grijs, R., Pooley, D., Stevens, I. R., & Portegies Zwart, S. F. 2007, *MNRAS*, 378, L44
Nakajima, H., Imanishi, K., Takagi, S.-I., Koyama, K., & Tsujimoto, M. 2003, *PASJ*, 55, 635
Nazé, Y., Rauw, G., & Manfroid, J. 2008, *A&A*, 483, 171
Osten, R. A. & Wolk, S. J. 2009, *ApJ*, 691, 1128
Ozawa, H., Grosso, N., & Montmerle, T. 2005, *A&A*, 438, 661
Pallavicini, R., Golub, L., Rosner, R., Vaiana, G. S., Ayres, T., & Linsky, J. L. 1981, *ApJ*, 248, 279
Poteet, C., Marchenko, S., Corcoran, M., & Andersen, M. 2004, *BAAS*, 36, 1380
Pravdo, S. H., Tsuboi, Y., Suzuki, Y., Thompson, T. J., & Rebull, L. 2009, *ApJ*, 690, 850
Preibisch, T. & Zinnecker, H. 2001, *AJ*, 122, 866
Preibisch, T. & Zinnecker, H. 2002, *AJ*, 123, 1613
Preibisch, T. 2003, *A&A*, 401, 543
Preibisch, T. 2003, *A&A*, 410, 951
Preibisch, T. & Zinnecker, H. 2004, *A&A*, 422, 1001
Prisinzano, L., Damiani, F., Micela, G., & Sciortino, S. 2005, *A&A*, 430, 941
Prisinzano, L., Damiani, F., Micela, G., & Pillitteri, I. 2007, *A&A*, 462, 123
Ramírez, S. V., Rebull, L., Stauffer, J., Hearty, T., Hillenbrand, L., Jones, B., Makidon, R., Pravdo, S., Strom, S., & Werner, M. 2004, *AJ*, 127, 2659
Ramírez, S. V., Rebull, L., Stauffer, J., Strom, S., Hillenbrand, L., Hearty, T., Kopan, E. L. Pravdo, S., Makidon, R., & Jones, B. 2004, *AJ*, 128, 787
Randich, S. 2000, *ASP Conf. Ser.*, 198, 401
Rauw, G., Nazé, Y.; Gosset, E., Stevens, I. R., Blomme, R., Corcoran, M. F., Pittard, J. M., & Runacres, M. C. 2002, *A&A*, 395, 499
Rauw, G., De Becker, M., Gosset, E., Pittard, J. M., & Stevens, I. R. 2003, *A&A*, 407, 925
Rebull, L. M., Stauffer, J. R., Ramírez, S. V., Flaccomio, E., Sciortino, S., Micela, G., Strom, S. E., & Wolff, S. C. 2006, *AJ*, 131, 2934
Rho, J., Ramírez, S. V., Corcoran, M. F., Hamaguchi, K., & Lefloch, B. 2004, *ApJ*, 607, 904
Robrade, J. & Schmitt, J. H. M. M. 2007, *A&A*, 461, 669
Sana, H., Rauw, G., Nazé, Y., Gosset, E., & Vreux, J.-M. 2006, *MNRAS*, 372, 661
Sana, H., Rauw, G., Sung, H., Gosset, E., & Vreux, J.-M. 2007, *MNRAS*, 377, 945
Scelsi, L., Sacco, G., Affer, L., Argiroffi, C., Pillitteri, I., Maggio, A., & Micela, G. 2008, *A&A*, 490, 601
Simon, T., Andrews, S. M., Rayner, J. T., & Drake, S. A. 2004, *ApJ*, 611, 940
Simon, T. & Dahm, S. E. 2005, *ApJ*, 618, 795
Simon, T. 2006, *AJ*, 131, 501

Simon, T. 2009, *ApJ*, 694, 425

Skinner, S., Gagné, M., & Belzer, E. 2003, *ApJ*, 598, 375

Skinner, S. L., Zhekov, S. A., Palla, F., & Barbosa, C. L. D. R. 2005, *MNRAS*, 361, 191

Skinner, S. L., Simmons, A. E., Zhekov, S. A., Teodoro, M., Damineli, A., & Palla, F. 2006, *ApJ* (Letters), 639, L35

Skinner, S. L., Simmons, A. E., Audard, M., & Güdel, M. 2007, *ApJ*, 658, 1144

Skinner, S. L., Sokal, K. R., Cohen, D. H., Gagné, M., Owocki, S. P., & Townsend, R. D. 2008, *ApJ*, 683, 796

Skinner, S. L., Sokal, K. R., Güdel, M., & Briggs, K. R. 2009, *ApJ*, 696, 766

Skinner, S. L., Sokal, K. R., Megeath, S. T., Güdel, M., Audard, M., Flaherty, K. M., Meyer, M. R., & Damineli, A. 2009, *ApJ*, 701, 710

Stelzer, B. & Schmitt, J. H. M. M. 2004, *A&A*, 418, 687

Stelzer, B., Micela, G., & Neuhäuser, R. 2004, *A&A*, 423, 1029

Stelzer, B. & Scholz, A. 2009, *A&A*, 507, 227

Sung, H., Bessell, M. S., & Lee, S.-W. 1998, *AJ*, 115, 734

Takagi, S.-I., Murakami, H., & Koyama, K. 2002, *ApJ*, 573, 275

Testa, P., Huenemoerder, D. P., Schulz, N. S., & Ishibashi, K. 2008, *ApJ*, 687, 579

Townsley, L. K., Feigelson, E. D., Montmerle, T., Broos, P. S., Chu, Y.-H., & Garmire, G. P. 2003, *ApJ*, 593, 874

Townsley, L. K., Broos, P. S., Feigelson, E. D., Brandl, B. R., Chu, Y.-H., Garmire, G. P., & Pavlov, G. G. 2006, *AJ*, 131, 2140

Townsley, L. K. 2006, in: Livio M. (ed.), *Massive Stars: From Population III and GRBs to the Milky Way*, in press (astro-ph/0608173)

Tsuboi, Y., Koyama, K., Hamaguchi, K., Tatematsu, K., Sekimoto, Y., Bally, J., & Reipurth, B. 2001, *ApJ*, 554, 734

Tsujimoto, M., Koyama, K., Tsuboi, Y., Goto, M., & Kobayashi, N. 2002, *ApJ*, 566, 974

Tsujimoto, M., Kobayashi, N., & Tsuboi, Y. 2005, *AJ*, 130, 2212

Tsujimoto, M., Hosokawa, T., Feigelson, E. D., Getman, K. V., & Broos, P. S. 2006, *ApJ*, 653, 409

Tsujimoto, M., *et al.* 2007, *ApJ*, 665, 719

Walter, F. M., Wolk, S. J., Freyberg, M., & Schmitt, J. H. M. M. 1997, *Mem. SAIt*, 68, 1081

Wang, Q. D., Dong, H., & Lang, C. 2006, *MNRAS*, 371, 38

Wang, J., Townsley, L. K., Feigelson, E. D., Getman, K. V., Broos, P. S., Garmire, G. P., & Tsujimoto, M. 2007, *ApJS*, 168, 100

Wang, J., Townsley, L. K., Feigelson, E. D., Broos, P. S., Getman, K. V., Román–Zúñiga, C. G., & Lada, E. 2008, *ApJ*, 675, 464

Wang, J., Feigelson, E. D., Townsley, L. K., Román–Zúñiga, C. G., Lada, E., & Garmire, G. 2009, *ApJ*, 696, 47

Winston, E., Megeath, S. T., Wolk, S. J., Muzerolle, J., Gutermuth, R., Hora, J. L., Allen, L. E., Spitzbart, B., Myers, P., & Fazio, G. G. 2007, *ApJ*, 669, 493

Wolk, S. J., Bourke, T. L., Smith, R. K., Spitzbart, B., & Alves, J. 2002, *ApJ* (Letters), 580, L161

Wolk, S. J., Spitzbart, B. D., Bourke, T. L., & Alves, J. 2006, *AJ*, 132, 1100

Wolk, S. J., Spitzbart, B. D., Bourke, T. L., Gutermuth, R. A., Vigil, M., & Comerón, F. 2008, *AJ*, 135, 693

Wolk, S. J. 2009, *AIP Conf. Ser.*, 1094, 959

Yusef–Zadeh, F., Law, C., Wardle, M., Wang, Q. D., Fruscione, A., Lang, C. C., & Cotera, A. 2002, *ApJ*, 570, 665

Star clusters: basic galactic building blocks
Proceedings IAU Symposium No. 266, 2009
R. de Grijs & J. R. D. Lépine, eds.

doi:10.1017/S1743921309991062

Infrared surveys of Galactic star clusters

Valentin D. Ivanov,[1] Maria Messineo,[2] Qingfeng Zhu,[3] Don Figer,[2] J. Borissova,[4] R. Kurtev[4] and G. R. Ivanov[5]

[1]ESO, Ave. Alonso de Cordova 3107, Casilla 19, Santiago 19001, Chile
email: `vivanov@eso.org`

[2]Chester F. Carlson Center for Imaging Science, Rochester Institute of Technology, 54 Lomb Memorial Drive, Rochester, NY 14623-5604, USA
email: `[messineo, figer]@cis.rit.edu`

[3]Key Laboratory for Research in Galaxies and Cosmology, The Chinese Academy of Sciences, University of Science and Technology of China, Hefei, Anhui, 230026, China
email: `zhuqf@ustc.edu.cn`

[4]Departamento de Física y Astronomía, Facultad de Ciencias, Universidad de Valparaíso, Ave. Gran Bretaña 1111, Playa Ancha, Casilla 53, Valparaíso, Chile
email: `[jura.borissova, radostin.kurtev]@uv.cl`

[5]Department of Astronomy, Sofia University, Bulgaria, 5 James Bourchier, 1164 Sofia, Bulgaria
email: `gr_ivanov@phys.uni-sofia.bg`

Abstract. Many attempts have been made to carry out a complete observational census of Milky Way star clusters based on recent near- and mid-infrared surveys. However, more clusters are still being discovered, indicating that existing catalogs are incomplete. We attempt to estimate the total number of supermassive (SM; $M_{\rm cl} \geqslant 10^4$ M$_\odot$) clusters in the Galaxy, and to improve the yield from the automated cluster searches. Assuming that the 'local' census of SM clusters is complete, and that their surface density accross the disk follows that of the stars, we predict that the Milky Way contains $\geqslant 81 \pm 21$ SM clusters. We apply a cluster-detection algorithm to the 2MASS Point Source Catalog after a preliminary color and/or magnitude selection of the point sources to improves the surface-density cluster-to-field contrast. Our algorithm identified 94 new candidates, and re-identified 34 known clusters. During the visual inspection, we detected an additional 41 new candidates, and re-identified 32 known objects. Preliminary characterization suggests that the new list may contain red-supergiant, open and globular clusters.

Keywords. globular clusters: general, open clusters and associations: general, Galaxy: bulge, Galaxy: disk, Galaxy: structure, infrared: general

1. Introduction

Lada & Lada (2003) estimated that 70–90% of the stars formed in giant molecular clouds are born in embedded clusters. The optical census has yielded large numbers of clusters (i.e., Phelps & Janes 1993, 1994), but most objects with the highest scientific impact (i.e., the Arches cluster; Cotera *et al.* 1992, 1996; Figer 1995; Nagata *et al.* 1996) have been discovered in the infrared (IR). It is not surprising that once the Two Micron All Sky Survey (2MASS; Skrutskie *et al.* 2006), the Deep Near-Infrared Southern Sky Survey (DENIS; Epchtein *et al.* 1997), the *Spitzer Space Telescope*'s Galactic Legacy Infrared Mid-Plane Survey Extraordinaire (GLIMPSE; Benjamin *et al.* 2003) and, most recently, the *UKIRT* Infrared Deep Sky Survey (UKIDSS; Lawrence *et al.* 2007) became available, a number of teams attempted to search for as yet unknown obscured clusters.

There were two basic strategies to identify new cluster candidates. The first consists of targeted searches around known IR and radio sources, or in areas of interest (e.g., the

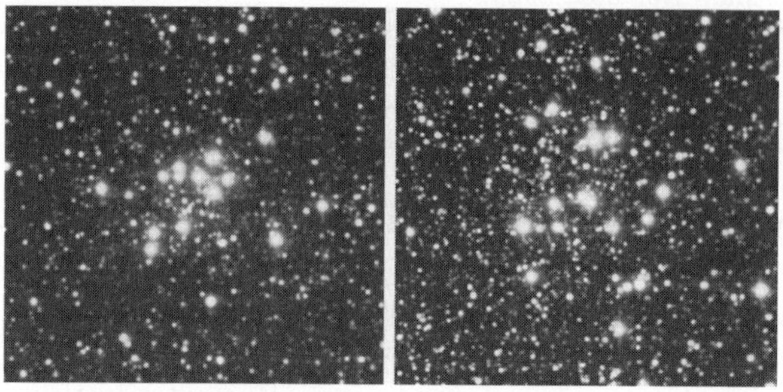

Figure 1. 2MASS three-color images of red-supergiant-containing clusters, RSGC 1 ([BDS2003] 122 or BDSB 122; *left*) and RSGC 2 (Stephenson 2; *right*). Blue, green and red correspond to J, H and K_s, respectively. The fields cover 6.7×6.7 arcmin^2. North is up, East to the left.

Galactic bulge or known nearby star-forming regions; i.e., Dutra & Bica 2000; Kronberger *et al.* 2006). The second is unbiased: it is based on automated techniques that apply sets of objective rules to identify overdensities in flux or stellar surface-number density in the pixel data or in the point-source catalogs, respectively. The morphology of some known supermassive clusters clearly demonstrates that they can be extremely difficult to detect because of their small angular size and compact structure. Both methods were used for the first time to find new Milky Way clusters in near-IR (NIR) surveys by Reylé & Robin (2002), using the DENIS data. Ivanov *et al.* (2002) and Borissova *et al.* (2003) reported lists of cluster candidates identified as number-density peaks in the 2MASS Point Source Catalog (PSC; Cutri *et al.* 2003). These studies were soon followed by others (i.e., Froebrich *et al.* 2007a, 2007b, 2008; Mercer *et al.* 2005; Davies *et al.* 2007), who found additional cluster candidates.

Our team pursued a rigorous campaign to determine the nature of the candidates and develop relevant diagnostic techniques (Borissova *et al.* 2005, 2006, 2008; Beletsky *et al.* 2009; Ivanov & Borissova 2002; Ivanov *et al.* 2002, 2005a, 2005b; Kurtev *et al.* 2007, 2008). Our NIR imaging of candidates selected through visual inspection suggests $\sim$ 40–50% contamination by multiple stars and extended sources. The visual inspection also shows that automated searches fail in the areas of highly variable extinction, i.e., they produce many spurious detections around dark dusty clouds or in areas of lower extinction, as also pointed by Froebrich *et al.* (2007a). The incompleteness of NIR-based cluster searches was also demonstrated in the mid-IR *Spitzer Space Telescope*-based work of Mercer *et al.* (2005). Most disturbingly, some of their mid-IR cluster candidates are clearly visible on 2MASS images but they were missed by both automated searches and visual inspection.

These considerations prompted us (i) to put in prospective the results from the current searches by estimating, at least approximately, the total number of detectable clusters in the Milky Way and (ii) to further improve our search algorithm. The visual inspection of three-color images suggest that most clusters stand out really well because cluster colors differ from those of foreground and background stars, because of either intrinsically different stellar populations or the dust reddening suffered by the clusters at greater distances than the foreground stars (Figure 1). The cluster-number estimate is based on the fundamental assumptions that the clusters follow the spatial distribution of the field stars and that we know the local density of the most massive clusters ($M_{\rm cl} \geqslant 10^4$ M$_\odot$), which we used as normalization.

Here we describe additional color and magnitude criteria to the 'normal' overdensity search, slightly modified with respect the description in Ivanov *et al.* (2002). We optimized our criteria for clusters containing red supergiants (RSGs), because many of the known supermassive Milky Way clusters host this type of star (i.e., Davies *et al.* 2007). The criteria can easily be modified to favor clusters containing other stellar types. We provide a list of new cluster candidates, and some re-identified clusters and cluster candidates. For all objects, we carry out a preliminary analysis and characterization based on the available 2MASS and GLIMPSE data.

2. Total number of Milky Way clusters

We attempt here to obtain a zeroth-order semi-empirical estimate of the total number of clusters in the Milky Way from simple scaling relations, using the following assumptions:

1. The surface density of clusters accross the Galactic X–Y plane follows the stellar surface density, described by an exponential law (de Vaucouleurs 1959) with the same scale length $R_{\rm eff}$ as for the stars.

2. All 'local' supermassive clusters with masses $M_{\rm cl} \geqslant 10^4\,{\rm M}_\odot$ in the solar vicinity are known, so we can scale the exponential law to match the measured surface density in the high-mass bin of the mass distribution.

3. The cluster mass distribution is described by a power law with an index of -2 (Piskunov *et al.* 2008) and cutoffs at 10^2 and $10^4\,{\rm M}_\odot$, so we can extrapolate the number of clusters to lower-mass bins.

The total number of clusters in the Milky Way is obtained by integrating over both the spatial extent of the Galaxy and the cluster mass range.

The main argument supporting the first assumption comes from the estimate of Lada & Lada (2003) that most stars formed in giant molecular clouds are born in embedded clusters. A word of caution is needed as regards the third assumption, however. Larsen (2006) argued for a truncated cluster initial mass function, which can be parametrized well using a Schechter function (Schechter 1976) as demonstrated by Gieles (2009). The tentative current upper-mass limit is $\sim 10^5\,{\rm M}_\odot$ for our Galaxy, well above the mass bin that we use for normalization, so it is likely to have a negligible effect on our estimates.

We adopt an exponential law with $R_{\rm eff} = 2.5\,{\rm kpc}$, starting at $R_{\rm in} = 2\,{\rm kpc}$ and with an outer cutoff at $R_{\rm max} = 20\,{\rm kpc}$ (Robin *et al.* 2003; Picaud & Robin 2004; Brand & Wouterloot 2007). The innermost disk profile is flat, with a constant surface density at a level equal to that of the exponential law at $R_{\rm in}$ (Freudenreich 1998; Lépine & Leroy 2000; López–Corredoira *et al.* 2001). The profile is shown in Figure 2 (left, top).

For the purpose of local normalization of the supermassive clusters, we considered the objects from table 4 in Messineo *et al.* (2009), with some updates discussed in detail in Ivanov *et al.* (2010). Up to which distance from the Sun is the supermassive cluster sample likely complete? The bottom right-hand panel of Figure 2 shows the K-band extinction, A_K, as a function of the distance, D, from the Sun. Some clusters, even at $D \sim 7$–$8\,{\rm kpc}$, suffer from low extinction: $A_K \sim 0.5$–$1.5\,{\rm mag}$, i.e., they are located behind 'holes' in the Galactic dust. None of the known clusters suffer from extinction much higher than $A_K \sim 3\,{\rm mag}$, regardless of distance, suggesting that extinction, and not distance, is the main obstacle that prevents us from detecting clusters in the Galactic plane. Indeed, the apparent magnitude of the stars in a cluster near the Galactic Center would be only about 0.6 mag fainter than that for a cluster at $D \sim 6\,{\rm kpc}$, if it was not for the effects of extinction. Bearing in mind the limited statistics, we conclude that the sample is likely to become incomplete for $A_K \geqslant 3\,{\rm mag}$, which for at least some clusters occurs at a distance $D \sim 7$–$8\,{\rm kpc}$, as can be seen in Figure 2. Indeed, the cumulative distribution (right, middle panel) deviates from the expected distribution for constant cluster surface

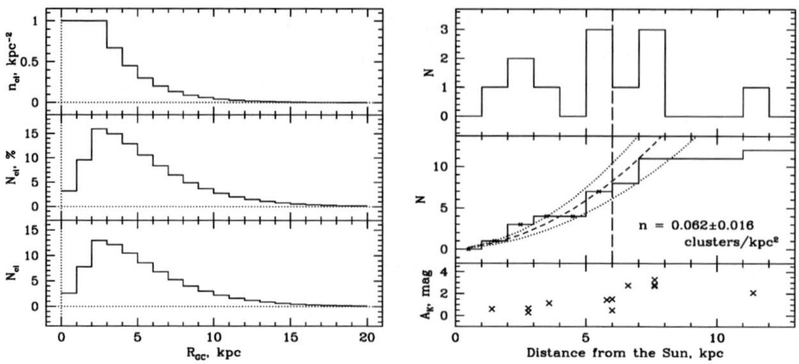

Figure 2. *(left)* Exponential profile of the cluster surface density, normalized at the peak value *(top)*, fractional and number distribution of the supermassive clusters as a function of Galactocentric distance (*middle* and *bottom*, respectively). See Section 2 for details. *(right)* Distribution of the supermassive clusters as a function of their distance to the Sun *(top)* and the corresponding cumulative histogram *(middle)*. The K-band absorption as function of distance is also shown *(bottom)*. The vertical long-dashed lines mark the maximum distance to which we consider the sample complete. The centers of the bins that fall bellow this limit are marked with crosses for clarity. The short-dashed line shows the predicted cumulative distribution for homogeneously distributed supermassive clusters with constant surface density, $n = 0.062 \pm 0.016$ kpc^{-2}. The dashed lines are the 1σ Poissonian uncertainties, based on the cluster numbers.

density (short-dashed line) for greater distances. Based on these considerations, we will make a somewhat arbitrary assumption that the sample is complete to $D \leqslant 6$ kpc, an upper limit consistent with the data, keeping in mind that the known clusters may reside in low-extinction windows throughout the disk. Seven clusters are located within the nearest 6 kpc. Averaging within this volume, we obtain a cluster surface density $n = 0.062 \pm 0.016\,\mathrm{kpc}^{-2}$. The error is random and is dominated by the small-number statistics. This calculation neglects the local density gradient.

Scaling the modified exponential radial profile to match the 'local' surface density, and integrating over the entire Galaxy, we estimate that it contains $\sim 81 \pm 21$ clusters with masses in the range $10^4 - 10^5\,\mathrm{M}_\odot$. We remind the reader that this number is a lower limit. What fraction of these clusters falls within the boundaries of our cluster search ($0 \leqslant l \leqslant 360$ deg, $-1.5 \leqslant b \leqslant 1.5$ deg)? Adopting a vertical cluster distribution scale height of $H_\mathrm{d} = 55$ pc (Phelps & Janes 1994) and a displacement of the Sun from the Galactic disk along the Z axis of 15 pc (Drimmel & Spergel 2001), we estimate that $\sim 79 \pm 20$ of the $\sim 81 \pm 21$ supermassive clusters in the Milky Way probably fall into the area covered by our search, of which $\sim 39 \pm 10$ are closer than the Galactic Center (i.e., observable), approximately three times exceeding the number of known objects.

3. Search algorithm with color/magnitude criteria

We imposed color/magnitude criteria on the stars in the 2MASS PSC to isolate the cluster members on unique sequences without counterparts on the field color–magnitude diagrams (CMDs). We create a two-step search algorithm, (i) pre-selection of stars from the 2MASS PSC that meet certain color and/or magnitude criteria and (ii) detection of statistically significant peaks in the density map constructed from this pre-selection.

The RSG stars in known RSG clusters (Figure 3) occupy relatively compact loci at $1.2 \leqslant J - K_\mathrm{s} \leqslant 2.2$ mag for RSGC 1 ([BDS2003] 122 or BDSB 122) and $0.2 \leqslant J - K_\mathrm{s} \leqslant 1.7$ mag for RSGC 2 (Stephenson 2). These limits were chosen to separate the

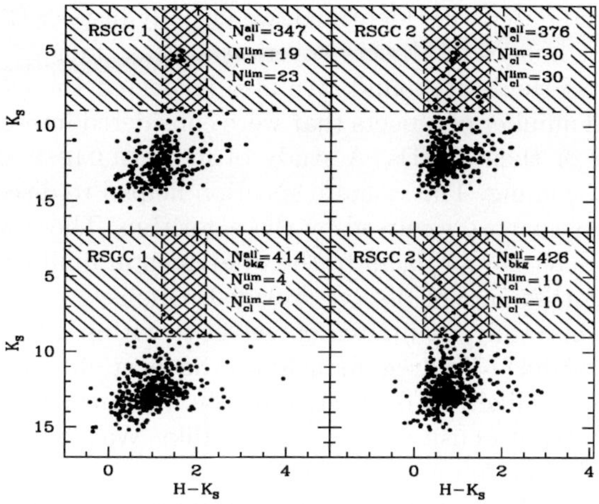

Figure 3. 2MASS color–magnitude diagrams of red-supergiant-containing clusters, RSGC 1 ([BDS2003] 122 or BDSB 122; *left*) and RSGC 2 (Stephenson 2; *right*) and their surroundings. Stars within $R_{\rm cl} \leqslant 2$ arcmin are shown in the top panels and stars within equal-area annuli, $7.0 \leqslant R \leqslant 7.28$ arcmin, are shown in the bottom panels. The double-hatched areas show the combination of the color and magnitude limits designed to select only RSG stars, and the single-hatched areas represent the adopted magnitude limits. The results from number counts are also given. They indicate that the total number of stars inside the 'cluster' is smaller than the total number of stars in the 'field' annulus: 347 versus 414 for RSGC 1 and 376 versus 426 for RSGC 2. This is caused by crowding and the presence of bright stars in the cluster area. It is probably the reason why unconstrained number counts used in Ivanov *et al.* (2002) and Froebrich *et al.* (2007a) failed to detect these clusters.

apparent regions of the most significant overdensities with respect to the field, but they can be limited further. The optimal color limits for each individual cluster depend on the distance, reddening and cluster age. To alleviate these complications, we adopted only an RSG magnitude limit of $K_{\rm s} \leqslant 9.0$ mag. For the RSGs in a 12 Myr cluster with absolute magnitudes of $M_K \sim -10$ to -12 mag (Davies *et al.* 2008), this corresponds to a maximum distance modulus of $(m - M)_K \sim 12$–21 mag, or a distance limit of $\sim$ 16–40 kpc for $A_K = 3$ mag, or $\sim$ 4–10 kpc for $A_K = 6$ mag. Therefore, unless the extinction significantly exceeds that for the Arches, Quintuplet or the Galactic Center cluster (i.e., more than one magnitude in A_K), the adopted magnitude limit should not exclude RSG-dominated clusters on the near side of the Milky Way's center.

The second step in the search is to build two-dimensional surface-density maps to detect the peaks. This is similar to the technique described by Ivanov *et al.* (2002), except—to make the search independent from a fixed grid—we calculated the surface density within a radius of $R_{\rm max} = 1.5$ arcmin around each pre-selected star, using only the other pre-selected stars. The surface densities suffer from the problem of not having been measured at equidistant points, which is inconvenient for further analysis, so we still introduce a grid, but it is finer than the $R_{\rm max}$: the bin size is 0.5×0.5 arcmin^2, effectively preserving the angular 'resolution' determined by $R_{\rm max}$. Next, we remove the background by subtracting from the grid a median-smoothed surface-density distribution, where the median-filter size was 21×21 bins (10.5×10.5 arcmin^2). The final step was to identify clusters as groups of neighboring bins, where neighbors are defined as bins separated by less than $R_{\rm neighbor} = 1.5$ arcmin. We found it advantageous to smooth the two-dimensional histogram with a 5×5 boxcar function before running the cluster-identification routine. This improved the detection of irregularly shaped candidates and objects that are larger than $R_{\rm neighbor}$.

Once the cluster candidates were identified, we visually inspected the 2MASS-generated CMDs and three-color images of objects covering at least four grid bins, with an overdensity of $\geqslant 10\sigma$ above the background. We also inspected the GLIMPSE survey, where available, for a limited number of objects that were considered interesting because of their apparent morphology or their CMDs. A study of more GLIMPSE and UKIDSS images of our candidates is forthcoming. The visual inspection helped to discover *serendipitously* a number of new clusters that were missed by the algorithm. These were usually compact, and we attribute the detection problem to the fundamental limitation of the 2MASS PCS (or any other catalog) that cannot resolve crowded regions.

For the purpose of this study, we limited ourselves to within ± 1.5 deg from the Galactic equator, an area of 1080 deg^2. This region includes the most obscured parts of the Galaxy, where the cluster census is likely to be least complete. Barring extinction effects, it covers the highest apparent stellar-density areas in the Milky Way, and therefore the location of most active cluster formation and disruption.

The list of identified objects, together with preliminary characterization from public data, will appear in Ivanov *et al.* (2010). All identified objects are concentrated within ~ 60 deg from the Galactic Center (i.e., roughly within the solar circle). There are more objects within the $0 \leqslant l \leqslant 60$ deg range than in $300 \leqslant l \leqslant 360$ deg (133 versus 68), for both automatically (91 versus 37) and visually (41 versus 31) detected cluster candidates, contrary to the samples of Froebrich *et al.* (2007b), Mercer *et al.* (2005) and the combined Bica and Dutra catalogs. This result may reflect a spatial variation in the global star-formation rate in the Galaxy, e.g., due to the Galactic bar. Our search may possibly be selecting more distant clusters. Incidentally, the three known RSG clusters are located in the overdense quadrant.

The automatically discovered cluster candidates are, on average, larger than those detected visually, with mean radii of 102 ± 45 versus 70 ± 40 arcsec and median radii of 100 versus 60 arcsec, respectively. The new automatically detected candidates have, on average, ~ 10 arcsec larger radii than the re-identified known clusters, which suggests that our algorithm finds objects that are more likely to have been missed because they are spread over large areas and have lower contrast with respect to the foreground and background populations. On the other hand, the new, visually detected candidates have on average ~ 20 arcsec smaller radii than the re-identified previously known objects. Most previous visual searches were pointed at known radio or mid-IR sources, and typically discovered young clusters with nebulosities. Our visually detected candidates are free from this bias, and we find many clusters without diffuse emission which are likely evolved to beyond the O-star stage.

We attempted to assess the nature of the candidates, even of the known but poorly studied clusters that we re-identified, using three sources:

1. Morphology of the 2MASS CMDs constructed from stars within the cluster radius and a comparison-field CMD based on a circular annulus around the object;
2. Morphology on $\sim 900 \times 900$ arcsec^2 three-color 2MASS images;
3. Literature data of sources within 4 arcmin from the candidate.

This is not always straightforward because of significant field contamination and possible random alignments of objects at different distances. Nevertheless, this assessment is useful as a first approximation and as basis for selecting objects for follow-up studies.

We found a total of 25 RSG cluster candidates, of which 21 among the automatically identified candidates (15 new and 6 re-identified previously known clusters). However, among the latter group only four are truly known to contain RSGs, the remaining two are known but poorly studied objects, and we suspect that they contain RSG stars based on their 2MASS CMD features. Finally, we have four more candidates among the new visually identified objects. We have 36 candidates for old open clusters or globulars, of which 20 were automatically (17 new and 3 previously known) and 16 were visually identified (9 new, 7 previously known). Most appear relatively poor, but this may be a combined effect due to distance, dust extinction and the shallow survey depth. There are 12 massive cluster candidates in our sample, 9 of which were automatically identified (2 new and 7 previously known objects) and 3 were visually identified (1 new and 2 previously known objects). Finally, we provide a list of 15 candidate cluster pairs with separations $\delta\theta \leqslant$ 5 arcmin. All but three of the latter show indications of physical connections.

4. Summary and conclusions

1. The Milky Way is likely to contain $\geqslant 81 \pm 21$ clusters with masses of $10^4 - 10^5$ $M_\odot$, estimated from the assumption that the cluster surface density in the disk is described by an exponentail law with the same scale length as for the stars, normalized at the 'local' density of such clusters. The error is random and reflects only the small number of the clusters used for the normalization. If half of these clusters are located on the far side of the Galactic Center and remain inaccessible because of high extinction, about 30 supermassive clusters remain to be discovered on the near side of our Galaxy.

2. We developed a new search algorithm that uses color/magnitude selection of stars, optimized to pre-select a certain type of clusters, but with higher contrast than a general, unconstrained algorithm. We applied it to the 2MASS PSC, searching mostly (but not exclusively) for potential RSG clusters. We discovered serendipitously a number of additional candidates, usually unresolved.

3. The existing Galactic cluster census is incomplete. We identified 135 new cluster candidates (94 automatically and 41 visually) in a ± 1.5 deg strip along the Galactic plane. We also re-identified 65 known clusters (or cluster candidates, since the cluster nature of many of them has not yet been confirmed; 34 automatically and 32 visually), including most of the well-known supermassive/RSG clusters, giving us confidence in the power of the new selection method.

4. A preliminary characterization is reported for 25 RSG cluster candidates, 36 globular or old open cluster candidates, and 12 massive cluster candidates. These classifications are based only on 2MASS, and (if available) literature information.

5. The number of massive cluster candidates is lower than predicted. This is expected because compact, supermassive clusters like the Arches cluster cannot be resolved by 2MASS. Some additional supermassive clusters similar to RSGC 1, 2 and 3 may hide among our RSG cluster candidates. However, we expect that many of our candidates, upon closer scrutiny, will turn out to be spurious detections, i.e. holes, in the dust.

6. Last but not least, we emphasize that the objects in our lists are candidates until their cluster nature is confirmed based on deeper imaging or spectroscopy.

References

Beletsky, Y., Carraro, G., & Ivanov, V. D. 2009, *A&A*, in press (arXiv:0908.2866)
Benjamin, R., *et al.* 2003, *PASP*, 115, 953
Borissova, J., *et al.* 2003, *A&A*, 411, 83
Borissova, J., Ivanov, V. D., Minniti, D., Geisler, D., & Stephens, A. W. 2005, *A&A*, 435, 95
Borissova, J., Ivanov, V. D., Minniti, D., & Geisler, D. 2006, *A&A*, 455, 923
Borissova, J., Ivanov, V. D., Hanson, M. M., Georgiev, L., Minniti, D., Kurtev, R., & Geisler, D. 2008, *A&A*, 488, 151
Brand, J. & Wouterloot, J. G. A. 2007, *A&A*, 464, 909

Cotera, A. S., Simpson, J. P., Erickson, E. F., Colgan, S. W. J., & Burton, M. G. 1995, *Bull. Am. Astron. Soc.*, 24, 1262

Cotera, A. S., Erickson, E. F., Colgan, S. W. J., Simpson, J. P., Allen, D. A., & Burton, M. G. 1996, *ApJ*, 461, 750

Cutri, R. M., Skrutskie, M. F., van Dyk, S., *et al.* 2003, *The IRSA 2MASS All-Sky Point Source Catalog*, NASA/IPAC Infrared Science Archive. `http://irsa.ipac.caltech.edu/applications/Gator/`

Davies, B., Figer, D. F., Kudritzki, R.-P., MacKenty, J., Najarro, F., & Herrero, A. 2007, *ApJ*, 671, 781

Davies, B., Figer, D. F., Law, C. J., Kudritzki, R.-P., Najarro, F., Herrero, A., & MacKenty, J. W. 2008, *ApJ*, 676, 1016

Drimmel, R. & Spergel, D. N. 2001, *ApJ*, 556, 181

de Vaucouleurs, G. 1959, in *Handbuch der Physik*, Vol. 53, S. Flügge (ed.), p. 311, Berlin: Springer–Verlag

Dutra, C. M. & Bica, E. 2000, *A&A* (Letters), 359, L9

Epchtein, N., *et al.* 1997, *ESO Messenger*, 87, 27

Figer, D. F. 1995, PhD Thesis, University of California, Los Angeles

Freudenreich, H. T. 1998, *ApJ*, 492, 495

Froebrich, D., Meusinger, H., & Scholz, A. 2007a, *MNRAS*, 377, 54

Froebrich, D., Scholz, A., & Raftery, C. L. 2007b, *MNRAS*, 374, 399

Froebrich, D., Meusinger, H., & Scholz, A. 2008, *MNRAS*, 390, 1598

Gieles, M. 2009, *MNRAS*, 394, 2113

Ivanov, V. D. & Borissova, J. 2002, *A&A*, 390, 937

Ivanov, V. D., Borissova, J., Pessev, P., Ivanov, G. R., & Kurtev, R. 2002, *A&A* (Letters), 394, L1

Ivanov, V. D., Borissova, J., Bresolin, F., & Pessev, P. 2005a, *A&A*, 435, 107

Ivanov, V. D., Kurtev, R., & Borissova, J. 2005b, *A&A*, 442, 195

Ivanov, V. D., Messineo, M., Zhu, Q., Figer, D., Borissova, J., Kurtev, R., & Ivanov, G. R. 2010, *A&A*, submitted

Kronberger, M., *et al.* 2006, *A&A*, 447, 921

Kurtev, R., Borissova, J., Georgiev, L., Ortolani, S., & Ivanov, V. D. 2007, *A&A*, 475, 209

Kurtev, R., Ivanov, V. D. Borissova, J., & Ortolani, S. 2008, *A&A*, 489, 583

Larsen, S. S. 2006, in: T. Richtler & S. S. Larsen (eds.), *Globular Clusters: Guides to Galaxies*, p. 95, Berlin: Springer

Lépine, J. R. D. & Leroy, P. 2000, *MNRAS*, 313, 263

Messineo, M., Davies, B., Ivanov, V. D., Figer, D. F., Schuller, F., Habing, H. J., Menten, K. M., & Petr-Gotzens, M. G. 2009, *ApJ*, 697, 701

Nagata, T., Woodward, C. E., Shure, M., & Kobayashi, N. 1995, *AJ*, 109, 1676

Lada, C. J. & Lada, E. A. 2003, *ARA&A*, 41, 57

Lawrence, A., *et al.* 2007, *MNRAS*, 379, 1599

López–Corredoira, M., Hammersley, P. L., Garzón, F., Cabrera–Lavers, A., Castro–Rodríguez, N., Schultheis, M., & Mahoney, T. J. 2001, *A&A*, 373, 139

Mercer, E., Clemens, D., & Meade, M. 2005, *ApJ*, 635, 560

Phelps, R. L. & Janes, K. A. 1993, *AJ*, 106, 1870

Phelps, R. L. & Janes, K. A. 1994, *ApJS*, 90, 31

Picaud, S. & Robin, A. C. 2004, *A&A*, 428, 891

Piskunov, A. E., Kharchenko, N. V., Schilbach, E., Röser, S., Scholz, R.-D., & Zinnecker, H. 2008, *A&A*, 487, 557

Reylé, C. & Robin, A. C. 2002, *A&A*, 348, 403

Robin, A. C., Reylé, C., Derriére, S., & Picaud, S. 2003, *A&A*, 409, 523

Schechter P. 1976, *ApJ*, 203, 297

Skrutskie, M. F., *et al.* 2006, *AJ*, 131, 1163

Session VI

Cluster dynamics

Star clusters: basic galactic building blocks
Proceedings IAU Symposium No. 266, 2009
R. de Grijs & J. R. D. Lépine, eds.

doi:10.1017/S1743921309991074

Stellar-mass black holes in star clusters: implications for gravitational-wave radiation

Sambaran Banerjee,[1,2] Holger Baumgardt[1] and Pavel Kroupa[1]

[1]Argelander-Institut für Astronomie, Auf dem Hügel 71, 53121, Bonn, Germany
email: `[sambaran, holger, pavel]@astro.uni-bonn.de`

[2]Alexander von Humboldt fellow

Abstract. We study the dynamics of stellar-mass black holes (BHs) in star clusters, with particular attention to the formation of BH–BH binaries, which are interesting as sources of gravitational waves (GWs). We examine the properties of these BH–BH binaries through direct N-body simulations of Plummer clusters of $N \leqslant 10^5$ low-mass stars with an initial population of stellar-mass BHs, using the NBODY6 code. We find that the stellar-mass BHs segregate rapidly into the cluster core and form a dense subcluster of BHs in which BH–BH binaries form through three-body encounters. While most BH binaries are ejected from the cluster by recoils due to superelastic encounters with the single BHs, we find that for clusters with $N \gtrsim 5 \times 10^4$, typically a few of them harden sufficiently so that they can merge via GW emission within the cluster. Also, for each of such clusters there are a few escaping BH binaries that merge within a Hubble time, with most merger times being within a few Gyr. These results imply that the intermediate-age massive clusters constitute the most important class of star cluster candidates that can produce dynamical BH–BH mergers at the present epoch. The BH–BH merger rates obtained from our computations imply a significant detection rate (~ 30 yr^{-1}) for the proposed *Advanced LIGO* GW detector.

Keywords. gravitational waves, black hole physics, stellar dynamics, scattering, methods: N-body simulations, methods: numerical, galaxies: star clusters, binaries: close

1. Introduction

Star clusters, e.g., globular clusters (henceforth GCs), young and intermediate-age massive clusters (henceforth IMCs) harbor a large overdensity of compact stellar remnants compared to the field by virtue of their mass segregation. These compact stars, which are neutron stars (henceforth NSs) and black holes (henceforth BHs), are produced by stellar evolution of the most massive stars of the cluster within the first ~ 50 Myr after cluster formation. Since they are heavier than the remaining low-mass stars in the cluster, they segregate quickly (within one or a few half-mass relaxation times) to the cluster core, forming a dense subcluster of compact stars. Such a compact stellar subcluster is of wide interest as it is efficient in producing compact-star binaries dynamically, which are of interest for a wide range of physical phenomena, e.g., X-ray emission (X-ray binaries), gravitational waves (BH–BH and NS–NS binaries) and short gamma-ray bursts (NS–NS and NS–BH binaries). In the present work, we investigate the dynamics of stellar-mass BHs in star clusters, which are formed through supernova explosions of $\gtrsim 18 \mathrm{M}_\odot$ stars, typically within the mass range $8\mathrm{M}_\odot \lesssim M_{\rm BH} \lesssim 12\mathrm{M}_\odot$ for solar-metallicity progenitors (Casares 2007). We are particularly interested in the dynamically formed BH–BH binaries, which are strong sources of GWs as they spiral in via GW radiation.

As studied earlier by several authors (e.g., Merritt *et al.* 2006; Mackey *et al.* 2007), BHs, formed through stellar evolution, segregate into the cluster core within ~ 0.3 pc and form a dynamically isolated subcluster of BHs, where the density of BHs is large enough that BH–BH binary formation through three-body encounters (Heggie & Hut 2003) becomes important. These dynamically formed BH binaries then 'harden' through repeated superelastic encounters with the surrounding BHs (Heggie's law; Heggie 1975). The binding energy of the BH binaries released is carried away by the single BHs and BH binaries involved in the encounters. This causes the BHs and BH binaries to get ejected from the BH core to larger radii in the cluster, and they deposit energy to the cluster stars while sinking back to the core through dynamical friction, thereby heating the cluster. As the BH binaries harden, the encounter-driven recoil becomes stronger and finally the recoil is large enough that the encountering single BH and/or the BH binary escapes from the cluster. This also results in cluster heating because of the associated mass loss from the cluster core that decreases its potential energy correspondingly. These heating mechanisms result in cluster expansion.

In the present work, we make a detailed study of the dynamics of BH–BH binaries formed in a BH subcluster. In particular, we investigate whether hard enough BH binaries that can merge via gravitational radiation in a Hubble time within the cluster or after getting ejected, can be formed in such a subcluster. To that end, we perform a series of N-body integrations of concentrated star clusters (half-mass radius $r_{\rm h} \leqslant 1$ pc) consisting of $\leqslant 10^5$ low-mass stars, to which a certain number of stellar-mass BHs is added, representing a star cluster with an evolved stellar population.

2. Computations

To study the dynamics of BHs in star clusters, we perform direct N-body simulations with star clusters in which a certain number of BHs are added initially, using the NBODY6 code on graphical-processing-unit (GPU) platforms (see Aarseth 2009). The initial cluster follows a Plummer model with half-mass radius $r_{\rm h} \leqslant 1$ pc, consisting of $N \leqslant 10^5$ low-mass main-sequence stars in the mass-range $0.5{\rm M}_\odot \leqslant m \leqslant 1.0{\rm M}_\odot$. A specified number of BHs are added to each cluster with the same initial distribution as the stars. The initial number of BHs are chosen according to the BH retention fraction, where we explore both full retention and the case where half of the BHs are ejected from the cluster by natal kicks. For simplicity, we consider only equal-mass BHs in the present work, with the representative value of $M_{\rm BH} = 10{\rm M}_\odot$. With such a cluster, we mimic the epoch at which the massive stars have already evolved to produce their remnant BHs.

To evolve the BH–BH binaries due to GW emission, the well-known Peters' formula (Peters 1964) is used in NBODY6, according to which the merger time $t_{\rm mrg}$ of an equal-mass BH–BH binary due to GW emission is given by

$$t_{\rm mrg} = 150\,{\rm Myr}\left(\frac{{\rm M}_\odot}{M_{\rm BH}}\right)^3 \left(\frac{a}{{\rm R}_\odot}\right)^4 (1-e^2)^{7/2}, \tag{2.1}$$

where a is the semimajor axis of the binary and e is its eccentricity. Numerical simulations of BH–BH mergers indicate that for unequal-mass BHs or even for equal-mass BHs with unequal spins, the merged BH product acquires a velocity kick of typically 100 km s^{-1} or more, due to asymmetry in momentum outflow from the system, associated with the GW emission. Therefore, in our computations, we provide an arbitrarily large velocity kick of 150 km s^{-1} immediately after a BH–BH merger, to make sure that the merged BH escapes. Also, we evolve the clusters isolated, i.e., in absence of a galactic tidal field as the formation of the BH core through mass segregation and its dynamics remain largely unaffected by the presence of a tidal field, which mainly affects stars near the

Table 1. Summary of the computations performed for isolated clusters and those with reflective boundary (see Section 2). The meaning of different columns is as follows: (1): Identity of the particular model; similar values with different names (ending with A, B, etc.) imply runs repeated with different random seeds. (2): Total number of stars, N. (3): Number of simulations, $N_{\rm sim}$, with the particular cluster. (4): Initial half-mass radius of the cluster, $r_{\rm h}(0)$ (isolated cluster), or radius of reflective sphere, $R_{\rm s}$. (5): Initial number of BHs, $N_{\rm BH}(0)$. (6): Total number of BH–BH binary mergers within the cluster, $N_{\rm mrg}$. (7): Times, $t_{\rm mrg}$, corresponding to the mergers. (8): Number of escaped BH pairs, $N_{\rm esc}$; the three values of $N_{\rm esc}$ are for $t_{\rm mrg} < 3$ Gyr, 1 Gyr and 100 Myr, respectively.

Model name	N	$N_{\rm sim}$	$r_{\rm h}(0)$ or $R_{\rm s}$ (pc)	$N_{\rm BH}(0)$	$N_{\rm mrg}$	$t_{\rm mrg}$ (Myr)	$N_{\rm esc}$
			Isolated clusters				
C10K20	10000	10	1.0	20	0	— —	— —
C25K50	25000	10	1.0	50	0	— —	3 1 1
C50K80	45000	1	1.0	80	1	698.3	3 1 0
C50K80.1	45000	1	0.5	80	2	217.1, 236.6	3 2 1
C50K40.1	45000	1	0.5	40	0	— —	1 1 1
C50K200	50000	1	1.0	200	2	100.8, 467.8	0 0 0
C65K110	65000	1	1.0	110	1	314.6	4 2 1
C65K110.1	65000	1	0.5	110	0	— —	4 3 1
C65K55.1	65000	1	0.5	55	1	160.5	1 0 0
C100K80	100000	1	1.0	80	2	219.4, 603.2	5 2 1
C100K200	100000	1	1.0	200	0	— —	5 4 4
			Reflective boundary				
R3K180	3000	1	0.4	180	1	1723.9	5 3 1
R4K180A	4000	1	0.4	180	1	3008.8	2 2 1
R4K180B	4000	1	0.4	180	2	100.2, 1966.5	2 1 0
R3K100	3000	1	0.4	100	2	3052.8, 3645.9	1 1 0
R4K100A	4000	1	0.4	100	2	104.4, 814.2	3 3 1
R4K100B	4000	1	0.4	100	1	1135.3	3 3 3

tidal boundary. Further, for simplicity, we do not take into account primordial binaries in this initial study.

2.1. *Simulation of a GC core: reflective boundary*

We also perform integrations with a smaller number of stars and BHs that are confined within a reflecting spherical boundary (Banerjee *et al.* 2009). With such a dynamical system, one can mimic the core of a massive cluster, where the BHs are concentrated after mass segregation. The advantage of this approach is that one can compute the evolution of a massive cluster with many fewer stars, allowing much faster computation. We integrate $N = 3000 - 4000$ stars, confined within 0.4 pc, which provides a stellar density of $\sim 10^4 {\rm M}_{\odot}$ pc^{-3}, appropriate for the core density of a massive cluster. However, stars and BHs faster than a pre-assigned speed, $v_{\rm esc} \approx 24$ km s^{-1}, representing the escape speed of the parent cluster, are allowed to escape through the reflective boundary.

3. Results

Table 1 summarizes the results of our computations. For demonstration purposes, we use one of the models, C50K80 (see Table 1). Other models generally possess similar properties. Figure 1 (left panel) demonstrates the mass segregation of the BHs in the cluster, which takes about 50 Myr. As the BHs segregate within about 0.3 pc of the

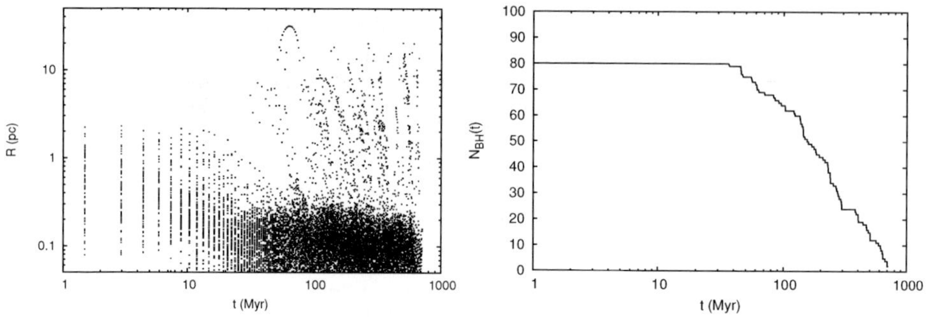

Figure 1. Mass segregation of BHs shown by the radial position R versus time t (left panel) for model C50K80. The BHs segregate in ~ 50 Myr, during which $N_{\rm BH}$ (the number of BHs bound to the cluster) remains unchanged (right panel). As the BH subcluster becomes dense enough that BH–BH binaries begin to form through three-body encounters, single BHs and BH binaries are ejected from the cluster, resulting in a decrease of $N_{\rm BH}$.

cluster core, the BH density of this subcluster becomes high enough to form BH–BH binaries through three-body encounters. Once BH–BH binaries start forming, single BHs and BH binaries begin to escape from the BH core due to encounter recoils (see Section 1). In Figure 1 (left panel), one can clearly distinguish the two phases of the BH subsystem, the initial segregation phase and BH-core formation, the radial positions of the BHs being scattered outwards in the latter phase due to the recoils. The resulting decrease of $N_{\rm BH}$ during this phase is also shown in Figure 1 (right panel).

3.1. *Mergers and escapers*

To study the possibility of BH–BH mergers, we consider the positions of the BH–BH binaries within the cluster in an a versus $(1 - e^2)$ plane, as shown in Figure 2, where each pair is represented by a different symbol and color coded with the evolution time in Myr. For each BH pair, a and e fluctuate over the plane, the changes occurring over a collision timescale, of order Myr. Since the orbital periods of the binaries corresponding to these points are much shorter (from $\sim$ days to years), these points generally represent binaries which are stable over many orbits. Overplotted in Figure 2 are lines of constant GW merger time $t_{\rm mrg}$, as given by Equation (2.1). While most points correspond to very large merger times, a few of them lie close to the $t_{\rm mrg} = 10$ Myr line. This indicates that these binaries are indeed hardened up to small enough a and/or acquire sufficient eccentricity that if they are left unperturbed, they can merge via GW emission within several Myr. However, these merging BH pairs can still be perturbed by further encounters on timescales of $\sim$ Myr, which can often prevent them from merging. In the particular example shown in Figure 2, only one of them could merge within the cluster. This feature is found to be generally true for all models reported here (see Table 1), i.e., each produces BH pairs that are capable of merging within several Myr. On the other hand, among the escaped BH binaries, all of them with GW merger times of less than a Hubble time are of interest, since they remain unperturbed afterwards. The positions of the escaped BH binaries from the model C65K110 are shown in Figure 3 as an interesting illustration.

Figure 4 (left panel) shows the distribution of merger times, $t_{\rm mrg}$, of BH binaries that occurred within the clusters, which indicates that $N_{\rm mrg}$ decreases with $t_{\rm mrg}$. This can be expected, as at later times $N_{\rm BH}$ in a cluster decreases, decreasing the hardening rate of BH binaries correspondingly. Figure 4 (right panel) shows the distribution of the merger times for the escapers, which also shows a decrease with time, as can be expected from Equation (2.1). The above merger-time distributions indicate that most mergers occur within the first few Gyr of cluster evolution.

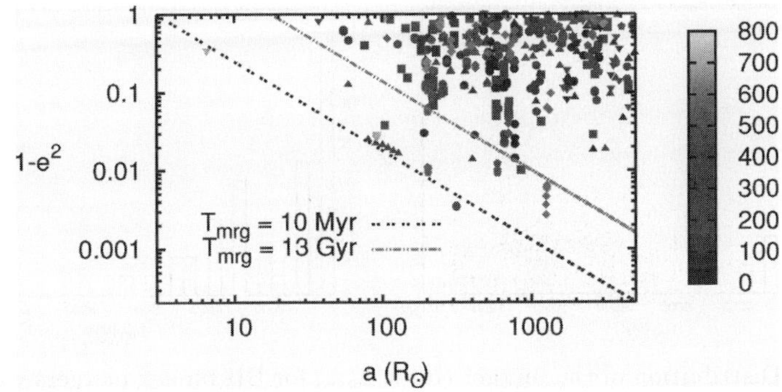

Figure 2. Positions of all BH binaries within the cluster on the $1 - e^2$ versus a plane, where different symbols are used to distinguish between different BH pairs (for model C50K80). The color coding of the points, as indicated by the color scale, represents the time (in Myr) at which they appear at a particular location in the above plane. Lines of constant GW-merger times $t_{\rm mrg}$ = constant are overplotted.

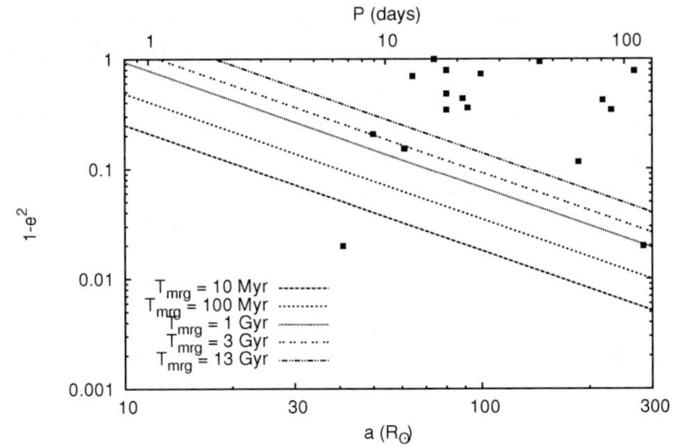

Figure 3. Positions of all escaped BH binaries in the $1 - e^2$ versus a plane for model C65K110. Lines of constant merger times are overplotted, as in Figure 2.

These results imply that an important class of candidates for dynamically forming BH binaries that merge at the present epoch are the star clusters with initial mass $M_{\rm cl} \gtrsim 3 \times 10^4 {\rm M}_\odot$, which are few Gyr old. Such clusters represent IMCs with initial masses close to the upper limit of the initial cluster mass function in spiral and starburst galaxies (Weidner *et al.* 2004; Gieles *et al.* 2006; Larsen 2009). Old GCs, which can be about 10 times more massive, are typically much older ($\sim$ 10 Gyr), so that they cannot contribute significantly to the present-day merger rate, as most of their mergers would have occurred much earlier. On the other hand, young massive clusters with ages younger than 50 Myr are generally too young to produce BH–BH mergers, as the segregation of BHs and the formation of the BH core itself takes longer. Hence, IMCs seem to be the most likely star clusters for dynamically producing present-day BH–BH mergers.

3.2. *Detection rate of BH–BH mergers*

Results from Table 1 give an average merger rate of $\approx$ 0.4 cluster^{-1} Gyr^{-1} for BHs within the clusters and $\approx$ 0.9 cluster^{-1} Gyr^{-1} for the escapers, totalling $\mathcal{R}_{\rm mrg} \approx 1.3$ cluster^{-1} Gyr^{-1}. To estimate the detection rate of BH–BH mergers from IMCs, we adopt the space density derived by Portegies Zwart & McMillan (2000) for young populous clusters, $\rho_{\rm cl} \approx 3.5h^3$ Mpc^{-3}, where $h = 0.73$ is the Hubble parameter. The above merger

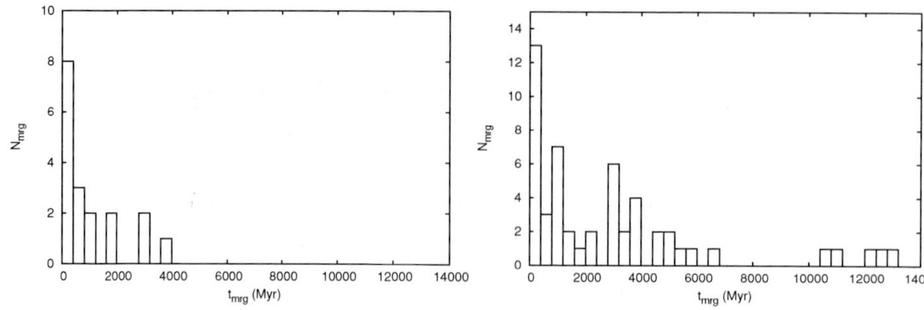

Figure 4. *(left)* Distribution of the merger times, $t_{\rm mrg}$, for BH binary mergers within the cluster for the models in Table 1. *(right)* Distribution of $t_{\rm mrg}$ for escaped BH binaries for the models in Table 1.

rate implies (Banerjee *et al.* 2009) that IMCs would yield a BH–BH-merger detection rate of $\mathcal{R}_{\rm AdLIGO} \approx 30$ yr^{-1} for the future *Advanced LIGO (AdLIGO)* detector (see, e.g., Belczynski *et al.* 2007). The present-day *LIGO* detection rate is, however, negligible, $\mathcal{R}_{\rm LIGO} \approx 4.6 \times 10^{-3}$ yr^{-1}. Interestingly, the above dynamical BH–BH-merger detection rate can be more than 10 times higher than that from individual primordial binaries (Belczynski *et al.* 2007). Our results imply that dynamically formed BH–BH binaries constitute the dominant contribution to BH–BH merger detection. Thus, the dynamical BH–BH inspirals from star clusters seem to be a promising channel for GW detection by *AdLIGO*, although their estimated detection rate with the present *LIGO* detector is negligible, conforming with the hitherto nondetection of GWs.

Acknowledgements

We thank the Symposium organizers for selecting this work for presentation. SB acknowledges support from the Alexander von Humboldt Foundation. HB acknowledges support from the German Science Foundation through a Heisenberg fellowship.

References

Aarseth, S. J. 2009, http://www.sverre.com/

Banerjee, S., Baumgardt, H., & Kroupa, P. 2009, *MNRAS*, in press (arXiv:0910.3954)

Belczynski, K., Taam, R. E., Kalogera, V., Rasio, F. A., & Bulik, T. 2007, *ApJ*, 662, 504

Casares, J. 2007, in: V. Karas & G. Matt. (eds.), *Black Holes from Stars to Galaxies: Across the Range of Masses*, IAU Symp, 238, p. 3

Gieles, M., Larsen, S. S., Scheepmaker, R. A., Bastian, N., Haas, M. R., & Lamers, H. J. G. L. M. 2006, *A&A*, 446, L9

Heggie, D. C. 1975, *MNRAS*, 173, 729

Heggie, D. C. & Hut, P. 2003, *The Gravitational Millon-Body Problem: A Multidisciplinary Approach to Star Cluster Dynamics* (Cambridge: Cambridge University Press)

Larsen, S. S. 2009, *A&A*, 494, 539

Mackey, A. D., Wilkinson, M. I., Davies, M. B., & Gilmore, G. F. 2007, *MNRAS*, 379, L40

Merritt, D., Piatek, S., Portegies Zwart, S., & Hemsendorf, M. 2004, *ApJ*, 608, L25

Peters, P. C. 1964, *Phys. Rev. B*, 136, 1224

Portegies Zwart, S. F. & McMillan, S. L. W., 2000, *ApJ* (Letters), 528, L17

Weidner, C., Kroupa, P., & Larsen, S. S. 2004, *MNRAS*, 350, 1503

Star clusters: basic galactic building blocks
Proceedings IAU Symposium No. 266, 2009
R. de Grijs & J. R. D. Lépine, eds.

doi:10.1017/S1743921309991086

Star cluster dynamics in galaxies

Kenji Bekki

School of Physics, University of New South Wales, Sydney, NSW, 2052, Australia
email: bekki@phys.unsw.edu.au

Abstract. I present a review of star cluster (SC) dynamics in galaxies, with special emphasis on the effects of global galactic dynamics on SC formation and evolution. I particularly discuss (i) dynamical friction processes affecting SCs in galaxies of different masses, (ii) formation of stellar galactic nuclei and massive globular clusters (GCs) through multiple merging of SCs, (iii) interactions between giant molecular clouds (GMCs) and SCs, (iv) SC destruction due to the strong tidal fields in galaxy mergers and (v) the formation of low-mass dwarfs from numerous SCs. I also discuss some recent observational results on SC mass functions in dwarf galaxies and physical properties of GC systems in luminous galaxies based on recent results of numerical simulations of SC dynamics in galaxies.

Keywords. globular clusters: general, open clusters and associations: general, galaxies: formation, galaxies: evolution

1. Introduction

The formation and evolution of star clusters (SCs) can be strongly influenced by the global dynamics of their host galaxies, such as by their strong tidal fields (e.g., Vesperini 2000; Baumgardt & Makino 2003), collisions between low-mass SCs and GMCs (e.g., Gieles *et al.* 2006) and dissipative gas dynamics (e.g., Bekki *et al.* 2002; Bournaud *et al.* 2008). Equally, basic galactic components such as stellar galactic nuclei and thin and thick disks can also be influenced dynamically by the evolution of SC systems (e.g., Kroupa 1998; Capuzzo–Dolcetta & Vicari 2005). Radial density profiles of the stars in individual SCs formed from GMCs may depend on the strength of the tidal field of their host galaxy (e.g., Hurley & Bekki 2008). Furthermore, mass functions (MFs) of old GCs and radial density profiles of GC systems (GCSs) in galaxies can be significantly changed owing to destruction of GCs with different masses within galaxies (e.g., Vesperini 2000).

Here, I focus on how the formation and evolution of SCs and SC systems (SCSs) in galaxies can be influenced by global dynamical processes. I do not intend to discuss *internal stellar dynamics of SCs*, mainly because a number of contributions at this conference (e.g., Boily, these proceedings) focus on internal SC evolution. I select five key topics and describe observational backgrounds, unresolved problems, results of previous theoretical studies and implications of the latest theoretical results. SCs in galaxy-scale simulations are usually represented by point-mass particles, so that internal evolution cannot be investigated properly (e.g., Fujii *et al.* 2008). I briefly discuss how to construct more self-consistent models for SC evolution in galaxies, based on more sophisticated numerical simulations.

2. Dynamical friction of SCs in galaxies

Tremaine *et al.* (1975) were among the first to discuss dynamical friction (DF) of massive GCs in galactic halos in the context of the formation of stellar galactic nuclei.

The timescale of DF ($t_{\rm fric}$) for typical GCs with masses ($M_{\rm gc}$) of $2\times10^5\,{\rm M}_\odot$ within galactic halos of luminous galaxies is, however, quite long (e.g., Binney & Tremaine 1987). For the Galaxy, with a circular velocity $v_{\rm c} = 220$ km s^{-1} and Coulomb logarithm $\ln\Lambda = 10$, $t_{\rm fric}$ can be estimated as

$$t_{\rm fric} = 1.2\times10^{11}\left(\frac{r_{\rm i}}{2\,{\rm kpc}}\right)^2\left(\frac{v_{\rm c}}{220{\rm km\,s^{-1}}}\right)\left(\frac{M_{\rm gc}}{2\times10^5\,{\rm M}_\odot}\right)^{-1}\,{\rm yr}, \tag{2.1}$$

where $r_{\rm i}$ is the distance of a given GC from the center of its host galaxy. For a galaxy with $v_{\rm c} = 70$ km s^{-1} and $\ln\Lambda = 10$,

$$t_{\rm fric} = 1.8\times10^{9}\left(\frac{r_{\rm i}}{1\,{\rm kpc}}\right)^2\left(\frac{v_{\rm c}}{70{\rm km\,s^{-1}}}\right)\left(\frac{M_{\rm gc}}{10^6\,{\rm M}_\odot}\right)^{-1}\,{\rm yr}. \tag{2.2}$$

Therefore, although DF cannot be important for the evolution of normal SCs and GCs within luminous galaxies like the Milky Way, it can significantly change the physical properties of SCs and SCSs in dwarf galaxies. For example, DF of SCs in the central regions of dwarf galaxies can be a key determinant for their transformation from nonnucleated to nucleated dwarf galaxies (e.g., Oh & Lin 2000; Bekki *et al.* 2004).

Recent theoretical studies have shown that the details of DF processes can be determined by the mass distributions of their host galaxies, such as by their dark-matter halos (e.g., Hernandez & Gilmore 1998; Goerdt *et al.* 2006; Inoue 2009). Orbital decay of GCs due to DF can significantly change the spatial distribution of GCs and thus increase their susceptibility to being destroyed by strong galactic tidal fields (e.g., Vesperini 2000). Although these previous studies contributed significantly to a better understanding of SC evolution within galaxies due to DF, their models are not fully self-consistent in the sense that they investigated DF of SCs *in galactic halos* but they did not include DF *caused by field stars in galactic bulges and disks.* Given that DF can depend strongly on the relative velocities between SCs and background components, SC DF can be quite

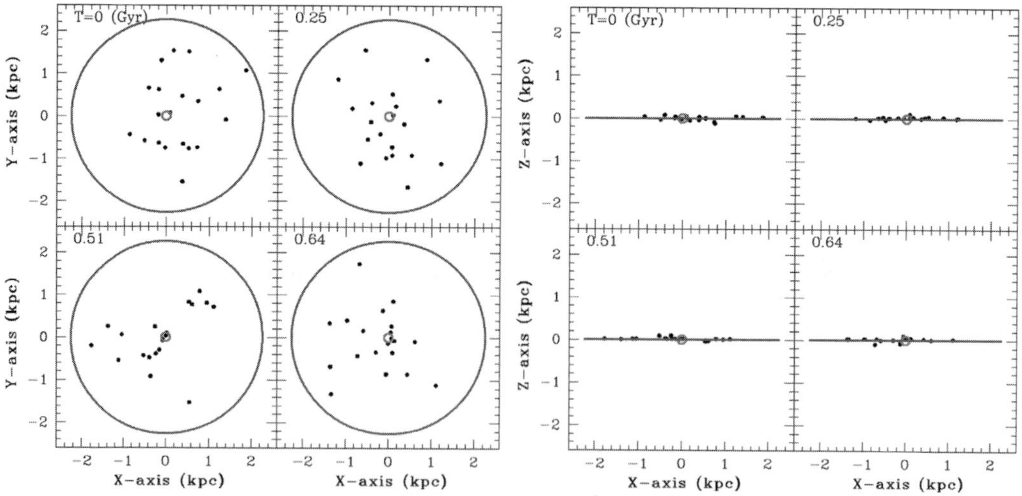

Figure 1. Spatial distributions of disk SCs (small black dots) in a disk galaxy for four different time steps in a model with $M_{\rm disk} = 10^9\,{\rm M}_\odot$. The left- and right-hand panels show the distributions of SCs projected onto the x–y and x–z planes, respectively. Time (T) in units of Gyr is given in each of the four panels. Large blue and small red circles represent the initial disk size of the galaxy and the nuclear region ($R = 100$ pc), respectively.

efficient in disk galaxies, where the relative velocities between SCs and disk field stars can be rather small.

Thus, here I discuss DF of SCs owing to disk field stars, based on the latest results of numerical simulations (primarily our own) of SC evolution due to DF within disk galaxies. DF processes of SCs caused by field stars in disk galaxies are important, first because recent cosmological hydrodynamical simulations of the possible formation sites of the first GCs (Kravtsov & Gnedin 2005) have shown that the present-day GCs can be formed within disk galaxies at high redshifts ($z \sim 3$), second because less luminous disk galaxies like the Large Magellanic Cloud (LMC) have disky GC systems in which the GCs have disky spatial distributions and rotational kinematics (e.g., Freeman *et al.* 1983; Olsen *et al.* 2004), and third because previous theoretical studies suggested that the present-day dwarf ellipticals (dEs) were previously less luminous disk galaxies with no or small bulges (Mastropietro *et al.* 2004). Therefore, DF of SCs due to disk field stars can be important for the evolution of the luminosity function (LFs) and MFs of SCs and GCs in dEs and the structural and kinematic properties of GCSs in less luminous disk galaxies.

Figure 1 shows the time evolution of the projected SCS distribution, composed of 20 SCs in a disk galaxy with an initial disk size ($R_{\rm disk}$) of 2.3 kpc and initial disk and halo masses ($M_{\rm disk}$ and $M_{\rm halo}$, respectively) of 10^9 and $10^{10}\,{\rm M}_\odot$, respectively. The canonical

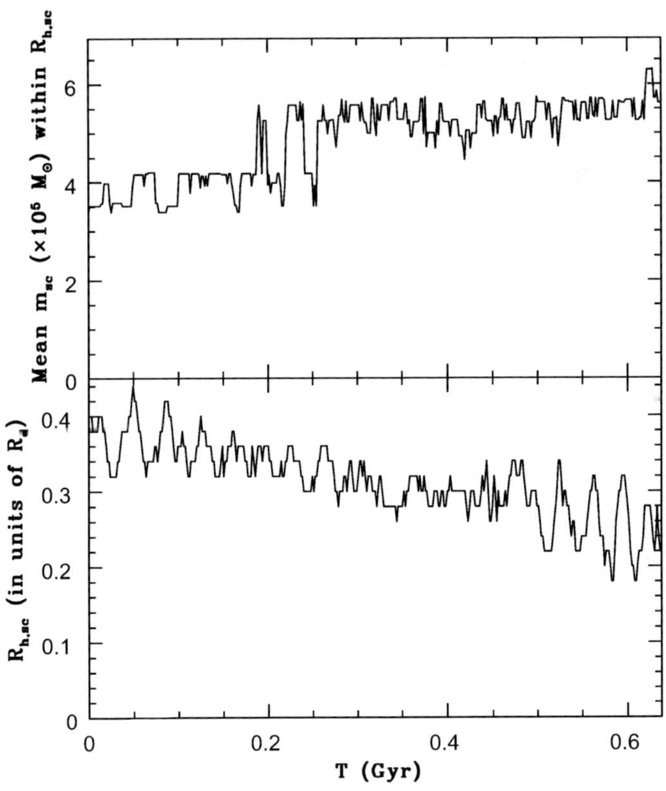

Figure 2. Time evolution of the half-number radius ($R_{\rm h,sc}$) of the SCS in a disk galaxy with $M_{\rm disk} = 10^9\,{\rm M}_\odot$ *(bottom)* and the mean SC mass within $R_{\rm h,sc}$ *(top)*. Note that both $R_{\rm h,sc}$ and the mean $m_{\rm sc}$ can change rapidly owing to orbital decay of SCs caused by DF, within much less than 1 Gyr.

MF of Galactic GCs (Harris 1991) is used for these SCs,

$$\Phi(M) \propto e^{-(M-M_0)^2/2\sigma_{\rm m}^2}, \tag{2.3}$$

where $M_0(V) = -7.23 \pm 0.23$ mag and $\sigma_{\rm m}$=1.25 mag. The minimum and maximum SC masses are set at 7×10^4 and $2 \times 10^6 {\rm M}_\odot$, respectively, for a reasonable mass-to-light (M/L) ratio (the details of the model are given in Bekki 2009). The SC models discussed in Sections 3 through 6 are the same. The SCs can be influenced not only by DF caused by disk field stars but also by the global dynamical action of spiral arms and bars in the disk. It is clear from Figure 1 that several massive SCs can pass through the nuclear region ($R \leqslant 100$ pc) at $T = 0.51$ Gyr and later, owing to orbital decay of these SCs caused by DF due to field stars in the disk galaxy. In this model, 30% of the SCs (6 out of 20) can be transferred to the nuclear region within 0.64 Gyr by DF.

Figure 2 shows that the half-number radius of the SCS ($R_{\rm h,sc}$) evolves from $0.38R_{\rm disk}$ (corresponding to = 0.86 kpc) to $0.30R_{\rm disk}$ (0.68 kpc) in 0.6 Gyr. The mean of the SC masses within $R_{\rm h,sc}$ becomes higher because DF is more efficient for more massive SCs: the number fraction of more massive SCs within $R_{\rm h,sc}$ becomes larger owing to the more rapid radial transfer of more massive SCs to the inner regions of the disk. The total number of SCs within $0.1R_{\rm disk}$ gradually increases, so that a multiple-SC system with a total mass of $4.3 \times 10^6 {\rm M}_\odot$ can form in the nuclear region of the disk. The x, y and z components of the velocity dispersions of the multiple SCs are 17.7, 23.6 and 3.2 km s^{-1}, respectively. As these velocity dispersions are comparable to the internal velocity

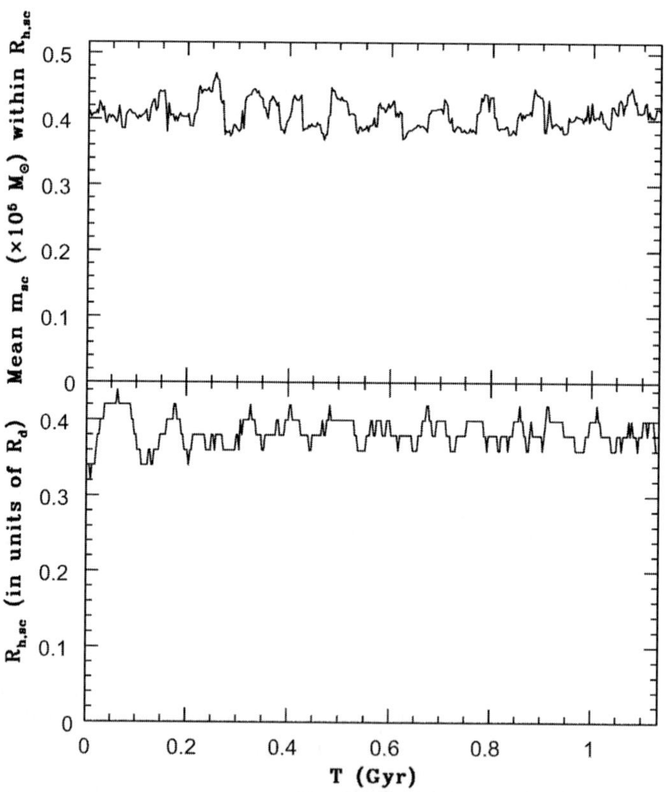

Figure 3. As Figure 2, but for a disk model with $M_{\rm disk} = 10^{10}{\rm M}_\odot$. Note that $R_{\rm h,sc}$ and the mean $m_{\rm sc}$ within $R_{\rm h,sc}$ in this high-mass disk model show little change in their time evolution in comparison with the low-mass disk simulation.

dispersions of the nuclear massive SCs, this suggests that the nuclear SCs may soon merge to form a single nucleus.

Thus, the timescale for DF of SCs in less luminous disk galaxies is rather short (less than 1 Gyr), so that DF can significantly change (i) the LFs and MFs of SCs and GCs and (ii) the structural and kinematic properties of SCSs and GCSs in these disk galaxies. One of the implications of these results is that the LFs and MFs of SCs and GCs can steepen owing to the consumption of more massive clusters in nuclear regions of galaxies (i.e., inward transfer and the resulting formation of stellar nuclei, which cannot be identified as a SC or GC). However, it should be stressed here that DF cannot be important for the evolution of SCSs in more luminous disk galaxies with $M_{\rm disk} \geqslant 10^{10}{\rm M}_\odot$. Figure 3 shows that neither $R_{\rm h,sc}$ nor the mean $m_{\rm sc}$ within $R_{\rm h,sc}$ evolve significantly with time for a disk galaxy with $M_{\rm disk} = 10^{10}{\rm M}_\odot$. For this more massive disk galaxy, DF cannot be effective within a few Gyr, not even for SCs with $m_{\rm sc} \sim 10^6{\rm M}_\odot$, so that the radial number-density profile of the SCS cannot change. The reason for this less effective DF is that relative velocities between disk field stars and SCs are so large that DF cannot work for these more massive disk galaxies. The results in Figures 3 and 4 thus imply that (i) DF affecting SCs in disk galaxies can only be effective in dwarf galaxies and thus (ii) the LFs and MFs of GCs can steepen more significantly due to DF in less luminous galaxies, as observed in dwarf galaxies (e.g., Jordán *et al.* 2006).

3. Formation of stellar galactic nuclei trough SC merging

Previous theoretical studies discussed the formation of stellar galactic nuclei through mergers of SCs (or GCs) in galaxies with different physical properties (e.g., Lotz *et al.* 2001; Capuzzo–Dolcetta & Vicari 2005). Recent, fully self-consistent N-body simulations have shown that mergers of SCs in the inner regions of galaxies lead to the formation of stellar galactic nuclei with projected radial density profiles similar to what is observed (e.g., Capuzzo–Dolcetta & Miocchi 2008). One of the remaining questions in the SC-merger scenario of nucleus formation is whether it can explain the observed relation between the masses of stellar nuclei and those of their host galaxies ($M_{\rm nuc} \sim 0.003 M_{\rm gal}$, where $M_{\rm gal}$ is the galaxy mass; Côté *et al.* 2006). Since the observational study of Côté

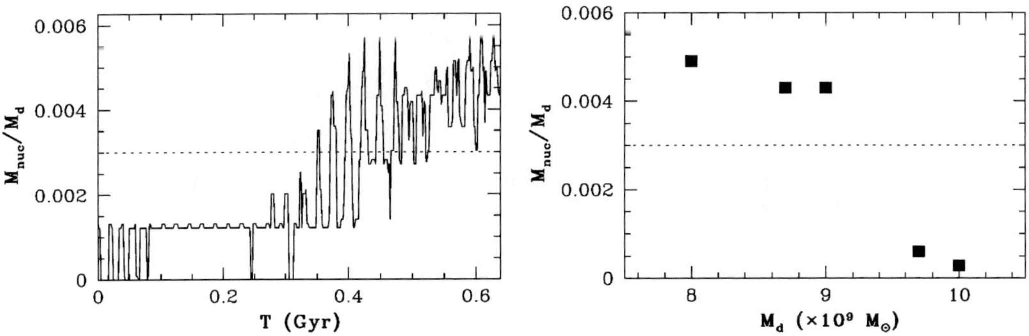

Figure 4. Time evolution of the nuclear mass ratio ($M_{\rm nuc}/M_{\rm disk}$) for a disk model with $M_{\rm disk} = 10^9{\rm M}_\odot$ *(left)* and as a function of disk mass, $M_{\rm disk}$. For both panels, the observed nuclear mass ratio (i.e., $M_{\rm nuc}/M_{\rm disk} = 0.003$) is indicated by a dotted line, for comparison. In this galaxy-scale simulation, the nuclear SCs are assumed to merge quickly and form a single stellar nucleus (Bekki *et al.* 2004). Therefore, $M_{\rm nuc}$ is assumed to be the total mass of SCs within the nuclear region of the galaxy; it is not literally the total mass of a single stellar nucleus. Thus, $M_{\rm nuc}/M_{\rm disk}$ can increase and decrease with time.

et al. (2006) derived a total mass of the luminous components (i.e., not including the extended dark-matter halos) of galaxies using reasonable M/L ratios, it is reasonable to use $M_{\rm disk}$ rather than $M_{\rm gal}$ for the purpose of discussing the observed $M_{\rm nuc}$ versus $M_{\rm gal}$ relation. I will now briefly discuss the origin of this relation based on the results of the galaxy-scale simulations with SCs discussed in the previous section.

Figure 4 shows the time evolution of the mass ratio of the stellar nucleus to the stellar disk in a galaxy with $M_{\rm disk} = 10^9 {\rm M}_{\odot}$. Here, the nuclear mass is the total mass of SCs within the nuclear region (i.e., $R = 0.1 R_{\rm disk}$) of the galaxy. It is not the total mass of the single stellar nucleus formed from multiple merging of SCs. The simulated nuclear mass ratio can rapidly become greater than 0.003 (in much less than 1 Gyr) because of orbital decay of massive SCs caused by DF of the clusters caused by the disk field stars. This result suggests that the formation of stellar galactic nuclei with masses similar to observed masses may be possible through SC mergers. However, the formation of stellar galactic nuclei through SC merging can proceed only in less luminous disk galaxies, because DF of SCs cannot be efficient for more luminous disk galaxies. Figure 4 shows that $M_{\rm nuc}/M_{\rm disk}$ can be as large as the observed value of 0.003, but *only for disk galaxies with* $M_{\rm disk} \leqslant 10^9$ ${\rm M}_{\odot}$. This result suggests that the observed $M_{\rm nuc}$ versus $M_{\rm disk}$ relation can be explained by SC mergers only for less luminous disk galaxies.

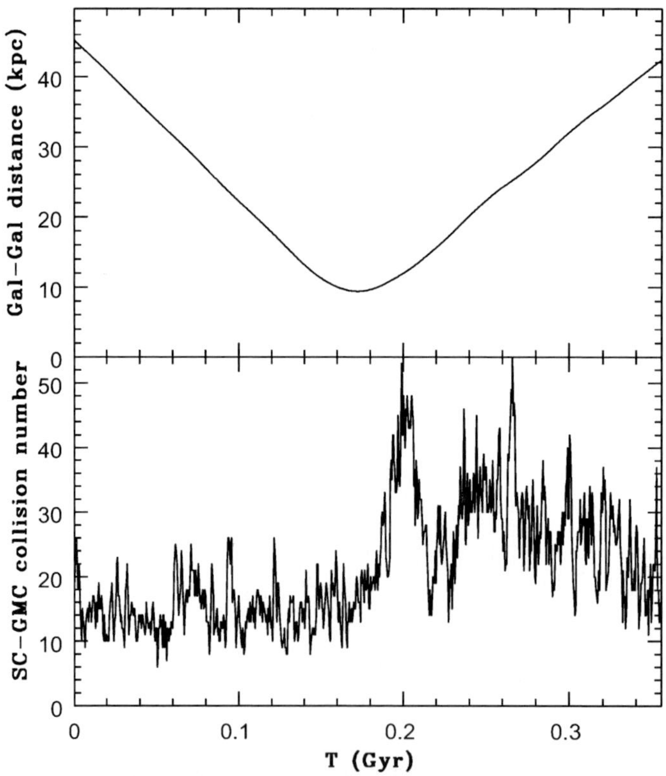

Figure 5. Time evolution of galaxy–galaxy distances *(top)* and SC–GMC collision number *(bottom)* for a galaxy-interaction model with a mass ratio of the two interacting galaxies of 2 and with a total mass of the primary galaxy of $4 \times 10^{10} {\rm M}_{\odot}$. If distances between SC–GMC pairs are less than 100 pc, they are regarded as colliding with each other. Initially, 100 SCs and 250 GMCs are distributed in the same way as the disk field stars, according to exponential radial distributions. Both SCs and GMCs have rotational kinematics similar to the disk field stars.

4. SC–GMC interaction in interacting galaxies

Strong dynamical interaction between SCs and GMCs and between different SCs is considered important for, e.g., the destruction of low-mass SCs (e.g., Gieles *et al.* 2006), the formation of rotationally flattened SCs through SC mergers (e.g., Makino *et al.* 1991) and the formation of binary SCs (e.g., Leon *et al.* 1999). It is, however, unclear how frequently such SC–GMC and SC–SC collisions/interactions can happen in galaxies owing to a lack of extensive galaxy-scale simulations that investigate directly and self-consistently such collisional events in galaxies. Thus, here I briefly discuss the SC–GMC and SC–SC collision rates in disk galaxies based on our galaxy-scale simulations with SCs and GMCs (Bekki *et al.*, in prep.). Since the parameter space for this numerical investigation is huge (i.e., regarding one's choice of different galaxy masses and structures), here I describe the results of the simulations for interacting disk galaxies with $M_{\rm disk} = 4 \times 10^9\,{\rm M}_\odot$ (corresponding to a total mass of $4 \times 10^{10}{\rm M}_\odot$), which is reasonable for the LMC and Magellanic-type interacting galaxies. The canonical MFs for SCs and GMCs observed for the Galaxy and the LMC are adopted for the numerical models described below.

Figure 5 shows that galaxy interactions can significantly (up to a factor of ~ 3) increase the SC–GMC collision rate during the pericenter passage of the two galaxies if their mass ratio is as high as 2. About 94% of all SCs can strongly interact with GMCs within mutual SC–GMC distances of less than 100 pc during the tidal interaction. Although the mean relative velocity ($V_{\rm rel}$) of SC–GMC collisions for mutual distances of less than 100 pc is 68 km s^{-1}, only 19% of SCs that are strongly interacting with GMCs show $V_{\rm rel} < 20$ km s^{-1}. For the latter, SC–GMC merging of more massive SCs would be possible owing to $V_{\rm rel}$ being comparable to the internal velocity dispersion of the SCs. About 73% of all SCs can strongly interact with other SCs with mutual SC–SC distances of less than 50 pc during the tidal interaction and only 2% of SCs have $V_{\rm rel} < 20$ km s^{-1}. This implies that formation of SC–SC pairs during galaxy interaction may be very rare.

Figure 6 shows that the mass ratios of GMC to SC masses for strongly interacting SC–GMC pairs in galaxy interactions can be variable for the adopted MFs for SCs and GMCs.

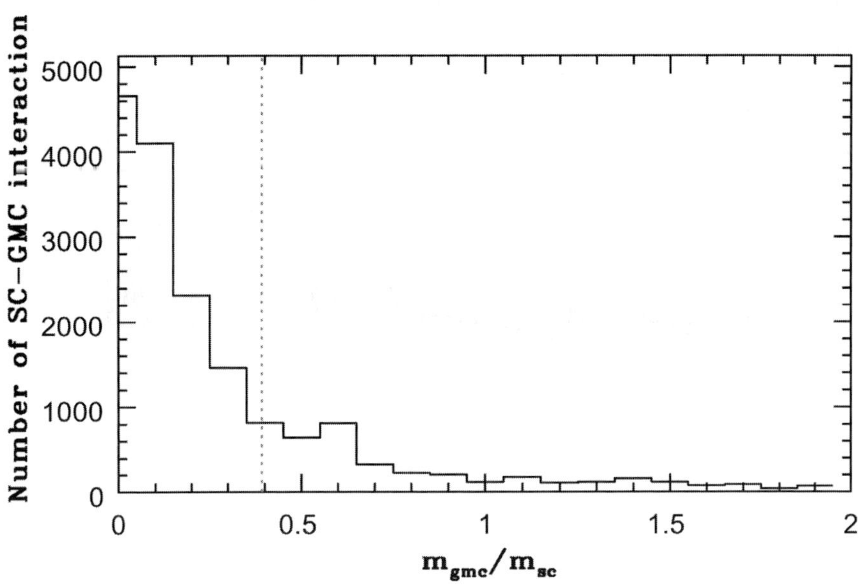

Figure 6. Distribution of the ratio of GMC to SC masses ($m_{\rm gmc}/m_{\rm sc}$) for colliding SC–GMC pairs in the same interaction model as shown in Figure 5. The mean value of $m_{\rm gmc}/m_{\rm sc}$ is indicated by the vertical dotted line, for comparison.

Most of the SC–GMC pairs show smaller $m_{\rm gmc}/m_{\rm sc}$ (< 1), with a mean of ~ 0.4. This suggests that some SCs, in particular those with smaller $m_{\rm gmc}/m_{\rm sc}$ ($0.3 - 0.5$), would accrete some gas during the SC–GMC interaction without being destroyed by GMCs owing to their smaller $m_{\rm gmc}/m_{\rm sc}$ (e.g., Bekki & Mackey 2009). If the gas transferred from GMCs to SCs during SC–GMC interactions can be used for further star formation in the SCs, the new stars can be identified as second-generation stars. There can be multiple stellar populations within individual SCs if the clusters experience strong SC–GMC interactions in interacting galaxies. One of the implications of these results is that the observed multiple stellar populations in intermediate-age GCs in the LMC (inferred from the double main-sequence turnoffs in their color–magnitude diagrams; e.g., Mackey *et al.* 2008) is due to SC–GMC collisions during the LMC–Galaxy interaction which occurred a few Gyr ago. Future, more sophisticated numerical simulations would need to be done to understand whether secondary star formation is really possible during SC–GMC collisions.

5. Tidal destruction of SCs in galaxy mergers

Although it is observationally clear that new SCs, some of which might eventually become GCs, are being formed in strongly interacting and merging galaxies like the Antennae system (Whitmore *et al.* 1995), formation and destruction processes of these new SCs during galaxy interactions/mergers are not so clear. Since SC 'infant morality' (i.e., disintegration of young SCs due largely to internal dynamical processes) in merging galaxies is extensively discussed in other contributions at this conference, here I focus

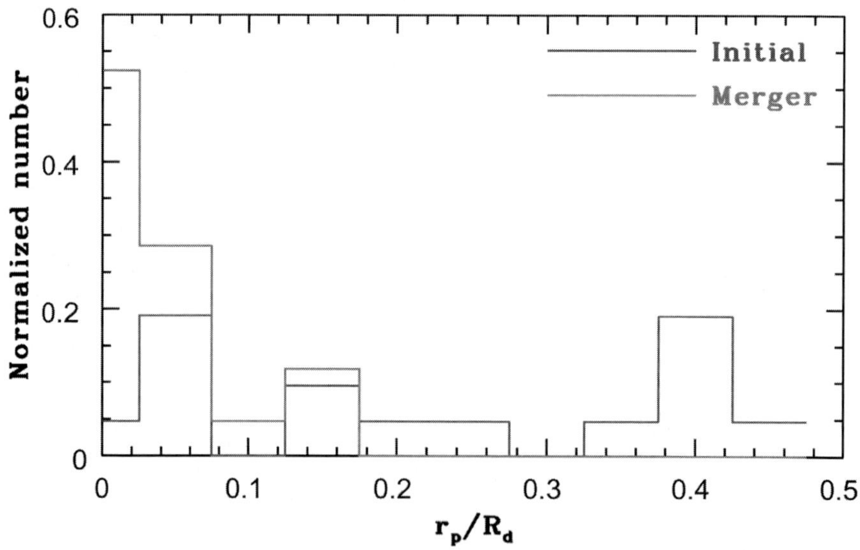

Figure 7. Normalized distribution of pericenter distances ($r_{\rm peri}$) in units of $R_{\rm disk}$ for orbits of 20 SCs from the initial disk (blue) and the galaxy merger (red) for a model with $M_{\rm disk} = 10^9 {\rm M}_{\odot}$. Here, $r_{\rm peri}$ for SCs in the initial disk are the initial distances of the SCs from the center of the galaxy, because initially SCs have almost circular orbits. Initially, these SCs are set to be within the merger progenitor disks. The nearly prograde–prograde merger of two disk galaxies with a pericenter distance of twice the disk size can eventually form a giant elliptical galaxy within 20 dynamical timescales. Note that most SCs in the merger show very small $r_{\rm peri}/R_{\rm disk}$ compared with those in the initial disk. This suggests that dynamics of major mergers can change circular orbits of the disk SCs into highly elongated orbits in the merger remnant.

on *destruction of initially bound SCs by strong tidal fields in galaxy mergers.* I base my discussion on the results of our numerical simulations (Bekki *et al.*, in prep.) of orbital evolution of SCs in mergers of two bulgeless disk galaxies with total masses of $M_{\rm disk} = 10^9 \rm M_\odot$ each.

Figure 7 shows the distribution of pericenter distances ($r_{\rm peri}$) in units of $R_{\rm disk}$ for the orbits of 20 SCs during a galaxy merger. It is clear that the SC orbits can become significantly elongated (i.e., the orbital eccentricity, $e_{\rm orb}$, becomes larger) during violent relaxation of galaxy mergers so that $r_{\rm p}$ can become rather small. For this model, the mean $r_{\rm p}$ and $e_{\rm orb}$ are $0.08 R_{\rm disk}$ and 0.8, respectively. Consequently, SCs with highly radial orbits can pass through the inner region of the merger (remnant) to feel the much stronger tidal field there. This implies that a significant fraction of SCs in a galaxy mergers can either lose a large amount of their initial mass or be destroyed completely by the strong tidal field. These results do not depend strongly on model parameters such as the orbital configurations of galaxy mergers or the masses of the progenitor disk galaxies. These results also imply that, although new GCs can be formed during galaxy mergers (Whitmore *et al.* 1995; Bekki *et al.* 2002; Bournaud *et al.* 2008), their specific frequencies (S_N) do not increase significantly after galaxy mergers because of the tidal destruction of newly formed SCs. The origin of the observed high S_N in giant elliptical

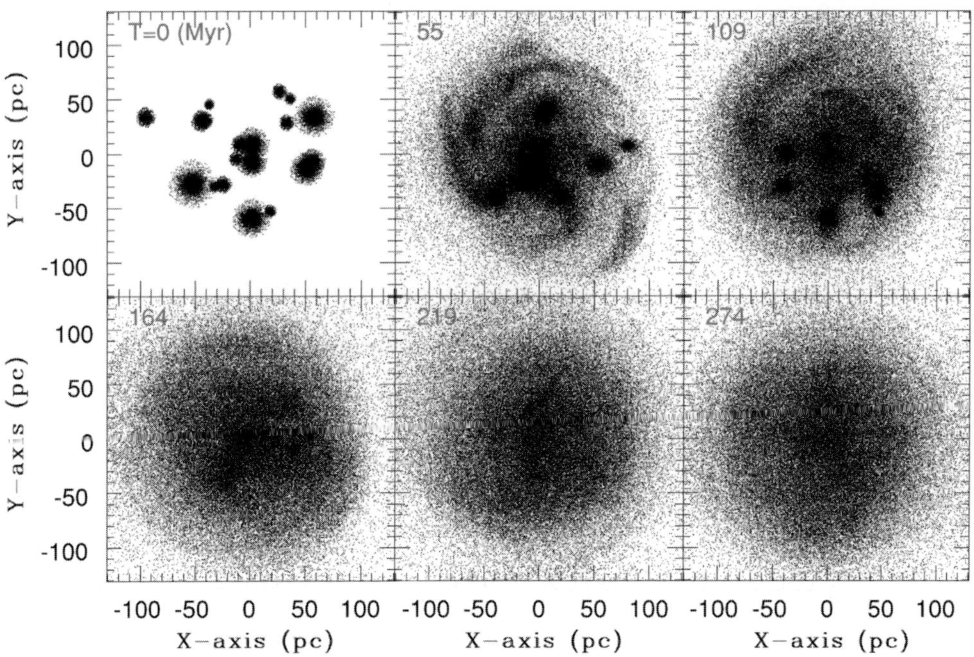

Figure 8. Time evolution of the distribution of SCs in a low-mass dark-matter halo projected onto the x–y plane. The 19 SCs are assumed to be unbound, in the sense that the total kinetic energy of a given SC ($T_{\rm kin}$) is four times greater than the absolute magnitude of the potential energy (W). This is done to mimic 50% mass loss of gas within newly formed SCs from GMCs due to energy feedback of massive OB stars and supernovae. The SC sizes and masses ($r_{\rm sc}$ and $m_{\rm sc}$, respectively) scale as $m_{\rm sc} \propto {r_{\rm sc}}^2$, which is similar to the scaling relation for GMCs. Owing to multiple-SC interactions and mergers and the strong tidal field of the dark-matter halo, most SCs eventually disintegrate to form a smooth distribution of field stars, which can be identified as a low-surface-brightness dwarf galaxy. If SCs are strongly bound initially (e.g., $2T_{\rm kin}/|W| = 1$; virial equilibrium), they cannot be destroyed as rapidly as in this model.

galaxies would not be so closely associated with the large number of new GCs formed during galaxy mergers that form giant elliptical galaxies.

6. Formation of low-mass dwarf galaxies from SCs

Given that the vast majority of field stars in galaxies originate from unbound or bound SCs (e.g., Lada & Lada 2003), it is reasonable to say that galaxies can be formed from SCs: they are fundamental building blocks of galaxies. If SCs are the first stellar objects formed in the very early phases of galaxy formation, one of the particularly interesting problems in terms of galaxy formation from multiple SCs relates to how the very first SCs, assumed to have formed in low-mass dark-matter halos, can disintegrate and then transform into a new galaxy dominated by field stars. Since this question has not been addressed by previous numerical simulations, here I discuss what can happen to multiple SCs in low-mass dark-matter halos when they interact and merge with one another and are influenced by the tidal fields of the halos. My analysis is based on our numerical simulations (Bekki *et al.*, in prep.).

Figure 8 shows that 19 initially unbound SCs with a total mass of $\sim 10^6 M_\odot$ in a dark-matter halo with a total mass of $\sim 10^9 M_\odot$ and with a Navarro, Frenk and White (NFW) profile can mutually interact with one another to form a diffuse dwarf galaxy in 274 Myr. The strong tidal field of the halo can play a vital role in this transformation from the initially clumpy distribution of SCs to the rather smooth field-star profile. As shown in Figure 9, the final galaxy has a very thick disk with a smaller ratio of maximum

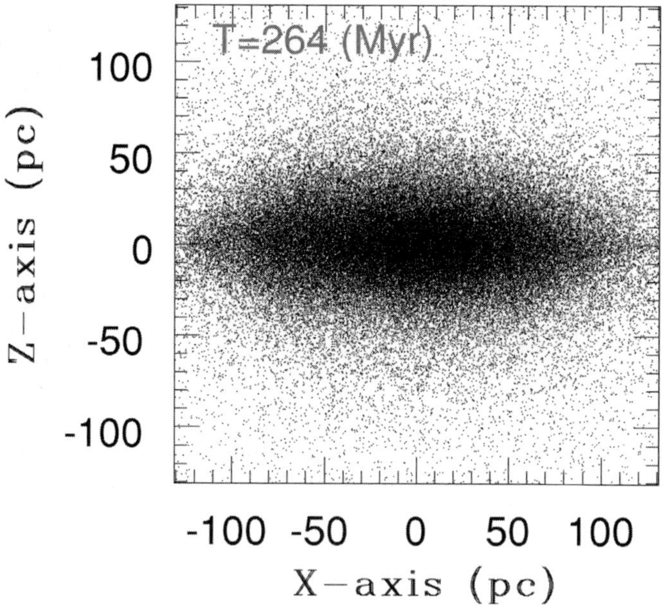

Figure 9. Final distribution of 19 SCs projected onto the x–z plane for the model shown in Figure 8. All unbound SCs completely disintegrate within 0.3 Gyr because of (i) the strong tidal field of their host dark-matter halo and (ii) mutual dynamical interactions between SCs. The field stars formed from disintegration of SCs can form a thick-disk component within the dark-matter halo. The initially unbound SCs can expand during their evolution because of rapid gas removal after their formation, and some fraction of the stars from the SCs can consequently form the stellar halo within the dark-matter halo. The formation of thick disks and stellar halos from disintegration of SCs within dark-matter halos is only possible in low-mass halos.

rotational velocity (V) to central velocity dispersion (σ) in the field stars ($V/\sigma \sim 3$). The thick-disk structure and kinematics are due largely to dynamical interactions between SCs and the dark-matter halo and between SCs and other SCs.

The final structural and kinematic properties of the simulated, very-low-mass dwarf galaxies depend strongly on the adopted initial MF of SCs and the total masses of the dark-matter halos. For example, the final disk-like structure of field stars in a dark-matter halo can be thicker in models with larger mean masses of SCs for a given initial mass of the halo. These results are based on rather idealized numerical models (e.g., fixed external tidal fields and a gravitational softening length for SC evolution), so that the internal dynamical evolution of individual SCs is not so self-consistently modeled. Thus, more self-consistent numerical simulations need to be done to gain a better understanding of how SCs form low-mass galaxies and how they evolve within them.

7. Conclusions

I have discussed five physical processes related to SCs, including (1) DF of SCs in disk galaxies, (2) multiple mergers of SCs in the central regions of galaxies, (3) collisions and mergers between SCs and GMCs and between different SCs, (4) tidal destruction of SCs in merging galaxies and (5) formation of low-mass dwarfs from multiple SCs in low-mass dark-matter halos. These processes are very important for a better understanding of the evolution of LFs and MFs of SCs and GCs, the origin of stellar galactic nuclei and their relationships with their host galaxies, formation of SCs with multiple stellar populations, evolution of S_N of GCs in early-type galaxies formed in major galaxy mergers and formation of very-low-mass galaxies at high redshifts. I have discussed these based mostly on galaxy-scale simulations in which internal evolution of SCs (e.g., two-body relaxation effects, binary formation and stellar evolution) is not properly modeled. Therefore, it is possible that future more sophisticated numerical simulations including *both galaxy- and cluster-scale internal dynamics* yield different results from those obtained in our simulations in terms of these five processes.

Ideally speaking, the internal dynamics of SCs in galaxies should be investigated using numerical codes by which we can investigate SC evolution in a *live (not a fixed) galactic potential*, represented by N-body particles. This is, first, because DF, which can not be investigated using models with fixed potentials, can significantly change the distances of SCs from their galactic centers (and thus the strengths of the galactic tidal fields, which are important for SC disintegration), and second because time-dependent global galactic structures (such as bars and spirals, which are hard to model as fixed, external gravitational forces) can dynamically influence SCs. Future sophisticated numerical simulations, including both cluster-scale (0.1–10 pc) internal and galaxy-scale (10 pc–10 kpc) global dynamics (e.g., Fujii *et al.* 2008) will allow us to investigate more self-consistently the formation and evolution of SCs within galaxies and thereby better understand not only the origin of the physical properties of SCs and SCSs, but also the formation of galaxies from their fundamental building blocks.

References

Baumgardt, H. & Makino, J. 2003, *MNRAS*, 340, 227
Bekki, K. 2009, *MNRAS*, submitted
Bekki, K., Forbes, D. A., Beasley, M. A., & Couch, W. J. 2002, *MNRAS*, 335, 1176
Bekki, K., Couch, W. J., Drinkwater, M. J., & Shioya, Y. 2004, *ApJ* (Letters), 610, L13
Bekki, K. & Mackey, A. D. 2008, *MNRAS* (Letters), 394, L124

Binney, J. & Tremaine, S. 1987, *Galactic Dynamics*, Princeton: Princeton University Press
Bournaud, F., Duc, P.-A., & Emsellem, E. 2008, *MNRAS* (Letters), 389, L8
Capuzzo–Dolcetta, R. & Miocchi, P. 2008, *MNRAS* (Letters), 388, L69
Capuzzo–Dolcetta, R. & Vicari, A. 2005, *MNRAS*, 356, 899
Côté, P., *et al.* 2006, *ApJS*, 165, 57
Freeman, K. C., Illingworth, G., & Oemler Jr., A. 1983, *ApJ*, 272, 488
Fujii, M., Iwasawa, M., Funato, Y., & Makino, J. 2008, *ApJ*, 686, 1082
Gieles, M., Portegies Zwart, S. F., Baumgardt, H., Athanassoula, E., Lamers, H. J. G. L. M., Sipior, M., & Leenaarts, J. 2006, *MNRAS*, 271, 793
Goerdt, T., Moore, B., Read, J. I., Stadel, J., & Zemp, M. 2006, *MNRAS*, 368, 1073
Harris, W. E. 1991, *ARA&A*, 29, 543
Hernandez, X. & Gilmore, G. 1998, *MNRAS*, 297, 517
Hurley, J. R. & Bekki, K. 2008, *MNRAS* (Letters), 389, L61
Inoue, S. 2009, *MNRAS*, 397, 701
Jordán, A., McLaughlin, D. E., Côté, P., Ferrarese, L., Peng, E. W., Blakeslee, J. P., Mei, S., Villegas, D., Merritt, D., Tonry, J. L., & West, M. J. 2006, *ApJ* (Letters), 651, L25
Kravtsov, A. V. & Gnedin, O. Y. 2005, *ApJ*, 623, 650
Kroupa, P. 1998, *MNRAS*, 300, 200
Lada, C. J. & Lada, E. A. 2003, *ARA&A*, 41, 57
Leon, S., Bergond, G., & Vallenari, A. 1999, *A&A*, 344, 450
Lotz, J. M., Telford, R., Ferguson, H. C., Miller, B. W., Stiavelli, M., & Mack, J. 2001, *ApJ*, 552, 572
Mackey, A. D., Broby Nielsen, P., Ferguson, A. M. N., & Richardson, J. C. 2008, *ApJ* (Letters), 681, L17
Makino, J., Akiyama, K., & Sugimoto, D. 1991, *ApSS*, 185, 63
Mastropietro, C., Moore, B., Mayer, L., Debattista, V. P., Piffaretti, R., & Stadel, J. 2005, *MNRAS*, 364, 607
Oh, K. S. & Lin, D. N. C. 2000, *ApJ*, 543, 620
Olsen, K. A. G., Miller, B. W., Suntzeff, N. B., Schommer, Robert A., & Bright, J. 2004, *AJ*, 127, 2674
Vesperini, E. 2000, *MNRAS*, 318, 841
Whitmore, B. C. & Schweizer, F. 1995, *AJ*, 109, 960

Star clusters: basic galactic building blocks
Proceedings IAU Symposium No. 266, 2009
R. de Grijs & J. R. D. Lépine, eds.

doi:10.1017/S1743921309991098

HST's hunt for intermediate-mass black holes in star clusters

Julio Chanamé,[1] Justice Bruursema,[2] Rupali Chandar,[3] Jay Anderson,[4] Roeland van der Marel[4] and Holland Ford[2]

[1]Department of Terrestrial Magnetism, Carnegie Institution of Washington, Washington, DC, USA
email: jchaname@dtm.ciw.edu

[2]Department of Physics and Astronomy, Johns Hopkins University, Baltimore, MD, USA

[3]Department of Physics and Astronomy, The University of Toledo, Toledo, OH, USA

[4]Space Telescope Science Institute, Baltimore, MD, USA

Abstract. Establishing or ruling out, either through solid mass measurements or upper limits, the presence of intermediate-mass black holes (IMBHs; with masses of $10^2 - 10^5$ $M_\odot$) at the centers of star clusters would profoundly impact our understanding of problems ranging from the formation and long-term dynamical evolution of stellar systems, to the nature of the seeds and the growth mechanisms of supermassive black holes. While there are sound theoretical arguments both for and against their presence in today's clusters, observational studies have so far not yielded truly conclusive IMBH detections nor upper limits. We argue that the most promising approach to solving this issue is provided by the combination of measurements of the proper motions of stars at the centers of Galactic globular clusters and dynamical models able to take full advantage of this type of data set. We present a program based on *HST* observations and recently developed tools for dynamical analysis designed to do just that.

Keywords. stellar dynamics, astrometry, stars: kinematics, globular clusters: general

1. Introduction

There is solid observational evidence for the existence of black holes (BHs) in two very different regimes. Stellar-mass BHs, with masses in the range 5 – 15 $M_\odot$, are the end products of normal high-mass stellar evolution. Solid mass estimates for these BHs come from measurements of the mass function in several tens of Galactic X-ray binaries. Supermassive BHs (SMBHs), at the other extreme, with masses between $10^{5.5}$ and $10^{9.5}$ $M_\odot$, have been reliably weighed in galactic centers based on the motions of their surrounding stars and gas. A tight correlation between the masses of SMBHs and the velocity dispersion of their host spheroid (the $M_{\rm BH} - \sigma$ relation) evidences that an intimate link must exist between these BHs and the processes relevant to galaxy formation, and highlights the importance of understanding the seeds and growth mechanisms of SMBHs. IMBHs, with masses in between those of the above two regimes, are likely to play a key role in these questions but they are yet to be convincingly detected.

Existing theoretical work provides arguments for and against the presence of IMBHs at the centers of globular clusters (GCs). Several possible channels of formation have been proposed (see van der Marel 2004 for a review), and, interestingly, all those mechanisms lead to IMBH masses consistent with an extrapolation of the $M_{\rm BH} - \sigma$ relation down to dispersions typical of GCs. On the other hand, it has been argued that it may be difficult for GCs, given their relatively shallow potential wells, to retain growing massive black holes at their centers for too long (Baker *et al.* 2008; Holley–Bockelmann *et al.* 2008; Moody & Sigurdsson 2009).

This current debate, however, should not prevent astronomers to set out plans to look for the existence of these objects. To the question of whether IMBHs are possible or not in GCs, the right attitude should be one that, although we do not know the answer at this moment, we will also act independently from theoretical prejudices and observationally probe for their presence anyway. Any tight upper limits to the masses of central compact objects in a number of GCs are equally important for the field as any solid mass measurements eventually confirming their existence. Certainly, any of these alternatives would clearly constitute an important improvement over today's situation.

2. Current state of the hunt for IMBHs in GCs

As in the case of SMBHs in active galactic nuclei, one could argue that the most unambiguous signature of the presence of a massive compact object would be the detection of central, unresolved radio and/or X-ray luminosity generated by the accretion of matter onto the central object. In the case of star clusters, however, one must make sure that this signature is not consistent with an accumulation of more common stellar-mass accreting objects (cataclysmic variables, X-ray binaries, stellar-mass black holes) or pulsars. This danger is well illustrated by the case of a GC associated with the giant elliptical galaxy NGC 4472 in the Virgo cluster, where Maccarone *et al.* (2007) reported a strong and highly variable X-ray signature "which rules out any object other than a black hole in such an old stellar population", then estimating about 400 $M_\odot$ for the mass of this object. Such a mass estimate, however, was necessarily based on several assumptions regarding highly uncertain variables of accretion physics (e.g., innermost stable orbit, accretion rate, disk and continuum model, unabsorbed luminosity, etc.), and can hardly be considered a mass measurement. Indeed, the same authors later determined, based on optical spectroscopy, that the object in question is most likely a stellar-mass black hole (Zepf *et al.* 2008).

As a consequence, while the observation of central, unresolved accretion signatures originating from star clusters certainly constitutes evidence for the possible presence of IMBHs, and, moreover, may be considered a prime way to search for IMBH candidates, it is nevertheless not adequate for the purposes of a solid, unambiguous mass measurement. The same considerations apply to the possibility of catching an event of stellar disruption by an IMBH (Irwin *et al.* 2009).

Therefore, for unambiguous mass measurements, we need to rely on dynamics, and there have been various observational efforts in this direction. Based on dynamical modeling of measured line-of-sight (LOS) kinematics, evidence for the presence of IMBHs has been claimed in three GCs: M15 and ω Cen in the Milky Way (Gerssen *et al.* 2002 and Noyola *et al.* 2008, respectively), and G1 in M31 (Gebhardt *et al.* 2002, 2005). In the cases of M15 and G1, however, the available data are also consistent with nonequilibrium models in which mass segregation occurs over time, but which do not possess an IMBH. (Baumgardt *et al.* 2003a,b). Moreover, those measurements are restricted to the luminous red giants in the clusters. Not being very numerous, any analysis based on the kinematics of the giants will be more affected by shot noise and projection effects (i.e., not knowing whether the tracers contributing to most of the velocity signal are actually probing the IMBH's sphere of influence, or whether they are instead located in front of or behind the cluster center) than if the data set were dominated by the more numerous, although fainter, main-sequence stars.

The cases of G1 and ω Cen are unique in their own way. We start by pointing out that, since the subject of this meeting is star clusters, and although the presence of an IMBH does not have anything to do with this fact, it is nevertheless appropriate to recall that ω Cen and G1 are both likely to be the nuclei of stripped dwarf galaxies rather than regular GCs. However, as this is yet to be considered an established result, we will discuss them here anyway and, instead of having to specify their uncertain nature by calling them 'stellar spheroids of low central velocity dispersion', we will refer to them loosely as clusters. With respect to G1, one can argue that this remains the best case so far for an IMBH in a GC. This is because, on top of the dynamical evidence quoted above, this cluster is also a source of radio and X-ray emission (Ulvestad *et al.* 2007; Kong 2007), as one would expect for an accreting compact object. The problem with this additional accretion evidence, however, is one of angular resolution, since both the radio and X-ray source positions are too uncertain and they could still be the result of a number of less exotic objects, such as X-ray binaries or CVs distributed throughout the cluster. Of course, this situation is expected to improve when performing higher-angular-resolution observations.† In the case of ω Cen, the problem has been again related to the effect of bright giants, in this case the effect that their shot noise has on the determination of the cluster center, a key step in the determination of light and velocity-dispersion profiles. Briefly, Anderson & van der Marel (2009) used high-angular-resolution *HST* images to measure the proper motions of cluster members, most still on the main sequence, and found a cluster center about 12 arcsec away from that determined by Noyola *et al.* (2008). The light and (2D) velocity-dispersion profiles computed with respect to the new center did not show the signatures that Noyola *et al.* (2008) atributed to a 40 000 $M_\odot$ IMBH. Next, van der Marel & Anderson (2009) constructed dynamical models with the proper-motion data and determined that models both with and without a black hole fit the data equally well, reporting a tight upper limit of 12 000 $M_\odot$ in case an IMBH is present.

3. Current theoretical understanding

Until just a few years ago, the generalized intuition indicated that the best places to look for IMBHs were the centers of very dense stellar systems such as the cores of Galactic globular clusters with very steep luminosity profiles. Solid theoretical basis for this expectation came from the seminal work of Bahcall & Wolf (1976, 1977), which predicted the development of stellar density cusps surrounding massive central BHs. In well-relaxed clusters, groups of stars of different mass formed cusps with correspondingly different slopes, reflecting the well-known phenomenon of mass segregation. This prompted the selection of Galactic globular clusters with dense, collapsed cores for observational studies that would look for the dynamical signature of an IMBH at their centers. However, drawing conclusions from the simulations as to what to expect observationally has never been an easy task, given that while the theoretical results are unambiguous regarding the mass of the particles, the observations can only track the luminous stars, and conversion from one to the other depends on many assumptions. More recently, it was realized that an IMBH at the center of a star cluster constitutes an additional source of heat acting against the rapid collapse of the cluster's core. Baumgardt *et al.* (2004a,b) found that the IMBH not only prevents the collapse of the core but also produces an expansion of the cluster, making it unlikely that dense core-collapsed clusters could harbor IMBHs. Therefore, a completely new picture of where it would be best to look for

† Kong *et al.* (2009) recently reported high-resolution observations with *Chandra*, locating the X-ray emission within the core radius of G1. Based on the ratio of X-ray to Eddington luminosity, the authors suggest the emission is more likely coming from low-mass X-ray binaries.

IMBHs has emerged: unlike collapsed cores and dense central regions, the most likely clusters to harbor an IMBH would be those whose projected surface brightness profile are well fit by regular King-like models with an extended core and intermediate concentrations ($W_0 \sim 7, c \sim 1.5$), with the surface brightness rising slightly only in the very inner regions of this core, forming a shallow, barely perceptible density cusp in the form of a power law with slope $\alpha = \mathrm{d}\log \mathrm{SB}/\mathrm{d}\log r \sim -0.25$ (Baumgardt *et al.* 2005; Miocchi 2007; Trenti *et al.* 2007).

4. Towards solid mass measurements or upper limits

In Section 2 we started by arguing that the observation of accretion signatures, while great for the detection of IMBH candidates, was not adequate for a reliable mass measurement of satisfactory precision, for which we had to rely on dynamical analyses. The current status of the evidence for IMBHs in GCs, however, proves that dynamics has its own problems. A large part of these, as summarized in Section 2, have to do with the limited amount of data available for those studies, which did not provide enough kinematic information to, say, distinguish the effect of a single massive central object from that of a more extended mass distribution. But this constitutes just one aspect of the potential problems associated with dynamics, the other side having to do with the subsequent handling of the available data. Indeed, in the process of the modeling of kinematic data, two of the most important dangers are (i) the inadequate exploitation of the full information content of a given dataset and (ii) the adoption of simplifying assumptions that are too restrictive and thus explore only a limited region of the space of possible solutions.

The management of these latter points can have profound implications for the inferred mass distribution (arguably the main goal of stellar dynamics), thus creating a delicate balance between the available observations for a given problem and the complexity of the models chosen to make use of such data. This balance was nicely exemplified by the initial debate regarding the dark-matter content of some intermediate-luminosity elliptical galaxies, during which LOS velocity measurements of planetary nebulae in the halos of these galaxies—showing a Keplerian-like decline of velocity dispersion with radius—were interpreted as evidence of the presence of little, if any, dark matter in those galactic halos (Romanowsky *et al.* 2003). Although this remains a subject of current discussion, it was later shown that such conclusions may well be the result of too restrictive assumptions regarding both the geometry of the underlying potential and the anisotropy of the orbits of the planetary nebulae around those galaxies (Dekel *et al.* 2005; de Lorenzi *et al.* 2009).

4.1. *The need for HST*

The key for an unambiguous assessment of the presence of IMBHs in GCs is availability of many stars with well-measured velocities probing the GC's core, and containing enough kinematic information to permit models to be very general (Section 4.2). This demands:

- *High spatial resolution:* stellar velocities need to be measured inside the sphere of influence of the putative BH, $r_{\rm BH} \simeq 0.39''(M_{\rm BH}/10^3\,\mathrm{M}_\odot)(10\,\mathrm{km\,s^{-1}}/\sigma)^2(10\,\mathrm{kpc}/D)$. Thus, for an IMBH of $3 \times 10^3\ \mathrm{M}_\odot$, many stars have to be well resolved inside $r_{\rm BH} \sim 1''$.
- *At least two velocity components:* a well-known degeneracy between mass and anisotropy (Binney & Mamon 1982) is the main obstacle for a reliable measure of the mass distribution in stellar systems based on line-of-sight velocities. Proper motions provide two components of the space velocity, allowing to actually measure the anisotropy of the stellar orbits instead of assuming it.

• *Optimization of observing (wavelength) window:* high levels of astrometric accuracy require minimizing as much as possible the red light from nearby, luminous giants and the unresolved background of low-mass stars populating the crowded cores of GCs. Thus, observations must be conducted at short enough wavelengths, from the blue to the near-ultraviolet.

Even with the largest-aperture telescopes and assisted by adaptive optics, ground-based observations are unable to produce radial velocities and/or proper motions for many stars in the inner arcsecond of GCs (e.g., Gebhardt *et al.* 2000). Built exactly to overcome these problems, *HST* is by design the *only* instrument that, by largely satisfying the three above requirements, will be able to resolve, within the next decade, the question of the existence of IMBHs in GCs.

4.2. *The need for specialized modeling*

Appropriate data represent only half of what is needed, and highly specialized models are required to avoid the shortcomings of more simplified methods that were developed for other circumstances. First, special machinery is necessary to handle data of a discrete nature (i.e., individual 2D velocities of a number of resolved stars). The most typical approaches for dynamical analysis (the Jeans equations and orbit-based models designed to work with observed velocity profiles from integrated light) must necessarily bin the data to produce averaged profiles of velocity dispersions (see figure 13 in van de Ven *et al.* 2006). Clearly, binning amounts to not exploiting the entire information content of the dataset, and, since only a limited number of stars are available inside the BH's sphere of influence, this can only have the effect of degrading and possibly erasing the signature of an IMBH.

Second, given that the signature of any central BH would be imprinted on the detailed orbital structure of the nearest stars, overly restrictive assumptions regarding the form of the stellar distribution function, most crucially the degree of isotropy of the orbits of the tracer stars, must be avoided. This point becomes obvious when considering that, for M15 for example, *HST* LOS velocities indicate that there is a clear increase in the net rotation of the stars in the inner arcseconds of the cluster (see figure 9 of Gerssen *et al.* 2002). This fact stands in stark contradiction to the standard expectation that relaxation should rapidly drive the velocity distribution at the centers of GCs towards isotropy, thus illustrating the necessity of using the most general models possible to fit these kinematics.

Third, the fact that there is flattening and substantial rotation seen in some GCs makes the use of axisymmetric models, rather than spherical ones, a must.

5. Our ongoing *HST* program

Starting around 2004, our team embarked on a comprehensive program to use accurate *HST* proper-motion measurements of the stars in the central regions of a number of Galactic GCs and search for evidence of IMBHs. Reflecting the theoretical expectations of that time, all GCs included in these initial *HST* programs were selected to be among those with high central densities (several of them being core-collapsed systems). Through the following years, target clusters spanning a variety of structural properties were included in the sample, and today we have accumulated data, from our own programs as well as by exploiting the *HST* Archive, for a total of nine GCs with two or more astrometric epochs, producing time baselines between 2 and 6 years. Reflecting the history of *HST* of the last few years, the data were obtained using those instruments providing the highest angular resolution possible at the time of the observations. Table 1 summarizes our

Table 1. Details of our *HST* program.

Target GC	Instruments Time baseline	Program ID PI
NGC 2808 NGC 6341 (M92) NGC 6752	HRC/HRC 2 years[1]	GTO-10335 GTO-11801 Ford
NGC 362 NGC 6624 NGC 6681 (M70) NGC 7078 (M15) NGC 7099 (M30)	HRC/WFPC2 4.5 years	GO-10401 GO-10841 GO-11988 Chandar
NGC 6266 (M62)	WFC/WFC3 6 years	GO-11609 Chanamé

[1]These clusters will have a third epoch with WFC3 in 2010, thus increasing the baseline to 5–6 years.

current dataset. We are currently in the process of reducing these data, and preliminary stellar proper-motion catalogs for some of these clusters are being generated (Bruursema *et al.*, in prep.). For a review of this intensive process, see the proper-motion catalog of ω Cen of Anderson & van der Marel (2009).

We plan to perform the dynamical analysis of all these data sets by making use of Schwarzschild models specifically designed to exploit data of a discrete nature, such as that of our proper-motion catalogs. Schwarzschild's technique is arguably the most developed and well-tested method available for constraining the detailed mass distribution of equilibrium stellar systems, and is based on the simple idea of finding the best combination (or superposition) of all possible orbits that, allowed by some previously specified potential, reproduces both the spatial distribution of the tracers (i.e., the light distribution) and the measured kinematics. The simplicity and success of the method therefore rely on two aspects: (i) that the overall stellar system can be considered to be in equilibrium (a safe assumption in the case of old Galactic GCs) and (ii) whether or not the set of orbits considered for the superposition is really comprehensive. As long as these two conditions are satisfied, the method is very general and free from most assumptions. As of today, the only orbit-based dynamical tool available that fulfills all requirements outlined in Section 4.2 is the discrete Schwarzschild code we recently developed and tested (Chanamé *et al.* 2008), and which we will be employing in this program.

Acknowledgements

JC acknowledges the support of the American Astronomical Society and the International Astronomical Union in the form of travel grants to attend the IAU General Assembly, as well as the support from NASA through Hubble Fellowship grant HF-01216.01-A, awarded by the Space Telescope Science Institute, which is operated by the Association of Universities for Research in Astronomy, Inc., under NASA contract NAS5-26555.

References

Anderson, J. & van der Marel, R. 2009, *ApJ*, submitted (arXiv:0905.0627)

Bahcall, J. N. & Wolf, R. A. 1976, *ApJ*, 209, 214

Bahcall, J. N. & Wolf, R. A. 1977, *ApJ*, 216, 883

Baker, J. G., Boggs, W. D., Centrella, J., Kelly, B. J., McWilliams, S. T., Miller, M. C., & van Meter, J. R. 2008, *ApJ* (Letters), 682, L29

Baumgardt, H., Hut, P., Makino, J., McMillan, S., & Portegies Zwart, S. 2003a, *ApJ* (Letters), 582, L21

Baumgardt, H., Makino, J., Hut, P., McMillan, S., & Portegies Zwart, S. 2003b, *ApJ* (Letters), 589, L25

Baumgardt, H., Makino, J., & Ebisuzaki, T. 2004a, *ApJ*, 613, 1133

Baumgardt, H., Makino, J., & Ebisuzaki, T. 2004b, *ApJ*, 613, 1143

Baumgardt, H., Makino, J., & Hut, P. 2005, *ApJ*, 620, 238

Binney, J. & Mamon, G. A. 1982, *MNRAS*, 200, 361

Chanamé, J., Kleyna, J., & van der Marel, R. 2008, *ApJ*, 682, 841

Dekel, A., Stoehr, F., Mamon, G. A., Cox, T. J., Novak, G. S., & Primack, J. R. 2005, *Nature*, 437, 707

de Lorenzi, F., *et al.* 2009, *MNRAS*, 395, 76

Gebhardt, K., Pryor, C., O'Connell, R. D., Williams, T. B., & Hesser, J. E. 2000, *AJ*, 119, 1268

Gebhardt, K., Rich, R. M., & Ho, L. C. 2002, *ApJ* (Letters), 578, L41

Gebhardt, K., Rich, R. M., & Ho, L. C. 2005, *ApJ*, 634, 1093

Gerssen, J., van der Marel, R. P., Gebhardt, K., Guhathakurta, P., Peterson, R. C., & Pryor, C. 2002, *AJ*, 124, 3270

Holley–Bockelmann, K., Gültekin, K., Shoemaker, D., & Yunes, N. 2008, *ApJ*, 686, 829

Irwin, J. A., Brink, T., Bregman, J. N., & Roberts, T. P. 2009, *ApJ* (Letters), submitted (arXiv:0908.1115)

Kong, A. K. H. 2007, *ApJ*, 661, 875

Kong, A. K. H., Heinke, C. O., Di Stefano, R., Barmby, P., Lewin, W. H. G., & Primini, F. A. 2009, *ApJ* (Letters), submitted (arXiv:0910.3944)

Maccarone, T. J., Kundu, A., Zepf, S. E., & Rhode, K. L. 2007, *Nature*, 445, 183

Miocchi, P. 2007, *MNRAS*, 381, 103

Moody, K. & Sigurdsson, S. 2009, *ApJ*, 690, 1370

Noyola, E., Gebhardt, K., & Bergmann, M. 2008, *ApJ*, 676, 1008

Romanowsky, A. J., Douglas, N. G., Arnaboldi, M., Kuijken, K., Merrifield, M. R., Napolitano, N. R., Capaccioli, M., & Freeman, K. C. 2003, *Science*, 301, 1696

Trenti, M., Ardi, E., Mineshige, S., & Hut, P. 2007, *MNRAS*, 374, 857

Ulvestad, J. S., Greene, J. E., & Ho, L. C. 2007, *ApJ* (Letters), 661, L151

van der Marel, R. P. 2004, in: L. C. Ho (ed.), *Coevolution of Black Holes and Galaxies*, Carnegie Obs. Astrophys. Ser., (Cambridge: Cambridge University Press), p. 37

van der Marel, R. P. & Anderson, J. 2009, *ApJ*, submitted (arXiv:0905.0638)

van de Ven, G., van den Bosch, R. C. E., Verolme, E. K., & de Zeeuw, P. T. 2006, *A&A*, 445, 513

Zepf, S. E., Maccarone, T. J., Bergond, G., Kundu, A., Rhode, K. L., & Salzer, J. J. 2007, *ApJ* (Letters), 669, L69

Star clusters: basic galactic building blocks
Proceedings IAU Symposium No. 266, 2009
R. de Grijs & J. R. D. Lépine, eds.

doi:10.1017/S1743921309991104

The internal dynamics of stellar clusters

Christian M. Boily

Observatoire astronomique de Strasbourg & CNRS UMR 7550
11, rue de l'université, Strasbourg F-67000, France
email: christian.boily@astro.unistra.fr

Abstract. In this brief review I summarise recent progress in the area of stellar dynamics, focusing on the dynamics of bound, self-gravitating stellar associations in isolation and (approximate) equilibrium. The basics of stellar dynamics are first outlined and the importance of stellar evolution is stressed. Subsequently, I argue that the evolution of anisotropic clusters of stars still holds solutions to current outstanding problems, such as the dynamics of galactic nuclei. I take a more personal standpoint when discussing the role of stellar evolution in the dynamics on relaxation timescales and draw from several recent models to underscore that a major step forward has been made in coupling stellar evolution and dynamics. I then briefly visit the issue of multiple stars and highlight some as yet unsolved problems.

Keywords. galaxies: star clusters, binaries: general

1. Introduction

It has been a long while since Newton wrote down the basic equations that govern the evolution of self-gravitating systems of point sources. That a general, exact solution to the problem of evolving a set of N mass elements over time will forever elude us must have come as a shock to the pioneers of stellar dynamics, who realised that progress in this field would come only through a statistical approach to constructing observables derived from theory. That long-ish statement can be rephrased as *nonlinearity rules* in Newtonian stellar dynamics, and so one must resort to statistical methods to solve for the dynamics of individual clusters (mainly drawn from a Fokker–Planck treatment of the equations) or, from averaging over large ensembles of models to smooth over singular solutions that could be attributed to the choice of initial conditions.

In this short review, I will take up the last point, but start with general considerations and the point-mass treatment of stellar clusters. I will then highlight what has been learned from evolving models with inclusion of a mass spectrum and stellar evolution. The single stellar evolution approach has been extended to multiple stars (binaries and hierarchical higher-order multiples), which are an essential ingredient to understand the demographics of exotic stellar remnants (pulsars, blue stragglers and the like). I will end with an overture to ongoing efforts by modellers to include ever more detailed physics in their calculation but with a view to 'packaging' software for the benefit of the broader community. The scope of this paper is narrow and not much will be found here by those seeking updates on observational campaigns and techniques: those aspects are covered in depth in the recent conference proceedings by Richtler & Larsen (2009) and Vazdekis & Peletier (2007). Constraints of time and space (and skills) mean that I have to select what instinctively seems more in tune with the spirit of the meeting in Rio. More substantial recent surveys of the field start with the well-balanced Meylan & Heggie (1997) review and include, among many others, several excellent contributions in Vesperini *et al.* (2008).

2. Basics

It will help to follow a thread of increasing complexity when discussing stellar cluster dynamics. Let me recall that the problem of retracing orbits in a self-gravitating system of identical point sources, satisfying the scalar virial theorem

$$\sigma^2 \equiv -\frac{GM}{r_{\rm g}} \tag{2.1}$$

(where σ is the average squared velocity dispersion, M the total mass and $r_{\rm g}$ the gravitational radius: not equal to the truncation radius, in general), boils down to two timescales: one measuring the time needed to cross a nominal volume enclosed by the truncation radius, and the other the time required for kinetic energy to diffuse effectively through the system by repeated two-body interactions. Since we are focussing on the internal cluster dynamics, it is convenient to introduce the (spherical) half-mass radius $r_{\rm h}$ and define the crossing time thus

$$t_{\rm cr} \equiv \frac{2r_{\rm h}}{\sigma} \tag{2.2}$$

and note that I make no distinction between the one- and three-dimensional velocity dispersion because the definition of $t_{\rm cr}$ is not unique and this issue is not important here. This is clear from the definition of the two-body relaxation time, $t_{\rm r}$, of Meylan & Heggie (1997), which features $r_{\rm g}$:

$$t_{\rm r} \equiv 0.138 \left(\frac{r_{\rm h}}{2r_{\rm g}}\right)^{3/2} \frac{N}{\ln(0.4N)} t_{\rm cr}. \tag{2.3}$$

In the latter equation, the numerical factor in front of $t_{\rm cr}$ is not fixed by mass (i.e., particle number) alone, but rather includes a dimensionless ratio $r_{\rm h}/r_{\rm g}$, which together with N encompasses the basic physics that drives analysis: whenever the ratio $t_{\rm r}/t_{\rm cr} \gg 1$, we can treat the system as being in near equilibrium and in the regime of slow evolution. It is remarkable that profiling of the mass in space only enters in the considerations through $r_{\rm h}/r_{\rm g}$. In the regime of slow evolution, one aims to solve the Boltzmann equations, suitably rewritten with Rosenbluth potentials to account for the collisional terms. This is the 'large-N' limit, when a distribution function is a sensible representation of the background mean potential. Formally, one writes

$$\frac{{\rm d}f(\boldsymbol{r},\boldsymbol{v},t)}{{\rm d}t} = \left(\frac{\partial f(\boldsymbol{r},\boldsymbol{v},t)}{\partial t}\right)_{\rm coll} \tag{2.4}$$

and expands the right-hand side using diffusion coefficients to evolve the distribution function, f, through phase space (e.g., Spitzer 1987). An appealing variant of this approach is to consider the diffusion coefficients as statistical weights for a two-body encounter leading to an exchange of kinetic energy: this is the road to Monte Carlo modelling. Now the core-halo structure of star clusters implies shorter dynamical times in the core and therefore more efficient diffusion there than in the outer envelope. This leads to core collapse, as the envelope drains potential energy from the core (by expanding) which loses mass and shrinks. Both Fokker–Planck and gas-dynamical solutions (treating the flow of mass in the continuous-fluid limit) have long confirmed theoretical expectations already developed in the 1960s of the irreversibility of this process, once the core density exceeds 709× the cluster average density (Cohn 1979; Lynden–Bell & Eggleton 1980). It may seem far-fetched to treat the decoupled collapsing core from a distribution-function perspective given that the dense core drops to zero size and mass in a finite time: clearly, methods based on distribution functions are aimed at massive, rich clusters: see Aarseth

& Heggie (1998) for a discussion. The potential mapped by a small number of stars confined to the core of a cluster becomes ever more granular and no longer amenable to a fluid approach. Indeed, small-N systems depart from spherical symmetry (through Poissonian noise) and so the fine details of the evolution of the cluster in the final stages of core collapse derived from Fokker–Planck modelling become more uncertain. This is where N-body modelling become essential, as it supplements other approaches to stellar dynamics through more general three-dimensional coding of the equations of motion. The caveat, of course, is that to date the largest-N calculations performed are limited to $N \sim 10^5$ (e.g., Hemsendorf & Merritt 2002; Baumgardt *et al.* 2003, 2008; Fujii *et al.* 2008) which is still below the average of $\simeq 4.3 \times 10^5$ for Galactic clusters (assuming a standard stellar mass function; Meylan & Heggie 1997) and further still from the large-N regime sought. It is reassuring to find that the simple problem of core collapse for equal-mass systems over a time $t \simeq 15 t_{\rm r}$ is well reproduced by completely different methods (e.g., Takahashi 1995; Makino 1996; Baumgardt *et al.* 2003; Spurzem *et al.* 2005).

Against this backdrop, let me chart the few issues that I will address in the remainder of this paper. Once more, my *leitmotiv* is the complexity of systems in terms of the physics that they include. I divide these into four broad classes of problems that are of interest here, which are organised according to whether or not one includes stellar evolution (SE) in the calculations and whether or not multiple stars are included in the initial conditions. Needless to say, the combination of multiple stars with full stellar evolution (including mass exchange, etc.) is the more challenging case to theorists. Below, I start with the point-mass dynamics of single stars.

3. Point-mass dynamics

3.1. *Single stars with anisotropy*

Surely, one should be content with the outline of Section 2 above, as this is a well-beaten path. Not quite. One problem still causing headaches is to deconvolve projected data while relaxing the hypothesis of isotropy. Recall that two-body relaxation tends to make the central region of clusters isotropic, while the outer expanding envelope is more and more anisotropic as a result of excess radial motion. Louis & Spurzem (1991), Takahashi (1995) and Drukier (1999) discussed the Fokker–Planck evolution towards core collapse of initially anisotropic clusters (see Giersz & Spurzem 1994 for an earlier similar study of gas-dynamical models). These models were all nonrotational and while evolution proceeded roughly as for the isotropic models, Takahashi (1995) reported delayed core collapse, from 15 to 17.6 $t_{\rm r}$, which may be a source of scatter in comparative photometric surveys of cuspy clusters (see, e.g., Noyola & Gebhardt 2006 for a sample of 38 clusters from *HST* archive data). The importance of anisotropy was stressed in an application to the G1 cluster in M31, where excellent fits to observed parameters (both kinematics and photometry) were recovered from merging two smaller clusters (Baumgardt *et al.* 2003). Such an interpretation has been challenged by Gebhardt *et al.* (2005) based on higher-resolution photometric data indicating a profile best adjusted with a $M_{\rm bh} \simeq 20\,000 {\rm M}_\odot$ central black hole. Pooley & Rappaport (2006) and Kong (2007) pointed out that X-ray data could provide confirmation of the presence of a black hole, if angular resolution could pin down the source well enough to rule out alternatives based on a hypothetical population of X-ray binary sources (for example). Whatever the ultimate outcome of these investigations, the point was not lost that anisotropy has to be factored in when exploring the dynamics of clusters at the most basic level.

But the yet more important effect attributable to anisotropy must be systematic streaming, which leads to net rotation and strong deviations from spherical symmetry.

That Galactic clusters such as ω Cen and 47 Tuc are rotating has been known for a long time and ideas were already put forward many years ago to include rotation in the models (e.g., Agekian 1958; Shapiro & Merchant 1976; Hachisu 1979). Still, it is only relatively recently that self-consistent two-dimensional Fokker–Planck modelling with distribution functions of the form $f(E, L_z)$ was introduced by Einsel & Spurzem (1999), who confirmed Hachisu's gravo-gyro catastrophe [see also Ernst *et al.* (2007) and Kim *et al.* (2008) for investigations of this problem with N-body integrations]. The inclusion of rotation speeds up core-collapse in single-mass cluster dynamics, by a multiplicative factor of up to ≈ 3 compared to isotropic models. Interestingly, projected quantities derived from these models have only been made available relatively recently by Fiestas *et al.* (2006),† who ran a grid of models with an improved version of Einsel's original Fokker–Planck code. As in the Baumgardt *et al.* work on anisotropic clusters resulting from the merging of two clusters, and hence quite possibly residual rotation from that event, one should hope that rotation will, in the future, become an integral part of the parameter space explored to fit data. Some work is underway to embed black holes in clusters with rotation and it will be interesting to monitor the outcome of such models in applications to clusters such as G1 or M15 (see Fiestas *et al.* 2008 for preliminary work).

3.2. *Chaos and sensitivity to the initial configuration*

The orbit integration of the collisional N-body problem is known to be sensitive to the initial configuration of the system, in the sense that small deviations in the coordinates of a body are amplified exponentially with time at the onset of integration. To track this issue, Miller (1971) set up a variational method to monitor the orbits of massless bodies, given a small shift in phase space with respect to actual massive bodies that define the potential. The game consists of measuring the distance Δ between orbits traced out by the massless bodies to that of the body nearest to them in the initial configuration. Linear perturbation analysis tells us that Δ grows exponentially with time. One's instinct is that the e-folding time should become longer with increasing N as one moves closer to the idealised limit of the smooth (and static) mean field for which orbits must satisfy the collisionless Boltzmann equation (CBE; see Binney & Tremaine 2008). Detailed analysis indicates instead that the e-folding time, t_e, *de*creases with N, a trend driven mainly by accrued small-scale two-body scatter (Goodman *et al.* 1993). Hemsendorf & Merritt (2002) confirmed this expectation with a series of N-body calculations with N running from 100 to $\approx 132\,000$, and obtained a rough fit for $N \gtrsim 32\,000$ of the form (see their figure 2)

$$t_e = \frac{0.12 t_{\rm cr}}{\ln 0.7 \ln N}, \tag{3.1}$$

with a notable weak dependence on N. Hut & Heggie (2002) argue that the exponential growth saturates at a few $\times t_e \propto t_{\rm cr}$. Their analytical approach, backed up by numerical experiments, leads to a turnover, from an exponential, to a power-law growth rate over a relaxation time. Since two diverging paths will have, sensibly, the same potential energy but different kinetic energy, this growth can be expressed as a relative shift in total energy,

$$\frac{\delta E}{E} \propto \left(\frac{t}{t_{\rm r}}\right)^{\frac{1}{2}}. \tag{3.2}$$

This near-saturation has two immediate consequences for the purpose of tracing orbits numerically: (i) it implies that the orbits can be known up to a finite precision, beyond

† See `http://www.ari.uni-heidelberg.de/clusterdata/`

which interpretation becomes unphysical and (ii) it sets a constraint on the level of fluctuations that can be 'spontaneously' transmitted from the background potential, to a multiple star (say, a binary or triple star).

Point (ii) is best left for Section 5. Point (i) is easier to understand if we carry out a *Gedankenexperiment* such that after a crossing time of evolution (say), the phase-space coordinates are synchronised and a restart performed. If the data were synchronised to arbitrary accuracy by interpolation of the known positions and velocities, that precision would be lost after $\sim t_{\rm cr}$ of integration through exponential growth of rounding errors. That remark can be turned on its head, by saying that when coordinates are made synchronous, the procedure can tolerate less-than-perfect accuracy and be equally valid for a restart (as long as one uses an integration time well exceeding a crossing time). Beyond issues of recovering accurate data for the sake of dissecting the orbits of a model cluster, one is confronted with the problem of data storage associated with a core-halo structure. In such a system, the dynamical time $t_{\rm cr}$ varies significantly: for a rich cluster, one may compute $t_{\rm cr} \sim 10^3$ years in the core, and $t_{\rm cr} \sim 10^6$ years in the halo. Clearly, resolving orbits everywhere requires a large throughput in the core, and a much reduced one in the halo. But synchronicity required over the whole system to construct snapshots translates to huge data sets. Practitioners have long realised that a self-respecting integrator would negotiate orbits more carefully where needed, by adopting a smaller time step (the block time step procedure in conjunction with the Hermite integration scheme is the standard of choice; McMillan 1986; Makino & Aarseth 1992; see Aarseth 2003 for more background). A major space-saving procedure is to dump phase-space coordinates at a fixed number of integration steps (instead of equal time intervals) and reconstruct the orbit using the same scheme for integration to interpolate to the desired time (e.g., Faber *et al.* 2009). It is not clear how to translate the precision achieved using this approach to requirements that meet the same level of tolerance as what is achieved by the divergence in energy over a crossing time (Hut & Heggie 2002), although one can guess that the optimal approach (to orbit reconstruction) factors in the uncertainty associated with the reference orbit.

I will now show that only statistics can save us from compiling heaps of data only to follow a single-mass cluster up to core collapse. The classical solution has the core radius going to zero at a time $t \equiv t_{\rm c}$. Prior to that instant, the core radius $\propto (t_{\rm c} - t)^{2/(6-\alpha)} \to 0$ as $t \to t_{\rm c}$ (the constant $\alpha \approx 2.21$; e.g., Lynden–Bell & Eggleton 1980; see Spitzer 1987). The mass fraction of stars confined to the core $M_{\rm c} \propto (t_{\rm c} - t)^{(6-2\alpha)/(6-\alpha)} \propto (t_{\rm c} - t)^{0.42}$ drops rapidly as $t \to t_{\rm c}$, however the central density $\rho_{\rm c} \propto (t_{\rm c} - t)^{-2\alpha/(6-\alpha)} \propto (t_{\rm c} - t)^{-1.16}$ diverges to the same limit, which in turn implies that the orbital periods of stars near the centre $\propto 1/\sqrt{G\rho_{\rm c}} \propto (t_{\rm c} - t)^{0.58}$ also drop to zero. But owing to the Lagrangian nature of the Hermite scheme, the volume of data stored over one revolution remains constant. As a result, the net volume of data scales with $\propto M_{\rm c} \times \sqrt{G\rho_{\rm c}} \propto (t_{\rm c} - t)^{-0.16}$ and therefore increases as a weak power of time. The singularity as $t \to t_{\rm c}$ will not be attained in practice owing to the formation of binary stars by three-body interactions (the probability $\propto \rho^2 \times \rho r_{\rm c}^2 \sigma \propto (t_{\rm c} - t)^{-2.92}$ is a steep function of a vanishing argument). Nevertheless, the trend with time will be real and may require a fix in the form of a coarser sampling in the later phases of a cluster's evolution.

3.3. *Single stars with a mass spectrum*

The dynamics of point-mass clusters becomes more entertaining once a spectrum of masses is included. The most immediate effect is the segregation of stars (by mass) through a process which Spitzer termed the asymptotic tendency towards equipartition of kinetic energy: since this is driven by two-body interactions, one finds that segregation operates on the relaxation time $t_{\rm r}$, but for a multiplicative factor that accounts for the spread in the mass spectrum of the point masses (Spitzer 1969; Meylan & Heggie 1997):

$$t_{\rm seg} \simeq \frac{\langle m \rangle}{\max\{m\}} \frac{\bar{\rho}}{\rho_{\rm h}} \left(\frac{r_{\rm h}}{r_{\rm g}}\right)^{3/2}, \tag{3.3}$$

where $\langle \ldots \rangle$ denotes ensemble averaging, and overbar mass-averaged quantities, and I have taken an estimate of $t_{\rm seg}$ at the half-mass radius where the density $\rho(r_{\rm h})$ is evaluated. Note that $t_{\rm seg}$ presumes that each mass element in the set $\{m_i\}$ is constant, so that one expects the initially more massive stars to orbit closer to the centre than less massive bodies. The hierarchy of masses would also define a hierarchy of $t_{\rm seg}$ times, using the same definition (3.3) but replacing the maximum mass by the one of interest.

In practice, an application of Equation (3.3) only makes sense if the segregation time is shorter than the lifetime of the zero-age main sequence stars of the same initial mass (metals are not taken into account for the moment). Several authors have bounced on this observation to deduce that massive stars converging near the centre of a cluster would inevitably dissipate tidal energy through their envelope or by a direct hit, coalesce and then possibly form a black hole (e.g., Portegies Zwart & McMillan 2002; Portegies Zwart *et al.* 2004; Gürkan *et al.* 2004; Freitag *et al.* 2006a,b; Ardi *et al.* 2008; Gaburov *et al.* 2008). Therefore, mass segregation would provide a natural setup to trigger the formation of very massive stars or even a black hole in any sufficiently rich and dense cluster at birth (with a full spectrum of stellar masses). The implications for current and future gravitational waves from inspiraling black holes and their detection are clear (see Fregeau *et al.* 2006; Amaro–Seoane *et al.* 2007).

There are several issues related to the formation of a black hole and the migration of stars in its neighbourhood. Spitzer (1969) realised that when two populations of stars cohabit in a cluster, the trend towards equipartition leads to the more massive component decoupling from the other, lighter component to form a (quasi-) self-bound core. This occurs in two-population systems of masses m_1, m_2, with $m_2 > m_1$, whenever the ratio of total mass in each group satisfies

$$\frac{M_2}{M_1} < \beta \left(\frac{m_1}{m_2}\right)^{3/2}, \tag{3.4}$$

where the dimensionless constant $\beta \approx 0.16$. This relation goes a step further than Equation (3.3), which does not take into consideration the self-gravity of the most segregated group. Now, in first instance one would like to know what degree of segregation a cluster can reach, and whether Equation (3.4) will take it back to the evolutionary track of single-mass clusters and core collapse (which involves the innermost region only). Gaburov *et al.* (2009) and Glebbeek *et al.* (2009) pointed out that a star orbiting in the core of a dense cluster may undergo *several* collisions, leading to fusion once dissipative effects are taken into account. Paramount to understanding how this will bear on observations is the state of equipartition of the system at the onset of evolution (be it numerical modelling or otherwise). Several observations point to clusters being segregated at birth (e.g., de Grijs *et al.* 2002; McCrady *et al.* 2005), while theoretical studies now seem to favour segregation taking place at the very formation of a cluster (mainly by scattering of substructures: McMillan *et al.* 2007; Allison *et al.* 2009; Vesperini *et al.* 2009). Baumgardt *et al.* (2008) and Decressin *et al.* (2008) explore some observational consequences of early mass segregation. A crucial result from theory is that mass segregation in clusters with rotation proceeds in a nontrivial way, on a timescale that optimises angular-momentum diffusion (see Kim *et al.* 2004 for details). Here, I mention an application to galactic nuclei where multimass models have drawn mainly from Monte Carlo simulations of dense environments (Fregeau *et al.* 2002; Freitag *et al.* 2006b), while N-body modelling has focussed on the orbital migration of black-hole binaries or the interplay between single-star populations and black-hole binaries (see Berentzen *et al.* 2009 for a recent update on this problem). An important distinction with respect to stellar clusters is the inclusion of a massive black hole in the calculations at rest at the centre of the system. The segregation time (Equation 3.3) may yet be shorter than a massive-star lifetime, τ, owing to the high density of light stars around the black hole (Freitag *et al.* 2006a). In that case, the massive stars will sink to the centre rapidly and hit the *stalling radius* (Merritt 2006), where the orbits are quasi-Keplerian. At that point, little evolution was expected from star–star scattering, an issue revisited recently by Alexander & Hopman (2009), who introduced the dimensional parameter Δ (rewritten in the notation of Equation 3.4),

$$\Delta \equiv 4N_2M_2^2/[N_1M_1^2(3 + M_2/M_1)].$$

$\Delta > 1$ for self-scattering among massive stars, whereas $\Delta < 1$ for a depleted sample of massive stars scattering off background light stars (here of mass m_1). An interpretation of this result is that massive stars may yet enhance their own scattering to reach below the stalling radius. It will be interesting to derive observational consequences of this possibility since the profile of massive bodies will differ from that of background stars: spatial gradients in spectroscopic indices would then indicate whether effectively $\Delta > 1$ in nuclei (see Alexander & Hopman 2009 for a discussion of this possibility). A most interesting test will be the Milky Way centre, where massive S stars are known to orbit within the stalling radius, a notoriously difficult puzzle to solve using the principle of stellar migration alone (see, e.g., Merritt *et al.* 2009 or Portegies Zwart *et al.* 2003 for discussions of the migration of clusters with intermediate-mass black holes in their midst).

4. Dynamics with stellar evolution

There have been great advances since 2001 in coupling dynamics with stellar evolution tracks, in large part driven by joint efforts such as MoDEST (Hut 2003) and MUSE (Portegies Zwart *et al.* 2009). Below. I revisit some outstanding issues and prospects for future developments.

4.1. *Chronicle of a dissolution foretold*

The basic problem with coupling stellar evolution and dynamics is that the stars lose mass at the same time as their luminosity varies rapidly, which makes the construction of observables a tad more tricky (all the more so with small-ish clusters for which broad-band photometry may be dominated by a few bright stars). At a more basic level, I note that stellar mass loss induces a steady drift in the potential not accounted for in the dynamics of fixed point-mass elements. To illustrate this, I write the fully dimensional Newtonian-dynamics equations for a body of mass m, which is an explicit function of time,

$$\frac{\mathrm{d}\boldsymbol{p}}{\mathrm{d}t} = m\dot{\boldsymbol{v}} + \dot{m}\boldsymbol{v} = -m\boldsymbol{\nabla}\phi(\boldsymbol{r}, t), \tag{4.1}$$

with $\boldsymbol{p}$ the linear momentum and other quantities assume their usual meaning. If I now specialise to the case of spherical symmetry, I can write for the density $\rho(r, t) \simeq m(t)/m(0)\rho(r, 0)$ if all stars have equal mass and the structure of the cluster has changed little over $t \sim t_{\mathrm{cr}}$, and $m(t)$ is some decreasing function of time. The same will be true of $\phi(r, t)$ in terms of $\phi(r, 0) \equiv \phi_o$ as for the density because of Poisson's equation. One may thus compute the rate of change of the binding energy of any star in the system as

$$\dot{E} = \dot{m}\phi + m\boldsymbol{v}\cdot\boldsymbol{\nabla}\phi + m\partial_t\phi + +m\boldsymbol{v}\cdot\dot{\boldsymbol{v}} + \frac{1}{2}\dot{m}v^2 \simeq -\dot{m}\left[\frac{1}{2}v^2 - 2\frac{m(t)}{m(0)}\phi_o\right], \qquad (4.2)$$

where the last step follows from the 'frozen potential' approximation. The key insight from Equation (4.2) is that $\dot{E} > 0$ for any initially bound orbit since $\dot{m} < 0$ by hypothesis. If we imagine that the cluster is so evolved that the self-similar core-collapse flow applies approximately to the inner region, then it is easy to show that the binding energy of the decoupled core $W_{\rm c} \propto GM_{\rm c}^2/r_{\rm c} \propto (t_{\rm c} - t)^{0.31}$ drops to zero (along with its mass) at a rate $\dot{W}_{\rm c} \propto (t_{\rm c} - t)^{-0.69}$. But since the net effect of mass loss through stellar winds is to pump in energy ($\mathrm{d}E/\mathrm{d}t > 0$), there must come a time at which the binding energy of the core is positive for a finite core mass and radius. This argument is correct for any orbit confined to a small region around the cluster centre when mass loss begins. Since these are generally the most bound orbits, one is drawn to the conclusion that stellar winds must severely disrupt the classical self-similar solution by applying brakes to core contraction. Whether this occurs before the formation of binaries by three-body encounters, which would also halt collapse, depends on the ratio $t_{\rm r}/\tau$ of cluster relaxation time to main-sequence lifetime, τ. For the problem at hand, we must understand that the answer is a strong function of the cluster's initial conditions and stellar spectral type. One thing is clear, though, the systematic inclusion of stellar evolution in the modelling has had a deep impact on our understanding of these objects: below I track two areas that have benefited much from this, mass estimation of extragalactic clusters, and dynamics of multiple star populations.

4.2. *Chronicles of a mass not so well foretold*

The virial theorem is a powerful tool when applied to averaged properties of a self-bound cluster (at least for clusters of age > 30 Myr or so; see Bastian & Goodwin 2006; de Grijs & Parmentier 2007). What is less clear is whether one may presume that the age of such young clusters inevitably is $\ll t_{\rm r}$ or indeed the equipartition timescale $t_{\rm seg}$. Theoretical modelling is a way of recovering a conversion factor (I will call it η) from mock spectrophotometric data of the projected models. Dimensional analysis alone gives

$$\eta = \frac{r_{\rm ph}\sigma_{\rm s}^2}{GM},$$

where $r_{\rm ph}$ is some photometric radius, $\sigma_{\rm s}$ a spectroscopic one-dimensional velocity dispersion and M the total mass of the model. Under virial equilibrium conditions, one finds $\eta \simeq 8$ for a good range of King models. The catch is that while η may remain constant for mass-weighted quantities, in reality flux-weighted quantities will evolve on the same or a similar timescale as the massive stars for several important cases (see Boily *et al.* 2005; Fleck *et al.* 2006 for a fuller exploration of single-star models). As a result, η may well vary by a factor of up to three in the case of single-star populations. Note that the idea of converting light to mass is used by observers, for instance see McCrady & Graham (2007) or Östlin *et al.* (2007), yet theorists have been slow to convert their model data into observable quantities. Some work has attempted to draw consequences for full globular cluster mass functions by deconvolving the mass of model clusters of a given initial mass and age (Fleck *et al.* 2006; Kruijssen & Portegies Zwart 2009), although much remains to be done in this area given the oversimplification of the stellar populations modelled. Yet, it is clear that η is a key parameter to evaluate as it may hold the answer to cluster survival and the true mass function (de Grijs & Parmentier 2007).

4.3. *Chronicled desperation, or dynamics with multiple stars...*

One of the main reasons for the poor situation with η is surely that an analytical treatment is impossible, and that high-resolution numerical models are expensive computationally, regardless of the method used. As a consequence, ensemble-averaged quantities are even more problematic to secure. A notable difficulty is that multiple stars complicate the issue even more.

Consider a cluster of mixed stellar populations, of single and binary stars. Surely, if the binary population represents only a small percentage of all stars, it can carry little of the total light and mass. In that case, one would treat binaries to zeroth order as (on average) stars twice as massive as the others and derive observables from that perspective. If, instead, it is the single stars that are a small minority, then a binary population with mass-independent parameters (periods, eccentricity, etc.) would produce the same segregation pattern as single-star models (since mean and maximum masses both double, except that $t_{\rm seg}$ would be even shorter for the same initial mass function as single stars). A little thinking and some tests with random pairing of stars leads to a fraction $\approx 40\%$ of binaries for an optimal bias in the parameter η over time. Kouwenhoven & de Grijs (2008), Kouwenhoven *et al.* (2009) and Weidner *et al.* (2009) have studied this issue quantitatively. These studies indicate that massive clusters (of mass $\sim 10^6$ $\mathrm{M}_\odot$) are robust to a bias in spectrophotometric measures because the velocity dispersion of the binary centres of mass are significantly larger on the mean than the internal dispersion of the binaries. This will break down for low-mass open clusters, since now the internal dispersion is larger than that caused by the cluster's self-gravity.

To pin down these issues more firmly will require a better understanding of cluster properties at birth. Considering that protostars may condense on a timescale not shorter than several $\times 10^5$ years, one cannot help but conclude that cluster and star formation censure each other, in the sense that protostars will trace out a significant number of orbits in a dense cluster over their formation timescale. That this will imprint the properties of the cluster (its mass profile and velocity dispersion) as well as the stellar population (binary periods, destruction of loose binaries and the like) is clear, as was shown by Wiersma *et al.* (2006), who concluded that pre-main-sequence stellar evolution in a cluster led to the depletion of the binary population as well as a shift in its period distribution. Because it is doubtful whether a generic formation scenario applies to all clusters (clumpy accretion for open clusters, e.g., McMillan *et al.* 2007; near-virial equilibrium throughout the phase of star formation for more massive clusters, e.g., Tan *et al.* 2006), one may have no choice but to cover as broad a range in initial conditions as possible but aimed at modelling target clusters. Spectacular examples of this are the study by Hurley *et al.* (2005) for the open cluster M67 using direct-integration N-body models, and Giersz & Heggie (2009) and Heggie & Giersz (2008) for Monte Carlo models of clusters M4 and NGC 6397. It is well worth stressing here that a winning strategy for treating multiple stars in clusters (binaries, triples, quadruples, and so forth) may be in the form of hybrid codes such as the Monte Carlo approach of Giersz & Spurzem (2000, 2004) which treats complex interactions with direct integration (no assumption on cross sections). The caveat is that at least a few realisations will be needed for a meaningful outcome. An important point which was made for single-star populations, that observables are best derived from light fluxes and not mass, applies here as well in the case of binary and multiple stars: see Hurley (2007) for an application to estimates of the core radius and a revised measure of expansion driven by binaries and intermediate-mass black holes (e.g., Mackey *et al.* 2008).

4.4. *Evolved stellar populations: millisecond pulsars, multiple main sequences*

The modelling of multiple stars is important as they open up a large array of collision events that may lead to exotic stellar remnants, such as millisecond pulsars (see, e.g., Ransom 2008 for a review). The demographics of binary and multiple stars have attracted much attention for a number of years, especially as regard their fraction in the dense cores of clusters. This is were physical collisions with stars are more likely to occur. Conventional wisdom would have primordial binaries ejected at high velocities through gravitational recoil and it is worth mentioning once more that stellar interactions would take place already in the protostellar stage of star formation, and that this phase ultimately bears on the statistics of multiple stars (see Wiersma *et al.* 2006 for binary stars). One would then infer a large fraction of primordial binaries so that at least a few percent survive long enough in the core to form an exotic stellar remnant. It is very exciting to find instead, with Hurley *et al.* (2007) using N-body simulations, that while binaries may be ejected from the core, they may well have centre-of-mass velocities below the cluster's escape velocity. These binaries may then sink back towards the cluster core efficiently, thanks to the shorter segregation time (Equation 3.3), and replenish the core. Overall, the core binary fraction of $\sim 20\%$ remains nearly constant over time (although the core itself will slowly shrink). I find comfort in the fact that this result was recently confirmed by Fregeau *et al.* (2009), who rectified their earlier statistics published in Ivanova *et al.* (2005) using an improved version of their Monte Carlo code. Thus, these two studies and methods independently provide insight into core binary dynamics. This point may appear banal to make, yet it is essential because one finds only partial overlap in the parameter space that the methods each can cover.

To conclude, I should mention another big challenge to dynamicists and stellar evolution experts, which illustrates perfectly the necessity of marrying these two fields and this is the observations of multiple stellar main sequences in rich clusters (for resent updates see, e.g., Piotto *et al.* 2007 [NGC 288] and Milone *et al.* 2009a [NGC 1851]). These different main-sequence threads arise from systematic differences in CNO abundances that are difficult to interpret through a universal stellar mass function and require self-enrichment with a flattened high-mass stellar distribution function (d'Antona & Caloi 2004; but see Pflamm–Altenburg & Kroupa 2009 for an alternative view). So far, the efficiency of self-enrichment was somewhat difficult to quantify, but recent developments in models of slow winds from rapidly rotating stars (Decressin *et al.* 2007; d'Erçole *et al.* 2008) have made it more likely that a standard stellar mass function but with added physics of rotating massive stars strongly suggest that the fine threads are, in fact, more common than could have been anticipated (Milone *et al.* 2009b), a situation that will keep cluster research buzzing for some time still.

Acknowledgements

It is a pleasure to thank the organisers of this fine meeting, and R. de Grijs and J. Lépine in particular for the invitation to attend, and both as editors for patiently waiting for this short contribution submitted late.

References

Aarseth, S. J. 2003, *Gravitational N-Body Simulations*, Cambridge: Cambridge University Press
Aarseth, S. J. & Heggie, D. C. 1998, *MNRAS*, 297, 794
Agekian, T. A. 1958, *Sov. Astron.*, 2, 22
Alexander, T. & Hopman, C. 2009, *ApJ*, 697, 1861
Allison, R. J., Goodwin, S. P., Parker, R. J., de Grijs, R., Portegies Zwart, S. F., & Kouwenhoven, M. B. N. 2009, *ApJ* (Letters), 700, L99

Amaro–Seoane, P., Gair, J. R., Freitag, M., *et al.* 2007, *Class. Quant. Grav.*, 24, 113

Ardi, E., Baumgardt, H., & Mineshige, S. 2008, *ApJ*, 682, 1195

Bastian, N. & Goodwin, S.P. 2006, *MNRAS* (Letters), 369, L9

Bastian, N., Saglia, R. P., Goudfrooij, P., Kissler–Patig, M., Maraston, C., Schweizer, F., & Zoccali, M. 2006, *A&A*, 448, 881

Baumgardt, H., De Marchi, G., & Kroupa, P. 2008, *ApJ*, 685, 247

Baumgardt, H., Heggie, D. C., Hut, P., & Makino, J. 2003, *MNRAS*, 341, 247

Baumgardt, H., Makino, J., Hut, P., McMillan, S., & Portegies Zwart, S. 2003, *ApJ* (Letters), 589, L25

Belleman, R. G., Bédorf, J., & Portegies Zwart, S. F. 2008, *NewA*, 13, 103

Berentzen, I., Preto, M., Berczik, P., *et al* 2009, *ApJ*, 695, 455

Binney, J. J. & Tremaine, S. D. 2008, *Galactic Dynamics*, 2nd ed., Princeton: Princeton University Press

Boily, C. M., Lançon, A., Deiters, S., & Heggie, D. C. 2005, *ApJ* (Letters), 620, L27

Cohn, H. N. 1979, *ApJ*, 234, 1056

D'Antona, F. & Caloi, V. 2004, *ApJ*, 611, 871

Decressin, T., Meynet, G., Charbonnel, C., Prantzos, N., & Ekström, S. 2007, *A&A*, 464, 1029

Decressin, T., Baumgardt, H., & Kroupa, P. 2008, *A&A*, 492, 101

de Grijs, R., Johnson, R. A., Gilmore, G. F., & Frayn, C. M. 2002, *MNRAS*, 331, 228

de Grijs, R. & Parmentier, G. 2007, *ChJA&A*, 7, 155

D'Ercole, A., Vesperini, E., d'Antona, F., McMillan, S. L. W., & Recchi, S. 2008, *MNRAS*, 391, 825

Drukier, G. A., Cohn, H. N., Lugger, P. M., & Yong, H. 1999, *ApJ*, 518, 233

Einsel, C. & Spurzem, R. 1999, *MNRAS*, 302, 81

Ernst, A., Glaschke, P., Fiestas, J., Just, A., & Spurzem, R. 2007, *MNRAS*, 377, 465

Faber, N. T., *et al.* 2009 *MNRAS* 2009, in press

Fiestas, J., Spurzem, R., & Kim, E. 2006, *MNRAS*, 373, 677

Fiestas, J., Porth, O., & Spurzem, R. 2008, in: E. Vesperini, M. Giersz, & A. Sills (eds), *Dynamical Evolution of Dense Stellar Systems*, Proc. IAU Symp. 246, p. 146

Fleck, J.-J., Boily, C. M., Lançon, A., Deiters, S. 2006, *MNRAS*, 369, 1392

Fregeau, J. M., Joshi, K. J., Portegies Zwart, S. F., & Rasio, F. A. 2002, *ApJ*, 570, 171

Fregeau, J. M., Larson, S. L., Miller, M. C., *et al.* 2006, *ApJ* (Letters), 646, L135

Fregeau, J. M., Ivanova, N., & Rasio, F. A. 2009, *ApJ*, submitted (arXiv:0907.4196)

Freitag, M., Rasio, F. A., & Baumgardt, H. 2006a, *MNRAS*, 368, 121

Freitag, M., Gürkan, M. A., & Rasio, F. A. 2006b, *MNRAS*, 368, 141

Fujii, M., Iwasawa, M., Funato, Y., & Makino, J. 2008, *ApJ*, 686, 1082

Gaburov, E., Gualandris, A., & Portegies Zwart, S. 2008, *MNRAS*, 384, 376

Gaburov, E., Harfst, S., & Portegies Zwart, S. 2009, *NewA*, 14, 630

Gebhardt, K., Rich, R. M., & Ho, L. C. 2005, *ApJ*, 634, 109

Giersz, M. & Heggie, D. C. 2009, *MNRAS*, 395, 1173

Giersz, M. & Spurzem, R. 1994, *MNRAS*, 269, 241

Giersz, M. & Spurzem, R. 2000, *MNRAS*, 317, 581

Giersz, M. & Spurzem, R. 2003, *MNRAS*, 343, 781

Glebbeek, E., Gaburov, E., de Mink, S. E., Pols, O. R., & Portegies Zwart, S. F. 2009, *A&A*, 497, 255

Goodman, J., Heggie, D. C., & Hut, P. 1993, *ApJ*, 415, 715

Gürkan, M. A., Freitag, M., & Rasio, F. A. 2004, *ApJ*, 604, 632

Hachisu, I. 1979, *PASJ*, 31, 523

Hemsendorf, M. & Merritt, D. 2002, *ApJ*, 580, 606

Hurley, J. R., Pols, O. R., Aarseth, S. J., & Tout, C. A. 2005, *MNRAS*, 363, 293

Hurley, J. R. 2007, *MNRAS*, 379, 93

Hurley, J. R., Aarseth, S. J., & Shara, M. M. 2007, *ApJ*, 665, 707

Hut, P. 2003, in: *Globular Clusters: Formation, Evolution and the Role of Compact Objects* (astro-ph/0309395)

Hut, P. & Heggie, D. C. 2002, *J. Stat. Phys.*, 109, 1017

Ivanova, N., Belczynski, K., Fregeau, J. M.,& Rasio, F. A. 2005, *MNRAS*, 358, 572

Kim, E., Einsel, C., Lee, H. M., Spurzem, R., & Lee, M. G. 2002, *MNRAS*, 334, 310

Kim, E., Lee, H. M., & Spurzem, R. 2004, *MNRAS*, 351, 220

Kim, E., Yoon, I., Lee, H. M., & Spurzem, R. 2008, *MNRAS*, 383, 2

Kong, A. K. H. 2007, *ApJ*, 661, 875

Kouwenhoven, M. B. N., Brown, A. G. A., Goodwin, S. P., Portegies Zwart, S. F. & Kaper, L. 2009, *A&A*, 493, 979

Kouwenhoven, M. B. N., & de Grijs, R. 2008, *A&A*, 480, 103

Louis, P. D. & Spurzem, R. 1991, *ApJ*, 251, 408

Lynden–Bell, D. & Eggleton, P. E. 1980, *MNRAS*, 191, 483

Mackey, A. D., Wilkinson, M. I., Davies, M. B., & Gilmore, G. F. 2008, *MNRAS*, 386, 65

Makino, J. 1996, *ApJ*, 471, 796

Makino, J. & Aarseth, S. J. 1992, *PASP*, 44, 141

McCrady, N. & Graham, J. R. 2007, *ApJ*, 663, 844

McCrady, N., Graham, J. R., & Vacca, W. D. 2005, *ApJ*, 621, 278

McMillan, S. L. W. 1986, *ApJ*, 307, 126

McMillan, S. L. W., Vesperini, E., & Portegies Zwart, S. F. 2007, *ApJ* (Letters), 655, L45

Merritt, D. 2006, *Rep. Prog. Phys.*, 69, 2513

Merritt, D., Gualandris, A., & Mikkola, S. 2009, *ApJ*, 693, 35

Meylan, G. & Heggie, D. C. 1997, *A&ARev.*, 8, 1

Miller, R. H. 1971, *J. Comput. Phys.*, 8, 449

Milone, A. P., Stetson, P. B., Piotto, G., Bedin, L. R.; Anderson, J., Cassisi, S., & Salaris, M. 2009b, *A&A*, 503, 755

Milone, A. P., Bedin, L. R., Piotto, G., & Anderson, J. 2009a, *A&A*, 497, 755

Noyola, E. & Gebhardt, K. 2006, *AJ*, 132, 447

Östlin, G., Cumming, R. J., & Bergvall, N. 2007, *A&A*, 461, 471

Pflamm–Altenburg, J. & Kroupa, P. 2009, *MNRAS*, 397, 488

Piotto, G., Bedin, L. R., Anderson, J., King, I. R., Cassisi, S., Milone, A. P., Villanova, S., Pietrinferni, A., & Renzini, A. 2007, *ApJ* (Letters), 661, L53

Pooley, D. & Rappaport, S. 2006, *ApJ* (Letters), 644, L45

Portegies Zwart, S. F. & McMillan, S. L. W. 2002, *ApJ*, 576, 899

Portegies Zwart, S. F., McMillan, S. L. W., & Gerhard, O. 2003, *ApJ*, 593, 352

Portegies Zwart, S. F., Baumgardt, H., Hut, P., *et al.* 2004, *Nature*, 428, 724

Portegies Zwart, S., McMillan, S., Harfst, S., *et al.* 2009, *NewA*, 14, 369

Ransom, S. M. 2008, in: E. Vesperini, A. Sills, & M. Giersz (eds.), *Dynamical Evolution of Dense Stellar Systems*, Proc. IAU Symp. 246, p. 291

Richtler, T. & Larsen, S. 2009, *Globular Clusters: Guides to Galaxies*, ESO Astrophys. Symp. Ser., Heidelberg: Springer

Schive, H.-Y., Chien, C.-H., Wong, S.-K., Tsai, Y.-C., & Chiueh, T. 2008, *NewA*, 13, 418

Shapiro, S. L. & Marchant, A. B. 1976, *ApJ*, 210, 757

Spitzer Jr., L. 1969, *ApJ* (Letters), 158, L139

Spitzer Jr., L. 1987, *Dynamics of Star Clusters*, Princeton: Princeton University Press

Spurzem, R., Giersz, M., Takahashi, K., & Ernst, A. 2005, *MNRAS*, 364, 948

Takahashi, K. 1995, *PASJ*, 47, 561

Tan, J. C., Krumholz, M. R., & McKee, C. F. 2006, *ApJ*, 641, L121

Vazdekis, A. & Peletier, R. F. 2007, *Stellar Populations as Building Blocks of Galaxies*, Proc. IAU Symp. 241 Cambridge: Cambridge University Press

Vesperini, E., McMillan, S. L. W., & Portegies Zwart, S.F. 2009, *ApJ*, 698, 615

Vesperini, E., Giersz, M., & Sills, A. 2008, *Dynamical Evolution of Dense Stellar Systems*, Proc. IAU Symp. 246 Cambridge: Cambridge University Press

Weidner, C., Kroupa, P., & Maschberger, T. 2009, *MNRAS*, 393, 663

Wiersma, R., Sills, A., & Portegies Zwart, S. F. 2006, *ApJ*, 637, 838

Star clusters: basic galactic building blocks
Proceedings IAU Symposium No. 266, 2009
R. de Grijs & J. R. D. Lépine, eds.

doi:10.1017/S1743921309991116

On the origin of globular cluster bimodality

Oleg Y. Gnedin

Department of Astronomy, University of Michigan, Ann Arbor, MI 48109, USA
ognedin@umich.edu

Abstract. Globular cluster systems in most large galaxies display bimodal color and metallicity distributions, which are frequently interpreted as indicating two distinct modes of cluster formation. The metal-rich (red) and metal-poor (blue) clusters have systematically different locations and kinematics in their host galaxies. However, the red and blue clusters have similar internal properties, such as their masses, sizes, and ages. It is therefore interesting to explore whether both metal-rich and metal-poor clusters could form by a common mechanism and still be consistent with the bimodal distribution. We show that if all globular clusters form only during mergers of massive, gas-rich protogalactic disks, their metallicity distribution could be statistically consistent with that of the Galactic globulars. We take the galaxy assembly history from cosmological dark-matter simulations and couple it with the observed scaling relations for the amount of cold gas available for star formation. In the best-fitting model, early mergers of smaller hosts create exclusively blue clusters, while subsequent mergers of progenitor galaxies with a range of masses create both red and blue clusters. Thus, bimodality arises naturally as the result of a small number of late, massive merger events. We calculate cluster mass loss, including the effects of two-body scattering and stellar evolution, and find that more blue than red clusters are disrupted by the present time because of their lower initial masses and older ages. The present-day mass function in the best-fitting model is consistent with the Galactic distribution. However, the spatial distribution of model clusters is much more extended than observed and is independent of the parameters of our model.

Keywords. galaxies: formation, galaxies: star clusters, globular clusters: general

1. Are the red and blue clusters really that different?

A self-consistent description of the formation of globular clusters remains a challenge to theorists. A particularly puzzling observation is the apparent bimodality, or even multimodality, of the color distribution of globular cluster systems in galaxies ranging from dwarf disks to giant ellipticals. This color bimodality likely translates into a bimodal distribution of the abundances of heavy elements such as iron. We know this to be the case in the Galaxy as well as in M31, where relatively accurate spectral measurements exist for a large fraction of the clusters. The two most frequently encountered modes are commonly called *blue* (metal poor) and *red* (metal rich).

Bimodality in the globular cluster metallicity distribution of luminous elliptical galaxies was proposed by Zepf & Ashman (1993), following a theoretical model by Ashman & Zepf (1992). The concept of cluster bimodality became universally accepted because the two populations also differ in other observed characteristics. The system of red clusters has a significant rotational velocity similar to the disk stars, while blue clusters have little rotational support in the three disk galaxies observed in detail, the Milky Way, M31, and M33. In elliptical galaxies, blue clusters have a larger velocity dispersion than red clusters, due to both a lack of rotation and their more extended spatial distribution. Red clusters are usually more spatially concentrated than blue clusters (Brodie & Strader 2006). All of these differences, however, are in the external properties (location and kinematics), which reflect *where* the clusters formed, but not *how*. The internal properties of the red and blue clusters are similar, i.e., their masses, sizes, and ages, with only slight differences. Even the metallicities themselves differ typically by a factor of 10 between the two modes, which is not enough to affect the dynamics of molecular clouds from which these clusters formed. Could it be, then, that both red and blue clusters form in a similar way on small scales, such as in giant molecular clouds, while the differences in their metallicities and spatial distributions reflect when and where such clouds assemble?

All scenarios proposed in the literature assume different formation mechanisms for the red and blue clusters, and most scenarios envision the stellar population of one mode to be tightly linked to that of the host galaxy (e.g., Forbes *et al.* 1997; Côté *et al.* 1998; Strader *et al.* 2005). The other mode is assumed to have formed differently, in some unspecified 'primordial' way. This assumption only pushed the problem back in time, but did not solve it. For example, Beasley *et al.* (2002) used a semi-analytic model of galaxy formation to study bimodality in luminous elliptical galaxies and needed two separate prescriptions for the blue and red clusters. In their model, red clusters formed in gas-rich mergers with a fixed efficiency of 0.007 relative to field stars, while blue clusters formed in quiescent disks with a different efficiency of 0.002. The formation of blue clusters also had to be artificially truncated at $z = 5$. Strader *et al.* (2005) and Rhode *et al.* (2005) suggested that the blue clusters could instead have formed in very small halos at $z > 10$, before cosmic reionization removed cold gas from such halos. This scenario requires high efficiency of cluster formation in the small halos and also places stringent constraints on the age spread of blue clusters, of less than 0.5 Gyr. Unfortunately, even the most recent measurements of the relative cluster ages in the Galaxy (De Angeli *et al.* 2005; Marín–Franch *et al.* 2009) cannot detect age differences smaller than 9%, or about 1 Gyr, and therefore cannot support or falsify the reionization scenario.

2. Globular clusters could form in protogalactic disks

We set out to test whether a common mechanism could explain the formation of both red and blue modes and produce an entire metallicity distribution consistent with the observations. We begin with a premise of hierarchical galaxy formation in a ΛCDM (cold dark matter) universe. Primordial density fluctuations in the early Universe, probed directly by anisotropies in the cosmic microwave background, set the seeds for structure formation. Cosmological numerical simulations study the growth of the fluctuations through gravitational instability and show us a history of galaxy assembly. The simulations begin with tiny deviations from the Hubble flow, whose amplitudes are set by the measured power spectrum, while the phases are assigned randomly. Therefore, each particular simulation provides only a statistical description of a representative part of the Universe, although current models successfully reproduce major features of observed galaxies.

Hubble Space Telescope observations have convincingly demonstrated one of the likely formation routes for massive star clusters today: in the mergers of gas-rich galaxies (e.g., Holtzman *et al.* 1992; O'Connell *et al.* 1995; Whitmore *et al.* 1999; Zepf *et al.* 1999). We adopt this single formation mechanism and assume that globular clusters form only during massive gas-rich mergers. We follow the merging process of progenitor galaxies in a Galaxy-sized environment using a set of cosmological N-body simulations from Kravtsov *et al.* (2004). We need to decide what type and how many clusters will form in each merger event. For this purpose, we use observed scaling relations to assign each dark-matter subhalo a certain amount of cold gas that will be available for star formation throughout cosmic time and an average metallicity of that gas. To keep the model transparent, we choose a parametrization of the cold gas mass that is as simple as

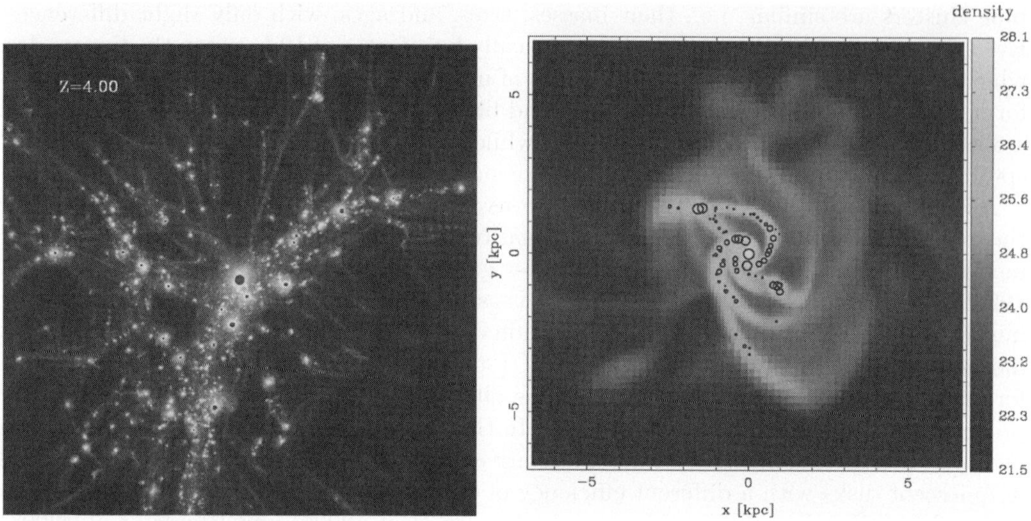

Figure 1. *(left)* Hierarchical buildup of a Milky Way-type galaxy in an adaptive-mesh simulation at $z = 4$. The view is centered on the largest progenitor galaxy in the simulation and shows a $1h^{-1}$ Mpc^2 region (comoving). Dark circles show protogalactic disks in the central regions of dark-matter halos. *(right)* A massive, gaseous disk with prominent spiral arms, seen face-on, in the process of active merging at $z = 4$. In Kravtsov & Gnedin (2005), model star clusters form in giant gas clouds, shown by circles with sizes corresponding to the cluster masses.

possible. Finally, we make the simplest assumption that the mass of all globular clusters formed in the merger is linearly proportional to the mass of this cold gas, $M_{\rm GC} \propto M_{\rm g}$. We discuss our results in Section 3.

Although such a model appears extremely simplistic, we have some confidence that it may capture the main elements of the formation of massive clusters. Kravtsov & Gnedin (2005) used a hydrodynamic simulation of the Galactic environment at high redshifts ($z > 3$) and found dense, massive gas clouds within the protogalactic clumps. These clouds assemble during gas-rich mergers of progenitor galaxies, when the cold gas forms a thin, self-gravitating disk. The disk develops strong spiral arms, which further fragment into separate molecular clouds located along the arms as beads on a string (see Figure 1). If the high-density regions of these clouds formed star clusters, the resulting distributions of cluster mass, size, and metallicity are consistent with those of Galactic metal-poor clusters. The high stellar density of Galactic clusters restricts their parent clouds to be in relatively massive progenitors, with a total mass of $M_{\rm h} > 10^9\ {\rm M}_\odot$. The mass of the molecular clouds increases with cosmic time, but the merger rate declines steadily. Therefore, the cluster formation efficiency peaks during an extended epoch, $5 < z < 3$, when the Universe is less than 2 Gyr old. The molecular clouds are massive enough to be shielded from extragalactic ultraviolet radiation, so that globular cluster formation is unaffected by the reionization of cosmic hydrogen. The mass function of model clusters is consistent with a power law $dN/dM \propto M^{-\alpha}$, where $\alpha = 2.0 \pm 0.1$, similar to that of the local young star clusters. The total mass of clusters formed in each progenitor is roughly proportional to the available gas supply and the total mass, $M_{\rm GC} \propto M_{\rm g} \propto M_{\rm h}$.

In Prieto & Gnedin (2008) we showed that subsequent mergers of the progenitor galaxies ensure that the present distribution of the globular cluster system is spheroidal, as observed. Since the hydrodynamic simulation was stopped at $z \approx 3.3$, we used the Kravtsov *et al.* (2004) N-body simulation to calculate cluster orbits to $z = 0$. We used the evolving properties of all progenitor halos, from the outputs with a time resolution of $\sim 10^8$ yr, to

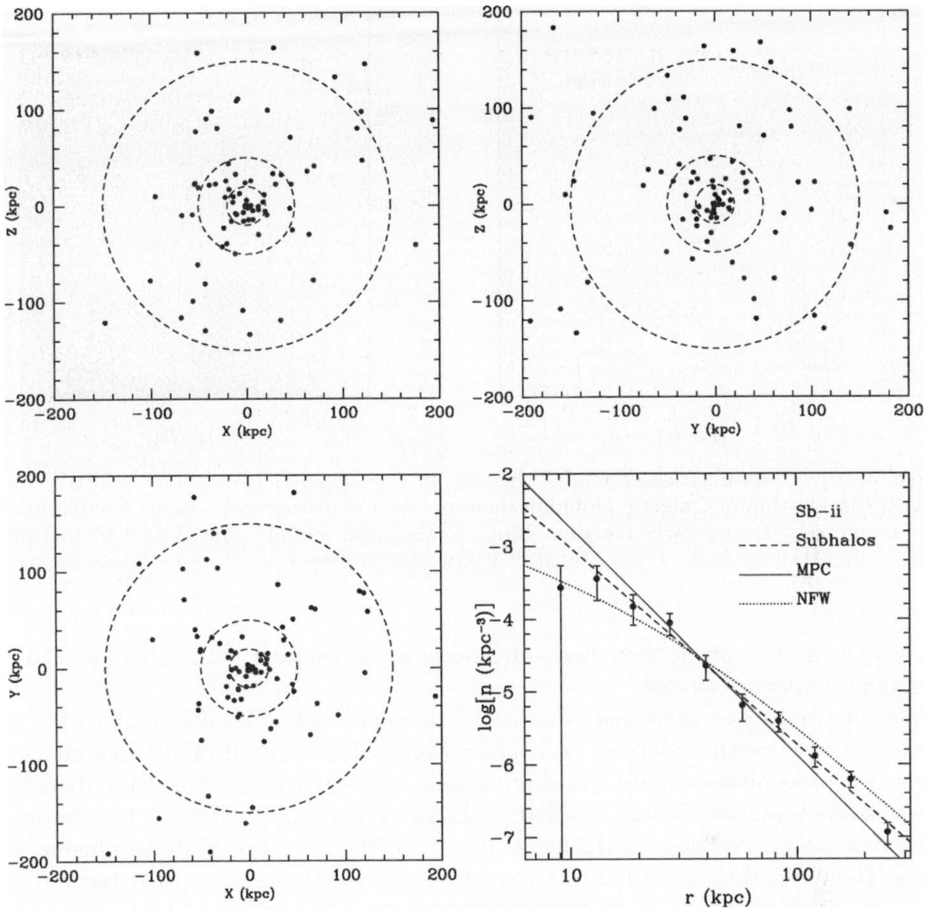

Figure 2. Spatial distribution of surviving model globular clusters, calculated by Prieto & Gnedin (2008). Dashed circles illustrate projected radii of 20, 50, and 150 kpc. The number–density profile *(bottom right)* can be fit by a power law, $n(r) \propto r^{-2.7}$. The distribution of model clusters is similar to that of surviving satellite halos (dashed line) and smooth dark matter (dotted line). It is also consistent with the observed slope of the metal-poor globular clusters in the Galaxy (solid line), plotted using data from the catalog of Harris (1996).

derive the gravitational potential in the entire computational volume at all epochs. We calculated the orbits of globular clusters in this potential from the time when their host galaxies accrete onto the main (most massive) galaxy. Using these orbits, we calculated the dynamical evolution of the model clusters, including the effects of stellar mass loss, two-body relaxation, tidal truncation, and tidal shocks.

In this model, all clusters form on nearly circular orbits within the protogalactic disks. Depending on the subsequent trajectories of the hosts, clusters form three main subsystems at the present time. *Disk clusters* formed in the most massive progenitor that eventually hosts the present Galactic disk. These clusters, found within the inner 10 kpc, do not actually stay on circular orbits but are instead scattered to eccentric orbits by perturbations from accreted galactic satellites. *Inner halo clusters*, found between 10 and 60 kpc, came from the now-disrupted satellite galaxies. Their orbits are inclined with respect to the Galactic disk and are fairly isotropic. *Outer halo clusters*, beyond 60 kpc from the center, are either still associated with the surviving satellite galaxies, or

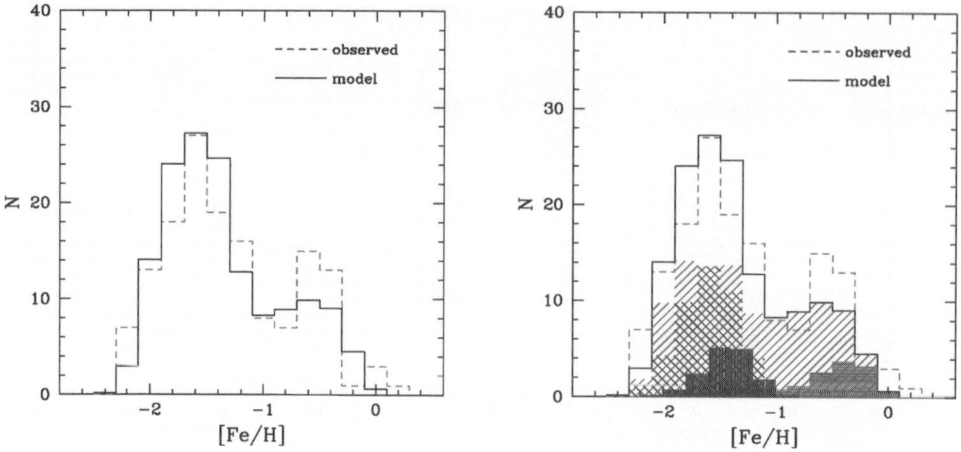

Figure 3. *(left)* Metallicities of model clusters at $z = 0$ (solid histogram) compared to the observed distribution of Galactic globular clusters (shaded histogram). *(right)* Contributions of major mergers (red) and early mergers (blue) to the total model count. Filled histograms are for the main Galactic disk. (From Muratov & Gnedin, in prep.)

were scattered away from their hosts during close encounters with other satellites and consequently appear isolated.

The azimuthally averaged space density of globular clusters is consistent with a power law, $n(r) \propto r^{-\gamma}$, with a slope $\gamma \approx 2.7$ (see Figure 2). Since all of the distant clusters originate in progenitor galaxies and share similar orbits with their hosts, the distribution of the clusters is almost identical to that of the surviving satellite halos. This power law is similar to the observed slope of the metal-poor ([Fe/H] < -1 dex) globular clusters in the Galaxy. However, the model clusters have a more extended spatial distribution (larger median distance) than observed. In the model it is largely determined by the orbits of the progenitor galaxies and the epoch of formation. Moore *et al.* (2006) showed that early-forming halos are more spatially concentrated and to match the Galactic distribution, globular clusters would need to form at $z \sim 12$. However, such an early formation is inconsistent with the requirement of high mass and high density of the parent molecular clouds. At present, we do not have an acceptable solution to this problem.

3. Metallicity bimodality as a natural outcome of hierarchical galaxy formation

Following the scenario outlined above, A. Muratov & Gnedin (in prep.) developed a semi-analytic model that aims to reproduce statistically the metallicity distribution of the Galactic globular clusters, as compiled by Harris (1996). The formation of clusters is triggered during a merger of gas-rich protogalaxies with a mass ratio of 1:5 or higher, and during very early mergers with any mass ratio when the cold-gas fraction in the progenitors is close to 100%. These criteria are applied at every time step of the simulation, every $\sim 10^8$ yr. In the best-fitting model, the mass of globular clusters formed in each event is

$$M_{\rm GC} \approx 7 \times 10^{-4}\ M_{\rm g} \approx 10^{-4}\ M_{\rm h}. \tag{3.1}$$

This imposes the minimum mass of a halo capable of forming a globular cluster. Based on dynamical-disruption arguments, we track only the clusters that are more massive

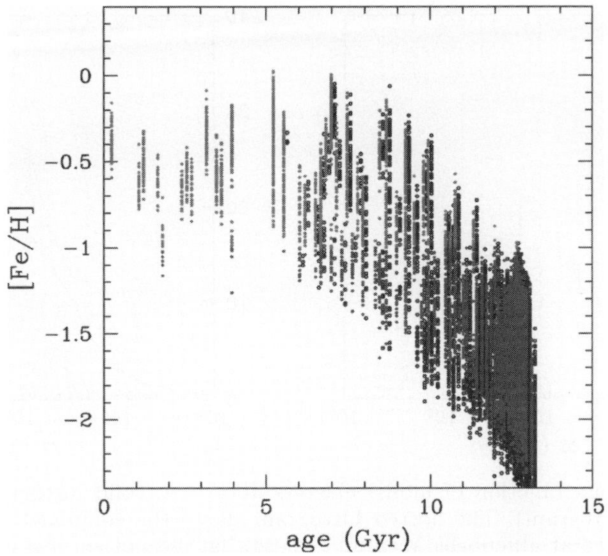

Figure 4. Age–metallicity relation of model clusters in 64 realizations of the Galactic sample. Filled (colored) circles show surviving clusters, open (black) circles show disrupted clusters. The buildup of massive halos drives the steep slope of this relation at early epochs. Notice an order-of-magnitude spread in metallicities of clusters forming at a given epoch.

than $M_{\rm min} = 10^5$ M$_\odot$. Therefore, to form even a single cluster with the minimum mass, the halo needs to be more massive than $\sim 10^9$ M$_\odot$. Individual cluster masses are drawn randomly from the assumed initial cluster mass function, $dN/dM \propto M^{-2}$, normalized to $M_{\rm GC}$. The clusters are assigned the mean metallicity [Fe/H] of their host protogalaxies, using the observed galaxy stellar-mass/metallicity relation. Overall, the model has five free parameters relating to the normalization of the cluster-formation rate and the merger mass ratio.

Figure 3 shows the metallicity distribution in the best-fitting model. The red peak is not as pronounced as in the observations but is noticeable. Interestingly, major mergers contribute both to the red and blue modes, in about equal proportions. Early mergers of low-mass progenitors contribute only blue clusters.

Note that the model has the same formation criteria for all clusters, without explicitly differentiating between the two modes. The only variables are the gradually changing amount of cold gas, the growth of protogalactic disks, and the merger rate. Yet, the model produces two peaks in the metallicity distribution, centered at [Fe/H] ≈ -1.6 and ≈ -0.6 dex, matching those of the Galactic globular clusters. The Kolmogorov–Smirnov probability of the model being consistent with the data is $P_{\rm KS} = 80\%$.

Our prescription links cluster metallicity to the average galaxy metallicity in a one-to-one relation, albeit with random scatter. Since the average galaxy metallicity grows monotonically with time, the cluster metallicity also grows with time. The model thus encodes an age–metallicity relation, in the sense that metal-rich clusters are younger than their metal-poor counterparts by several Gyr. However, Figure 4 shows that clusters of the same age may differ in metallicity by as much as a factor of 10, as they formed in progenitors of different mass. Observations of Galactic globular clusters do not show a clear age–metallicity relation, but instead indicate an age spread that increases with metallicity (De Angeli *et al.* 2005; Marín–Franch *et al.* 2009). Our model does not appear to obviously contradict this trend.

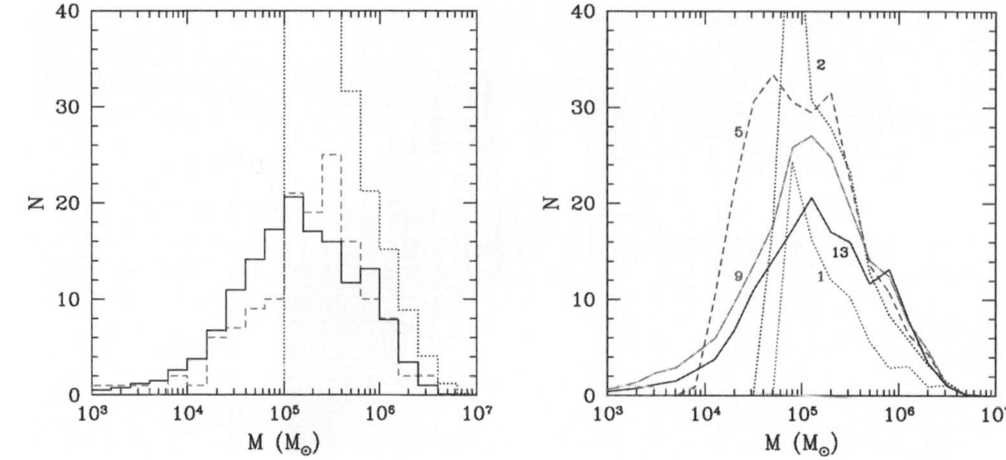

Figure 5. *(left)* Mass function of model clusters at $z = 0$ (solid histogram) versus Galactic clusters (dashed histogram). The dotted histogram shows the combined initial masses of the model clusters formed at all epochs, including those that did not survive until the present. We do not follow clusters with initial masses below 10^5 M$_\odot$. *(right)* Actual mass function at cosmic times of 1 Gyr ($z \approx 5.7$; dotted), 2 Gyr ($z \approx 3.2$; dotted), 5 Gyr ($z \approx 1.3$; dashed), 9 Gyr ($z \approx 0.5$; dot-dashed), and 13 Gyr ($z \approx 0$; solid).

Some of the old and low-mass clusters will be disrupted by the gradual escape of stars and will not appear in the observed sample. We calculated the effects of the dynamical evolution of the model clusters, including stellar mass loss and two-body evaporation, but ignored tidal shocks for simplicity. Figure 5 compares the resulting model mass function at $z = 0$ with the observed Galactic distribution. Since the model parameters were tuned to reproduce the metallicity distribution, the mass functions do not match as well but are still consistent at the level of $P_{\rm KS} = 7\%$. The majority of the disrupted clusters were blue clusters that formed in early low-mass progenitors.

The right panel of Figure 5 illustrates the evolution of the mass function as an interplay between the continuous buildup of massive clusters ($M > 10^5$ M$_\odot$) and the dynamical erosion of the low-mass clusters ($M < 10^5$ M$_\odot$). Expecting that the clusters below $M_{\rm min}$ would eventually be disrupted, we did not track their formation in the model. Instead, the low end of the mass function was built by the gradual evaporation of more massive clusters. Note that most of the clusters were not formed until the Universe was 2 Gyr old, corresponding to $z \approx 3$. The fraction of clusters formed before $z \approx 6$, when cosmic hydrogen was reionized, is small.

In calculating the rate of two-body relaxation, we assumed a standard result for the evaporation time, $t_{\rm d}(M) \approx 10^{10}(M/2\times 10^5\ {\rm M}_\odot)$ yr. We also used an alternative rate with the weaker mass dependence, $t_{\rm d}(M) \approx 10^{10}(M/2 \times 10^5\ {\rm M}_\odot)$ yr, suggested by the recent results of Gieles & Baumgardt (2008). While this prescription leads to slower disruption of the low-mass clusters, we find that the resulting distribution at $z = 0$ is still consistent with observations at the level of $P_{\rm KS}$ of a few percent.

In our scenario, bimodality results from the history of galaxy assembly (rate of mergers) and the amount of cold gas in protogalactic disks. Early mergers are very frequent but involve relatively low-mass protogalaxies, which produce preferentially blue clusters. Late mergers are infrequent but typically involve more massive galaxies. As the number of clusters formed in each merger increases with progenitor mass, just a few late, supermassive mergers can produce a significant number of red clusters. The concurrent growth of the average metallicity of galaxies between the late mergers leads to an apparent 'gap' between the red and blue clusters.

We expect that our formalism could be applied to other galactic environments, such as those of elliptical galaxies with larger samples of globular clusters. For example, Peng *et al.* (2008) showed that the fraction of red clusters increases from 10 to 50% with increasing luminosity of elliptical galaxies in the Virgo cluster. Compared to the Galaxy, giant ellipticals are expected to experience more massive mergers, which would produce similar numbers of red and blue clusters according to Figure 3. Thus, a red-cluster fraction peaking at ~50% is a natural outcome of hierarchical formation.

References

Ashman, K. M. & Zepf, S. E. 1992, *ApJ*, 384, 50

Beasley, M. A., Baugh, C. M., Forbes, D. A., Sharples, R. M., & Frenk, C. S. 2002, *MNRAS*, 333, 383

Brodie, J. P. & Strader, J. 2006, *ARA&A*, 44, 193

Côté, P., Marzke, R. O., & West, M. J. 1998, *ApJ*, 501, 554

De Angeli, F., Piotto, G., Cassisi, S., Busso, G., Recio–Blanco, A., Salaris, M., Aparicio, A., & Rosenberg, A. 2005, *AJ*, 130, 116

Forbes, D. A., Brodie, J. P., & Grillmair, C. J. 1997, *AJ*, 113, 1652

Gieles, M. & Baumgardt, H. 2008, *MNRAS* (Letters), 389, L28

Harris, W. E. 1996, *AJ*, 112, 1487

Holtzman, J. A., Faber, S. M., Shaya, E. J., Lauer, T. R., Groth, J., Hunter, D. A., Baum, W. A., Ewald, S. P., Hester, J. J., Light, R. M., Lynds, C. R., O'Neil, E. J., & Westphal, J. A. 1992, *AJ*, 103, 691

Kravtsov, A. V. & Gnedin, O. Y. 2005, *ApJ*, 623, 650

Kravtsov, A. V., Gnedin, O. Y., & Klypin, A. A. 2004, *ApJ*, 609, 482

Marín–Franch, A., Aparicio, A., Piotto, G., Rosenberg, A., Chaboyer, B., Sarajedini, A., Siegel, M., Anderson, J., Bedin, L. R., Dotter, A., Hempel, M., King, I., Majewski, S., Milone, A. P., Paust, N., & Reid, I. N. 2009, *ApJ*, 694, 1498

Moore, B., Diemand, J., Madau, P., Zemp, M., & Stadel, J. 2006, *MNRAS*, 368, 563

O'Connell, R. W., Gallagher, J. S., Hunter, D. A., & Colley, W. N. 1995, *ApJ* (Letters), 446, L1

Peng, E. W., Jordán, A., Côté, P., Takamiya, M., West, M. J., Blakeslee, J. P., Chen, C.-W., Ferrarese, L., Mei, S., Tonry, J. L., & West, A. A. 2008, *ApJ*, 681, 197

Prieto, J. L. & Gnedin, O. Y. 2008, *ApJ*, 689, 919

Rhode, K. L., Zepf, S. E., & Santos, M. R. 2005, *ApJ* (Letters), 630, L21

Strader, J., Brodie, J. P., Cenarro, A. J., Beasley, M. A., & Forbes, D. A. 2005, *AJ*, 130, 1315

Whitmore, B. C., Zhang, Q., Leitherer, C., Fall, S. M., Schweizer, F., & Miller, B. W. 1999, *AJ*, 118, 1551

Zepf, S. E. & Ashman, K. M. 1993, *MNRAS*, 264, 611

Zepf, S. E., Ashman, K. M., English, J., Freeman, K. C., & Sharples, R. M. 1999, *AJ*, 118, 752

Star clusters: basic galactic building blocks
Proceedings IAU Symposium No. 266, 2009
R. de Grijs & J. R. D. Lépine, eds.

doi:10.1017/S1743921309991128

The progeny of stellar dynamics and stellar evolution within an N-body model of NGC 188

Aaron M. Geller,[1] Jarrod R. Hurley[2] and Robert D. Mathieu[1]

[1]University of Wisconsin–Madison, 475 N. Charter Street, Madison WI 53706, USA

[2]Swinburne University of Technology, P.O. Box 218, Hawthorn, VIC 3122, Australia

Abstract. We present a direct N-body simulation modeling the evolution of the old (7 Gyr) open cluster NGC 188. This is the first N-body open cluster simulation whose initial binary population is directly defined by observations of a specific open cluster: M35 (150 Myr). We compare the simulated color–magnitude diagram at 7 Gyr to that of NGC 188, and discuss the blue stragglers produced in the simulation. We compare the solar-type main-sequence binary period and eccentricity distributions of the simulation to detailed observations of similar binaries in NGC 188. These results demonstrate the importance of detailed observations in guiding N-body open cluster simulations. Finally, we discuss the implications of a few discrepancies between the NGC 188 model and observations and suggest a few methods for bringing N-body open cluster simulations into better agreement with observations.

Keywords. open clusters and associations: individual (NGC 188), binaries: spectroscopic, blue stragglers, N-body simulations

1. Introduction

Recently, sophisticated N-body simulations have incorporated stellar dynamics and stellar evolution self-consistently, gaining the ability to accurately model open-cluster-sized populations ($N \sim 10^4$) of single and binary stars through many billions of years. (e.g., NBODY6, Aarseth 2003; with stellar and binary evolution included by Hurley *et al.* (2000, 2002). Concurrently, large observational surveys of open clusters, like the *WIYN* Open Cluster Study (WOCS; Mathieu 2000), are coming to maturity, providing comprehensive databases of open cluster characteristics, including detailed information on their binary populations. These observations, and specifically those of the binaries, provide important tests and guidance for N-body simulations. In this paper, we describe our first steps in using this wealth of data to help guide an N-body simulation with the aim of creating a realistic model of NGC 188.

2. N-body model of NGC 188

The groundbreaking Hurley *et al.* (2005) model of M67 is a clear example of the potential for N-body simulations to model real open clusters, and thereby address long-standing questions about the evolution of a binary population, origins of blue stragglers (BSs), and specific observed characteristics of a given open cluster. Hurley *et al.* had very limited knowledge of the real binary population in M67, and thus aimed to match other important observables like the number of BSs, the spatial profile, etc. Today, through the WOCS survey, work at the Harvard–Smithsonian Center for Astrophysics (e.g., Latham 2006) and others, we know a great deal about the M67 binaries. It is now clear that, although Hurley *et al.* (2005) were successful in matching many observed characteristics

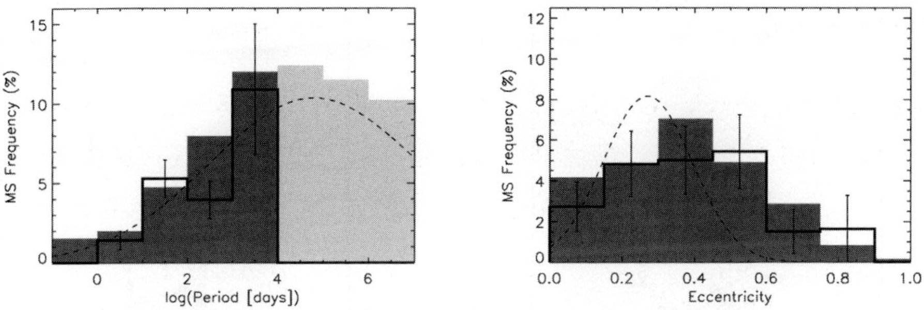

Figure 1. Binary period *(left)* and eccentricity *(right)* frequency distributions, comparing the primordial binary population in the simulation (filled histograms) to the observed main-sequence binary population of M35 (150 Myr; solid black lines). The M35 observed distributions have been corrected for incompleteness to show all binaries with periods, $P < 10^4$ days (excluding those with periods below the circularization period in the eccentricity distribution plot). The primordial binary distributions used in the simulation are plotted with dark gray for observationally detectable binaries ($P < 10^4$ days), and light gray for longer-period binaries. We also show the Duquennoy & Mayor (1991) Galactic-field binary frequency distributions as the dashed lines for comparison.

of M67, the binary frequency at 4 Gyr was too high and the simulation contained too many short-period binaries throughout the evolution of the cluster (see Geller & Mathieu, in prep.). The binary population has a significant effect on the formation mechanisms and rates of BSs as well as the dynamical evolution of the entire cluster. Therefore, correctly modeling the binary population is one of the most important ingredients in creating a realistic open cluster model.

The old (7 Gyr) open cluster NGC 188, like M67 (4 Gyr), has a rich and well-studied binary population (e.g., Geller *et al.* 2009a; Geller & Mathieu, in prep.), a large population of BSs (Mathieu & Geller 2009), and a wealth of observations from the WOCS group and others (see Geller *et al.* 2008). Indeed, NGC 188 is quickly becoming one of the most best-studied and most important open clusters for understanding the origins of BSs, the characteristics of an evolved binary population and the dynamical state of an evolved open cluster. These data provide ideal and much-needed guidance for an N-body simulation of NGC 188.

2.1. *Simulation method*

All simulations were run using the NBODY6 code (Aarseth 2000), with slight modifications to define the initial binary population and output format. All stellar evolution is incorporated as in Hurley et al. (2000) and binary evolution, including tides, mass transfer, etc., is treated as in Hurley *et al.* (2002). BSs are modeled in the same manner as in Hurley *et al.* (2005). The simulations were executed on the supercomputer at the Swinburne Centre for Astrophysics and Supercomputing.

The NGC 188 simulation was initialized with a mass of 23 870 $M_\odot$ ($N = 24\,000$ stars) chosen from a Kroupa (2001) initial mass function and distributed according to a Plummer profile. We assume solar metallicity for the stars in the model (Sarajedini et al. 1999) and place the cluster in orbit around a point-mass Galactic potential according to the observed orbit of NGC 188 (Carraro & Chiosi 1994). In the following, we discuss the binary population of the model in some detail, focusing on the period and eccentricity distributions and leaving more thorough discussions of methods and results for future papers.

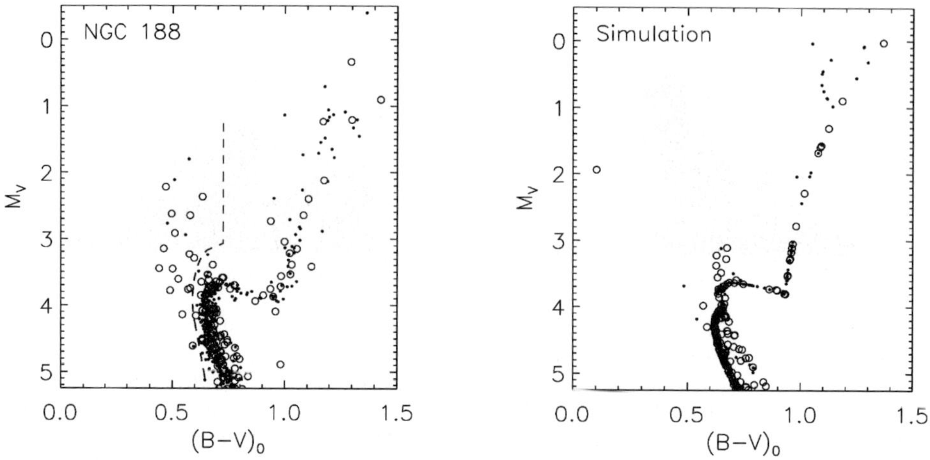

Figure 2. Color–magnitude diagrams of NGC 188 *(left)* and the NGC 188 simulation at 7 Gyr *(right)*. The single stars are plotted as small black points, and the detected binaries are plotted using larger open circles. The NGC 188 BSs are identified as being to the left of the dashed line (as in Geller *et al.* 2008), and within the simulation BSs are identified as being main-sequence stars with masses greater than the turnoff mass (and reside as a similar region to observed BSs in NGC 188). The simulation succeeds as reproducing the overall form of the observed single- and binary-star sequence, but has not created the large number of observed BSs.

2.2. *Primordial binary population and observations of M35*

The initial parameters defining a binary population have long been derived from theoretical analyses and predictions, since we lacked sufficiently detailed observations of young open cluster binary populations to help guide these choices. The WOCS survey of M35 (Geller *et al.* 2009c) now allows for a major step forward. We can use these binaries as a proxy for the primordial binary population in this simulation. There is no a priori reason to presume that the M35 binaries will evolve into the NGC 188 binaries. Still, the detailed observations of the M35 binaries allow us to create the first N-body model whose initial binary population is *directly defined by observations* and to study the evolution of these binaries for the entire age of NGC 188 (7 Gyr).

M35 is observed to have a main-sequence (MS) binary frequency of $22 \pm 3\%$ out to periods of 10^4 days. We have chosen to initialize the simulation with a binary frequency of 25% out to the same period range. In Figure 1 we show the M35 incompleteness-corrected binary period and eccentricity distributions (black line), compared to the initial conditions used in the NGC 188 model (filled histograms). As we only detect binaries with periods $< 10^4$ days, we use the Galactic-field binaries (Duquennoy & Mayor 1991) to extend the period distribution to longer periods. Thus, we input a primordial binary population with a lognormal period distribution and a Gaussian eccentricity distribution.

2.3. *Comparing to observations of NGC 188*

The WOCS observations of NGC 188 (Geller *et al.* 2008, 2009a; Geller & Mathieu, in prep.) allow for a detailed comparison with the NGC 188 model. In the following analysis, we limit the output of our simulation to only include stars within the Geller *et al.* observable domain (in both mass and binary period). We first focus on the color–magnitude diagram (CMD) at 7 Gyr shown in Figure 2. Note that the stellar evolution incorporated into the simulation does an excellent job of reproducing the 'normal' single- and binary-star sequence compared to the observed CMD. However, the simulation is notably lacking in BSs: there are 21 observed BSs in NGC 188 and only 5 at 7 Gyr in the simulation.

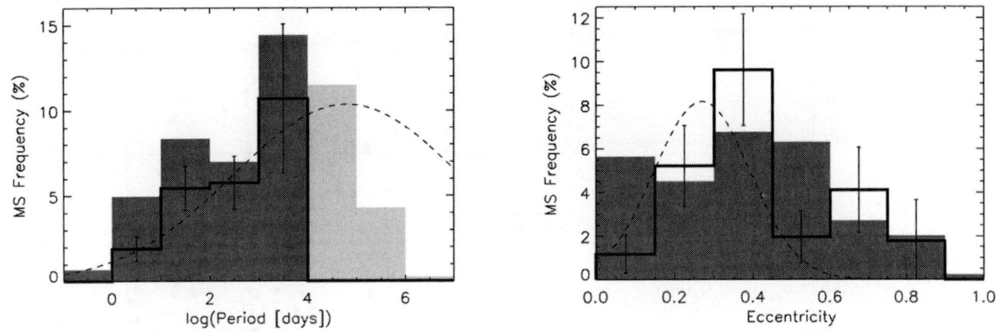

Figure 3. Period *(left)* and eccentricity *(right)* MS-binary frequency distributions at 7 Gyr. The NGC 188 observed distributions are plotted in the solid black lines and have been corrected for incompleteness in the same manner as in Figure 1, showing all binaries with $P < 10^4$ days and excluding those with periods below the circularization period in the eccentricity distribution plot. The simulated binary distributions are plotted in the filled histograms with the same gray-scale coding as in Figure 1. Again, we also show the Duquennoy & Mayor (1991) Galactic-field binary frequency distributions in the dashed lines for comparison. Note the close agreement between the simulated and observed binaries in both distributions and binary frequency.

This low number of BSs stays relatively constant throughout the simulation, and is likely due to the initial binary population. We will discuss the implications of this result in Section 3.

We find the solar-type MS binary population in the simulation at 7 Gyr to correspond remarkably well to observations of similar binaries in NGC 188. In Figure 3 we compare the observed incompleteness-corrected NGC 188 binaries (black lines) to those of the simulation at 7 Gyr (filled histograms) in both period and eccentricity. The form of both the simulated distributions, as well as the binary frequency agree well with the observations of NGC 188 as well as the Galactic field (dashed lines; Duquennoy & Mayor 1991). This close agreement is a testament to the power of using detailed observations to define the initial binary population in an N-body simulation.

Still, there are a few discrepancies between the simulated binary population and the observations. The circularization period in the model is significantly shorter than the observed value for NGC 188 (Figure 4). Meibom & Mathieu (2005) find a circularization period of $14.5^{+1.4}_{-2.2}$ days, while a similar analysis of the model shows a period of 2.5 days. Interestingly, we also see *too many* circular MS binaries in the simulation at 7 Gyr (first bin in Figure 3), most having periods greater than the circularization period of both the simulation and observations. We discuss both of these results in the following section.

3. Discussion and future directions

By using the observed M35 binaries as a proxy for our primordial binary population, we have managed to nearly reproduce the NGC 188 binaries within our N-body simulation at 7 Gyr. However, as noted in Section 2.3, the current NGC 188 model is deficient in BSs. Interestingly, the Hurley *et al.* (2005) model of M67 was able to produce the large number of observed BSs (though not their binary frequency), but was unable to reproduce the observed binary population. The quantity of BSs produced in the M67 model was largely dependent on the initial binary population. Hurley *et al.* found that a large fraction of short-period binaries was necessary to create the observed numbers of BSs in M67 (and NGC 188). These short-period binaries not only have a higher likelihood of creating BSs in isolation (e.g., through mass transfer resulting from Roche-lobe overflow),

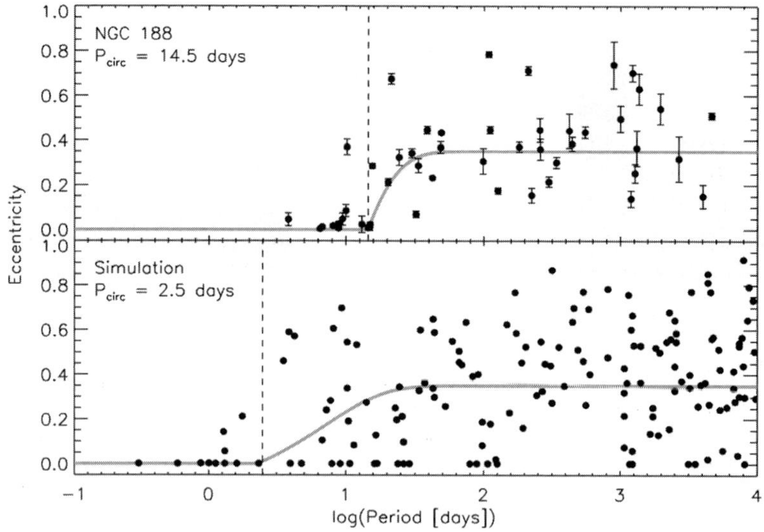

Figure 4. Eccentricity plotted against the logarithm of the period ($e - \log P$) at 7 Gyr for the observed *(top)* and simulated *(bottom)* solar-type MS NGC 188 binaries. The solid gray lines represent the best-fit circularization functions from Meibom & Mathieu (2005) for each sample, respectively. The tidal processes in the simulation result in a circularization period that is significantly lower than in NGC 188. Also, the simulation produces a population of circular binaries with $P \gg P_{\rm circ}$ that is not observed in NGC 188 (nor in the Galactic field). Most of these circular binaries in the model have previously gone through a stage of mass transfer.

but dynamical interactions involving short-period binaries have a higher probability of resulting in collisions and mergers (Fregeau *et al.* 2004). Observations of M35 binaries are very different from the flat period distribution and large binary frequency used in the M67 model (e.g., Figure 1). The important next step will be to match both the binary population and the BS population within one cohesive simulation. We are currently exploring a number of possible methods for addressing this issue, the most promising of which appears to be the inclusion of primordial triples (e.g., Kozai-induced mergers; Perets & Fabrycky 2009, larger cross section for dynamical interactions, etc.).

The tidal circularization period is one of our best observational tools to study the effects of tides on a binary population. In Figure 4 and Section 2.3 we show that the circularization period at 7 Gyr in the simulation is significantly lower than what is observed in NGC 188. This suggests that the tidal-energy-dissipation rate may be *underestimated* in the N-body model (at least for solar-type MS binaries). We may also (or alternatively) require a form of pre-MS tidal circularization (e.g., Kroupa 1995). As pointed out by Mathieu (2008), tidal processes have a significant impact on dynamical interactions between single and binary stars, and the energy dissipation rates can be the deciding factor between a close fly-by, tidal capture or even a merger or collision that could result in a BS. WOCS data (e.g., Meibom & Mathieu 2005) provide the ideal foundation from which we will improve the tidal physics within the N-body model.

Finally, the simulation has an excess of circular binaries with periods beyond the circularization period. The majority of these have undergone some form of mass transfer, which has quickly circularized the orbit. However, similar populations of circular binaries at $P \gg P_{\rm circ}$ are not observed in open clusters nor in the Galactic field (e.g., Duquennoy & Mayor 1991; Meibom & Mathieu 2005). This may suggest that mass transfer, especially in initially eccentric binaries, does not necessarily lead to circular binaries (e.g., Sepinsky *et al.* 2007; Bonačić Marinović *et al.* 2008), or that these cases of mass transfer happen much more frequently in the simulation than in reality. Correctly modeling the products of mass transfer and common-envelope evolution is critical to accurately reproduce BSs as well as the binary population within an N-body simulation.

We will present further details about this NGC 188 model in future papers. Additionally, we will explore the various methods mentioned above for bringing the NGC 188 model, and N-body simulations in general, in better agreement with observations. This synergy between observations and simulations will continue to refine the N-body method and reveal further insights into the origins of BSs, the evolution of a binary population and the dynamical evolution of star clusters.

Acknowledgements

This work was funded in part by a National Science Foundation (NSF) East Asia and Pacific Summer Institutes (EAPSI) fellowship, the Wisconsin Space Grant Consortium and NSF grant AST-0406615.

References

Aarseth, S. J. 2000, in: V. G. Gurzadyan & R. Ruffini (eds.), *The Chaotic Universe*, Proc. Second ICRA Network Workshop, Adv. Ser. Astrophys. Cosmol., vol. 10, p. 286
Aarseth, S. J. 2003, *Gravitational N-Body Simulations*, Cambridge: Cambridge University Press
Bonačić Marinović, A. A., Glebbeek E., & Pols, O. R. 2008, *A&A*, 480, 797
Carraro, G. & Chiosi, C. 1994, *A&A*, 288, 751
Duquennoy, A. & Mayor, M. 1991, *AA*, 248, 485
Fregeau, J. M., Cheung, P., Portegies Zwart, S. F., & Rasio, F. A. 2004, *MNRAS*, 352, 1
Geller, A. M., Mathieu, R. D., Braden, E. K., Meibom, S., Platais, I. & Dolan, C. 2009b, *AJ*, submitted
Geller, A. M., Mathieu, R. D., Harris, H. C., & McClure, R. D. 2008, *AJ*, 135, 2264
Geller, A. M., Mathieu, R. D., Harris, H. C., & McClure, R. D. 2009a, *AJ*, 137, 3743
Hurley, J. R., Pols, O. R., Aarseth, S. J., & Tout, C. A. 2005, *MNRAS*, 363, 293
Hurley, J. R., Pols, O. R., & Tout, C. A. 2002, *MNRAS*, 329, 897
Hurley, J. R., Tout, C. A., & Pols, O. R. 2000, *MNRAS*, 315, 543
Kroupa, P. 1995, *MNRAS*, 277, 1507
Kroupa, P. 2001, *MNRAS*, 322, 231
Latham, D. W. 2006, in: K. A. van der Hucht (ed.) *Highlights Astron.* vol. 14, p. 444
Mathieu, R. D. 2000, *ASP Conf. Ser.*, 198, 517
Mathieu, R. D. 2008, in: E. Vesperini, M. Giersz & A. Sills (eds.), *Dynamical Evolution of Dense Stellar Systems*, Proc. IAU Symp. No. 246, 79
Mathieu, R. D. & Geller, A. M. 2009, *Nature*, in press
Meibom, S. & Mathieu, R. D. 2005, *ApJ*, 620, 970
Mermilliod, J.-C., Rosvick, J. M., Duquennoy, A., & Mayor, M. 1992, *A&A*, 265, 513
Perets, H. B. & Fabrycky, D. C. 2009, *ApJ*, 697, 1048
Sarajedini, A., von Hippel, T., Kozhurina–Platais, V., & Demarque, P. 1999, *AJ*, 118, 2894
Sepinsky, J. F., Willems, B., Kalogera, V., & Rasio, F. A. 2007, *ApJ*, 667, 1170

Star clusters: basic galactic building blocks
Proceedings IAU Symposium No. 266, 2009
R. de Grijs & J. R. D. Lépine, eds.

doi:10.1017/S174392130999113X

The formation of brown dwarfs

Ant Whitworth,[1] Dimitri Stamatellos,[1] Steffi Walch,[1] Murat Kaplan,[1] Simon Goodwin,[2] David Hubber[2] and Richard Parker[2]

[1]School of Physics & Astronomy, Cardiff University, Cardiff CF24 3AA, UK
email: ant@astro.cf.ac.uk

[2]Department of Physics & Astronomy, University of Sheffield, Sheffield S3 7RH, UK

Abstract. We argue that brown dwarfs (BDs) and planemos form by the same mechanisms as low-mass hydrogen-burning stars, but that as one moves to lower and lower masses, an increasing fraction of these objects is formed by fragmentation of the outer parts ($R \gtrsim 100\,\mathrm{AU}$) of protostellar accretion discs around more massive primary protostars, which in turn formed in their own very-low-mass prestellar cores. Numerical simulations of disc fragmentation with realistic thermodynamics show that low-mass objects are readily formed by fragmentation of short-lived massive, extended protostellar accretion discs. Such objects tend subsequently to be liberated into the field at low speed, due to mutual interactions with the primary protostar. Many ($\sim$20%) are in low-mass ($M_1 + M_2 < 0.2\,\mathrm{M}_\odot$) binary systems with semi-major axes $a \sim 1$ to $2\,\mathrm{AU}$ or $\sim 200\,\mathrm{AU}$ and mass ratios $q \equiv M_2/M_1 \gtrsim 0.7$. Most of the brown dwarfs have sufficiently large attendant discs to sustain accretion and outflows. Most of the BDs that remain bound to the primary protostar have wide orbits (i.e., there is a BD desert), and these BDs also have a significantly higher probability of being in a BD/BD binary system than do the brown dwarfs that are liberated into the field (just as observed). In this picture, the multiplicity statistics and velocity dispersion of brown dwarfs are largely determined by the *eigen evolution* of a small-N system, born from a single prestellar core, rather than the larger-scale dynamics of the parent cluster. Consequently, many of the statistical properties of brown dwarfs should not differ very much from one star-formation region to another.

Keywords. accretion, accretion disks, gravitation, hydrodynamics, radiative transfer, stellar dynamics, binaries: general, stars: formation, stars: low-mass, brown dwarfs

---

1. Introduction

The appearance of the Universe we live in, and the environment in which we ourselves have prospered, are both strongly conditioned by the fact that almost all seriously dense (say, $\rho \gtrsim 1\,\mathrm{g\,cm^{-3}}$) baryonic matter in the Universe is in objects with a rather small logarithmic range of mass (say, 0.01 to $100\,\mathrm{M}_\odot$), i.e., stars. The study of star formation is largely concerned with understanding why this is so. There are two main approaches to answering this question. One approach argues that the thermodynamics of the interstellar medium delivers a more or less universal mean Jeans mass $\sim 0.3\,\mathrm{M}_\odot$, and then the range of stellar masses simply reflects the variances of all factors determining the final stellar mass in a particular location. This is the basis of turbulence theories. The second approach argues that it becomes increasingly hard to form stars as one goes to higher masses (because self-gravitating gas has to go through ever more complicated contortions to get past radiation pressure) or lower masses (because opacity makes it increasingly hard for self-gravitating gas to lose entropy on a dynamical timescale), so that there is naturally a concentration of masses between these limits. The full solution may actually involve all three elements (mean Jeans mass, radiation pressure and opacity). Here we will be most concerned with understanding the formation of low-mass stars, i.e., star formation close to the opacity limit, and what constraints can be inferred from the observed properties of such stars.

In the context of star formation, it is appropriate to describe brown dwarfs (and planemos) as stars. This is because all available evidence (e.g., Luhman *et al.* 2007) suggests that brown dwarfs form just like low-mass hydrogen-burning stars, i.e., by gravitational instability, on a dynamical timescale†, normally (possibly always) in clusters, and with an initially uniform elemental composition (the same as the elemental composition of the interstellar medium out of which they condense). There is no compelling evidence for a discontinuity in stellar statistical properties across the hydrogen-burning limit at $M_H \simeq 0.075 \pm 0.005\,M_\odot$, at least, not in those properties that might be expected to reflect the formation process of the objects concerned (as distinct from the nuclear processes which contribute later to their intrinsic luminosity and long-term evolution). With this definition of a star, the opacity limit, which is variously estimated to be between 0.003 and $0.007\,M_\odot$, is the minimum mass for star formation. (It makes no sense to call it the minimum mass for planemo formation, because it evidently is not!)

In contrast, planets form by accumulation of a rocky core, on a much longer timescale, $\gtrsim 10^7$ yr, with subsequent acquisition of a gaseous envelope if the circumstances allow this, and with an initially fractionated elemental composition. With these definitions of stars and planets, based on their formation mechanisms, there is probably an interval in the mass range 0.003 to $0.010\,M_\odot$ which hosts both stars and planets. It may be quite hard to distinguish the two without information on their internal compositions.

In Section 2, we describe briefly the different formation mechanisms that have been proposed for brown dwarfs. In Section 3, we rehearse the observational constraints on brown dwarf formation. In Section 4, we explore the possibility that brown dwarfs form by disc fragmentation, and show that this mechanism reproduces the observed statistical properties of brown dwarfs rather well. In Section 5, we summarise our conclusions.

2. How do brown dwarfs form?

Turbulence: Brown dwarfs and low-mass hydrogen-burning stars might form by turbulent fragmentation (e.g., Klessen & Burkert 2002; Padoan & Nordlund 2002; Hennebelle & Chabrier 2008). In this picture, it is envisaged that turbulent energy is injected into the interstellar medium on large scales (possibly by the gravitational and/or thermal instabilities involved in the formation of molecular clouds, or by galactic shear), and then cascades to smaller and denser scales, until it creates self-gravitating cores which condense to form stars. Allowing for the fact that a single pre-stellar core is not expected to put all its mass into a single star, this requires that occasionally, not very often, turbulence can deliver self-gravitating cores with masses $\lesssim 0.01\,M_\odot$. This has not yet been demonstrated to happen in a numerical simulation. As presently developed, the theory also has the following outstanding issues. (i) It seems to require some fine tuning, perhaps by as yet unidentified thermodynamic processes, to keep the ratio of brown dwarfs to hydrogen-burning stars roughly constant, as observed across widely differing star-formation regions. (ii) The mechanism seems to imply that more massive stars form in less dense, more extended cores, and hence in the outer parts of clusters. This may be difficult to reconcile with the growing evidence for primordial mass segregation. (iii) The mechanism does not yet make any firm predicitons about binary statistics.

† We note that the dynamical timescale for low-mass stars condensing out of gas with isothermal sound speed $c \simeq 0.2\,\mathrm{km\,s^{-1}}$ (i.e., H_2 at $\sim$10 K) is given by $t_{\rm dyn} \simeq GM_\star/c^3 \simeq 600\,\mathrm{years}\,(M_\star/M_{\rm Jupiter})$ and is therefore very short for low-mass objects.

Dynamical fragmentation and ejection: Brown dwarfs and low-mass hydrogen-burning stars might also form by dynamical fragmentation and ejection. This mechanism starts with a pre-stellar core (perhaps formed by turbulent fragmentation), but the core is sufficiently massive and turbulent that it collapses and undergoes dynamical fragmentation, thereby producing a small ensemble of protostellar embryos at the centre of an infalling envelope. Some of these protostellar embryos are then ejected from the core by dynamical interactions amongst themselves, and consequently never acquire sufficient mass to ignite hydrogen-burning. Those that remain in the core compete to accrete the remaining mass, and become hydrogen-burning stars with a range of masses. This mechanism was first proposed by Reipurth & Clarke (2000), and has been demonstrated on the basis of numerical simulations by Bate *et al.* (2003) and Goodwin *et al.* (2004). It has two main problems. First the brown dwarfs tend to be ejected with rather large velocities, and therefore there should be a diaspora of brown dwarfs visible round young clusters. Despite concerted observational efforts to detect it, such a diaspora has not been observed. Second, this mechanism does not seem *either* to produce the tight BD/BD binary systems in which about ∼20% of brown dwarfs are found, *or* to explain the brown-dwarf desert (the paucity of close brown-dwarf companions to Sun-like stars).

Photo-erosion of pre-existing cores: Brown dwarfs and low-mass hydrogen-burning stars might also form where a pre-existing core is overrun by an HII region, as first proposed by Hester (1996). In this case, there is a competition between core collapse and the boiling off of its outer layers by ionisation. Whitworth & Zinnecker (2004) have shown that this is a very robust mechanism, in the sense that brown dwarfs are produced for a wide range of likely initial conditions. Moreover, evaporating gaseous globules (EGGs) are seen in M16, so the mechanism evidently operates. However, it is a very inefficient mechanism. It requires quite a massive core to make a single brown dwarf. In addition, it can only operate in clusters that produce OB stars, so it cannot explain the brown dwarfs observed in lower-mass star-formation regions like Taurus. Finally, it makes no predictions about the binary statistics of brown dwarfs. It is—at best—a viable way to produce some single brown dwarfs in star-formation regions like the Orion Nebula Cluster.

Disc fragmentation and liberation: In the remainder of this contribution, we shall promote the idea that a large fraction of brown dwarfs and some low-mass hydrogen-burning stars form by fragmentation of accretion discs around more massive primary protostars. In this picture, the low-angular-momentum material in a pre-stellar core collapses first to form the primary protostar (hereafter the *ur* star). Much of the remaining material in the core then forms an accretion disc round the ur star. Once this disc becomes sufficiently massive ($M_{\rm disc} \gtrsim 0.3 M_\star$) and extended ($R \gtrsim 100\,{\rm AU}\,(M_\star/{\rm M}_\odot)^{1/3}$), it fragments to form lower-mass secondaries in its outer parts, primarily brown dwarfs and very-low-mass hydrogen-burning stars, but also a few planemos. Gravitational interactions between the secondaries and the ur star then liberate most of the secondaries, but a few remain bound to the ur star. More specifically, we argue that, as one proceeds to lower masses, from low-mass hydrogen-burning stars through brown dwarfs to planemos, an increasing fraction of stars is formed as secondaries by disc fragmentation, rather than being formed as primaries by the monolithic collapse of a core, or by dynamical fragmentation during the collapse of a core. By means of numerical simulations, we show that this formation mechanism seems to be able to deliver all observed statistical properties of brown dwarfs. One advantage of this model is that most statistical properties of brown dwarfs and low-mass hydrogen-burning stars are determined by the *eigen evolution* of a small-N system formed within a single core, so they are largely independent of environmental factors, or of the overall dynamics of the parent cluster. André *et al.* (2007) recently presented evidence that the prestellar cores in Ophiuchus are sufficiently far apart, and move sufficiently slowly (relative to one another), that their internal evolution will occur more or less in isolation.

3. Constraints on brown dwarf formation

The stellar initial mass function (IMF): The distribution of stellar masses—at least for contemporary, local star formation—is approximately lognormal (e.g., Kroupa 2001; Chabrier 2003), and similar in shape to the observed prestellar-core mass function. However, the core-mass function peaks at a mass about three times higher than the stellar IMF. Thus, to reproduce the masses and binary statistics of stars, a typical prestellar core must convert between 30 and 60% of its mass into between two and five stars. Since (a) the core-mass function is not significantly narrower than the stellar IMF and (b) the masses of stars produced by a single core are not identical, the mapping from core-mass function to stellar IMF cannot be self-similar. Rather, lower-mass cores must, on average, produce fewer stars and/or convert a higher fraction of their mass into stars. Hence, the decline in the stellar IMF at low masses reflects both the decline in the core-mass function at low masses, and the difficulty of producing low-mass secondaries in such cores. In fact, in a low-mass, low-turbulence core, disc fragmentation is particularly effective at producing secondaries, because material can be parked in a disc, losing entropy, whilst it slowly builds up towards gravitational instability.

Accretion and outflows: A significant fraction of young brown dwarfs show evidence of accretion discs, magnetospheric accretion, bipolar outflows, etc. The inference is that their formation is simply a scaled-down version of the formation of higher-mass stars. For example, mass-accretion rates appear to scale approximately as the stellar mass squared, $\dot{M}_\star \propto M_\star^2$, for both hydrogen-burning stars and brown dwarfs. However, we should be mindful that these observations simply require young brown dwarfs to be attended by significant accretion discs, i.e., that they did not form from material with negligible angular momentum, and that they have not suffered a sufficiently violent perturbation to separate them from their accretion discs. A brown dwarf of mass $m_{\rm bd}$, which is in orbit round an ur star of mass $M_{\rm ur}$ at radius $R_{\rm orbit}$, will, if subsequently ejected by dynamical interactions with other stars, retain a disc of radius $r_{\rm disc} \sim R_{\rm orbit}\,(m_{\rm bd}/M_{\rm ur})^{1/2}$. We will show that brown dwarfs formed by disc fragmentation are formed at such large distances from the ur star that they retain discs several tens of AU in extent, in accordance with observations.

Maps and SEDs: If brown dwarfs are formed by disc fragmentation, one might hope to see evidence for massive extended discs in continuum maps and SEDs of Class 0 and I protostars. The problem here is that the process of disc fragmentation is very fast, and the fragmenting disc is likely to be deeply embedded in the core envelope. Therefore, the disc is probably already largely dissipated by the time that the protostar reaches the Class I phase.

Spatial and velocity distribution in clusters: The distributions of brown dwarfs in young and evolved clusters should hold clues to the way in which brown dwarfs form. However, these clues have proved hard to decipher. Setting aside the issue of selection effects (brown dwarfs become increasingly hard to detect as they age and fade), the problem is that one has to disentangle primordial mass segregation (for example, any tendency that brown dwarfs might have to form preferentially in the outer parts of a cluster), from subsequent segregation due to systematic differences in the velocity dispersions of stars of different mass. This second effect works on three levels, the first being the *eigen evolution* of the stars formed in a single core (which determines the local velocity dispersion of the individual stars relative to the centre of mass of the core), the second being the overall dynamics of the cluster (which determines the velocity dispersion of the individual cores relative to one another, and may involve violent relaxation), and the third being dynamical mass segregation (more quasistatic relaxation driven by equipartition of energy).

Binary statistics: Arguably the most discriminating constraints on brown-dwarf formation derive from the binary statistics of brown dwarfs. As one looks at stars of decreasing mass, the probability of their being the primary in a binary system decreases [for brown dwarfs, $f_{\rm bin} \sim 0.20 \pm 0.05$], and—for those that are primaries—the mean semi-major axis decreases [for brown dwarfs, $\bar{a} \sim 5 \pm 2\,{\rm AU}$], the (logarithmic) spread of semi-major axes decreases [for brown dwarfs, $\sigma_{\log_{10}(a)} \sim 0.4 \pm 0.1$] and the mean mass ratio increases [for brown dwarfs, $\bar{q} \sim 0.8 \pm 0.1$]. Two further aspects are remarkable. (i) When Sun-like stars have brown-dwarf companions, the semi-major axis is normally large, say, $a \sim 200\,{\rm AU}$. Close systems are very rare (this is termed the Brown Dwarf Desert; e.g., Burgasser *et al.* 2007). (ii) Those brown dwarfs that are companions to Sun-like stars (orbiting at $a \sim 200\,{\rm AU}$) are much more likely to be in a close ($a \sim 5\,{\rm AU}$) binary system with another brown dwarf than brown dwarfs in the field (Burgasser *et al.* 2005).

4. Disc fragmentation

Suppose that there exists a Keplerian disc with surface density $\Sigma(R)$, sound speed $a(R)$ and orbital angular speed $\Omega(R) = \left(GM_{\rm ur}/R^3\right)^{1/2}$, where r is the radial distance from the central ur star. Now consider a small circular proto-fragment with radius r, and hence mass $m = \pi r^2 \Sigma(R)$. The equation controlling its radial excursions is

$$\ddot{r} \simeq -\frac{Gm}{r^2} + \frac{1}{\rho}\frac{{\rm d}P}{{\rm d}r} + \frac{\Omega^2(R)r}{4} \tag{4.1}$$

$$\simeq -\pi G\Sigma(R) + \frac{a^2(R)}{r} + \frac{\Omega^2(R)r}{4}. \tag{4.2}$$

Hence, the timescale on which the proto-fragment condenses out is given by

$$t_{\rm cond} \simeq \left(\frac{r}{-\ddot{r}}\right)^{1/2} \simeq \left(\frac{\pi G\Sigma(R)}{r} - \frac{a^2(R)}{r^2} - \frac{\Omega^2(R)}{4}\right)^{-1/2} \tag{4.3}$$

and the fastest-condensing proto-fragment has radius and growth time given by

$$r_{\rm fast} \simeq \frac{2a^2(R)}{\pi G\Sigma(R)}, \tag{4.4}$$

$$t_{\rm fast} \simeq 2\left(\left(\frac{\pi G\Sigma(R)}{a(R)}\right)^2 - \Omega^2(R)\right)^{-1/2}. \tag{4.5}$$

Evidently, the growth time is only real if

$$\Sigma(R) \gtrsim \frac{a(R)\Omega(R)}{\pi G}, \tag{4.6}$$

which is the condition for the disc to be unstable to gravitational fragmentation (Toomre 1964; Goldreich & Lynden–Bell 1965). In an equilibrium disc that is too massive to be Keplerian, Ω should be replaced by the epicyclic frequency, and in a nonequilibrium disc by twice the local vorticity, but otherwise the condition takes the same form.

Gammie (2001) has shown that such a disc will only fragment if a proto-fragment can cool on a dynamical timescale, otherwise the fragment undergoes an adiabatic bounce and is sheared apart. For a marginally unstable disc, the dynamical timescale is

$$t_{\rm dyn} \simeq \frac{2a(R)}{\pi G \Sigma(R)}. \tag{4.7}$$

The cooling time is given by

$$t_{\rm cool} \simeq \frac{\Sigma(R) a^2(R)}{2\sigma_{\rm sb} T^4(R)/\Sigma(R)\bar{\kappa}_{\rm R}(R)} = \frac{15c^2 h^3 \Sigma^2(R) \bar{\kappa}_{\rm R}(R)}{2^2 \pi^5 \bar{m}^4 a^6(R)}. \tag{4.8}$$

In the first expression, the numerator, Σa^2, is the thermal energy per unit area and the denominator, $2\sigma_{\rm sb} T^4(R)/\Sigma(R)\bar{\kappa}_{\rm R}(R)$, is the blackbody flux from the two sides of the disc, divided by the optical depth, $\tau = \Sigma \bar{\kappa}_{\rm R}$. Here, $\bar{\kappa}_{\rm R}$ is the Rosseland mean opacity and we are assuming that the disc is optically thick (which can be confirmed retrospectively). To obtain the second expression, we have substituted $\sigma_{\rm sb} T^4 = 2\pi^5 \bar{m}^4 a^8/15c^2 h^3$, where $\bar{m}$ is the mean gas-particle mass ($\sim 4 \times 10^{-24}$ g for molecular gas). Combining Equations (4.7) and (4.8), Gammie's condition, $t_{\rm cool} < t_{\rm dyn}$, reduces to

$$\Sigma(R) \lesssim \left(\frac{2^3 \pi^4 \bar{m}^4 a^7(R)}{15 G c^2 h^3 \bar{\kappa}_{\rm R}(R)} \right)^{1/3}. \tag{4.9}$$

We shall assume that in cool parts of the disc, the opacity is dominated by ice-coated grains with emission efficiency proportional to frequency squared. Hence, the Rosseland mean opacity is proportional to the temperature squared, or the sound speed to the fourth power. For standard interstellar dust, this gives $\bar{\kappa}_{\rm R} \simeq \kappa_{\rm O} a^4$ with $\kappa_{\rm O} \simeq 3 \times 10^{-19}\,{\rm s^4\,cm^{-2}\,g^{-1}}$. With this expression for $\bar{\kappa}_{\rm R}$, Eqn. (4.9) becomes

$$\Sigma(R) \lesssim \left(\frac{2^3 \pi^4 \bar{m}^4}{15 G c^2 h^3 \kappa_{\rm O}} \right)^{1/3} a(R). \tag{4.10}$$

Since the lower limit on Σ in Equation (4.6) and the upper limit in Equation (4.10) are both linear in a, when we combine them, the dependence on a drops out, and we are left with a constraint on Ω:

$$\Omega(R) \lesssim \left(\frac{2^3 \pi^7 G^3 \bar{m}^4}{15 c^2 h^3 \kappa_{\rm O}} \right)^{1/3}, \tag{4.11}$$

or equivalently (for a Keplerian disc)

$$R \gtrsim R_{\rm min}(M_{\rm ur}) \simeq \left(\frac{15^2 c^4 h^6 \kappa_{\rm O}^2 M_{\rm ur}^3}{2^6 \pi^{14} G \bar{m}^8} \right)^{1/9} \simeq 100\,{\rm AU} \left(\frac{M_{\rm ur}}{{\rm M}_\odot} \right)^{1/3}. \tag{4.12}$$

This is a rather robust explanation for the brown-dwarf desert. Only the outer parts of a circumstellar disc can be both massive enough, and at the same time able to cool fast enough, to fragment. Moreover, the masses of the secondaries that form there are typically in the range 0.003 to $0.1\,{\rm M}_\odot$ (see Whitworth & Stamatellos 2006 for details). Parenthetically, we note that the dependence on the dust emissivity is very weak: first, $R_{\rm min} \propto \kappa_{\rm O}^{2/9}$; second, even if the dependence on a does not cancel out exactly (i.e., $\bar{\kappa}_{\rm R}$ is not $\propto a^4$), the dependence of a on R is so weak compared with that of Ω (typically, $a \propto R^{-1/4}$ as opposed to $\Omega \propto R^{-3/2}$) that the result is unlikely to be changed by much. Thus, even if the dust properties in a protostellar disc evolve away from those in the general interstellar medium, this is unlikely to have a large effect.

We note that to form a disc of this extent requires the material trying to accrete onto the ur star to have angular momentum

$$h \gtrsim h_{\rm min} \simeq 3 \times 10^{20}\,{\rm cm^2\,s^{-1}} \left(\frac{M_{\rm ur}}{\rm M_\odot}\right)^{2/3} \equiv \left(0.1\,{\rm km\,s^{-1}}\right)(0.01\,{\rm pc}) \left(\frac{M_{\rm ur}}{\rm M_\odot}\right)^{2/3}. \quad (4.13)$$

Alternatively, the original core must have

$$\beta \equiv \frac{U_{\rm rot}}{|U_{\rm grav}|} \gtrsim 0.005 \left(\frac{M_{\rm ur}}{\rm M_\odot}\right)^{4/3}. \quad (4.14)$$

These are very modest requirements compared with the observed properties of prestellar cores.

We have simulated disc fragmentation using smoothed-particle hydrodynamics, solving the energy equation with a scheme that treats realistically the equation of state of protostellar gas and the transport of cooling radiation against dust opacity (for details see Stamatellos *et al.* 2007), and introducing sink particles only at very high densities ($\gtrsim 10^{-3}\,{\rm g\,cm^{-3}}$; Stamatellos *et al.* 2007; Stamatellos & Whitworth 2009a,b). We find that fragmentation is limited to the outer parts of a protostellar disc, and forms multiple secondaries with masses in the range 0.003 to $0.200\,{\rm M_\odot}$. Dynamical interactions amongst these secondaries have two main effects. First, a significant fraction end up in tight ($a \sim 1$ to 2 AU), low-mass ($M_1 + M_2 \lesssim 0.2\,{\rm M_\odot}$) binary systems. Second, the higher-mass secondaries tend to be scattered or migrate into tighter orbits (relative to the ur star), whilst the lower-mass secondaries tend to be liberated into the field. And the intermediate-mass secondaries remain at large distances from the ur star. Thus, the properties of the brown-dwarf desert are reproduced. In addition, (i) both bound and

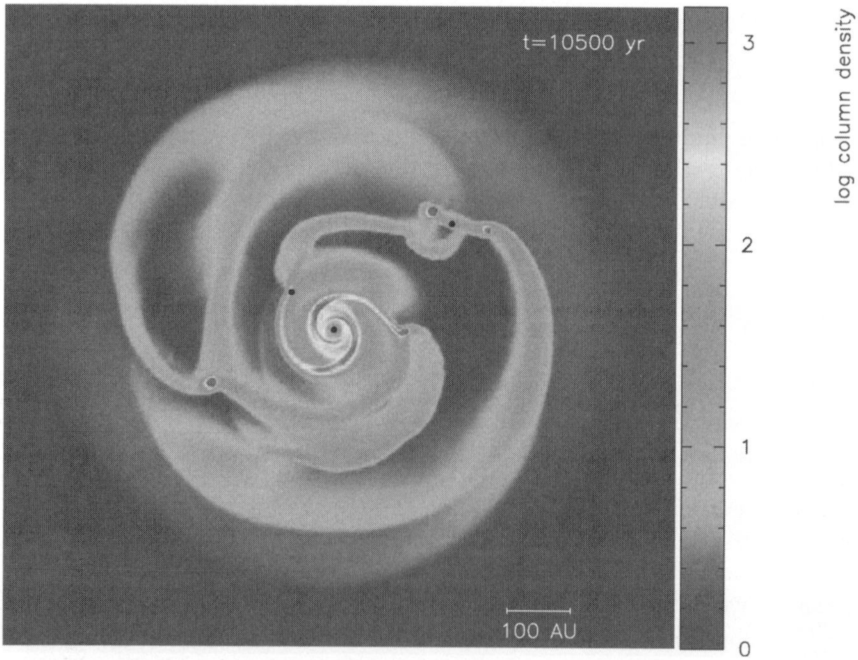

Figure 1. Simulation of the fragmentation of an $0.7\,{\rm M_\odot}$ disc round an $0.7\,{\rm M_\odot}$ star. Note that no fragmentation takes place in the inner 70 AU of the disc, although some low-mass secondaries are subsequently scattered into the inner region. Two low-mass secondaries have already formed, and three or four more are about to form.

liberated brown dwarfs retain discs, with masses $\lesssim 0.01\,\mathrm{M}_\odot$ and radii $\lesssim 40\,\mathrm{AU}$, (ii) brown dwarfs that remain bound to the ur star are about four times more likely to be in a tight binary system with another brown dwarf than those which are liberated, (iii) BD/BD binaries tend to have high mass ratios ($\bar{q} \sim 0.85$), (iv) very few planemos remain bound to the ur star, but some end up in tight binary systems with brown dwarfs and (v) wide BD/BD binaries ($a \sim 200\,\mathrm{AU}$) are also formed.

5. Conclusions

Disc fragmentation and the subsequent eigen evolution of the resulting small-N system explain (a) the brown-dwarf desert, (b) the statistics of BD/BD and BD/planemo binaries (frequency, location and orbital parameters), (c) the properties of discs around young brown dwarfs and (d) the existence of free-floating planemos. This scenario also predicts (e) the low velocity dispersion of brown dwarfs and hence their spatial distribution in clusters and (f) that planets cannot form on a dynamical timescale, even with convective cooling. In addition, disc fragmentation has the advantage that many of the statistical properties of brown dwarfs and low-mass hydrogen-burning stars are determined by the eigen evolution of a small-N system formed within a single protostellar core, so they are largely independent of the environment or the overall dynamics of the parent cluster.

References

André, P., Belloche, A., Motte, F., & Peretto, N. 2007, *A&A*, 472, 519
Bate, M. R., Bonnell, I. A., & Bromm, V. 2003, *MNRAS*, 339, 577
Burgasser, A. J., Kirkpatrick, J. D., & Lowrance, P. J. 2005, *AJ*, 129, 2849
Burgasser, A. J., Reid, I. N., Siegler, N., Close, L., Allen, P., Lowrance, P. J., & Gizis, J. 2007, in: B. Reipurth, D. Jewitt & K. Keil (eds.), *Protostars & Planets V*, p. 427, Tucson: University of Arizona Press
Chabrier, G. 2003, *PASP*, 115, 763
Gammie, C. A. 2001, *ApJ*, 553, 174
Goldreich, P. & Lynden–Bell, D. 1965, *MNRAS*, 130, 97
Goodwin, S. P., Whitworth, A. P., & Ward–Thompson, D. 2004, *A&A*, 414, 633
Hennebelle, P. & Chabrier, G. 2008, *ApJ*, 684, 395
Hester, J. J., *et al.* 1996, *AJ*, 111, 2349
Klessen, R. & Burkert, A. 2002, *ApJS*, 128, 287
Kroupa, P. 2002, *Science*, 295, 82
Luhman, K., Joergens, V., Lada, C., Muzerolle, J., Pascucci, I., & White, R. 2007, in: B. Reipurth, D. Jewitt & K. Keil (eds.), *Protostars & Planets V*, p. 443, Tucson: University of Arizona Press
Padoan, P. & Nordlund, Å 2002, *ApJ*, 576, 870
Reipurth, B. & Clarke, C. J. 2001, *AJ*, 122, 432
Stamatellos, D., Whitworth, A. P., Bisbas, T., & Goodwin, S. P. 2007, *A&A*, 475, 37
Stamatellos, D., Hubber, D. A., & Whitworth, A. P. 2007, *MNRAS* (Letters), 382, L30
Stamatellos, D. & Whitworth, A. P. 2009a, *MNRAS*, 392, 413
Stamatellos, D. & Whitworth, A. P. 2009b, *MNRAS*, in press (arXiv:0908.2247)
Toomre, A. 1964, *ApJ*, 139, 1217
Whitworth, A. P., Bate, M. R., Nordlund, Å., Reipurth, B., & Zinnecker, H. 2007, in: B. Reipurth, D. Jewitt & K. Keil (eds.), *Protostars & Planets V*, p. 459, Tucson: University of Arizona Press
Whitworth, A. P. & Stamatellos, D. 2006, *A&A*, 458, 817
Whitworth, A. P. & Zinnecker, H. 2004, *A&A*, 427, 299

Star clusters: basic galactic building blocks
Proceedings IAU Symposium No. 266, 2009
R. de Grijs & J. R. D. Lépine, eds.

doi:10.1017/S1743921309991141

On the origin of the hypervelocity runaway star HD 271791

V. V. Gvaramadze

Sternberg Astronomical Institute, Moscow State University, Universitetskij Pr. 13, Moscow 119992, Russia
email: vgvaram@mx.iki.rssi.ru

Abstract. We discuss the origin of the early-B-type runaway star HD 271791 and show that its extremely high velocity ($\simeq$530 − 920 km s^{-1}) cannot be explained within the framework of the binary–supernova ejection scenario. Instead, we suggest that HD 271791 attained its peculiar velocity in the course of a strong dynamical encounter between two hard, massive binaries or through an exchange encounter between a hard, massive binary and a very massive star, formed through runaway mergers of ordinary massive stars in the dense core of a young massive star cluster.

Keywords. binaries: close, stars: individual (HD 271791), stars: kinematics, galaxies: star clusters

1. Introduction

HD 271791 is a massive (11 ± 1 M$_\odot$; Przybilla *et al.* 2008), extremely high-velocity runaway star with a Galactic rest-frame velocity of $\simeq$530 − 920 km s^{-1} (Heber *et al.* 2008), typical of the so-called hypervelocity stars (HVSs), i.e., the ordinary stars moving with peculiar velocities exceeding the escape velocity of our Galaxy (Brown *et al.* 2005; Edelmann *et al.* 2005; Hirsch *et al.* 2005). The existence of HVSs was predicted by Hills (1988), who showed that a close encounter between a tight binary system and the supermassive black hole (BH) in the Galactic Centre could be responsible for ejection of one of the binary components with a velocity of up to several 1000 km s^{-1}. Yu & Tremaine (2003) proposed an additional mechanism for HVS production based on the interaction between a single star and a putative binary BH in the Galactic Centre. It is therefore plausible that some HVSs were produced in that way (Gualandris *et al.* 2005; Baumgardt *et al.* 2006; Levin 2006; Sesana *et al.* 2006; Ginsburg & Loeb 2006; Lu *et al.* 2007; Löckmann & Baumgardt 2008). At present, however, proper-motion measurements are not available for all but one of the known HVSs, so that it is impossible to unambiguously associate their birthplace with the Galactic Centre. HD 271791 is the only HVS with measured proper motion and all measurements show that this star was ejected from the periphery of the Galactic disc (at a galactocentric distance of $\geqslant$15 kpc; Heber *et al.* 2008).

There are two possible alternative explanations for the origin of HVSs. The first suggests that the HVSs attain their high peculiar velocities in the course of strong dynamical three- or four-body encounters in young and dense star clusters located in the Galactic disc (Gvaramadze 2006a, 2007; Gvaramadze *et al.* 2008, 2009) or in the Large Magellanic Cloud (Gualandris & Portegies Zwart 2007). The second was proposed by Abadi *et al.* (2009). According to these authors, some HVSs could originate from tidal disruption of dwarf galaxies during their close passage near the Milky Way. The young age of HD 271791, 25 ± 5 Myr (Przybilla *et al.* 2008), is inconsistent with the second possibility since there are no indications of a recent encounter between a dwarf satellite and the Milky Way. So, we are left with the first possibility. Before discussing it in Section 4, we consider a proposal by Przybilla *et al.* (2008) that HD 271791 attained its peculiar velocity in the course of disintegration of a close massive binary system following a supernova (SN) explosion.

2. HD 271791: former secondary in a massive binary system

The spectral analysis of HD 271791 by Przybilla *et al.* (2008) revealed that the Fe abundance in its atmosphere is subsolar and that the α-process elements are enhanced. The first finding is consistent with the origin of HD 271791 in the metal-poor outskirts of the Galactic disc, while the second suggests that this star was a secondary component of a massive tight binary and that its surface was polluted by nucleosynthetic products after the primary star exploded as a SN. Przybilla *et al.* (2008) believe that the binary–SN explosion could be responsible not only for the α enhancement in HD 271791 but also for the extremely high space velocity of the star. Below, we outline their scenario.

The large separation of HD 271791 from the Galactic plane ($\simeq$10 kpc), along with the proper-motion measurements (Heber *et al.* 2008), implies that the time of flight of this B2III star is comparable to its evolutionary age, which in turn implies that the star was ejected within several Myr after its birth in the Galactic disc. The ejection event was connected with disruption of a massive tight binary following the SN explosion. The original binary was composed of a primary star of mass of $\gtrsim 55\,M_\odot$ and an early-B-type secondary (HD 271791), so that the SN explosion and binary disruption occurred early in the lifetime of HD 271791. The system was close enough to go through the common-envelope phase before the primary exploded as a SN. During the common-envelope phase, the primary star lost most of its hydrogen envelope and the binary became a tight system composed of a Wolf–Rayet star and an early-B-type main-sequence star. At the moment of SN explosion, the mass of the primary star was $\leqslant 20\,M_\odot$ and the binary semimajor axis was $\sim 14\,R_\odot$ (corresponding to an orbital velocity of the secondary of $\leqslant 420\,\mathrm{km\,s^{-1}}$). The exploded star expelled $\simeq 10\,M_\odot$ of its mass while the remaining mass collapsed to a $\leqslant 10\,M_\odot$ BH. The SN explosion was asymmetric enough to disrupt the system. Przybilla *et al.* (2008) assumed that HD 271791 was released at its orbital velocity and that, at the time of binary disruption, the vector of the orbital velocity was directed by chance along the Galactic-rotation direction. The first assumption is based on the widespread erroneous belief that runaways produced from a SN in a binary system have peculiar velocities comparable to their pre-SN orbital velocities. The second assumption is required to explain the difference between the assumed space velocity from the binary disruption and the Galactic rest-frame velocity of HD 271791 (provided that the latter is on the low end of the observed range, $530 - 920\,\mathrm{km\,s^{-1}}$).

In the next section, we discuss the conditions under which the secondary star could be launched into free flight at a velocity equal to its pre-SN orbital value.

3. Binary–supernova ejection scenario

One of two basic mechanisms producing runaway stars is based on a SN explosion in a massive, tight binary system (Blaauw 1961). After the primary star exploded as a SN, the binary system could disintegrate if the system lost more than half of its pre-SN mass (Boersma 1961) and/or the SN explosion was asymmetric (so that the stellar remnant, either a neutron star, NS, or a BH, received at birth a kick velocity exceeding the escape velocity from the system; Stone 1982; Tauris & Takens 1998).

In the case of binary disruption following a symmetric SN explosion, the stellar remnant is released at its orbital velocity, while the space velocity of the secondary star, $V_{\rm sec}$, is given by (Boersma 1961; Radhakrishnan & Shukre 1985; Tauris & Takens 1998)

$$V_{\rm sec} = \sqrt{1 - 2\frac{m_1 + m_2}{m_1^2}}\, V_{\rm orb}, \tag{3.1}$$

where $m_1 > 2 + m_2, m_1 = M_1/M_{\rm co}, m_2 = M_2/M_{\rm co}$, M_1 and M_2 are the pre-SN masses of the primary and the secondary stars, $M_{\rm co}$ is the mass of the compact object formed in the SN explosion, $V_{\rm orb} = [GM_1^2/(M_1 + M_2)a]^{1/2}$ is the orbital velocity of the secondary star, G is the gravitational constant and a is the binary semimajor axis. It follows from Equation (2.1) that $V_{\rm sec} \simeq V_{\rm orb} \simeq (GM_1/a)^{1/2}$ if $M_1 \gg M_2, M_{\rm co}$.

The above consideration shows that the secondary star could achieve a high peculiar speed if one adopts a large pre-SN mass for the primary (i.e., $M_1 \gg M_{\rm co}$). However, stellar evolutionary models suggest that the pre-SN masses of stars with initial masses, from 12 to 120 $M_\odot$ do not exceed $\sim$10 – 17 $M_\odot$ (Vanbeveren *et al.* 1998; Meynet & Maeder 2003). Using these figures and assuming that the pre-SN binary is as tight as possible (i.e., the secondary main-sequence star is close to filling its Roche lobe), one can estimate the maximum possible velocity achieved by a runaway star in the process of binary disruption following a symmetric SN explosion. Assuming that the SN explosion left behind a NS (i.e., $M_{\rm co} = 1.4\,M_\odot$) and adopting $M_1 = 10 - 17\,M_\odot$, one finds that a

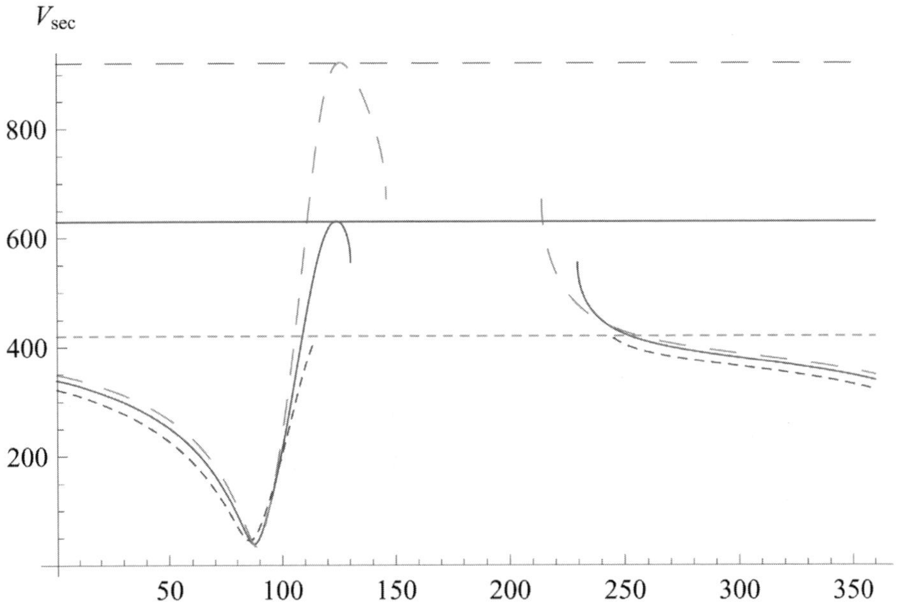

Figure 1. Peculiar velocity of the secondary star (HD 271791) as a function of the angle, θ, between the kick vector and the direction of motion of the exploding star and the magnitude of the kick [750 km s^{-1} (short-dashed line), 1000 km s^{-1} (solid line), 1200 km s^{-1} (long-dashed line)]. The discontinuities in the curves correspond to a range of angles θ for which the system remains bound. The horizontal solid line indicates the Galactic rest-frame velocity of HD 271791 of 630 km s^{-1}, corresponding to the 'best' proper motion given in Heber *et al.* (2008). The horizontal short- and long-dashed lines indicate, respectively, the orbital velocity of HD 271791 of 420 km s^{-1} [suggested by the scenario of Przybilla *et al.* (2008)] and the maximum possible Galactic rest-frame velocity of HD 271791 of 920 km s^{-1} (Heber *et al.* 2008). See text for details.

$3\,M_\odot$ secondary star could attain a peculiar velocity of $\simeq 300-500\,\mathrm{km\,s^{-1}}$, while a $10\,M_\odot$ star could be ejected with a speed of $\simeq 350\,\mathrm{km\,s^{-1}}$.

Note that the pre-SN mass is maximum for stars with initial mass $\simeq 20-25\,M_\odot$ and $\geqslant 80\,M_\odot$ [see figure 6 of Meynet & Maeder (2003)]. In the first case, the SN explosion leaves behind a NS, while in the second the stellar SN remnant is a BH of mass $M_{co} \geqslant 5\,M_\odot$ (e.g., Woosley *et al.* 2002). The large separation of HD 271791 from the Galactic plane implies that this massive star was ejected very soon after its birth in the Galactic disc. From this, it follows that to explain the space velocity of HD 271791 within the framework of the binary–SN scenario, one should assume that the primary was a short-lived very massive star (Przybilla *et al.* 2008). In this case, the stellar SN remnant is a BH and the SN ejecta are not massive enough to cause disruption of the binary system.

According to Przybilla *et al.* (2008), the pre-SN mass of the primary star was $\leqslant 20\,M_\odot$ (i.e., somewhat higher than the maximum mass predicted by stellar evolutionary models; see above) and the SN explosion left behind a BH of mass $\leqslant 10\,M_\odot$ (comparable to the mass of the secondary star, HD 271791), i.e., the system lost less than a half of its mass. Thus, to disrupt the binary, the SN explosion should be asymmetric. In this case, the space velocities of the BH and HD 271791 depend on the magnitude and direction of the kick imparted to the BH at birth (Tauris & Takens 1998). To estimate V_{sec}, one can use equations (44)–(47) and (54)–(56) of Tauris & Takens (1998). It follows from these equations that V_{sec} is maximal if the vector of the kick velocity does not strongly deviate from the orbital plane of the binary and is directed nearly towards the secondary, i.e., the angle, θ, between the kick vector and the direction of motion of the exploding star is $\sim\theta_* = \arccos(-v/w)$, where $v = [G(M_1 + M_2)/a]^{1/2}$ is the relative orbital velocity and w is the kick velocity (see Gvaramadze 2006b).

Figure 1 shows how the direction and the magnitude of the kick affect V_{sec}. The three solid lines represent V_{sec} calculated for the binary parameters suggested by Przybilla *et al.* (2008) and three kick magnitudes of $750\,\mathrm{km\,s^{-1}}$ (short-dashed), $1000\,\mathrm{km\,s^{-1}}$ (solid line) and $1200\,\mathrm{km\,s^{-1}}$ (long-dashed line). One can see that to launch HD 271791 at its pre-SN orbital velocity $V_{orb} \simeq 420\,\mathrm{km\,s^{-1}}$, the kick imparted to the BH should be at least as large as $750\,\mathrm{km\,s^{-1}}$. In fact, the kick magnitude should be much larger since for kicks $\simeq 750\,\mathrm{km\,s^{-1}}$ the kick direction must be carefully tuned (see Figure 1), i.e., θ should be either $\simeq 115°$ or $\simeq 245°$ (note that for $115° \leqslant \theta \leqslant 245°$ the binary system remains bound). The even larger kicks of $\gtrsim 1000$ and $\gtrsim 1200\,\mathrm{km\,s^{-1}}$ are required to explain the Galactic rest-frame velocities of HD 271791 of 630 and $920\,\mathrm{km\,s^{-1}}$ [corresponding, respectively, to the 'best' and the maximum proper motions given in Heber *et al.* (2008); see also Przybilla *et al.* (2008)]. Although one cannot exclude the possibility that BHs attain a kick at birth, we note that there is no evidence that the kick magnitude could be as large as required by the above considerations (e.g., Nelemans *et al.* 1999).

Thus, we found that to explain the peculiar velocity of HD 271791 the magnitude of the kick attained by the $\leqslant 10\,M_\odot$ BH must be unrealistically large ($\gtrsim 750\,\mathrm{km\,s^{-1}}$), which makes the binary–SN ejection scenario highly unlikely (cf. Gvaramadze 2007; Gvaramadze & Bomans 2008). Nevertheless, some authors still believe that "HD 271791 is a plausible runaway star produced by a supernova explosion in a massive binary system" (Bromley *et al.* 2009).

4. Dynamical-ejection scenario

The second basic mechanism responsible for the origin of runaway stars is based on dynamical three- or four-body interactions in dense stellar systems (Poveda *et al.* 1967;

Aarseth 1974; Gies & Bolton 1986). Below, we discuss two possible channels for producing high-velocity runaways within the framework of the dynamical-ejection scenario (cf. Gvaramadze 2009).

The first possibility is that the high-velocity runaways originate from the interaction between two massive, hard binaries (Mikkola 1983; Leonard & Duncan 1990). The runaways produced in binary–binary encounters are frequently ejected at velocities comparable to the orbital velocities of the binary components (Leonard & Duncan 1990) and occasionally they can attain a velocity as high as the escape velocity, $V_{\rm esc} = (2GM_*/R_*)^{-0.5}$, from the surface of the most massive star in the binaries (Leonard 1991). For upper-main-sequence stars with the mass–radius relationship (Habets & Heintze 1981), $R_* = 0.8(M_*/{\rm M}_\odot)^{0.7}\,{\rm R}_\odot$, where R_* and M_* are the stellar radius and mass, respectively, one has $V_{\rm esc} \simeq 700\,{\rm km\,s}^{-1}(M_*/{\rm M}_\odot)^{0.15}$, so that the ejection velocity could, in principle, be as large as $\simeq 1100 - 1200\,{\rm km\,s}^{-1}$ if the binaries contain at least one star of mass of $20 - 40\,{\rm M}_\odot$.

To reconcile this ejection scenario with the presence of nucleosynthetic products in the atmosphere of HD 271791, one should assume that (i) HD 271791 was a secondary component of one of the binaries involved in the encounter and (ii) by the moment of the encounter, the binary containing HD 271791 has experienced SN explosion and remained bound [i.e., the stellar SN remnant (BH) received a small or no kick at birth]. The requirement that HD 271791 was a member of a post-SN binary should also be fulfilled in the second dynamical process discussed below.

The second possibility is that the high-velocity runaway stars attain their peculiar velocities in the course of close encounters between massive, hard binaries and a very massive star (Gvaramadze 2007; Gvaramadze *et al.* 2009), formed through runaway collisions of ordinary massive stars in dense star clusters (e.g., Portegies Zwart *et al.* 1999). A close encounter with the very massive star results in tidal breakup of the binary, after which one of the binary components becomes bound to the very massive star while the second recoils with a high velocity, given by (Hills 1988)

$$V_\infty \simeq 500\,{\rm km\,s}^{-1}\,(M_{\rm VMS}/100\,{\rm M}_\odot)^{1/6}(a'/30\,{\rm R}_\odot)^{-1/2}(M_1/10\,{\rm M}_\odot)^{1/3}, \qquad (4.1)$$

where $M_{\rm VMS}$ is the mass of the very massive star and a' is the post-SN binary semimajor axis. It follows from Equation (4.1) that, to explain the peculiar velocity of HD 271791 of $\simeq 400 - 600\,{\rm km\,s}^{-1}$, the mass of the very massive star should be $\gtrsim 100 - 300\,{\rm M}_\odot$ [the first figure corresponds to the mass of the most massive star formed in a 'normal' way in a cluster with a mass $M_{\rm cl} \simeq 10^4\,{\rm M}_\odot$ (Weidner *et al.* 2009)]†. The above estimates can be supported by the results of three-body scattering experiments, which showed that $\gtrsim 3$ percent of encounters between hard, massive binaries and a very massive star of mass of $200 - 300\,{\rm M}_\odot$ produce runaways with $V_\infty \gtrsim 500 - 600\,{\rm km\,s}^{-1}$ (Gvaramadze *et al.* 2009).

Note that the requirement that HD 271791 was a member of a post-SN binary does not contradict our proposal that this star can attain its high speed via a three-body encounter with a very massive star (i.e., with the star more massive than the primary star in the original binary). Merging of ordinary stars results in effective rejuvenation of the collision product (e.g., Meurs & van den Heuvel 1989) so that the very massive star could still be on the main sequence when the most massive ordinary stars start to explode as SNe (e.g., Portegies Zwart *et al.* 1999).

† Simple estimates show that our Galaxy can currently host about 100 star clusters with masses $M_{\rm cl} \gtrsim 10^4\,{\rm M}_\odot$ (Gvaramadze *et al.* 2008).

Acknowledgements

I am grateful to L. R. Yungelson for useful discussions, to the Russian Foundation for Basic Research and the International Astronomical Union for travel grants and to the Deutsche Forschungsgemeinschaft for partial financial support.

References

Aarseth S. J. 1974, *A&A*, 35, 237
Abadi, M. G., Navarro, J. F., & Steinmetz, M. 2009, *ApJ* (Letters), 691, L63
Baumgardt, H., Gualandris, A., & Portegies Zwart, S. 2006, *MNRAS*, 372, 174
Blaauw, A. 1961, *Bull. Astron. Inst. Netherlands*, 15, 265
Boersma, J., 1961, *Bull. Astron. Inst. Netherlands*, 15, 291
Bromley, B. C., Kenyon, S. J., Brown, W. R., & Geller, M. J. 2009, *ApJ*, 706, 925
Brown W. R., Geller M. J., Kenyon S. J., & Kurtz M. J. 2005, *ApJ* (Letters), 622, L33
Edelmann, H., Napiwotzki, R., Heber, U., Christlieb, N., & Reimers, D. 2005, *ApJ* (Letters), 634, L181
Gies, D. R. & Bolton, C. T. 1986, *ApJS*, 61, 419
Ginsburg, I. & Loeb, A. 2006, *MNRAS*, 368, 221
Gualandris, A. & Portegies Zwart, S. 2007, *MNRAS* (Letters), 376, L29
Gualandris, A., Portegies Zwart, S., & Sipior, M. S. 2005, *MNRAS*, 363, 223
Gvaramadze, V. V. 2006a, in *On the Present and Future of Pulsar Astronomy*, IAU JD06 (IAU 26th General Assembly), # 25
Gvaramadze, V. V. 2006b, *A&A*, 454, 239
Gvaramadze, V. V. 2007, *A&A* (Letters), 470, L9
Gvaramadze, V. V. 2009, *MNRAS* (Letters), 395, L85
Gvaramadze, V. V. & Bomans, D.J. 2008, *A&A*, 485, L29
Gvaramadze, V. V., Gualandris, A., & Portegies Zwart, S. 2008, *MNRAS*, 385, 929
Gvaramadze, V.V., Gualandris, A., & Portegies Zwart, S. 2009, *MNRAS*, 396, 570
Habets, G. M. H. J., & Heintze, J. R. W. 1981, *A&AS*, 46, 193
Heber, U., Edelmann, H., Napiwotzki, R., Altmann, M., & Scholz, R.-D. 2008, *A&A*, 483, L21
Hills, J.G. 1988, *Nature*, 331, 687
Hirsch, H. A., Heber, U., O'Toole, S. J., & Bresolin, F. 2005, *A&A* (Letters), 444, L61
Leonard, P. J. T. 1991, *AJ*, 101, 562
Leonard, P. J. T., & Duncan, M. J. 1990, *AJ*, 99, 608
Levin, Y. 2006, *ApJ*, 653, 1203
Löckmann, U. & Baumgardt, H. 2008, *MNRAS*, 384, 323
Lu, Y., Yu, Q., Lin, D. N. C. 2007, *ApJ* (Letters), 666, L89
Meurs, E. J. A., van den Heuvel, E. P. J. 1989, *A&A*, 226, 88
Meynet, G. & Maeder, A. 2003, *A&A*, 404, 975
Mikkola, S. 1983, *MNRAS*, 203, 1107
Nelemans, G., Tauris, T. M., & van den Heuvel, E. P. J. 1999, *A&A* (Letters), 352, L87
Portegies Zwart, S. F., Makino, J., McMillan, S. L. W., & Hut, P. 1999, *A&A*, 348, 117
Poveda, A., Ruiz, J., & Allen, C. 1967, *Bol. Obs. Tonantzintla Tacubaya*, 4, 86
Przybilla, N., Nieva, M. F., Heber, U., & Butler, K. 2008, *ApJ* (Letters), 684, L103
Radhakrishnan, V. & Shukre, C. S. 1985, in: G. Srinivasan & V. Radhakrishnan (eds.), *Supernovae, Their Progenitors and Remnants*, p. 155, Bangalore: Indian Academy of Sciences
Sesana, A., Haardt, F., & Madau, P. 2006, *ApJ*, 651, 392
Stone R. C. 1982, *AJ*, 87, 90
Tauris, T. M. & Takens, R. J. 1998, *A&A*, 330, 1047
Vanbeveren, D., De Loore, C., & Van Rensbergen, W. 1998, *A&A Rev.*, 9, 63
Weidner, C., Kroupa, P., & Bonnell, I. 2009, *MNRAS*, in press (arXiv:0909.1555)
Woosley, S. E., Heger, A., & Weaver, T. A. 2002, *Rev. Mod. Phys.*, 74, 1015
Yu, Q. & Tremaine, S. 2003, *ApJ*, 599, 1129

Session VII

Star clusters as laboratories for stellar evolution

Star clusters: basic galactic building blocks
Proceedings IAU Symposium No. 266, 2009
R. de Grijs & J. R. D. Lépine, eds.

doi:10.1017/S1743921309991153

Globular clusters as laboratories for stellar evolution

Márcio Catelan,[1] Aldo A. R. Valcarce,[1] and Allen V. Sweigart[2]

[1]Pontificia Universidad Católica de Chile, Departamento de Astronomía y Astrofísica, Av. Vicuña Mackenna 4860, 782-0436 Macul, Santiago, Chile
email: [mcatelan, avalcarc]@astro.puc.cl

[2]NASA Goddard Space Flight Center, Exploration of the Universe Division, Code 667, Greenbelt, MD 20771, USA
email: allen.v.sweigart@nasa.gov

Abstract. Globular clusters have long been considered the closest approximation to a physicist's laboratory in astrophysics, and as such a near-ideal laboratory for (low-mass) stellar evolution. However, recent observations have cast a shadow on this long-standing paradigm, suggesting the presence of multiple populations with widely different abundance patterns, and—crucially—with widely different helium abundances as well. In this review we discuss which features of the Hertzsprung–Russel diagram may be used as helium-abundance indicators, and present an overview of available constraints on the helium abundance in globular clusters.

Keywords. Hertzsprung–Russell diagram, stars: abundances, stars: evolution, stars: Population II, globular clusters: general

1. Introduction

In the words of Moehler (2001), "globular clusters [GCs] are the closest approximation to a physicist's laboratory in astrophysics." Not only do these spheroidal stellar agglomerations contain up to several million stars, all at the same distance from us, but their stars have also been thought to have all been born at the same instant, from a cloud with homogeneous chemical composition. According to this canonical paradigm, GCs represent the best examples of a simple stellar population. As a result, GCs have been used extensively to place constraints on key ingredients of canonical stellar evolution models, such as the mixing-length parameter of convective energy transport theory (e.g., Palmieri *et al.* 2002; Ferraro *et al.* 2006). In addition, GC studies have played an important role in the field of particle physics, with measured color–magnitude-diagram (CMD) properties being used to place some of the strongest constraints available on several dark-matter candidates and other particle-physics parameters that play a key role in the physics beyond the so-called 'standard model' of particle physics, including, e.g., the neutrino magnetic moment, the mass of the axion and the cross sections of weakly interacting massive particles (e.g., Raffelt 1996, 2000, 2008; and references therein).

However, recent observations suggest that GCs may not play as reliable a role as astrophysical laboratories as previously believed. For one, large variations in some light-element abundances, particularly O, Na, Al and Mg (but, importantly, not in the iron-peak or other α-capture elements), have been found in virtually all GCs for which suitable spectroscopic data have been obtained (e.g., Carretta *et al.* 2009; and references therein), unlike what is seen among metal-poor field stars (Gratton *et al.* 2000). For another, recent *Hubble Space Telescope* observations (e.g., Piotto 2009; and references therein) have revealed the presence of multiple stellar populations in at least some of the more massive GCs in our Galaxy, including multimodal main sequences (MSs) and subgiant branches (SGBs), possibly connected with multimodal horizontal branches (HBs), the latter having been known since much earlier (see Catelan 2008 for a recent review).

While at least some GC CMDs still reveal exquisitely tight sequences down to the bottom of the MS, with no evidence of multiple populations (or even for sizeable samples of binary stars; Davis *et al.* 2008), the fact that several GCs show split sequences strongly suggests that GCs do not all represent the simple stellar populations that they were once thought to. Instead, they appear to be chemically complex entities whose stars did not form all at the same time, undergoing instead several bursts of star formation which progressively contaminate the medium before the formation of the next stellar generation in the cluster.

Of particular importance, in this regard, is the possibility that each stellar generation may increase the helium content, Y, of the medium before the next generation forms. Helium is the second most abundant element in the Universe, and as such, changes in Y may dramatically change the stellar evolutionary paths and associated timescales. In this sense, while helium enrichment is a consistent prediction of different scenarios for the abundance variations that are seen in GCs, the amount of He enhancement differs markedly from one model to the next (e.g., Marcolini *et al.* 2009). Importantly, while the observed MS splits in some GCs indicate very large levels of He enrichment, with the He abundance in the most extreme populations reaching values as high as about 40% by mass (Norris 2004; D'Antona *et al.* 2005; Piotto *et al.* 2005, 2007), it is not straightforward to produce such large levels of He enrichment on the basis of available chemical-enrichment scenarios. In this sense, and given that He enhancements among GC stars have now been suggested to be the rule rather than the exception (with some GCs—those with purely blue HBs—possibly lacking first-generation, non-He-enriched stars altogether; e.g., D'Antona & Caloi 2008), it is important to check all possible signatures of He enhancement, based on the observed properties of GC stars, to constrain the He-enrichment scenario. This is a crucial task to help establish the extent to which at least some GCs may still be safely used as laboratories of low-mass stellar evolution. On the other hand, the heavily contaminated GCs will continue to play an important role, although as laboratories of the evolution (and ejecta produced by) more massive stars.

Since He-abundance measurements are only possible for hot stars, which are generally lacking in GCs, most of the available Y estimates use indirect techniques, based on their CMD properties. In this sense, in the next section we first review empirical determinations of the He abundance in GC stars and then discuss the impact of He-abundance variations among several GC observables, based on a new set of evolutionary tracks for canonical and He-enhanced compositions (Valcarce 2010), along with the results that have been obtained using some of these observables. Finally, in Section 3, we present our conclusions.

2. Helium-abundance measurements for GCs

2.1. *Direct methods*

Most GCs lack sufficiently hot stars for direct He-abundance measurements.† Even for those that contain sufficiently hot stars, primarily on the blue HB, the results of direct He-abundance measurements can be strongly affected by diffusion effects, which can dramatically lower the photospheric He abundance for HB stars hotter than 11 500 K (e.g., Behr 2003; Moehler *et al.* 2003).‡ Conversely, at the hot end of the HB, very high

† To be sure, a He I line is found in near-infrared spectra of cool red giants at 10 830 Å, but this is due to chromospheric emission and is thus a much better indicator of mass loss than of He abundance (e.g., Dupree *et al.* 2009; and references therein).

‡ This corresponds to the so-called 'Grundahl jump' (Grundahl *et al.* 1999).

photospheric He abundances have indeed been measured, which several authors have interpreted in terms of the 'late-flasher' scenario for the origin of these hot stars. In this scenario, stars that lose too much mass while on the red-giant branch (RGB) may ignite He not at the RGB tip, but rather as they evolve towards the white-dwarf region of the CMD, undergoing extensive mixing in the process. Their photospheres may thus become dramatically enriched in He, to levels that can be much higher than even the highest levels that have been suggested under the 'primordial' (multiple-population) scenario (e.g., Moehler *et al.* 2004, 2007; Cassisi *et al.* 2009; and references therein).

This leaves us with blue HB stars situated between the blue edge of the instability strip, at around 7200 K, and the Grundahl jump, at 11 500 K. Very recently, Villanova *et al.* (2009) carried out the first detailed spectroscopic He-abundance measurement for HB stars in this temperature range. Based on *VLT*/UVES spectra of cool blue HB stars in NGC 6752, they found that "... all our targets... have a homogeneous He content with a mean value $Y = 0.245 \pm 0.012$, compatible with the most recent measurements of the primordial He content of the Universe." Note that NGC 6752, as a cluster whose HB is comprised solely of blue HB stars, might have been viewed as an object lacking first-generation (i.e., non-He-enriched) stars, but these new measurements show that even clusters having entirely blue HBs likely have at least some stars with primordial He.

2.2. *Indirect Methods*

In Figure 1, we show a comparison between two isochrones for an old (12 Gyr) and metal-poor ($Z = 0.0005$) population, for two different He abundances on the zero-age MS: $Y = 0.24$ (which should be representative of the 'first generation' of GC stars) and $Y = 0.34$ (which is even lower than the amount of He enhancement that has been suggested for some GCs; see, e.g., Table 1 in D'Antona & Caloi 2008). These models were computed using an updated version of the Schwarzschild stellar evolution code, with updated input physics as described in Valcarce (2010). The tracks assume a solar-scaled mix (and thus $[\mathrm{M/H}] = -1.54$ for these models), but the scaling relation of Salaris *et al.* (1993) can be straightforwardly used to apply our tracks to the α-enhanced case as well.

As can be seen from Figure 1, these isochrones reveal that not only is the MS affected by an increase in the He abundance, but so are several other key evolutionary stages in the lives of low-mass stars. This is summarized in Tables 1 (for $Y = 0.24$) and 2 (for $Y = 0.34$), where the properties of the following evolutionary stages are indicated: (i) MS, at a temperature $T_{\rm eff} = 5000$ K (corresponding to a color $B-V \simeq 0.8$ mag), (ii) MS, at a luminosity $\log(L/\mathrm{L}_\odot) = -0.5$, (iii) MS turnoff (TO) point, (iv) SGB, corresponding to a point on the isochrone that is cooler than the MS TO by $\Delta \log T_{\rm eff} = 0.03$, (v) SGB, at a temperature $\log T_{\rm eff} = 3.78$, (vi) base of the RGB, corresponding to a point on the isochrone that is brighter than the MS TO by $\Delta \log(L/\mathrm{L}_\odot) = +0.5$, (vii) base of the RGB, at a luminosity $\log(L/\mathrm{L}_\odot) = +1.0$, (viii) RGB luminosity-function 'bump,' (ix) RGB tip, (x) zero-age HB (ZAHB), at a temperature $\log T_{\rm eff} = 3.85$ (which is close to the blue edge of the RR Lyrae instability strip) and (xi) asymptotic giant branch (AGB) 'clump.' The differences between the predicted values for the two different He abundances are given in Table 3 and the implied slopes in Table 4. In what follows, we address some of the outstanding results of this analysis (see also Salaris *et al.* 2006).

MS. An expanded view of Figure 1 around the MS and SGB regions is given in Figure 2, clearly confirming that He-enhanced models are hotter and/or less luminous than their low-Y counterparts. Indeed, detections of MS splits, as mentioned in Section 1, have so far been the main indicator of variations in the He abundance within individual GCs.

MS TO. According to our results, the MS TO of a He-enhanced population is both hotter and fainter than for low Y, as can also be seen from Figure 2.

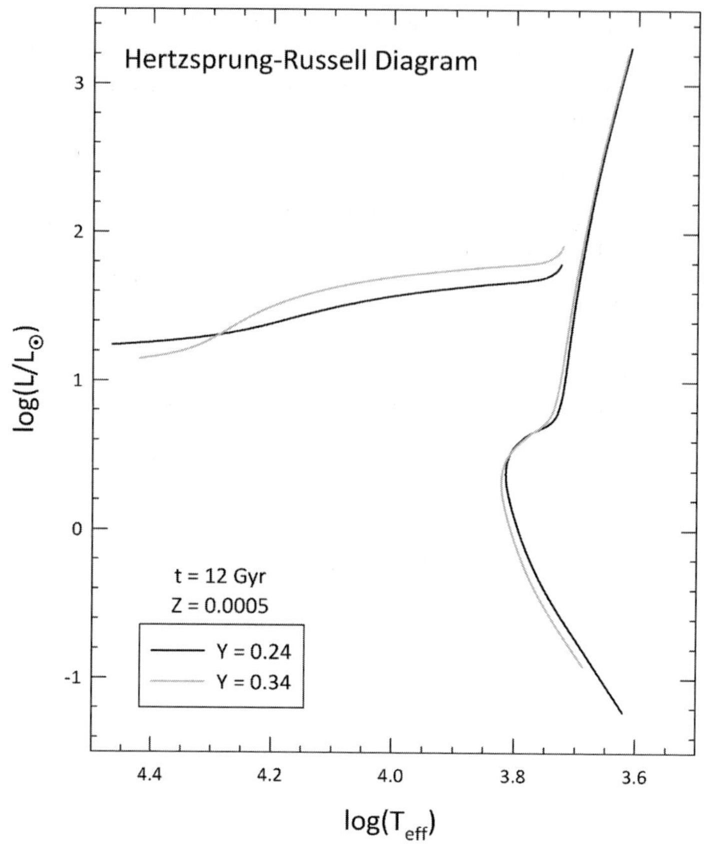

Figure 1. Isochrones for an age of 12 Gyr, metallicity $Z = 0.0005$ and two different initial helium abundances, $Y = 0.24$ (black lines) and $Y = 0.34$ (gray lines).

SGB. Our calculations reveal that the position of the SGB is *not* sensitive to Y (see also Figure 2). Differences in He abundance are thus not a viable candidate to explain the SGB splits that have been found in the literature, and changes in the abundances of other abundant species, such as the CNO elements, are accordingly also required to reproduce these splits (see also Cassisi *et al.* 2008; Salaris *et al.* 2008; Ventura *et al.* 2009).
Base of the RGB. Figure 2 shows that the base of the RGB of He-enhanced models is hotter (at a given luminosity) than their low-Y counterparts. The detected difference corresponds to $\mathrm{d}\Delta(B-V)_{\mathrm{RGB}}/\mathrm{d}Y \approx \mathrm{d}\Delta(V-I)_{\mathrm{RGB}}/\mathrm{d}Y \approx -0.35$, implying $\mathrm{d}\Delta(B-I)_{\mathrm{RGB}}/\mathrm{d}Y \approx -0.7$. This is not a negligible effect. As a matter of fact, Table 3 shows that the temperature split is predicted to be about half as large as the MS separation. Our calculations further reveal that such a difference in temperature, hence color, persists until the RGB tip, as can also be seen from Figures 3 and 4, although the size of the predicted split decreases progressively towards the RGB tip, where the presence of AGB stars may also complicate any empirical tests (see also Caloi & D'Antona 2005; Salaris *et al.* 2006; Pietrinferni *et al.* 2009). Still, since photometry of RGB stars is nowadays often precise to a level much better than 0.01 mag, large He enhancements might also manifest themselves as RGB splits in well-populated CMDs. On the other hand, one should also keep in mind the possibility that differences in color transformations between He-enriched and non-He-enriched GC stars might mask, at least in part, this effect (but see Girardi *et al.* 2007). Indeed, such RGB splits appear to have so far been reported

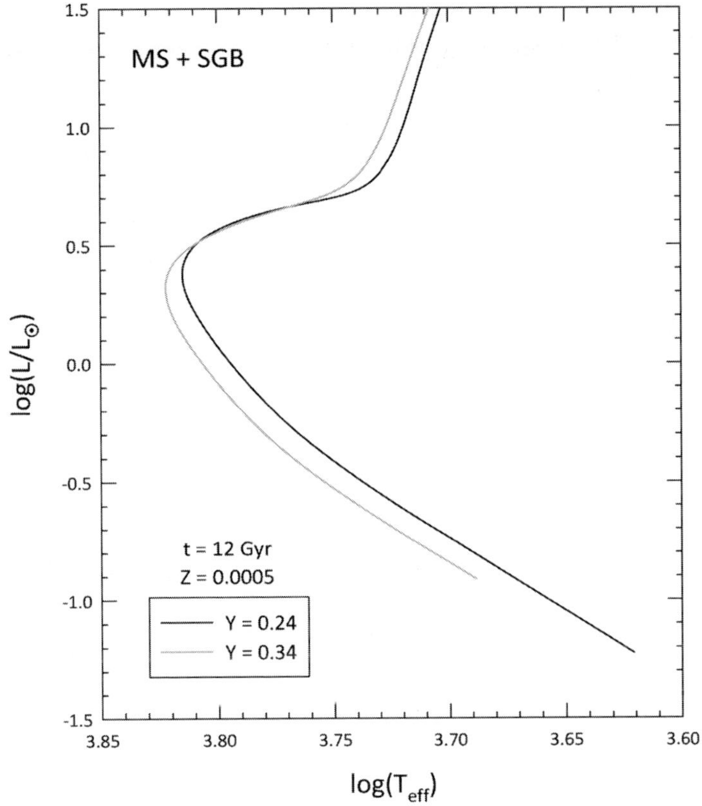

Figure 2. As Figure 1, but zooming in around the MS–SGB region.

only in the cases of M4 (NGC 6121; Marino *et al.* 2008) and NGC 1851 (Lee *et al.* 2009), and even in those cases, only using bluer passbands and the Strömgren filter system, respectively. This suggests that these RGB splits may, in reality, be due to the effect of abundance variations on the color transformations, rather than to a He enhancement proper (see also Yong *et al.* 2008, 2009), but further studies would be of much interest.

RGB bump. Our calculations confirm many previous indications that both the position and shape of the RGB bump depend sensitively on the He abundance (e.g., Salaris *et al.* 2006). This is particularly clear from Figure 3, which shows that the higher-Y models have a brighter bump than their lower-Y counterparts. Accordingly, multiple generations with widely different He abundances should also manifest themselves in the form of multiple RGB bumps. A search for these multiple bumps has been conducted for ω Cen (Sollima *et al.* 2005; F. R. Ferraro 2009, priv. comm.), with an indication that the implied range in Y is smaller than indicated by analyses of the MS splits. However, analysis of this cluster is complicated because it contains a large spread in metallicities and ages, both of which also affect the position of the RGB bump. Therefore, further analysis of this and other (monometallic) GCs would certainly prove of interest (see also Riello *et al.* 2003; Caloi & D'Antona 2005; Carretta *et al.* 2007).

Bono *et al.* (2001) carried out a comparison between the number of stars around and below the detected bump in a large and homogeneous sample of GCs and the prediction of He-enriched models. Their results, as summarized in their figure 2, again do not reveal statistically significant evidence for large He-abundance variations, except perhaps for NGC 6441 (which, however, they suggest may be an artifact of differential reddening,

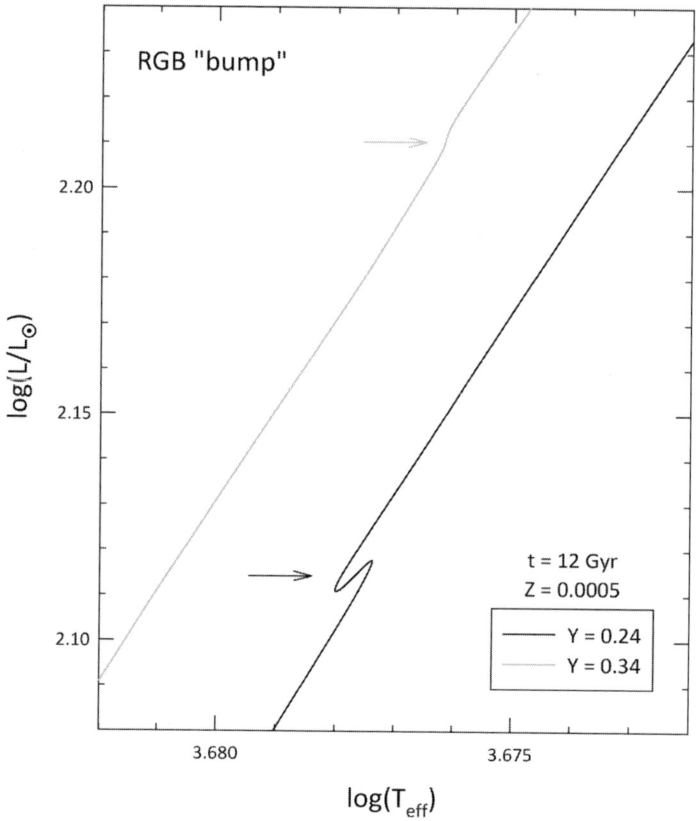

Figure 3. As Figure 1, but zooming in around the RGB-bump region.

noting that NGC 6441's 'twin,' NGC 6388, appears entirely consistent with the expectations for a 'normal' Y). Interestingly, Bono *et al.* also call attention to a possible increase in Y in the blue-HB cluster M13 (NGC 6205), but the measurements for this cluster are still consistent with a 'normal' Y, within the error bars.

RGB tip. As is clear from Figure 4, the luminosity of the RGB tip, where He ignition occurs, also depends on the He abundance, by an amount that is slightly less than 50% of that seen (at a fixed $T_{\rm eff}$) on the MS. Therefore, GCs that lack first-generation stars and possess only second- and third-generation stars, as has been suggested for GCs with completely blue HBs (e.g., D'Antona & Caloi 2008), should have fainter RGB tips, by of order 0.11 mag for a difference in Y of 0.1 (Table 3). While evolution close to the RGB tip is quite fast, and therefore not many stars are usually found in that region of the CMD (which, in addition, may also contain AGB stars), careful analysis based on large samples might give indications of whether fainter RGB tips indeed tend to be more common, statistically speaking, in blue-HB GCs. In this sense, some authors have used the RGB-tip luminosity to place constraints on the He core mass at the He flash, $M_{\rm c}^{\rm HeF}$, which is of strong interest for particle physicists in particular. However, such studies have so far revealed evidence for large increases in neither $M_{\rm c}^{\rm HeF}$ nor Y with respect to canonical values. On the contrary, available analyses have tended to favor a somewhat low He abundance see, e.g., figure 2 in Raffelt; and references therein), even though more detailed analysis of individual clusters, which has not been the focus of most such studies, would certainly prove of interest.

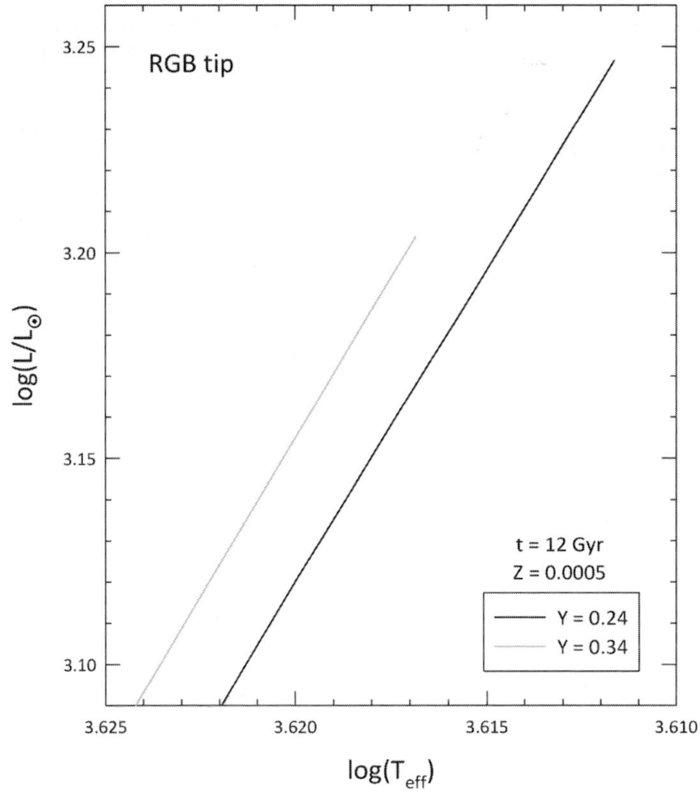

Figure 4. As Figure 1, but zooming in around the RGB-tip region.

As we have seen, contrary to the RGB tip, the RGB bump becomes *more* luminous with an increase in Y. Therefore, the difference in magnitude between the RGB tip and the RGB bump should be more sensitive to Y than either of these features alone. Based on Tables 1 and 2, we find $\mathrm{d}\Delta M_{\rm bol}(\text{RGB bump} - \text{tip})/\mathrm{d}Y = -3.6$, which is a stronger Y dependence than for many of the individual indicators listed in Table 4.

ZAHB. As well known, the ZAHB luminosity is strongly sensitive to Y (see, e.g., Sweigart 1987; and also our Figures 1, 5 and 6). This strong dependence constitutes the basis for a number of methods that are often used to infer the He abundance in GCs.

One such method is the 'A method' of Caputo & Castellani (1975), which uses the fact that the periods of RR Lyrae stars depend strongly on their luminosities, and the latter on Y, to infer the He abundance on the basis of period and temperature measurements for RR Lyrae variables. Related to this are the so-called 'period-shift techniques,' in which differences in period, at a given temperature, are taken as evidence for variations in luminosity, and hence Y. Based on these techniques, an overluminosity, and hence evidence for an increase in Y, could not be detected among either ω Cen's (Sollima *et al.* 2006) or NGC 1851's (Lee *et al.* 2009) RR Lyrae stars. Similarly, Sandquist (2000) could not confirm the presence of significant He-abundance variations in his analysis of a large number of GCs, although he found a difference between metal-poor and moderately metal-rich objects, concluding however that "it is unlikely that the difference in $\langle A \rangle$ between the two groups is due to a difference in helium abundance." On the other hand, large overluminosities, possibly related to He enhancement, have been confirmed for the metal-rich bulge GCs NGC 6388 and NGC 6441, which are known to contain large

Table 1. Evolutionary predictions for $Y = 0.24$.[a]

Stage	$M/M_\odot$	$\log T_{eff}$	$\log(L/L_\odot)$
MS[b]	0.606	3.699	−0.744
MS[c]	0.661	3.738	−0.500
TO	0.815	3.815	0.379
SGB[d]	0.832	3.785	0.622
SGB[e]	0.833	3.780	0.636
Base RGB[f]	0.838	3.725	0.879
Base RGB[g]	0.839	3.719	1.000
RGB bump	0.843	3.679	2.112
RGB tip	0.844	3.611	3.247
ZAHB[h]	0.663	3.850	1.650
AGB clump	0.663	3.708	2.012

NOTES:
[a] For an assumed $Z = 0.0005$, $t = 12$ Gyr.
[b] For a temperature $T_{eff} = 5000$ K.
[c] For a luminosity $\log(L/L_\odot) = -0.50$.
[d] At a temperature cooler than the TO by $\Delta \log T_{eff} = 0.03$.
[e] At a temperature $\log T_{eff} = 3.780$.
[f] At a level more luminous than the TO by $\Delta \log(L/L_\odot) = 0.5$.
[g] At a luminosity $\log(L/L_\odot) = 1.0$.
[h] At a temperature $\log T_{eff} = 3.850$.

Table 2. Evolutionary predictions for $Y = 0.34$.[a]

Stage	$M/M_\odot$	$\log T_{eff}$	$\log(L/L_\odot)$
MS[b]	0.513	3.699	−0.849
MS[c]	0.577	3.754	−0.500
TO	0.687	3.822	0.321
SGB[d]	0.699	3.792	0.591
SGB[e]	0.700	3.780	0.630
Base RGB[f]	0.704	3.737	0.821
Base RGB[g]	0.705	3.727	1.000
RGB bump	0.708	3.676	2.211
RGB tip	0.708	3.617	3.203
ZAHB[h]	0.625	3.850	1.770
AGB clump	0.625	3.724	2.157

NOTES:
[a] For an assumed $Z = 0.0005$, $t = 12$ Gyr.
[b] For a temperature $T_{eff} = 5000$ K.
[c] For a luminosity $\log(L/L_\odot) = -0.50$.
[d] At a temperature cooler than the TO by $\Delta \log T_{eff} = 0.03$.
[e] At a temperature $\log T_{eff} = 3.780$.
[f] At a level more luminous than the TO by $\Delta \log(L/L_\odot) = 0.5$.
[g] At a luminosity $\log(L/L_\odot) = 1.0$.
[h] At a temperature $\log T_{eff} = 3.850$.

Table 3. Differences between evolutionary predictions.[a]

Stage	$\Delta(M/M_\odot)$	$\Delta \log T_{eff}$	$\Delta \log(L/L_\odot)$
MS[b]	−0.093	—	−0.105
MS[c]	−0.084	+0.016	—
TO	−0.128	+0.007	−0.058
SGB[d]	−0.133	+0.007	−0.033
SGB[e]	−0.133	—	−0.006
Base RGB[f]	−0.134	+0.012	−0.058
Base RGB[g]	−0.134	+0.008	—
RGB bump	−0.135	−0.003	+0.099
RGB tip	−0.136	+0.006	−0.044
ZAHB[h]	−0.038	—	+0.120
AGB clump	−0.038	+0.016	+0.145

NOTES:
[a] In the sense $Y = 0.34$ minus $Y = 0.24$, for an assumed $Z = 0.0005$, $t = 12$ Gyr.
[b] For a temperature $T_{eff} = 5000$ K.
[c] For a luminosity $\log(L/L_\odot) = -0.50$.
[d] At a temperature cooler than the TO by $\Delta \log T_{eff} = 0.03$.
[e] At a temperature $\log T_{eff} = 3.780$.
[f] At a level more luminous than the TO by $\Delta \log(L/L_\odot) = 0.5$.
[g] At a luminosity $\log(L/L_\odot) = 1.0$.
[h] At a temperature $\log T_{eff} = 3.850$.

Table 4. Predicted slopes.[a]

Stage	$\frac{d \log T_{eff}}{dY}$	$\frac{d \log(L/L_\odot)}{dY}$	$\frac{dM_{bol}}{dY}$
MS[b]	—	−1.05	+2.63
MS[c]	+0.16	—	—
TO	+0.07	−0.58	+1.45
SGB[d]	+0.07	−0.33	+0.83
SGB[e]	—	−0.06	+0.15
Base RGB[f]	+0.12	−0.06	+0.15
Base RGB[g]	+0.08	—	—
RGB bump	−0.03	+0.99	−2.48
RGB tip	+0.06	−0.44	+1.10
ZAHB[h]	—	+1.20	+3.00
AGB clump	+0.16	+1.45	+3.63

NOTES:
[1] In the sense $Y = 0.34$ minus $Y = 0.24$, for an assumed $Z = 0.0005$, $t = 12$ Gyr.
[2] For a temperature $T_{eff} = 5000$ K.
[3] For a luminosity $\log(L/L_\odot) = -0.50$.
[4] At a temperature cooler than the TO by $\Delta \log T_{eff} = 0.03$.
[5] At a temperature $\log T_{eff} = 3.780$.
[6] At a level more luminous than the TO by $\Delta \log(L/L_\odot) = 0.5$.
[7] At a luminosity $\log(L/L_\odot) = 1.0$.
[8] At a temperature $\log T_{eff} = 3.850$.

populations of RR Lyrae and blue HB stars, an uncommon feature in the metal-rich domain, which by itself may also point to the need for He enhancement (e.g., Sweigart & Catelan 1998; Pritzl *et al.* 2002; Caloi & D'Antona 2007; Busso *et al.* 2007).

Another method which exploits the strong dependence of HB luminosities on Y is the 'Δ method' of Caputo *et al.* (1983), which uses the difference in magnitude between the HB and the MS as a He indicator. This is a particularly strong Y indicator because, while the HB luminosity increases with Y, the MS luminosity decreases. Based on Tables 1 and

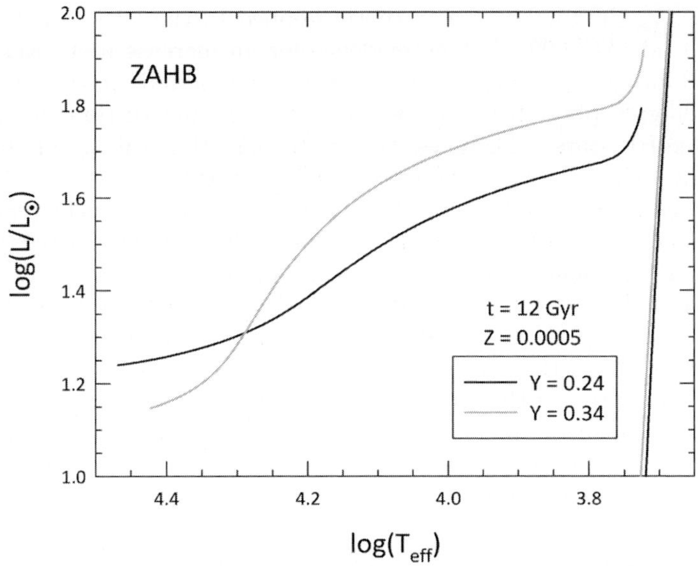

Figure 5. As Figure 1, but zooming in around the ZAHB region.

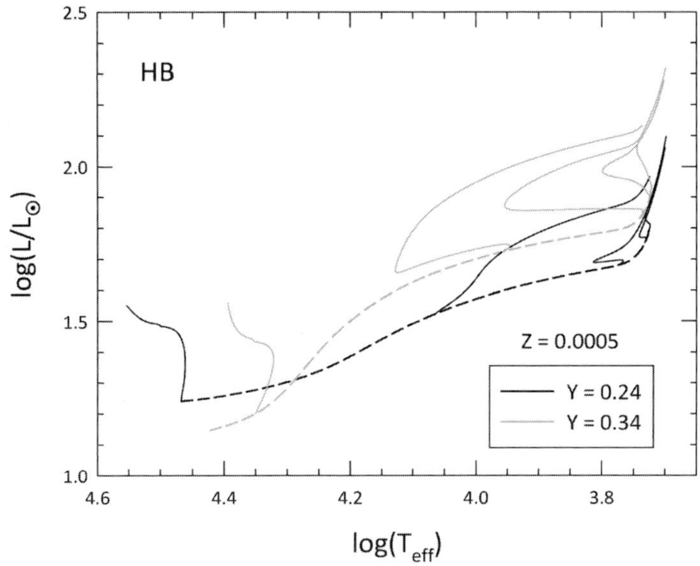

Figure 6. As Figure 5, but including evolutionary tracks for the following mass values (from left to right): $0.49, 0.60, 0.70, 0.80$ and $0.90\,M_\odot$. The same ZAHB sequences as in the previous figure are shown as dashed lines.

2, we find $d\Delta M_{\rm bol}(\rm MS - ZAHB)/dY = 5.6$. Sandquist (2000) applied this method to a large sample of GCs, but did not find evidence for significant variations in Y.

Finally, the increased luminosities of He-enhanced HB stars also lead to decreased surface gravities at a given $T_{\rm eff}$. Therefore, and as suggested by several authors (see, e.g., Catelan 2009; and references therein), spectroscopic $\log g$ measurements can also be of interest in constraining the amount of He enhancement, particularly among blue-HB stars cooler than 11 500 K, whose atmospheres are not affected by the strong diffusion effects that make it extremely difficult to properly interpret the results for hotter HB stars. In this sense, the method was recently applied to the GCs M3 (NGC 5272) and M13 by Catelan *et al.* (2009), but no evidence for an increase in Y over the canonical value was found for either cluster. In fact, these authors found, on the basis of a detailed analysis of Strömgren photometry for M3, that the amount of He enhancement among the cluster's blue-HB stars is likely less than 0.01 in Y, thus ruling out the much higher He enhancements that have been proposed in the literature.

HB. The color distribution of HB stars has been widely used in recent years to study the level of He enhancement in GCs (e.g., D'Antona & Caloi 2008; and references therein). The main reason why high-Y GCs of a given age and metallicity are expected to have a bluer HB morphology is the fact that their more compact progenitors evolve faster. Thus, at a given age, high-Y GCs feed stars of lower mass (and thus bluer; see Figure 6) onto the HB phase than their lower-Y counterparts. However, the colors of HB stars are sensitive not only to the He abundance, but also to many other parameters, such as age, CNO and α-element abundances, rotation and—crucially—mass loss on the RGB (see Catelan 2009 for a recent review). It is thus important to use indicators other than the color distribution of HB stars to establish whether, and to what level, He may be enhanced among HB stars. In this sense, the *luminosity distribution* of HB stars also plays a very important role that should not be ignored (e.g., Crocker *et al.* 1988; Fusi Pecci *et al.* 1996; Sweigart & Catelan 1998; Salaris *et al.* 2006, 2008; Catelan *et al.* 2009).

Examination of Figure 6 reveals that not only ZAHB luminosities (and their run with $T_{\rm eff}$), but also other HB features, are sensitive to the He abundance. Note, in particular, that high-Y evolutionary tracks tend to evolve much more significantly in luminosity than at lower Y, which leads to an increase in the HB 'thickness' (or luminosity width) with increasing Y. Similarly, an increase in Y leads to much more pronounced 'blueward loops,' which also has important observational consequences (e.g., Sweigart & Catelan 1998; Pritzl *et al.* 2002; Catelan 2009). In addition, the predicted separation in luminosity between extreme-HB (EHB) stars and post-EHB stars is also a strong function of Y (Brown *et al.* 2008), which could, at least in principle, provide a test of He enhancement for those clusters for which extensive ultraviolet photometry is available.

Another key He abundance indicator for GCs that is based on properties of HB stars is the number ratio R of HB stars and RGB stars brighter than the HB (Iben 1968). In brief, not only does the HB luminosity increase markedly with increasing Y, but also HB lifetimes—and thus the number of HB stars relative to RGB stars brighter than the HB—increase markedly with increasing Y. Indeed, for HB stars beginning their evolution close to the instability strip, our models give $dR/dY = 8.3$,† thus confirming the strong sensitivity of the method to He-abundance variations.

The most recent application of this method to a large sample of GCs is the one by Salaris *et al.* (2004). They find that "the mean... $Y = 0.250 \pm 0.006$. An instrinsic dispersion with a firm upper limit of 0.019 around this value... is a priori possible given the observational errors." Interestingly, the authors find some evidence for higher R values for clusters with blue HBs, as would be expected in the He-enhancement scenario. However, they point out that this result may be more straightforwardly explained in terms of an increase in HB lifetimes for bluer, less massive HB stars (see also Zoccali *et al.* 2000).

AGB clump. According to our results (e.g., Table 3), the AGB clump, which is seen in some GCs at the base of the AGB, becomes significantly brighter with increasing Y. The

† This decreases to $dR/dY = 5.8$ if we assume that the HB luminosity taken as reference for RGB number counts is the same for both He-rich and He-poor models, as might be more appropriate for clusters for which stars on the 'horizontal' part of the HB are not He-enhanced, for which, in addition, one should consider that a fraction of the RGB stars might also be poor in He (see also Riello *et al.* 2003; Salaris *et al.* 2004).

difference in luminosity between the AGB clump and the ZAHB is, however, much less strongly dependent on Y, with $\mathrm{d}M_{\mathrm{bol}}(\mathrm{ZAHB} - \mathrm{AGB\ clump})/\mathrm{d}Y \approx 0.3$ (see also Bono *et al.* 1995). In addition, AGB clumps are only found in GCs with predominantly red HBs, which are not expected to be the ones displaying the largest levels of He enhancement.

3. Conclusions

Nature has conspired to make reliable measurements of the He abundance of GC stars an extremely difficult task. Still, many features of the CMD, besides MS and HB colors, present at least some sensitivity to Y. These can be best exploited, depending on the specific CMD location, in different wavelength regimes, using a variety of photometric and spectroscopic techniques. It is thus important to establish, on a cluster-by-cluster basis, which features are consistent (or not), to the limit of our empirical and theoretical knowledge, with the He-enhancement levels that have been suggested. Only then will we be in a good position to firmly establish the extent to which such He enhancements occur in nature. This is a crucial task to establish the role played by GCs as laboratories of stellar evolution and indeed, more generally, of astrophysics.

Acknowledgements

Support for M.C. is provided by Proyecto Basal PFB-06/2007, FONDAP Centro de Astrofísica 15010003, Proyecto FONDECYT Regular #1071002 and a John Simon Guggenheim Memorial Foundation Fellowship. Support for A.V. is provided by the IAU, CONICYT, SOCHIAS, MECESUP and ALMA.

References

Behr, B. B. 2003, *ApJS*, 149, 67
Bono, G., Cassisi, S., Zoccali, M., & Piotto, G. 2001, *ApJ* (Letters), 546, L109
Bono, G., Castellani, V., Degl'Innocenti, S., & Pulone, L. 1995, *A&A*, 297, 115
Brown, T. M., Smith, E., Ferguson, H. C., Sweigart, A. V., Kimble, R. A., & Bowers, C. W. 2008, *ApJ*, 682, 319
Busso, G., *et al.* 2007, *A&A*, 474, 105
Caloi, V. & D'Antona, F. 2005, *A&A*, 435, 987
Caloi, V. & D'Antona, F. 2007, *A&A*, 463, 949
Caputo, F. & Castellani, V. 1975, *Ap&SS*, 38, 39
Caputo, F., Cayrel, R., & Cayrel de Strobel, G. 1983, *A&A*, 123, 135
Carretta, E., Bragaglia, A., Gratton, R., & Lucatello, S. 2009, *A&A*, in press (arXiv:0909.2941)
Carretta, E., *et al.* 2007, *A&A*, 464, 939
Cassisi, S., Salaris, M., Anderson, J., Piotto, G., Pietrinferni, A., Milone, A., Bellini, A., & Bedin, L. R. 2009, *ApJ*, 702, 1530
Cassisi, S., Salaris, M., Pietrinferni, A., Piotto, G., Milone, A. P., Bedin, L. R., & Anderson, J. 2008, *ApJ* (Letters), 672, L115
Catelan, M. 2008, *MemSAIt*, 79, 388
Catelan, M. 2009, *Ap&SS*, 320, 261
Catelan, M., Grundahl, F., Sweigart, A. V., Valcarce, A. A. R., & Cortés, C. 2009, *ApJ* (Letters), 695, L97
Crocker, D. A., Rood, R. T., & O'Connell, R. W. 1988, *ApJ*, 332, 236
D'Antona, F., Bellazzini, M., Caloi, V., Fusi Pecci, F., Galleti, S., & Rood, R. T. 2005, *ApJ*, 631, 868
D'Antona, F. & Caloi, V. 2008, *MNRAS*, 390, 693
Davis, D. S., Richer, H. B., Anderson, J., Brewer, J., Hurley, J., Kalirai, J. S., Rich, R. M., & Stetson, P. B. 2008, *AJ*, 135, 2155

Dupree, A. K., Smith, G. H., & Strader, J. 2009, *AJ*, 138, 1485

Ferraro, F. R., Valenti, E., Straniero, O., & Origlia, L. 2006, *ApJ*, 642, 225

Fusi Pecci, F., Bellazzini, M., Ferraro, F. R., Buonanno, R., & Corsi, C. E. 1996, in: H. Morrison & A. Sarajedini (eds.) *Formation of the Galactic Halo .... Inside and Out*, ASP Conf. Ser., Vol. 92, p. 221 (San Francisco: ASP)

Girardi, L., Castelli, F., Bertelli, G., & Nasi, E. 2007, *A&A*, 468, 657

Gratton, R. G., Carretta, E., Matteucci, F., & Sneden, C. 2000, *A&A*, 358, 671

Grundahl, F., Catelan, M., Landsman, W. B., Stetson, P. B., & Andersen, M. I. 1999, *ApJ*, 524, 242

Iben Jr., I. 1968, *Nature*, 220, 143

Lee, J.-W., Lee, J., Kang, Y.-W., Lee, Y.-W., Han, S.-I., Joo, S.-J., Rey, S.-C., & Yong, D. 2009, *ApJ* (Letters), 695, L78

Marcolini, A., Gibson, B. K., Karakas, A. I., & Sánchez–Blázquez, P. 2009, *MNRAS*, 395, 719

Marino, A. F., Villanova, S., Piotto, G., Milone, A. P., Momany, Y., Bedin, L. R., & Medling, A. M. 2008, *A&A*, 490, 625

Moehler, S. 2001, *PASP*, 113, 1162

Moehler, S., Dreizler, S., Lanz, T., Bono, G., Sweigart, A. V., Calamida, A., Monelli, M., & Nonino, M. 2007, *A&A* (Letters), 475, L5

Moehler, S., Landsman, W. B., Sweigart, A. V., & Grundahl, F. 2003, *A&A*, 405, 135

Moehler, S., Sweigart, A. V., Landsman, W. B., & Dreizler, S. 2009, *Ap&SS*, 291, 231

Norris, J. E. 2004, *ApJ* (Letters), 612, L25

Palmieri, R., Piotto, G., Saviane, I., Girardi, L., & Castellani, V. 2002, *A&A*, 392, 115

Pietrinferni, A., Cassisi, S., Salaris, M., Percival, S., & Ferguson, J. W. 2009, *ApJ*, 697, 275

Piotto, G. 2009, *Proc. Int'l Astron. Union*, 4, 233

Piotto, G., *et al.* 2005, *ApJ*, 621, 777

Piotto, G., *et al.* 2007, *ApJ* (Letters), 661, L53

Pritzl, B. J., Smith, H. A., Catelan, M., & Sweigart, A. V. 2002, *AJ*, 124, 949

Raffelt, G. G. 1996, Stars as Laboratories for Fundamental Physics: The Astrophysics of Neutrinos, Axions, and Other Weakly Interacting Particles (Chicago: Univ. of Chicago Press)

Raffelt, G. G. 2000, *Phys. Rep.*, 333, 593

Raffelt, G. G. 2008, *Lect. Notes Phys.*, 741, 51

Riello, M., *et al.* 2003, *A&A*, 410, 553

Salaris, M., Cassisi, S., & Pietrinferni, A. 2008, *ApJ* (Letters), 678, L25

Salaris, M., Chieffi, A., & Straniero, O. 1993, *ApJ*, 414, 580

Salaris, M., Riello, M., Cassisi, S., & Piotto, G. 2004, *A&A*, 420, 911

Salaris, M., Weiss, A., Ferguson, J. W., & Fusilier, D. J. 2006, *ApJ*, 645, 1131

Sandquist, E. L. 2000, *MNRAS*, 313, 571

Sollima, A., Borissova, J., Catelan, M., Smith, H. A., Minniti, D., Cacciari, C., & Ferraro, F. R. 2006, *ApJ* (Letters), 640, L43

Sollima, A., Ferraro, F. R., Pancino, E., & Bellazzini, M. 2005, *MNRAS*, 357, 265

Sweigart, A. V. 1987, *ApJS*, 65, 95

Sweigart, A. V. & Catelan, M. 1998, *ApJ*, 501, L63

Valcarce, A. A. R. 2010, Ph.D. Thesis, Pontificia Universidad Católica de Chile (in prep.)

Ventura, P., Caloi, V., D'Antona, F., Ferguson, J., Milone, A., & Piotto, G. P. 2009, *MNRAS*, 399, 934

Villanova, S., Piotto, G., & Gratton, R. 2009, *A&A*, 499, 755

Yong, D., Grundahl, F., D'Antona, F., Karakas, A. I., Lattanzio, J. C., & Norris, J. E. 2009, *ApJ* (Letters), 695, L62

Yong, D., Grundahl, F., Johnson, J. A., & Asplund, M. 2008, *ApJ*, 684, 1159

Yoon, S.-J., Joo, S.-J., Ree, C. H., Han, S.-I., Kim, D.-G., & Lee, Y.-W. 2008, *ApJ*, 677, 1080

Zoccali, M., Cassisi, S., Bono, G., Piotto, G., Rich, R. M., & Djorgovski, S. G. 2000, *ApJ*, 538, 289

Star clusters: basic galactic building blocks
Proceedings IAU Symposium No. 266, 2009
R. de Grijs & J. R. D. Lépine, eds.

doi:10.1017/S1743921309991165

Close pairs: keys to comprehension of star cluster evolution

Dany Vanbeveren

Astrophysical Institute, Vrije Universiteit Brussel, Belgium
email: dvbevere@vub.ac.be

GroepT Leuven Engineering College, Association KU Leuven, Belgium
email: dany.vanbeveren@groept.be

Abstract. In this review I first summarize why binaries are key objects in the study of stellar populations, to understand the evolution of star clusters and galaxies, and thus to understand the universe. I then focus on four specific topics:

(i) the formation (through binaries) and evolution of very massive stars in dense clusters and the importance of stellar-wind mass loss. I discuss preliminary computations of wind mass-loss rates of very massive stars performed with the Munich hydrodynamical code and the influence of these new rates on the possible formation of an intermediate-mass black hole in the cluster MGG 11 in M82;

(ii) the evolution of intermediate-mass binaries in a starburst with special emphasis on the variation of the supernova (SN) Ia rate (i.e., on the delayed time distribution of SNe Ia). A comparison with SN Ia rates in elliptical galaxies may provide important clues to SN Ia models as well as to the evolution of SN Ia progenitors;

(iii) the evolution of double-neutron-star mergers in a starburst (i.e., the delayed time distribution of these mergers) and what this tells us about the suggestion that these mergers may be important production sites of r-process elements;

(iv) the possible effect of massive binaries on the self-enrichment of globular clusters.

Keywords. stellar dynamics, binaries: close, stars: evolution

1. Introduction

The question of whether or not binaries are important in population studies in general and cluster studies in particular is obviously rhetoric but sometimes it is useful to summarize why this is so. In section 2, I give a personal selection of observational and theoretical facts that illustrate the importance of binaries. In sections 3–5, I will highlight three topics related to binaries in clusters/starbursts,

1. one of the most spectacular events related to cluster stellar dynamics is a physical collision of two or more stars. The most probable scenario for this collision process is as follows. A primordial binary or a binary that is formed dynamically interacts with a third object (a single star or another binary). This may result in the formation of unconventionally formed objects (UFOs; Vanbeveren 2007), where binary components are exchanged or where a new binary originates with a component which is a merger of two or more stars. Some standard massive binaries such as the Wolf–Rayet (WR) binary γ^2 Velorum may have been formed this way (see also Vanbeveren 2007). When the interaction results in a merger of two or more stars, the product becomes more massive and attracts other stars. This may initiate a runaway collision process, which in turn may result in the formation of an intermediate-mass black hole. The latter will be discussed in Section 3;

2. the delay-time distribution of SNe Ia in elliptical galaxies (considered to be remnants of super starbursts) and what this tells us about the binary formation mechanism of these SNe (Section 4);

3. the delay-time distribution of merging double-neutron-star binaries (NS+NS) or merging neutron-star and black-hole binaries (NS+BH) and the link with the formation of r-process elements (Section 5).

The present population of low-mass stars in globular clusters (GCs) shows clearly the effects of chemical enrichment of a population of intermediate-mass and/or massive stars that formed at an earlier evolutionary phase during the GC's life (see various contributions in these proceedings). Studies aiming at explaining this self-enrichment have mainly focused on single stars. In Section 6, I offer a few suggestions as to how massive binaries could have affected the chemical self-enrichment of globular clusters (see also de Mink *et al.*, these proceedings).

2. The role of binaries in population synthesis

The influence of binaries on population number/spectral synthesis has been studied in numerous papers the last three decades (e.g., Van Bever & Vanbeveren 2000, 2003; De Donder & Vanbeveren 2004; Belczynski *et al.* 2008; and references therein). It is clear that a discussion of the evolution of binaries is essential here, especially focusing on uncertainties in all physical processes that govern binary evolution and how they affect population predictions. Let me mention stellar-wind mass loss during core hydrogen and core helium burning of massive stars, rotation, Roche-lobe overflow and mass and angular-momentum accretion in Case A and Case Br binaries, common-envelope evolution in Case Bc and Case C binaries, the spiral-in process in binaries with an extreme mass ratio, the distribution of kick velocities of a compact SN remnant and last but not least, the evolution of binary mergers. I discussed some of these uncertainties in a recent review (Vanbeveren 2009). Below, I list a number of facts resulting from all of these studies (a personal and, therefore, probably a somewhat subjective selection).

• Many (most) of the massive and intermediate-mass stars in clusters are binary members (Kouwenhoven 2006; Mason *et al.* 2009), but the binary frequency may vary from cluster to cluster, possibly depending on the cluster's stellar density. About half of these binaries are interacting binaries (with orbital periods less than 10 years). Note that the binary frequency among solar-type stars may be lower (Zinnecker & Yorke 2007).

• During binary evolution, both components may merge and form a single star. When the primary of a massive binary explodes, due to the asymmetry of the SN explosion most binaries are disrupted. Due to stellar dynamics in dense clusters, binaries may be disrupted. Although binaries may also be formed due to N-body processes in dense clusters, the net result of all this is that the presently observed binary frequency (at least in the massive and intermediate-mass range) is lower than that at birth. This also means that a fraction of the observed single stars may have had a binary evolutionary past. To illustrate this, the single star ζ Pup in the solar neighborhood is a runaway star, which means that it most probably has had a binary history. In the past, ζ Pup has frequently been considered a prototype massive single star, but it is probably not a typical massive single star at all.

• About 10% of O-type stars are runaways (defined as stars with a peculiar space velocity $\geqslant$30 km s^{-1}; Gies 1987). We know of two processes that can produce such stars, close encounters in dense clusters of a binary and another object (either a single star or another binary; Lada *et al.* 1984) and a SN explosion in massive close binaries (Blaauw

1961). Interestingly, in the close-encounter scenario, many runaway stars are expected to be mergers of at least two stars (see, for example, the close-encounter scenario for ζ Pup in Vanbeveren *et al.* 2009). At present, there are too many unknown (cluster) parameters to determine the frequency of runaways formed through the close-encounter scenario. De Donder *et al.* (1997) proceeded as follows. Massive close binaries are a fact, SN explosions in these binaries are a fact and thus runaways formed through the Blaauw scenario are a fact. Using a binary population synthesis code, it is possible to predict this frequency. We concluded that ~5–8% of O-type stars are expected to be Blaauw-type runaways (less than one third of them have a compact companion, either a NS or a BH). In other words, we expect that ~50–80% of O-type runaways are formed through a SN explosion in binaries, and this means that 20–50% of O-type runaways may be formed through the close-encounter process.

• The influence of recent stellar-wind mass-loss-rate formalisms (which account for the wind inhomogeneities, i.e., clumping) of massive core-hydrogen-burning stars (pre-luminous blue variables; LBVs) on massive-star evolution is very moderate and the uncertainties in these formalisms do not imply large uncertainties in massive-star population synthesis on the whole. However, the mass-loss rates during a possible LBV phase, the red-supergiant and WR phases (i.e., the hydrogen-deficient core-helium-burning phase of massive stars) are very important for stellar evolutionary prediction. The uncertainties in both the observed and the theoretically predicted rates are still a factor of 2 or more and unfortunately this uncertainty critically affects massive-star evolution and therefore also massive-star population synthesis predictions.

• The rotational-velocity distribution of O-type stars in the solar neighborhood reveals two important features. (i) A significant fraction are relatively slow rotators with an average velocity of 100 km s^{-1}. When this is translated to an average initial velocity, one arrives at the conclusion that a significant fraction of O-type stars are born with an average velocity of ~200 km s^{-1}. The evolutionary calculations of the Geneva group reveal then that the effect of rotation on the pre-SN explosion of this group of massive stars is rather moderate, comparable to the effect of moderate convective-core overshooting (note that rotation implies rotational mixing of CNO elements up to the stellar surface, but this hardly affects the overall evolutionary behavior of massive stars). (ii) The distribution has an extended tail towards rapid rotators. However, a fraction of these rapid rotators are runaway stars, indicating that they may have had a binary past. Mokiem *et al.* (2006) studied 21 OB dwarfs in the Small Magellanic Cloud (SMC) and concluded that their average rotational velocity is ~160–190 km s^{-1}. Since massive dwarfs are very young stars close to the zero-age main sequence, the latter value should be representative for the average rotational velocity of massive stars at birth. Note then that this SMC value is very similar to the initial value of Galactic O stars while also in the SMC the most rapid rotators seem to be runaway stars, as in the Galaxy.

• Rotating single-star evolutionary models have difficulties to explain atmospheric CNO abundance anomalies in the observed massive-star sample (Hunter *et al.* 2008). However, the combination of rotation and binaries provides a much better correspondence (Langer et al. 2008). I think that in this discussion the massive binary HD 163181 deserves some special attention. It is an eclipsing binary with a period of 12 days, accompanied by a nitrogen-enriched BN0.5Ia primary. Hutchings (1975) derived masses for the components of 13 and 22 $M_\odot$ (for the primary and secondary, respectively). The primary is 1.5–2 mag brighter than the secondary, which indicates that it is a core-helium-burning star that lost most of its hydrogen-rich layers through Roche-lobe overflow. This binary is therefore an illustration of the process where the atmospheric N enhancement is caused by binary-type mass loss rather than rotational mixing.

• The observed overluminosity of the optical components of some of the standard massive X-ray binaries can be explained as caused by rotational mixing in spun-up mass gainers of massive binaries (Vanbeveren & De Loore 1994).

• The effect of tides in short-period binaries on the rotation of massive binary components in low-metallicity regions (where stellar-wind mass loss is small) explains, in a straightforward way, long γ-ray bursts (Detmers *et al.* 2008).

• Rapidly rotating stars are formed through mass transfer in binaries or through binary mergers (due to either common-envelope evolution or dynamics in dense stellar systems). This means that a cluster in which the initial population consists mainly of slow rotators, but with a significant population of binaries, will become populated with rapid rotators as a result of binary evolution or due to the interplay of binaries and cluster dynamics. At least part of the cluster Be-type star population is expected to be formed this way (Pols & Marinus 1994; Van Bever & Vanbeveren 1998).

• One of the hot topics in stellar evolution research is the formation and evolution of stellar mergers. Mergers result as a consequence of canonical binary evolution (because of a nonconservative Roche-lobe-overflow process and/or common-envelope evolution) or of close encounters in star clusters where (in most cases) at least one of the players is a binary. It may therefore be expected that mergers are rapid rotators. Below, I list six different kinds of mergers which have been studied in literature, but one may think of more combinations. (i) The merger of two main-sequence (MS) stars. Smoothed-particle hydrodynamics simulations reveal that, during the merger process, large-scale mixing occurs (this means that in a cluster they will show up as blue stragglers) and mass is lost. As an example, Suzuki *et al.* (2007) calculated the merging of two massive MS stars (an 88+88 and an 88+28 $M_\odot$ merger). After the merger, the new star is largely homogenized (which means that this star shows the products of CNO burning in its atmosphere) while during the merger, $\sim$10 $M_\odot$ is lost. It is tempting to link these results to the η Car event in the 19th century. (ii) The merger of a WR and a MS star. No detailed models have been calculated but it can be expected that the resulting star may be quite spectacular. (iii) The merger of a NS (BH?) with a MS star (a Thorne–Zytkow object; Thorne & Zytkow 1977). Calculations by Canon *et al.* (1992) indicate that such objects may show up as red supergiants. Their subsequent evolution is uncertain. (iv) The merger of a white dwarf (WD) and a MS star. Population synthesis of intermediate-mass binaries reveals that many WD+MS binaries merge during the common-envelope phase when the MS star fills its Roche lobe (De Donder & Vanbeveren 2004). The subsequent evolution of these mergers is also uncertain, but I guess that it will be a rapid rotator and may show up as a Be star. (v) Double-neutron-star mergers, rapid rotators and a favorite model for short γ-ray bursts and possible sites of r-process element production and ejection (Dessart *et al.* 2009; see also Section 5). (vi) The merger of two WDs, a valuable model to explain SNe Ia if one accounts properly for the effects of rotation during the merger (Piersanti *et al.* 2003; see also Section 4).

• Accounting for the previous six points, the following statement is worth considering. "The effect of rotation is important for the evolution of some massive stars but perhaps mainly in the framework of binaries or in the framework of binaries in combination with stellar dynamics in dense clusters."

• Binaries are an essential ingredient in population number/spectral synthesis. Examples include the evolution of massive-star spectral features in starburst galaxies (Van Bever & Vanbeveren 1998, 2003; Belkus *et al.* 2003; Brinchmann *et al.* 2008), the ultraviolet upturn in elliptical galaxies (Han *et al.* 2007), the X-ray binary population in galaxies (Van Bever & Vanbeveren 2000), the population of double pulsars (De Donder & Vanbeveren 1998, 2003; Belczynski *et al.* 2002), the population of carbon-enhanced metal-poor stars (Pols *et al.* 2009), the population of short γ-ray bursts (i.e., merging of double-neutron-star binaries; De Donder & Vanbeveren 1998; O'Shaughnessy *et al.* 2009); the population of long γ-ray bursts (Detmers *et al.* 2008) and, last but not least, the SNe Ia (see Section 4) which are responsible for some 70% of all iron in the Universe.

• The discovery of the double pulsar J0737–3039 and recent population synthesis models of massive binaries have reopened the discussion on the origin of r-process elements (De Donder & Vanbeveren 2003; Section 5).

• Massive Population III binaries, where the primary is a very massive star that ends its life in a pair-instability SN, may be important sites of primary nitrogen (Vanbeveren & De Donder 2006).

• Most of the theoretical models that aim at explaining the chemical evolution of galaxies intrinsically assume that all stars are single (e.g., only single-star yields are used). However, most of massive stars are born as binary components and De Donder & Vanbeveren (2004) showed that the integrated chemical yields of a population of massive binaries differs by a factor of 2–3 from the integrated chemical yields of a population of massive single stars. I think that it is time to include binaries in all Galactic chemical codes.

3. The ultraluminous X-ray source in the young, dense cluster MGG11

Ultraluminous X-ray sources (ULXs) are point sources with X-ray luminosities of up to 10^{42} erg s^{-1}. MGG 11 is a young, dense star cluster of solar metallicity ~200 pc from the center of the starburst galaxy M82 (McCrady *et al.* 2003). A ULX is associated with the cluster. If the X-rays are due to Eddington-limited mass accretion onto a BH, it is straightforward to show that the mass of the BH has to be at least 1000 $M_\odot$. However, how to form a star with solar metallicity and a mass greater than 1000 $M_\odot$? Mass segregation in a dense, young cluster associated with core collapse and the formation of a runaway stellar-collision process was promoted by Portegies Zwart *et al.* (2004). Note that the latter authors mainly addressed the dynamical evolution of a dense cluster, but the evolution of the very massive stellar-collision product was poorly described.

The evolution of very massive stars has been studied in detail by Belkus *et al.* (2007). It was concluded that stellar-wind mass loss during core hydrogen and core helium burning is very important. A convenient evolutionary recipe for such very massive stars was presented, which can easily be implemented in an N-body dynamical code. Our N body code, which includes this recipe, has been described in Belkus (2008) and Vanbeveren *et al.* (2009), and was applied to simulate the evolution of MGG 11. Our main conclusion was that stellar-wind mass loss of massive and very massive stars does not prevent occurrence of a runaway collision event and the formation of a very massive star in a cluster like MGG 11. However, after this event stellar-wind mass loss during the remaining core-hydrogen-burning phase is large enough to reduce the mass again and the formation of a BH with a mass greater than ~75 $M_\odot$ is rather unlikely.

These calculations and conclusion depend critically on the adopted stellar-wind mass-loss formalism for very massive stars. We used a formalism proposed by Kudritzki (2002) and this requires some discussion. Kudritzki (2002) published rates for stars with luminosities up to $\log L/L_\odot = 7$, but we extrapolated his formalism for stars with mass greater than 1000 $M_\odot$, corresponding to $\log L/L_\odot \geqslant 7.5$. Furthermore, the Kudritzki results are calculated using the Munich stellar wind code from 2000 when the line force had not yet been calculated consistently with the line blocking/blanketing NLTE (non-local thermal equilibrium) computations (Pauldrach 2009, priv. comm.). Using the new, more consistent version of the Munich code (for a description see Pauldrach *et al.* 2003)

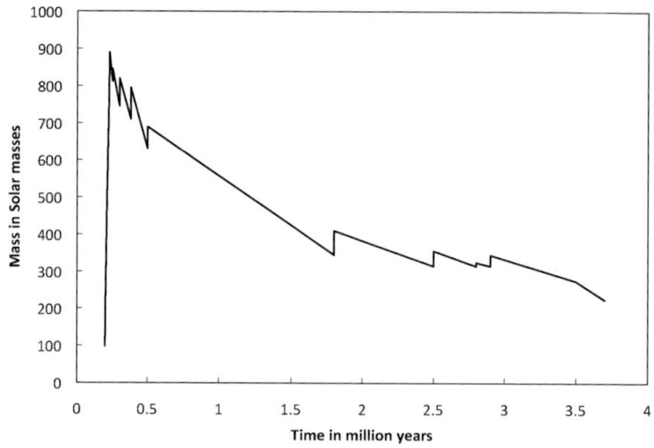

Figure 1. Mass evolution of the runaway merger in the cluster MGG 11 predicted by an N-body code that self-consistently includes stellar evolution of massive and very massive stars, using the stellar-wind mass formalism of Pauldrach & Vanbeveren (in prep.).

stellar-wind mass-loss rates were determined for very massive stars with masses up to 3000 $M_\odot$ using the properties of these stars predicted by evolution (Pauldrach & Vanbeveren, in prep.). A main result is that the new mass-loss rates are significantly smaller than the Kudritzki values. To illustrate this, for a star with $T_{\rm eff} = 50\,000$ K, $M/M_\odot = 250$ and $\log L/L_\odot = 6.9$, the old mass-loss rate was 2.5×10^{-4} $M_\odot$ yr^{-1}, while the new value is 5.2×10^{-5} $M_\odot$ yr^{-1}, or about a factor of five smaller.

We implemented the new rates in our N-body code (note that our code consistently combines the effects of dynamics and stellar evolution of very massive stars using a very efficient evolutionary algorithm) and we recalculated the dynamical evolution of MGG 11 using the same initial cluster conditions as in Vanbeveren *et al.* (2009). During core helium burning (the WR phase) of the very massive stars, we used the same WR wind formalism as in Vanbeveren *et al.* (2009). The new mass evolution of the runaway merger is illustrated in Figure 1. We notice a similar behavior as in the old simulations, e.g., rapid mass growth that is not prevented by stellar-wind mass loss and significant mass loss after the main collision event but, as expected, the final mass is significantly higher. The formation of an intermediate-mass BH in MGG 11 with a mass between 200 and 300 $M_\odot$ cannot be excluded. On the question of what the maximum mass is of the BH formed dynamically in a cluster like MGG 11, the final answer, my friend, is clearly blowing in the stellar-wind mass loss of very massive stars.

4. The delay-time distribution of SNe Ia

It is generally believed that SNe Ia are thermonuclear explosions of WDs that exceed their Chandrasekhar limit. Two main scenarios explain how this can happen. Both involve the evolution of interacting intermediate-mass binaries, i.e., the single-degenerate (SD) and double-degenerate (DD) scenarios. In the former, the WD has a MS or a red giant companion that is filling its Roche lobe. Hydrogen-rich matter is transferred towards the WD and this pushes the WD mass over the Chandrasekhar limit. The DD model involves the formation of a double CO WD binary. Due to gravitational-wave radiation, both WDs spiral in and merge. When the mass of the merger reaches 1.4 $M_\odot$, a SN Ia may happen (Webbink 1984). It has been argued that the merger of two WDs will lead to the formation of a NS and a SN Ia will not happen (e.g., Saio and Nomoto 1998; and references therein). However, these counterarguments do not account for the consequences of rotation and

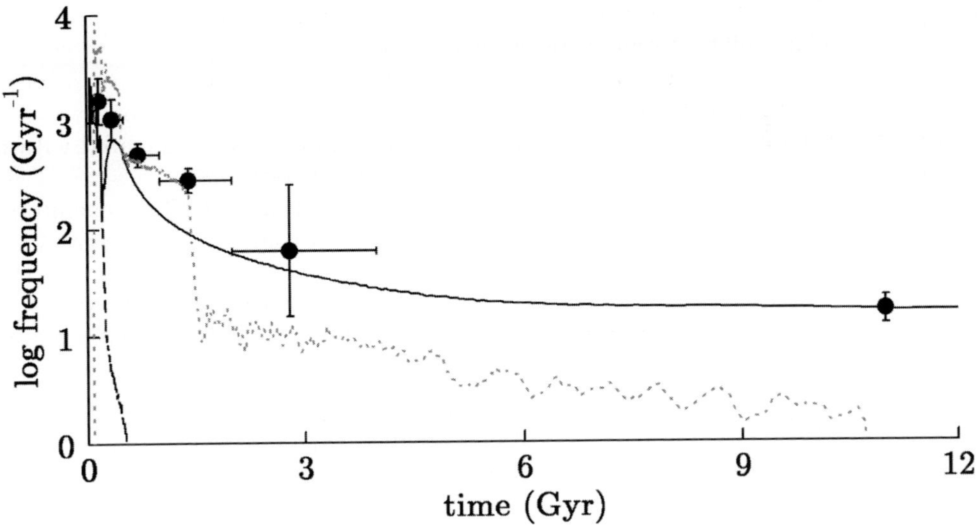

Figure 2. Delay-time distributions (DTDs) for $\beta = 1$ for the double-degenerate (DD; solid black) and single-degenerate (SD; dotted grey) scenarios as well as DTD for $\beta = 0$ for the DD (dashed black) scenario. (Observational data points from Totani *et al.* 2008.)

angular-momentum transport during the merger. This effect was investigated by Piersanti *et al.* (2003; see also more recent papers published by this group) and they showed that the rotating-DD model naturally produces a hydrogenless SN Ia.

To solve the question as to which of the two scenarios is the dominant contributor to the SN Ia rate, population synthesis of starburst regions may be very useful and we focus on the delay-time distribution (DTD) of SNe Ia. The DTD is defined as the number of SN Ia events as a function of time in a starburst. By observing elliptical galaxies, which are for this purpose equivalent to starburst galaxies, at similar metallicity but at different redshifts, one can construct an observational DTD. This can then be compared to DTDs for starburst galaxies predicted by population synthesis. Such studies have been performed by Yungelson & Livio (2000; only DD models), Han & Podsiadlowski (2004; only SD models) and Ruiter *et al.* (2008; both SD and DD models). We recently performed a similar study (Mennekens et al. 2009) and I summarize some of the results here.

The theoretical DTDs are calculated using an updated version of the population code of De Donder & Vanbeveren (2003) and we compare with the observational DTD of Totani *et al.* (2008). We obviously account for the common-envelope process in binary evolution, but we also focus on the evolution of Case A and Case Br binaries that evolve through a canonical Roche-lobe overflow (RLOF) process, mass transfer and mass accretion (as observed in Algol systems). Binary population synthesis results depend on many parameters and after a detailed parameter study we conclude that double WD progenitors experience two RLOF phases. The RLOF may result in the formation of a common envelope. However, our population predictions reveal that the first RLOF of most DD SN Ia progenitors is a canonical RLOF with mass transfer and mass accretion (the second RLOF when the original primary is already a WD is obviously a common-envelope process).

An important consequence of this conclusion is that any analytical formalism aimed at describing the DTD of DDs that is based on the assumption that the progenitors went through two common-envelope phases is wrong.

Figure 2 shows the DTD prediction of the DD model if the first RLOF of the progenitor binaries is assumed to be conservative (all mass lost by the loser is accreted by the gainer) and nonconservative (all mass lost by the loser leaves the system through the second Lagrangian point). Comparison with the observed DTD allows us to conclude that only a population model where the first RLOF is (quasi-)conservative results in a reasonable match with observations. Note that this conclusion corresponds with the situation when population synthesis predictions of Algol binaries is compared with the observed properties of Algol binaries in the solar neighborhood.

In Figure 2 we also show the DTD of the SD model using the SD progenitors of Hachisu *et al.* (2008). In this paper, these progenitors are identified as contours for different WD masses as a function of binary period and companion mass. A SD SN Ia is assumed to result if the evolutionary track of the progenitor system traverses this contour. As can be noticed, especially the late-time behavior of the DTD poorly matches the observations. The predicted DTD with SD SN Ia progenitors only does not reproduce the late time behavior of the observed DTD. Using DD SN Ia progenitors (eventually in combination with SD progenitors) and (quasi-)conservative RLOF in Case A and Case Br intermediate-mass binaries results in the best match.

5. The delay-time distribution of NS+NS and NS+BH mergers

It is generally accepted that the rapid neutron-capture process (r-process) is responsible for the existence of the heaviest elements in the Universe. There are two sites where the physical conditions are such that the r-process can happen, i.e., in SN explosions of massive stars and in binary-neutron-star mergers (NS+NS and possibly also NS+BH) (Qian & Woosley 1996; Rosswog *et al.* 2001; Dessart *et al.* 2009).

One of the major arguments against the binary-neutron-star merger was published by Matthews *et al.* (1992). He considered the observed periods and eccentricities of the binary neutron stars known at that time and calculated the expected merger-rate timescale using the theory of gravitational-wave radiation. He implemented this in a Galactic evolutionary model and when a comparison was made between the merger rate and the observed Galactic evolution of the r-process element europium, he concluded that compared to Eu double-compact-star mergers appear too late (see the dashed line in Figure 3) and thus, neutron-star-binary mergers cannot be major r-process production sites. However, the discovery of the short-period (2.4 hr) double pulsar J0737–3039 (Lyne *et al.*

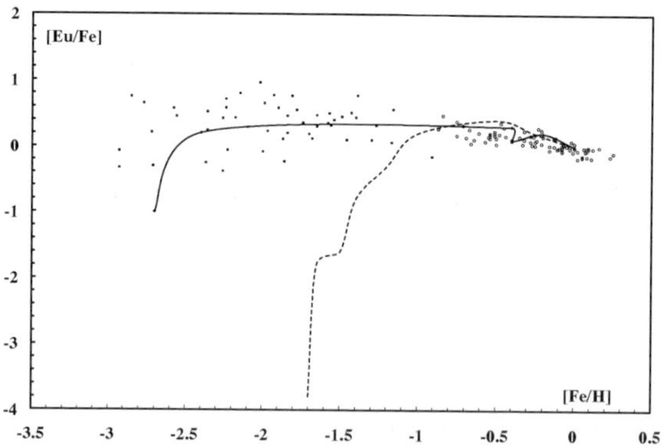

Figure 3. Temporal evolution of the predicted double-neutron-star merger rate in the solar neighborhood. The dotted curve corresponds to the prediction of Matthews (1992), while the solid line is that of De Donder & Vanbeveren (2003, 2004). The observations are from various sources discussed in those two papers.

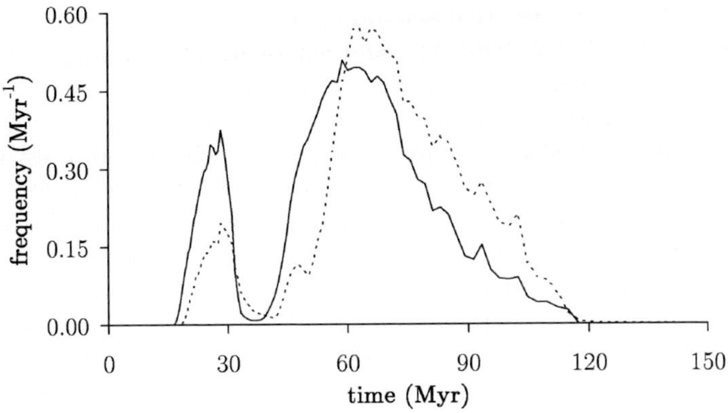

Figure 4. Delay-time distributions of double-neutron-star mergers (DTDmergers) using standard population synthesis parameters (see text). The solid and dotted curves correspond to $Z = 0.02$ and 0.002, respectively.

2004) reopened the discussion. Moreover, since 1992 our physics view of the evolution of binaries has improved significantly. Ivanova *et al.* (2003) and Dewi & Pols (2003) studied the Case BB evolution of binaries with a hydrogen-deficient helium-burning component and a normal MS companion. They proposed a scenario leading to the formation of ultracompact double-neutron-star binaries which can merge on a 1 million year timescale. De Donder & Vanbeveren (2003, 2004) included this scenario in a binary population synthesis model and calculated the DTD of these mergers (DTDmerger), defined as the evolution of double-neutron-star mergers in a starburst. The various evolutionary phases of massive binaries prior to double-neutron-star binary formation depend on a number of uncertain parameters and the DTDmerger also depends on these parameters. A typical distribution is shown in Figure 4 for $Z = 0.02$ and 0.002 (for a Salpeter-type initial mass function, flat binary mass-ratio distribution, flat binary period distribution in the log, common-envelope evolution efficiency $\alpha = 1$, mass-transfer efficiency during the canonical RLOF phase in Case Br binaries $\beta = 1$ and our preferred kick-velocity distribution during the SN explosion, with an average $v_{\rm kick} = 450\,{\rm km\,s^{-1}}$). As can be seen, double-neutron-star mergers appear very early. De Donder & Vanbeveren included this model in a Galactic chemical code and Figure 3 shows that the temporal evolution of the average merger rate now follows the observed evolution of Eu in the solar neighborhood.

6. The role of massive binaries in the self-enrichment of globular clusters

It is generally accepted that a large fraction of the low-mass stars GCs are formed from material that was enriched with hydrogen-burning products produced in more massive stars: the self-enrichment process in GCs.

The first self-enrichment scenario that was proposed in the literature suggested that the enrichment was due to low-metallicity (Z) intermediate-mass stars (Cottrell & Da Costa 1981). Detailed intermediate-mass evolutionary calculations manage to explain the observed abundance patterns in GCs but fine tuning of the evolutionary processes is required, especially of those that operate during the AGB phase (e.g., Decressin *et al.* 2009; Ventura & D'Antona 2009). Note that if the GC initially had an intermediate-mass binary population similar to the one in other aggregates (like in the association Sco OB2; Kouwenhoven 2006), and if the DD scenario (Webbink 1984) is responsible for the SN Ia events, population synthesis reveals that many SNe Ia should have occured in the past. To illustrate this, using the Brussels population code, a starburst simulation of 10^6 $M_\odot$ in stars with initial masses $\geqslant 0.8$ $M_\odot$ and where 30% of the intermediate-mass stars are primaries of binaries with an orbital period $\leqslant 10$ years, predicts 10 000–15 000 SN Ia events. To avoid a conflict with the observed Z of GCs, it is clear that most of the matter ejected during the SNe Ia must have left the cluster. Interestingly, if the SD scenario applies we do not expect many SNe Ia at all. The reason for this is that the number of SNe Ia predicted by the SD scenario strongly depends on Z, and low Z implies very few SNe Ia (Hachisu *et al.* 2008).

A second self-enrichment scenario is based on the assumption that prior to the formation of the low-mass stars, a population of low-Z massive stars was present. This model may work if a process is available to remove hydrogen-burning products at low velocity (below the escape velocity of the cluster) from the low-Z massive stars. Decressin *et al.* (2007) propose the *Winds of Fast Rotating Massive Stars* scenario. For this scenario to work, fast means really very fast (equatorial velocities of order 800–1000 $\mathrm{km\,s^{-1}}$) and it remains to be demonstrated whether such high average values are realistic in low-metallicity regions. However, if a massive star is a binary member, it may lose its CNO-processed layers naturally through the RLOF/common-envelope/spiral-in process and fast rotation is not needed (remember that most of the massive stars in our Galaxy are observed as members of binaries; Section 2). Therefore, I propose the *RLOF/common-envelope/spiral-in Mass Loss in Massive Binaries* scenario.

The DTD of merging double-neutron-star binaries was discussed in Section 5 and it was concluded that the Galactic temporal evolution of these mergers follows the observed temporal evolution of Eu in the solar neighborhood, which may be an indication that they are production sites of r-process elements that cannot be neglected. I suggest that these binaries may also be important sites of r-process self-enrichment of GCs.

References

Belczynski, K., Kalogera, V., Rasio, F. A., Taam, R. E., Zezas, A., Bulik, T., Maccarone, T. J., & Ivanova, N. 2008, *ApJS*, 174, 223

Belczynski, K., Bulik, T., & Kalogera, V. 2002, *ApJ*, 571, 147

Belkus, H. 2008, PhD Thesis, Vrije Universiteit, Brussels

Belkus, H., Van Bever, J., & Vanbeveren, D. 2007, *ApJ*, 659, 1576

Belkus, H., Van Bever, J., Vanbeveren, D., & Van Rensbergen, W. 2003, *A&A*, 400, 429

Blaauw, A. 1967, *Bull. Astr. Inst. Netherlands*, 15, 265

Brinchmann, J., Kunth, D., & Durret, F. 2008, *A&A*, 485, 657

Canon, R., Eggleton, P. P., Zytkow, A. N., & Podsiadlowski, P. 1992, *ApJ*, 386, 206

Cottrell, P. L. & Da Costa, G. S. 1981, *ApJ*, 245, 79

De Donder, E. & Vanbeveren, D. 2003, *NewA*, 9, 1

De Donder, E. & Vanbeveren, D. 2004, *NewAR*, 48, 861

De Donder, E. & Vanbeveren, D. 1998, *A&A*, 333, 557

De Donder, E. Vanbeveren, D., & Van Bever, J. 1997, *A&A*, 318, 812

Decressin, T., Charbonnel, C., Siess, L., Palacios, A., Meynet, G., & Georgy, C. 2009, *A&A*, 505, 727

Decressin, T., Meynet, G., Charbonnel, C., Prantzos, N., & Ekström, S. 2007, *A&A*, 464, 1029

Dessart, L., Ott, C.D., Burrows, A., Rosswog, S., & Livne, E. 2009, *ApJ*, 690, 1681

Detmers, R. G., Langer, N., Podsiadlowski, P., & Izzard, R. G. 2008, *A&A*, 484, 831

Dewi, J. D. M. & Pols, O. R. 2003, *MNRAS*, 344, 629

Gies, D. R. 1987, *ApJS*, 64, 545

Hachisu, I., Kato, M., & Nomoto, K. 2008, *ApJ*, 679, 1390

Han, Z. & Podsiadlowski, P. 2004, *MNRAS*, 350, 1301

Han, Z., Podsiadlowski, P., & Lynas–Gray, A. E. 2007, *MNRAS*, 380, 1098

Hunter, I., Lennon, D. J., Dufton, P. L., Trundle, C., Simón–Díaz, S., Smartt, S. J., Ryans, R. S. I., & Evans, C. J. 2008, *A&A*, 479, 541

Hutchings, J. B. 1975, *ApJ*, 200, 122

Ivanova, N., Belczynski, K., Kalogera, V., Rasio, F. A., & Taam, R. E. 2003, *ApJ*, 592, 475

Kouwenhoven, M. B. N. 2006, PhD Thesis, University of Amsterdam

Kudritzki, R. P. 2002, *ApJ*, 577, 389

Lada, C. J., Margulis, M., & Dearborn, D. 1984, *ApJ*, 285, 141

Langer, N., Cantiello, M., Yoon, S.-C., Hunter, I., Brott, I., Lennon, D., de Mink, S., & Verheijdt, M. 2008, *Proc. IAU Symp.* No. 250, 167

Lyne, A. G., *et al.* 2004, *Science*, 303, 1153

Mason, B. D., Hartkopf, W. I., Gies, D. R., Henry, T. J., & Helsel, J. W. 2009, *AJ*, 137, 3358

Matthews, G. J., Bazan, G., & Cowan, J. J. 1992, *ApJ*, 391, 719

McCrady, N., Gilbert, A. M., & Graham, J. R. 2003, *ApJ.*, 596, 240

Mennekens, N., Vanbeveren, D., De Greve, J. P., & De Donder, E. 2009, *A&A*, submitted

Mokiem, M. R., *et al.* 2006, *A&A 456*, 1131

O'Shaughnessy, R., Kalogera, V., & Belczynski, K. 2009, *ApJ*, submitted (arXiv:0908.3635)

Piersanti, L., Gagliardi, S., Iben Jr., I., & Tornambe, A. 2003, *ApJ*, 583, 885

Pols, O. R., Izzard, R. G., Lugaro, M., & de Mink, S. E. 2009, *Proc. IAU Symp.* No. 252, 383

Pols, O. R. & Marinus, M. 1994, *A&A*, 288, 475

Portegies Zwart, S.F., Baumgardt, H., Hut, P., Makino, J., & McMillan, S. L. W. 2004, *Nature*, 428, 724

Qian, Y.-Z. & Woosley, S. E. 1996, *ApJ*, 471, 331

Rosswog, S., Freiburghaus, C., Thielemann, F.-K., Davies, M. B. 2001, in: J. C. Wheeler & H. Martel, *20th Texas Symp. on relativistic astrophysics*, AIP Conf. Proc. No. 586, p. 343

Ruiter, A., Belczynski, K., & Fryer, C. 2009, *ApJ*, 699, 2026

Saio, H. & Nomoto, K. 1998, *ApJ*, 500, 388

Suzuki, T. K., Nakasato, N., Baumgardt, H., Ibukiyama, A., Makino, J., & Ebisuzaki, T. 2007, *ApJ*, 668, 435

Thorne, K. S. & Zytkow, A. N. 1977, *ApJ*, 212, 832

Totani, T., Morokuma, T., Oda, T., Doi, M., & Yasuda, N. 2008, *PASJ*, 60, 1327

Van Bever, J. & Vanbeveren, D. 1998, *A&A*, 334, 21

Van Bever, J. & Vanbeveren, D. 2000, *A&A*, 358, 462

Van Bever, J. & Vanbeveren, D. 2003, *A&A*, 400, 63

Vanbeveren, D. 2007, in: N. St-Louis & A. F. J. Moffat, *Massive Stars in Interacting Binaries*, ASP Conf. Ser., 367, p. 13, San Francisco: Astron. Soc. Pac.

Vanbeveren, D. 2009, *NewAR*, 53, 27

Vanbeveren, D., Belkus, H., Van Bever, J., & Mennekens, N. 2009, in: E. Pérez, R. de Grijs & R. M. González Delgada, *Young massive star clusters – Initial conditions and environments*, *ApSS*, 324, 271

Vanbeveren, D. & De Donder, E. 2006, *NewA*, 12, 95

Vanbeveren, D. & De Loore, C. 1994, *A&A*, 290, 129

Ventura, P. & D'Antona, F. 2009, *A&A*, 499, 835

Webbink, R. 1984, *ApJ*, 277, 355

Yungelson, L. & Livio, M. 2000, *ApJ*, 528, 108

Zinnecker, H. & Yorke, H. W. 2007, *ARA&A*, 45, 481

Star clusters: basic galactic building blocks
Proceedings IAU Symposium No. 266, 2009
R. de Grijs & J. R. D. Lépine, eds.

doi:10.1017/S1743921309991177

Star clusters and single stellar populations

Licai Deng[1] and Yu Xin[2]

[1]National Astronomical Observatories, Chinese Academy of Sciences,
Beijing 100012, China
email: licai@bao.ac.cn

[2]Argelander Institute for Astronomy, Unversity of Bonn,
53121 Bonn, Germany
email: yxin@astro.uni-bonn.de

Abstract. Star clusters are ideal laboratories to test the theory of stellar evolution and provide very tight constraints on the concept of single stellar poputions (SSPs). Observations show that some stars fail to conform to the theoretical evolutionary scenario applicable to single stars. These special objects, particularly blue stragglers, present a challenge to our current theory of stellar evolution. They may be very important in the context of the integrated spectral properties of clusters. Here, we review the construction of SSP models, both empirically using star clusters and theoretically based on binary-interaction theory.

Keywords. blue stragglers, open clusters and associations: general, Galaxy: stellar content

---

1. Introduction

Simple stellar population (SSP) theory provides a very effective way to study the stellar content in complicated, unresolved stellar systems. By definition, an SSP is a group of stars born at the same time from the same gas cloud, so that they have the same chemical composition at the time of starbirth and the same age. The masses of the SSP's member stars are given by a certain power-law mass distribution at the time of their birth. By implementing the theory of stellar evolution (for single stars) and stellar atmosphere models, the population's behavior during evolution and its observable quantities such as the integrated spectrum, photometric colors, mass-to-light ratio, etc., can be predicted. This technique is known as evolutionary population synthesis (EPS; Bressan *et al.* 1994; Bertelli *et al.* 1994; Worthy 1994; Bruzual & Charlot 2003, hereafter BC03). However, the central engine driving EPS, the theory of stellar evolution, does not cover all observational evidence, even for those stellar populations resolved on color–magnitude diagrams (CMDs) of star clusters. The blue stragglers (BSs), first discovered in the globular cluster M3 (Sandage 1953) in the Galaxy, and later proved common in all resolved stellar systems, for instance, cannot be explained directly. As they are thoroughly studied in the well-known Galactic clusters, BSs cannot be excluded from the clusters' stellar populations, but they cannot be descendants of single stars. Therefore, they have to be accounted for in some way other than through single-star evolution theory. BSs are now understood as the remnants of stellar interactions, including mass exchange (Tian *et al.* 2006; Chen & Han 2009), coalescence of close binary systems and collisional encounters in dense environments (Sills *et al.* 2005; Ferraro *et al.* 2003). These processes happen in the real world and certainly affect the general properties of SSPs. They therefore introduce some uncertainties into applications of EPS.

SSP models can also be built by direct observations of star clusters. This technique is referred to as 'optimized population synthesis' (e.g., Bica & Alloin 1986a,b). It optimizes the problems of the stellar initial mass function and stellar atmospheres. However, due to limitations of parameter-space coverage and issues to do with observations such as sampling uncertainties, it has fewer applications to studies of galaxies compared with EPS.

Assessment of BSs in terms of their contributions to their host population can be realized either theoretically or empirically. In principle, by theoretically collating all physics relevant to stellar interactions and assembling the results into the conventional SSP framework, one could construct more realistic SSP models. However, because the various combinations of binary parameters and interacting processes are usually too complicated, the results are not sufficiently convincing. In fact, the formation of BSs is much more complicated and stochastic than can be dealt with by existing theoretical models. This is evident from the distributions of BSs in star cluster CMDs. Therefore, empirical treatment seems a better choice. By building up synthetic SSP models based on star clusters covering large grids in both age and metallicity, one can make a model database available for EPS similar to that of the conventional SSP grids. Empirical models constructed this way can be used as an observational constraint to theoretical studies of SSP models and the stellar-interaction processes included in the models.

Here, we analyze the uncertainties in the conventional SSP models due to BSs, using empirical models based on our Galactic open cluster (OC) sample (Deng *et al.* 1999; Xin & Deng 2005—hereafter XD05—Xin *et al.* 2006, hereafter XDH06), and then compare the results with theoretical binary-synthesis models.

2. Observational facts

Since the first identification of BSs by Sandage (1953) in his work on the globular cluster M3, BSs have been found to be a common ingredient of resolved stellar systems on various scales, from star clusters to galaxies. BSs in star clusters can be clearly identified in cluster CMDs based on photometric observations. As BSs do not obey the best-fitting cluster isochrones in a given CMD, determining whether or not they are cluster members is crucial. For most Galactic globular clusters, membership is not of special concern, simply because field contamination is not a big problem. For OCs located in the disk and therefore surrounded by diffusely distributed field stars, however, membership is an important issue. This problem can be handled with membership information from proper-motion and/or radial velocity observations. For our analysis, we will rely on cataloged clusters with relatively robust information on their stellar membership (Ahumada & Lapasset 1995; Dias *et al.* 2002).

In principle, Galactic GCs can also be used to study SSP models. The BS contents in these clusters have been made available using *HST* observations by Piotto *et al.* (2004). However, considering the general (diffuse) stellar content of galaxies, the conditions related to the formation mechanisms of BSs (e.g., binary fractions, binary parameter distributions, dynamical evolution of the system, overall stellar density and collision rate, and its evolution within the system, etc.) will likely be more like those in OCs than in GCs. For this reason, we will focus on OCs.

3. Results from open clusters

Based on the photometric catalog of BSs in OCs (Ahamada & Lapasset 1995), we have constructed synthetic integrated spectral-energy distributions (ISEDs) for selected clusters in a series of recent papers (Deng *et al.* 1999; XD05; XDH06). The underlying ideas, assumptions and our approach are as follows.

• Assuming that OCs can be approximated as natural SSPs, we first restore the contributions of the evaporated low-mass stars (not important in terms of their integrated light, but still considered for reasons of consistency). This can be realized by substituting the observed stellar component by a conventional SSP model isochrone. All stars in the population obey a single initial mass function (assumed to be Salpeter-like), including stars that eventually evolve into BSs through any formation channel.

• We assume that the formation of BSs will not change the mass function of the cluster. Considering the number of BSs (usually of order 10) and the total number of stars (of order 1000), this is a good approximation.

• We also assume that BSs are basically hydrogen-burning stars, so that their fundamental physical parameters can be determined by fitting them using main-sequence models. According to our present understanding of BSs, this assumption is quite secure. This can also be justified by spectrophotometric observations of BSs, as done in, e.g., NGC 2682 (M67; Liu *et al.* 2008).

• All other faint 'straggling' stars, including BSs below the turnoff and underlopers below the subgiant branch, are neglected in constructing the synthetic SSP model. This is justified by the fact that the total number of such objects is usually very small, and they are not important when considering the integrated light.

After taking all these assumptions into account, the synthetic SSP model for a cluster can be constructed based on two components. The first is composed of all stars represented by the best-fitting isochrone in the CMD and all other stars evaporated dynamically during the past evolutionary history of the cluster. This is nothing but the conventional SSP model. The second component includes the BSs; bright stragglers at redder colors (i.e., yellow or red stragglers) are also taken into account if they exist in the catalog. For the second component (the 'BS component'), the spectral properties are derived by fitting single-star models to the individual objects using the algorithm presented in XD05. The sum of all light from the members making up this component provides us with the final modifications to the conventional SSP model.

As a result, the synthetic ISEDs of stellar populations are drastically different from the conventional SSP models for a given age and metallicity. In terms of the shape of the spectral properties, the synthetic ISED of clusters (regarded as the real SSP) are enhanced towards shorter wavelengths compared with the conventional models. This implies that a stellar population may look younger (bluer colors) or more metal-poor than the conventional SSP. In other words, if conventional SSP models are used to fit the ISED and measure the age or metallicity of a real stellar population, one or both parameters will be underestimated compared to the true parameters, thus posing a problem for our current understanding of galaxies.

We will show that the synthetic ISED can be fitted very well with conventional SSP models, but the uncertainties are significant. Shown in Figure 1 is an ISED fit to NGC 2420 using conventional SSP models, by either fixing the metallicity to obtain the age (left), or fixing the age to derive the metallicity (right). It is clear that good-quality fits (much better than for unresolved observations) yield nevertheless very different ages or metallicities compared to the true parameters. This technique actually measures the uncertainties in age or metallicity in EPS.

3.1. *Uncertainty in age*

For a given set of parameters of age and metallicity, the synthesized ISED of a cluster is enhanced towards shorter wavelengths with respect to the conventional SSP models because of the inclusion of BSs. This enhancement is the cause of incorrect age determinations on the basis of conventional SSP model fits (XD05; XDH06).

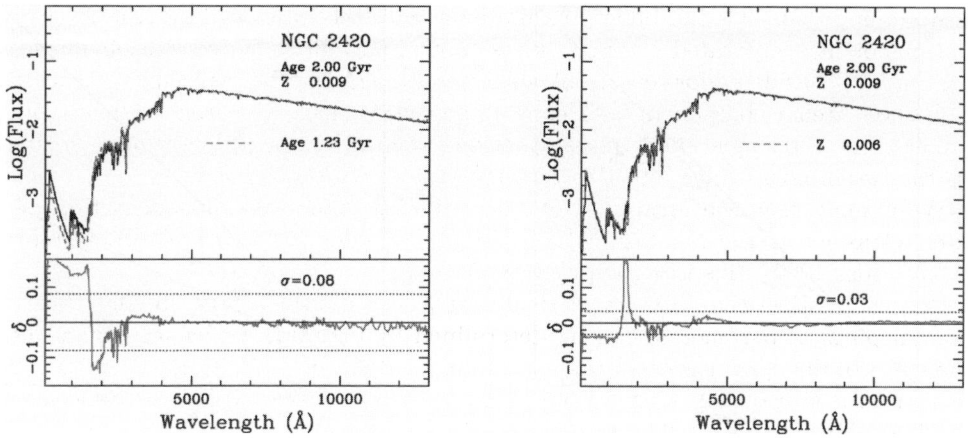

Figure 1. Examples of fits to the synthetic ISEDs of clusters using conventional SSP models. *(left)* Adopting the correct metallicity and leaving age as a free parameter. The original parameters of the clusters are given in the upper right-hand corner in each panel. The solid line is the synthetic ISED, while the dashed line represents the conventional SSP model. The lower part shows the fit errors. *(right)* As for the left panel, but for metallicity as a free parameter.

For clusters with relatively robust parameter determinations, the true age values versus those from conventional SSP fits are shown in the left panel of Figure 1. The distance from the diagonal represents the uncertainties in age for each cluster, shown as dots in the plot.

3.2. *Uncertainty in metallicity*

Figure 1 (right panel) shows fits of the synthetic ISEDs for the same clusters as in previous subsection using the conventional SSP models. In each panel, the original parameters of the clusters are included. The solid lines in the plots are for the synthetic ISED, while the dashed lines are for the fitted conventional SSPs (whose metallicities are also indicated). The lower part of each panel shows the fitting accuracy. As can be seen, the fits are very good. Across the entire range of optical bands, the difference is generally below a few percent, better than what observational photometry can actually distinguish. The only sizable deviation is found in the ultraviolet.

It is clear from Figure 2 that, when compared with age fits, the scatter in metallicity is much larger. The linear fit only shows the trend. One of the possible reasons for this would be the limited grid coverage in metallicity. There are only five metallicities in the library, which makes interpolation much less accurate than for our age fits.

3.3. *Uncertainty in mass-to-light ratio*

Due to the BS content in general stellar populations, the normal way of estimating their mass using photometric data will be affected. This can be shown on the basis of a comparison between the mass-to-light ratios (MLRs) of the synthetic with respect to the conventional SSP models. The mass of an SSP will not be affected by BS formation, because most BS formation processes conserve mass, at least approximately. Therefore, alterations to the MLR are due to changes in the luminosity of the SSP. In general, the formation of BSs will contribute more light to the system, since BSs form a more massive star from low-mass progenitors. However, this needs more investigation.

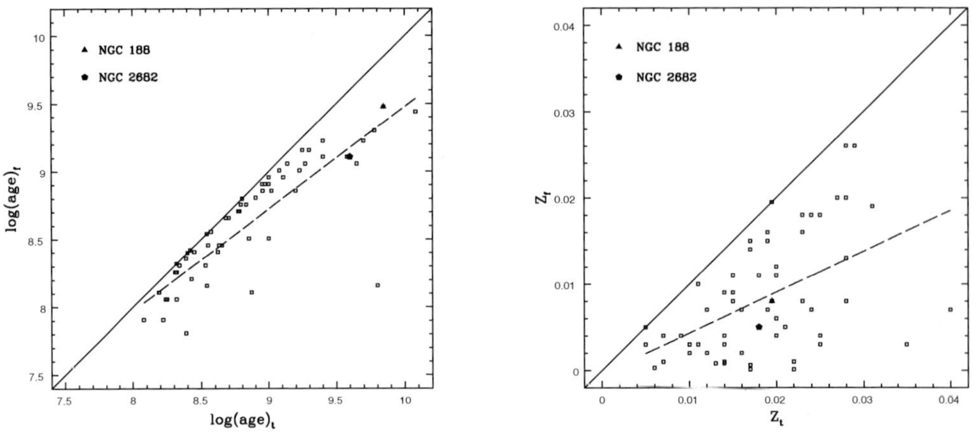

Figure 2. Synthetic age *(left)* and metallicity *(right)* of open clusters (vertical axis) versus isochrone-fitted values (horizontal axis). The two well-studied and relatively well-populated clusters M67 (NGC 2682) and NGC 188 are indicated by heavy dots. A linear fit to the sample clusters is given by the dotted line.

4. Comparison between cluster ISEDs and binary synthesis

Because the empirical method discussed above is limited by both age and metallicity coverage of the sample of stars clusters, constructing a complete library of revised SSP models is not possible at the present time. Further applications of observationally constrained SSP models to EPS need more effort and a larger, more diverse sample. Therefore, a theoretical approach to the problem becomes important if we want to reach a perfect solution. This will need to rely on statistical studies of the detailed physical processes, modeling of stellar evolution with all processes included, and hence will need a lot of effort.

Interacting binary systems are believed to be one of the major ways of BS formation. There are two channels of BS formation related to interactions in close binaries, mass exchange between the components (Zhang *et al.* 2004, 2005; Tian *et al.* 2006; Lu *et al.* 2009) and coalescence (Chen & Han 2009), which will eventually happen in some close binary systems. Thanks to recent advances in studies of binary evolution, population synthesis involving binaries ('binary synthesis') can be done. This provides another way of getting a more realistic picture of SSPs than the conventional scenario (Hurley *et al.* 2005; Tian *et al.* 2006; Zhang *et al.* 2005) and can be regarded as a partial solution to the current problem.

To explore the capability of binary evolutionary models and the binary synthesis technique, we compare the results of Zhang *et al.* (2005) to the conventional SSP models from BC03, in terms of broad-band colors derived from SEDs of both datasets. Figure 3 (left) shows that, due to the formation of BSs with the assumptions made for the binary parameters in a population, the binary population has bluer colors for almost all ages (except for part of $V - I$) with respect to BC03. Just like the synthetic ISED of a star cluster, the ISED of a binary population is enhanced at shorter wavelengths compared with the conventional SSP model, as shown in Figure 3 (left). This enhancement is systematic in the model grid (see Zhang *et al.* 2004).

A comparison with our empirical approach is shown in the right-hand panel of Figure 3. If we derive the age using the above-mentioned technique, both binary synthesis (dashed line) and the empirical ISED (open dots; size proportional to fitting error) predict a

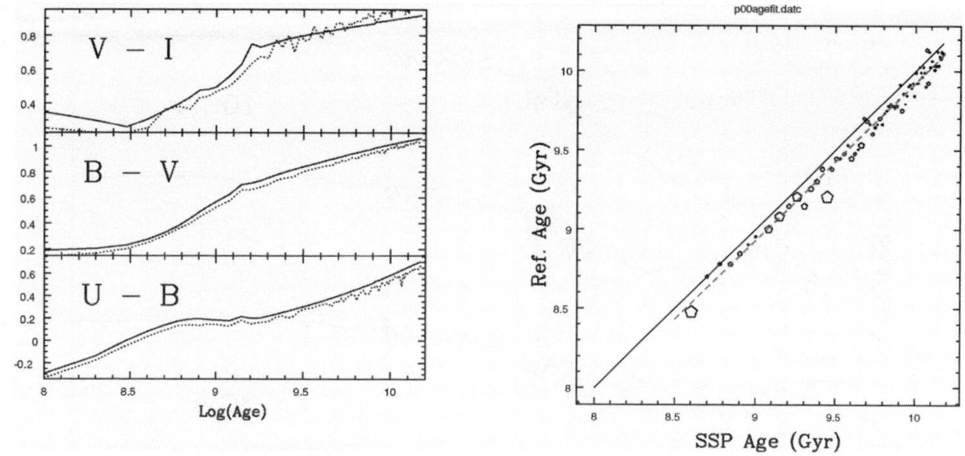

Figure 3. *(left)* Broad-band colors derived from binary synthesis (dotted lines) compared with the conventional BC03 model (solid lines). The ages predicted by binary synthesis are systematically younger than resulting from the conventional model *(right)* (dashed line) due to the bluer colors. The synthetic ISEDs of the clusters are shown as open symbols (size proportional to fitting errors), showing not only a large scatter due to stochasticity, but also younger ages, on average, than predicted by binary synthesis.

younger age. It is clear that the theoretical results from binary population models cannot fully explain the age discrepancy due to the existence of BSs in star clusters. We anticipate that, instead of considering binary mass exchange alone, a full consideration of all BS formation channels in clusters may reach closure with regard to this problem. Although not perfect in covering parameter space and all possible evolutionary phases, stellar population models incorporating close binary systems can, at least in part, reveal the effects of remnants (mostly BSs) on conventional SSP approaches. Bearing in mind that collisional encounters and mergers are not involved, the results presented here on the basis of binary synthesis should be regarded as a lower limit to the modifications to the conventional SSP models.

5. Discussion and concluding remarks

We have reviewed the role of BSs in star clusters and the contribution of BSs to the integrated light of the host stellar populations based on a sample of Galactic open clusters.

Considered as a general case for stellar populations covered by the parameter range of our OC sample, the uncertainties in the conventional SSP model due to neglecting stellar interactions are measured in terms of the basic parameters. It is shown that the results of EPS for the populations represented by the OC sample differ significantly from the correct values. If age and metallicity are both uncertain because of other problems (Charlot *et al.* 1996), the MLR will also be highly uncertain. If one uses the conventional SSP models, the mass of a population estimated based on V-band data, can be overestimated by as much as 20% based on our OC analysis. However, considering binary population synthesis, deriving the correct MLR is more complicated. BS formation in a stellar population from binary mass exchange usually involves a star near the turnoff. The high-luminosity red giant-branch (RGB) phase following the single-star evolutionary path will thus be avoided. As a result, the population gains light in the blue (BS contribution) and loses

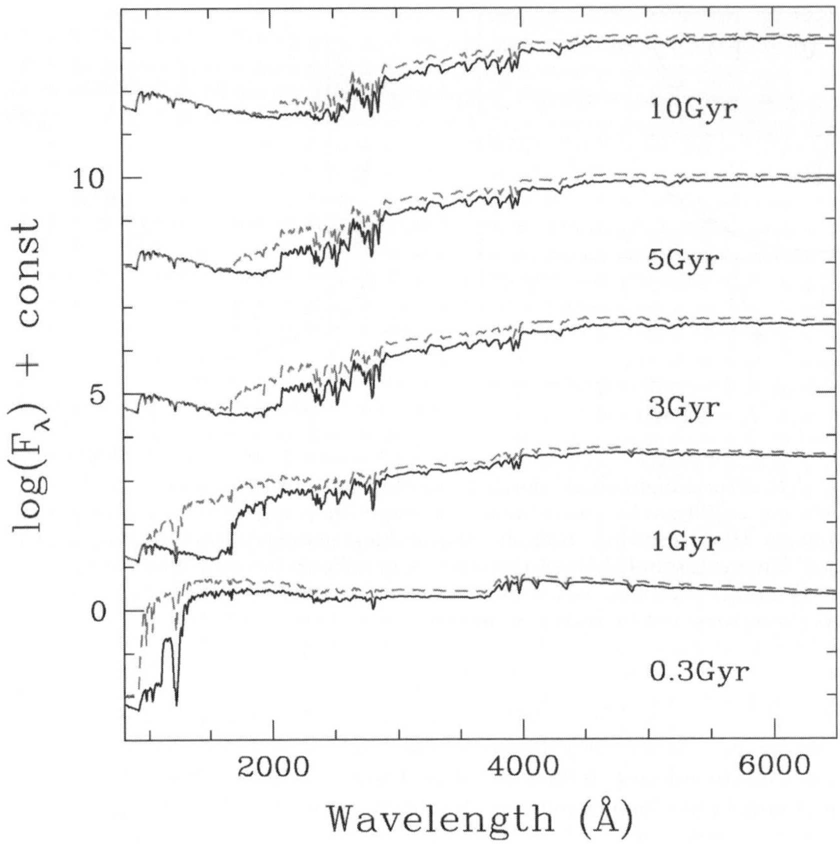

Figure 4. Comparison between BS03 ISEDs of SSP models (lower solid spectra) and the empirically revised SSP models with the same ages (indicated for each pair of ISEDs) for solar metallicity.

light in the red (avoiding the RGB phase). Therefore, the bolometric luminosity of a population may be reduced because of BS formation (Chen & Han 2009).

Using photometric observations of Galactic OCs (Ahumada & Lapasset 1995), we have tested the modifications to the main properties of the conventional SSP model and its uncertainties in applications of population analysis. We have shown that, limited to the age and metallicity parameters of the OC sample, the conventional SSP models are seriously affected, and this may have severe implications for galaxy population studies using EPS. To correct for these errors, studies of SSPs using star clusters of a larger sample and using the binary synthesis method with all interacting processes taken into account are both needed. Work to generate modified SSP models based on our current results is in progress (see for preliminary results Figure 4).

Acknowledgements

We thank the Chinese National Science Foundation for partial support through grants 10973015, 10573022 and 10778719, and the Ministry of Science and Technology of China through grant 2007CB815406. We would like to thank Z. Han, F. H. Zhang and X. F. Chen for providing binary population synthesis models.

References

Ahumada, J. & Lapasset, E. 1995, *A&AS*, 109, 375 (AL95)
Bertelli, G., Bressan, A., Chiosi, C., Fagotto, F., & Nasi, E. 1994, *A&AS*, 106, 275
Bica, E. & Alloin, D. 1986a, *A&AS*, 66, 171
Bica, E. & Alloin, D. 1986b, *A&A*, 162, 21
Bressan, A., Chiosi, C., & Fagotto, F. 1994, *ApJS*, 94, 63
Bruzual G. & Charlot, S. 2003, *MNRAS*, 344, 1000 (BC03)
Charlot, S. Worthy, G., & Bressan, A. 1996, *ApJ*, 457, 625
Chen, X., & Han, Z.W. 2009, *MNRAS*, 395, 1822
Deng, L., Chen, R., Liu, X. S., & Chen, J. S. 1999, *ApJ*, 524, 824
Dias, W. S., Alessi, B. S., Moitinho, A., & Lépine, J. R. D. 2002, *A&A*, 389, 871
Ferraro, F. R., Sills, A., Rood, R. T., Paltrinieri, B., & Buonanno, R. 2003, *ApJ*, 588, 464
Hurley, J. R., Pols, O. R., Aarseth, S. J., & Tout, C. A. 2005, *MNRAS*, 363, 293
Liu, G. Q., Deng, L., Chávez, M. D., Bertone, E., Herrero, D. A., & Mata-Chávez, M. D. 2008, *MNRAS*, 390, 665
Piotto, G., De Angeli, F., King, I. R., Djorgovski, S. G., Bono, G., Cassisi, S., Meylan, G., Recio–Blanco, A., Rich, R. M., & Davies, M. B. 2004, *ApJ* (Letters), 604, L109
Sandage, A. R. 1953, *AJ*, 58, 61
Sills, A. I., Adams, T., & Davies, M. B. 2005, *MNRAS*, 358, 716
Worthey, G. 1994, *ApJS*, 95, 107
Xin, Y. & Deng, L. 2005, *ApJ*, 619, 824 (XD05)
Xin, Y., Deng, L., & Z. Han 2007, *ApJ*, 660, 319 (XDH06)
Zhang, F., Han, Z., Li, L., & Hurley, J. R. 2004, *A&A*, 415, 117
Zhang, F., Li, L., & Han, Z. 2005, *MNRAS*, 364, 503

Star clusters: basic galactic building blocks
Proceedings IAU Symposium No. 266, 2009
R. de Grijs & J. R. D. Lépine, eds.

doi:10.1017/S1743921309991189

A new method to measure stellar mass-loss rates

Jason S. Kalirai

Space Telescope Science Institute, 3700 San Martin Drive, Baltimore, MD 21218, USA
email: jkalirai@stsci.edu

Abstract. The total stellar mass loss that a star suffers through post-main-sequence evolution is of vital importance to understand its subsequent evolution. The mass-loss rate along the first-ascent red-giant branch alone determines the upper red-giant-branch luminosity function and horizontal-branch morphology. The distribution of stars in these phases directly affects our interpretation of the integrated colors of distant galaxies, and is therefore of fundamental importance for galaxy formation and evolution studies in the higher-redshift Universe. Yet, these mass-loss rates, especially as a function of age and metallicity, are very poorly constrained in current models. I present new constraints on this field based on imaging and spectroscopic observations of the end products from this evolution, white dwarfs. By studying the mass distribution of these dead stars in nearby star clusters with a range of (known) ages and metallicities, we can directly constrain the mass-loss rates of stars across a range of environments. These observations directly impact several fields in astrophysics, including our knowledge of the enrichment of the interstellar medium, our ability to construct population synthesis models to interpret galaxy colors and the general interpretation of the sources and processes responsible for the observed ultraviolet upturn in elliptical galaxies.

Keywords. stars: mass loss, techniques: photometric, techniques: spectroscopic, white dwarfs

1. The importance of measuring stellar mass loss

Correctly characterizing the amount of mass loss that stars suffer through post-main-sequence evolution represents one of the most important and fundamental goals of stellar astrophysics. The mass-loss rate on the first-ascent red-giant branch alone can drastically affect the eventual fate of an evolving star. For example, the total integrated red-giant-branch mass loss can affect the location that a star leaves the red-giant branch for low-mass stars (e.g., Castellani & Castellani 1993; D'Cruz *et al.* 1996; Kalirai *et al.* 2007), and therefore alters the upper red-giant-branch luminosity function. Equally important, higher rates of mass loss lead to hotter exposed stars, which will occupy a bluer position on the subsequent core helium-burning horizontal branch (Rood 1973). Extremely high levels of mass loss on the red-giant branch can also lead to stars with very thin hydrogen envelopes that bypass both of these phases and evolve directly to the white-dwarf cooling sequence with helium cores (e.g., Hansen 2005; Kalirai *et al.* 2007), or those that experience late flashes (e.g., see D'Cruz *et al.* 1996). Clearly, the initial phases of mass loss after the main sequence can have widespread implications for the subsequent evolution of stars.

Unfortunately, theoretical predictions of post-main-sequence stellar mass loss are difficult to calculate because of our insufficient understanding of mass-loss mechanisms on the red-giant branch and at the helium flash, and also the unknown number of thermal pulses on the asymptotic giant branch (e.g., Habing 1996; Weidemann 2000). The mass-loss rates depend on the assumed composition of the dust grains, the dust-to-gas ratio, the expansion velocity of the stellar envelope and the temperature/luminosity of the star (e.g., Groenewegen 2006). Given our lack of knowledge of these basic stellar properties, variations in mass-loss recipes are large, leading to large uncertainties in our basic predictions from stellar evolution models and therefore our interpretation of the properties of stars in these phases today (see, e.g., Catelan 2007). We note that the different models especially disagree with one another at metallicities above solar.

A combination of photometric and spectroscopic efforts have recently led to the construction of an initial–final mass relation over a wide mass range. The relation simply illustrates the connection between the final remnant white-dwarf mass, and its initial progenitor mass, and therefore directly probes stellar mass loss. Kalirai *et al.* (2008) present the first results on the low-mass end of the relation, by undertaking a new spectroscopic campaign targeting white dwarfs in old open star clusters. Kalirai *et al.* (2007) and Kalirai *et al.* (2009) also extend the relation to both a single metal-rich and metal-poor cluster, and therefore a preliminary estimate of the dependences of mass loss on metallicity can be calculated. Given that the white dwarfs are end products of both the red-giant and asymptotic giant-branch phases, we cannot strictly use our observations to constrain the red-giant-branch mass loss, but rather the total integrated mass loss through all of the post-main-sequence evolutionary phases.

2. The data

A large number of observations over the past 30 years have successfully established the initial–final mass relation for solar metallicity and $M_{\rm initial} \gtrsim 3$ M$_\odot$ (a complete summary is provided in Kalirai *et al.* 2009). The first step in this process involves imaging rich star clusters down to sufficient depth to yield white-dwarf candidates (e.g., the *CFHT* Open Cluster Survey: Kalirai *et al.* 2001a,b,c, 2003). These stars are then followed with spectroscopy to (i) verify their white-dwarf nature and (ii) build up enough signal-to-noise ratio to accurately characterize the higher-order Balmer lines (e.g., Hϵ and H8). Recently, the bulk of these observations have been carried out using the *Keck*/LRIS multiobject spectrograph on Mauna Kea (e.g., Williams, et al. 2004, 2009; Williams & Bolte 2007; Kalirai *et al.* 2005, 2007, 2008, 2009). Multiobject capabilities essentially allow sampling of fainter white dwarfs since their spectra are simultaneously obtained on the same spectroscopic mask, in the same exposures (e.g., 20–30 white-dwarf candidates are observed together). As mentioned above, the more recent observations of the initial–final mass relation presented in Kalirai *et al.* (2008) provide the first mass measurements of white dwarfs in old open clusters.

The white-dwarf spectra are analyzed with atmosphere models to yield the fundamental properties of each star, e.g., temperature and gravity (Bergeron *et al.* 1992, 1995). With these in hand, the radius (and therefore mass) of the star can be easily calculated. As the stars belong to a star cluster, the main-sequence turnoff mass of their parent cluster is also known, and a small correction can be applied to yield the progenitor mass of each of the white dwarfs (e.g., the main-sequence lifetime of the progenitor is simply the difference between the cluster age and the white-dwarf cooling age of each individual star).

3. Characterizing stellar mass loss from white-dwarf spectroscopy

3.1. *The solar-metallicity initial–final mass relation*

The most recent initial–final mass relation, calculated using the entire set of observations up to 2009, is presented in Kalirai *et al.* (2008). This relation requires a small correction to reflect the new Stark broadening calculations presented in Tremblay & Bergeron (2009),

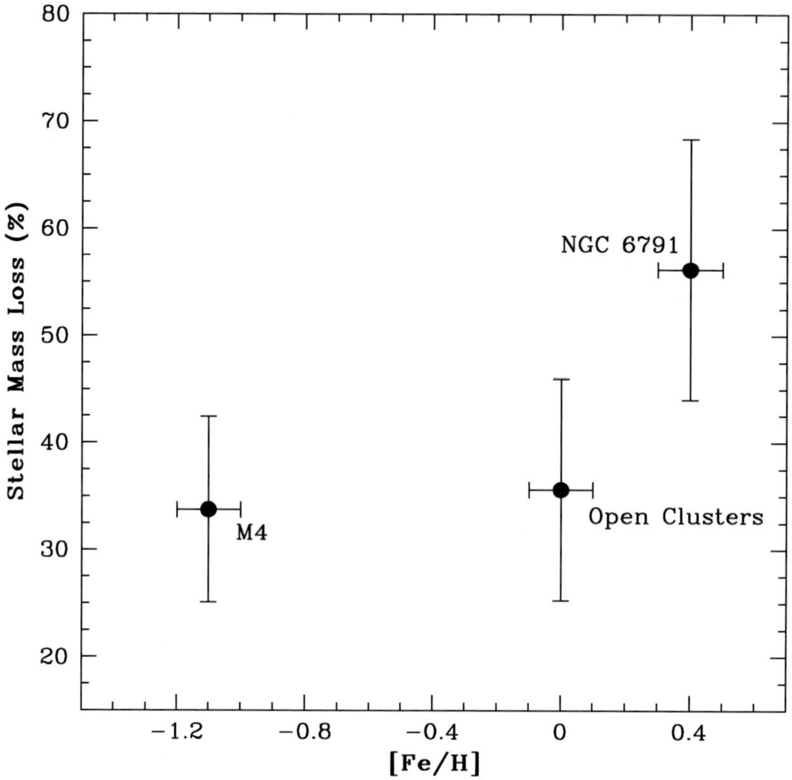

Figure 1. The total integrated mass loss through post-main-sequence evolution is shown for three different regimes, all measured by comparing the mass distribution of white dwarfs in star clusters to their progenitor mass. For both a single metal-poor globular cluster and for many rich open clusters, the mass loss is measured to be ~35%. At supersolar metallicities, our spectroscopic measurements of the mass of NGC 6791's white dwarfs suggest high levels of mass loss.

which results in a shift in all spectroscopic white-dwarf mass measurements of $+0.03$ M$_\odot$. The general form of the relation is such that more massive progenitor stars produce more massive white dwarfs. The data are well fit by a simple linear relation, $M_{\rm final} = (0.109 \pm 0.007)\ M_{\rm initial} + (0.428 \pm 0.025)$ M$_\odot$. We can therefore estimate that solar-metallicity stars with $M_{\rm initial} = 0.80 \pm 0.05$ M$_\odot$ will form white dwarfs with $M_{\rm final} = 0.52 \pm 0.03$ M$_\odot$. These stars will therefore lose 35% of their mass through post-main-sequence stellar evolution, as shown in Figure 1.

3.2. *Mass loss in metal-poor and metal-rich stars*

Kalirai *et al.* (2009) provided the first data point on the initial–final mass relation that comes from a population II system. Their *Keck*/LRIS spectroscopic study of white dwarfs in M4, [Fe/H] = −1.1 (Marino *et al.* 2008), indicates that 0.8 M$_\odot$ progenitors form 0.53 M$_\odot$ white dwarfs, an integrated mass loss of 34%. This therefore indicates that the evolution of low-mass stars that are chemically anemic leads to white dwarfs with approximately the same mass as in the solar-metallicity case. So, the mass-loss/metallicity relation may, in fact, be relatively flat between these metallicity regimes.

Unlike the findings above, Kalirai *et al.* (2007) found evidence for enhanced mass loss at extremely high metallicities. By studying the white-dwarf mass distribution in the supersolar-metallicity star cluster NGC 6791 ([Fe/H] = +0.40; Origlia *et al.* 2006;

Gratton *et al.* 2006), they showed that not only is the mass loss stochastic, but can also approach values as high as 55–60%. Formally, correcting for the +0.03 $M_\odot$ offset in Tremblay & Bergeron (2009), Kalirai *et al.* (2007) measure the mean white-dwarf mass to be $M_{\rm final} = 0.46 \pm 0.06$ $M_\odot$ (the cluster main-sequence turnoff mass is $\sim$1.05 $M_\odot$). As they discuss, such evolution likely leads to the formation of helium-core white dwarfs in this cluster.

Data points for these two regimes are also shown in Figure 1 and suggest that the form of the mass-loss/metallicity relation is not linear over a large range in metallicity. As Catelan (2007) show, some theoretical relations predict an almost exponential-type behavior of mass loss as a function of metallicity at very high metallicities [e.g., the Goldberg (1979) and Judge & Stencel (1991) relations]. Our preliminary results from this analysis, from the study of just one metal-rich and one metal-poor cluster, are qualitatively similar to such relations.

4. Summary and implications

The relationship between stellar mass loss and metallicity is fundamentally important to understand the evolution of stars beyond the main sequence. Characterizing the amount of mass loss that stars suffer, as a function of their initial mass and metallicity (as well as rotation, magnetism and other properties), leads to the eventual distribution of stars along post-main-sequence evolutionary phases such as the horizontal and asymptotic giant branches. Unlike nearby stellar systems such as Galactic star clusters, we have no direct means to separate the unresolved light that we measure from distant galaxies into these evolutionary sequences. Yet, given the brightness of stars in these phases of stellar evolution, the ultraviolet and blue light is dominated by hot horizontal-branch stars and the red light by asymptotic and red-giant-branch stars.

An excellent example of the influence of stellar evolution in shaping an unexpected spectral response can be seen in most elliptical galaxies. Despite these systems being old, metal-rich galaxies, the spectral-energy distribution for many ellipticals shows a rise in ultraviolet light (Burstein *et al.* 2008; Bressan *et al.* 1994; O'Connell 1999). In the absence of star formation and hot, young stars, the most natural source that can emit appreciably in the ultraviolet is an extremely hot horizontal-branch star (Greggio & Renzini 1990). The formation of such stars almost certainly results from enhanced mass loss along the red-giant branch, either from binary evolution or from wind-driven mass loss (Han *et al.* 2007; Kalirai *et al.* 2007). In its basic properties, the open star cluster NGC 6791 is similar to an elliptical galaxy; it is an old, metal-rich system. To explain the ultraviolet excess, Dorman, O'Connell, & Rood (1995) estimate that 15–20% of a population's evolving stars would need to populate the extremely blue horizontal branch, which is much lower than the fraction of stars actually seen in this phase in the [Fe/H] = +0.4 cluster NGC 6791.

Recently, the metallicity distribution function (MDF) of the Galactic bulge has been probed with dwarfs for the first time. Given the need for high-resolution, high signal-to-noise spectra, previous studies have been forced to measure the MDF with red giants (e.g., Zoccali *et al.* 2003). The new dwarf measurements result from stars that are microlensed, and therefore have amplified fluxes. With quick spectroscopic follow up, Cohen *et al.* (2008, 2009) and Johnson *et al.* (2008) showed that the few dwarfs in the Galactic bulge with accurate metallicity measurements are extremely metal-rich and inconsistent with the MDF from red giants. A possible explanation for this disparity may be related to biases in the giant MDF, for example, if the metal-rich dwarfs 'peel' away from the red-giant branch prior to reaching the tip, given enhanced mass loss (e.g., Castellani

& Castellani 1993). As the present sample only includes half a dozen dwarfs, it will be important to target more microlensed stars and build upon the statistics in this comparison.†

As Kilic *et al.* (2007) demonstrate, production of low-mass white dwarfs through enhanced mass loss appears to be a general occurrence in our Galaxy. In addition to the studies above, such a relation between mass loss and metallicity may also impact our understanding of planetary systems around white dwarfs and the varying characteristics of supernovae in spiral versus elliptical galaxies. More concrete conclusions will require verifying the hints that we have thus far seen by probing white-dwarf properties in systems spanning a wide range in metallicity. Fortunately, many more star clusters can be subjected to these types of studies with current 8–10m-class telescopes, as well as with the next-generation 30 m telescopes.

References

Bensby, T., *et al.* 2009, *IAU Symp.* No. 265, in press (arXiv:0908.2779)
Bergeron, P., Saffer, R. A., & Liebert, J. 1992, *ApJ*, 394, 228
Bergeron, P., Liebert, J., & Fulbright, M. S. 1995, *AJ*, 444, 810
Bressan, A., Chiosi, C., & Fagotto, F. 1994, *ApJS*, 94 63
Burstein, D., Bertola, F., Buson, L. M., Faber, S. M., & Lauer, T. R. 1988, *ApJ*, 328, 440
Castellani, M. & Castellani, V. 1993, *ApJ*, 407, 649
Catelan, M. 2007, in: M. Chavez, E. Bertone, D. Rosa-Gonzalez, and L. H. Rodríguez–Merino (eds.), *New Quests in Stellar Astrophysics II: The Ultraviolet Properties of Evolved Stellar Populations*, p. 175, Springer
Cohen, J. G., Huang, W., Udalski, A., Gould, A., & Johnson, J. A. 2008, *ApJ*, 682, 1029
Cohen, J. G., Thompson, I. B., Sumi, T., Bond, I., Gould, A., Johnson, J. A., Huang, W., & Burley, G. 2009, *ApJ*, 699, 66
D'Cruz, N. L., Dorman, B., Rood, R. T., & O'Connell, R. W. 1996, *ApJ*, 466, 359
Dorman, B., O'Connell, R. W, & Rood, R. T. 1995, *ApJ*, 442, 105
Goldberg, L. 1979, *QJRAS*, 20, 361
Gratton, R., Bragaglia, A., Carretta, E., & Tosi, M. 2006, *ApJ*, 642, 462
Greggio, L. & Renzini, A. 1990, *ApJ*, 364, 35
Groenewegen, M. A. T. 2006, *A&A*, 448, 181
Habing, H. J. 1996, *ARA&A*, 7, 97
Han, Z., Podsiadlowski, P., & Lynas–Gray, A. E. 2007, *MNRAS*, 380, 1098
Hansen, B. M. S. 2005, *ApJ*, 635, 522
Johnson, J. A., Gaudi, B. S., Sumi, T., Bond, I. A., & Gould, A. 2008, *ApJ*, 685, 508
Judge, P. G. & Stencel, R. E. *1991*, ApJ, 371, 357
Kalirai, J. S., Richer, H. B., Fahlman, G. G., Cuillandre, J., Ventura, P., D'Antona, F., Bertin, E., Marconi, G. & Durrell, P. 2001a, *AJ*, 122, 257
Kalirai, J. S., Richer, H. B., Fahlman, G. G., Cuillandre, J., Ventura, P., D'Antona, F., Bertin, E., Marconi, G., & Durrell, P. 2001b, *AJ*, 122, 266
Kalirai, J. S., Ventura, P., Richer, H. B., Fahlman, G. G., D'Antona, F. & Marconi, G. 2001c, *AJ*, 122, 3239
Kalirai, J. S., Fahlman, G. G., Richer, H. B., & Ventura, P. 2003, *AJ*, 126, 1402
Kalirai, J. S., Richer, H. B., Reitzel, D., Hansen, B. M. S., Rich, R. M., Fahlman, G. G., Gibson, B. K., & von Hippel, T. 2005, *ApJ* (Letters), 618, L123
Kalirai, J. S., Bergeron, P., Hansen, B. M. S., Kelson, D. D., Reitzel, D. B., Rich, R. M., & Richer, H. B. 2007, *ApJ*, 671, 748
Kalirai, J. S., Hansen, B. M. S., Kelson, D. D., Reitzel, D. B., Rich, R. M., & Richer, H. B. 2008, *ApJ*, 676, 594

† A recent preprint by Bensby *et al.* (2009) shows that the disparity in the MDF between the dwarfs and giants is significantly reduced based on a larger set of observations.

Kalirai, J. S., Davis, D. S., Richer, H. B., Bergeron, P., Catelan, M., Hansen, B. M. S., & Rich, R. M. 2009, *ApJ*, 705, 408
Kilic, M., Stanek, K. Z., & Pinsonneault, M. H. 2007, *ApJ*, 671, 761
Marino, A. F., Villanova, S., Piotto, G., Milone, A. P., Momany, Y., Bedin, L. R., & Medling, A. M. 2008, *A&A*, 490, 625
O'Connell, R. W. 1999, *ARA&A*, 37, 603
Origlia, L., Valenti, E., Rich, R. M., & Ferraro, F. R. 2006, *ApJ*, 646, 499
Rood, R. T. 1973, *ApJ*, 184, 815
Tremblay, P.-E. & Bergeron, P. 2009, *ApJ*, 696, 1755
Weidemann, V. 2000, *A&A*, 363, 647
Williams, K. A., Bolte, M., & Koester, D. 2004, *ApJ* (Letters), 615, L49
Williams, K. A & Bolte, M. 2007, *AJ*, 133, 1490
Williams, K. A., Bolte, M., & Koester, D. 2009, *ApJ*, 693, 355
Zoccali, M., *et al.* 2003, *A&A*, 399, 931

Star clusters: basic galactic building blocks
Proceedings IAU Symposium No. 266, 2009
R. de Grijs & J. R. D. Lépine, eds.

doi:10.1017/S1743921309991190

The dynamical effect of white dwarf kicks in star clusters

Harvey B. Richer

Department of Physics and Astronomy, University of British Columbia,
6224 Agricultural Road, Vancouver, British Columbia, Canada V6T 1Z1
email: richer@astro.ubc.ca

Abstract. Globular star clusters generally have large cores, i.e., r_c/r_h (the ratio of core to half-light radii) exceeds 0.3 for more than 50% of the Galactic globular clusters. In the absence of a central heating source, dynamical models suggest that massive clusters will contract, typically on a timescale shorter than a Hubble time, and exhibit a compact core. To explain the disagreement between observations and theory, intermediate-mass mass black holes have been invoked to explain the core structure. Recent observations, however, have failed to definitively prove their existence in clusters. A new scenario, involving a natal kick given to white dwarfs may provide the required heating and help clusters avoid or delay core collapse.

Keywords. globular clusters: general, white dwarfs, black hole physics, stellar dynamics

1. Introduction

An examination of the distribution of core to half-light radii (r_c/r_h) in globular star clusters indicates that many clusters have large core radii relative to their half-light radii, fully 50% have this ratio >0.3. So, most clusters do not currently appear to be in a core-collapsed (or binary-burning) phase of evolution. They may still be in the standard initial contraction phase or they may have had their core sizes enhanced through some excess energy source in the cluster. N-body or Monte Carlo models (e.g., Heggie *et al.* 2006) demonstrate that, with zero binaries, a cluster rapidly goes into core collapse and only with an appreciable primordial binary fraction (>10%) could core collapse be avoided or delayed for a time comparable to the age of the Universe. Thus, putting an appreciable binary fraction into the models helps bring the theoretical structural picture of globular clusters closer to the real data, although even with 100% binaries, very large core radii could not be achieved in the dynamical simulations. Thus the structure of the bulk of the globular star clusters cannot be explained simply by assuming a large binary fraction: the cluster must either be dynamically young or have some new energy source.

2. Intermediate-mass black holes in star clusters

With binaries clearly not the solution to large cluster cores, the role of intermediate-mass black holes (IMBHs) was examined. These are black holes in the mass range from roughly 100 to a few times 10 000 $M_\odot$. N-body calculations by numerous authors (e.g., Gill *et al.* 2008; Trenti *et al.* 2007) showed that large cores could easily be obtained with the presence of an IMBH with a mass a few percent of that of the total cluster. Some authors even made statements to the effect that an IMBH was required in the cores of those clusters exhibiting very large core radii. An observational signature of the presence of an IMBH in a star cluster is an increase in the cluster velocity dispersion near the core (Baumgardt *et al.* 2005). Several heroic attempts have been made to search for this effect in a few clusters and while none of these cases was rock solid, they were certainly suggestive of the presence of IMBHs in some clusters. However, recently two proper-motion studies at the centers of loose clusters (Ruberg 2008; Anderson & van der Marel 2009) failed to yield any evidence for the presence of IMBHs.

3. A new heat source: kicked white dwarfs

A new paradigm seems required to explain the core dynamics in globular star clusters. Davis *et al.* (2008) showed that in the globular cluster NGC 6397 the young white dwarfs appear to have been given a small 'kick' when they were born, on the order of 3–5 km s^{-1}. Including such a kick in dynamical models demonstrated that large core radii could easily be obtained in globular clusters and that core collapse could be delayed for a period longer than a Hubble time (Fregeau *et al.* 2009). How does this explain the large cluster cores? A kicked white dwarf will lose energy to its surrounding stars as it is moving too fast for its mass in equipartition. In this sense, it supplies heat to the cluster, puffing up its core (Heyl 2008).

The physical mechanism for a white-dwarf kick is not immediately apparent. Asymmetric mass loss in the late asymptotic giant branch or planetary-nebula phase seems most obvious, but a kick at the core helium flash may also supply the energy input needed. We are currently investigating this observationally by exploring the radial distributions of stars in clusters in all evolutionary phases beyond the main-sequence turnoff. In addition, we are examining the location of the white dwarf in planetary nebulae to determine whether it is located off center. This could be evidence for a kick late in the evolution of a star.

References

Anderson, J. & van der Marel, R. P. 2009, *ApJ*, submitted (arXiv:0905.0627)

Baumgardt, H., Makino, J., & Hut, P. 2005, *ApJ*, 620, 238

Davis, D. S., Richer, H. B., King, I. R., Anderson, J., Coffey, J., Fahlman, G. G., Hurley, J., & Kalirai, J. S. 2008, *MNRAS* (Letters), 383, L20

Fregeau, J. M., Richer, H. B., Rasio, F. A., & Hurley, J. R. 2009, *ApJ* (Letters), 695, L20

Gill, M., Trenti, M., Miller, M. C., van der Marel, R., Hamilton, D., & Stiavelli, M. 2008, *ApJ*, 686, 303

Heggie, D. C., Trenti, M., & Hut, P. 2006, *MNRAS*, 368, 677

Heyl, J. 2008, *MNRAS*, 390, 622

Ruberg, A. 2008, MSc Thesis, University of British Columbia

Trenti, M., Ardi, E., Mineshige, S., & Hut, P. 2007, *MNRAS*, 374, 857

Star clusters: basic galactic building blocks
Proceedings IAU Symposium No. 266, 2009
R. de Grijs & J. R. D. Lépine, eds.

doi:10.1017/S1743921309991207

Star clusters with dual red clumps

Léo Girardi,[1] Stefano Rubele[1,2] and Leandro Kerber[3]

[1]Osservatorio Astronomico di Padova, Italy
[2]Dipartimento di Astronomia, Università di Padova, Italy
[3]Universidade Estadual de Santa Cruz, Ilhéus, Brazil

Abstract. A few star clusters in the Magellanic Clouds exhibit composite structures in the red-clump region of their colour–magnitude diagrams. The most striking case is NGC 419 in the Small Magellanic Cloud (SMC), where the red clump is composed of a main blob as well as a distinct secondary feature. This structure is demonstrated to be real and corresponds to the simultaneous presence of stars which passed through electron degeneracy after central-hydrogen exhaustion and those that did not. This rare occurrence in a single cluster allows us to set stringent constraints on its age and on the efficiency of convective-core overshooting during main-sequence evolution. We present a more detailed analysis of NGC 419, together with a first look at other populous Large Magellanic Cloud clusters which are apparently in the same phase: NGC 1751, NGC 1783, NGC 1806, NGC 1846, NGC 1852 and NGC 1917. We also compare these Magellanic Cloud cases with their Galactic counterparts, NGC 752 and NGC 7789. We emphasise the extraordinary potential of these clusters as *absolute* calibration marks on the age scale of stellar populations.

Keywords. galaxies: star clusters, Magellanic Clouds, stars: evolution, Hertzsprung–Russell diagram

1. NGC 419 and its dual red clump

NGC 419 is a populous star cluster located to the east of the Small Magellanic Cloud (SMC)'s bar in a region relatively devoid of dust and free from contamination by the SMC field. There are two wonderful pairs of *HST* images of this cluster, taken in the F555W and F815W filters, originally obtained as part of programme GO-10396 (PI J. S. Gallagher) and now retrievable from the *HST* archive. They are shown in Figure 1. While the ACS/WFC images reveal the overall cluster structure, the ACS/HRC observations show the details of the cluster core.

The HRC images allow us to perform photometry of an impressive quality, and reveal at least two surprises: the presence of multiple main-sequence turnoffs (MMSTOs), and a composite red clump which contains a pronounced faint extension (Figure 2). Both features were noticed by Glatt *et al.* (2008). They tentatively interpreted the faint red-clump structure as being caused by SMC field contamination. However, this explanation does not stand up to a simple star count in the neighbouring field selected from the ACS/WFC image (see Girardi *et al.* 2009): the expected number of field red-clump stars in the ACS/HRC area is just 4.5, while the faint red clump contains at least 50 objects (Figure 2). Assuming a Poissonian distribution for the field stars, the probability (P) that these 50 stars are drawn from the SMC field alone is virtually zero ($P < 10^{-9}$).

Detached binaries cannot give rise to this faint red clump either, since any combination of single stars will cause an extension of the red clump to brighter magnitudes.

What, then, is the dual red clump of NGC 419 made of? Girardi *et al.* (2009) show that it is caused by the presence of stars following two very different evolutionary paths.

In the following, we will expand their arguments a little. There are, in fact, a few different ways of looking at dual red clumps, as detailed below.

2. Electron degeneracy on the scene

The basic fact behind dual red clumps is the dilemma facing low- to intermediate-mass stars after exhausting their central hydrogen and departing from the main sequence: *to RGB or not to RGB, that is the question!* (RGB: red-giant branch; adapted from Shakespeare 1600). Stars slightly more massive than 2 $M_{\odot}$ simply contract their hydrogen-exhausted cores until a temperature of about 10^8 K is reached. This causes core-helium ignition and the overall expansion of the star, which quickly settles in the red clump and

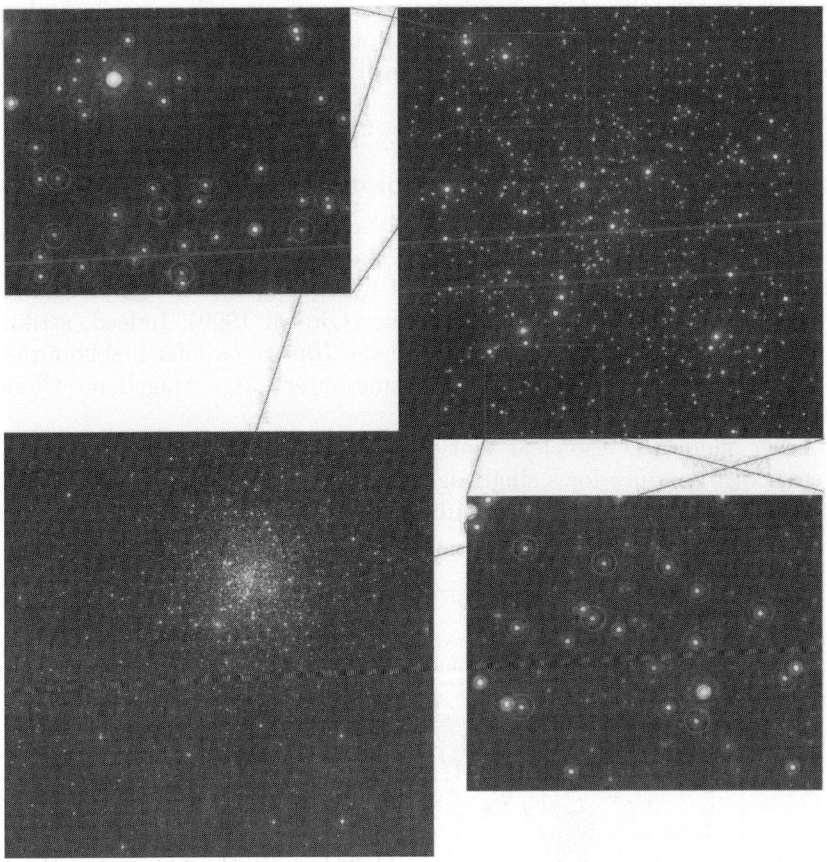

Figure 1. False-colour images (in the electronic version) of the Small Magellanic Cloud star cluster NGC 419, derived from ACS/WFC *(bottom left)* and HRC *(top right)* images in the F555W and F814W filters. The top left and bottom right panels zoom in onto the HRC image. Red-clump stars are marked with circles. At first glance, it is evident that they are quite homogeneous in their colours and luminosities, as expected for red-clump stars. However, when comparing their first Airy rings, one notices some quite subtle and *systematic* differences in their brightnesses. Indeed, there are two kinds of red-clump stars: the most numerous and brighter, and a subsample (about 15% of the total) of fainter 'secondary red-clump stars' (marked with red and green circles, respectively, in the electronic version). The luminosity difference between these groups is about 0.4 mag. There is also a ~0.04 mag difference in the mean colour which, however, is too small to be noticeable in the figure. The two kinds of red-clump stars are uniformly distributed across the HRC image.

quietly burns helium thereafter. Stars slightly less massive instead suffer from electron degeneracy in the core before helium can be ignited. Electron pressure stops core contraction and the core becomes nearly isothermal. Subsequently, neutrino losses cool its very centre. The result is that He ignition is prevented for a long time. These stars have to climb all the way up to the RGB tip, where helium ignites in a flash, finally freeing their cores from electron degeneracy.

The difference between these two evolutionary paths is a quite abrupt function of initial mass (see Sweigart *et al.* 1990; Girardi 1999). In an interval of order 0.2 $M_\odot$, the core mass at He ignition changes from 0.46 to 0.33 $M_\odot$, depending on the presence or absence of electron degeneracy. This core-mass difference alone causes a ~0.4 mag difference in the initial luminosity of He-burning stars (with the red clump of the younger population being fainter), which is adequate to explain the two clumps of NGC 419.

The quantitative details of this explanation, based on stellar tracks and isochrones including the complete He-burning phase and computed with a fine mass resolution, can be found in Girardi *et al.* (2009; see also Girardi 1999).

3. Just another manifestation of prolonged star formation

From the discussion above, one may already obtain a hint that dual red clumps are a common feature in galaxy fields containing stars of all initial masses (ages), provided that such fine CMD features are not blurred by effects such as differential reddening, photometric errors and metallicity dispersions (Girardi 1999). Indeed, striking examples of faint secondary clumps are provided in the *Hipparcos* solar-neighbourhood CMD (Girardi *et al.* 1998) and by the CMDs of some outer Large Magellanic Cloud (LMC) fields (Bica *et al.* 1998; Piatti *et al.* 1998).

On the other hand, it is also clear that dual red clumps were *not expected* to be found in single star clusters, since for a single age the dispersion in turnoff masses is negligible and much smaller than the ~0.2 $M_\odot$ required to explain the presence of dual red clumps.

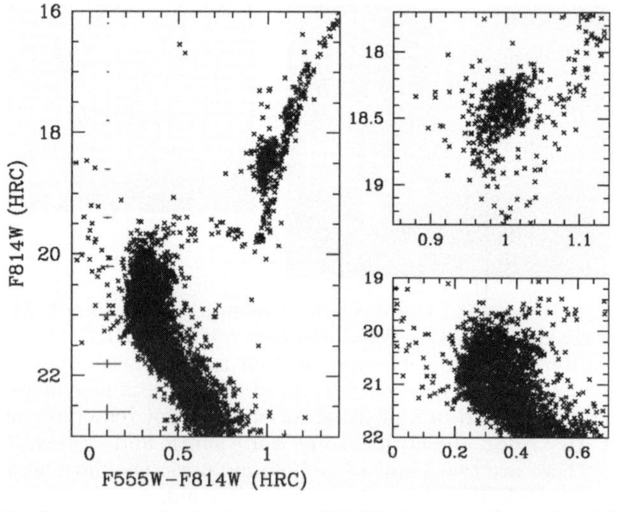

Figure 2. NGC 419 colour–magnitude diagram (CMD; *left panel*), derived from the ACS/HRC data (Girardi *et al.* 2009). The error bars in the left panel are upper limits to the photometric errors. The two right-hand panels detail the red-clump and main-sequence turnoff, clearly showing their composite structure.

This was concluded by Girardi *et al.* (2000) in their study of the open clusters with dual red clumps, NGC 752 and NGC 7789.

However, we now have convincing evidence that star clusters are not single-aged stellar populations. In addition to the old globular clusters and NGC 419 itself, many intermediate-age clusters in the LMC show MMSTOs. The simplest explanation of MMSTOs is based on prolonged periods of star formation (Bertelli *et al.* 2003; Baume *et al.* 2007; Mackey *et al.* 2008; Milone *et al.* 2009).

Bastian & de Mink (2009) recently provided an alternative explanation for the presence of MMSTOs, based on the $T_{\rm eff}$/colour spread of main-sequence stars with different rotation rates. The idea is, no doubt, very interesting and worth of further exploration. One crucial point to be clarified is whether rotation may cause the small spread in core mass at the end of the main sequence that is *required* to explain the dual red clump in NGC 419. In other words, the different rotation velocities on the main sequence have to produce a direct effect in the mass of H-exhausted cores, otherwise Bastian & de Mink's (2009) explanation cannot be applied to NGC 419.

Rubele *et al.* (in prep.) go a little further into this question. They apply to NGC 419 the classical method of star-formation-history recovery through CMD reconstruction (cf. Kerber *et al.* 2009), deriving a star-formation rate which lasts for at least 500 Myr (Figure 3). The spread in turnoff masses turns out to be as large as 0.18 $M_\odot$. A χ^2-like statistic reveals that their best-fitting solutions are *excellent* representations of the data. It remains to be clarified whether the rotation advocated by Bastian & de Mink (2009) will produce similarly good descriptions of the CMDs of clusters with MMSTOs, keeping at the same time the number of free parameters in the models at an acceptable level.

4. A new way of weighing stellar cores

The above interpretation implies that stars leaving the NGC 419 main sequence have an H-exhausted core of almost exactly 0.33 $M_\odot$. The uncertainty in this quantity is as small as $\sim$0.01 $M_\odot$. It only depends on basic principles of stellar structure (e.g., Schönberg &

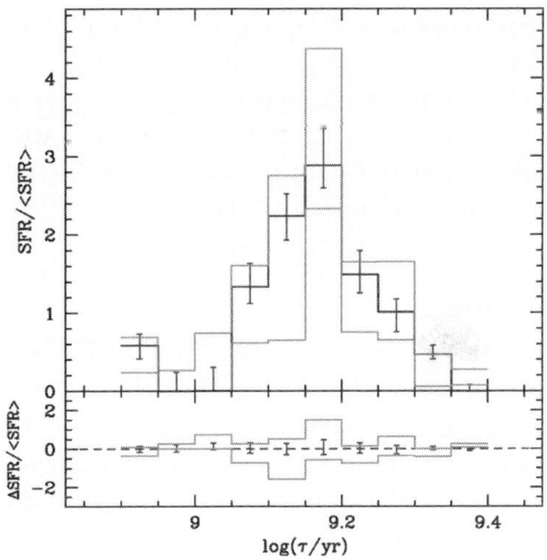

Figure 3. Star-formation history of NGC 419, as determined based on the classical method of CMD reconstruction (Rubele *et al.*, in prep.). The heavy line with error bars is the best solution with its 68% confidence levels, reflecting random errors only. The top and bottom histograms instead indicate the total systematic errors derived from the full range of possible distances, reddenings and metallicities of the cluster.

Chandrasekhar 1942) and on the well-known physics of partially degenerate matter. One may consider NGC 419 observations as providing the most precise (although indirect) determinations of the H-exhausted core mass of living stars.

Asteroseismology will eventually provide direct measurements of the core mass of individual red giants in the solar vicinity (see, e.g., Dupret *et al.* 2009), but at the cost of considerable investments in terms of dedicated telescopes and data analysis. The NGC 419 results, in contrast, come almost for free.

5. An absolute mark on the age scale of stellar populations

A direct consequence of knowing the core mass of a star at the end of its main-sequence evolution, is that one has a way of measuring the extent of convective-core overshooting, a phenomenum that has plagued age determinations of star clusters for decades. One just has to combine the red-clump observations (indicative of 0.33 $M_\odot$ H-exhausted cores) with some indicator of the main-sequence mass, which can be, very simply, the main-sequence location in the CMD. And once convective-core overshooting is constrained, age determinations become much more solid.

Girardi *et al.* (2009) pursued this idea by simultaneously fitting the red-clump morphology and the colour difference between the main-sequence termination and the red clump in NGC 419. The best simultaneous fit was found for a cluster mean age of $t = 1.35^{+0.11}_{-0.04}$ Gyr and an efficiency of convective overshooting of $\Lambda_c = 0.47^{+0.14}_{-0.04}$ pressure scale heights (see Bressan *et al.* 1993 for the formalism). It is worth mentioning that our results allow us to exclude the classical Schwarzschild criterion for convective boundaries with very high confidence.

It is remarkable that the above-mentioned age determination is quite independent of distance, reddening, binary fraction *and* overshooting (see Girardi *et al.* 2009). Hence, star clusters with dual red clumps have great potential of becoming *absolute marks* on the age scale of stellar populations.

We recall that absolute age marks are extremely rare in stellar astrophysics. Accurate absolute ages can be derived for young associations on the basis of their kinematics (for ages up to ~4 Myr; e.g., Brown *et al.* 1997), and for old globular clusters once their upper age limit is constrained by cosmological measurements (e.g., $<$13.73 Gyr; Spergel *et al.* 2007). Inside this very wide age interval, ages of stellar aggregates depend somewhat on the extension of convective boundaries, which are notoriously uncertain from a theoretical point of view. Clusters with dual red clumps may represent the exception to this rule, providing absolute marks at ages of about 1.3 Gyr.

6. NGC 419 is not alone

The analysis in Girardi *et al.* (2009) is about to be repeated with up-to-date stellar models (Bressan *et al.*, in prep.), and removing the assumption of a constant star-formation rate during the period of formation of the MMSTOs. The error bars in the age determination will inevitably increase compared to Girardi *et al.* (2009). Moreover, we have recently realised that the uncertainty in the metal content represents a good fraction of the error budget for NGC 419 (Rubele *et al.*, in prep.). More clusters with dual red clumps, with as far as possible good [Fe/H] determinations, are necessary to obtain tighter constraints on their ages and overshooting efficiency.

Such star clusters exist and can even be numerous in the Magellanic Clouds. A quick look at the *HST*/ACS CMDs from Milone *et al.* (2009) reveals that dual red clumps are present in the LMC star clusters NGC 1751, NGC 1783, NGC 1806, NGC 1846,

NGC 1852 and NGC 1917 (Girardi *et al.* 2009). All exhibit MMSTOs at a magnitude which is compatible with the lower limit for the settling of electron degeneracy. We are now examining their archival *HST* data, checking accurately how much of the faint clump features can be attributed to contamination from the LMC field, and then repeating the same analysis as for NGC 419.

In our Galaxy, the open clusters NGC 752 and NGC 7789 contain, among their radial-velocity members, a handful of stars which apparently belong to a faint secondary clump (Mermilliod *et al.* 1998; Girardi *et al.* 2000). Other candidates are NGC 2660, NGC 2204 (Girardi *et al.* 2000) and Tr 20 (Seleznev *et al.* 2009). In all these cases, the small numbers of red-clump stars may hamper the derivation of clear constraints to the stellar models and cluster ages. The situation with the populous Magellanic Cloud clusters is clearly different, and much more exciting.

Acknowledgements

The *HST* data illustrated in this talk were obtained from the Multimission Archive at the Space Telescope Science Institute (STScI). STScI is operated by the Association of Universities for Research in Astronomy, Inc., under NASA contract NAS5-26555. We thank the IAU 266 Scientific Organising Committee for the invited talk slot, I. Platais, P. Marigo, J. D. do Nascimento, G. Carraro and A. Miglio for their comments and acknowledge the support from INAF/PRIN07 CRA 1.06.10.03, contract ASI-INAF I/016/07/0, and the Brazilian agencies CNPq and FAPESP.

References

Bastian, N. & de Mink, S. E. 2009, *MNRAS* (Letters), 398, L11
Baume, G., Carraro, G., Costa, E., Méndez, R. A., & Girardi, L. 2007, *MNRAS*, 375, 1077
Bertelli, G., Nasi, E., Girardi, L., Chiosi, C., Zoccali, M., & Gallart, C. 2003, *AJ*, 125, 770
Bica, E., Geisler, D., Dottori, H., Clariá, J. J., Piatti, A. E., & Santos, Jr., J. F. C. 1998, *AJ*, 116, 723
Bressan, A., Fagotto, F., Bertelli, G., & Chiosi C. 1993, *A&AS*, 100, 647
Brown, A. G. A., Dekker, G., & de Zeeuw, P. T. 1997, *MNRAS*, 285, 479
Dupret M.-A., *et al.* 2009, *A&A*, 506, 57
Girardi, L. 1999, *MNRAS*, 308, 818
Girardi, L., Mermilliod, J.-C., & Carraro, G. 2000, *A&A*, 354, 892
Girardi, L., Groenewegen, M. A. T., Weiss, A., & Salaris M. 1998, *MNRAS*, 301, 149
Girardi, L., Rubele, S., & Kerber, L. 2009, *MNRAS* (Letters), 394, L74
Glatt, K., *et al.* 2008, *AJ*, 136, 1703
Kerber, L. O., Girardi, L., Rubele, S., & Cioni M.-R. 2009, *A&A*, 499, 697
Mackey, A. D., Broby Nielsen, P., Ferguson, A. M. N., & Richardson J. C. 2008, *ApJ* (Letters), 681, L17
Mermilliod, J.-C., Mathieu, R. D., Latham, D. W., & Mayor, M. 1998, *A&A*, 339, 423
Milone, A. P., Bedin, L. R., Piotto, G., & Anderson J. 2009, *A&A*, 497, 755
Piatti, A. E., Geisler, D., Bica, E., Clariá, J. J., Santos Jr., J. F. C., Sarajedini, A., & Dottori, H. 1999, *AJ*, 118, 2865
Schönberg, M. & Chandrasekhar, S. 1942, *ApJ*, 96, 161
Seleznev, A. F., Carraro, G., Costa, E., Loktin, A. V. 2010, *NewA*, 15, 61
Shakespeare, W., 1600, in *Hamlet*
Spergel D. N., *et al.* 2007, *ApJS*, 170, 377
Sweigart, A. V., Greggio, L., & Renzini A. 1990, *ApJ*, 364, 527

Star clusters: basic galactic building blocks
Proceedings IAU Symposium No. 266, 2009
R. de Grijs & J. R. D. Lépine, eds.

doi:10.1017/S1743921309991219

The multipopulation phenomenon in Galactic globular clusters: M4 and M22

S. Villanova,[1] G. Piotto,[2] A. F. Marino,[2] A. P. Milone,[2] A. Bellini,[2] L. R. Bedin,[3] Y. Momany[4] and A. Renzini[4]

[1]Departamento de Astronomía, Universidad de Concepción, Casilla 160-C, Concepción, Chile
email: svillanova@astro-udec.cl
[2]Dipartimento di Astronomia, Università di Padova, Vicolo dell'Osservatorio 3, Padova, I-35122, Italy
email: [giampaolo.piotto, anna.marino, antonino.milone, andrea.bellini]@unipd.it
[3]Space Telescope Science Institute, 3700 San Martin Drive, Baltimore, MD 21218, USA
email: bedin@stsci.edu
[4]Osservatorio Astronomico di Padova, Vicolo dell'Osservatorio 5, 35122 Padova, Italy
email:[yazan.momany, alvio.renzini]@oapd.inaf.it

Abstract. We present an abundance analysis based on high-resolution spectra of red-giant-branch (RGB) stars in the Galactic globular clusters NGC 6121 (M4) and NGC 6656 (M22). Our aim was to study their stellar population in the context of the multipopulation phenomenon recently discovered to affect some globular clusters. Analysis was performed for the following elements: O, Na, Mg, Al, Ca, Fe, Y, and Ba. Spectroscopic data were completed by high-precision wide-field $UBVI_C$ ground-based photometry and *HST*/ACS observations. For M4, we find a well-defined Na–O anticorrelation composed of two distinct groups of stars with significantly different Na and O content. The two groups of Na-rich and Na-poor stars populate two different regions along the RGB. As regards M22, Na and O follow the well-known anticorrelation found in many other GCs. However, at odds with M4, it appears to be continuous without any hint of clumpiness. On the other hand, we identified two clearly separated groups of stars with significantly different abundances of the s-process elements Y, Zr and Ba. The relative numbers of the members of both groups are very similar to the ratio of the stars in the two subgiant branches of M22 recently found by Piotto (2009). The s-element-rich stars are also richer in iron and have higher Ca abundances. This makes M22 the second cluster after ω Centauri where an intrinsic spread in Fe was found. Both spectroscopic and photometric results imply the presence of two stellar populations in M4 and M22, even if both clusters have completely different characteristics.

Keywords. globular clusters: individual (M 4, M 22)

1. Introduction

Globular clusters (GCs) are generally chemically homogeneous in their Fe-peak elements, while they show star-to-star abundance variations in light elements such as C, N, O, Na, Mg, and Al, among others. In some cases, these chemical inhomogeneities result in well-defined anticorrelations. For example, all GCs for which Na and O abundances have been measured show a well-defined Na–O anticorrelation (see Carretta *et al.* 2006, 2008 for an update). The origin of these variations is not yet well understood, since both a primordial and an evolutionary explanation, or a combination of both have been proposed (see Gratton *et al.* 2004 for a recent review).

Interestingly enough, abundance anomalies have been found also among stars on the lower part of the red-giant branch (RGB) or even on the main sequence (MS). For instance, Cannon *et al.* (1998) found a CN bi-modality in MS and subgiant-branch (SGB) stars in 47 Tuc, and the Na–O anti-correlation was found at the level of the MS turnoff (TO) and SGB in M 13 (Cohen & Meléndez 2005), NGC 6397 and NGC 6752 (Carretta *et al.* 2005; Gratton *et al.* 2001), NGC 6838 (Ramírez & Cohen 2002). On their own, these results suggest a possible primordial origin for the abundance inhomogeneities.

More recent results, made possible by a significant improvement of the photometric precision of *HST* images, indicate the presence of multimodal sequences in the color–magnitude diagrams (CMDs) of some GCs (Bedin *et al.* 2004; Piotto *et al.* 2007; Milone *et al.* 2008), which further confirms that at least in some GCs there is more than one generation of stars formed from material chemically contaminated by previous generations.

Among the clusters with multiple stellar populations, ω Centauri is the most complex and interesting case. This object is the only GC for which variations in iron-peak elements have been identified unambiguously (Freeman & Rodgers 1975; Norris *et al.* 1996; Suntzeff & Kraft 1996). The Fe multimodal distribution is at least in part responsible for the multiple RGB (Lee *et al.* 1999; Pancino *et al.* 2000) of ω Cen. Its MS also splits into three sequences, as shown by Bedin *et al.* (2004). Recently, Villanova *et al.* (2007) showed that the SGB also splits into at least four branches with a large age difference (greater than 1 Gyr) among the different populations.

NGC 2808 is the second cluster in which a MS split into three branches was found (Piotto *et al.* 2007). The MS multimodality was associated with the observed multimodal distribution of the stars along the Na–O anticorrelation, where three groups of stars with different O (and Na) content were found by Carretta *et al.* (2006).

In NGC 1851, Yong & Grundahl (2008) found abundance variations in various elements by studying a sample of eight RGB stars. Sodium and oxygen follow the Na–O anticorrelation, and evidence of the presence of two groups of stars with different s-process element content, as well as two groups of stars with different CN strength were also found, probably related to the two populations of stars photometrically identified by Milone *et al.* (2008) along the SGB.

2. The case of M4

As far as we know, M4 shows no evidence of multiple stellar populations in its CMD, and its mass ($\log M/M_{\odot} = 4.8$; Mandushov *et al.* 1991) is much lower than the mass of the clusters exhibiting photometric peculiarities discussed above. Chemical abundances of M4 RGB stars have already been measured by different groups. A study by Norris (1981) showed a CN bimodality in M4, e.g., stars of very similar magnitudes and colors have a bimodal distribution of CN band strengths. More recently, Smith *et al.* (2005), studying the fluorine abundance in seven RGB stars in M4, found a large variation in F, which is anticorrelated with the Na and Al abundances. In these previous studies, both an evolutionary and a primordial scenario were taken into account to explain the light-element variations and the CN bimodality in M4.

Variations in light-element abundances are common among GCs and are also present in M4. According to our results based on high-resolution high signal-to-noise ratio (S/N) spectra collected with UVES+FLAMES, sodium and oxygen have very large dispersions and show the typical anticorrelation found in many other GCs (Gratton *et al.* 2004), as shown in Figure 1 (left panel). Our most interesting result is that, thanks to the large number of stars in our sample, we can show that the [Na/Fe] distribution is bimodal (Figure 1, right panel). Adopting an arbitrary separation between the two peaks in Figure 1 (right panel) at ~ 0.2 dex, we obtain for the first group of Na-rich stars a mean

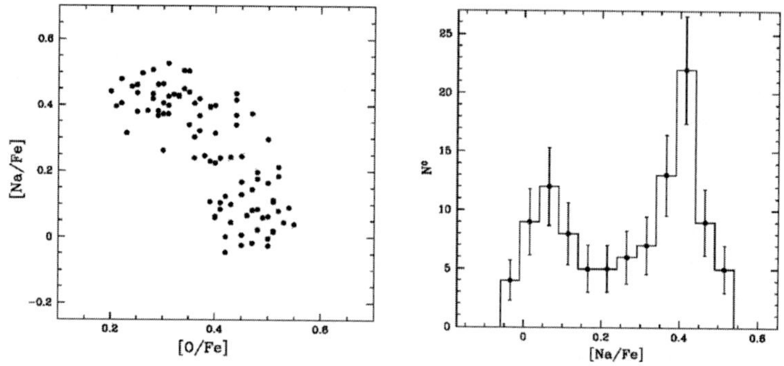

Figure 1. *(left)* [Na/Fe] versus [O/Fe] abundance ratios. *(right)* Histogram of the distribution of the [Na/Fe] abundances.

content of [Na/Fe] $= 0.38 \pm 0.01$ and [O/Fe] ~ 0.30 dex, while for the second group of Na-poor stars we get [Na/Fe] $= 0.07 \pm 0.01$ and [O/Fe] ~ 0.47 dex, with a few stars having intermediate sodium and oxygen abundances.

Previous studies also revealed a trend of Na and Al with CN-band strength. The CN-band strengths of red giants in M4 have been measured by many authors, and Smith & Briley (2005) have homogenized all available data for CN-band strengths in terms of the S(3839) index, i.e., the ratio of the flux intensities of the cyanogen band near 3839 Å and the nearby continuum. Figure 2 (left panel) shows the [Na/Fe], [Al/Fe], [O/Fe], [Mg/Fe], and [Ca/Fe] values we obtained as a function of S(3839) for the stars in common with Smith & Briley (2005). In this figure, CN-strong objects are represented by filled circles, CN-weak by open circles, while open squares are stars in the middle. We note that the CN-strong objects clearly show significantly enhanced Na and depleted O abundances, while the CN-weak stars have lower Na and higher O. There is also a systematic difference in Al between CN-weak and CN-strong stars, with CN-strong stars richer in Al content. Figure 2 shows some differences in [Mg/Fe] between the two groups, while no difference is found for Ca.

It is instructive to look at the position of the stars belonging to these two Na groups in the U versus $(U-B)$ CMD (Figure 2, right panel). The Na-rich and Na-poor stars are

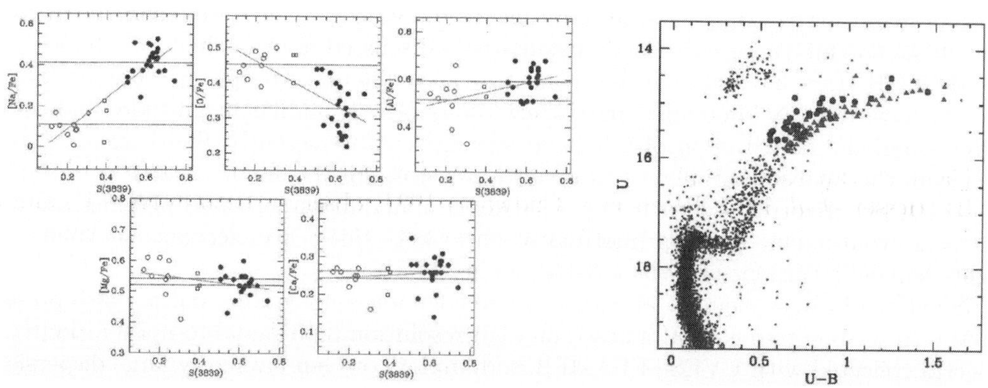

Figure 2. *(left)* Abundance ratios [Na/Fe], [O/Fe], [Al/Fe], [Mg/Fe], and [Ca/Fe] are plotted as a function of the CN index S(3839). *(right)* U versus $(U-B)$ CMD from WFI photometry. The stars belonging to the two different Na groups are represented by two different symbols: red triangles and blue circles represent the stars with [Na/Fe] $\geqslant 0.2$ and [Na/Fe] < 0.2 dex, respectively.

represented by red triangles and blue circles, respectively. Interestingly, the two groups of Na-rich and Na-poor stars form two distinct branches on the RGB. Na-rich stars define a narrow sequence on the red side of the RGB, while the Na-poor sample populates the blue, more dispersed portion of the RGB. Even more interestingly, the anomalous broadening of the RGB is visible down to the base of the RGB, at $U \sim 17.5$ mag, indicating that the two abundance groups are present all over the RGB, even well below the RGB bump, where no deep mixing is expected. This evidence further strengthens the idea that the bimodal Na, O, and CN distribution must have been present in the material from which the stars we presently observe in M4 originated.

The conclusion is that M4 hosts two groups of stars identified both spectroscopically and photometrically that are due to the presence of two distinct populations. We did not find any dependence of the chemical-abundance distribution on the evolutionary status (along the RGB) of the target stars, from below the RGB bump to the RGB tip. This means that the abundance anomalies are very likely due to primordial variations in the chemical content of the material from which M4 stars formed, and not to different evolutionary paths of the present stellar population of M4. This is surprising because of the relatively low mass of M4, an order of magnitude lower than the masses of ω Cen, NGC 2808, and NGC 1851, the other clusters in which multiple stellar generations have been confirmed. Where did the gas which polluted the material for the CN–Na-rich stars come from? Has it been ejected from a first generation of stars? How could it stay within the shallow gravitational potential of M4? The idea that M4 hosts two generations of stars makes the multipopulation phenomenon in GCs even more puzzling than originally thought. It becomes harder and harder to accept the idea that the phenomenon can be totally internal to the cluster, unless this object is what remains of a much larger system (a larger GC or the nucleus of a dwarf galaxy?). Surely, because its orbit involves frequent passages at high inclination through the Galactic disk, always at distances from the Galactic center of less than 5 kpc (see Dinescu *et al.* 1999), M4 must have been strongly affected by tidal shocks, and therefore it might have been much more massive in the past.

3. The case of M22

Located at a distance of ~ 3.2 kpc (Harris 1996), NGC 6656 (M22) is a particularly interesting GC, because a large number of photometric and spectroscopic studies have suggested a complex metalicity spread, similar to, albeit significantly smaller than, that found in ω Centauri. In particular, it has often been suggested, although never convincingly confirmed, that M22 may also have a spread in the content of its iron-peak elements.

The first evidence for a spread in metalicity comes from the significant spread along the RGB (Hesser *et al.* 1977; Peterson & Cudworth 1994) observed both in $(B-V)$ and in Strömgren colors. However, it is still uncertain whether this spread can be attributed to a metalicity spread or to reddening variations. Spectroscopic studies are divided between those which conclude that no significant metalicity variations are present in M22 (Cohen 1981, based on three stars; Gratton 1982, four stars) and studies claiming a spread in iron, with $-1.4 < [\mathrm{Fe/H}] < -1.9$ dex (Pilachowsky *et al.* 1984, six stars; Lehnert *et al.* 1991, four stars).

Particularly interesting are the findings regarding the CN-band strengths. Norris & Freeman (1983) showed that CN variations in M22 were correlated with Ca H and K line variations, similar to those in ω Cen. By studying a sample of four stars, Lehnert *et al.* (1991) found Ca and Fe variations that also correlated with variations in CH

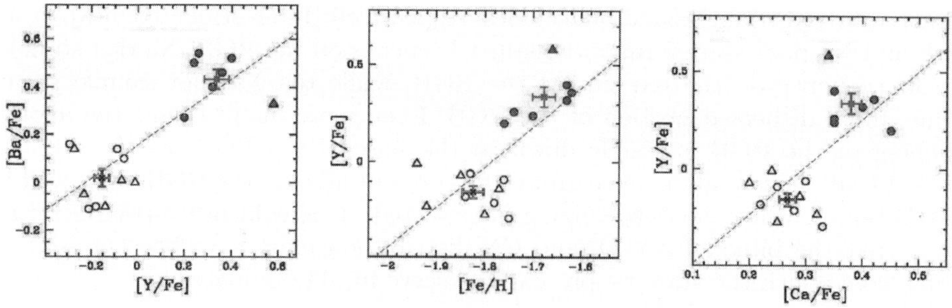

Figure 3. From left to right: [Ba/Fe] versus [Y/Fe], [Y/Fe] versus [Fe/H], [Y/Fe] versus [Ca/Fe] abundance ratios.

and CN-band strength. However, Brown & Wallerstein (1992) found no Ca abundance differences between CN-strong and CN-weak stars, although they observed differences in [Fe/H] correlating with the CN strength. More recently, Kayser *et al.* (2008) found some indications of a CN–CH anticorrelation in SGB stars, perhaps diluted by large uncertainties introduced by differential reddening.

For those reasons, we decided to focus our attention on this enigmatic cluster. As for M4, results we obtained are based on high-resolution high-S/N spectra collected with UVES+FLAMES. We found that M22 shows the typical NaO anticorrelation common to all the GCs where Na and O have been measured to date (see Carretta *et al.* 2006). But the most interesting result comes from the heavier elements. We have measured abundance for the three s-process elements yttrium, zirconium, and barium. All span a wide range of abundance values. The maximum variations of [Y/Fe], [Zr/Fe], and [Ba/Fe] have amplitudes of 0.87, 0.73, and 0.65 dex respectively, much greater than the internal error ($\sigma \leqslant 0.1$ dex).

In Figure 3 (left panel), we show [Ba/Fe] as a function of [Y/Fe]. A clear correlation between the elements is evident. The s-process elements clearly show a bimodal distribution: one group is overabundant in s elements with average values of [Y/Fe] $= +0.34 \pm 0.05$, [Zr/Fe]$= +0.60 \pm 0.04$, and [Ba/Fe] $= +0.43 \pm 0.04$. This group contains seven out of seventeen stars (i.e., $\sim 40\%$ of the full sample studied in this paper). The remaining ten stars have [Y/Fe] $= -0.16 \pm 0.03$, [Zr/Fe] $= +0.20 \pm 0.03$, and [Ba/Fe] $= +0.02 \pm 0.04$. Stars rich in s-process elements are represented by red filled symbols, while s-poor stars are shown as black open symbols. Circles are RGB stars, while triangles are asymptotic giant branch stars. Crosses with error bars indicate the average abundances of the stars in each group.

On the other hand, there is only a partial correlation between Y, Ba, and Zr abundances with Na, O or Al. However, it seems that stars enriched in s-process elements are all Na and Al rich, while the group of stars with lower s-process element abundances spans almost all values of Na and Al abundance.

This scenario is not limited to s-process elements. If we compare the iron abundances with the s-process element abundances, we find a strong correlation, with s-process element-rich stars having systematically higher [Fe/H], as shown in Figure 3 (central panel). The two s-process element-rich and poor groups have average $[\mathrm{Fe/H}]_{\mathrm{s-rich}} = -1.68 \pm 0.02$ and $[\mathrm{Fe/H}]_{\mathrm{s-poor}} = -1.82 \pm 0.02$ dex, respectively (δ[Fe/H] $= 0.14$ dex). This makes M22 the second cluster after ω Cen showing an intrinsic Fe spread. As shown in Figure 3 (right panel), the Ca abundance is also well correlated with s-process elements. All this seems to suggest that core-collapse supernovae (CCSNe) are the best candidates to produce the iron excess of the second-generation stars. Indeed, would iron

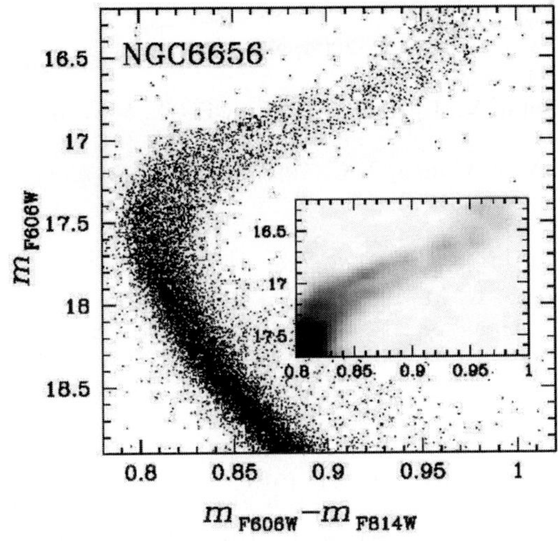

Figure 4. The M22 SGB observed with *HST*.

be produced by Type Ia SNe, we would expect a lower α-element content in s-process-rich stars with respect to the first generation of (s-process element-poor) stars. In fact, SNIa events are selectively enriched in iron. On the other hand, CCSNe along with iron, also produce α elements in higher quantities compared with SNIa.

The bi-modality in s-process elements in M22 resembles the case of NGC 1851. In NGC 1851, Yong & Grundahl (2008) noted that the abundances of the s-process elements Zr and Ba appear to cluster around two distinct values. They suggested that the two corresponding groups of stars may be related to the two stellar populations photometrically observed by Milone *et al.* (2008) along the SGB. At variance with M22, in NGC 1851 the s-element abundance appears to correlate with Na, Al, and O abundance.

From the similarity between our results and those of Yong & Grundahl (2008) for NGC 1851, it is possible to associate the presence of the two groups of stars with different s-process element content with the two populations of stars isolated along the SGB of M22 by Piotto (2009; see also Figure 4). The fraction of stars on the bright SGB corresponds to $62 \pm 5\%$ of the total SGB population, while the faint SGB includes the remaining $38 \pm 5\%$ of the SGB stars. In the stellar sample of the present paper, the fraction of Ba-strong, Y-strong, and Zr-strong stars is $\sim 41\%$. It is therefore tempting to connect the s-process element-poor sample to the bright SGB stars, while the faint SGB stars could have enhanced s-process elements. This second (younger) generation may have formed from material which was also enriched by CCSNe ejecta, as indicated by the higher iron and α-element abundances.

References

Bedin, L. R., Piotto, G., Anderson, J., Cassisi, S., King, I. R., Momany, Y., & Carraro, G. 2004, *ApJ*, 605, 125

Brown, J. A. & Wallerstein, G. 1992, *AJ*, 104, 1818

Cannon, R. D., Croke, B. F. W., Bell, R. A., Hesser, J. E., & Stathakis, R. A. 1998, *MNRAS*, 298, 601

Carretta, E., Gratton, R. G., Lucatello, S., Bragaglia, A., & Bonifacio, P. 2005, *A&A*, 433, 597

Carretta, E., *et al.* 2009, *A&A*, 505, 117

Cohen, J. & Meléndez, J. 2005, *AJ*, 129, 303

Dinescu, D. I., Girard, T. M., & van Altena, W. F. 1999, *AJ*, 117, 1792

Gratton, R. G. 1982, *A&A*, 115, 171
Gratton, R., Sneden, C., & Carretta, E. 2004, *ARA&A*, 42, 385
Harris, W. E. 1996, *AJ*, 112, 1487
Lee, Y. W., Joo, J. M., Sohn, Y. J., Rey, S.-C., Lee, H.-C., & Walker, A. R. 1999, *Nature*, 402, 55
Lehnert, M. D., Bell, R. A., & Cohen, J. G. 1991, *ApJ*, 367, 514
Milone, A. P., *et al.* 2008, *ApJ*, 673, 241
Norris, J. 1981, *ApJ*, 248, 177
Norris, J. & Freeman, K. C. 1983, *ApJ*, 266, 130
Peterson, R. C. & Cudworth, K. M. 1994, *ApJ*, 420, 612
Piotto, G., Bedin, L. R., Anderson, J., King, I. R., Cassisi, S., Milone, A. P., Villanova, S., Pietrinferni, A., & Renzini, A. 2007, *ApJ*, 661, 53
Smith, V. V., Cunha, K., Ivans, I. I., Lattanzio, J. C., Campbell, S., & Hinkle, K. H. 2005, *ApJ*, 633, 392
Smith, G. H. & Briley, M. M. 2005, *PASP*, 117, 895
Suntzeff, N. B. & Kraft, R. P. 1996, *AJ*, 111, 1913
Villanova, S., *et al.* 2007, *ApJ*, 663, 296
Yong, D. & Grundahl, F. 2008, *ApJ*, 672, 29

Star clusters: basic galactic building blocks
Proceedings IAU Symposium No. 266, 2009
R. de Grijs & J. R. D. Lépine, eds.

doi:10.1017/S1743921309991220

Primordial binary evolution and blue stragglers

Xuefei Chen and Zhanwen Han

National Astronomical Observatories/Yunnan Observatory, Chinese Aacademy of Sciences, Kunming, 650011, P. R. China
email: [xuefeichen717, zhanwenhan]@hotmail.com

Abstract. Blue stragglers have been found in all populations. They are important in both stellar evolution and stellar population synthesis. Much evidence shows that blue stragglers are relevant to primordial binaries. Here, we summarize the links between binary evolution and blue stragglers, describe the characteristics of blue stragglers originating from different binary evolutionary channels and show their consequences for binary population synthesis, such as for the integrated spectral-energy distribution, the colour–magnitude diagram, their specific frequency, and their influence on colours, etc.

Keywords. binaries: close, stars: evolution, blue stragglers

1. Introduction

Blue stragglers (BSs) are an important population component in stellar evolution as well as in star clusters. These objects have remained on the main sequence for a time exceeding that expected from standard stellar evolution theory, and they may affect the integrated spectra of their host clusters by contributing excess spectral energy in blue and ultraviolet bands. Many mechanisms, including single-star and binary models, have been proposed to account for the existence of BSs (see the review by Stryker 1993). At present, it is widely believed that more than one mechanism plays a role in the production of BSs in clusters and that binaries are important or even dominant for the production of BSs in open clusters and in the field (Lanzoni *et al.* 2007; Dalessandro *et al.* 2008; Sollima *et al.* 2008). Binaries may produce BSs through mass transfer, coalescence of the two components, binary–binary and binary–single star collisions. The collision of binary–binary or binary–single stars may lead binaries to be tighter or farther apart, which is relevant for the dynamics and the environment in the host cluster. In this contribution, we are only concerned with BSs resulting from the evolution of primordial binaries, i.e., mass transfer and coalescence of the two components.

2. Evolutionary channels

Before we describe the details of the evolutionary channels to BSs relevant to binary evolution, we first introduce the critical mass ratio, q_c, for dynamically unstable mass transfer, which is crucial to determine the fate of a binary during mass transfer. The value of q_c differs for different evolutionary stages of the primary at the onset of mass transfer, and it has been well studied through polytropic models (Hjellming & Webbink 1987; Han *et al.* 2001) and detailed binary evolution (Han *et al.* 2002; Chen & Han 2008a). Figure 1 shows two examples of these studies. We see obvious differences for q_c between polytropic models and detailed binary evolution, which in turn lead to differences in the products, including BSs, after mass transfer.

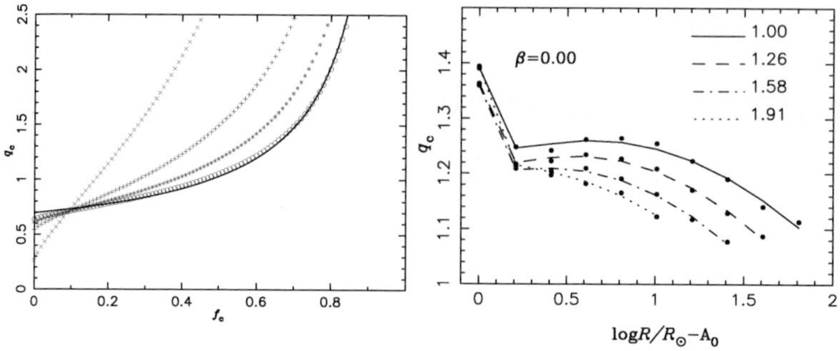

Figure 1. Critical mass ratio, q_c, for dynamically unstable mass transfer. *(left)* Polytropic model. f_c is the core mass fraction of the mass donor. The crosses, pluses, asterisks and circles are for mass-transfer efficiencies $\beta = 0.25$, 0.5, 0.75 and 1.0, respectively. The lost mass is assumed to carry away the same specific angular momentum as pertains to the mass donor. The solid line is from Webbink (1988) for conservative mass transfer (see also Han *et al.* 2001). *(right)* Detailed binary evolution for low-mass binaries between a first-ascent giant-branch and a main-sequence star. The giant star has a mass of 1.00, 1.26, 1.60 and $1.90 M_\odot$, respectively. The lines are from the fitting formulae of equations (4) to (6) in Chen & Han (2008a).

2.1. *Mass transfer*

In general, if the mass ratio $q = M_1/M_2$ (donor/accretor mass) at the onset of mass transfer is lower than q_c, mass transfer is stable, and the accretor evolves upwards along the main sequence in response to accretion, if it is a main-sequence star, and becomes a BS when it is more massive than the turnoff of the host cluster.

This channel may produce binary BSs with various orbital periods (Chen & Han 2004, 2008a). If the primary is on the main sequence or in the Hertzsprung gap at the onset of mass transfer, the products have short or relatively short orbital periods. These include Algol systems, which are still in a slow stage of mass transfer. Mass transfer between giant stars and main-sequence companions may produce long-orbital-period BSs. The value of q_c here significantly affects the BS number and orbital periods (Chen & Han 2008a).

Since the accreted material may be originating from the nuclear region of the mass donor, it is rich in helium compared to the accretor's surface and has a higher mean molecular weight, resulting in secular instability during or after accretion. Thermohaline mixing will occur in this case. This mixing was once believed to cause the surface-abundance abnormality of the accretor to be invisible. However, the study of Chen & Han (2004) showed no distinction in surface composition between the models with and without thermohaline mixing during mass transfer, although their evolutionary tracks diverge. After mass transfer, CNO abundance abnormalities may exist for about 10^8 yr, comparable to the lifetime of a typical BS. Thus, we could observe CNO abundance abnormalities in BSs this way.

2.2. *Coalescence of contact binaries*

During mass transfer, the accretor likely fills its Roche lobe and the system becomes a contact binary. This contact binary eventually coalesces into a single star (Webbink 1976; Eggleton 2000; Li *et al.* 2005). If both components are on the main sequence, their remnant is also a main-sequence star and evolves similarly to a normal star with that mass. So, the remnant may be a BS if it is more massive than the turnoff.

By assuming that the matter from the secondary mixes homogeneously with the envelope of the primary and that no mass is lost from the system during the merger process, Chen & Han (2008b) constructed mergers of contact binaries and studied their characteristics. Their study shows that some mergers are located on the left of the zero-age main sequence defined by normal surface composition (i.e., helium content $Y = 0.28$ and metallicity $Z = 0.02$ for Population I) on a colour–magnitude diagram, because of enhanced surface-helium content. In addition, the central hydrogen content of the mergers is independent of mass. Thus, the concentration towards the blue side of the main sequence with decreasing mass predicted by Sandquist & Shetrone (2003), does not appear in their models. In fact, there is no evidence for such a concentration from observations.

In old clusters, angular-momentum loss (AML) of low-mass binaries induced by magnetic braking is a main factor leading to contact binaries and final merging. A simple estimation (Chen & Han 2008b) shows that, in old clusters, BSs resulting from AML are much more numerous than those originating from evolutionary effects only, indicating that AML in low-mass binaries makes a major contribution to BSs in old clusters such as NGC 188, NGC 2682, etc. In clusters of intermediate age, e.g., in NGC 2660, the models of Chen & Han (2008b) can account for several BSs, although in the most likely region in the colour–magnitude diagram of NGC 2660 no BSs have as yet been observed. About 0.5 $M_\odot$ of mass loss in the merger process is necessary to resolve this conflict.

2.3. *Coalescence due to dynamically unstable mass transfer*

If the mass ratio, q, is greater than q_c, mass transfer is dynamically unstable, and a common envelope (CE) is formed. The CE may be ejected if the orbital energy deposited into the envelope overcomes its binding energy, or else the binary will merge into a single star. If the two components are main-sequence stars, the remnant of coalescence will be on the main sequence, and it becomes a BS if its mass is higher than the turn-off mass of the host cluster.

In this case, the core of the secondary spirals in quickly and remains in the centre of the merger. The merger then has a chemical composition similar to that of the primary, resembling the result of smoothed-particle-hydrodynamics calculations (Lombardi *et al* 1996; Sills & Lombardi 1997; Sills *et al.* 2001). Binary coalescence of a contact binary or dynamically unstable mass transfer is a popular hypothesis for the formation of single BSs (Mateo *et al.* 1990; Pols & Marinus 1994; Andronov *et al.* 2006; Chen & Han 2008b).

3. Binary population synthesis

We performed five sets of simulations for a Population I composition ($X = 0.70, Y = 0.28$ and $Z = 0.02$) to systematically investigate BS formation from primordial binary evolution. The age ranges from 0.1 to 20 Gyr. The mass-transfer efficiency, β, defined as the mass fraction of the matter lost from the primary and accreted by the secondary, is an important parameter affecting the final results. Generally, we set $\beta = 1$ when the mass donor is on the main sequence and $\beta = 0.5$ otherwise. However this value is very uncertain, and it is likely higher than 0.5 when the mass donor is in the Hertzsprung gap and less than 0.5 when the mass donor is on the first giant branch or the AGB. We therefore also studied the case of $\beta = 1$, when mass transfer begins in the Hertzsprung gap (Chen & Han 2009).

3.1. *Distribution in colour–magnitude diagrams*

Figure 2 shows colour–magnitude diagrams of a population with an age of 4.3 Gyr for various β values when the primary is in the Hertzsprung gap at the onset of Roche-lobe overflow. We see that BSs from $\beta = 1$ may be more massive than those for $\beta = 0.5$, indicating that a high value of β may produce BSs far away from the turnoff. In particular, we may obtain BSs with masses higher than twice the turnoff mass, even though these objects have very short lifetimes. It is appropriate to assume a high value of β for old

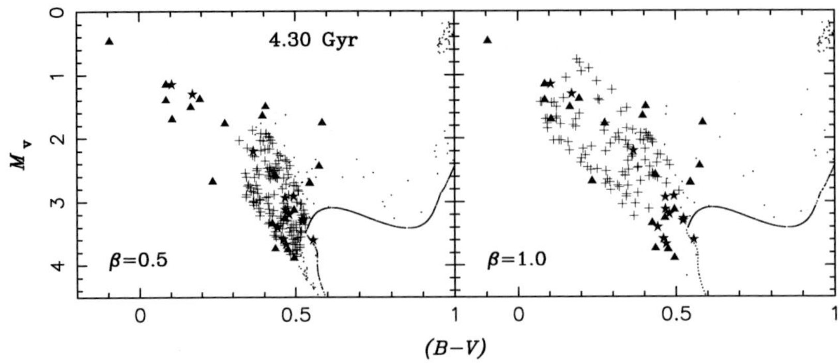

Figure 2. Colour–magnitude diagrams for a 4.3 Gyr-old population. Here, β is the mass fraction of the matter lost from the primary and accreted by the secondary when the primary is in the Hertzsprung gap at the onset of mass transfer. The crosses and stars are for BSs from mass transfer and binary coalescence, respectively, and the small dots are for other objects in the population. The triangles are observed BSs in M67 from Sandquist & Shetrone (2003).

clusters, since the binaries contributing to BSs are less massive than those in young clusters.

We also see in the figure that many BSs are below but bluer than the turnoff. They may extend into the region 1 mag below the turnoff. These objects are mainly formed through mass transfer and binary coalescence of dynamically unstable mass transfer. They are less evolved than the turnoff and may contribute more to the flux in the V band in the subsequent evolution. Most BSs from mass transfer and binary coalescence of dynamically unstable mass transfer are within 1.5 mag of the turnoff, while some from binary coalescence of contact binaries may have magnitudes about 2.3 mag above the turnoff.

3.2. *The specific frequency*

The BS number from the simulations, $N_{\rm BS}$, depends slightly on the population's age, while the specific frequency, $\log F_{\rm BSS} (\equiv \log(N_{\rm BS}/N_2)$, where N_2 is the number of stars within 2 mag below the main-sequence turnoff), depends heavily on age due to the related increase of N_2. In all of the five sets, $\log F_{\rm BSS}$ first decreases with time and then increases when the age becomes older than 10 Gyr. The decrease of $\log F_{\rm BSS}$ before 1.5 Gyr comes from the increase of N_2. Subsequently, the number of potential binaries which may contribute to BSs decreases, leading to the formation of fewer BSs. On the other hand, N_2 continues to increase. Thus, $\log F_{\rm BSS}$ continues to decrease. Over time, the primaries in long-orbital-period binaries gradually enter the AGB phase, expand dramatically and some may fill their Roche lobe and start mass transfer. Due to strong stellar winds in the AGB phase, these mass donors at the onset of mass transfer are probably much less massive than before and this mass transfer is easily stabilized, resulting in some long-orbital-period BSs. As a consequence, $\log F_{\rm BSS}$ begins to increase when their age is older than 10 Gyr. AML becomes more and more important for BS formation with time and exceeds that from primordial binary evolution when the age is older than 2.5 Gyr.

When we investigated the role of primordial binary evolution for BS formation in Galactic open clusters, we found a serious problem that the BS specific frequency obtained in our simulations is much lower (only about 20%) than observed, which may result from the following aspects. (i) Observational errors, such as N_2 counted in clusters, the cluster ages and the BS sample, etc. For example, the study of Carraro (2008) showed that a large fraction of earlier BSs are actually field stars. (ii) The adopted Monte Carlo simulation

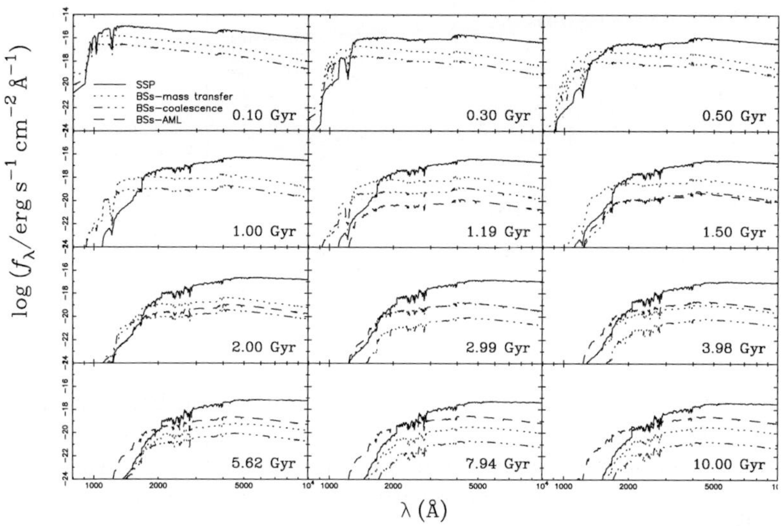

Figure 3. Integrated rest-frame intrinsic spectral-energy distributions for a stellar population with a mass of 1 $M_\odot$ at a distance of 10 kpc with $\beta = 0.5$ when the primary is in the Hertzsprung gap at the onset of mass transfer. The solid lines are for the SSP (a population without binary interactions) and the other lines are the contributions of BSs from different evolutionary channels.

parameters, including the initial mass function, distributions of the initial ratio and the initial separation. All these parameters in our simulation are for field stars and it is very likely that they are different in clusters. For instance, recent studies show that the initial mass function might be quite different for disk and halo stars in the Galaxy (Pols *et al.* 2008). (iii) Other channels, e.g., a more recent era of star formation and dynamical interaction may produce BSs in open clusters in addition to primordial binary evolution and AML.

3.3. *Contribution to the integrated spectral-energy distributions*

We plotted the integrated spectral-energy distributions (ISEDs) of a population for various cases in Figure 3, where SSP refers to a population without binary interaction. This figure shows that BSs resulting from binary evolution are dominant contributors to the ISED in ultraviolet and blue bands for ages between 0.3 and 2.0 Gyr. The BSs and SSP have comparable energy in the ultraviolet and blue between 2 and 4 Gyr. The contribution from AML becomes more important with time, and exceeds that from primordial binary evolution for a population older than $\sim$ 3 Gyr. Thus, primordial binaries are important contributors to BSs over the whole age range. Since BSs from a high β have higher masses, their contributions to the ISED become more important in ultraviolet and blue bands (see figure 5 in Chen & Han 2009).

3.4. *Influence on colours and ages*

Obviously, the excess spectral energy due to BSs in ultraviolet and blue bands inevitably results in some changes in colours involving these bands. The results are shown in Figure 4, from which we see that the colours are affected from about 0.32 Gyr (all five colours shown in this figure) to older than 10 Gyr (i.e., F185W–F336W and F218W–F336W), and the maximum difference may be up to 1.5 mag for F170W–F336W.

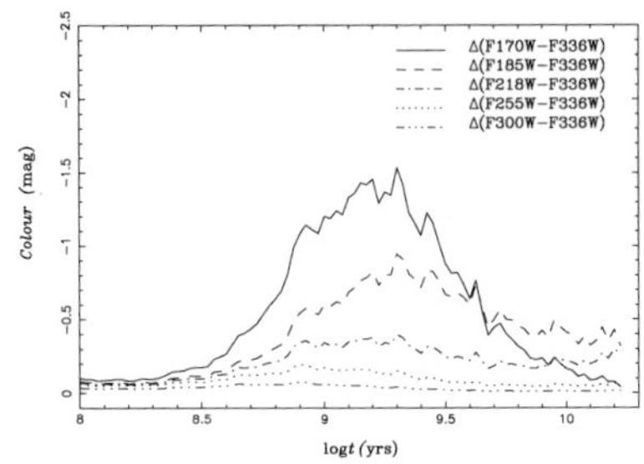

Figure 4. Colour differences versus age for a population with and without BSs.

4. Conclusions

We have summarized the links between binary evolution and BSs, and the relevant BS characteristics. We also showed binary population synthesis results of BSs from primordial binary evolution, such as their distribution in the colour–magnitude diagram, their contribution to the ISED, the specific frequency and their influence on colours.

Acknowledgements

This work was in part supported by the Chinese National Science Foundation (grants 10973036, 10603013, 10821061 and 2007CB815406) and the Yunnan National Science Foundation (grant 08YJ041001).

References

Andronov, N., Pinsonneault, M. H., & Terndrup, D. M. 2006, *ApJ*, 646, 1160
Carraro, G., 2008, *A&A*, 482, 777
Chen, X. & Han, Z. 2004, *MNRAS*, 355, 1182
Chen, X. & Han, Z. 2008a, *MNRAS*, 387, 1416
Chen, X. & Han, Z. 2008b, *MNRAS*, 384, 1263
Chen, X. & Han, Z. 2009, *MNRAS*, 395, 1822
Dalessandro, E., Lanzoni, B., Ferraro, F. R., Rood, R. T., Milone, A., Piotto, G., & Valenti, E. 2008, *ApJ*, 677, 1069
Eggleton, P. P. 2000, *NewAR*, 44, 111
Han, Z., Eggleton, P. P., Podsiadlowski, P., Tout C. A., & Webbink, R. F. 2001, *ASP Conf. Ser.*, 229, 205
Han, Z., Podsiadlowski, P., Maxted, P. F. L., Marsh, T. R., & Ivanova, N. 2002, *MNRAS*, 336, 449
Hjellming, M. S. & Webbink, R. F. 1987, *ApJ*, 318, 794
Lanzoni, B., Dalessandro, E., Perina, S., Ferraro, F. R., Rood, R. T., & Sollima, A. 2007, *ApJ*, 670, 1065
Li, L., Han, Z., & Zhang, F. 2005, *MNRAS*, 360, 272
Lombardi Jr., J. C., Rasio, F. A., & Shapiro, S. L. 1996, *ApJ*, 468, 797
Mateo, M., Harris, H., Nemec, J., & Olszewski, E. 1990, *AJ*, 100, 469
Pols, O. R., Izzard, R. G., Lugaro, M., & de Mink, S. E. 2008, *IAU Symp.*, 252, 383
Pols, O. R. & Marnus, M. 1994, *A&A*, 288, 475
Sandquist, E. L. & Shetrone, M. D. 2003, *AJ*, 125, 2187
Sills, A. & Lombardi Jr., J. C. 1997, *ApJ*, 105, 1081
Sills, A., *et al.* 2001, *ApJ*, 105, 1081
Sollima, A., Lanzoni, B., Beccari, G., Ferraro, F. R., & Fusi Pecci, F. 2008, *A&A*, 481, 701
Stryker, L. L. 1993, *PASP*, 105, 1081
Webbink, R. F. 1976, *ApJ*, 209, 829
Webbink, R. F. 1988, in: Mikolajewska, J., Friedjung, M., Kenyon, S. J., & Viotti, R., *The Symbiotic Phenomenon*, p. 311 (Dordrecht: Kluwer)

Summary

Star clusters: basic galactic building blocks
Proceedings IAU Symposium No. 266, 2009
R. de Grijs & J. R. D. Lépine, eds.

doi:10.1017/S1743921309991232

IAU Symposium 266: Summary

Giovanni Carraro

ESO, Alonso de Cordova 3107, 19001, Santiago de Chile, Chile
email: gcarraro@eso.org

Dipartimento di Astronomia, Università di Padova, Vicolo Osservatorio 5, I-35122 Padova, Italy
email: giovanni.carraro@unipd.it

Abstract. A review is given of the topics discussed during IAU Symposium 266 on star clusters, and of the major results presented. The current state of the art is discussed, together with potential prospects which may help to direct and generate future advances.

Keywords. globular clusters: general, open clusters and associations: general, Galaxy: halo, Galaxy: disk, Galaxy: bulge, stars: low-mass, brown dwarfs, Hertzsprung–Russell diagram, stars: horizontal-branch, binaries: general, stars: abundances

1. Introduction

IAU Symposium 266, *Star clusters: basic galactic building blocks throughout time and space*, hosted 47 talks including the plenary review and summary talk. As the title emphasizes, star cluster studies currently cover a large range of topics in Galactic and extragalactic astronomy. In fact, star clusters are laboratories of stellar evolution and dynamics, as well as building blocks facilitating studies of the properties of external galaxies, including their cosmological formation and assembly, star-formation history (SFH) and chemical evolution. In this summary, I report on the diverse and rich material discussed at the meeting, grouping together the various talks where appropriate.

2. Star clusters and star formation

Star clusters are ideal sites to study the formation of stars in a statistical way. Researchers at the meeting agreed that stars form from dense CO cores inside molecular clouds. In addition, (almost) all stars form in star clusters and tend to group together in hierarchical structures. This model emerges and acquires further confirmation from recent and ongoing extensive submillimeter, infrared, optical and X-ray surveys, both photometric and spectroscopic, which were presented during the meeting. The major role that the *Atacama Large Millimeter Array* will play in this context was emphasized, particularly its unprecedented spatial resolution at these wavelengths.

The study of very young clusters or associations can help us probe the first phases of the star-formation process and the complex interplay between stellar evolution, feedback and interactions. A large variety of stellar systems exist, from nuclear star clusters to associations, and from embedded clusters to super star clusters, such that the star-formation process is actually studied in very different environments. From detailed X-ray studies, evidence was presented of clusters having a sequential SFH, while others show prolonged SFHs.

At some point, we should correlate these findings with the dynamical status of clusters and their location in the Galaxy, to see whether star formation is an environmentally dependent process.

3. Star clusters and stellar evolution

This is a classic topic in astronomy and, as usual, observations offer interesting news and challenges, which in turn motivate continuous revision of the theory. For instance, the discovery and characterization of the intriguing multiple stellar populations in Galactic and Large/Small Magellanic Cloud globular clusters (the commonly believed standard example of single stellar populations) through extensive photometric and spectroscopic campaigns is boosting major theoretical efforts.

We need to understand how this complicated mix of populations formed, in essence how clusters could retain processed material from first-generation stars to produce second- or even third-generation stars. The competing roles of rotating massive stars and evolved stars on the asymptotic giant branch were discussed in this context, but without reaching clear-cut conclusions. Alongside this discussion, a new scenario in which the gas necessary for the second generation is expelled from first-generation binary systems was proposed and attracted much interest.

The enhanced/nonenhanced helium-abundance explanation for secondary blue sequences in the color–magnitude diagrams is another hot topic, and a clear example that there must be some missing ingredient in our current understanding of stellar evolution. Thus far, researchers have mainly concentrated on the internal evolution of star clusters, but it might be worth to take a look at completely different scenarios, such as external perturbation and minor mergers. At any rate, we need to pose the fundamental question: *Can we still consider globulars as simple stellar populations?*

Studies of brown dwarfs in young clusters, helium-burning stars in the horizontal-branch phase, blue stragglers and white dwarfs in intermediate-age and old systems were presented. These will potentially modify several fundamental aspects in our understanding of stellar structure and evolution (e.g., mass limits, lifetimes, the role of binaries) in the near future.

4. Star clusters in external galaxies

The study of star clusters in external galaxies seems quite lively. Here, star clusters are mostly considered test particles, and their integrated properties are used to deduce general properties of galaxies. Quantities such as luminosity and mass functions, age–mass relationships and color distribution are derived to determine the initial mass function (IMF), the SFH and (more generally) the cosmological formation scenario (either monolithic or hierarchical) of the host galaxy.

Critical issues related to measuring these quantities were discussed, like limitations of the integrated photometry, proper filter selection and binning techniques, as they might interfere with the objectivity of the basic measurements and the resulting conclusions. This is illustrated by the intriguing situation of using completely different approaches for deriving star cluster luminosity functions which produce basically the same results! It was argued that better communication among researchers and exchanges of processed data can help solve many of these conflicts.

In various models, star clusters are the building blocks of galaxies and the interaction between clusters and their environment is investigated to study both cluster evolution and the formation of structures such as bulges. The role of the tidal field, disk shocking and dynamical friction was discussed in unprecedented detail.

5. Star cluster formation and internal dynamics

Many models of the formation, early evolution and dissolution of star clusters were discussed during the conference. Feedback from massive stars and their interaction with the environment, giant molecular clouds and the host galaxy's potential well determine how long a cluster with a given initial mass can survive, and possibly when mass segregation occurs. This has been observed to take place at different times in different clusters, but it is not yet clear whether or not mass segregation occurs at the very beginning of star cluster evolution.

The role of the IMF and its universality was discussed in this context, together with the never-sufficiently-stressed importance of binary stars, white dwarfs and black holes. Dedicated proper-motion and radial-velocity studies allow us to pin down actual cluster members and provide more solid estimates. Discrepant results were shown, casting doubts on the reality of black holes in clusters. The IMF was shown to be the same in very old and very young clusters, supporting is possible universality.

N-body models are getting more and more sophisticated, and we heard that the coupling of stellar evolution with dynamics making rapid progress, although significant differences in the results from different codes are still present. However, N-body and stellar-evolution models combined already show an impressive level of detail and a remarkable similarity to actual observations for some individual clusters (notably M67 and NGC 188).

6. Star clusters in the Milky Way

The structure and chemical evolution of the Milky Way and its components have traditionally been investigated using star clusters. The parallel IAU Symposium 265 on chemical abundances discussed many of these classical topics, which had only minor coverage during our meeting.

Globular clusters are used to probe the formation of the Galactic halo, and very detailed chemical studies attempt to clarify whether stars floating in the halo originate from globular-cluster mass loss. Isotopic abundances are leading to completely different opinions, which hopefully will converge when more data come available. Opposite claims are, in fact, based only on a limited number of clusters.

The origin, age, shape and metallicity distribution of the Galactic bulge was also addressed using globulars, in spite of the difficulties that the high and patchy reddening poses. The Galactic-disk spiral structure and chemical evolution is another traditional topic, where progress will benefit from the intense hunt for highly obscured, overlooked star clusters. Still, new spiral features have recently been discovered in the outer disk, mostly in the third Galactic quadrant. Hopefully, some other massive young clusters – so rare in the Milky Way – will be discovered exploiting data from UKIDSS and the upcoming *VISTA* surveys.

7. Conclusions

All participants agreed that the Symposium was interesting and successful. The results presented during the conference boosted lively and important discussions, which will surely result in new collaborations and, in turn, in a greater interest in the field of star cluster research.

Posters

Star clusters: basic galactic building blocks
Proceedings IAU Symposium No. 266, 2009
R. de Grijs & J. R. D. Lépine, eds.

doi:10.1017/S1743921309991244

Further developments in star cluster spectral libraries†

Andrea V. Ahumada,[1,2,3] Juan J. Clariá,[2,3] Eduardo Bica,[4] Andrés E. Piatti,[2,5] João F. C. Santos Jr.,[6] M. Lorena Talavera,[7] Tali Palma,[2,3] Daniela B. Pavani,[8] M. Celeste Parisi,[2,3] M. Cristina Torres[3] and Carlos M. Dutra[9]

[1]European Southern Observatory, Alonso de Córdova 3107, Santiago, Chile
email: aahumada@eso.org
[2]CONICET (Argentina)
[3]Observatorio Astronómico de la Universidad Nacional de Córdoba, Laprida 854, 5000 Córdoba, Argentina
email: [claria, tali, celeste]@oac.uncor.edu, crico@emprendinet.com.ar
[4]Universidade Federal do Rio Grande do Sul, Departamento de Astronomia CP 15051, RS, Porto Alegre 91501-970, Brazil
email: bica@if.ufrgs.br
[5]Instituto de Astronomía y Física del Espacio, CC 67, Suc. 28, 1428, Ciudad de Buenos Aires, Argentina
email: andres@iafe.uba.ar
[6]Departamento de Fisica, ICEx, UFMG, CP 702, 30123-901, Belo Horizonte, MG, Brazil
email: jsantos@fisica.ufmg.br
[7]Observatorio Astronómico Centroamericano de Suyapa, Cdad. Univ., Boulevard Suyapa, Ap. Postal 4432, Tegucigalpa M. D. C., Honduras
email: loren.briones@gmail.com
[8]Universidade Federal de Pelotas, Instituto de Física e Matemática, Dep. de Física, Campus Universitário, s/n, Caixa Postal 354, 96010-900 Pelotas, RS, Brazil
email: dpavani@if.ufrgs.br
[9]Universidade Federal do Pampa, BR 472 km 592, CEP:97500-970, Uruguaiana, RS, Brazil
email: cmdutra@gmail.com

Abstract. We present flux-calibrated integrated spectra in the optical spectral range of Galactic open clusters (GOCs) and Magellanic Cloud (MC) stellar clusters (SCs) obtained at *CASLEO* (Argentina). The SC parameters were derived using the equivalent-width (EW) method and the template-matching procedure by comparing the line strengths and continuum distribution of the cluster spectra with those of template spectra with known parameters. MC cluster reddening values were also estimated by interpolation between the available extinction maps. The derived ages for the GOCs range from 3 Myr to 4 Gyr, while those of the MC SCs vary from 3 Myr to 7 Gyr. $E(B-V)$ colour-excess values in the MCs appear to be all lower than 0.17 mag, while those of the GOCs range from 0.00 to 2.40 mag. The present data led us to upgrade the spectral libraries of reference spectra or templates of solar and MC metallicities.

Keywords. galaxies: star clusters, techniques: spectroscopic

---

1. Introduction

It is well known that star clusters (SCs) are building blocks of galaxies, in the form of young associations, open and globular clusters. SCs are, in general, powerful tracers of

† The full poster (in pdf format) is available at http://www.astro.iag.usp.br/~iaus266/Posters/pAhumada.pdf.

the formation and evolution of their parent galaxies. This work is an ongoing systematic spectroscopic survey of Galactic open clusters (GOCs) and Magellanic Cloud (MC) SCs. The main goal of this long-term study is to collect and analyse a large sample of SC integrated spectra with the aim of deriving their fundamental parameters and making them available as template spectra to complement previous spectral libraries.

2. Galactic open clusters

GOCs have always played a prominent role in the derivation of both the chemical and dynamical evolution of the Galactic disk. This is due to the fact that their fundamental parameters may be determined more easily and more accurately than those for isolated stars. However ~50% of the 1787 OCs known to exist in the Milky Way (Dias *et al.* 2002) are unstudied. Integrated spectra of small-angular-size OCs have proved very useful tools in increasing the number of OCs studied and are a valuable means to provide independent information about reddenings, ages and, in some cases, metallicities. Several years ago, Bica & Alloin (1986a) published a pioneering study of integrated spectroscopy applied to composite stellar populations. Compact SCs are the most suitable objects to carry out integrated spectroscopic observations with a middle-sized telescope. Our sample was selected using the WEBDA database (Paunzen & Mermilliod 2009), where most of the selected OCs are compact clusters. On this basis, we have observed and studied 71 GOCs in the last 10 years by means of integrated spectroscopy. Table 1 shows the GOC sample (see the full electronic poster version).

3. Observations

All observations were carried out with the 2.15m telescope at the *Complejo Astronómico El Leoncito* (*CASLEO*, Argentina) during several runs. We employed a CCD camera with a chip of 1024 × 1024 pixels attached to a REOSC spectrograph (simple mode). The observations were performed by scanning the slit across the objects in the north–south direction. A grating of 300 grooves mm^{-1} was used. The spectral coverage was in the visible range: 3600–6800 Å, with an average dispersion in the observed region of ~140 Å mm^{-1} (3.46 Å pix^{-1}). The slit width was 4.2″, which provided a resolution of ~14 Å.

4. Reddening and age determination

Age and foreground-reddening values of the selected clusters were derived simultaneously by means of a template-matching method. This was done by achieving the best possible match between the continuum and the lines of the analysed cluster's integrated spectrum and those from a template of integrated spectra with known properties. To apply the template-matching method, a direct, reddening-independent age estimate was first obtained from the EWs of the Balmer absorption lines in each spectrum. This was done by interpolating the EW values from the age calibration of Bica & Alloin (1986b). Based on this age estimate, we selected, among the available template libraries, a subset of templates to compare with the observed spectrum. This selection allowed us to constrain our search of the most appropriate template, as those selected were within a certain age range. The next step was to determine the final age of the spectrum by varying reddening and template values until the best match of continuum, Balmer and metal lines to that of the best-fitting template was obtained. Results of the study of the full sample are shown in Table 1 (poster version), where the columns show the first cluster designation, the derived $E(B-V)$ and the age adopted for the cluster, taking into account the Balmer lines and the template age.

5. GOC template libraries

Different authors, including Bica & Alloin (1986a), Bica (1988), Bonatto *et al.* (1995), Santos *et al.* (1995) and more recently Schiavon *et al.* (2005) struggled to create reference spectra of SCs in different spectral ranges and to define grids of composite stellar populations. Because SCs may be considered stellar population units of a given age and metallicity, their spectral properties have been used for stellar population synthesis and for the interpretation of galaxy spectra. We collected part of the OCs spectra presented here and averaged them into templates, together with previously available spectra of well-studied OCs, with a view of obtaining the highest possible signal-to-noise ratio and higher time resolution to study their integrated-light evolution (Piatti *et al.* 2002; Ahumada *et al.* 2007). Figure 1 shows the spectral evolution up to intermediate ages (Ia and Ib) of stellar populations with nearly solar metallicity. For the younger populations, we constructed different templates (Ya and Yb), considering the foreground, the mean (Ya2 and Yb2) and the internal reddening. At such young ages, luminous stars play an important role in the resulting slope of the continuum energy distribution, which was also taken into account (Y*–WR–SG).

6. Magellanic Cloud SCs

The study of extragalactic stellar systems provides relevant information about star-formation and chemical-enrichment histories of the host galaxies. Despite the multiple observational and theoretical projects undertaken in the past, our current knowledge of these processes is rather imcomplete, even for the galaxies in the Local Group. Therefore,

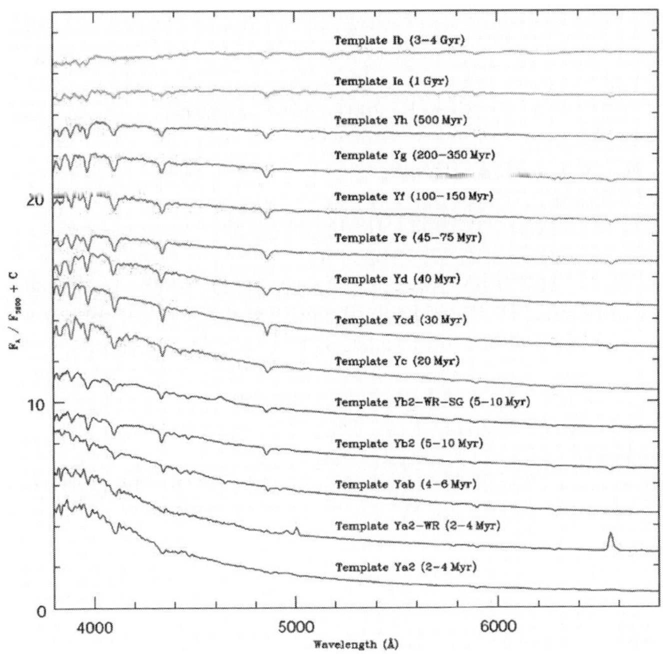

Figure 1. GOC template spectra for the young (Y) and intermediate (I) age groups.

the SCs of the MCs may furnish us with the ideal basis to conduct a detailed examination of these processes because of their proximity, richness and variety. One of the goals of the present study is to collect and analyse a large sample of MC SC spectra with the aim of studying the integrated-light properties of such metal-deficient clusters, deriving their fundamental parameters and making them available as template spectra which can complement previous subsolar-metallicity libraries. The main denomination of the 42 observed SMC SCs are shown in Table 2 (poster version). Some of the SCs are very-well-known objects and they were observed to have SMC template spectra at the ages of those clusters. Table 3 (poster version) shows the denomination of the 54 LMC SCs selected for this study. Most of the LMC clusters were classified as type II (30–70 Myr), according to Searle *et al.* (1980). In addition to using the template-matching method, we determined ages using diagnostic diagrams involving the sum of the EWs of selected spectral lines combined with the calibrations of age and metallicity by Santos & Piatti (2004). Based on the Burstein & Heiles (1982) maps we determined another $E(B-V)$ value, which turned out to be very similar to the one derived from the template matching method. $E(B-V)$ colour-excess values in the MCs appear all to be lower than 0.17 mag. Tables 2 and 3 show the ages determined for SMC and LMC SCs, respectively (poster version).

7. Future work

Among some previously defined consecutive spectral groups, we are working on defining new GOC template spectra with the aim to refine the spectral library of Piatti *et al.* (2002) and to improve its temporal resolution. The present MC data constitute part of the elements to enhance the spectral libraries at the metallicities of the SMC and LMC SCs. We are planning, in the near future, to extend the existing libraries to the near-infrared. We will also create new libraries for other metallicities, which will be very useful to study the stellar populations in more distant galaxies.

References

Ahumada, A. V., Clariá, J. J., & Bica, E. 2007, *A&A*, 473, 437

Bica, E. 1988, *A&A*, 195, 76

Bica, E. & Alloin, D. 1986a, *A&A*, 162, 21

Bica, E. & Alloin, D. 1986b, *A&AS*, 66, 171

Bonatto, C., Bica, E., & Alloin, D. 1995, *A&AS*, 112, 71

Burstein, D. & Heiles, C. 1982, *AJ*, 87, 1165

Dias, W. S., Alessi, B. S., Moitinho, A., & Lépine, J. R. D. 2002, *A&A*, 389, 871

Paunzen, E. & Mermilliod, J.-C. 2009, `http://www.univie.ac.at/webda/`

Piatti, A. E., Bica, E., Clariá, J. J., Santos Jr., J. F. C., & Ahumada, A. V. 2002, *MNRAS*, 335, 233

Santos Jr., J. F. C., & Piatti, A. E. 2004, *A&A*, 428, 79

Santos Jr., J. F. C., Bica, E., Clariá, J. J., Piatti, A. E., Girardi, L. A., & Dottori, H. 1995, *MNRAS*, 276, 1155

Schiavon, R. P., Rose, J. A., Courteau, S., & MacArthur, L. A. 2005, *ApJS*, 160, 163

Searle, L., Wilkinson, A., & Bagnuolo, W. G. 1980, *ApJ*, 239, 803

Star clusters: basic galactic building blocks
Proceedings IAU Symposium No. 266, 2009
R. de Grijs & J. R. D. Lépine, eds.

doi:10.1017/S1743921309991256

Measuring the GC luminosity function up to $z \sim 0.2$

K. A. Alamo–Martínez,[1] R. A. González–Lópezlira[1] and J. Blakeslee[2]

[1]Centro de Radioastronomía y Astrofísica, Universidad Nacional Autónoma de México, Campus Morelia, Michoacán CP 58190, Mexico

[2]Dominion Astrophysical Observatory, Herzberg Institute of Astrophysics, National Research Council of Canada, Victoria, BC V9E 2E7, Canada

Abstract. Globular clusters (GCs) are stellar systems ($\sim 10^6$ $M_\odot$) with very regular symmetry, single age, and single metallicity. Spectroscopic studies have revealed very old ages, suggesting that GCs were formed in the earliest stages of galaxy formation and assembly. The aim of this work is to find out how far we can measure the GC luminosity function, specific frequency, and radial distribution, applying the surface-brightness-fluctuations (SBF) technique to deep ACS images. To this end, we apply the effects caused by higher redshift to *HST*/ACS images (in two optical bands, F606W and F814W) of M87, an elliptical galaxy with a very well-studied GC system. The effects involved are: (i) evolution, (ii) inverse k correction, (iii) binning of the image to smaller angular size, (iv) cosmological dimming of surface brightness, and (v) noise addition to account for different exposure times. After processing the images we detect the brightest GCs through direct photometry (e.g., with SExtractor), whereas the unresolved clusters are measured through SBFs. The above treatment is repeated for $z = 0.05, 0.1, 0.14$, and 0.18, and the results are compared to the measurements at $z = 0$ to estimate biases and incompleteness.

Keywords. techniques: photometric, globular clusters: general, galaxies: star clusters

---

The full poster (in pdf format) is available at
`http://www.astro.iag.usp.br/~iaus266/Posters/pAlamo-Martinez.pdf`.

Star clusters: basic galactic building blocks
Proceedings IAU Symposium No. 266, 2009
R. de Grijs & J. R. D. Lépine, eds.

doi:10.1017/S1743921309991268

Determination of astrophysical parameters of open clusters using classical and semi-automated isochrone-fitting methods

Virginia Mello Alves,[1] Eduardo Bica[2] and Daniela Borges Pavani[1]

[1]Universidade Federal de Pelotas (UFPel), Brazil

[2]Universidade Federal do Rio Grande do Sul (UFRGS), Brazil

Abstract. When we use optical isochrone-fitting solutions from the literature to 2MASS CMDs, they are often not the best solutions in the infrared domain. We analysed 10 open clusters with 2MASS, nine of them previously studied with optical photometry (NGC 1245, NGC 1342, NGC 1502, NGC 2104, NGC 2204, NGC 2243, NGC 2281, NGC 6709 and NGC 744) and one using integrated spectroscopy (BH 132). The study involved the classical (by eye) and a semi-automated method of isochrone fitting. We used the solutions of the first method as input for the second, looking for refined solutions. The semi-automated method uses a synthetic color–magnitude diagram (CMD), based on different Padova isochrones, to compare with the observed CMDs by means of likelihood statistics. The derived astrophysical parameters are age, distance and reddening values. The present results show better fits than those implied by the optical values. We also show that the semi-automated method decreases the parameter uncertainties.

Keywords. open clusters and associations: general, infrared: stars

---

The full poster (in pdf format) is available at
`http://www.astro.iag.usp.br/~iaus266/Posters/pAlves.pdf`.

Star clusters: basic galactic building blocks
Proceedings IAU Symposium No. 266, 2009
R. de Grijs & J. R. D. Lépine, eds.

doi:10.1017/S174392130999127X

Star clusters as building blocks for dSph galaxy formation

P. Assmann,[1] M. Fellhauer[1] and M. I. Wilkinson[2]

[1]Departamento de Astronomía, Universidad de Concepción, Casilla 160-C, Concepción, Chile
email: [passmann, mfellhauer]@astro-udec.cl

[2]Department of Physics & Astronomy, University of Leicester, Leicester, UK
email: miw6@astro.le.ac.uk

Abstract. We study numerically the formation of dSph galaxies. Intense starbursts, e.g., in gas-rich environments, typically produce a few to a few hundred young star clusters within a region of just a few hundred pc. The dynamical evolution of these star clusters may explain the formation of the luminous component of dwarf spheroidal (dSph) galaxies. Here, we perform a numerical experiment to show that the evolution of star cluster complexes in dark-matter haloes can explain the formation of the luminous components of dSph galaxies.

Keywords. methods: N-body simulations, galaxies: dwarf, galaxies: halos

1. Introduction

Understanding the formation and evolution of galaxies is a fundamental goal in astronomy. Galaxies are believed (in the standard cosmological scenario) to build up by the merging of smaller constituents, namely dwarf galaxies. This scenario makes dwarf galaxies the basic building blocks of the structure formation in the Universe (e.g., Springel *et al.* 2005). Therefore, to understand galaxy formation and evolution on all scales, it is crucial to understand the formation and evolution of the smallest constituents first. Dwarf galaxies come in different shapes and dynamical states, e.g., dwarf disc galaxies, dwarf elliptical galaxies (with or without a nucleus), dwarf irregulars, ultracompact dwarfs and dwarf spheroidals (dSphs), to name a few. According to the standard scenario, all galaxies reside in an extended dark-matter (DM) halo, whereby dwarf galaxies should reside in the highest-M/L (mass-to-light ratio) DM haloes (relative to their luminous, baryonic mass) of all galaxies.

We investigate numerically the initial conditions for the formation of dSph galaxies. Our model is based on the assumption that stars form in starbursts (Lada & Lada 2003), producing star clusters within a region of just a few hundred pc. The dynamical evolution of these star clusters, i.e., their dissolution due to gas expulsion and subsequent merging in the centre of the DM halo, may explain the formation of dSph galaxies. Our numerical simulations are carried out using the particle mesh code SUPERBOX (Fellhauer *et al.* 2000), which allows us to keep track of many objects in one simulation. The distribution of the star clusters inside the central area of the DM halo follows a Plummer distribution, while we take two distinct types of DM haloes into account: a cusped Navarro, Frenk & White (NFW; Navarro *et al.* 1997) and a cored Plummer profile (Plummer 1911). The setup of our simulation is as follows: the single star clusters are represented by Plummer spheres with a Plummer radius of 4 pc and a cutoff radius of 25 pc. Each cluster has an initial mass of 10^5 $M_\odot$ and an initial crossing time of 2.4 Myr, and is represented by 100 000 particles. Fifteen of these clusters are placed in a star cluster (SC) complex that is modelled again as a Plummer distribution (i.e., positions and velocities according to the

Plummer distribution function), now with the star clusters as particles. This Plummer distribution is given a Plummer radius of 25 pc and a cutoff radius of 150 pc. The DM haloes have a scale length of 500 pc and the mass enclosed within one scale length is 1.01×10^{8} $M_{\odot}$, using 1 000 000 particles. For the DM-halo NFW profile (Plummer profile), we used a cutoff radius of 16 kpc (2.5 kpc).

2. Experimental results

In Figure 1 top (bottom), we show a time sequence of surface-density contours of the star clusters travelling through the central area of the cored (cusped) DM halo, while expanding due to gas expulsion, collisions, etc. This leads to subsequent dissolution of the SCs and the distribution of their stars in the central area of the DM halo, building up the luminous component of the dSph galaxy.

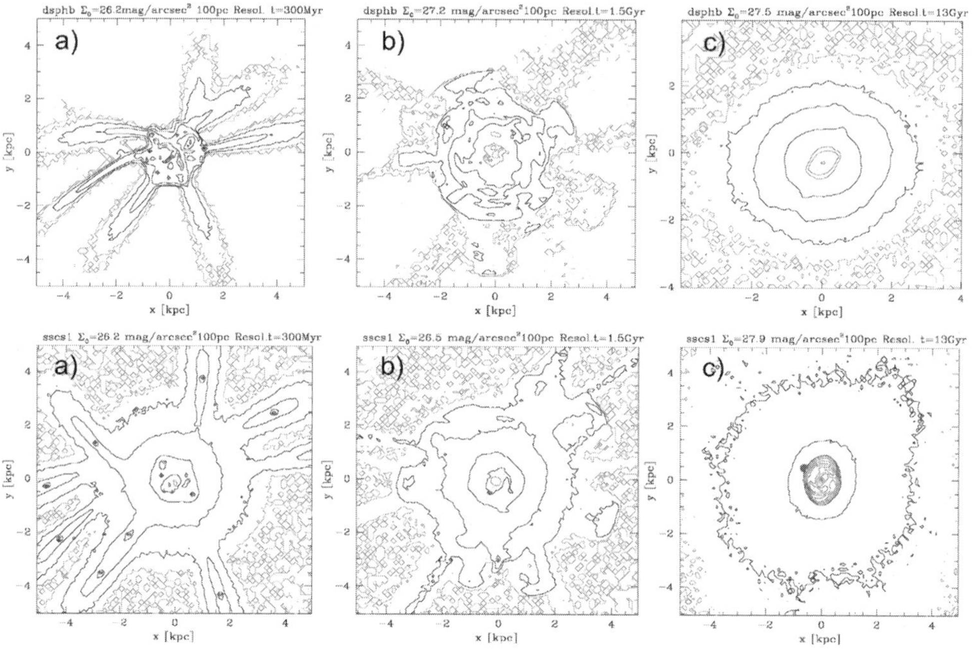

Figure 1. *(top)* dSph fomation based on star cluster dissolution in a cored DM halo. The surface density is shown at *(a)* 300 Myr, *(b)* 1.5 Gyr and *(c)* 13 Gyr. *(bottom)* dSph fomation based on star cluster dissolution in a cusped DM halo. The surface density is shown at the same time steps as in the top panel.

In Figure 2 for both scenarios of DM haloes, the outer contours of our objects are rather smooth, as has been observed in most of the larger dwarf galaxies close to the Milky Way. However, it is interesting to note that there are still some fluctuations visible in the central parts, and observations of these fluctuations might corroborate our model for dSph galaxy formation. We note that in the cusped simulation, the stars are more centrally concentrated. In the middle panels, we can see that the velocity dispersion is still about 50% higher than expected considering the observations of Draco or the Ursa Minor dSph, but focusing on the core of our simulated objects, in the cusped simulation the dispersion velocities are lower than the cored simulation. The mean velocities per pixel, shown in the right panels, show areas which differ by $\sim$ 9 km s^{-1} in the cored and $\sim$ 13 km s^{-1} in the cusped simulation from the average, which should be zero. We call these 'streams' fossil remnants of the formation process. With the advent of new techniques for astronomical observations, these streams should be detectable and can be used to test our formation scenario.

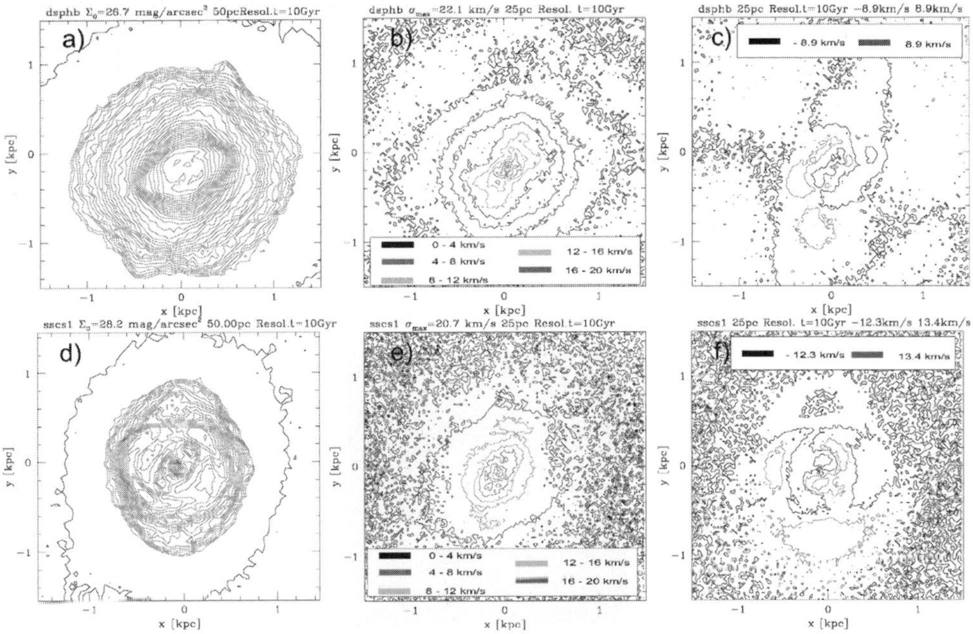

Figure 2. Projection of the dSph onto the xy plane. In *(a)* and *(d)* the surface density is shown with 50 pc resolution for the Plummer and NFW profiles, in *(b)* [*(e)*] the velocity dispersion with 25 pc resolution for the Plummer [NFW] profile is shown. In *(c)* and *(e)* we show the mean velocities of the stars for the Plummer and NFW profiles.

The surface-brightness profile for the cored simulation after 10 Gyr of evolution is shown in Figure 3 (left panel). We fit our model with King, Sérsic and Dehnen profiles. Our results are best described by the King profile. There, the core radius is 720 pc and the central surface brightness is 26 mag arcsec^{-2}. In the right panel of Figure 3 we show the line-of-sight velocity dispersion within a radius of 500 pc. The central velocity dispersion is 14.5 km s^{-1}.

In Figure 4 we show some star clusters after 10 Gyr of evolution for the cusped simulation. The two star clusters shown represent the two main types of stellar streams we see. One from a radial orbit has spread its stars uniformly across a spherical region, while the other one, on a more circular orbit, has stars in a torus-like configuration. These torus-like 'streams' are responsible for an artificially raised velocity dispersion in projection. One also sees two surviving star clusters in the outskirts of the galaxy similar to the star clsuters in Sagittarius and Fornax.

3. Conclusion and outlook

We study numerically the formation of dSph galaxies from merging star clusters, considering two distinct types of profiles for the dark-matter halo. We compare the results of both simulations. We see that in the cusped simulation, stars are more centrally concentrated but have lower central velocity dispersions. In both simulations there are streams of stars (fossil remnants of dissolved star clusters) which survive more than 10 Gyr of evolution, raising the velocity dispersion artificially. With the advent of new techniques

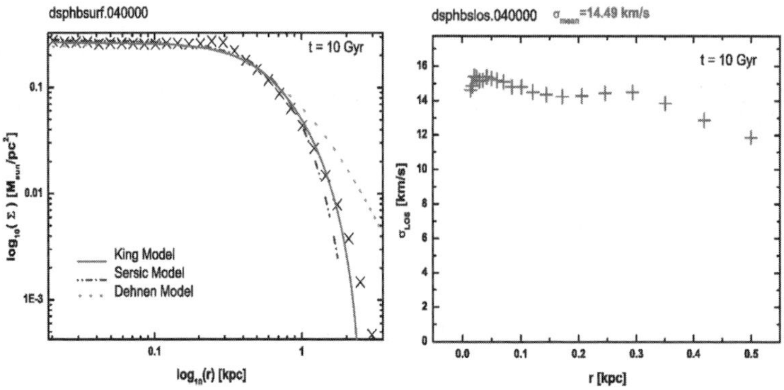

Figure 3. *(a)* Surface-density distribution fitted with different profiles. *(b)* Line-of-sight velocity-dispersion profile.

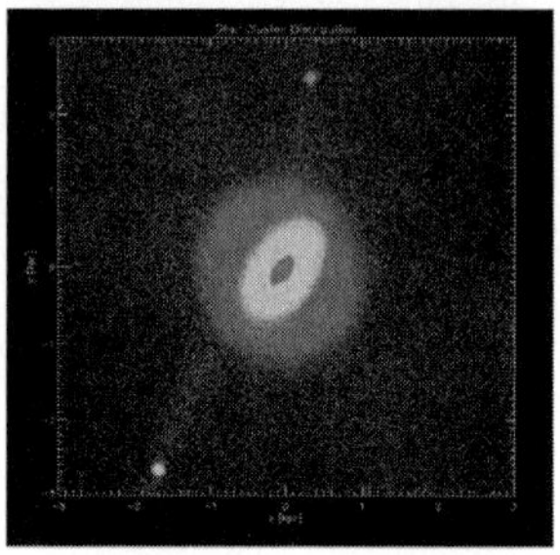

Figure 4. The red and yellow points represent two types of streams of stars at 10 Gyr in the central region of the dSph. We can also see two small star clusters orbiting the galaxy.

for astronomical observations, these streams should soon be detectable and can test our formation scenario. Our initial setup was chosen to produce a stellar component with a line-of-sight velocity dispersion of 10 km s^{-1}, similar to the larger dSph galaxies of the Milky Way. But due to the stellar 'streams' we see an artificially raised dispersion in both models. These streams can only produce such an effect in a low-density environment like that of a dSph galaxy.

We will investigate this effect further and if we find that it is persistent substructure, we will explore the potential impact of undetected streams on estimates of the DM content of the Milky Way's dSph galaxies (based on measurements of the velocity dispersion).

References

Fellhauer, M., Kroupa, P., Baumgardt, H., Bien, R., Boily, C. M., Spurzem, R., & Wassmer, N. 2000, *NewA*, 5, 305
Lada, C. J. & Lada, E. A. 2003, *ARA&A*, 41, 57
Navarro, J. F., Frenk, C. S., & White, S. D. M. 1997, *ApJ*, 490, 493
Plummer, H. C. 1911, *MNRAS*, 71, 460
Springel, V., *et al.* 2005, *Nature*, 435, 629

Star clusters: basic galactic building blocks
Proceedings IAU Symposium No. 266, 2009
R. de Grijs & J. R. D. Lépine, eds.

doi:10.1017/S1743921309991281

The globular cluster NGC 6642: evidence for a depleted mass function

Eduardo Balbinot, Basílio X. Santiago, Eduardo Bica and Charles Bonatto

Departamento de Astronomia, Universidade Federal do Rio Grande do Sul,
Porto Alegre, Brazil
email: balbinot@if.ufrgs.br

Abstract. We present photometry for the globular cluster NGC 6642 using the F606W and F814W filters with the ACS/WFC third-generation camera aboard the *Hubble Space Telescope*. The colour–magnitude diagram shows sources reaching ≈ 6 mag below the turnoff in m_{F606W}. Theoretical isochrone fitting was performed and evolutionary parameters were obtained, including the metallicity $[Fe/H] = -1.80 \pm 0.2$ dex and age, $\log(\text{age/yr}) = 10.14 \pm 0.05$. We confirm that NGC 6642 is located in the Galactic bulge, at a distance of $d_\odot = 8.05 \pm 0.66$ kpc and suffers from a reddening of $E(B-V) = 0.46 \pm 0.02$ mag. These values are in general agreement with those of previous authors. Completeness-corrected luminosity and mass functions were obtained for different annuli centred on NGC 6642. Their spatial variation indicates the existence of mass segregation and depletion of low-mass stars. Most striking is the inverted shape of the mass function itself, with an increase in stellar numbers as a function of increasing mass. This has been observed previously in other globular clusters and is also the result of N-body simulations of stellar systems which have reached $\simeq 90\%$ of their lifetime and are subjected to strong tidal effects. We thus conclude that NGC 6642 is a very old, highly evolved globular cluster. Its current location close to perigalacticon, at only 1.4 kpc from the Galactic Centre, may contribute to this high level of dynamical evolution and stellar depletion.

Keywords. globular clusters: general, globular clusters: individual (NGC 6642), Galaxy: structure

---

1. Introduction

The bulge is known to host old but fairly metal rich stellar populations, with high $[\alpha/Fe]$ ratios, indicating that it formed on a short timescale early in the Galactic history (Origlia *et al.* 2005). The bulge is thus an appropriate region to look for ancient GCs with distinct metallicities from those in the outer stellar halo. The bulge environment may also cause extreme dynamical evolution in star clusters due to bulge shocking and strong tidal effects (Aguilar *et al.* 1988; Shin *et al.* 2008).

Analyses of the structure and dynamical evolution of bulge GCs require high-resolution imaging to resolve individual stars in their cores and to effectively subtract contaminant field stars. The Advanced Camera for Surveys' Wide Field Channel (ACS/WFC) on board of the *Hubble Space Telescope (HST)* has provided a significant leap in the amount of such data on GCs, not only in the bulge, but also elsewhere (Sarajedini *et al.* 2007; Richer *et al.* 2008; Paust *et al.* 2009).

2. Data

NGC 6642 is projected towards the Galactic bulge, thus superimposed onto bulge and inner halo stellar populations. Dynamical studies of this GC, as well as other bulge

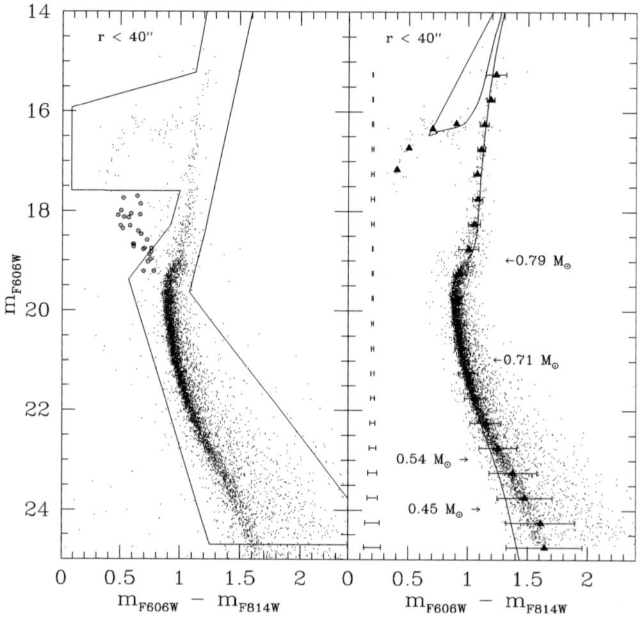

Figure 1. *(left)* CMD of all sources located within $r \leqslant 40''$ from the centre of NGC 6642. We also show the region used to select the cluster's evolutionary sequences. The blue-straggler candidates are marked with open circles. *(right)* Same CMD as in the left panel, but cut according to the selected region. The solid triangles show the CMD fiducial line and corresponding dispersion. The mean photometric error is shown in the extreme left of this panel. Some mass values are given along the main sequence. The best-fit isochrone is overplotted and has the following parameters: $\log(\text{age/yr}) = 10.14$, $[\text{Fe/H}] = -1.80$ dex, $E(B-V) = 0.46$ mag and $(m - M) = 14.55$ mag.

clusters, have been extremely difficult due to limitations of ground-based images. Here, we use *HST*/ACS data to overcome this difficulty.

The images were retrieved from the Space Telescope Science Institute (STScI)'s data archive (ID GO-9799) and were automatically reduced by the STScI pipeline. Two exposures were taken in the F606W and F814W filters, one short (10s) and one long (340s).

The photometry was done using the DOLPHOT software (Dolphin *et al.* 2000), which essentially uses *HST*-optimized PSF fitting. DOLPHOT was employed again in artificial-star mode. Based on fake-star experiments, we determined the completeness as a function of magnitude, colour and position. For more details on the photometry see Balbinot *et al.* (2009).

3. Colour–magnitude diagrams

Figure 1 shows the colour–magnitude diagram (CMD) for the stars located within $r \leqslant 40''$ ($\leqslant 1.6$ pc) from the nominal centre of NGC 6642. By restricting the image region, we strongly reduce field contamination.

We can see a well-defined structure in the main sequence (MS) spanning ~ 6 mag below the MS turnoff. The subgiant, red-giant and horizontal branches are also clearly seen. Some candidate asymptotic giant branch stars are also present. The solid contour in the left panel indicates our selected limits for these stellar evolutionary loci. The chosen limits also avoid effects of saturation and incompleteness. Open circles in the left panel indicate blue-straggler candidates.

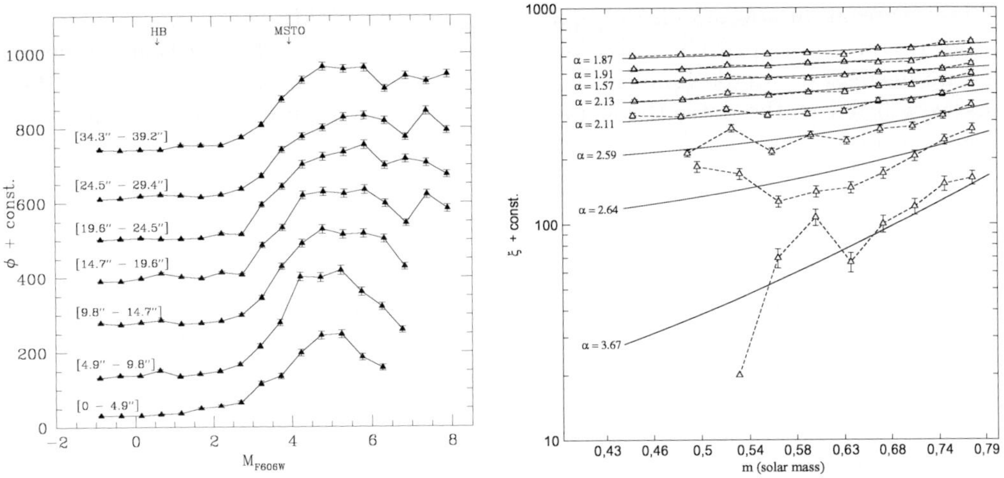

Figure 2. *(left)* Luminosity function of NGC 6642 stars at different radii, as indicated next to each curve. The positions of the MS turnoff (MSTO) and horizontal branch (HB) are indicated at the top. An offset was added to each curve to avoid overlap. In the inner regions, the completeness function falls off more rapidly as a function of magnitude, yielding a brighter cutoff limit. *(right)* Present-day mass functions (PDMFs) for different annuli around the centre of NGC 6642. Distance from cluster centre increases upwards. We use the same radial bins as in the left-hand panel. A constant offset was added to the star counts to avoid confusion. Fits to power-law PDMFs are also shown and the corresponding slopes given.

In the right panel of Figure 2, the GC fiducial line is shown on top of the selected stars. We also show the dispersion around the fiducial points plus the mean photometric error for each magnitude bin.

The NGC 6642 inner-field CMD is used to visually fit isochrones. We use the isochrone grid computed by Girardi *et al.* (2000) for that purpose. To determine parameter uncertainties and to test our visual fit, we generated model cluster fiducial lines and compared them with the observations using a simple algorithm described in Balbinot *et al.* (2009). The best-fit isochrone is shown in the right panel of Figure 2. The corresponding parameters are $\log(\text{age/yr}) = 10.14 \pm 0.05$ and $[\text{Fe/H}] = -1.80 \pm 0.2$ dex, $E(B-V) = 0.46 \pm 0.02$ mag and $(m - M) = 14.55 \pm 0.18$ mag.

4. Mass and luminosity functions

We now study the distribution of stars as a function of luminosity and mass. These are important tools to assess dynamical effects, such as mass segregation and stellar evaporation, that take place throughout a GC's lifetime.

Figure 2a shows the resulting luminosity functions (LFs) for different radial annuli. A 0.5 absolute-magnitude bin size was adopted. A constant offset was added to the curves to avoid cluttering. Poissonian error bars are also shown. Figure 2 includes evolved as well as MS stars. The most striking variation in the LF shapes occurs in the MS domain ($M_{\rm F606W} > 3.5$ mag), where the LF is clearly depleted of low-luminosity stars and shows a clear peak in the inner cluster regions. In contrast, the outer regions display flatter LFs all the way to the detection limit.

Using the best-fit Padova isochrone, we convert absolute magnitudes to masses. Only stars below masses ~ 0.8 M$_\odot$ are considered, since according to the model, mass loss is more pronounced at higher masses. Figure 2b shows the resulting present-day mass function (PDMF).

Mass segregation now stands out more clearly, as the PDMF is systematically steeper in the central regions of NGC 6642. More striking, however, is the fact that the number of stars decreases towards lower masses, in contrast to most observed PDMFs. Power laws were fitted to the data; the slopes are indicated in Figure 2b. The annuli are the same as in Figure 2a.

5. Conclusion

The CMD of NGC 6642 yields an age of 13.8 ± 1.6 Gyr and $[Fe/H] = -1.80 \pm 0.20$ dex. We also derived a colour excess of $E(B-V) = 0.46 \pm 0.02$ mag towards NGC 6642 and a distance comparable to that to the Galactic centre, $d_{\odot} = 8.05 \pm 0.66$ kpc. Given the direction of NGC 6642 on the sky, and assuming that the distance from the Sun to the Galactic centre is 8.0 kpc, we conclude that NGC 6642 currently occupies a very central position in the Galaxy, $d_{GC} = 1.4$ kpc.

We point out that this inverted PDMF slope has also been observed in other recent studies (De Marchi & Pulone 2007; Paust *et al.* 2009). N-body GC simulations show that in tidal fields the preferential depletion of low-mass stars leads to a PDMF with inverted slope when the cluster has reached 90% or more of its associated dissolution time (Baumgardt & Makino 2003). The timescale associated with slope inversion is related to the time when the compact stellar remnants start to dominate the cluster mass. Since NGC 6642 is a very old GC that resides in a central region of the Galaxy, the shape of the PDMF can easily be understood in the context of tidal interactions. However, alternative explanations may be important, such as the gas-expusion model of Marks *et al.* (2009).

We thus conclude from an independent perspective that NGC 6642 is a very old, highly evolved, core-collapsed globular cluster.

References

Aguilar, L., Hut, P., & Ostriker, J. 1988, *ApJ*, 335, 720
Balbinot, E., Santiago, B., Bica, E., & Bonatto C. 2009, *MNRAS*, 396, 1596
Baumgardt, H. & Makino, J. 2003, *MNRAS*, 340, 227
De Marchi, G. & Pulone, L. 2007, *A&A*, 467, 107
Dolphin, A. E. 2000, *PASP*, 112, 1383
Girardi, L., Bressan, A., Bertelli, G., & Chiosi, C. 2000, *A&AS*, 141, 371
Marks, M., Kroupa, P., & Baumgardt, H. 2008, *MNRAS*, 386, 2047
Paust, N., *et al.* 2009, *AJ*, 137, 246.
Origlia, L., Valenti, E., Rich, R., & Ferraro, F. 2005, *MNRAS*, 363, 897
Richer, H., *et al.* 2008, *AJ*, 135, 2141
Sarajedini, A., *et al.* 2007, *AJ*, 133, 1658
Shin, J., Kim, S., & Takahashi, K. 2008, *MNRAS* (Letters), 386, L67

Star clusters: basic galactic building blocks
Proceedings IAU Symposium No. 266, 2009
R. de Grijs & J. R. D. Lépine, eds.

doi:10.1017/S1743921309991293

Multicolor CCD photometry of the open cluster NGC 752

Stanislava Bartašiūtė,[1] **Robert Janusz,**[2,3] **Richard P. Boyle,**[3] **A. G. Davis Philip**[4] **and Viktoras Deveikis**[1]

[1]Astronomical Observatory of Vilnius University, Čiurlionio 29, LT-03100, Vilnius, Lithuania
email: `stanislava.bartasiute@ff.vu.lt`

[2]University School Ignatianum, ul. Kopernika 26, 31-501, Kraków, Poland
email: `rj@jezuici.krakow.pl`

[3]Vatican Observatory Research Group, Steward Observatory, Tucson, AZ 85721, USA
email: `boyle@ricci.as.arizona.edu`

[4]ISO and Union College, 1125 Oxford Place, Schenectady, NY 12308, USA
email: `agdp@union.edu`

Abstract. We obtained CCD observations of the open cluster NGC 752 with the 1.8m *Vatican Advanced Technology Telescope* (Mt. Graham, Arizona) with a 4K CCD camera and eight intermediate-band filters of the *Stromvil* (Strömgren + Vilnius) system. Four $12' \times 12'$ fields were observed, covering the central part of the cluster. The good-quality multicolor data made it possible to obtain precise estimates of distance moduli, metallicity and foreground reddening for individual stars down to the limiting magnitude, $V = 17.5$, enabling photometric identification of faint cluster members. The new observations provide an extension of the lower main sequence to three magnitudes beyond the previous (photographic) limit. A relatively small number of photometric members identified at fainter magnitudes seems to be indicative of actual dissolution of the cluster from the low-mass end.

Keywords. techniques: photometric, open clusters and associations: individual (NGC 752)

1. Introduction

The Galactic open cluster NGC 752 [$\alpha(2000) = 01^{\rm h}57.7^{\rm m}$, $\delta(2000) = +37°47'$; $\ell = 137.1°$, $b = -23.3°$], at a distance of 450 pc from the Sun, is one of the very few clusters in the solar neighborhood with an age between 1 and 2 Gyr. With its relatively small number of stars (~ 130), which are observed out to $40'$ from the cluster center, NGC 752 has been the subject of some controversy regarding its lower main sequence: the great majority of the stars are found in the region of the turnoff, whereas the lower part of the main sequence appears to be very sparsely populated. From the available photometric, proper-motion and radial-velocity data the cluster's main sequence is known at best to $M_V \sim +6$ mag. This may be due to both a lack of observations at magnitudes fainter than the present (photographic) limit and dynamic escape of low-mass stars from the cluster. To define the extension of the cluster's main sequence towards fainter magnitudes, we obtained CCD observations of several $12' \times 12'$ fields in the central part of the cluster.

2. Observations and data reductions

Observations were taken in November 2007 and 2008 at the 1.8m *Vatican Advanced Technology Telescope (VATT)* on Mt. Graham, Arizona, with a 4K CCD camera and eight intermediate-band filters of the *Stromvil* (Strömgren + Vilnius) system. The scale

Figure 1. NGC 752. Marked are four VATT 4K CCD fields of 12′ × 12′ each.

on the chip is 0.37″ per pixel, yielding a field of view of 12′ × 12′. Both long and short exposures were obtained in four slightly overlapping fields, A, B, C and D (Figure 1).

CCD frames were processed using the CommandPhot software (Janusz 2007). Flat-field corrections and color transformations were derived from observations of the standard cluster M67. Photoelectric standards in the field of NGC 752 (Bartašiūtė *et al.* 2007) were also used, either for calibrations or as a check on the external accuracy of our photometry. A total of 440 stars down to 18.0 mag have been measured in the four fields (Figure 2).

3. Results

To identify photometric members of the cluster, we determined for individual stars the absolute magnitudes, M_V, interstellar reddening values, $E(Y-V)$, and metallicities, [Fe/H]. Accuracies of ±0.2 mag, ±0.02 mag and ±0.15 dex for these three parameters, respectively, were regularly obtained for stars brighter than magnitude 16.5.

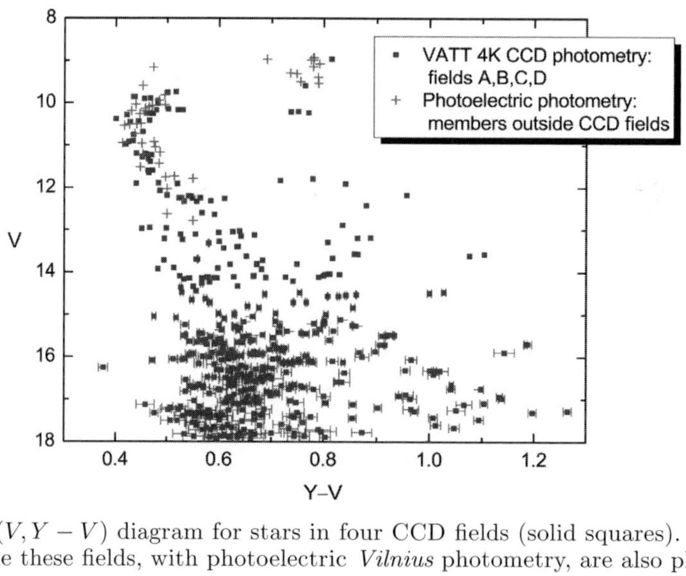

Figure 2. $(V, Y-V)$ diagram for stars in four CCD fields (solid squares). The cluster members outside these fields, with photoelectric *Vilnius* photometry, are also plotted (crosses).

Photometric members were selected according to the values of the cluster parameters determined earlier from photoelectric *Vilnius* photometry of 62 cluster stars (Bartašiūtė *et al.* 2007)—$(m - M)_V = 8.38 \pm 0.14$ mag, $E(Y - V) = +0.03 \pm 0.01$ mag and $[\mathrm{Fe/H}] = -0.11 \pm 0.08$ dex—by adopting the 3σ criterion to allow for the effect of classification errors and taking into account the photometric effect of unknown binaries.

Seventy of the 440 stars classified have cluster membership status known from proper-motion and/or radial-velocity data (Daniel *et al.* 1994). Therefore, we were able to check the validity of our photometric membership estimates. For 65 such stars, photometric and astrometric results show a remarkable coincidence, and only five stars, or less than 10%, show the opposite. The overwhelming majority of background stars (the crowding below $V \approx 15$ mag in the the diagram of Figure 2) are classified as dwarfs of spectral types from late F to early K, having, in most cases, a mild metal deficiency.

The photometric members are shown in Figure 3, together with the best-fitting isochrone (1.5 Gyr) and model binary trajectories. The new observations provide an extension of the lower main sequence to four magnitudes beyond the photoelectric limit and to three magnitudes beyond the proper-motion (photographic) limit. The color–magnitude diagram (CMD) clearly shows an inadequate number of low-mass stars with respect to the expected main sequence, at least in the central, observed part of the cluster. In this context, it is interesting to note that the red giants in the cluster, regardless of their higher initial masses, seem to be amongst the least centrally concentrated cluster stars.

Within the cluster radius, the existing data give a more or less complete census of the member stars only for the upper part of the CMD. Therefore, we can make only a cursory comparison of the number of observed stars with that following from the cluster's initial mass (Figure 4). To obtain the mass of a cluster containing, like NGC 752, 15 red giant stars observed at an age of 1.5 Gyr, we assumed a Kroupa (2001) stellar initial mass function (IMF) and generated with the SimClust tools (Deveikis *et al.* 2008) 1000 artificial clusters in the mass interval from 1000 to 5000 $M_\odot$. Our simulations, which included stellar evolution for single stars, give an initial mass of 2200 $M_\odot$. If we extrapolate the number of identified photometric members across the entire diameter of

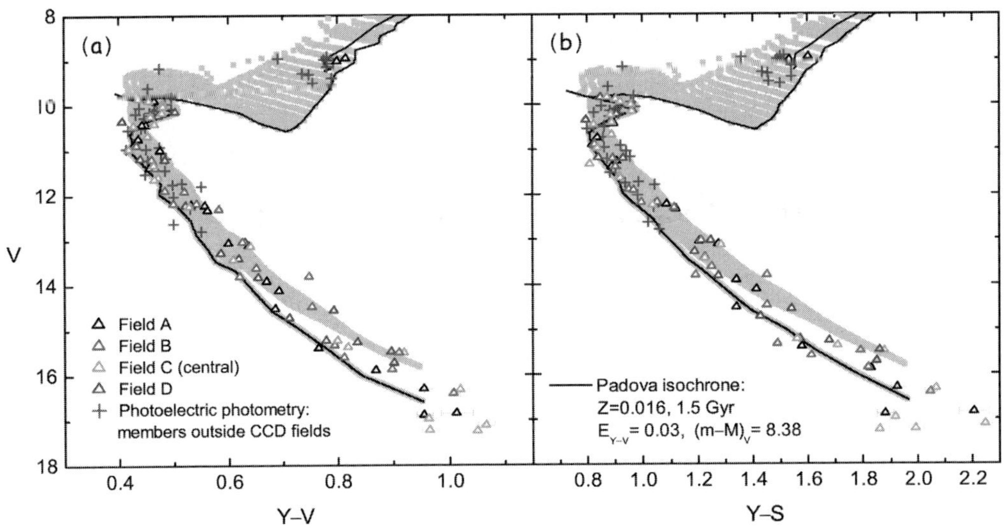

Figure 3. CMDs of the photometric members of NGC 752. The shaded area denotes model binary trajectories outlined by the best-fitting isochrone.

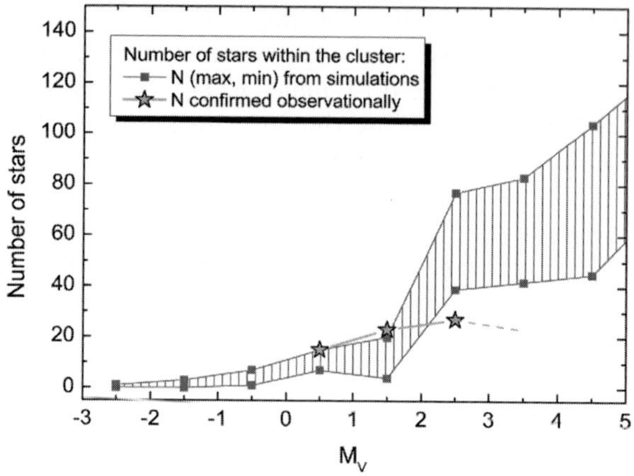

Figure 4. Present luminosity function based on Monte Carlo simulations.

the cluster, we would still have a fairly small number of lower-mass stars relative to the initial mass spectrum.

According to the predicted dissolution times for clusters in the solar neighborhood given by Lamers & Gieles (2006), the survival time of a cluster with an initial mass of 2200 $M_\odot$ should be 0.6 Gyr, i.e., much shorter than the present age of NGC 752. The discrepancy can be reconciled by assuming for this cluster a larger initial mass. As Figure 4 suggests (see the observed number of stars in the range $0 < M_V < 2$ mag), the cluster mass is likely underestimated. There are at least two reasons for such a possibility. First, the actual number of red giants, on which the basic point of our simulations rests, can be incomplete. Indeed, a radial-velocity survey by Mermilliod *et al.* (1998) has revealed an additional two red giants at nearly 100′ from the cluster center, suspected to be corona or escaping members. Second, the result is highly dependent on the assumed stellar IMF: assuming a Salpeter IMF, the initial mass of the cluster should be twice that simulated with a Kroupa-type IMF.

In summary, we confirm the status of NGC 752 as a dissolving cluster. To discern the escaping low-mass members, high-precision photometric, astrometric and radial-velocity observations secured in the outer parts of the cluster would be most helpful.

References

Bartašiūtė, S., Deveikis, V., Straižys, V., & Bogdanovičius, A. 2007, *BaltA*, 16, 199
Daniel, S. A., Latham, D. W., Mathieu, R. D., & Twarog, B. A. 1994, *PASP*, 106, 281
Deveikis, V., Narbutis, D., Stonkutė, R., Bridžius, A., & Vansevičius, V. 2008, *BaltA*, 17, 351
Janusz, R. 2007, *BaltA*, 16, 467
Kroupa, P. 2001, *MNRAS*, 322, 231
Lamers, H. J. G. L. M. & Gieles, M. 2006, *A&A* (Letters), 455, L17
Mermilliod, J.-C., Mathieu, R. D., Latham, D. W., & Mayor, M. 1998, *A&A*, 339, 423

Star clusters: basic galactic building blocks
Proceedings IAU Symposium No. 266, 2009
R. de Grijs & J. R. D. Lépine, eds.

doi:10.1017/S174392130999130X

Limits on the dark matter content of globular clusters

H. Baumgardt,[1] P. Côté, M. Hilker, M. Rejkuba, S. Mieske, S. G. Djorgovski and P. Stetson

[1]Argelander-Institut für Astronomie, Universität Bonn, Auf dem Hügel 71, 53121 Bonn, Germany

Abstract. We have measured the velocity dispersion of the Galactic globular cluster NGC 2419 to determine if a substantial amount of dark matter is present in this cluster. NGC 2419 is one of the best globular clusters to look for dark matter due to its large mass, long relaxation time and large Galactocentric distance, which makes tidal stripping of dark matter unlikely. Our results can be summarized as follows. (i) We found a global velocity dispersion of 4.14 ± 0.48 km s^{-1}, which leads to a total cluster mass of $(9.02 \pm 2.22) \times 10^5$ M$_\odot$ and implies a global mass-to-light ratio of 2.05 ± 0.50 M$_\odot$/L$_\odot$. (ii) Our derived mass-to-light ratio is completely consistent with the mass-to-light ratio of a standard stellar population at the metallicity and age of NGC 2419. In addition, the mass-to-light ratio of NGC 2419 does not increase towards the outer cluster parts. (iii) We can therefore rule out the presence of a dark-matter halo with a central density greater than about 0.02 M$_\odot$ pc^{-3}. Similar limits are found for other halo globular clusters, like Pal 14. These observations therefore indicate that NGC 2419 and other halo globular clusters did not form at the centers of dark-matter halos similar to those surrounding dwarf galaxies. Instead, an origin driven by gas-dynamical processes during mergers between galaxies or proto-galactic fragments seems to be the more likely explanation for the formation of even the lowest-metallicity globular clusters.

Keywords. globular clusters: individual (NGC 2419), dark matter

---

The full poster (in pdf format) is available at
`http://www.astro.iag.usp.br/~iaus266/Posters/pBaumgardt.pdf`.

Star clusters: basic galactic building blocks
Proceedings IAU Symposium No. 266, 2009
R. de Grijs & J. R. D. Lépine, eds.

doi:10.1017/S1743921309991311

Hidden infrared star clusters in our Galaxy: follow-up observations

Jura Borissova,[1] Radostin Kurtev,[1] Margaret M. Hanson,[2] Leonid Georgiev,[3] Valentin Ivanov[4] and Sergio Ortolani[5]

[1]Departamento de Física y Astronomía, Facultad de Ciencias, University of Valparaíso, Chile
[2]Department of Physics, University of Cincinnati, Cincinnati, OH, USA
[3]UNAM, Mexico [4]European Southern Observatory, Santiago 19001, Chile
[5]Dipartimento di Astronomia, University of Padova, Padova, Italy

Abstract. We are reporting some recent results from our long-term program aimed at characterizing the obscured present-day star cluster population in the Galaxy. Our goal is to expand the current census of the Milky Way's inner stellar disk to guide models seeking to understand the structure and recent star-formation history of our Galaxy. The immediate goal is to derive accurate cluster physical parameters using precise infrared photometry and spectroscopy. So far, we observed approximately 60 star cluster candidates selected from different infrared catalogs. Their nature, reddening, distance, age and mass are analyzed. Two of them, Mercer 3 and Mercer 5, are new obscured Milky Way globular clusters. Among the newly identified open clusters, the objects [DBS2003] 179, Mercer 23, Mercer 30, Mercer 70, and [DBS2003] 106 are particularly interesting because they contain massive young OB and Wolf–Rayet stars with strong emission lines.

Keywords. open clusters and associations: general, open clusters and associations: individual ([DBS2003]179, [DBS2003]106, [MCM2005b]23, [MCM2005b]30, [MCM2005b]70), globular clusters: individual ([MCM2005b]3, [MCM2005b]5), stars: early-type, stars: Wolf–Rayet

---

The full poster (in pdf format) is available at
`http://www.astro.iag.usp.br/~iaus266/Posters/pBorissova.pdf`.

Star clusters: basic galactic building blocks
Proceedings IAU Symposium No. 266, 2009
R. de Grijs & J. R. D. Lépine, eds.

doi:10.1017/S1743921309991323

UBVRI CCD photometric study of southern open star clusters observed at Pico dos Dias Observatory (LNA/Brazil)

Thiago Costa Caetano and Wilton S. Dias
Universidade Federal de Itajubá (UNIFEI), MG, Brazil

Abstract. Open clusters have long been recognized as important objects to investigate aspects of Galactic structure such as the location of spiral arms, Galactic dynamics or even the chemical-abundance gradients in the Galactic disk. Great effort has been dedicated to studies of these objects throughout the past few years to determine parameters like age, distance, reddening, metallicity as well as kinematic information in a systematic and consistent manner. In this work, we present results on several open clusters selected from the DAML02 catalog that have been observed and investigated for the first time. CCD *UBVRI* (Johnsons–Cousins system) observations were carried out in 2009 using the 60 cm telescope of the Pico dos Dias Observatory (LNA/Brazil). The color–color and color–magnitude diagrams were analyzed for each cluster with the help of a program written by our group, which allows handling photometric and astrometric data simultaneously. We determined the fundamental parameters (reddening, distance and age) using the main-sequence fitting method, taking into account kinematic information of individual stars in the clusters. The mean proper motion and radial velocity were estimated using a set of stars selected photometrically.

Keywords. Hertzsprung–Russell diagram, open clusters and associations: general

---

The full poster (in pdf format) is available at
`http://www.astro.iag.usp.br/~iaus266/Posters/pCosta-Caetano.pdf`.

Star clusters: basic galactic building blocks
Proceedings IAU Symposium No. 266, 2009
R. de Grijs & J. R. D. Lépine, eds.

doi:10.1017/S1743921309991335

Lithium survey in evolved stars observed in the SACY project

B. L. Canto Martins,[1] S. Vieira,[2], C. A. O. Torres,[2] G. R. Quast,[2] L. da Silva,[3] R. de La Reza,[3] C. H. F. de Melo[4] and J. R. de Medeiros[1]

[1]DFTE/UFRN, Natal, RN, Brazil

[2]Laboratório Nacional de Astrofísica, Itajubá, MG, Brazil

[3]Observatório Nacional, Rio de Janeiro, RJ, Brazil

[4]European Southern Observatory, La Silla, Chile

Abstract. The primary goal of the SACY project (Search for Associations Containing Young Stars) was to identify possible associations of stars younger than the Pleiades association among optical counterparts of *ROSAT* X-ray-bright sources. The study of the chemical abundance in stars located in regions of stellar formation is extremely important to understand stellar nucleosynthesis, the physical mechanisms controlling mixing in stellar interiors, and chemical enrichment in the Galaxy. The present work highights the first results of a chemical-abundance study of evolved stars identified in the SACY survey. For this, we performed a detailed spectroscopic analysis for the determination of atmospheric parameters and Li abundance for a sample of giant and subgiant stars. The observations were carried out with high resolution using the FEROS ($R = 48\,000$) échelle spectrograph. We measured the stellar parameters ($T_{\rm eff}, \log g, v_{\rm mic}$, [Fe/H]) from LTE analysis in the complete range of $420 - 1100$ nm. Li abundance was derived from the region around the lithium line at 6707.78 Å for the entire sample of stars.

Keywords. stars: abundances, stars: fundamental parameters, Hertzsprung–Russell diagram, open clusters and associations: general

---

The full poster (in pdf format) is available at
`http://www.astro.iag.usp.br/~iaus266/Posters/pCanto-Martins.pdf`.

Star clusters: basic galactic building blocks
Proceedings IAU Symposium No. 266, 2009
R. de Grijs & J. R. D. Lépine, eds.

doi:10.1017/S1743921309991347

Building stable stellar systems with ρ as the only datum

D. D. Carpintero
Facultad de Ciencias Astronómicas y Geofísicas, UNLP, Argentina
Instituto de Astrofísica de La Plata, CONICET–UNLP, Argentina

Abstract. Schwarzschild's method to set up a model galaxy is most useful when the distribution in velocity space is unknown. Nevertheless, one has to know beforehand which kind of orbits are spawned by the potential of the model. Moreover, although the system thus generated is in equilibrium, it is not necessarily stable. Here, we present a new method that allows to build up a stable stellar system without any previous knowledge of its distribution in velocity space.

Keywords. methods: numerical, galaxies: kinematics and dynamics

---

The full poster (in pdf format) is available at
`http://www.astro.iag.usp.br/~iaus266/Posters/pCarpintero.pdf`.

Star clusters: basic galactic building blocks
Proceedings IAU Symposium No. 266, 2009
R. de Grijs & J. R. D. Lépine, eds.

doi:10.1017/S1743921309991359

Membership probability via control-field colour–magnitude decontamination†

Wagner J. B. Corradi,[1] Francisco F. S. Maia[2] and João F. C. Santos Jr.[3]
Universidade Federal de Minas Gerais, ICEx, Dept. de Física CP 702, 30123-970,
Av. Antônio Carlos, 6627, Belo Horizonte, Brazil
email: `[wag, jsantos]@fisica.ufmg.br, ffsmaia@ufmg.br`

Abstract. The fundamental physical parameters of open clusters are important tools to understand the formation and evolution of the Galactic disk and to test star-formation and evolution models. However, only a small fraction of the known open clusters in the Milky Way have precise determinations of distance, reddening, age, metallicity, radial velocity and proper motion. One of the major problems in determining these parameters lies in the difficulty to separate cluster members from field stars and to assign membership. We propose a decontamination method by employing 2MASS data in the regions around the clusters NGC 1981, NGC 2516, NGC 6494 and M11. We present decontaminated colour–magnitude diagrams of these objects showing the membership probabilities and structural parameters as derived from King-profile fitting.

Keywords. open clusters and associations: general, methods: data analysis

1. Introduction

Despite the large number of known open clusters, many have only been discovered recently and have not been studied in detail or do not have well-determined parameters (Mermilliod 1995; DAML02: Dias *et al.* 2002; Kharchenko *et al.* 2005). Sky surveys like 2MASS (Skrutskie et al. 2006) have produced large amounts of near-infrared data and contributed to the discovery of even more objects.

The determination of the fundamental parameters of open clusters provides important tools for investigations of the Galactic disk and star-formation models. However, contamination by field stars and significant reddening severely hamper the detection and characterization of clusters, especially towards the Galactic plane or the Galactic Center.

We devised a method to remove the field-star contamination on colour–magnitude diagrams (CMDs) by sampling the population from a nearby control field and then removing it from the cluster's CMD. The 2MASS catalog was selected because of its full-sky coverage and our ability to extract data from spatially unlimited regions. Also, near-infrared wavelengths are particularly sensitive to discriminate cluster stars from the contaminating field for young stellar systems (e.g., Santos *et al.* 2005).

2. Data

VIZIER (`http://vizier.u-strasbg.fr/viz-bin/VizieR`) was used to extract near-infrared photometric data from 2MASS in circular fields centered on NGC 1981, NGC 2516, NGC 6494 and M 11, with the clusters' center coordinates taken from DAML02. The data comply with the 2MASS Level 1 requirement (i.e., $J < 15.8$, $H < 15.1$ and $K < 14.3$ mag) and was extracted within 30 arcmin of the clusters' center coordinates.

† The full poster (in pdf format) is available at
`http://www.astro.iag.usp.br/~iaus266/Posters/pCorradi.pdf`.

Comparison fields were selected for the targets with the same area of the cluster data and similar reddening, as deduced from near-infrared (2MASS) and mid-infrared (*IRAS*) images. The NGC 1981 comparison field was extracted 1 deg to the northwest of the cluster due to the southward nebulosity associated with NGC 1977. For the remaining targets, circular annular fields centered on the cluster were used. Figure 1 compares spatial-density profiles for the targets constructed from raw data and the decontaminated subsample, as discussed below.

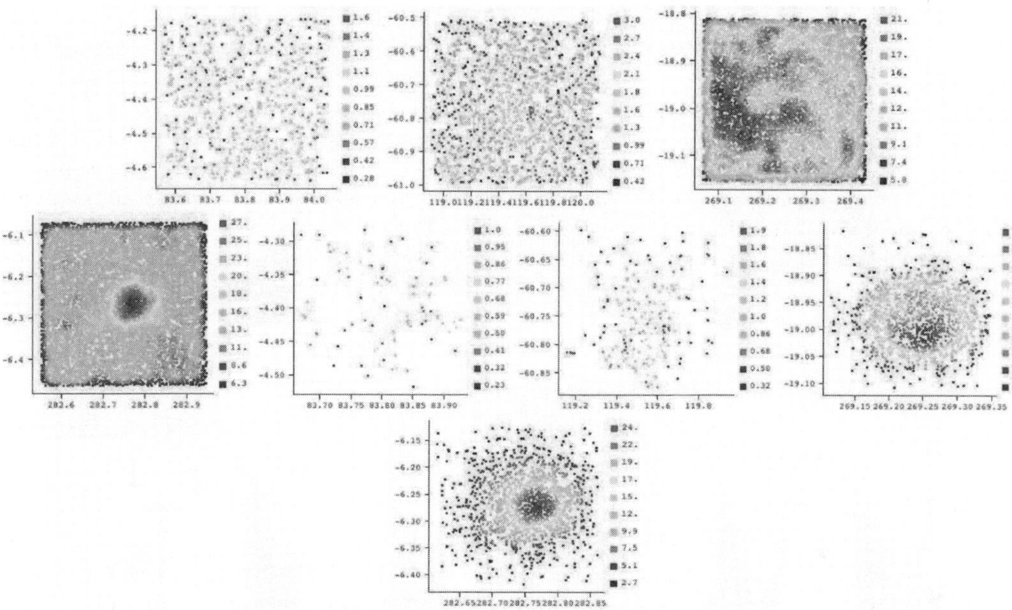

Figure 1. Spatial-density profiles for targets (left to right) NGC 1981, NGC 2516, NGC 6494 and M 11 using raw *(top)* and decontaminated data *(bottom)*. Color bars on the right represent stellar density.

3. New center determination

Precise determination of the clusters' central coordinates is an essential step before a reasonable radial density profile can be obtained. Catalogued center values are mainly intended for identification purposes and generally imprecise. Therefore, we estimated the centers of the targets by first selecting a region around the coordinates given by DAML02. In addition, the selected region was divided into bins of right ascension (RA) and declination (Dec) and star counts were obtained inside each bin. We used these star counts to construct spatial profiles and fit a Gaussian function to the stellar distributions in RA and Dec. The center coordinates were initially taken as the center of the fitted Gaussian.

This procedure was applied to different bin sizes and the corresponding center coordinates were used to create histograms showing the most recurrent RA and Dec. No trends were found between the coordinates and bin size and we adopted these most recurrent values as the new center coordinates of the clusters. Table 1 shows our results.

Table 1. (*left*) Center determination and their difference to the cluster center in DAML02 catalog. (*right*) Determined King parameters and uncertainties.

	Center determination			King-profile parameters		
Cluster	$\alpha_{\rm new}$	$\delta_{\rm new}$	$\Delta_{\rm lit}$	σ_0	$R_{\rm c}$	$\sigma_{\rm bg}$
NGC 1981	$+05^h\ 35^m\ 07.9^s$	-04° 20′ 34.8″	3.57′	2.61 (0.3)	1.0 (0.2)	0.51 (0.05)
NGC 2516	$+07^h\ 57^m\ 42.0^s$	-60° 43′ 22.8″	3.35′	14.9 (2.0)	0.5 (0.1)	0.95 (0.07)
NGC 6494	$+17^h\ 56^m\ 37.7^s$	-19° 00′ 32.4″	5.49′	17.8 (0.8)	2.9 (0.2)	10.7 (0.4)
M11	$+18^h\ 51^m\ 02.2^s$	-06° 15′ 32.4″	0.71′	42.9 (2.5)	2.1 (0.2)	4.0 (1.3)

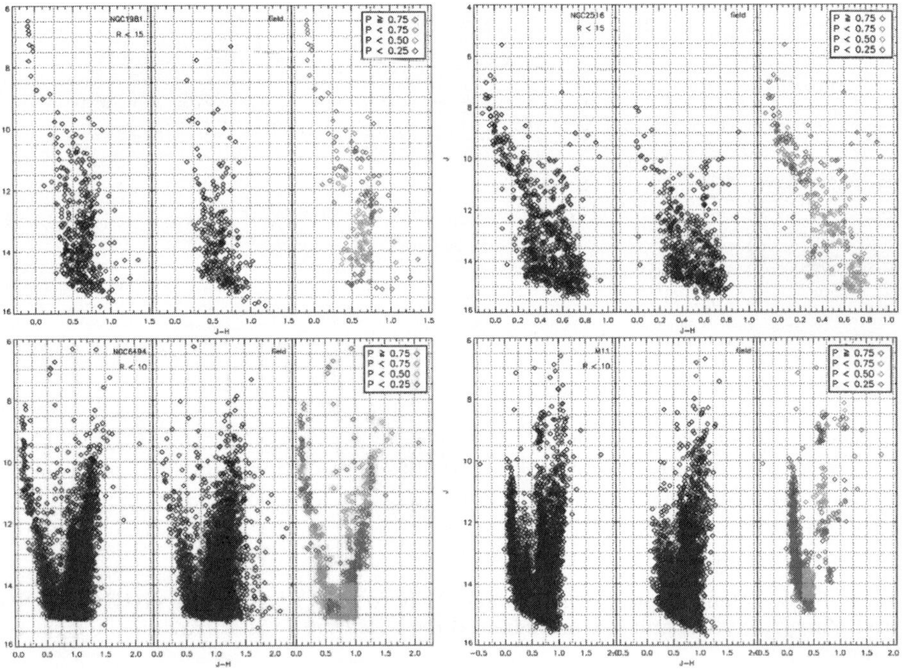

Figure 2. Cluster and control-field CMDs. From top left to bottom right: NGC 1981, NGC 2516, NGC 6494 and M11. Raw cluster data *(left)*, control field *(middle)*, cleaned subsample *(right)*. The color-coded inset shows the assigned membership probability.

4. Membership-probability assignment

For membership assignment we constructed CMDs for both cluster and field stars and divided the diagrams into small cells in J and $J-H$. The cells are small enough to detect local variations of field-star contamination of the various sequences in the CMD, but large enough to accommodate a significant number of stars. Typical cell sizes are $\Delta J = 0.5$ mag and $\Delta(J-H) = 0.25$ mag, with dense (sparse) clusters requiring smaller (larger) cells. Membership probabilities are assigned to stars within each cell based on the overdensity of cluster stars in relation to field stars, according to the relation $P = (N_{\rm clu} - N_{\rm fld})/N_{\rm clu}$. Zero probability was assigned whenever an excess of field stars over cluster stars occurred in a given cell.

For CMD decontamination, a subset of the original cluster sample was created by removing from each cell in the cluster's CMD, the expected number of field stars as measured in the control-field CMD. Stars were removed based on their distance to the center of the cluster, and cells without cluster overdensity had all stars inside their limits removed. We will refer to this final subsample as the 'clean' subsample. Figure 1 shows

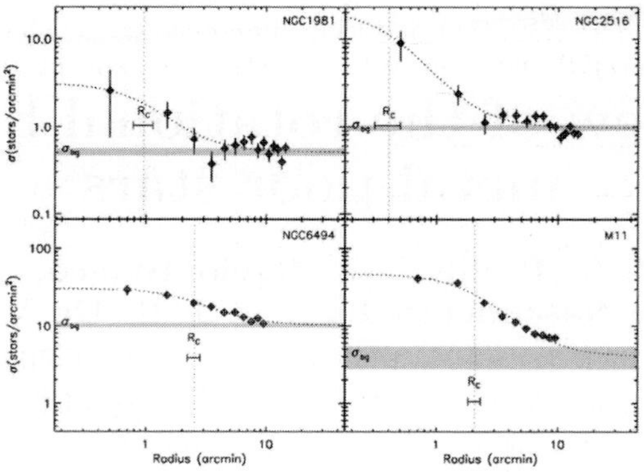

Figure 3. King-profile fits to the target clusters. From top left to bottom right: NGC 1981, NGC 2516, NGC 6494 and M11. Core radii and background densities are indicated.

the spatial-density profiles of the targets using the raw and clean data. Figure 2 shows the cluster, field and clean CMDs for the target clusters.

Radial density profile

To probe the spatial profile and derive structural properties of the targets, radial density profiles (RDPs) were constructed by counting stars inside rings of one arcmin width, centered on the cluster, and then dividing by the area of the ring. King-profile fitting used the modified function $\sigma(R) = \sigma_{\rm bg} + \sigma_0/(1+(R/R_{\rm c})^2)$, as introduced in King (1962). Figure 3 shows RDPs around the new center coordinates for the target clusters using the clean subsample.

Table 1 shows the central surface density, σ_0, core radius, $R_{\rm c}$ and background surface density, $\sigma_{\rm bg}$, with their respective uncertainties, for the decontaminated samples. The values found for M11 agree, within the uncertainties, with those determined by Santos *et al.* (2005).

5. Conclusion

By removing stars similar to the field population in the cluster region, the method effectively leaves the final subsample of stars less contaminated by the background. Additionally, it provides membership probabilities for all cluster stars, removing stars with zero probability from the sample. This subsample allows for much better defined cluster sequences in the CMD, providing good conditions for subsequent isochrone fitting and more accurate values for the clusters' parameters, as can be seen clearly in Figures 1 and 2.

References

Bonatto, C., Santos Jr., J. F. C., & Bica, E. 2006, *A&A*, 445, 567
Dias, W. S., Alessi, B. S., Moitinho, A., & Lépine, J. R. D. 2002, *A&A*, 389, 871
Kharchenko, N. V., Piskunov, A. E., Röser, S., Schilbach, E., & Scholz, R.-D. 2005, *A&A*, 438, 1163
King, I. 1962, *AJ*, 67, 471
Mermilliod, J.-C. 1995, *Inform. On-line Data Astron.*, 203, 127
Santos Jr., J. F. C., Bonatto, C., & Bica, E. 2005, *A&A*, 442, 201
Skrutskie, M. F., *et al.* 2006, *AJ*, 131, 1163

Star clusters: basic galactic building blocks
Proceedings IAU Symposium No. 266, 2009
R. de Grijs & J. R. D. Lépine, eds.

doi:10.1017/S1743921309991360

An overview of the rotational behavior of metal-poor stars

C. Cortés,[1] J. R. P. Silva,[2] A. Recio–Blanco,[3] M. Catelan,[4] J. D. Do Nascimento Jr.[1] and J. R. De Medeiros[1]

[1]Universidade Federal do Rio Grande do Norte (UFRN), Brazil

[2]Universidade Estadual do Rio Grande do Norte (UERN), Brazil

[3]Observatoire de la Côte d'Azur, France

[4]Pontifícia Universidad Católica, Chile

Abstract. We describe the behavior of the rotational velocity in metal-poor stars ([Fe/H] ~ -0.5 dex) at different evolutionary stages, based on $v \sin i$ values from the literature. Our sample is composed of stars in the field and in some Galactic globular clusters, including stars on the main sequence (MS), red-giant branch (RGB), and horizontal branch (HB). The metal-poor stars are mainly slow rotators, and their $v \sin i$ distribution along the Hertzsprung–Russell diagram is quite homogeneous. Nevertheless, a few moderate to high values of $v \sin i$ are found for stars located on the MS and the HB. We show that the overall distribution of $v \sin i$ values is basically independent of metallicity for the stars in our sample. In particular, the fast-rotating MS stars in our sample exhibit similar rotation rates as their metal-rich counterparts, suggesting that some may actually be fairly young, in spite of their low metallicity, or else that at least some would be better classified as blue straggler stars. We do not find significant evidence of evolution in $v \sin i$ values as a function of position on the RGB. In particular, we do not confirm previous suggestions that stars close to the RGB tip rotate faster than their less-evolved counterparts. While the presence of fast rotators among moderately cool blue-HB stars has been suggested as due to angular-momentum transport from a stellar core that has retained significant angular momentum during its prior evolution, we find that any such transport mechanisms must likely operate very fast as the star arrives on the zero-age HB (ZAHB), since we do not find a link between evolution off the ZAHB and $v \sin i$ values.

Keywords. stars: evolution, stars: fundamental parameters, stars: Population II, stars: rotation, stars: statistics

---

The full poster (in pdf format) is available at
`http://www.astro.iag.usp.br/~iaus266/Posters/pCortes.pdf`.

Star clusters: basic galactic building blocks
Proceedings IAU Symposium No. 266, 2009
R. de Grijs & J. R. D. Lépine, eds.

doi:10.1017/S1743921309991372

The fragmentation of expanding shells – limitations of the thin-shell model

Jim Dale,[1] Richard Wünsch,[1] Jan Palouš[1] and Ant Whitworth[2]

[1]Astronomical Institute, AVCR, Prague, Czech Republic

[2]School of Physics and Astronomy, Cardiff University, Cardiff, UK

Abstract. We study the fragmentation of expanding shells in the context of the linear thin-shell analysis. We simulate shell fragmentation using the FLASH AMR code and a variant of the Benz SPH code.

Keywords. stars: formation, ISM: bubbles

The full poster (in pdf format) is available at
`http://www.astro.iag.usp.br/~iaus266/Posters/pDale.pdf`.

Star clusters: basic galactic building blocks
Proceedings IAU Symposium No. 266, 2009
R. de Grijs & J. R. D. Lépine, eds.

doi:10.1017/S1743921309991384

Clusters of PMS stars in the Magellanic Clouds: the case of LH95

Nicola Da Rio
Max Planck Institute for Astronomy, Heidelberg, Germany

Abstract. We present a study of the pre-main-sequence (PMS) population in the stellar association LH95 in the LMC, based on the deepest *HST*/ACS photometry ever taken in a region of the Magellanic Clouds. This association hosts $\sim$ 2500 PMS stars, distributed in three main subclusters. We isolate this population by subtracting the LMC field, study the reddening distribution which peaks at $A_V \sim 0.5$ mag, and assign masses, down to 0.2 $M_\odot$, to each member using a newly derived conversion of evolutionary models for the LMC metallicity and the ACS photometric system. We derive the first IMF in the subsolar regime ever measured outside our Galaxy, complete down to 0.4 $M_\odot$. It presents a flattening at 1 $M_\odot$, changing its slope from $x = 2.05$ for intermediate-mass stars to $x = 1.05$ in the subsolar regime. Correcting for unresolved binarity, it is compatible with the Galactic IMF. We do not find evidence of spatial variations of the IMF in the region. We study the ages of the PMS members using maximum-likelihood methods based on the location of the stars in the color–magnitude diagram with respect to 2D isochrones obtained by applying intrinsic and observational biases (differential extinction, variability, confusion, unresolved binarity) to modeled simple stellar populations. We find a most likely global age of $\sim$ 10 Myr, and demonstrate that neglecting the aforementioned biases leads to an underestimation of the cluster age of up to 50%. We find that the observed luminosity spread is more than twice larger than can be explained by such biases, confirming that the star formation in LH95 lasted a few Myr.

Keywords. Magellanic Clouds, open clusters and associations: individual (LH95), stars: pre–main-sequence, stars: formation, stars: luminosity function, mass function

---

The full poster (in pdf format) is available at
`http://www.astro.iag.usp.br/~iaus266/Posters/pDaRio.pdf`.

Star clusters: basic galactic building blocks
Proceedings IAU Symposium No. 266, 2009
R. de Grijs & J. R. D. Lépine, eds.

doi:10.1017/S1743921309991396

Hierarchical stellar clusters in molecular clouds

Srabani Datta

Jodrell Bank Centre for Astrophysics, Turing Building, Upper Brook Street, Manchester M13 9PL, UK
email: Srabani.Datta@manchester.ac.uk

Abstract. Observations show that massive stars always form in clusters or associations, and that the most massive stars form in the dense cores of large clusters. This suggests that accretion processes in cluster cores may be responsible for the formation of stars. In addition, young stellar clusters have been found to contain subclusters, so that star formation can be seen to be a hierarchical process that involves clustering on a range of scales. In this paper, we propose a fractal model of the parental molecular cloud, namely that of the Julia set given by $f(z) = z^2 + c$, where z and c are complex numbers and $c = -0.745430 + 0.113008i$, to explain this phenomenon and the associated complex structures seen in star-forming regions.

Keywords. stars: luminosity function, mass function, ISM: clouds, open clusters and associations: general

1. Introduction

Observations show that massive molecular clouds are sources of recent or ongoing star formation, and that the most massive stars tend to form in the dense cores of large clusters. Lower-mass clouds contain sites of low-mass stellar systems, as in Taurus and Ophiuchus (Williams & McKee 1997). Many young stellar clusters have been found to contain subclusters and, on the smallest scales, most stars are in multiple or binary systems. Star formation can, therefore, be seen to be a hierarchical process that involves clustering on a range of scales. A very good example of this type of clustering can be seen in the Orion Nebula, where one such cluster is embedded in the Kleinmann–Low infrared nebula and the other is in the Trapezium cluster.

Stars form in a hierarchy of clusters and subclusters in a molecular cloud, depending on the amount of material available for star formation. This is borne out by the fact that the mass of the most massive star present increases systematically with the mass of the associated molecular cloud (Larson 1995). In addition, the mass of the most massive star in a cluster is proportional to the total mass of stars and inversely proportional to the number of stars in that cluster. Each level in a stellar hierarchy corresponds to a fixed logarithmic interval in mass, with the number of stars in that hierarchy inversely proportional to their mass.

Giant star–gas complexes (SGC), with sizes ranging from 170 to 700 pc, contain 90% of star clusters with ages from 2 to 3×10^7 years, as well as molecular clouds with masses from 10^5 to 10^6 $M_\odot$. The larger ones contain 10 or more stellar groupings, with the largest ones still having H I cloud remnants. OB associations in SGCs are found to be from 0.3 to 1.2×10^7 years old, with the older ones located towards the outer egdes. In the inner regions of the SGCs are found OB associations with ages of less than 6.5×10^6

years, HII regions, cloud clumps, clusters of early spectral types and young water masers. An example is SCG 2, located in the Sagittarius arm (Sitnik 1990).

2. The initial mass function

In 1955, Salpeter showed that the number of stars, N, per unit logarithmic mass interval in the solar neighbourhood followed the relation

$$\frac{\mathrm{d}N}{\mathrm{d}\log M} \propto M^{-1.35}, \tag{2.1}$$

where M is the stellar mass (Salpeter 1955). This stellar initial mass function (IMF) was later modified by Silk (1995), to include molecular clouds, into

$$\frac{\mathrm{d}N(M)}{\mathrm{d}M} = AM^{-(1+\gamma)}, \tag{2.2}$$

where $\mathrm{d}N(M)$ is the number of stars or cloud cores in the mass interval $\mathrm{d}M$, $\gamma = 1.5 \pm 0.3$, a range which includes the Salpeter IMF and A is a constant. Observational support for the Salpeter IMF and its connection with its parental cloud has come from *HST* data (Maíz Apellániz 2001) and submillimetre observations (Clark *et al.* 2007).

3. The IMF and fractal molecular clouds

Old models of the IMF arising from molecular-cloud fragmentation, ambipolar diffusion, stellar winds and collisions had to be abandoned because of insufficient evidence (Chabrier 2003), leaving turbulence in the parental molecular cloud (Elmegreen & Scalo 2004) to be the only plausible explanation. In addition, a hierarchical stellar structure is possible only from a fractal parental molecular cloud created from chaotic dynamics. Evidence for the fractal structure of molecular clouds has been found by various authors (e.g., Larson 1995; Elmegreen & Falgarone 1996; Kramer *et al.* 1998). It has been further shown (Datta 2001, 2003, 2007) that fractal molecular-cloud structure is in the shape of a Julia set given by

$$f(z) = z^2 + c, \tag{3.1}$$

where $c = -0.745430 + 0.113008i$, of average box dimension 1.67.

4. Relation between the IMF and fractal dimension

We now propose that there exists a connection between the box dimension of the molecular cloud and the IMF (Equation 2.2) for molecular clouds. To make that connection, we look at the later stages of molecular-cloud evolution in the presence of strong gravitational forces. At this stage, the cloud IMF is related to the core-mass function (CMF; Goodwin *et al.* 2008) and our assumption of cloud cores to be spheres is reasonable. In a large collection of such cores, as in a fractal, we can use the mean mass, M_c, to represent core masses in the range $M_\mathrm{c} \pm \mathrm{d}M_\mathrm{c}$ (law of large numbers). Replacing the frequency by the number of cores, $N(M_\mathrm{c})$, in the numerator of Equation (2.2) and taking logarithms, we get

$$\log N(M_\mathrm{c})/\log M_\mathrm{c} = \log A/\log M_\mathrm{c} - \gamma. \tag{4.1}$$

We also know that the mass of the core is a function of the core radius, R_c, so substituting this we get

$$\frac{\log N(R_c)}{\log 4/3\pi\rho R_c^3} = \frac{\log A}{\log 4/3\pi\rho R_c^3} - \gamma. \quad (4.2)$$

Taking limits as R_c tends to zero and since $4/3\pi\rho$ is a constant, we get

$$\lim_{R_c \to 0} \frac{N(R_c)}{\log 1/R_c} = 3\gamma. \quad (4.3)$$

From the definition of Kolmogorov (or Minkowski) dimension (Falconer 1997) given by

$$D = \lim_{\epsilon \to 0} \sup \frac{\log N(\epsilon)}{\log 1/\epsilon}, \quad (4.4)$$

where $N(\epsilon)$ is the minimum number of open balls of radius ϵ centred on points of a nonempty compact subset of a metric space X covering the set and 'sup' is the supremum (Datta 2003). Equation (4.4) is equivalent to the box dimension of the set or physical object, as in our case. Comparing Equations (4.3) and (4.4), we get

$$D \equiv 3\gamma. \quad (4.5)$$

Hence, the result.

References

Chabrier, G. 2003, *PASP*, 115, 763

Clark, P. C., Klessen, R. S., & Bonnell, I. A. 2007, *MNRAS*, 379, 57

Datta, S. 2001 in: E. D. Feigelson & G. J. Babu (eds.), *Statistical Challenges in Astronomy III*, Springer, p. 409

Datta, S. 2003, *A&A*, 401, 193

Datta, S. 2007, PhD thesis, University of Calcutta

Elmegreen, B. G. & Scalo, J. 2004, *ARA&A*, 42, 211

Elmegreen, B. G. & Falgarone, E. 1996, *ApJ*, 471, 816

Falconer, K. 1997, *Fractal Geometry*, Wiley

Goodwin, S. P. 2008, *A&A*, 477, 823

Kramer, C., Stutzki, J., Rohrig, R., & Corneliussen, U. 1998, *A&A*, 329, 249

Larson, R. B. 1995,*MNRAS*, 272, 213

Maíz Appellániz, J. 2001, *ApJ*, 563, 151

Salpeter, E. E. 1995, *ApJ*, 121, 161

Silk, J. 1995, *ApJ* (Letters), 438, L41

Sitnik, T. G. 1990 in: E. Falgarone & F. Boulanger (eds.), *Fragmentation of Molecular Clouds and Star Formation*, IAU Symp. 180, p. 21

Williams, J. P. & McKee, C. F. 1997, *ApJ*, 476, 166

Star clusters: basic galactic building blocks
Proceedings IAU Symposium No. 266, 2009
R. de Grijs & J. R. D. Lépine, eds.

doi:10.1017/S1743921309991402

No relation between the vertical velocity component and the absolute magnitude among globular clusters

E. A. De Souza–Rossetto and H. J. Rocha–Pinto

Observatório do Valongo, Universidade Federal do Rio de Janeiro, Ladeira Pedro Antônio, 43, Saúde, Rio de Janeiro, Brazil
email: [erika, helio]@astro.ufrj.br

Abstract. The globular cluster luminosity function distribution shows a peak at $M_V \approx -7.5$ mag. There are some indications that the kinematic parameters are correlated with luminosity. In particular, Alfaro *et al.* (2001) have studied the properties of the Galactic globular cluster system and they found a correlation between spatial-velocity component and globular cluster absolute magnitude. The authors assumed that the globular clusters can be separated into two groups. The first is composed of globular clusters with $M_V < -7.5$ mag and moving preferentially towards the north Galactic pole, while the faintest globular clusters, composing the second group, move towards the Galactic disk. We have selected a sample of globular clusters using the same criteria as Alfaro *et al.* (2001) and have checked that this apparent relation indeed exists. Nevertheless, we decided to investigate whether it could be a fortuitous relation or an intrinsic property by checking its validity for eight different epochs at past and future times. The orbital parameters for the globular clusters at these eight epochs were found by orbital integration using a typical Galactic potential. We show that this relation between the vertical velocity component and the absolute magnitude among globular clusters is not coherent with time and the velocity distribution does not support the hypothesis of Alfaro *et al.* for the existence of two dynamical groups of globular clusters.

Keywords. globular clusters: general

1. Introduction

The Milky Way hosts around 160 globular clusters (GCs; Harris 1996). The GC system is often divided into subgroups according to parameters like metalicity, spatial distribution, dynamics, kinematics, etc. For example, Hertz & Grendaly (1983) showed that the GC luminosity function is bimodal. Moreover, Mclaughlin & Pudritz (1996) have indicated that the GC formation process implies a luminosity function with a characteristic peak for each galaxy type.

van den Bergh (1985) showed that the Galactic GC luminosity function is similar to that of M31, pointing out that the processes that lead to peaks in the distribution should be similar in both galaxies. Also, Capuzzo–Dolcetta (1993) concluded that the evolution of the M31 GC system was strongly affected by dynamical friction and tidal disruption of the GCs near the center.

Based on these assertions, Alfaro *et al.* (2001) investigated whether it is possible to establish relations between kinematic properties and the luminosity function and, if this were the case, how this could be reflected in the velocity space of GCs. They selected a sample of GCs with proper-motion measurements within a Galactocentric radius of 20 kpc. They separated the GCs into two groups: those with M_V brighter and fainter

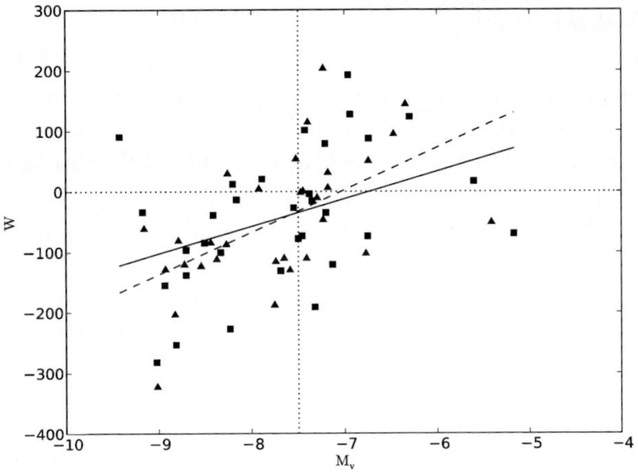

Figure 1. Relation between W and M_V. The squares and solid line are, respectively, our data and the linear best fit, while the triangles are data from Alfaro *et al.* (2001) and the dashed line is the best fit to their data.

than −7.5 mag, respectively. The authors found an apparent relation between the vertical velocity component (W) and absolute magnitude (M_V). Also, the brightest GCs in their sample, on average, are moving outwards from the Galactic plane and towards the north Galactic pole. The faintest GCs are characterized by negative motion. To test whether this relation was a selection effect, they estimated the W velocity component and found $\langle W \rangle = -32 \pm 6$ and 46 ± 6 km s^{-1} for the brightest and faintest clusters, respectively. Based on this, they rejected the selection-effect hypothesis and concluded that the $M_V - W$ relation could be primordial. They also found a probability of $< 1\%$ that the two groups of GCs come from the same population.

The goal of this paper is to investigate whether this relation is fortuitous and what its behaviour over the evolution of the Galaxy might be.

2. A reanalysis of the $M_V - W$ relation

To investigate the relation proposed by Alfaro *et al.* (2001), we selected a sample of GCs, applying the same criteria used by those authors. We looked for globular clusters with measurements available in the literature of their proper motion, right ascension, declination, distance and radial velocity. We adopted the M_V values compiled by Harris (1996). Our final sample contains 44 GCs, of which 32 are located within a Galactocentric radius 20 kpc. We used the equations of Johnson & Soderblom (1987) to calculate the spatial velocity components, U, V and W in a left-handed system, with the U component pointing from the Sun to the Galactic Center. We corrected for solar motion using $(u, v, w)_{\odot} = (-9, 11, 6)$ and $\Theta_0 = 220$ km s^{-1}.

In Figure 1 we show the $W - M_V$ plot for our (squares) and Alfaro *et al.* (2001)'s (triangles) samples. The solid and dashed lines are, respectively, our and Alfaro *et al.* (2001)'s linear fits. Our data show that this relation is much less defined than in Alfaro *et al.* (2001), to the point of rendering it doubtful. Take, for instance, the much larger numbers of bright GCs characterized by negative W that were not seen in Alfaro *et al.* (2001). Nevertheless, a relation does indeed exist.

3. Temporal evolution of the $M_V - W$ relation

Despite having found a similar relation as Alfaro *et al.* (2001), we decided to investigate whether it remains coherent over time. For this purpose, we used the Galactic potential proposed by Flynn *et al.* (1996). We integrated this potential and calculated the W velocities for eight different epochs from 8 Gyr ago until 8 Gyr in the future, at intervals of 2 Gyr. For each specific epoch, we analysed the relation between W and M_V. Moreover, we checked whether the brightest and faintest globular clusters are separated into two kinematical groups using by a K–S test. For the current epoch, providing support to Alfaro *et al.* (2001)'s claim, we find a probability of 0.074 that the GCs are from the same sample. However, for others epochs, this probability does not support this interpretation. In Figure 2 we plot the $W - M_V$ diagrams for the eight epochs analysed. The panels in this figure show that the GCs mix and the relationship between W and M_V changes considerably. Analysing the evolution of the relation, we conclude that the present-day $W - M_V$ relation is an artifact of the present-day kinematical configuration of the GC system.

4. Conclusions

We have found a similar relation between W and M_V for the current epoch among GCs as proposed by Alfaro *et al.* (2001). However, analysing the past and the future analogues of this relation, we conclude that the hypothesis presented by these authors is not valid. In others words, there is no physical relation between W and M_V for GCs and the GCs cannot be separated into two distinct groups according to their W velocity distribution.

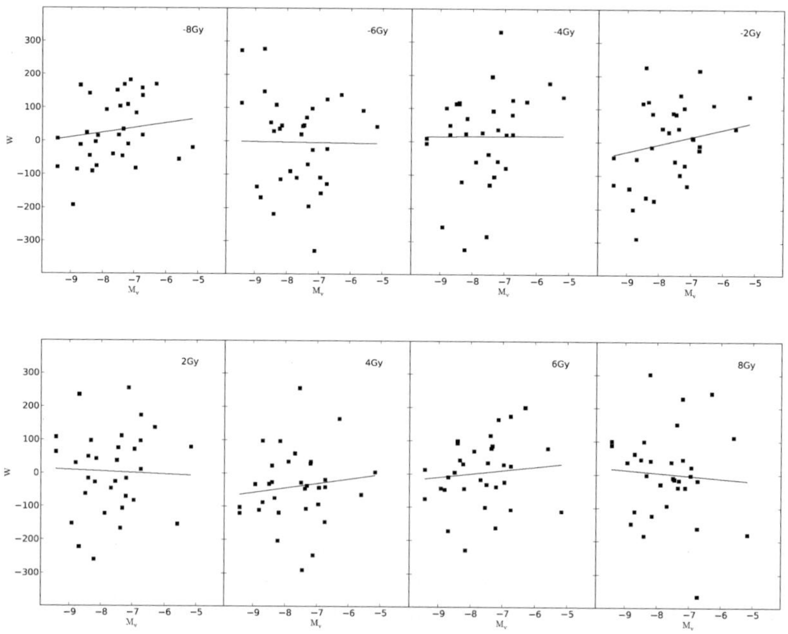

Figure 2. Relation between W and M_V for eight different epochs. We see that the GCs are mixing randomly. As a consequence, the relation between W and M_V does not fluctuate periodically, so that the relation is not primordial as proposed by Alfaro *et al.* (2001).

References

Alfaro, E. J., Delgado, A. J., Gómes–Flechoso, M. A., Ferrini, F., & Castro, I. 2001, *A&A*, 370, 45
Capuzzo–Dolcetta, R. 1993, *ApJ*, 415, 616
Flynn, C., Sommer–Larsen, J., & Christensen, P. R. 1996, *MNRAS*, 281
Johnson, D. R. H. & Soderblom, D. R. 1987, *AJ*, 93, 864
Hertz, P. & Grendaly, J. 1983, *ApJ*, 275, 105
Mclaughlin, D. E. & Pudritz, R. E. 1996, *ApJ*, 475, 578
van den Bergh, S. 1985, *ApJ*, 297, 361
Harris, W. E. 1996, *AJ*, 112, 1487

Star clusters: basic galactic building blocks
Proceedings IAU Symposium No. 266, 2009
R. de Grijs & J. R. D. Lépine, eds.

doi:10.1017/S1743921309991414

An unusual abundance of T Tauri stars? NIR study of the southern high-mass star-forming region RCW 34

H. M. de Villiers and D. J. Van der Walt

Unit for Space Physics, North West University, Potchefstroom Campus, South Africa

Abstract. RCW 34 is a special star-forming region with a few stars showing an infrared excess, but seems as if it has an uncommon abundance in T Tauri stars. This possibility was confirmed by (i) clustering of classical T Tauri (CTT) stars in the two-color diagram, (ii) clustering of pre-main-sequence stars in the color–magnitude diagram, as well as (iii) the significant clustering at dimensions larger than the image-frame size, indicated by the second 'bump' in the two-point correlation analysis. A possible explanation for the above features could be the existence of an underlying wide CTT cluster with a smaller cluster centered around the massive star. The K_s-band luminosity function ($\alpha = 0.31$) of RCW 34 shows that it is indeed a region of low stellar masses with an age of about of 1 Myr. Spectroscopic confirmation of such an underlying T Tauri cluster is necessary before any definite conclusions can be reached.

Keywords. methods: data analysis, techniques: image processing, techniques: photometric, stars: formation, stars: fundamental parameters, stars: luminosity function, mass function, stars: pre–main-sequence, infrared: stars

---

The full poster (in pdf format) is available at
`http://www.astro.iag.usp.br/~iaus266/Posters/pdeVilliers.pdf`.

Star clusters: basic galactic building blocks
Proceedings IAU Symposium No. 266, 2009
R. de Grijs & J. R. D. Lépine, eds.

doi:10.1017/S1743921309991426

Open clusters: a program to analyse the color–magnitude diagram with the help of stellar kinematic data

Wilton S. Dias and Thiago Costa Caetano
Universidade Federal de Itajubá (UNIFEI), MG, Brazil

Abstract. We constructed a program that allows one to use simultaneously and interactively photometric and astrometric stellar data (proper motion) to analyse color–color and color–magnitude diagrams. With this program, we are able to determine, based on photometric membership, the distance and age as well as the mean proper motion and radial velocity of several open clusters which had very uncertain parameters in previous analyses.

Keywords. methods: data analysis, Hertzsprung–Russell diagram, open clusters and associations: general

---

The full poster (in pdf format) is available at
`http://www.astro.iag.usp.br/~iaus266/Posters/pDias.pdf`.

Star clusters: basic galactic building blocks
Proceedings IAU Symposium No. 266, 2009
R. de Grijs & J. R. D. Lépine, eds.

doi:10.1017/S1743921309991438

The dynamics of an extremely young cluster probed for the first time

Christine Ducourant,[1,2] Sylvain Bontemps,[1,2] Didier Despois,[1,2] Philippe André,[3] Ramachrisna Teixeira,[4] Estelle Moraux,[5] Jerome Bouvier,[5] Alberto Krone–Martins[1,2,4] and Dimitris Sinachopoulos[6]

[1]Université de Bordeaux, Observatoire Aquitain des Sciences de l'Univers, 2 rue de l'Observatoire, BP 89, F-33271 Floirac Cedex, France

[2]CNRS, UMR 5804, Laboratoire d'Astrophysique de Bordeaux, 2 rue de l'Observatoire, BP 89, F-33271 Floirac Cedex, France
email: ducourant@obs.u-bordeaux1.fr

[3]CEA, DSM, IRFU, Service d'Astrophysique, 91191 Gif-sur-Yvette, France; Laboratoire AIM, CEA/DSM - CNRS - Université Paris Diderot, IRFU/Service d'Astrophysique, CEA–Saclay, 91191 Gif-sur-Yvette, France,

[4]Instituto de Astronomia, Geofísica e Ciências Atmosféricas, Universidade de São Paulo, Rua do Matão 1226, Cidade Universitária, São Paulo, Brazil

[5]Laboratoire d'Astrophysique de l'Observatoire de Grenoble, 414, Rue de la Piscine, Domaine Universitaire, BP 53, 38041 Grenoble Cedex 09, France

[6]National Observatory of Athens, P. O. Box 20048, Thissio 11810, Athens, Greece

Abstract. We present a proper-motion program that we have developed at ESO to measure, for the first time, proper motions of members of the nearby (125 pc), very young (4×10^5 yr) infrared ρ Ophiuchi cluster. Repeated imaging over an extended period will allow us to measure the global proper motion of the cluster and its velocity dispersion and will provide access to the ~ 1 km s^{-1} velocities for objects as faint as $K = 15.7$ mag. Access to the velocity field of such an extremely young cluster is of crucial importance to investigate the origin and early evolution of open cluster systems.

Keywords. stars: distances, stars: kinematics, stars: evolution, stars: formation

1. Introduction

Stars are believed to form predominantly in associations, and young stars are observed in infrared (IR) clusters still embedded in nearby molecular clouds (e.g., Lada *et al.* 1991; André *et al.* 2007). These extremely young clusters (1–2 Myr) appear to be the precursors of the visible open clusters and/or of dispersed Galactic stellar streams. While OB associations have recently been surveyed by *Hipparcos* for detailed proper-motion studies (de Zeeuw *et al.* 1999), no IR clusters have yet been studied in this sense.

Studying proper motions in IR clusters is the only way to determine if these clusters are gravitationally bound and therefore stable, or unbound, i.e., already in the process of dispersing. Indeed, embedded clusters are young enough to be still associated with a large amount of molecular gas, which may keep the system of young stars and gas still bound. Proper-motion investigation is also an excellent and classical way to significantly improve the completeness of the known cluster members and can be applied to IR clusters. They are young enough to not yet have lost any stars and therefore constitute excellent cases to investigate the fundamental issue of the shape and origin of the initial mass function (IMF).

For these reasons, searching for proper motions in embedded IR clusters is unique and should shed light on two crucial issues in star-formation studies, i.e., the origin/early evolution of young stellar clusters and the IMF observed directly at the stellar birth place.

The ρ Ophiuchi IR cluster is unique in its proximity (125 pc), its richness in young stars and protostars, and its extreme youth (the youngest cluster known, ~ 0.5 Myr old; André *et al.* 2007). It is certainly the best target presently accessible from the ground to measure, for the first time, internal stellar motions in an embedded IR cluster. In addition, we can use various surveys that have studied this cluster (Bontemps *et al.* 1998; Duchêne *et al.* 2004; Stamatellos *et al.* 2007), which have achieved a census of the population of young stars with IR excesses. This allowed these authors to discuss, for the first time, the mass function in such a young cluster down to masses of less than 0.1 $M_\odot$.

2. The project

2.1. *Selection of the field*

We have selected (from Bontemps *et al.* 1998) the subcluster ρ Ophiuchi F, which exhibits a clear association of young stars with the Ophiuchi F molecular dense core (see Motte *et al.* 1998; André *et al.* 2007). This dense core/IR cluster is roughly a 0.1 pc-radius sphere containing $15 - 30$ $M_\odot$ (Motte *et al.* 1998) of gas and $5 - 10$ $M_\odot$ of stars. The typical Keplerian velocity in such a system is then $v_K = (2GM_{tot}/R)^{1/2}$, or 1.3 and 1.7 km s^{-1} for, respectively, 20 and 40 $M_\odot$ spheres with a radius of $R = 0.1$ pc. The typical differential velocity between two stars can reach 3 km s^{-1}, i.e., 5 mas yr^{-1} at a distance of 125 pc. The systematic proper motion of the ρ Ophiuchi IR cluster can be estimated from the proper motion of the associated, nearby Upper Scorpius OB association recently measured by de Zeeuw *et al.* (1999) with *Hipparcos*. They found a global proper motion of 25.8 ± 0.1 mas yr^{-1}. Such a large proper motion with respect to the background field stars can easily be measured with our observational technique. We plan to investigate the proper motion of stars down to $K_s = 15.7$ (signal-to-noise ratio $= 33$ for individual images), which corresponds to a 0.02 $L_\odot$ (for $A_V = 30$), or 0.045 $M_\odot$ i.e., a young brown dwarf. We will be able to recognize all members, irrespective of their nature (protostars, young stars with or without IR excess, brown dwarfs). Our completeness will therefore be limited only by our ability to recognize systematic proper motions of the members (which, in principle, is not significantly flux dependent). This is extremely important for addressing the issue of the shape and origin of the IMF in this cluster, since we will get a membership census based on an unbiased technique. Additional spectroscopic observations of kinematic members will confirm their nature.

2.2. *Binarity*

We have investigated, through Monte Carlo simulations, the effects of binarity. The data for the simulation were taken from Bontemps et al. (1998; IMF), D'Antona & Mazzitelli (1994; *L–M* relation), Duquennoy & Mayor (1991; binary-period distribution). Orbital inclinations were taken as random and component masses considered uncorrelated. Several thousand systems were simulated and analysed in terms of proper motion. We normalize the results, for presentation purposes, to 100 binary systems (assuming that all young stars belong to binary systems). The simulation results in 65 unresolved and 35 resolved couples for 0.7 arcsec seeing. We thus have 135 'objects' (65 unresolved binaries and 2×35 resolved-system components). Only nine unresolved binaries have a proper motion of their photocenter due to orbital motions in excess of 0.5 mas yr^{-1}. These stars (7% of the simulated population) will affect the results slightly. Twenty six components of

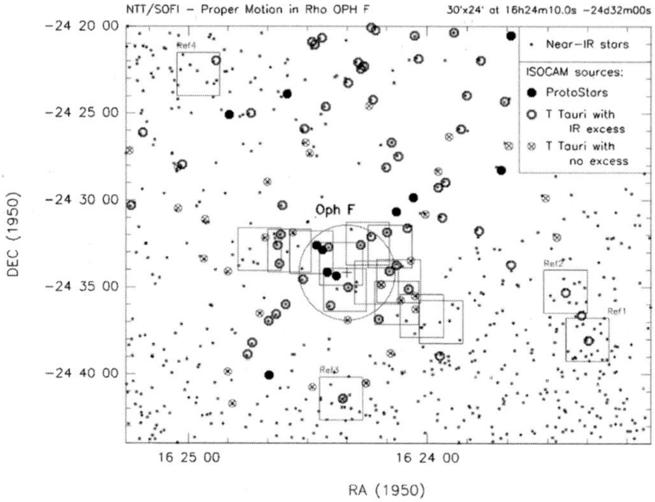

Figure 1. *NTT*/SOFI selected fields in the ρ Ophiuchi E–F dark core.

resolved binary systems also exhibit a measurable proper motion due to orbital motion. However, the separation of these 26 binaries with proper motions $\mu > 0.5$ mas yr^{-1} appears to be very small and essentially always less than 2 arcsec. Since the typical separation of young stars is much greater than 2 arcsec (on the order of 1 arcmin), this ensures that we will always recognize these pairs. They will be put apart for special analysis. It is therefore clear that binarity does not significantly affect our first goal, determination of the internal velocity dispersion of the forming cluster. It does not hamper either our second important goal, i.e., using proper motion as a cluster-membership criterion. This motion is much larger than any orbital motions and internal velocity dispersion (25 mas yr^{-1}; from *Hipparcos*).

2.3. *Astrometry*

The astrometric objective of this program is to measure individual proper motions with an accuracy of $\sigma_\mu = 0.5 - 1$ mas yr^{-1} in a young IR cluster.

The first observational epochs were acquired in April 2001, 2002, 2003 and 2004. Figure 1 presents the observed dense core with the 10 overlapping fields. Data have been measured and reduced using software developed in Bordeaux (see Ducourant *et al.* 2007). We have derived preliminary proper motions for 89 stars in the zone covered by our observations with an accuracy of $\sigma_\mu = 1 - 3$ mas yr^{-1} (Figure 2).

The known problem of field distortions of the SOFI instrument had to be addressed. Much effort was dedicated to correct our measurements for these distortions. To minimize residual effects, we also included (in the global reduction) photometric images (SOFI large field) that suffer from less distortion, to average the effects.

The combined problem of field distortion, which has a significant impact on the astrometry, and the poor numbers of stars present in each field led us to extend the program, acquiring a new set of observations in May 2009. With the addition of an additional epoch 8 years after the first observations, the astrometric precision should be improved by a factor of 2–3, i.e., enabling a precision of $\sigma_\mu = 0.5 - 1$ mas yr^{-1} and thus allowing us to measure the largest expected differential velocities within the cloud and the expected atypical velocities greater than that of the cloud (due to, e.g., binary ejection, ...). Data are being reduced and will be presented soon.

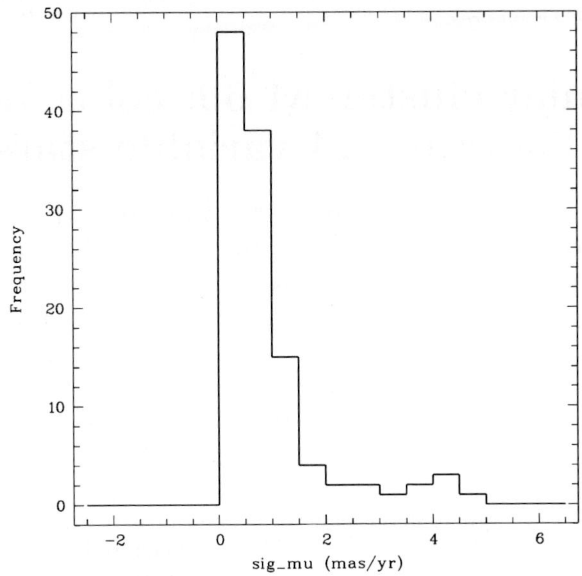

Figure 2. Histogram of the precisions of proper motions σ_μ derived from a preliminary global overlap solution of the 10 overlapping fields including the first four observational epochs.

References

André, P., Belloche, A., Motte, F., & Peretto, N. 2007, *A&A*, 472, 519

André, P., Bouwman, J., Belloche, A., & Hennebelle, P. 2004, *ApSS*, 292, 325

Barsony, M., Ressler, M. E., & Marsh, K. A. 2005, *ApJ*, 630, 381

Bontemps, S., *et al.* 1998, in: J. Yun & R. Liseau (eds.), *Star Formation with the Infrared Space Observatory*, ASP Conf. Ser., 132, p. 141

D'Antona, F. & Mazzitelli, I. 1994, *ApJS*, 90, 467

Duchêne, G., Bouvier, J., Bontemps, S., André, P., & Motte, F. 2004, *A&A*, 427, 651

Ducourant, C., Teixeira, R., Chauvin, G., Daigne, G., Le Campion, J.-F., Song, I., & Zuckerman, B. 2008, *A&A* (Letters), 477, L1,

Ducourant, C., Teixeira, R., Hambly, N. C., Oppenheimer, B. R., Hawkins, M. R. S., Rapaport, M., Modolo, J., & Lecampion, J. F. 2007, *A&A*, 470, 387

Duquennoy, A. & Mayor, M. 1991, *A&A*, 248, 485

Lada, E. A., Bally, J., & Stark, A. A. 1991, *ApJ*, 368, 432

Lazorenko, P. F., Mayor, M., Dominik, M., Pepe, F., Segransan, D., & Udry, S. 2007, *A&A*, 471, 1057

Stamatellos, D., Ward-Thompson, D., Whitworth, A. P., & Bontemps, S. 2007, *A&A*, 462, 677

Teixeira, R., Ducourant, C., Chauvin, G., Krone-Martins, A., Song, I., & Zuckerman, B. 2008, *A&A*, 489, 825

de Zeeuw. P. T., Hoogerwerf, R., de Bruijne, J. H. J., Brown, A. G. A., & Blaauw, A. 1999, *AJ*, 117, 354

Star clusters: basic galactic building blocks
Proceedings IAU Symposium No. 266, 2009
R. de Grijs & J. R. D. Lépine, eds.

doi:10.1017/S174392130999144X

The globular cluster M 69: color–magnitude diagram and variable stars

M. E. Escobar,[1] M. Catelan,[1] M. Zoccali,[1] H. A. Smith,[2] B. J. Pritzl,[3] A. Layden,[4] J. Gregorsok,[4] D. L. Welch[5] and T. Webb[6]

[1]Departamento de Astronomía y Astrofísica, Pontifícia Universidad Católica, Santiago, Chile
[2]Department of Physics and Astronomy, Michigan State University, East Lansing, MI, USA
[3]Department of Physics and Astronomy, Macalester College, Saint Paul, MN, USA
[4]Bowling Green State University, Bowling Green, OH USA
[5]McMaster University, Hamilton, ON, Canada
[6]Sterrewacht, Leiden University, Leiden, the Netherlands

Abstract. We present *BV* photometry and the results of a search for stellar variability in the Galactic globular cluster M 69. The resulting color–magnitude diagram shows significant contamination by field stars. In our variability search we found 62 variable stars, 54 of which were new discoveries.

Keywords. Hertzsprung–Russell diagram, stars: variables: other, globular clusters: individual (M 69)

The full poster (in pdf format) is available at
`http://www.astro.iag.usp.br/~iaus266/Posters/pEscobar.pdf`.

Star clusters: basic galactic building blocks
Proceedings IAU Symposium No. 266, 2009
R. de Grijs & J. R. D. Lépine, eds.

doi:10.1017/S1743921309991451

A deep dive into NGC 604 with *Gemini*/NIRI imaging

Cecilia Fariña,[1,2] Guillermo L. Bosch[1,2] and Rodolfo R. Barbá[3,4]

[1]Facultad de Ciencias Astronómicas y Geofísicas, Universidad Nacional de La Plata, Paseo de Bosque s/n, 1900 La Plata, Argentina
email: `ceciliaf@fcaglp.unlp.edu.ar`

[2]IALP, CONICET, Argentina

[3]ICATE, CONICET, Argentina

[4]Departamento de Física, Universidad de La Serena, Benavente 980, La Serena, Chile

Abstract. The giant HII region NGC 604 constitutes a complex and rich population to study in detail many aspects of massive star formation, such as their environments and physical conditions, the evolutionary processes involved, the initial mass function for massive stars and star-formation rates, among many others. Here, we present our first results of a near-infrared study of NGC 604 performed with NIRI images obtained with *Gemini North*. Based on deep *JHK* photometry, 164 sources showing infrared excess were detected, pointing to the places where we should look for star-formation processes currently taking place. In addition, the color–color diagram reveals a great number of objects that could be giant/supergiant stars or unresolved, small, tight clusters. An extinction map obtained based on narrow-band images is also shown.

Keywords. ISM: individual (NGC 604), HII regions, galaxies: individual (M 33), techniques: photometric, stars: formation, stars: early-type, infrared: stars

1. Introduction

NGC 604 is a giant HII region (GHR) located in an outer spiral arm of M33, at a distance of 840 kpc. It is the second most luminous HII region in the Local Group, after 30 Doradus in the LMC. Both are nearby examples of giant star-forming regions whose individual objects can be spatially resolved for further study. NGC 604 has been the target of many studies during the past few decades. A brief summary of known facts about NGC 604 includes the following. This GHR is ionized by a massive, young cluster, with at least 200 O stars (some as early as O3–O4). The cluster does not exhibit a central core distribution. Instead, the stars are widely spread over its projected area in a structure called 'scaled OB association' (SOBA; Hunter et al. 1996; Maíz Apellániz *et al.* 2004; Bruhweiler *et al.* 2003). Wolf–Rayet stars, a confirmed and many candidate red supergiant stars, a luminous blue variable and a supernova remnant are all part of NGC 604's stellar population (Conti & Massey 1981; D'Odorico & Rosa 1981; Drissen *et al.* 1993; Díaz *et al.* 1996; Churchwell & Goss 1999; Terlevich *et al.* 1996; Barbá *et al.* 2009). The age of the central ionizing cluster has been determined by different authors as between 3 and 5 Myr (González–Delgado & Pérez 2000; Bruhweiler *et al.* 2003; Díaz *et al.* 1996; Hunter *et al.* 1996; Relaño & Kennicut 2009).

The interstellar medium reveals a complex structure with a high-excitation central region (made up of multiple two-dimensional structures), asymmetrically surrounded by a low-excitation halo. The whole region shows a very complex geometry of cavities, expanding shells and filaments, as well as dense molecular regions. All of these structures

Table 1. Main characteristics of broad-band and narrow-band filters used in our *Gemini*/NIRI observations.

Broad bands			Narrow bands		
Filter	Central λ (μm)	Coverage (μm)	Filter	Central λ (μm)	Coverage (μm)
J	1.25	0.97–1.07	Paβ	1.282	~ 0.1
H	1.65	1.49–1.78	Brγ	2.16	~ 0.1
K_s	2.15	1.99–2.30	H_2(2–1)	2.24	~ 0.1

show different kinematic behaviour (Maíz Apellániz *et al.* 2004; Tenorio–Tagle *et al.* 2000; Sabalisck *et al.* 1995; Relaño & Kennicut 2009).

Aiming to characterize the youngest stellar population and its environment, we performed near-infrared (NIR) photometry (JHK) and analysed narrow-band images in Paβ, Brγ and H_2(2–1). Taking into account that NIR observations are less affected by dust extinction characteristic in star-fomation environments, we can take a deep dive into NGC 604 to study those very young objects which are still immersed in their parental clouds at the sites of current star formation.

2. Images and data processing

The images were obtained with the Near InfraRed Imager and Spectrometer (NIRI) at *Gemini North*. The resulting plate scale is 0.117 arcsec pixel^{-1}, with a field of view of 120×120 arcsec^2. The filters used and their main characteristics are listed in Table 1.

The images were taken under excellent seeing conditions, on average $\sim 0.35''$ in the J, H and K_s images. A set of approximately 10 individual exposures was taken in each band to combine into the final image. Data reduction and processing were performed with specific tasks using the GEMINI-NIRI IRAF package. Images were sky subtracted and flat fielded and short dark exposures were used to identify bad pixels. Stellar magnitudes were obtained by point-spread-function (PSF) fitting in crowded fields using the DAOPHOT software (Stetson 1987) in IRAF. Although a standard procedure, PSF construction and fitting involves an iterative and careful process in which several tries were made to get the best results. The effective area covered by our photometry is $\sim 107 \times 107$ arcsec^2 ($\sim 430 \times 430$ pc^2 at the distance of M33). The average photometric errors are 0.09, 0.11 and 0.21 mags in the H, J and K_s filters, respectively, and the completeness limits are 22 mag in J and 21 mag in H and K_s. Magnitudes in the individual filters were matched in a unique list containing 5566 objects in the field in which all three J, H and K_s magnitudes were measured. Astrometry was derived using 35 objects in common in our field of NGC 604 and the GSC–II Catalog, Version 2.3.2 (2006) in the ICRS, Equinox J2000.0.

3. Results and discussion

The resulting color–magnitude (CM) and color–color (CC) diagrams are shown in Figures 1 and 2, respectively. We have included ~ 2000 selected objects, located within a radius of 48 arcsec (~ 200 pc) and centered on NGC 604, meeting a certain photometric quality level [magnitude error $\leqslant$ median(error) + 1.0$\times$ standard deviation(error)]. Red symbols in both plots are objects that lie on the right side of the reddening line for a O6–O8 V star. For each of these objects, the error in $(H - K_s)$ color is smaller than its distance to the reddening line, so that we can ensure that they undoubtedly show an IR excess.

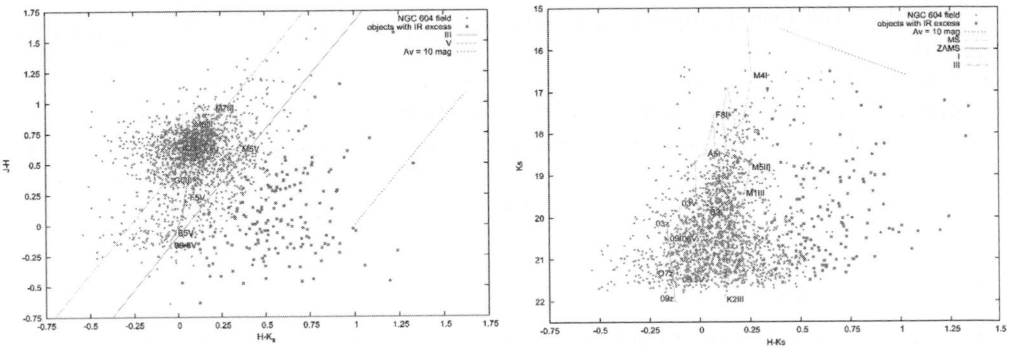

Figure 1. Color–color *(left)* and color–magnitude *(right)* diagrams for objects observed in a circular area centered on NGC 604. Red squares are objects that show an IR excess.

Among the objects with intrinsic IR excesses in GHRs we expect to find Wolf–Rayet stars, early supergiants and Of stars, and massive young stellar objects (MYSOs). Many of these types of objects (or candidates) have already been found in NGC 604. Based on our deep *JHK* photometry we found a total of 164 objects that show IR excesses. As can be expected, $\sim$ 70% of these lie in a small area near the region's center and a large proportion are tightly grouped in regions coincident with the radio-continuum peaks at 8.4 GHz from Churchwell & Goss (1999), as shown in Figure 3 (left panel). These results are in complete agreement with the analysis of Barbá *et al.* (2009) based on *HST*/NICMOS NIR images, where MYSO candidates also appeared aligned with the radio peak structures.

What we found supports the idea mentioned by many authors, that those areas may be embedded star-forming regions, also taking into account that regions coincident with the radio knots show conditions of high temperature and density (Tosaki *et al.* 2007), and in two of them Maíz Apellániz *et al.* (2004) identified compact Hii regions. Most objects with IR excesses will be the targets for further observing programs (making use of the integral-field spectroscopic facilities at *Gemini North*) to elucidate their nature and derive accurate properties by means of spectroscopic study. Also, those objects that

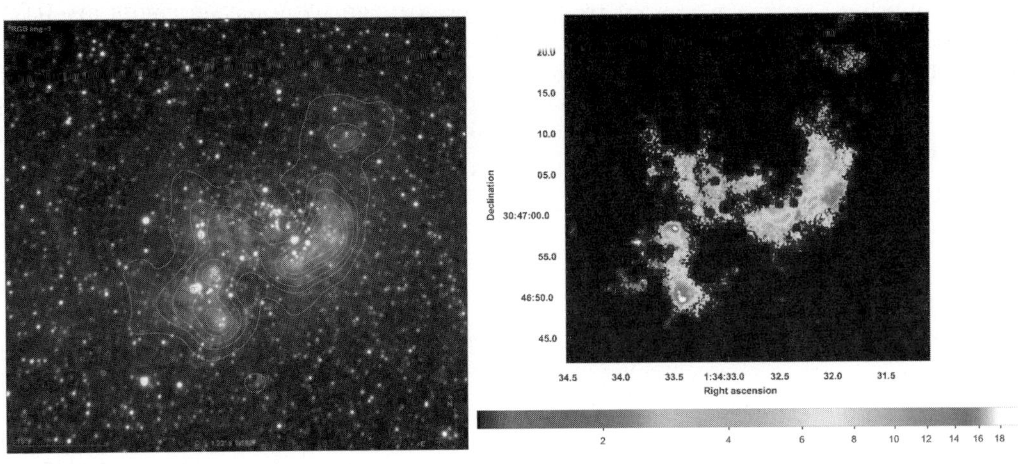

Figure 2. *(left)* Three–color (J, H and K_s) composite image of NGC 604 with 8.4 GHz radio–continuum countours overlaid (adapted from Churchwell & Goss 1999). Red circles are objects that show IR excess in our photometry. *(right)* Extinction map from Brγ/Hα, regions with higher extinction are red/white.

are located at the bright, massive end on the CM diagram deserve further observation and study.

On the basis of our narrow-band images we generated an extinction map, derived from the observed variations of the Brγ to Hα ratio (with the Hα image taken from Bosch *et al.* 2002). The color scale used in the map shows regions with higher extinction in red/white. The present and future results of our NIR study will be placed in the context of previous studies of NGC 604 to complete the picture of the overall formation and evolution scenario of this GHR.

Acknowledgements

RB acknowledges partial support from the Universidad de La Serena, project DIULS CD08102.

References

Barbá, R. H., Maíz Apellániz, J., Pérez, E., Rubio, M., Bolatto, A., Fariña, C., Bosch, G., & Walborn, N. R. 2009, in: E. Pérez, R. de Grijs & R. M. González–Delgado (eds.), *Young massive star clusters: initial conditions and environments*, *ApSS*, 324, 309

Bosch, G., Terlevich, E., & Terlevich, R. 2002, *MNRAS*, 329, 481

Bruhweiler, F. C., Miskey, C. L., & Smith Neubig, M. 2003, *AJ*, 125, 3082

Churchwell, E. & Goss, W. M. 1999, *ApJ*, 514, 188

Conti, P. S. & Massey, P. 1999, *ApJ*, 249, 471

Díaz, A. I., Terlevich, E., Terlevich, R., González–Delgado, R. M., Pérez, E., & García–Vargas, M. L. 1996, in: C. Leitherer, U. Fritze–von Alvensleben & J. Huchra (eds.), *From Stars to Galaxies: The Impact of Stellar Physics on Galaxy Evolution*, ASP Conf. Ser., 98, 399

D'Odorico, S. & Rosa, M. 1981, *ApJ*, 248, 1015

Drissen, L., Moffat, A. F. J., & Shara, M. M. 1993, *AJ*, 105, 1400

González Delgado, R. M. & Pérez, E. 2000, *MNRAS*, 317, 64

Hunter, D. A., Baum, W. A., O'Neil Jr., E. J., & Lynds, R. 1996, *ApJ*, 456, 174

Maí Apellániz, J., Pérez, E., & Mas–Hesse, J. M. 2004, *AJ*, 128, 1196

Relaño, M. & Kennicutt, R. C. 2009, *ApJ*, 699, 1125

Sabalisck, N. S. P., Tenorio–Tagle, G., Castaneda, H. O., & Muñoz–Tuñón, C. 1995, *ApJ*, 444, 200

Stetson, P. B. 1987, *PASP*, 99, 191

Tenorio–Tagle, G., Muñoz–Tuñón, C., Pérez, E., Maíz Apellániz, J., & Medina–Tanco, G. 2000, *ApJ*, 541, 720

Terlevich, E., Díaz, A. I., Terlevich, R., González–Delgado, R. M., Pŕez, E., & García–Vargas, M. L. 1996, *MNRAS*, 279, 1219

Tosaki, T., Miura, R., Sawada, T., Kuno, N., Nakanishi, K., Kohno, K., Okumura, S. K., & Kawabe, R. 2007, *ApJ*, 664, 27

Star clusters: basic galactic building blocks
Proceedings IAU Symposium No. 266, 2009
R. de Grijs & J. R. D. Lépine, eds.

doi:10.1017/S1743921309991463

Determination of kinematic distances of pre-main-sequence stars in star-forming regions

Phillip A. B. Galli,[1] Ramachrisna Teixeira,[1] Christine Ducourant[2] and Claude Bertout[3]

[1]Instituto de Astronomia, Geofísica e Ciências Atmosféricas, Universidade de São Paulo, Rua do Matão 1226, Cidade Universitária, São Paulo, Brazil
email: galli@astro.iag.usp.br

[2]Laboratoire d'Astrophysique de Bordeaux, Université de Bordeaux I, 2 Rue de l'Observatoire, 33270 Floirac, France

[3]Institut d'Astrophysique de Paris, 98bis Bd. Arago, 75014, Paris, France

Abstract. Many studies of star-forming regions have been carried out since the discovery of compact HII regions in the late 1960s. The kinematic properties of young stars in the nearest regions with ongoing and recent star formation provide essential tests of their formation mechanisms. The detection of coeval moving groups allows determination of individual distances through the convergent-point method. As a result, the main physical properties of these stars and their early evolutionary stages can be determined if we know how distant they are.

Keywords. stars: distances, stars: kinematics, stars: evolution, stars: formation

1. Introduction

A precise determination of distances is required to accurately determine the main physical parameters of stars (mass, age, luminosity and temperature) and to trace the evolution of young stars. Kinematic studies of young stars based on distances, proper motions and radial-velocity information are fundamental to investigate their formation mechanisms. Young stellar objects, including pre-main-sequence (PMS) stars, are supposed to be very close to their birth places and have velocities still very similar to the initial values, so that one can get clear constraints on the birth mechanisms. Many PMS stars were not observed by the *Hipparcos* satellite because of their faint magnitudes nor do they have any trigonometric parallaxes measured from the ground due to their distances. In this context, knowledge of the proper motions of young stars is essential because they allow detection of structures such as moving groups and associations to which these stars belong. The members of a moving group, which share the same space motion, allow us to obtain the convergent point from which we can determine the kinematic distance if the radial velocity is known.

2. The sample of pre-main-sequence stars

We have a sample of 1250 PMS stars and 104 candidates, spread over major star-forming regions (SFRs), as shown in Figure 1, taken from the proper-motion catalogue of PMS stars Ducourant *et al.* (2005). An exhaustive search in the literature for PMS stars in the proper-motion catalogue with known radial velocities has been conducted to develop our strategy. So far, we have found radial-velocity values for 410 stars, corresponding to 30% of our sample. Likewise, we searched for PMS stars with known parallaxes to compare with our future results. We found parallax values for 58 stars, corresponding to 4% of our sample. The main source for parallaxes is the *Hipparcos* catalogue.

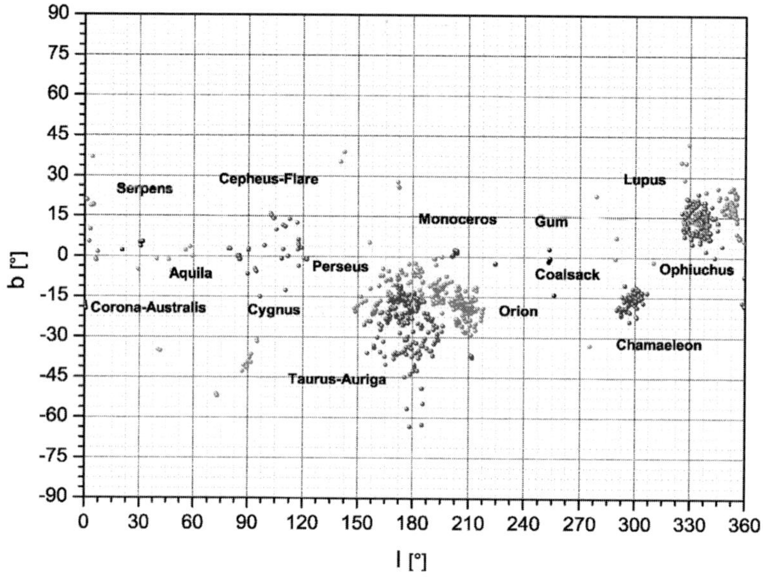

Figure 1. Sky distribution of the SFRs in the proper-motion catalogue of PMS stars.

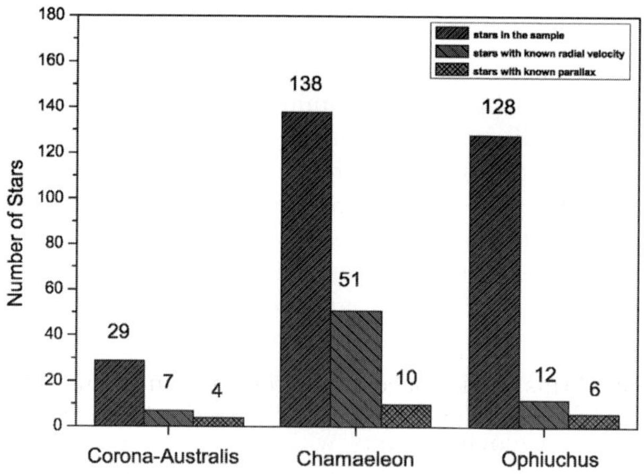

Figure 2. Brief census of the Corona Australis, Chamaeleon and Ophiuchus star-forming regions. We present the total number of stars in each star-forming region compared with the number of stars that have radial-velocity and parallax information.

This work will focus on a kinematic study of three SFRs and the same strategy developed here may be applied to the other regions in the future. Our priority list includes the following SFRs: Corona Australis, Chamaeleon and Ophiuchus. A brief census of each SFR considered is presented in Figure 2.

3. Convergent-point analysis

Although average distances for a moving group provide valuable information, what we really need for evolutionary models are the distances to individual stars in the group. The procedure that we use here to derive individual parallaxes is based on a knowledge of radial velocities and on the convergent-point method which makes full use of directly observed astrometric data, i.e., positions and proper motions. We developed our first convergent-point analysis based on Mamajek (2005). In this implementation, we approximately follow the convergent-point grid technique of Jones (1971), using the definition of t^2 and velocity dispersion of de Bruijne (1999a). The convergent-point code has been applied to the TW Hydrae association for a check of our results. A comparison among convergent-point analyses from different authors is shown in Figure 3. We use the TW Hydrae association only as a check of our data processing. The whole strategy developed here will be applied to the PMS stars in the Corona Australis, Chamaeleon and Ophiuchus star-forming regions.

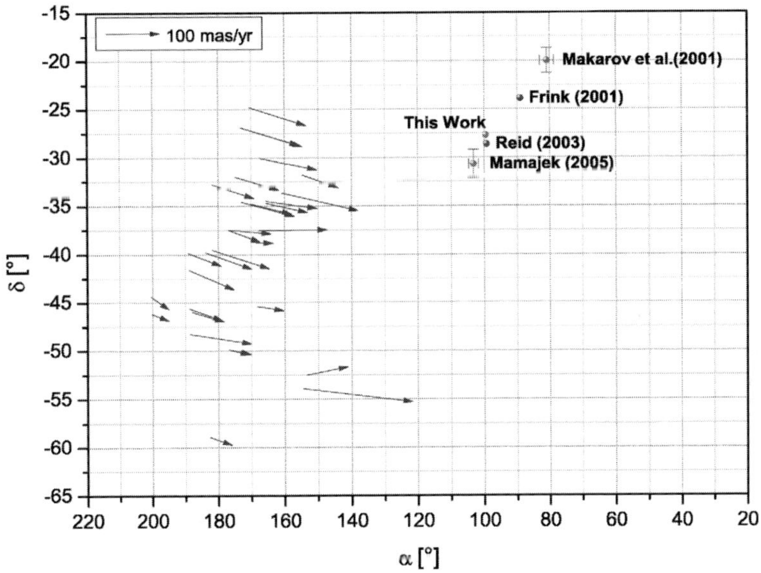

Figure 3. Comparison of convergent-point analysis for the TW Hydrae association.

The current stage of this work concerns the evaluation of the convergent point's uncertainty, of which the determination is not clear in the literature. As soon as we have finished the implementation of our own convergent-point strategy, we will be able to evaluate the kinematic distance to each group member considering that its radial velocity is known:

$$\Pi_{\rm ind} = \frac{4.74\mu}{V_{\rm rad} \tan \lambda} \tag{3.1}$$

The next step is dedicated to application of the distances derived here and analysis of the kinematic properties of each SFR. This work is the initial part of a more complete project which represents an effort to better constrain the early stages of stellar evolution as regards the formation mechanism, disk lifetime, timescale of planet formation and kinematics of young stellar objects.

References

Bertout, C. & Genova, F. 2006, *A&A*, 460, 499

Bertout, C., Siess, L., & Cabrit, S. 2007, *A&A* (Letters), 473, L21

de Bruijne, J. H. J. 1999, *MNRAS*, 306, 381

Ducourant, C., Teixeira, R., Périé, J. P., Lecampion, J. F., Guibert, J., & Sartori, M. J. 2005, *A&A*, 438, 769

Frink, S. 2001, in: R. Jayawardhana & T. Greene (eds.), *Young Stars Near Earth: Progress and Prospects*, in ASP Conf. Ser., 244, p. 16

Jones, D. H. P. 1971, *MNRAS*, 152, 231

Makarov, V. V. & Fabricius, C. 2001, *A&A*, 368, 866

Mamajek, E. E. 2005, *ApJ*, 634, 1385

Reid, N. 2003, *MNRAS*, 342, 837

Teixeira, R., Ducourant, C., Sartori, M. J., Camargo, J. I. B., Périé, J. P., Lépine, J. R. D., & Benevides–Soares, P. 2000, *A&A*, 361, 1143

Star clusters: basic galactic building blocks
Proceedings IAU Symposium No. 266, 2009
R. de Grijs & J. R. D. Lépine, eds.

doi:10.1017/S1743921309991475

Gas trapping in extremely massive clusters in NGC 1365

Emmanuel Galliano[1] **and Danielle Alloin**[2]
[1]Observatório Nacional, MCT, Rio de Janeiro, Brazil
[2]CEA, Saclay, France

Abstract. Three radio cm sources in the central star-forming kpc of the closeby barred Seyfert galaxy NGC 1365 are observed. The complete dataset includes *VLT* infrared J, K, L and N images and spectra. The main observed features include (i) a rising dust continuum towards the mid-infrared (MIR), (ii) bright H emission lines and PAH (polycyclic aromatic hydrocarbon) features, (iii) a bright MIR [NeII] line, and (iv) no detection of MIR [SIV] or [ArIII] lines. For a typical cluster star-formation history and initial mass function, the [ArIII]/[NeII] and [SIV]/[NeII] ratios exhibit an abrupt decrease of several orders of magnitude at an age of $\sim$ 7 Myr. The nondetection of [ArIII] and [SIV] indicates that the clusters are of approximately this age.

Keywords. galaxies: starburst, galaxies: star clusters, infrared: galaxies

---

The full poster (in pdf format) is available at
`http://www.astro.iag.usp.br/~iaus266/Posters/pGalliano.pdf`.

Star clusters: basic galactic building blocks
Proceedings IAU Symposium No. 266, 2009
R. de Grijs & J. R. D. Lépine, eds.

doi:10.1017/S1743921309991487

Search for cluster candidates in the UKIDSS database

Mario Gennaro,[1] Wolfgang Brandner,[1] Thomas Henning[1] and Andrea Stolte[2]

[1]Max-Planck-Institut für Astronomie, Königstuhl 17, 69117 Heidelberg, Germany

[2]I. Physikalishes Institut, Universität Köln, Zülpicher Str. 77, 50937 Köln, Germany

Abstract. UKIDSS GPS is a deep and high-angular-resolution survey of the Galactic plane in the J, H, and K bands. Using this survey, we searched the inner region of the Galaxy ($|l| < 10°$) for young cluster candidates associated with molecular bubbles (MBs). The latter are observed in the mid-infrared survey GLIMPSE II. The underlying idea is that MBs may be blown out by powerful sources, such as young O and B stars; some of the MBs in the sample are indeed associated with HII regions, a hint of the presence of young stars. One of the major observational problems in the Galactic plane is the high degree of interstellar absorption and its spatial variability. A significant effort is also devoted to quantifying this issue.

Keywords. surveys, open clusters and associations: general, infrared: stars

---

The full poster (in pdf format) is available at
`http://www.astro.iag.usp.br/~iaus266/Posters/pGennaro.pdf`.

Star clusters: basic galactic building blocks
Proceedings IAU Symposium No. 266, 2009
R. de Grijs & J. R. D. Lépine, eds.

doi:10.1017/S1743921309991499

Nuclear clusters in dwarf irregular galaxies and their connection to massive galactic clusters

Iskren Y. Georgiev,[1,2] Michael Hilker,[3] Thomas H. Puzia,[4] Paul Goudfrooij[2] and Holger Baumgardt[1]

[1]Argelander-Institut für Astronomie, Universität Bonn, Auf dem Hügel 71, 53121 Bonn, Germany

[2]Space Telescope Science Institute, Baltimore, MD, USA

[3]ESO, Garching, Germany

[4]Dominion Astrophysical Observatory, Herzberg Institute of Astrophysics, National Research Council of Canada, Victoria, BC, Canada

Abstract. We compare nuclear globular clusters (nGCs) in dwarf galaxies and Galactic GCs with extended (hot) horizontal branches (EHB–GCs) to test the suggested external origin of the latter and the conditions under which GC self-enrichment can operate. We show that the present-day escape velocity ($v_{\rm esc}$) of stellar ejecta to reach the cluster tidal radius compares with those of EHB–GCs. For EHB–GCs, we find a correlation between the present-day $v_{\rm esc}$ and their metallicity as well as $(V-I)$ colour. The similar $v_{\rm esc}$ and $(V-I)$ distribution of nGCs and EHB–GCs implies that nGCs could also have complex stellar populations. The $v_{\rm esc}$–[Fe/H] relation could reflect the known relation of increasing stellar-wind velocity with metallicity, which in turn explains why more metal-poor clusters typically show more peculiarities in their stellar population than more metal-rich clusters of the same mass.

Keywords. galaxies: star clusters, galaxies: dwarf, galaxies: nuclei

---

The full poster (in pdf format) is available at
`http://www.astro.iag.usp.br/~iaus266/Posters/pGeorgiev.pdf`.

Star clusters: basic galactic building blocks
Proceedings IAU Symposium No. 266, 2009
R. de Grijs & J. R. D. Lépine, eds.

doi:10.1017/S1743921309991505

Investigation of star clusters found in the 2MASS catalog

Elena Glushkova,[1] Ramakant Yadav[2] and Sergey Koposov[3,4]

[1]Sternberg Astronomical Institute, Universitetsky pr. 13,
Moscow 119992, Russia
email: elena@sai.msu.ru

[2]Aryabhatta Research Institute of Observational Sciences,
Manora Peak, Nainital 263129, India
email: rkant@aries.res.in

[3]Max Planck Institute for Astronomy, Königsstuhl 17,
Heidelberg 69117, Germany
email: koposov@mpia-hd.mpg.de

[4]Institute of Astronomy, University of Cambridge, Madingley Road, Cambridge CB3 0HA, UK

Abstract. We have launched a program of *UBVR* CCD photometric investigations of new clusters detected in the 2MASS catalog based on an automatic detection method, which uses a convolution with a density function. Three new clusters, Koposov 12, Koposov 53 and Koposov 77 were observed in *UBVI* with the 104 cm *Sampurnanand* telescope located at ARIES, Manora Peak, India. Their ages, distances, color excesses and metallicities have been derived using photometric data.

Keywords. open clusters and associations: general, open clusters and associations: individual (Koposov 12, Koposov 53, Koposov 77)

The full poster (in pdf format) is available at
`http://www.astro.iag.usp.br/~iaus266/Posters/pGlushkova.pdf`.

Star clusters: basic galactic building blocks
Proceedings IAU Symposium No. 266, 2009
R. de Grijs & J. R. D. Lépine, eds.

doi:10.1017/S1743921309991517

Star clusters: age, metallicity and extinction from integrated spectra

Rosa M. González Delgado[1] and Roberto Cid Fernandes[2]

[1]Instituto de Astrofísica de Andalucía–CSIC, Granada, Spain

[2]Departamento de Física, Universidade Federal de Santa Catarina, Florianopolis, Brazil

Abstract. Integrated optical spectra of star clusters in the Magellanic Clouds and a few Galactic globular clusters are fitted using high-resolution spectral models for single stellar populations. The goal is to estimate the age, metallicity and extinction of the clusters, and evaluate the degeneracies among these parameters. Several sets of evolutionary models that were computed with recent high-spectral-resolution stellar libraries (MILES, GRANADA, STELIB), are used as inputs to the STARLIGHT code to perform the fits. The comparison of the results derived from this method and previous estimates available in the literature allow us to evaluate the pros and cons of each set of models to determine star cluster properties. In addition, we quantify the uncertainties associated with the age, metallicity and extinction determinations resulting from variance in the ingredients for the analysis.

Keywords. galaxies: star clusters, galaxies: stellar content, Magellanic Clouds

1. Introduction

Star clusters (SCs) are one of the main modes of star formation in galaxies. They are powerful tracers of the formation and evolution history of galaxies because they retain information about the parent galaxy. In this respect, the age, metallicity and mass of cluster populations are key information to test different galaxy-formation scenarios.

Evolutionary synthesis models are powerful tools to infer the properties of extragalactic clusters, since their integrated light is easily modelled by a single-age and single-metallicity stellar population (SSP). For this reason, they are very useful objects to check the capabilities and uncertainties of the stellar population models.

In this contribution, we estimate the age, metallicity and extinction of a sample of clusters in the Magellanic Clouds (Large and Small: LMC and SMC) and Milky Way by fitting their optical spectra. We use several sets of high-resolution spectral stellar population models to quantify the accuracies and uncertainties associated with the age, extinction and metallicity determinations of stellar clusters that results from variance in the model ingredients.

2. The sample and data

The sample analysed here consists of 27 SCs from Leonardi & Rose (2003; LR03), 20 of which belong to the LMC, three to SMC and four are Galactic globular clusters (GCs). The spectra cover 3650–4600 Å, with a final resolution of 5.7 Å.

With respect to the literature data, we have compiled the ages and metallicities derived by LR03 for the MC clusters, and for the GC the values obtained by Koleva *et al.* (2008). LR03 adopted ages based on CMDs and/or derived from a secondary indicator, the S parameter, based on the $U-B$ and $B-V$ integrated colors (calibrated by Girardi *et al.* 1995). For the clusters' metallicities, we have taken the values from Olszewski *et al.* (1991), obtained by measuring the CaII triplet in individual stars in the SCs.

3. Set of evolutionary models and method

The stellar library is one of the most important ingredients in evolutionary synthesis models. In the last few years, several groups have spent a lot of effort to build complete stellar libraries at high/intermediate spectral resolution at optical wavelengths. As a result, several sets of models are currently available for spectral synthesis (see, e.g., the review by González Delgado 2009). Here, we have used seven versions of the most up-to-date high-resolution spectral models. They are:

• **Galaxev and STELIB**: the models (Bruzual & Charlot 2003, BC03) are computed with the STELIB library (Le Borgne *et al.* 2003), Salpeter (BC94s, BC00s) and Chabrier (BC94c, BC00c) initial mass functions (IMFs), and the evolutionary tracks from the Padova group, (1994 isochrones: BC94s, BC94c; 2000 isochrones: BC00s, BC00c). STELIB contains 249 stellar spectra in the 3200–9500 Å range at a resolution of 3 Å.

• **SED@ and GRANADA**: the GRANADA stellar library (Martins *et al.* 2005) contains 1654 synthetic spectra with a sampling of 0.3 Å, covering the range 3000–7000 Å, for a wide range of effective temperatures (3000–55000 K), gravity (−0.5 to 5.0), and metallicity (solar, half and 0.2 solar). The library was implemented in the SED@ code (Cerviño & Luridiana 2006) to calculate SSP models (González Delgado *et al.* 2005) using the Padova 2000 isochrones and a Salpeter IMF (RG00s).

• **Vazdekis and MILES**: the MILES library (Sánchez–Blázquez *et al.* 2006) contains a thousand stars spanning a large range of stellar parameters covering the 3500–7500 Å range at a resolution of 2.3 Å. The library was incorporated into the code by Vazdekis *et al.* (2009). They compute SSP spectra with the Padova 2000 isochrones and a Salpeter IMF (V00s).

• **Galaxev and MILES/GRANADA**: these models are analogous to those in BC03. The STELIB library was replaced by a combination of the MILES and GRANADA stellar libraries (Charlot & Bruzual, in prep.) (CB94c).

To determine the stellar population properties for each cluster, we fit the entire observed spectrum using the STARLIGHT code (Cid Fernandes *et al.* 2005) and the set of models described above. The version of STARLIGHT used here provides the following results:

• The best fit: for each of the SSP model suites, the best SSP is obtained as that characterized by the minimum χ^2.

• Multi-population fit: the code combines several SSPs and provides for each set of a given metallicity a population vector that represents the fractional contribution of each SSP of a given age to the model flux at a given wavelength.

• Bayesian estimates: following the Bayes theorem, the posterior probability for age, metallicity and extinction is given for each set of models.

4. Results and conclusions

Table 1 presents the Bayesian age, metallicity and extinction for all clusters. Figures 1 and 2 show some of the results obtained. A more extensive discussion of the quality of the fits and a comparison with literature data is included in Cid Fernandes & González Delgado (2009) and González Delgado & Cid Fernandes (2009).

Table 1. The Bayesian age, metallicity and extinction for all clusters obtained with the models by Charlot & Bruzual (2009; CB94c: columns 2–4), González Delgado *et al.* (2005; RG00s: columns 6–7), Vazdekis *et al.* (2009; V00s: columns 8–10)

Name	log(age/yr)	log $Z/Z_{\odot}$	A_V	log(age/yr)	log $Z/Z_{\odot}$	A_V	log(age/yr)	log $Z/Z_{\odot}$	A_V
M 411	9.26±0.00	−0.68±0.02	0.17±0.04	9.13±0.07	−0.52±0.17	0.06±0.06	9.13±0.10	−0.68±0.20	0.41±0.11
M 416	9.53±0.07	−0.71±0.17	0.02±0.08	9.28±0.02	−0.38±0.00	0.01±0.01	9.90±0.01	−1.28±0.00	0.22±0.02
M 419	9.24±0.08	−0.64±0.17	0.15±0.04	9.06±0.03	−0.39±0.07	0.09±0.04	9.10±0.00	−0.67±0.06	0.46±0.05
NGC1651	9.40±0.02	−0.68±0.03	0.14±0.09	9.24±0.04	−0.50±0.15	0.04±0.04	9.58±0.19	−1.10±0.34	0.36±0.19
NGC1754	9.94±0.00	−1.68±0.00	0.29±0.05	9.22±0.03	−0.39±0.06	0.02±0.02	9.81±0.03	−1.28±0.04	0.02±0.02
NGC1783	9.11±0.17	−0.29±0.35	0.14±0.13	9.19±0.05	−0.49±0.15	0.03±0.03	9.25±0.06	−0.69±0.13	0.26±0.10
NGC1795	9.41±0.04	−0.68±0.05	0.11±0.09	9.25±0.04	−0.53±0.15	0.05±0.05	9.66±0.16	−1.21±0.25	0.38±0.17
NGC1806	9.33±0.01	−0.68±0.00	0.42±0.04	9.25±0.02	−0.65±0.08	0.10±0.05	9.26±0.02	−0.67±0.03	0.34±0.08
NGC1818	7.41±0.07	+0.35±0.16	0.48±0.19	7.39±0.02	−0.38±0.03	0.45±0.05	8.00±0.00	−1.68±0.00	0.02±0.02
NGC1831	8.61±0.07	−0.38±0.01	0.15±0.14	8.75±0.01	−0.68±0.00	0.03±0.03	8.73±0.03	−0.40±0.08	0.03±0.04
NGC1846	9.31±0.00	−0.68±0.01	0.25±0.07	9.24±0.03	−0.65±0.08	0.04±0.04	9.23±0.07	−0.70±0.13	0.27±0.12
NGC1866	8.01±0.00	+0.42±0.03	0.56±0.04	8.20±0.00	0.00±0.00	0.29±0.03	8.05±0.03	+0.19±0.07	0.53±0.07
NGC1978	9.54±0.02	−0.68±0.00	0.15±0.04	9.29±0.02	−0.38±0.02	0.03±0.03	9.42±0.11	−0.69±0.10	0.11±0.09
NGC2010	8.05±0.03	+0.41±0.07	0.13±0.08	8.20±0.01	0.00±0.00	0.03±0.03	8.14±0.04	+0.14±0.09	0.09±0.07
NGC2121	9.57±0.06	−0.68±0.01	0.17±0.10	9.33±0.05	−0.46±0.14	0.11±0.09	9.99±0.19	−1.23±0.17	0.55±0.19
NGC2133	8.38±0.03	+0.02±0.10	0.19±0.11	8.35±0.01	0.00±0.02	0.26±0.05	8.27±0.03	+0.11±0.10	0.33±0.09
NGC2134	8.87±0.22	−1.94±0.81	0.06±0.07	8.39±0.07	−0.24±0.18	0.14±0.11	8.41±0.08	−0.29±0.19	0.09±0.10
NGC2136	7.90±0.03	+0.41±0.07	0.60±0.08	7.95±0.05	−0.41±0.10	0.62±0.10	8.12±0.11	−0.47±0.19	0.40±0.19
NGC2203	9.39±0.02	−0.67±0.04	0.42±0.09	9.27±0.04	−0.55±0.15	0.12±0.07	9.35±0.05	−0.69±0.09	0.25±0.10
NGC2210	9.86±0.09	−1.69±0.09	0.20±0.05	9.20±0.02	−0.41±0.10	0.02±0.02	9.80±0.00	−1.68±0.00	0.10±0.03
NGC2213	9.37±0.02	−0.68±0.02	0.32±0.06	9.27±0.03	−0.65±0.08	0.04±0.03	9.30±0.00	−0.68±0.01	0.13±0.05
NGC2214	7.85±0.01	+0.42±0.00	0.15±0.04	7.90±0.01	−0.38±0.04	0.14±0.03	8.04±0.01	−0.38±0.01	0.02±0.02
NGC2249	8.96±0.00	−0.68±0.01	0.01±0.020	8.79±0.05	−0.68±0.00	0.06±0.14	8.84±0.02	−0.67±0.04	0.15±0.06
47 Tuc	9.98±0.03	−0.37±0.03	0.02±0.020	9.90±0.08	−0.60±0.13	0.06±0.06	9.98±0.05	−0.37±0.04	0.02±0.02
M 15	9.85±0.00	−1.68±0.01	0.05±0.030	9.14±0.27	−0.41±0.14	0.06±0.20	9.80±0.00	−1.68±0.00	0.02±0.02
M 79	9.93±0.02	−1.68±0.00	0.01±0.010	9.21±0.03	−0.43±0.11	0.02±0.02	9.84±0.11	−1.68±0.00	0.01±0.01
NGC1851	9.01±0.05	−0.68±0.00	0.01±0.010	9.31±0.02	−0.38±0.01	0.02±0.02	10.2±0.04	−1.28±0.00	0.11±0.04

Our main conclusions are:

• All models can be used to derive ages that are in good agreement both with each other and with literature data, but they are slightly older than those based on the *S*–CMD method (see Figure 1).

• There is less agreement between the models for metallicity and extinction. In particular, Galaxev/STELIB models underestimate the metallicity by 0.6 dex (see Figure 2), while the extinction is overestimated by 0.1 mag.

• The age–metallicity relation for LMC clusters is well reproduced by the spectral fitting method, except for Galaxev/STELIB models.

• Accuracies of about 0.1 dex in age and 0.3 dex in metallicity can be achieved with the GRANADA and MILES libraries as long as the models are not extrapolated beyond their expected range of validity.

Acknowledgements

We acknowledge support from a joint CNPq–CSIC bilateral collaboration grant, and from grant AYA2007-64712.

References

Bruzual, G. & Charlot, S. 2003, *MNRAS*, 344, 1000

Cerviño, M. & Luridiana, V. 2006, *A&A*, 413, 145

Cid Fernandes, R., Mateus, A., Sodré, L., Stasińska, G., & Gomes, J. M. 2005, *MNRAS*, 358, 363

Cid Fernandes, R. & González Delgado, R. M. 2009, *MNRAS*, submitted

Girardi, L., Chiosi, C., Bertelli, G., & Bressan, A. 1995, *A&A*, 298, 87

González Delgado, R. M., Cerviño, M., Martins, L. P., Leitherer, C., & Hauschildt, P. H. 2005, *MNRAS*, 357, 945

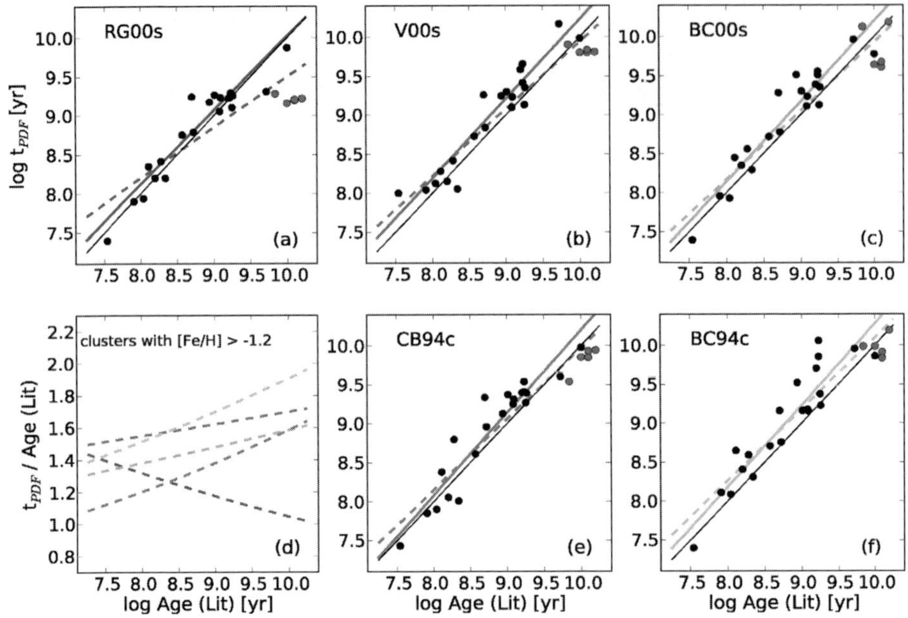

Figure 1. Panels *a, b, c, e* and *f*: Correlation between the age obtained with STARLIGHT and ages in the literature derived using the *S*–CMD calibration. Each panel represents the result for each set of models. Metal-poor clusters ($\log(Z/Z_\odot) \leqslant -1.2$) are marked with magenta circles. The black line is the one-to-one relationship, while the dashed line is a linear fit to all points and the solid line is a linear fit excluding the metal-poor clusters. The bottom left-hand panel shows the deviation of the linear fit obtained for each set of models with respect to the one-to-one relationship, expressed as the ratio between the spectral fit and the literature ages.

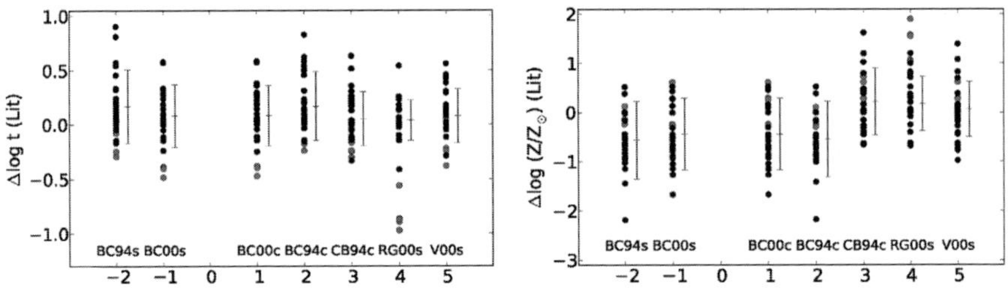

Figure 2. Logarithmic difference between the spectral-fit age *(left)* or metallicity *(right)* obtained for each set of models and the literature values. Magenta circles mark the metal-poor clusters ($\log(Z/Z_\odot) \leqslant -1.2$). Error bars mark the mean and standard deviation for each set of models.

González Delgado, R. M. 2009, in: E. Pérez, R. de Grijs & R. M. González–Delgado (eds.), *Young massive star clusters: initial conditions and environments*, *ApSS*, 324, 231

González Delgado, R. M. & Cid Fernandes, R. 2009, *MNRAS*, submitted

Le Borgne J. -F., Bruzual, G., Pelló, R., Lançon, A., Rocca–Volmerange, B., Sanahuja, B., Schaerer, D., Soubiran, C., & Vílchez–Gómez, R. 2003, *A&A*, 402, 433

Leonardi, A. J. & Rose J. A. 2003, *ApJ*, 126, 1811

Koleva, M., Prugniel, P., Ocvirk, P., Le Borgne, D., & Soubiran, C. 2008, *MNRAS*, 385, 1998

Martins, L. P., González Delgado, R. M., Leitherer, C., Cerviño, M., & Hauschildt, P. H. 2005, *MNRAS*, 358, 49

Olszewski, E. W., Schommer, R. A., Suntzeff, N. B., & Harris, H. C. 1991, *AJ*, 101, 515

Vazdekis, A., *et al.* 2009, *MNRAS*, submitted

Star clusters: basic galactic building blocks
Proceedings IAU Symposium No. 266, 2009
R. de Grijs & J. R. D. Lépine, eds.

doi:10.1017/S1743921309991529

Lithium abundances of main-sequence and subgiant stars in the globular cluster NGC 6397

J. I. González Hernández,[1,2]† P. Bonifacio,[1,2,3] E. Caffau,[1] M. Steffen,[4] H.-G. Ludwig,[1,2] N. Behara,[1,2] L. Sbordone,[1,2] R. Cayrel[1] and S. Zaggia[5]

[1]GEPI, Observatoire de Paris, CNRS, Université Paris Diderot, Place Jules Janssen, 92190 Meudon, France
email: `Jonay.Gonzalez-Hernandez@obspm.fr`

[2]Cosmological Impact of the First Stars (CIFIST) Marie Curie Excellence Team

[3]Istituto Nazionale di Astrofisica, Observatorio Astronomico di Trieste, Italy

[4]Astrophysikalisches Institut Potsdam, An der Sternwarte 16, D-14482 Potsdam, Germany

[5]INAF, Osservatorio Astronomico di Padova, Vicolo dell'Osservatorio 5, Padua 35122, Italy

Abstract. We present FLAMES/GIRAFFE spectroscopy obtained with the *Very Large Telescope (VLT)*. Using these observations, we have been able (for the first time) to observe the Li I doublet in the main-sequence (MS) stars of a globular cluster. We also observed Li in a sample of subgiant (SG) stars of the same $B - V$ colour. Our final sample is composed of 84 SG and 79 MS stars. In spite of the fact that SG and MS stars span the same temperature range, we find that the equivalent widths of the Li I doublet in SG stars are systematically greater than in MS stars, suggesting a higher Li content among SG stars. This is confirmed by our quantitative analysis, which makes use of both 1D and 3D model atmospheres. We find that SG stars show, on average, a higher Li abundance, by 0.1 dex, than MS stars. We also detect a positive slope of Li abundance with effective temperature: the higher the temperature the higher the Li abundance, both for SG and MS stars, although the slope is slightly steeper for MS stars. These results provide unambiguous evidence that the Li abundance changes with evolutionary state. The physical mechanisms that contribute to this are not yet clear, since none of the proposed models seem to describe accurately the observations. Whether such a mechanism can explain the cosmological lithium problem is still an open question.

Keywords. stars: abundances, stars: atmospheres, stars: fundamental parameters, stars: Population II, globular clusters: individual (NGC 6397)

---

1. Introduction

The primordial Li abundance inferred from fluctuations in the cosmic-microwave background measured by the *WMAP* satellite (Spergel *et al.* 2007; Cyburt *et al.* 2008) is $\log(\mathrm{Li/H}) + 12 = 2.72 \pm 0.06$ dex, approximately 0.3–0.5 dex higher than the Li abundance determined in metal-poor stars in the Galactic halo.

Many studies have tried to explain this difference: (i) Piau *et al.* (2006) propose that the first generation of stars, Population III stars, could have processed some fraction of the halo gas, thus lowering the lithium abundance, while (ii) other authors suggest that the primordial Li abundance has been uniformly depleted in the atmospheres of metal-poor dwarfs by some physical mechanism (e.g., turbulent diffusion, as in Richard

† Present address: Dpto. de Astrofísica y Ciencias de la Atmósfera, Facultad de Física, Universidad Complutense de Madrid, E-28040 Madrid, Spain, email: `jonay@astrax.fis.ucm.es`

et al. 2005 and Korn *et al.* 2006; gravitational waves, as in Charbonnel & Talon 2005, etc.). (iii) Finally, it has also been suggested that the standard Big Bang nucleosynthesis (SBBN) calculations should be revised, possibly by introducing new physics as in, e.g., Jedamzik (2004), Jedamzik (2006), Jittoh *et al.* (2008) and Hisano *et al.* (2009).

Here, we present the Li abundances of subgiant (SG) and main-sequence (MS) stars in NGC 6397. This study represents the first Li-abundance measurements in MS stars in a globular cluster.

2. Observations

Spectroscopic observations of the globular cluster NGC 6397 were carried out with the multi-object spectrograph FLAMES/GIRAFFE at the *VLT* in 2007 April, May, June and July, covering the spectral range $\lambda\lambda$6400–6800 Å at a resolving power of $\lambda/\delta\lambda \sim 17\,000$.

We selected subgiant and dwarf stars in the colour range $B - V = 0.6 \pm 0.03$ mag, which ensures that both sets of stars fall in a similar and narrow effective-temperature range (see figure 3 online in González Hernández *et al.* 2009).

The spectra were reduced with the ESO pipeline and later on treated within MIDAS. We corrected the spectra for sky lines and barycentric and radial velocities. We typically combine 17 spectra of dwarfs and 4 spectra of subgiants to achieve a similar signal-to-noise ratio. The mean radial velocity of the cluster stars is $v_{\rm r} = 18.5$ km s^{-1}.

3. Stellar parameters

We derived the effective temperature by fitting the observed Hα line profiles with synthetic profiles, using 3D hydrodynamical model atmospheres computed with the CO^{5}BOLD code (Freytag *et al.* 2002; Wedemeyer *et al.* 2004). The ability of 3D models to reproduce Balmer-line profiles has been shown by Behara *et al.* (2009), where the Hα profiles of the Sun and the metal-poor stars HD 84937, HD 74000 and HD 140283 were investigated. See Ludwig *et al.* (2009) for further details.

The effective temperatures of MS and SG stars were also derived adopting 1D ATLAS 9 model atmospheres (see Kurucz 2005) and using the same fitting procedure.

Fixed values for the surface gravity were adopted for both subgiant and dwarf stars, according to the values that best match the position of the stars on a 12 Gyr isochrone (Straniero *et al.* 1997). The adopted values were $\log(g/\text{cm s}^2) = 4.40$ and 3.85 for MS and SG stars, respectively.

4. Li abundances

We measure the equivalent width (EW) of the Li I 6708 Å line in SG and MS stars by fitting synthetic spectra of known EW to the observed Li profiles. The SG stars show on average larger EWs than MS stars. This is clearly seen in figure 1 of González Hernández *et al.* (2009), where the histograms of the EWs measured in SG and MS stars in this cluster are displayed. Although the $B-V$ colour is sensitive to surface gravity, this result was not expected a priori and suggests that subgiants in this cluster have actually higher Li abundances than dwarfs.

The Li abundances were derived using 3D model atmospheres. Line formation of Li was treated in nonlocal thermodynamical equilibrium (NLTE) using the same code and model atoms used in Cayrel *et al.* (2007). The analysis was also done with 1D model atmospheres, providing essentially the same picture, although $T_{\rm eff}$ in 1D yields lower

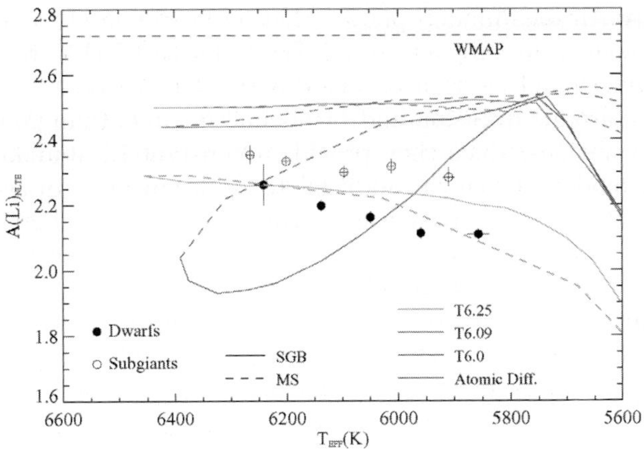

Figure 1. 1D NLTE (nonlocal thermodynamical equilibrium) Li abundances versus 1D effective temperatures of the observed stars and Li isochrones for different turbulent diffusion models. Black solid and open circles correspond to dwarf and subgiant stars, respectively. Different turbulent diffusion model isochrones of lithium are displayed for different levels of turbulence. In each isochrone, the dashed section of the line shows the Li abundance in dwarf stars, while the solid line shows that in subgiant stars. The horizontal black dashed line represents the cosmological Li abundance.

values (see figure 6 online in González Hernández *et al.* 2009). In the 1D case, we use the Carlsson *et al.* (1994) NLTE corrections.

5. Discussion and conclusions

In Figure 1 we display the derived 1D-NLTE Li abundances versus the 1D effective temperatures of dwarf and subgiant stars in the globular cluster NGC 6397 [see figure 2 in González Hernández *et al.* (2009) for a similar picture but with $T_{\rm eff}$ and Li abundances computed using 3D models]. The stars have been divided into five bins of effective temperature. The error bar in the Li abundance is the dispersion for each bin, divided by the square root of the number of stars in the bin. The Li abundance decreases with decreasing temperature. This lithium-abundance pattern is different from that found among field stars (Meléndez & Ramírez 2004; Bonifacio *et al.* 2007; González Hernández *et al.* 2008).

The difference in the Li abundance of dwarfs and subgiants is $\sim$ 0.14 dex if one computes the mean 1D A(Li) and the standard deviation of the mean for the two samples. For subgiants and dwarfs we find mean 1D Li abundances of 2.31 ± 0.01 and 2.17 ± 0.01 dex, respectively. The 3D models provide similar results, although slightly higher Li abundances. Lind *et al.* (2009) also find such differences between the mean Li abundances in MSs and SGs, but only by 0.03 dex, although this is still signicant at 1σ. However, this result is partly affected by the very narrow range of $T_{\rm eff}$ for MS stars derived by Lind *et al.* (2009) ($\sim$ 80 K) compared to the wide range ($\sim$ 450 K) for the SGs (see figure 7 online in González Hernández *et al.* 2009).

Our results imply that the Li surface abundance depends on the evolutionary status of the star. In Figure 1 we show the Li isochrones for different turbulent-diffusion models (Richard *et al.* 2005). These models have been shifted up by 0.14 dex in Li abundance to make the initial abundance of the models, log(Li/H) = 2.58 dex, coincide with the primordial Li abundance predicted from fluctuations of the microwave background measured by the *WMAP* satellite (Cyburt *et al.* 2008). The models assuming pure atomic diffusion, and, among those including turbulent mixing, T6.0 and T6.09, are ruled out by our observations. All such models predict that in dwarf stars Li should be either more abundant or the same as in subgiant stars. The only model that predicts a Li pattern which is qualitatively similar to that observed is the T6.25 model.

Models including atomic diffusion and tachocline mixing (Piau 2008) do not seem to reproduce our observations, since they provide a constant Li abundance up to 5500 K. The sophisticated models which, in addition to diffusion and rotation, also take into account the effects of internal gravity waves (Talon & Charbonnel 2004) seem to predict accurately the Li abundance pattern in solar-type stars, at solar metallicity (Charbonnel & Talon 2005), but models at low metallicity are still needed.

The cosmological lithium discrepancy still needs to be solved. Given that none of the existing models of Li evolution in stellar atmospheres match the observations, we hope our results will prompt further new theorctical investigations.

References

Behara, N. T., Ludwig, H.-G., Steffen, M., & Bonifacio, P. 2009, *AIP Conf. Ser.*, 1094, 784

Bonifacio, P., *et al.* 2007, *A&A*, 462, 851

Carlsson, M., Rutten, R. J., Bruls, J. H. M. J., & Shchukina, N. G. 1994, *A&A*, 288, 860

Cayrel, R., Steffen, M., Chand, H., Bonifacio, P., Spite, M., Spite, F., Petitjean, P., Ludwig, H.-G., & Caffau, E. 2007, *A&A* (Letters), 473, L37

Charbonnel, C. & Talon, S. 2005, *Science*, 309, 2189

Cyburt, R. H., Fields, B. D., & Olive, K. A. 2008, *J. Cosmol. Astro-Part. Phys.*, 11, 12

Freytag, B., Steffen, M., & Dorch, B. 2002, *AN*, 323, 213

González Hernández, J. I., *et al.* 2008, *A&A*, 480, 233

González Hernández, J. I., *et al.* 2009, *A&A* (Letters), 505, L13

Hisano, J., Kawasaki, M., Kohri, K., & Nakayama, K. 2009, *Phys. Rev. D*, 79, 063514

Jedamzik, K. 2004, *Phys. Rev. D*, 70, 083510

Jedamzik, K. 2006, *Phys. Rev. D*, 74, 103509

Jittoh, T., Kohri, K., Koike, M., Sato, J., Shimomura, T., & Yamanaka, M. 2008, *Phys. Rev. D*, 78, 055007

Kurucz, R. L. 2005, *Mem. Soc. Astron. Ital. Suppl.*, 8, 14

Korn, A. J., Grundahl, F., Richard, O., Barklem, P. S., Mashonkina, L., Collet, R., Piskunov, N., & Gustafsson, B. 2006, *Nature*, 442, 657

Lind, K., Primas, F., Charbonnel, C., Grundahl, F., & Asplund, M. 2009, *A&A*, 503, 545

Ludwig, H.-G., Behara, N. T., Steffen, M., & Bonifacio, P. 2009, *A&A* (Letters), 502, L1

Meléndez, J. & Ramírez, I. 2004, *ApJ* (Letters), 615, L33

Piau, L., Beers, T. C., Balsara, D. S., Sivarani, T., Truran, J. W., & Ferguson, J. W. 2006, *ApJ*, 653, 300

Piau, L. 2008, *ApJ*, 689, 1279

Richard, O., Michaud, G., & Richer, J. 2005, *ApJ*, 619, 538

Spergel, D. N., *et al.* 2007, *ApJS*, 170, 377

Straniero, O., Chieffi, A., & Limongi, M. 1997, *ApJ*, 490, 425

Talon, S. & Charbonnel, C. 2004, *A&A*, 418, 1051

Wedemeyer, S., Freytag, B., Steffen, M., Ludwig, H.-G., & Holweger, H. 2004, *A&A*, 414, 1121

Star clusters: basic galactic building blocks
Proceedings IAU Symposium No. 266, 2009
R. de Grijs & J. R. D. Lépine, eds.

doi:10.1017/S1743921309991530

Looking for building blocks of the Galactic halo: variable stars in the Fornax, Bootes I, Canes Venatici II dwarfs and in NGC 2419

Claudia Greco,[1] Gisella Clementini,[1] E. V. Held,[2] E. Poretti,[3] M. Catelan,[4] L. Federici,[1] M. Maio,[1] M. Gullieuszik,[2] V. Ripepi,[5] M. Dall'Ora,[5] L. Di Fabrizio,[6] K. Kinemuchi,[7,8] M. Di Crescienzo,[9] M. Marconi,[5] I. Musella,[5] B. Pritzl,[10] A. Rest,[11] N. De Lee[8] and H. Smith[12]

[1]INAF – Bologna, Italy, [2]INAF – Padova, Italy [3]INAF – Brera, Italy, [4]Pontifícia Universidad Católica, Chile, [5]INAF – Naples, Italy [6]INAF – La Palma, Italy [7]Universidad de Concepción, Chile, [8]University of Florida, USA, [9]INAF – Rome, Italy, [10]University of Wisconsin, Oshkosh, USA, [11]Harvard–Smithsonian Center for Astrophysics, USA, [12]Michigan State University, USA

Abstract. Λ cold-dark-matter hierarchical models of galaxy formation suggest that the halo of the Milky Way (MW) has been assembled, at least in part, through accretion of protogalactic fragments partially resembling the present-day dwarf spheroidal (dSph) satellites of the MW. Investigation of the stellar populations of the MW's globular clusters (GCs) and dSph companions can thus provide excellent tests to infer the dominant Galaxy-formation scenario, whether merger/accretion or cloud collapse. Pulsating variable stars offer a very powerful tool in this context, since variables of different types allow tracing the different stellar generations in a galaxy and to reconstruct the galaxy's star-formation history and assembly back to the first epochs of galaxy formation. In particular, the RR Lyrae stars, belonging to the old population ($t > 10$ Gyr), witnessed the epoch of halo formation, and thus hold a crucial role to identify the MW satellites that may have contributed to build up the Galactic halo. In the MW, most GCs with an RR Lyrae population sharply divide into two distinct groups (Oosterhoff types I and II) based on the mean periods and relative proportion of fundamental-mode (RRab) and first-overtone (RRc) RR Lyrae stars. On the other hand, the Galactic-halo field RR Lyrae stars show a dominance of Oosterhoff I properties. Here, we investigate the Oosterhoff properties of a number of different stellar systems, starting from relatively undisturbed dwarf galaxies (the Fornax dSph and its globular clusters), through distorted and tidally disrupting ones (the Bootes and Canes Venatici II dSphs), to possible final relics of the disruption process (the Galactic globular cluster NGC 2419). We are addressing the crucial question of whether the RR Lyrae pulsation properties in these systems conform to the Oosterhoff dichotomy characterizing the MW variables. If they do not, the Galaxy's halo cannot have been assembled by dSph-like protogalactic fragments resembling the present-day dSph companions of the MW. We have reduced and combined long time series from different telescopes, both ground- and space-based. Variable stars have been detected with image-subtraction techniques using the package ISIS2.1. Periods, amplitudes and Oosterhoff type for all variable stars, as well as color–magnitude diagrams of the stellar populations are discussed for each stellar cluster analyzed.

Keywords. stars: variables: other, Galaxy: halo, galaxies: dwarf, Local Group

---

The full poster (in pdf format) is available at
`http://www.astro.iag.usp.br/~iaus266/Posters/pGreco.pdf`.

Star clusters: basic galactic building blocks
Proceedings IAU Symposium No. 266, 2009
R. de Grijs & J. R. D. Lépine, eds.

doi:10.1017/S1743921309991542

Distribution of very young stellar clusters in grand-design spirals

Preben Grosbøl[1] and Horacio Dottori[2]

[1]European Southern Observatory, Garching, Germany

[2]Instituto de Física, Porto Alegre, Brazil

Abstract. Many grand-design spiral galaxies display strings of knots along their arms on K-band images. Near-infrared (NIR) spectra and broad-band colours of such knots have identified them as very young, massive stellar complexes. The low absorption in the NIR makes it possible to derive complete statistics of such complexes and thereby estimate the associated star-formation rate. We have obtained deep NIR maps of eight grand-design spirals using *VLT*/HAWK-I and identified massive complexes with ages of < 10 Myr using NIR colour–colour diagrams. The youngest, most massive complexes are well-aligned and concentrated in the arm regions of the grand-design galaxies with strong spiral perturbations. Their absolute magnitudes have a bright tail reaching almost $M_K = -16$ mag. Both the fraction of young to old sources and the ratio of diffuse to more compact objects suggest a dependence on the strength of the spiral pattern in the host galaxy.

Keywords. galaxies: spiral, galaxies: star clusters, infrared: galaxies

---

The full poster (in pdf format) is available at
`http://www.astro.iag.usp.br/~iaus266/Posters/pGrosbol.pdf`.

Star clusters: basic galactic building blocks
Proceedings IAU Symposium No. 266, 2009
R. de Grijs & J. R. D. Lépine, eds.

doi:10.1017/S1743921309991554

High-velocity runaway stars from three-body encounters

V. V. Gvaramadze,[1] A. Gualandris[2] and S. Portegies Zwart[3]

[1]Sternberg Astronomical Institute, Moscow State University, Universitetskij Pr. 13, Moscow 119992, Russia
email: vgvaram@mx.iki.rssi.ru

[2]Center for Computational Relativity and Gravitation, Rochester Institute of Technology, 85 Lomb Memorial Drive, Rochester NY 14623, USA
email: alessiag@astro.rit.edu

[3]Astronomical Institute 'Anton Pannekoek' and Section Computational Science, University of Amsterdam, Kruislaan 403, 1098 SJ Amsterdam, the Netherlands
email: spz@science.uva.nl

Abstract. We performed numerical simulations of dynamical encounters between hard, massive binaries and a very massive star (VMS; formed through runaway mergers of ordinary stars in the dense core of a young massive star cluster) to explore the hypothesis that this dynamical process could be responsible for the origin of high-velocity ($\geqslant 200 - 400$ km s^{-1}) early or late B-type stars. We estimated the typical velocities produced in encounters between very tight massive binaries and VMSs (of mass of $\geqslant 200$ M$_\odot$) and found that about 3 – 4% of all encounters produce velocities $\geqslant 400$ km s^{-1}, while in about 2% of encounters the escapers attain velocities exceeding the Milky Ways's escape velocity. We therefore argue that the origin of high-velocity ($\geqslant 200 - 400$ km s^{-1}) runaway stars and at least some so-called hypervelocity stars could be associated with dynamical encounters between the tightest massive binaries and VMSs formed in the cores of star clusters. We also simulated dynamical encounters between tight massive binaries and single ordinary 50 – 100 M$_\odot$ stars. We found that from 1 to $\simeq 4$% of these encounters can produce runaway stars with velocities of $\geqslant 300 - 400$ km s^{-1} (typical of the bound population of high-velocity halo B-type stars) and occasionally (in less than 1% of encounters) produce hypervelocity ($\geqslant 700$ km s^{-1}) late B-type escapers.

Keywords. stellar dynamics, methods: N-body simulations, binaries: general, stars: neutron

1. Introduction

The origin of high-velocity runaway stars can be attributed to two basic processes, (i) disruption of a tight, massive binary following the (asymmetric) supernova explosion of one of the binary components (Blaauw 1961) and (ii) dynamical three- or four-body encounters in dense stellar systems (Poveda *et al.* 1967). In the first process, the maximum velocity attained by runaway stars depends on the magnitude of the kick imparted to the stellar supernova remnant [either a neutron star (NS) or a black hole (BH)] and for reasonable values of this magnitude, the runaway velocity does not exceed ~ 200 km s^{-1} (e.g., Portegies Zwart 2000; cf. Gvaramadze 2009). In the second process, the ejection velocity could be higher. For example, the maximum velocity that a runaway star can attain in binary–binary encounters is equal to the escape velocity from the surface of the most massive star in the binaries and could be as large as ~ 1400 km s^{-1} (Leonard 1991).

The recent discovery of the so-called hypervelocity stars (HVSs; e.g., Brown *et al.* 2005)—ordinary stars moving with velocities exceeding the Milky Way's escape velocity—attracted attention to dynamical processes involving the supermassive BH in the Galactic

Centre (e.g., Gualandris *et al.* 2005; Baumgardt *et al.* 2006). These processes [originally proposed by Hills (1988) and Yu & Tremaine (2003)] can result in ejection velocities of several 1000 $\mathrm{km\,s^{-1}}$. Similar processes, but acting in the cores of young massive star clusters (YMSCs) and involving dynamical encounters with intermediate-mass ($\sim 100 - 1000\ \mathrm{M_\odot}$) BHs (IMBHs), were considered by Gvaramadze *et al.* (2008) to explain the origin of extremely-high-velocity ($\sim 1000\ \mathrm{km\,s^{-1}}$) NSs (e.g., Chatterjee *et al.* 2005) and HVSs. Gualandris & Portegies Zwart (2007) proposed that exchange encounters between hard binaries and an IMBH formed in the core of a YMSC in the Large Magellanic Cloud could be responsible for the origin of the HVS HE 0437−5439. Strong support for the possibility that at least some HVSs originate in star clusters rather than in the Galactic Centre comes from the proper-motion measurements of the HVS HD 271791, which constrain the birth place of this star to the outer parts of the Galactic disc (Heber *et al.* 2008).

2. High-velocity runaway stars from three-body encounters

In this paper, we explore the hypothesis (Gvaramadze 2007) that some high-velocity runaway stars could attain their peculiar velocities in the course of strong dynamical encounters between hard, massive binaries and a very massive ($\gtrsim 100 - 150\ \mathrm{M_\odot}$) star (VMS), formed in the core of a YMSC through collisions and mergers of ordinary massive stars (e.g., Portegies Zwart & McMillan 2002). In this process, one of the binary components is replaced by the VMS, while the second is ejected (a so-called exchange encounter), sometimes with a high velocity. Our goal is to check whether or not this process can produce early B-type stars (the progenitors of NSs) with velocities of $\gtrsim 200 - 400\ \mathrm{km\,s^{-1}}$ (typical of pulsars) and $3 - 4\ \mathrm{M_\odot}$ stars with velocities of $\gtrsim 300 - 400\ \mathrm{km\,s^{-1}}$ (typical of some late B-type halo stars).

In our study, we proceed from the similarity between encounters involving a VMS and an IMBH (the latter process is already known to be able to produce high-velocity runaways; Gvaramadze *et al.* 2008) and the fact that the radii of VMSs are smaller than the tidal radii of the intruders (the massive binaries), so that tidal breakup and ejection can occur before the binary components merge with the VMS (Gvaramadze 2007; Gvaramadze *et al.* 2009). Our study is motivated by the recent evolutionary models of VMSs (Belkus *et al.* 2007; Yungelson *et al.* 2008), which suggest that VMSs can lose most of their mass through copious winds and leave behind IMBHs with masses of $\lesssim 70\ \mathrm{M_\odot}$, which are not high enough to contribute significantly to the production of high-velocity runaway stars (see Gvaramadze *et al.* 2008). We therefore explore the possibility that a VMS could produce high-velocity escapers (either early or late B-type stars) before it finishes its life in a supernova and forms a BH. To check this possibility, we performed numerical simulations of three-body exchange encounters using the SIGMA3 package, which is part of the STARLAB software environment (McMillan & Hut 1996; Portegies Zwart *et al.* 2001). For a detailed description of the simulations, see Gvaramadze *et al.* (2009).

First, we focus on exchange encounters producing high-velocity early B-type stars and simulate encounters between a binary consisting of two main-sequence stars with masses $m_1 = 40\ \mathrm{M_\odot}$ and $m_2 = 8\ \mathrm{M_\odot}$, and a massive compact body, either a VMS or an ordinary star of mass of $50 - 100\ \mathrm{M_\odot}$ (the most massive ordinary stars formed in clusters with mass of $\simeq 10^3 - 10^4\ \mathrm{M_\odot}$; e.g., Weidner *et al.* 2009). To maximize the ejection speed, we assume that the binary system is very tight, e.g., formed by tidal capture. In this case, $a \simeq 3\,r_1 \simeq 0.15$ AU, where r_1 is the radius of the primary star (see Gvaramadze *et al.* 2009). Figure 1 (left panel) shows the velocity distribution for 8 $\mathrm{M_\odot}$

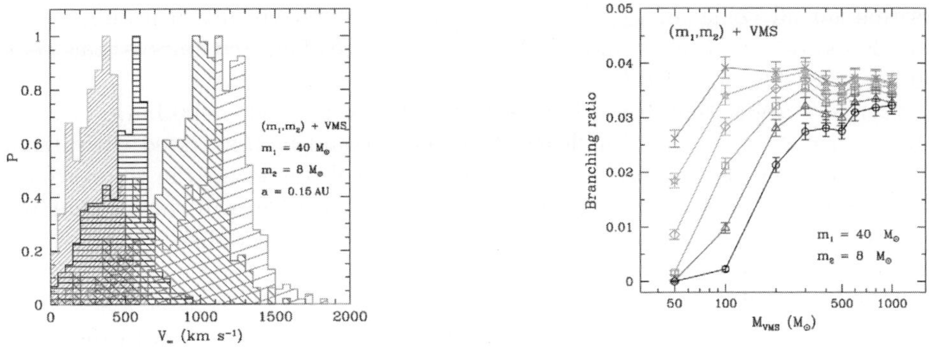

Figure 1. *(left)* Velocity distributions at infinity for escaping (8 $M_\odot$) stars in encounters between a binary consisting of a primary star with mass $m_1 = 40\ M_\odot$ and a secondary star with mass $m_2 = 8\ M_\odot$, and a single VMS of mass $M_{\rm VMS} = 50\ M_\odot, 100\ M_\odot, 500\ M_\odot$ and $1000\ M_\odot$ (from left to right). The binary semi-major axis is $a = 0.15$ AU. *(right)* Probability of exchange encounters between a $(40, 8)\ M_\odot$ binary (with $a = 0.15$ AU) and a single VMS resulting in ejection of the 8 $M_\odot$ star with a velocity from 200 to 700 $\rm km\,s^{-1}$ (top to bottom).

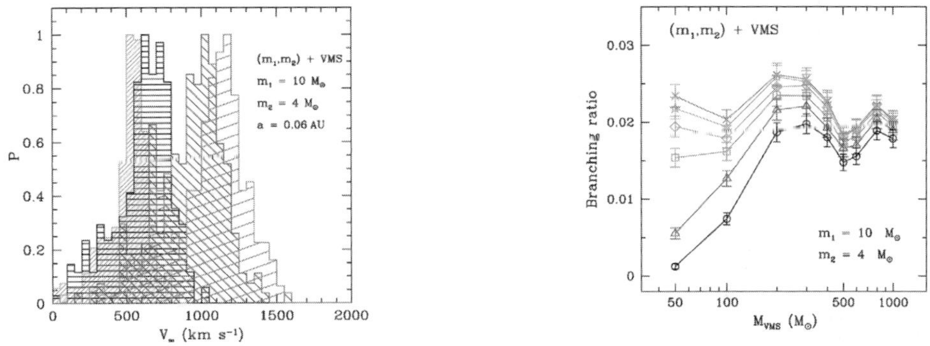

Figure 2. *(left)* Velocity distributions at infinity for escaping stars in encounters between a binary consisting of a primary star with mass $m_1 = 10\ M_\odot$ and a secondary star with mass $m_2 = 4\ M_\odot$, and a single VMS of mass $M_{\rm VMS} = 50\ M_\odot, 100\ M_\odot, 500\ M_\odot$ and $1000\ M_\odot$ (from left to right). The binary semi-major axis is $a = 0.06$ AU. *(right)* Probability of exchange encounters between a $(10, 4)\ M_\odot$ binary (with $a = 0.06$ AU) and a single VMS resulting in ejection of the 4 $M_\odot$ star with a velocity from 200 to 700 $\rm km\,s^{-1}$ (top to bottom).

escapers. As expected, the more massive VMSs are more likely to eject stars with high velocities. In the right panel of Figure 1 we show the probability of exchange encounters resulting in ejection of the 8 $M_\odot$ binary component with a velocity from 200 to 700 $\rm km\,s^{-1}$ (top to bottom). For $M_{\rm VMS} \geqslant 100\ M_\odot$, about 3% of all encounters produce runaways with peculiar velocities $\geqslant$ 400 $\rm km\,s^{-1}$. It can be seen that even an ordinary star of mass of 50 $M_\odot$ can occasionally (in $\sim$ 1% of encounters) produce an escape velocity $\geqslant$ 400 $\rm km\,s^{-1}$. To produce escapers with velocities typical of HVSs ($V_\infty \geqslant 700\ \rm km\,s^{-1}$), a VMS of several hundred solar masses is required. For a $200 - 300\ M_\odot$ VMS, $\geqslant$ 2% of all encounters result in an escape velocity of $\geqslant$ 700 $\rm km\,s^{-1}$. This fraction gradually increases to $\geqslant$ 3% for the more massive VMSs.

Next, we consider exchange encounters producing high-velocity late B-type stars and simulate encounters between a very tight binary ($m_1 = 10\ M_\odot, m_2 = 4\ M_\odot$ and $a \simeq 3r_1 \simeq 0.06$ AU) and a (very) massive star of mass of $50 - 1000\ M_\odot$. The velocity distributions for 4 $M_\odot$ escapers are shown in Figure 2 (left panel) for four different values of the VMS mass: $M_{\rm VMS} = 50, 100, 500$ and $1000\ M_\odot$. The right-hand panel of Figure 2 shows that for all values of $M_{\rm VMS}$ about 2% of encounters result in peculiar velocities $\geqslant 300-400$ km s^{-1} and that about the same percentage of ejected stars attains velocities $\geqslant 700$ km s^{-1} if $M_{\rm VMS} \geqslant 200$ M$_{\odot}$.

We therefore argue that the origin of high-velocity ($\geqslant 200 - 400$ km s^{-1}) early and late B-type runaway stars and at least some HVSs could be associated with dynamical encounters between the tightest massive binaries and VMSs formed in the cores of YMSCs. Future proper-motion measurements of HVSs with *Gaia* will reveal what fraction of these extremely-high-velocity stars originated in the Galactic disk.

Acknowledgements

We are grateful to L. R. Yungelson for useful discussions. VVG acknowledges the Russian Foundation for Basic Research and the International Astronomical Union for travel grants and the Deutsche Forschungsgemeinschaft for partial financial support. AG is supported by NASA grant NNX07AH15G. SPZ acknowledges support from the Netherlands Organization for Scientific Research (NWO; grant 643.200.503) and the Netherlands Research School for Astronomy (NOVA).

References

Baumgardt, H., Gualandris, A., & Portegies Zwart, S. 2006, *MNRAS*, 372, 174
Belkus, H., Van Bever, J., & Vanbeveren, D. 2007, *ApJ*, 659, 1576
Blaauw, A. 1961, *Bull. Astron. Inst. Neth.*, 15, 265
Brown, W. R., Geller, M. J., Kenyon, S. J., & Kurtz, M. J. 2005, *ApJ* (Letters), 622, L33
Chatterjee, S., Vlemmings, W. H. T., Brisken, W. F., Lazio, T. J. W., Cordes, J. M., Goss, W. M., Thorsett, S. E., Fomalont, E. B., Lyne, A. G., & Kramer, M. 2005, *ApJ* (Letters), 630, L61
Gualandris, A. & Portegies Zwart, S. 2007, *MNRAS* (Letters), 376, L29
Gualandris, A., Portegies Zwart, S., & Sipior, M. S. 2005, *MNRAS*, 363, 223
Gvaramadze, V. V. 2007, *A&A* (Letters), 470, L9
Gvaramadze, V. V. 2009, *MNRAS* (Letters), 395, L85
Gvaramadze, V. V., Gualandris, A., & Portegies Zwart, S. 2008, *MNRAS*, 385, 929
Gvaramadze, V. V., Gualandris, A., & Portegies Zwart, S. 2009, *MNRAS*, 396, 570
Heber, U., Edelmann, H., Napiwotzki, R., Altmann, M., & Scholz, R.-D. 2008, *A&A* (Letters), 483, L21
Hills, J. G. 1988, *Nature*, 331, 687
Leonard, P. J. T. 1991, *AJ*, 101, 562
McMillan, S. L. W. & Hut P. 1996, *ApJ*, 467, 348
Portegies Zwart, S. F. 2000, *ApJ*, 544, 437
Portegies Zwart, S. F. & McMillan, S. L. W. 2002, *ApJ*, 576, 899
Portegies Zwart, S. F., McMillan, S. L. W., Hut, P., & Makino, J. 2001, *MNRAS*, 321, 199
Poveda, A., Ruiz, J., & Allen, C. 1967, *Bol. Obs. Tonantzintla Tacubaya*, 4, 86
Weidner, C., Kroupa, P., & Bonnell, I. 2009, *MNRAS*, in press (arXiv:0909.1555)
Yu, Q. & Tremaine, S. 2003, *ApJ*, 599, 1129
Yungelson, L. R., van den Heuvel, E. P. J., Vink, J. S., Portegies Zwart, S. F., & de Koter, A. 2008, *A&A*, 477, 223

Star clusters: basic galactic building blocks
Proceedings IAU Symposium No. 266, 2009
R. de Grijs & J. R. D. Lépine, eds.

doi:10.1017/S1743921309991566

Galactic consequences of clustered star formation

M. R. Haas[1] and P. Anders[2]

[1]Leiden Observatory, Leiden University, Postbus 9513, 2300 RA, Leiden, the Netherlands
email: haas@strw.leidenuniv.nl

[2]Sterrenkundig Instituut, Utrecht University, Princetonplein 5, 3584 CC Utrecht, the Netherlands
email: p.anders@uu.nl

Abstract. If all stars form in clusters and both stars and clusters follow a power-law distribution which favours the creation of low-mass objects, the numerous low-mass clusters will be deficient in high-mass stars. Therefore, the stellar mass function integrated over the entire galaxy (the integrated galactic initial mass function; IGIMF) will be steeper at the high-mass end than the underlying stellar IMF. We show how the steepness of the IGIMF depends on the sampling method and on the assumptions made regarding the star cluster mass function. We also investigate the O-star content, integrated photometry and chemical enrichment of galaxies that result from several IGIMFs compared to more standard IMFs.

Keywords. stars: luminosity function, mass function, galaxies: stellar content, galaxies: abundances, galaxies: fundamental parameters, methods: statistical

1. Introduction

In recent years, a series of papers (Kroupa & Weidner 2003; Weidner & Kroupa 2004; Weidner & Kroupa 2005; Weidner & Kroupa 2006: WK06) have proposed that the stellar content of an entire galaxy may not be well described by the same initial mass function (IMF) that describes the distribution of stellar masses in the star clusters, where these stars form. The reason for this is that star clusters also form with a cluster mass function (CMF), which is a power law with an index of ~ -2. If the lower cluster mass limit is very low (in the mass range that is also covered by single stars, i.e., below 100 $M_\odot$, and under the assumption that the most massive star in a cluster cannot be more massive than its host cluster, M_{cl}), the low-mass clusters will be deficient in high-mass stars. Therefore, if the stellar content of all clusters is added up, resulting in the integrated galactic IMF (IGIMF), the distribution of stellar masses may be steeper at the high-mass end, depending on the exact shape of the CMF.

The exact form of the IGIMF could have profound implications for the integrated properties of galaxies. High-mass stars are important, both in terms of their energy output (massive stars are bright and end their lifes in supernova explosions) and in terms of chemical enrichment. Assuming an IGIMF instead of an IMF, the supernova rate and chemical enrichment of galaxies are studied by Goodwin & Pagel (2005) and the relation between Hα and ultraviolet luminosity in star-forming disks by Pflamm–Altenburg & Kroupa (2008). Recchi *et al.* (2009) investigates [α/ Fe] abundance ratios for IGIMFs.

In most of these studies, the assumed CMF and sampling method are kept the same. For sampling they use the 'sorted sampling' method, which was proposed by WK06 as they found it to best fit the relation between the most massive star in a cluster and the cluster mass. The CMF was assumed to be a pure power law with an index of −2.2 and a lower cluster mass limit of 5 $M_\odot$.

In this contribution, we vary the sampling method and CMF parameters and investigate the resulting IGIMFs. We will also estimate the observability of the effects of sampling-method and CMF variations through number counts of O-type stars, as will be observed by *Gaia*, the integrated photometry of distant galaxies and the chemical enrichment of galaxies (in terms of their total gas-phase metallicity).

2. Constructing IGIMFs

As default choice for the CMF we will use the same CMF as used in the past,

$$dN/dM_{\rm cl} \propto M_{\rm cl}^{-2.2}, \tag{2.1}$$

with a lower mass limit of 5 $M_\odot$. For computational simplicity we use a Salpeter IMF for stars between 0.1 and 100 $M_\odot$.

We test several sampling methods (*italic phrases* refer to the legend in Figure 1):

1. Random sampling until the cluster mass is reached. As this will never reach exactly the cluster mass aimed for, we either include or exclude the last star (*stop after* versus *stop before*), only exclude it if that brings the total mass in stars closer to the cluster mass aimed for than by including it (*stop nearest*) or exclude the last star (which first exceeded the cluster mass) with a 50% probability (*stop 50/50*).
2. Sampling a specific number of stars, estimated from the total cluster mass and average mass in the IMF, with the maximum stellar mass either limited to the cluster mass or unlimited (*number* and *number unlimited*).
3. The *sorted sampling* method of WK06.
4. *Analytical sampling*, in which we assume that the IMF in a cluster is always sampled analytically, thus excluding stochastic effects.

For a more detailed description, see Haas & Anders (2009). We use a Monte Carlo approach to reduce the numerical scatter to less than the difference between IGIMFs. We investigate the dependence of the results on the underlying IMF (we test for a Kroupa-type IMF) and the results are virtually indistinguishable.

As the lack of high-mass stars is mainly caused by the low-mass clusters, the behaviour of the CMF at the low-mass end is very important. The default choice is an extrapolation of the CMF observed at higher masses ($M_{\rm cl} \gtrsim 10^3$ $M_\odot$) down to the mass of the smallest observed star-forming clumps in the Taurus–Auriga region. We therefore test, in addition to the default choice given above, also lower limits of 1 and 50 $M_\odot$ for the clusters and power-law indices of −1.8, −3.2 and −4.2 (effectively changing the number of low-mass with respect to high-mass clusters).

3. Resulting IGIMFs

In Figure 1 we show the resulting IGIMFs from different sampling methods (left) and CMFs (right). The value of the IGIMF is divided by the value of the underlying IMF to enhance the visibility of the differences. Dotted lines indicate the slopes the different IGIMFs would have had.

It can be clearly seen that the different sampling methods give different IGIMFs. The recovered CMFs after filling the clusters with stars are indistinguishable from the input CMF. Depending on the method, the steep end has a power-law indix of about −2.6, compared to the input value of −2.35. The 'sorted sampling' and 'analytical sampling' methods yield even slightly steeper slopes and deviate strongly from a power-law shape.

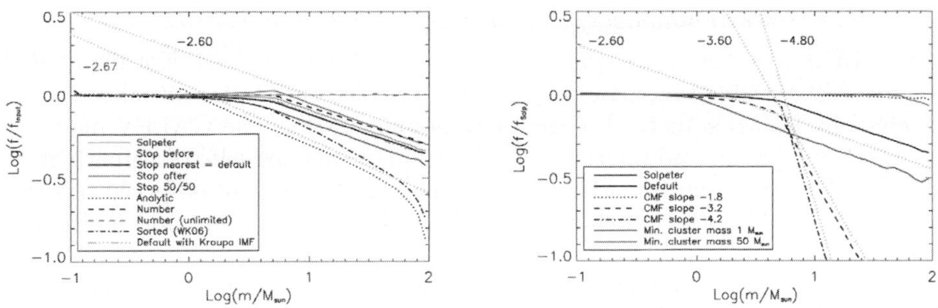

Figure 1. Resulting IGIMFs, normalized to the input IMF for different sampling methods *(left)* and CMFs *(right)*. The indications correspond to the italicized descriptions in the text. The dotted lines with indices indicate the slope the respective line would have in an IGIMF.

From the Monte Carlo realisations using different CMFs, it can be seen that the slope of the CMF determines the high-mass slope of the IGIMF, while the minimum cluster mass sets the stellar mass at which the steepening sets in. If clusters do not typically go down in mass all the way to the mass range of stars, sampling effects in the IGIMF will be negligible. The low-mass end of the CMF is poorly constrained. The IGIMF could be obtained observationally, giving an indirect measure of the low-mass end of the CMF (assuming a sampling method and an underlying IMF).

4. Galaxy properties from IGIMFs

Using the IGIMFs as IMFs in galaxy evolution models, we can investigate the integrated properties of galaxies if their total stellar initial mass distribution is given by an IGIMF rather than a more standard IMF. The high-mass end of the stellar-mass distribution is important for galaxies, as their light in many passbands is dominated by the most massive stars for a wide range of star-formation histories. Their chemical composition depends on the metals expelled by the massive stars and the interstellar medium is strongly affected by the energy output of the massive stars exploding as supernovae.

4.1. *The number of O stars in the Galaxy*

As the most pronounced effect occurs for the most massive stars, we first estimate the difference in massive-star content of the galaxies. Specifically, we estimate the number of O stars (with masses > 17 $M_\odot$) that *Gaia* will be able to observe. Assuming that *GAIA* will observe 10% of all O stars in the Galaxy, that they live for 10 Myr and that in the past 10 Myr the average star-formation rate of the Galaxy was 1 $M_\odot$ yr^{-1}, *GAIA* will be able to rule out several of the very extreme IGIMFs (resulting from, e.g., steep CMFs and sorted sampling) with very high significance. Judging the several different sampling methods is difficult, as the differences are on the order of 1σ, assuming purely Poissonian errors on the number counts.

If our assumptions regarding the observed fraction of O stars, the star-formation rate and/or the lifetimes of O stars are inaccurate, the resulting number of O stars varies linearly with either. The errors on the numbers scale with the square root of the numbers. As the numbers we calculate are on the order of 1000 to 2000, and the differences an order of magnitude smaller, the combined effect of our assumptions may be off by a factor of a few and the differences between all IGIMFs will still be of more than 2σ significance.

4.2. *Photometry and chemical evolution of galaxies*

Using the GALEV evolutionary synthesis models (Bicker *et al.* 2004; Kotulla *et al.* 2009), we can assess observable properties of galaxies, in which the stellar initial mass distribution is given by IGIMFs, instead of the more popular IMFs. The GALEV models follow the chemical enrichment and photometric evolution of galaxies self-consistently in closed-box models of galaxy evolution. They can match a wide range of observed properties of galaxies of all Hubble types, such as stellar masses, star formation and multi-passband photometry.

The ingredients of the models are star-formation histories, which depend on the gas mass present (no in- or outflows are assumed, nor are they necessary), and Hubble type, spectral synthesis, and yield tables. Subsequent generations of stars are formed from the gas enriched by earlier generations.

For all Hubble types, the difference in integrated photometry in any band is smaller than the observed galaxy-to-galaxy scatter in the HyperLeda database† (Paturel *et al.* 2003). Therefore, integrated photometry of galaxies will not be a discriminant between IMFs and/or IGIMFs.

Many heavy elements are mainly produced in supernova explosions of Type II (including Ib,c) and in the winds of massive asymptotic giant branch stars. With a deficiency in massive stars, as expected from the IGIMF, less production of metals is to be expected. The GALEV models do not follow the chemical enrichment on an element-by-element basis, but rather follow the total metallicity. Element abundances can then be obtained by assuming solar abundance ratios.

With IGIMFs as determined above, metallicities of galaxies vary by half a dex at a given gas-mass fraction. For the IGIMFs resulting from extremely steep CMFs, the difference can be up to two orders of magnitude. At given gas mass fractions, the HyperLeda database gives a scatter of about half a dex as well. Therefore, the most extreme CMFs can be ruled out by the observed gas-phase metallicities. To differentiate between several IGIMFs, more precise measurements of metallicities are necessary. The difference between sampling methods is comparable to, or slightly larger than, the spread that results from using individual metallicity measurements for the same galaxy. It is thus a possible discriminant between IGIMFs, provided that absolute determinations of gas-phase metallicities are reliable.

References

Bicker, J., Fritze–v. Alvensleben, U., Möller, C. S., & Fricke, K. J. 2004, *A&A*, 413, 37

Goodwin, S. P. & Pagel, B. E. J. 2005, *MNRAS*, 359, 707

Haas, M. R. & Anders, P. 2009, *A&A*, submitted

Kotulla, R., Fritze, U., Weilbacher, P., & Anders, P. 2009, *MNRAS*, 396, 462

Kroupa, P. 2001, *MNRAS*, 322, 231

Kroupa, P. & Weidner, C. 2003, *ApJ*, 598, 1076

Paturel, G., Petit, C., Prugniel, P., Theureau, G., Rousseau, J., Brouty, M., Dubois, P., & Cambrésy, L. 2003, *A&A*, 412, 45

Pflamm–Altenburg, J. & Kroupa, P. 2008, *Nature*, 455, 641

Recchi, S., Calura, F., & Kroupa, P. 2009, *A&A*, 499, 711

Salpeter, E. E. 1955, *ApJ*, 121, 161

Weidner, C. & Kroupa, P. 2004, *MNRAS*, 348, 187

Weidner, C. & Kroupa, P. 2005, *ApJ*, 625, 754

Weidner, C. & Kroupa, P. 2006, *MNRAS*, 365, 1333

† `http://leda.univ-lyon1.fr/`

Star clusters: basic galactic building blocks
Proceedings IAU Symposium No. 266, 2009
R. de Grijs & J. R. D. Lépine, eds.

doi:10.1017/S1743921309991578

A high-spectral-resolution study of abundances and isotopic composition of BHB stars in NGC 6397 and NGC 6752

S. Hubrig,[1] F. Castelli,[2] G. De Silva,[3] J. F. González,[4] Y. Momany,[3] M. Netopil,[5] S. Moehler[3]

[1]Astrophysikalisches Institut Potsdam, An der Sternwarte 16, 14482 Potsdam, Germany

[2]Instituto Nazionale di Astrofisica–Osservatorio Astronomico di Trieste, via Tiepolo 11, 34131 Trieste, Italy

[3]European Southern Observatory, Karl-Schwarzschild-Str. 2, 85748 Garching, Germany

[4]Complejo Astronómico El Leoncito, Casilla 467, 5400 San Juan, Argentina

[5]Institut für Astronomie der Universität Wien, Türkenschanzstr. 17, 1180 Vienna, Austria

Abstract. Large abundance anomalies have previously been detected in horizontal-branch B-type stars. We present the first high-resolution study of isotopic anomalies and chemical abundances in six horizontal-branch B-type stars in the globular clusters NGC 6397 and NGC 6752, carried out with UVES on the *VLT* and compare them to those observed in chemically peculiar main-sequence stars.

Keywords. diffusion, stars: abundances, stars: chemically peculiar, stars: horizontal-branch, stars: fundamental parameters, stars: Population II, globular clusters: individual (NGC 6397, NGC 6752)

The full poster (in pdf format) is available at
`http://www.astro.iag.usp.br/~iaus266/Posters/pHubrig.pdf`.

Star clusters: basic galactic building blocks
Proceedings IAU Symposium No. 266, 2009
R. de Grijs & J. R. D. Lépine, eds.

doi:10.1017/S174392130999158X

Proper-motion measurements in the Quintuplet cluster

Benjamin Hußmann,[1] **Andrea Stolte**[1] **and Wolfgang Brandner**[2]

[1]I. Physikalisches Institut, University of Cologne, Zülpicher Str. 77, 50937 Köln, Germany

[2]Max-Planck-Institut für Astronomie, Königstuhl 17, 69117 Heidelberg, Germany

Abstract. The three young, massive star clusters found in the Galactic Centre region (Young Nuclear Cluster, the Arches and Quintuplet clusters) are among the six most massive star clusters in our Galaxy, with masses similar to low-mass, extragalactic starburst clusters. The conditions for star formation in this region are extreme and likely comparable to those found in the HII regions in starburst galaxies and tidal-interaction zones of mergers. As the inner Galactic star clusters can be resolved, they can serve as templates for extragalactic starburst clusters. With knowledge of the spectral types, masses and ages of the individual stars, their stellar population can be studied in detail, allowing derivation of their present-day mass function (PDMF). The Quintuplet cluster, with an age of about 4 Myr, is the most extended of the three clusters and also displays a lower spatial density. To determine its mass function correctly, the distinction between cluster and field stars is therefore of particular importance. We present the first determination of a proper-motion-membership sample for the Quintuplet cluster. By comparing two high-precision astrometric *VLT*/NACO data sets with a time baseline of 5 years, the displacement of the Quintuplet cluster relative to the field population was measured and a selection of the proper-motion cluster members could be established, from which the PDMF can be derived.

Keywords. Hertzsprung–Russell diagram, open clusters and associations: individual (Quintuplet)

The full poster (in pdf format) is available at
`http://www.astro.iag.usp.br/~iaus266/Posters/pHussmann.pdf`.

Star clusters: basic galactic building blocks
Proceedings IAU Symposium No. 266, 2009
R. de Grijs & J. R. D. Lépine, eds.

doi:10.1017/S1743921309991591

Tracing star cluster formation in the interacting galaxy M51

Narae Hwang[1,2] and Myung Gyoon Lee[2]

[1]National Astronomical Observatory of Japan, 2-21-1 Osawa, Mitaka, Tokyo 181-8588, Japan
email: narae.hwang@nao.ac.jp

[2]Department of Physics and Astronomy, Seoul National University, 599 Gwanak-ro, Gwanak-gu, Seoul 151-747, Republic of Korea
email: mglee@astro.snu.ac.kr

Abstract. We present a study of star clusters in the interacting galaxy M51 using a star cluster catalog that includes about 3600 star clusters with $m_{\mathrm{F555W}} < 23$ mag, compiled by Hwang & Lee (2008). Combined with m_{F336W}-band imaging data taken with the *Hubble Space Telescope (HST)*'s WFPC2 camera, we have derived the ages and masses of star clusters in M51 using theoretical population synthesis models. The cluster age distribution displays multiple peaks that correspond to the epochs of dynamical encounters predicted by theoretical model studies and the cluster-formation rate appears to increase around the same epochs.

Keywords. galaxies: individual (M 51, NGC 5194, NGC 5195), galaxies: star clusters, galaxies: evolution, galaxies: interactions

1. Introduction

Star clusters in M51, an interacting galaxy system composed of the Sbc-type spiral galaxy NGC 5194 and the SB0 galaxy NGC 5195, provide a unique opportunity to understand star cluster formation and evolution in the context of dynamical interactions of their host galaxies. There are some theoretical (Toomre & Toomre 1972; Salo & Laurikainen 2000) and observational (Durrell *et al.* 2003) studies suggesting that the two galaxies may have experienced a single or multiple encounters a few hundred Myr ago. M51 is located at a distance of 8.4 Mpc (Feldmeier *et al.* 1997) so that, with the resolving power of *HST*, it is possible to marginally resolve individual star clusters.

We have carried out star cluster surveys using the deep and wide-field image data of M51 taken with *HST*/ACS and have compiled a catalog of about 3600 star clusters in M51 (Hwang & Lee 2008). The same image data have been used in several studies of M51 star clusters, such as of faint fuzzy clusters (Hwang & Lee 2006), the cluster size distribution (Scheepmaker *et al.* 2007), the cluster luminosity function (Haas *et al.* 2008) and the radial distribution of cluster formation (Scheepmaker *et al.* 2009). We have also analyzed m_{F336W}-band image data of M51 taken with *HST*/WFPC2 and have combined the m_{F336W}-band photometry with the photometric catalog of Hwang & Lee (2008) to estimate the ages and masses of star clusters. The theoretical population synthesis models of Bruzual & Charlot (2003) were used for age and mass estimation. The total number of star clusters with derived ages and masses is about 2000. For further information, see Hwang & Lee (2009).

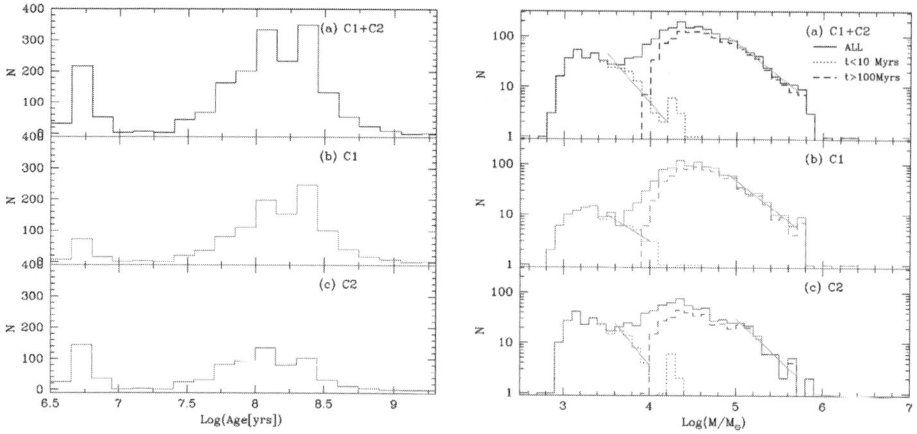

Figure 1. Age *(left)* and mass distribution *(right)* of the M51 Class 1 (C1) and Class 2 (C2) star clusters. There are prominent age peaks at about 100 and 250 Myr. The mass function is exponential and the slope of old component is relatively shallow compared with that of the young component.

2. Age and mass distributions

The age and mass distributions of the M51 star clusters are shown in Figure 1. The age distribution displays two overdense epochs, one at $\log(t/\text{yr}) < 7$ and the other at $7.5 < \log(t/\text{yr}) < 9.0$. The distribution in the latter epoch exhibits two peaks at about 100 and 250 Myr. This is true for both Class 1 (circular shape and isolated) and Class 2 (irregular shape and/or having neighbors) clusters. The large number of clusters with $7.5 < \log(t/\text{yr}) < 9.0$ is consistent with Lee *et al.* (2005), in which the number of star clusters with ages of 100–400 Myr was found to be larger in M51 than in the late-type galaxy M101.

The derived star cluster masses range from 10^3 to 10^6 $M_\odot$ and the mass distribution is composed of two components with different ages. These two components can be represented by a power law with a slope of $\alpha = -2.23 \pm 0.34$ for $t < 10$ Myr and $\alpha = -1.37 \pm 0.11$ for $t > 100$ Myr. The slope of the young cluster mass distribution is in agreement with the initial mass function slope $\alpha \approx -2.0$ of star clusters in late-type galaxies (e.g., Gieles *et al.* 2006).

Figure 2 displays the spatial distribution of star clusters with $\log(t/\text{yr}) < 7$ (young and less massive) and $8.0 < \log(t/\text{yr}) < 8.6$ (intermediate age and more massive) along with the distribution of very luminous HII regions with $L_{\text{H}\alpha} > 10^{37.5}$ erg s^{-1} (J. H. Lee *et al.*, in prep.). It shows a subtle difference in the spatial distribution of young and intermediate-age clusters. Although star clusters are mostly found along the spiral arms of M51, young clusters are preferentially concentrated in the regions where luminous HII regions are found, while intermediate-age clusters are distributed over the entire spiral-arm structure.

3. Star cluster formation rate

We have derived star cluster formation rate using star clusters with $4.6 < \log(M/M_\odot) < 5.0$ and $\log(M/M_\odot) > 5.0$. Only star clusters with $m_{\text{F336W}} < 22$ mag were selected for the cluster-formation-rate calculation to reduce the effects of incompleteness which can be very severe for clusters of $\log(t/\text{yr}) > 8.5$. The result is shown in Figure 3, where solid (dashed) lines display the cluster-formation rate derived from the observed (extrapolated

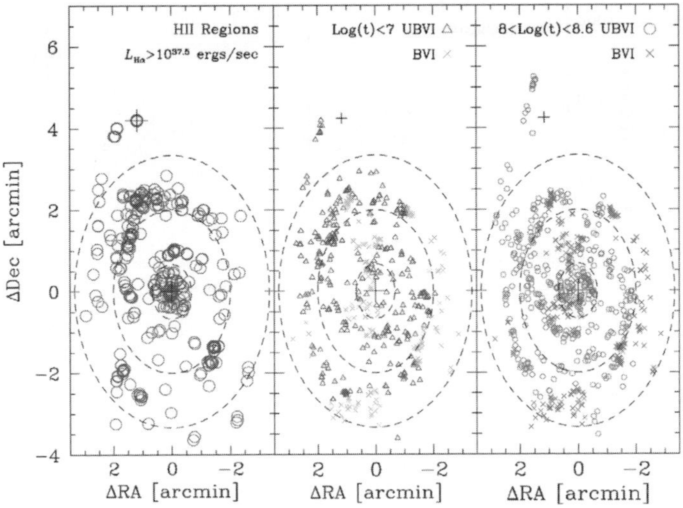

Figure 2. Spatial distribution of HII regions with $L_{\mathrm{H}\alpha} > 10^{37.5}$ erg s^{-1} *(left)* compared with the spatial distribution of the young ($\log(t/\mathrm{yr})$ <7; *middle*) and intermediate-age ($8 < \log(t/\mathrm{yr}) < 8.6$; *right*) star clusters in M51.

model) cluster age distribution. The extrapolated model age distribution was constructed by fitting the mass distribution of star clusters to a power law and then adding up the derived power-law distribution in the corresponding age bin.

Figure 3 suggests that the cluster-formation rate significantly increased at about 250 Myr and decreased thereafter. This can be seen from star clusters with $4.6 < \log(M/\mathrm{M}_\odot) < 5.0$ and $\log(M/\mathrm{M}_\odot) > 5.0$, respectively. Interestingly, this period is consistent with the theoretically expected interaction epoch of M51 (Toomre & Toomre 1972; Salo & Laurikainen 2000; Durrell *et al.* 2003). Combined with the age distribution which displays a peak at 250 Myr, this result of an enhanced cluster-formation rate during the same period suggests that the dynamical interaction expected to happen around this period may have significantly induced the formation of star clusters.

4. Implication

Dynamical galaxy interactions are believed to have a very strong impact on the formation and disruption of star clusters. However, studies on the dynamical evolution of galaxies and their star cluster populations have been carried out rather independently. With the available theoretical studies on dynamical interactions, M51 provides a unique opportunity for us to compare the evolution of dynamical events and the stellar population in the corresponding galaxies over a certain time interval.

We have shown that the number and the formation rate of star clusters in M51 significantly increased around a certain period. This period is consistent with the theoretically expected epoch of dynamical interactions between NGC 5194 and NGC 5195. However, to fully understand the mechanism and correlation of star cluster formation and galaxy dynamical interaction, we need spectroscopic observations of the star clusters. When we can compare the velocity distribution of star clusters with theoretical models of galaxy interactions, we will be able to understand exactly when and how effectively star clusters are formed and where star clusters are dispersed thereafter. Based on these studies, we

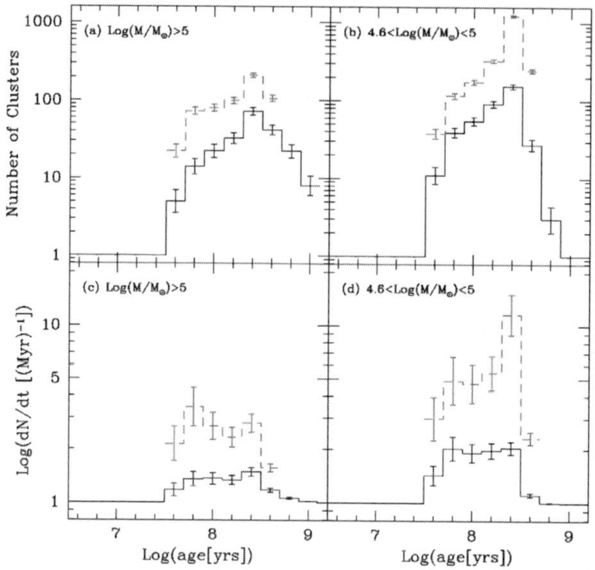

Figure 3. Number distribution and cluster-formation rate for clusters with $\log(M/\mathrm{M}_\odot) > 5.0$ *(left)* and $4.6 < \log(M/\mathrm{M}_\odot) < 5.0$ *(right)*. The black solid line shows the observed data, while the red dashed line shows the extrapolated data from the best-fit mass function in the corresponding age bins.

would be able to test and improve the existing theoretical models for galaxy interactions that can be applied to other general classes of galaxies.

Acknowledgements

NH acknowledges support from a Grant-in-Aid for JSPS Fellows, No. 20-08325, and travel support from the IAU. MGL is supported in part by a grant (R01-2007-000-20336-0) from the Basic Research Program of the Korea Science and Engineering Foundation.

References

Bruzual, G. & Charlot, S. 2003 *MNRAS*, 344, 1000
Durrell, P. R., Mihos, J. C., Feldmeier, J. J., Jacoby, G. H., & Ciardullo, R. 2002, *ApJ*, 582, 170
Feldmeier, J. J., Ciardullo, R., & Jacoby, G. H. 1997, *ApJ*, 479, 231
Gieles, M., Larsen, S. S., Bastian, N., & Stein, I. T. 2006, *A&A*, 450, 129
Haas, M. R., Gieles, M., Scheepmaker, R. A., Larsen, S. S., & Lamers, H. J. G. L. M. 2008, *A&A*, 487, 937
Hwang, N. & Lee, M. G. 2006, *ApJ* (Letters), 638, L79
Hwang, N. & Lee, M. G. 2008, *AJ*, 135, 1567
Hwang, N. & Lee, M. G. 2009, *ApJ*, in press (arXiv:0912.0321)
Lee, M. G., Chandar, R., & Whitmore, B. W. 2005, *AJ*, 130, 2128
Salo, H. & Laurikainen, E. 2000, *MNRAS*, 319, 377
Scheepmaker, R. A., Haas, M. R., Gieles, M., Bastian, N., Larsen, S. S., & Lamers, H. J. G. L. M. 2007, *A&A*, 469, 925
Scheepmaker, R. A., Lamers, H. J. G. L. M., Lamers, Anders, P., & Larsen, S. S. 2009, *A&A*, 494, 81
Toomre, R. P. J. & Toomre, J. 1972, *ApJ*, 178, 623

Star clusters: basic galactic building blocks
Proceedings IAU Symposium No. 266, 2009
R. de Grijs & J. R. D. Lépine, eds.

doi:10.1017/S1743921309991608

The field of Lacerta OB1†

Nadia T. Kaltcheva

Department of Physics and Astronomy, University Wisconsin–Oshkosh, 800 Algoma Blvd., Oshkosh, WI 54901, USA
email: kaltchev@uwosh.edu

Abstract. We present a new analysis of the Galactic structure in the field of Lacerta. The investigation uses precise $uvby\beta$ photometry and the recalculated *Hipparcos* parallaxes to provide an improved homogeneous distance scale to the main stellar groups in the field. The OB association Lac OB1 (Lac OB1b) is reliably identified as a very compact young group at an average distance of 520 ± 80 pc. The characteristics of the older Lac OB1a group and the open cluster NGC 7243 are extensively studied based on photometric diagrams and the inferred distances. A good agreement between the photometry-derived distances and the recalculated *Hipparcos* parallaxes is found for the Lac OB1 group, but in general discrepancies still exist for this particular field.

Keywords. stars: distances, open clusters and associations: individual (Lac OB1), stars: individual (10 Lac)

---

1. Introduction

Harboring one of the nearest OB associations, the field of Lacerta presents a unique opportunity to study the influence of massive stars on the interstellar medium. A reliable reconstruction of the star-forming history of the field should be possible once the spatial distribution of the young stars has been determined reliably. Despite extensive efforts to clarify the basic structural parameters of and distances to the young stellar groups in Lacerta, discrepancies still remain in published studies.

The field considered in this contribution extends between 90° and 110° Galactic longitude and from −25° to −5° in Galactic latitude. All available $uvby\beta$ photometry of O and B stars within this coordinate range was extracted from the catalog of Hauck & Mermilliod (1998). The sample contains 55 stars with complete photometry and homogeneously derived color excesses and distances. The color excesses were obtained using the calibration of Crawford (1978) for luminosity III, IV, and V, and based on the calibration of Kilkenny & Whittet (1985) for luminosity classes II, Ib, Iab, and Ia. The calibration of Balona & Shobbrook (1984) was used for all stars to derive absolute magnitudes. The photometry-derived stellar parameters are used to study the stellar content, the distribution of interstellar absorption and the structure of the field to reveal spatially coherent groups and layers, and provide precise stellar distances to the main apparent groupings in this direction (Kaltcheva 2009).

2. Results

Based on the derived photometric distances used in conjunction with the photometric diagrams, Lac OB1 (Lac OB1b; Blaauw 1958) is identified as a compact group of 12 low-reddening main-sequence stars located at a distance of 520 ± 20 pc in the direction

† The full poster (in pdf format) is available at
http://www.astro.iag.usp.br/~iaus266/Posters/pKaltcheva.pdf.

$l = 96.4°, b = -16.6°$. The available radial-velocity and proper-motion measurements support the impression that this is a real group. The recalculated *Hipparcos* parallaxes (van Leeuwen 2007) provide a much better agreement with the photometric parallaxes of the stars of the Lac OB1 group than the original *Hipparcos* values.

The stars in the northeastern part of the field (Lac OB1a group; Blaauw 1958) do not form a physical group, but are rather located in a layer between 140 and 400 pc, with some more distant stars towards $l \sim 104°$. For this subsample, a disagreement between the photometric and the recalculated *Hipparcos* parallaxes is still present. Based on the brightest members, an estimate of 678 ± 50 pc is obtained for the distance to the open cluster NGC 7243, which is in good agreement with that obtained by Jilinski *et al.* (2003).

The photometric distance to the O9V star 10 Lac (HD 214680) is estimated at 715^{+107}_{-92} pc. Although there is agreement within the errors, this photometric distance is considerably larger than the estimate of 529^{+70}_{-50} pc based on the new *Hipparcos* parralax (van Leeuwen 2007). However, the agreement with the recalculated parallax is much better in comparison to the old *Hipparcos* estimate of 325^{+82}_{-55} pc. This star is a standard star for Hβ photometry, with numerous $uvby\beta$ measurements collected. The star does not show peculiar behavior in the photometric diagrams. Since the $uvby\beta$ photometric distance in general is not influenced by rotation (Kaltcheva & Knude 1998), the photometric distance of 715 pc should be considered reliable. It is difficult, however, to conclude whether or not Lac 10 is a member of the Lac OB1 association. Photometrically, the star is significantly more distant than the Lac OB1 group. On the other hand, the radial velocity for 10 Lac suggests that the star might indeed be a member of the association.

Acknowledgements

This work is supported by the National Science Foundation (grant AST-0708950).

References

Balona, L. A. & Shobbrook, R. R. 1984, *MNRAS*, 210, 375
Blaauw, A. 1995, *AJ*, 63, 187
Crawford, D. L. 1978, *AJ*, 83, 48
Hauck, B. & Mermilliod, M. 1998, *A&AS*, 129, 143
Jilinski, E. G., Frolov, V. N., Ananjevskaja, J. K., Straume, J., & Drake, N. A. 2003, *A&A*, 401, 531
Kaltcheva, N. 2009, *PASP*, 121, 1045
Kaltcheva, N. & Knude, J. 1998, *A&A*, 337, 178
Kilkenny, D. & Whittet, D. C. B. 1985, *MNRAS*, 216, 127
van Leeuwen, F. 2007, *Hipparcos, the new reduction of the raw data*, Astrophys. Space Sci. Libr., vol. 350, Dordrecht: Springer

Star clusters: basic galactic building blocks
Proceedings IAU Symposium No. 266, 2009
R. de Grijs & J. R. D. Lépine, eds.

doi:10.1017/S174392130999161X

First results of the Southern Open Cluster Study

K. Kinemuchi,[1,2] A. Sarajedini,[2] D. Geisler,[1] F. Mauro,[1] G. Carraro,[3] S. Kafka[4] and E. Jeffery[5]

[1]Departamento de Astronomía, Universidad de Concepción, Concepción, Chile
email: [kkinemuchi, dgeisler, fmauro]@astro-udec.cl

[2]Department of Astronomy, University of Florida, 211 Bryant Space Science Center, Gainesville, FL 32611, USA
email: [kinemuchi, ata]@astro.ufl.edu

[3]European Southern Observatory
email: gcarraro@eso.org

[4]SSC/Caltech, Pasadena, CA 91125, USA
email: stella@caltech.edu

[5]Department of Astronomy, University of Texas at Austin, Austin, TX 78712, USA
email: ejeffery@astro.as.utexas.edu

Abstract. Galactic open clusters are excellent laboratories to study stellar populations that are coeval and have a common chemical composition. These clusters allow for the investigation of stellar and chemical abundance evolution, binary star systems, mass and luminosity functions, Galactic metallicity gradient, and disk structure, among other things. The Southern Open Cluster Study (SOCS) is a database of a select set of 24 open clusters with a broad range of ages and metallicities. It is an extension of the northern *WIYN* Open Cluster Study (WOCS). The main goal of both studies is to obtain comprehensive photometric, spectroscopic, and astrometric information of these key open clusters. Here, we provide an overview of the SOCS survey and present results of the wide-field photometry on two SOCS clusters, NGC 3532 and Tombaugh 2.

Keywords. surveys, open clusters and associations: individual (NGC 3532, Tombaugh 2)

1. Introduction

The Southern Open Cluster Study (SOCS) is a comprehensive study of 24 carefully selected southern open clusters. This survey is the southern extension of the northern survey, the *WIYN* Open Cluster Study (WOCS; Mathieu 2000; Sarajedini *et al.* 2003). The SOCS survey has similar science goals as the WOCS study, primarily to provide a complete and detailed analysis of the clusters through photometry and spectroscopy. This database will be an excellent resource for other studies related to open clusters and their stellar populations.

The primary goals of the SOCS survey can be described by three aspects: photometry, spectroscopy, and astrometry. Presently, our photometric campaign is nearly complete. We have achieved both wide-field and deep pencil-beam photometry for color–magnitude diagrams in a variety of optical bandpasses. For the spectroscopy of the SOCS clusters, we will obtain chemical abundances and stellar radial velocities. Combining all available kinematic resources, we will determine precise positions and proper-motion membership of the cluster stars. We will improve the astrometric information for all our SOCS clusters.

The open clusters selected for SOCS have a wide range in age (0.04–8.9 Gyr) and metallicity ($-0.5 < [Fe/H] < +0.1$). A list of the SOCS clusters is presented in Table 1.

Table 1. SOCS clusters and their parameters.

Name	RA (2000)	Dec (2000)	$(m - M)$ (mag)	[Fe/H] (dex)	Age (Gyr)
NGC 2204	06:15:33	−18:39:54	12.4	−0.3	0.8
NGC 2243	06:29:34	−31:17:00	13.4	−0.4	4.5
NGC 2287	06:46:01	−20:45:24	9.3	+0.04	1.9
NGC 2360	07:17:43	−15:38:30	11.7	−0.2	0.6
Tombaugh 2	07:03:05	−20:49:00	15.9	−0.4	1.0
Melotte 66	07:26:23	−47:40:00	13.6	−0.4	2.8
NGC 2423	07:37:06	−13:52:18	...	...	...
NGC 2451	07:44:27	−37:40:00	7.6	−0.5	0.04
Berkeley 39	07:46:42	−04:36:00	13.8	−0.2	7.9
NGC 2477	07:52:10	−38:31:48	11.3	+0.01	0.7
NGC 2516	07:58:04	−60:45:12	8.4	+0.06	0.1
NGC 2506	08:00:01	−10:46:12	13.0	−0.4	1.1
NGC 2547	08:10:09	−49:12:54	8.4	−0.16	0.36
NGC 2539	08:10:37	−12:49:06	10.9	+0.13	0.37
NGC 2546	08:12:15	−37:34:42	10.2	+0.12	0.74
IC 2395	08:42:30	−48:09:00	9.5	...	0.16
NGC 2660	08:42:38	−47:12:00	13.2	−0.2	1.1
NGC 3532	11:05:39	−58:45:12	8.6	−0.02	0.3
NGC 3680	11:25:38	−43:14:36	10.1	−0.1	1.2
NGC 3960	11:50:33	−55:42:24	12.7	−0.2	0.7
NGC 4103	12:06:40	−61:15:00	12.0	...	0.27
CR 261	12:37:57	−68:22:00	12.5	−0.1	8.9
NGC 5822	15:04:21	−54:23:48	10.3	−0.02	0.7
IC 4651	17:24:49	−49:56:00	10.1	+0.1	1.1

To demonstrate the spatial distribution of the open clusters, Figure 1 shows where the clusters are located in an Aitoff plot with Galactic coordinates. Figure 2 shows the metallicity distribution of the SOCS clusters with respect to log(age).

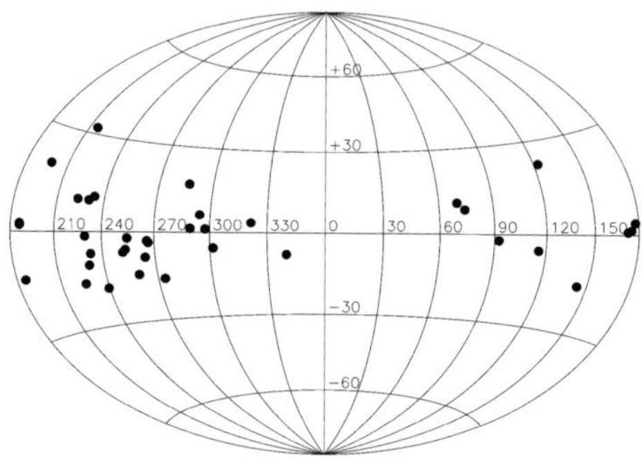

Figure 1. Aitoff plot in Galactic coordinates of the WOCS/SOCS clusters.

2. Data acquisition for photometry

For the wide-field photometry, all data were obtained at the *Warsaw 1.3m telescope* at Las Campanas Observatory and the *1.0m telescope* at Cerro Tololo Inter-American Observatory (CTIO), Chile. The open clusters were observed in 2005 and 2007 during four separate observing runs. The *Warsaw telescope* had a MOSAIC CCD camera with a field of view of $34' \times 34'$. The open clusters were observed in the Johnson $BVRI$ filters

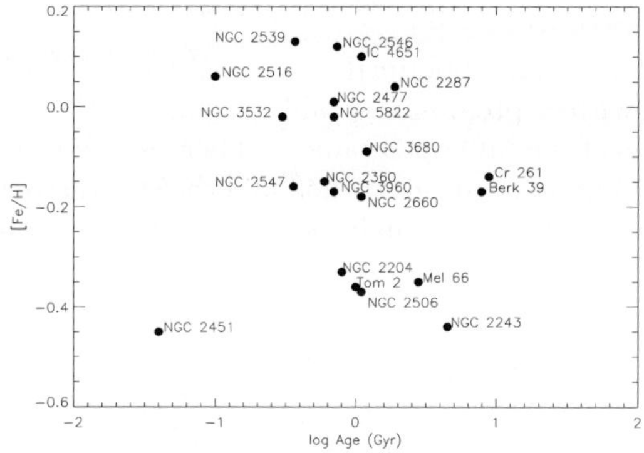

Figure 2. Distribution of the SOCS clusters with respect to metallicity and log(age).

with three different exposure times. Six open clusters were observed in *UBVRI* filters at the *CTIO 1.0m*, which covered a smaller region ($20' \times 20'$).

To perform point-spread-function photometry on the images, we developed a pipeline which interfaced with the stand-alone version of DAOPHOT IV (Stetson 1987, 1994). All data from the *Warsaw 1.3m telescope* were processed and each open cluster is being systematically photometered. In these proceedings, we present our preliminary results of two SOCS clusters from our photometric campaign.

3. Wide-field photometry results

We present color–magnitude diagrams (CMDs) of the SOCS open clusters NGC 3532 and Tombaugh 2. We provide our results with a preliminary photometric calibration of the CMDs. Further work is underway to determine the final standard calibration and to perform a careful analysis of the stellar populations.

3.1. *NGC 3532*

For this relatively metal-rich open cluster, we present in Figure 3 the CMD. Even with the strong field contamination, we can clearly see the cluster's main sequence and a hint of the turnoff. Data for this CMD was culled primarily from the inner three MOSAIC chips of CCD camera at the *Warsaw telescope*.

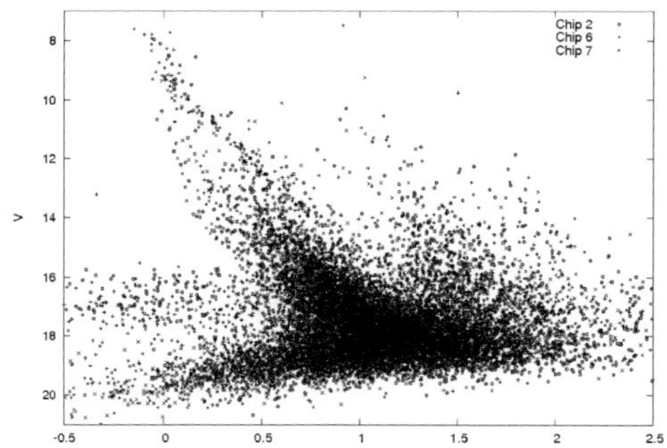

Figure 3. Color–magnitude diagram of NGC 3532.
(Data from the three inner MOSAIC chips.)

3.2. *Tombaugh 2*

Data for Tombaugh 2 were also obtained at the *Warsaw 1.3m* in 2007. The CMD in Figure 4 is also done with a preliminary calibration, but the diagram is composed of data from all eight chips of the MOSAIC camera. There is still heavy field contamination. However, one can see the main-sequence turnoff and the stars associated with the cluster's red clump. Standard calibration and analysis of Tombaugh 2 is currently underway.

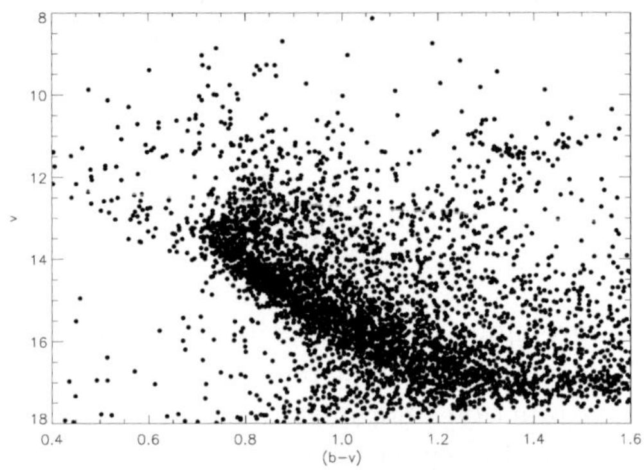

Figure 4. CMD of Tombaugh 2.

4. Future work

The SOCS project is now finishing up the optical-photometry portion but has delved into near-infrared (NIR) photometry. We have NIR data for two SOCS clusters taken with the ISPI instrument at the CTIO *Blanco 4.0m telescope*. The NIR data will be particularly useful in understanding the binary-star population in these open clusters (see Robitaille *et al.* 2007; Frinchaboy & Nielsen 2008). Projects are planned for the spectroscopy of the SOCS clusters, mainly to investigate their chemical abundances in detail.

Acknowledgements

KK, AS, and DG gratefully acknowledge support from the National Science Foundation through grant AST-0606703.

References

Frinchaboy, P. M. & Nielsen, D. 2008, in: E. Vesperini, M. Giersz & A. Sills (eds.), *Dynamical Evolution of Dense Stellar Systems*, Proc. IAU Symp. 246, p. 109

Mathieu, R. D. 2000, in: R. Pallavicini, G. Micela, & S. Sciortino (eds.), *Stellar Clusters and Associations: Convection, Rotation, and Dynamos*, ASP Conf. Ser., 198, p. 517, San Francisco: ASP

Robitaille, T. P., Whitney, B. A., Indebetouw, R., & Wood, K. 2007, *ApJS*, 169, 328

Sarajedini, A., Mathieu, R. D., & Platais, I. 2003, in: T. D. Oswalt (ed.), *The Future of Small Telescopes In The New Millennium. III. Science in the Shadows of Giants*, *ASSL*, 289, 257

Stetson, P. B. 1987, *PASP*, 99, 191

Stetson, P. B. 1994, *PASP*, 106, 250

Star clusters: basic galactic building blocks
Proceedings IAU Symposium No. 266, 2009
R. de Grijs & J. R. D. Lépine, eds.

doi:10.1017/S1743921309991621

Constraining star cluster disruption mechanisms

I. S. Konstantopoulos,[1,2,*] N. Bastian,[3] M. Gieles[2] and H. J. G. L. M. Lamers[4]

[1]Department of Physics & Astronomy, University College London, Gower Street, London, WC1E 6BT, UK

[2]European Southern Observatory, Casilla 19001, Santiago 19, Chile

[3]Institute of Astronomy, University of Cambridge, Madingley Road, Cambridge, CB3 0HA, UK

[4]Astronomical Institute, Utrecht University, Princetonplein 5, 3584 CC Utrecht, the Netherlands

*Currently at Penn State University; email: iraklis@astro.psu.edu

Abstract. Star clusters are found in all sorts of environments, and their formation and evolution is inextricably linked to the star-formation process. Their eventual destruction can result from a number of factors at different times, but the process can be investigated as a whole through the study of cluster age distributions. Observations of populous cluster samples reveal a distribution following a power law of index approximately −1. In this work, we use M33 as a test case to examine the age distribution of an archetypal cluster population and show that it is, in fact, the evolving shape of the mass detection limit that defines this trend. That is to say, any *magnitude-limited* sample will appear to follow a $dN/d\tau = \tau^{-1}$ relation, while cutting the sample according to mass gives rise to a composite structure, perhaps implying a dependence of the cluster disruption process on mass. In the context of this framework, we examine different models of cluster disruption from both theoretical and observational perspectives.

Keywords. galaxies: star clusters, galaxies: individual (M 33)

1. Introduction

Star clusters are commonly used to trace the stellar content and star-formation history of their host systems. The main limitations to this approach are the finite lifetime of a cluster (disruption) and evolutionary fading. From a theoretical point of view, the lifetime of a cluster should depend on its initial mass and the properties of the environment in which it evolves (Spitzer 1958†; Baumgardt & Makino 2003). Observations have, however, produced *two* empirical disruption laws:

- Mass-dependent disruption (MDD; Boutloukos & Lamers 2003: BL03);
- Mass-independent disruption (MID; Fall *et al.* 2005).

In the MID model, 'survivors' are selected on a purely random basis and a constant fraction is destroyed every age dex. Intriguing as it may be, this model clashes with several principles of cluster dynamics that would need to be revised considerably to accommodate it. In this contribution, we test cluster dissolution and attempt to disentangle it from incompleteness and the statistical biases that have (in the past) limited such studies. We then compare theory to observations of the M33 cluster system.

† This work showed that the lifetime due to encounters with interstellar clouds depends on the cluster density. Observations of young clusters have shown the radius to scatter tightly about 3.5 pc and can thus be considered a constant for any practical formulation. This immediately establishes the cluster mass as the main deciding factor for the lifetime of a cluster.

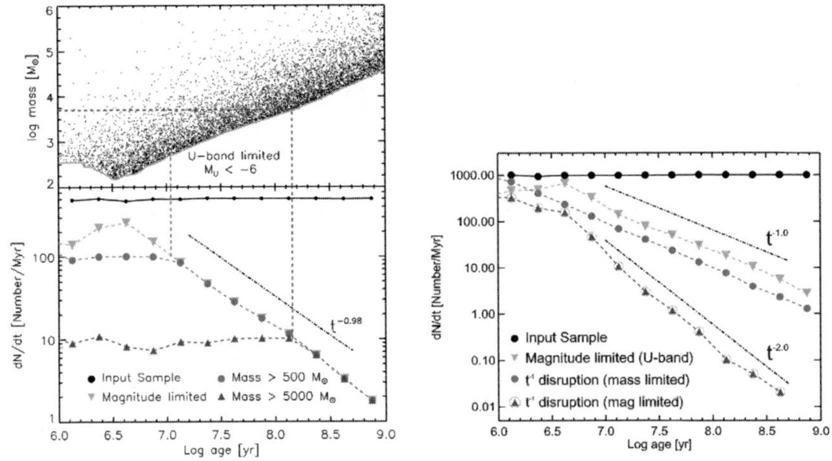

Figure 1. *(left)* Analysis of an artificial cluster population at the distance of M33. The population contains 5×10^5 clusters, sampled from a power-law mass function with index -2, and assuming a constant cluster-formation rate. The top panel shows the mass distribution for the observable part of the population. The detection limit is denoted by the solid line. This takes into account evolutionary fading and the sample is assumed to be limited by the U band, as is most commonly the case in extragalactic studies. The bottom panel shows the number of clusters per logarithmic age bin ($\mathrm{d}N/\mathrm{d}\tau$). The dark solid line corresponds to all clusters 'formed' (i.e., the entire simulated set), while the dashed green line tracks 'observed' clusters, i.e., those above the detection limit. The theoretical expectation is for a magnitude-limited sample to follow a simple power-law distribution, while the introduction of a mass limit segments the distribution into a flat and a power-law part (the transition is dictated by the mass cut and the point where incompleteness sets in). *(right)* A similar population, according to the predictions of MID. This presents the age distribution of a cluster population with a constant cluster-formation rate and an imposed a τ^{-1} disruption law.

2. A theoretical preamble

Before providing a treatment of an observed data set, let us try to understand the 'tools of the trade', the statistical distributions used to study and characterise the cluster disruption process. Therefore, we have created an artificial population with a constant cluster-formation history (CFH) and masses drawn from a power-law distribution with index -2, although we note that the cluster initial mass function was recently found to be better described by a Schechter function (Gieles 2009; Larsen 2009). This population is presented in the left-hand panel of Figure 1.

2.1. *Age–mass diagram*

This plot simply shows the number of clusters with increasing age. Logarithmic bins accentuate the increase of the number of high-mass clusters with increasing age. This arises because the sample grows with time: a 'size-of-sample' effect. As the sample increases, so does the likelihood of producing a massive cluster.

2.2. *Detection limit*

Unlike the top envelope of the age–mass diagram, the line along the bottom of the distribution is not caused by statistics, but it is the detection limit of our modelled observations. As clusters fade with time, this constant limiting flux translates into an increasing limiting mass.

In most cases, observed cluster samples will be luminosity limited. As the population ages, clusters of higher mass fade to below the detection limit. We have denoted this by a green line, representing the mass of a cluster with $M_U = -6$ mag (adapted to the study of M33 that will follow) at a given age. The U band is normally the limiting filter in observational studies and we emulate the detection limit in this plot.

2.3. *Age distribution: the* dN/dτ *plot*

The lower panel shows the main diagnostic used in the study of cluster disruption. The vertical axis gives the number of clusters our imaginary galaxy formed in each log(age) bin, normalised to unit time (in this case, 1 Myr). Here, we present the input sample as a solid line; a constant CFH gives rise to a flat line. This will not, however, be observed, due to the detection limit.

2.4. *A magnitude-limited sample*

The green line shows all clusters that lie above the detection limit in each log(age) bin and displays the characteristic power-law shape found in all observational studies. The interpretation of this plot is far from trivial. This τ^{-1} shape means that 90% of the population is lost with each age dex. MID interprets this as 90% of clusters dissolving each age dex (e.g., in M33; Sarajedini & Mancone 2007). In this example, however, this is due to detection incompleteness. Thus, a model has to treat *fading* and *dissolution* as two separate and concurrent causes of cluster disappearance.

2.5. *Taking mass cuts*

Having established the above, we can now perform a simple test for the mass dependence of cluster disruption: cutting the sample according to mass. In the dN/dτ plot we present two mass cuts as a red and blue line, respectively. Both cuts split the power-law distribution into a composite shape: a flat part, i.e., a constant cluster-formation rate, and a power-law decline, due to clusters fading to below the detection limit.

3. Observational study

After understanding the caveats inherent in the study of the dN/dτ plot, we can proceed to plot the distribution of an observed cluster sample. We obtained the age and mass of a sample of $\sim$ 350 bona fide clusters in M33 (covering the entire surface of the galaxy, an unbiased, *HST*-selected sample; Sarajedini & Mancone 2007; San Roman et al. 2009) using *UBVI* imaging from the Local Group Survey (Massey et al. 2006). The star-formation rate is known to have been constant for $\sim$ 1 Gyr in M33, making it ideal for our study. Figure 2 shows the age distribution of the magnitude-limited sample (left) and a mass cut (right). It exhibits the same shape as the dN/dτ plot of Figure 1 (left), with a flat part leading to a power law.

We also provide a direct test of MID in Figure 1 (right panel), where we create a population that destroys 90% of its clusters each age dex *and* is subject to evolutionary fading.

3.1. *Observed* dN/dτ *plot*

To dampen the effect of local overdensities in the dN/dτ plot (caused by uncertainties in the simple stellar population models), in Figure 2 we represent each age by a Gaussian, where the wings are defined by the uncertainty. The result is consistent with a τ^{-1} power law for a magnitude-limited sample. Crucially, as predicted by the models of Figure 1 (left), taking a mass cut produces a composite structure, with a flat initial part and a power-law decline at older ages. This demonstrates that an observed sample needs to be interpreted as the result of fading and dissolution on different timescales.

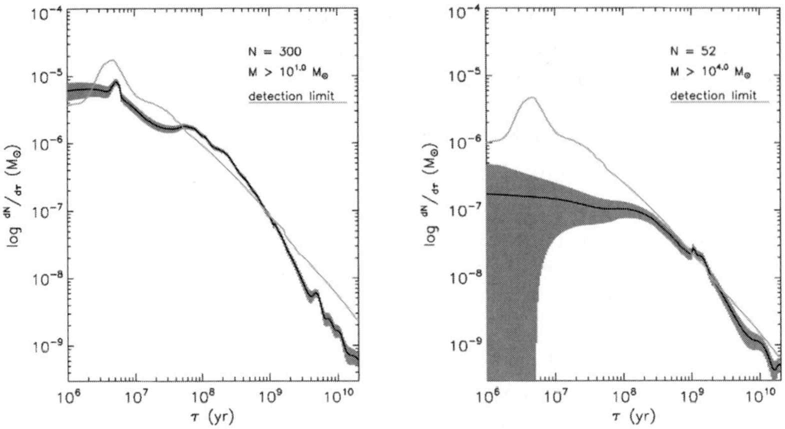

Figure 2. *(left)* A 'smoothed' $\mathrm{d}N/\mathrm{d}\tau$ plot. We represent each measurement as a Gaussian, where the peak occurs at the best-fit value (from 3DEF) and the wings are shaped after the uncertainty in the age fit (defined as the extrema calculated by 3DEF). This minimises the effect of uncertain measurements on the shape of the distribution, as an erratic measurement will be represented by a flat Gaussian. The solid green line is a prediction of the $\mathrm{d}N/\mathrm{d}\tau$ of a luminosity-limited sample. It is the detection limit of Figure 1 *(left)*, inverted to reflect the maximum expected number of detections, given the detection limit (due to the complex scaling of the $\mathrm{d}N/\mathrm{d}\tau$ plot we have normalised the two lines at 1 Gyr). *(right)* The same plot with an imposed mass cut at 10^4 $M_\odot$. The obvious flattening with respect to the magnitude-limited sample on the left strongly supports a mass dependence in the cluster disruption process (consistent with BL03).

3.2. *Mass-independent dissolution plus fading*

We have argued so far that the observed $\mathrm{d}N/\mathrm{d}\tau$ plot will show a combination of disruption and fading. The right-hand panel of Figure 1 shows the combined effect of 90% dissolution (the MID hypothesis) *and* 90% disappearance due to fading. This results in a very steep power-law decline (blue line) that is *inconsistent* with the observations of Figure 2.

This result stands firmly against the MID cluster disruption model and implies the existence of a mass dependence.

4. Summary

Sample incompleteness causes a τ^{-1} decline in the observed age distribution of cluster populations. Size-of-sample effects and thorough statistical methods can help to interpret the disruption process at play. The fundamental difference between magnitude- and mass-limited samples can then be used to discover the underlying physics. Our results based on the nearby population of M33 clusters are inconsistent with mass-independent disruption of a fixed fraction of clusters per age dex.

Acknowledgements

ISK gratefully acknowledges the support of an ESO studentship for the undertaking of this work and would like to thank the Vitacura staff and students for their hospitality. Thanks are also due to the organising committee of the General Assembly and Symposium 266. *Obrigado Rio!*

References

Baumgardt, H. & Makino, J. 2003, *MNRAS*, 340, 227
Boutloukos, S. G. & Lamers, H. J. G. L. M. 2003, *MNRAS*, 338, 717

Fall, S. M., Chandar, R., & Whitmore, B. C. 2005, *ApJ* (Letters), 631, L133
Larsen, S. S. 2009, *A&A*, 503, 467
Gieles, M. 2009, *MNRAS*, 394, 2113
Sarajedini, A. & Mancone, C. L. 2007, *AJ*, 134, 447
San Roman, I., Sarajedini, A., Garnett, D. R., & Holtzman, J. A. 2009, *ApJ*, 699, 839
Massey, P., Olsen, K. A. G., Hodge, P. W., Strong, S. B., Jacoby, G. H., Schlingman, W., & Smith, R. C. 2006, *AJ*, 131, 2478
Spitzer Jr., L. J. 1958, *ApJ*, 127, 17

Star clusters: basic galactic building blocks
Proceedings IAU Symposium No. 266, 2009
R. de Grijs & J. R. D. Lépine, eds.

doi:10.1017/S1743921309991633

The origin of very wide binary systems

M. B. N. Kouwenhoven,[1,2] S. P. Goodwin,[2] Richard J. Parker,[2] M. B. Davies,[3] D. Malmberg[3] and P. Kroupa[4]

[1]Kavli Institute for Astronomy and Astrophysics, Peking University, Yi He Yuan Lu 5, Hai Dian District, Beijing 100871, P. R. China
email: kouwenhoven@kiaa.pku.edu.cn

[2]University of Sheffield, Hicks Building, Hounsfield Road, Sheffield S3 7RH, UK

[3]Lund Observatory, Box 43, SE-221 00, Lund, Sweden

[4]Argelander Institute for Astronomy, University of Bonn, Auf dem Hügel 71, 53121, Bonn, Germany

Abstract. The majority of stars in the Galactic field and halo are part of binary or multiple systems. A significant fraction of these systems have orbital separations in excess of thousands of astronomical units, and systems wider than a parsec have been identified in the Galactic halo. These binary systems cannot have formed through the 'normal' star-formation process, nor by capture processes in the Galactic field. We propose that these wide systems were formed during the dissolution phase of young star clusters. We test this hypothesis using N-body simulations of evolving star clusters and find wide binary fractions of $1-30\%$, depending on initial conditions. Moreover, given that most stars form as part of a binary system, our theory predicts that a large fraction of the known wide 'binaries' are, in fact, multiple systems.

Keywords. binaries: general, stars: formation, stellar dynamics, methods: N-body simulations

1. Observations and formation of very wide binary systems

Numerous wide (10^3 AU$-$0.1 pc) binary systems are known to exist in the Galactic field and halo (e.g., Close *et al.* 1990; Chanamé & Gould 2004). These binaries are usually detected as common-proper-motion pairs (e.g., Wasserman & Weinberg 1991; Lépine & Bongiorno 2007) or through statistical methods (e.g., Bahcall & Soneira 1981; Garnavich 1988). Approximately 15% of the wide binaries in the field have a semi-major axis in the range 10^3 AU $< a <$ 0.1 pc (Duquennoy & Mayor 1991; see also Poveda *et al.* 2007). The origin of these wide binaries cannot be explained by star formation or by dynamical interactions in the Galactic field, and has long been a mystery.

The majority of stars are known to form in embedded clusters (Lada & Lada 2003). Wide binary systems cannot have formed through clustered star formation, simply because their semi-major axis is comparable to the size of a typical embedded cluster. Even if it were possible to form wide binaries in star clusters, they would immediately be destroyed by dynamical interactions. Although most stars are believed to have formed in star clusters, a fraction ($< 10-30\%$) may form through diffuse (isolated) star formation. However, this fraction makes it difficult to account for the $\sim$ 15% wide binaries in the Galactic field. If wide binaries are formed through diffuse star formation, then the majority of stars formed in this process must form a wide binary, implying a star-formation process that is fundamentally different from that of clustered star formation.

Binary systems of any separation may also form through dynamical capture processes. To form a binary system from two interacting stars, a third star should be present to serve as energy sink. Each of the three stars should be at the same location at the same time,

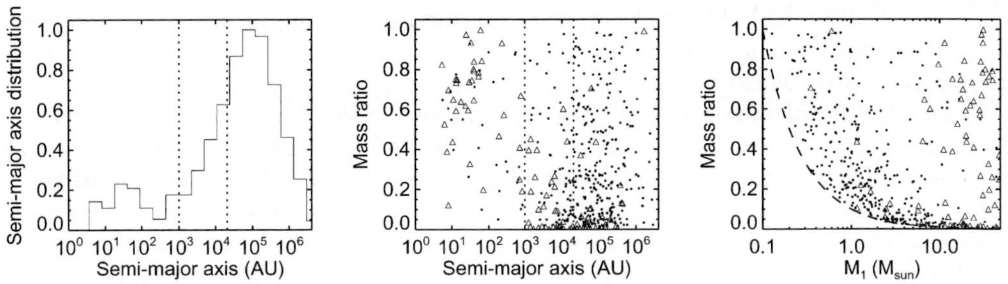

Figure 1. The binary population resulting from a Plummer model with $N = 1000$ and a virial radius of 0.1 pc (50 runs), showing the semi-major-axis distribution *(left)*, the correlation between mass ratio q and semi-major axis a *(middle)* and between primary mass and mass ratio (right). Binary and multiple systems are indicated with circles and triangles, respectively. The vertical dashed lines indicate $a = 10^3$ AU and $a = 0.1$ pc, respectively, the semi-major-axis range we consider in our analysis. The dashed curve in the right-hand panel indicates the minimum mass ratio $q_{\rm min}(M_1) = M_{\rm min}/M_1$ (Kouwenhoven *et al.* 2009).

and should have precisely those velocities and impact angles that allow the formation of a new binary system. These conditions are extremely rare in the Galactic field, resulting only in a handful of binaries over the lifetime of the Galaxy (Goodman & Hut 1993). Note that binaries can form through the capture process in the cores of star clusters through this process. However, wide binary systems cannot form in these environments, simply because they have a semi-major axis comparable to (or even larger than) star cluster cores.

We propose that wide binaries are formed during the dissolution of young star clusters (Kouwenhoven *et al.*, in prep). Young clusters generally have a short lifetime and dissolve into the field stellar population within typically 20 Myr (e.g., Mengel *et al.* 2005; Bastian *et al.* 2005; de Grijs & Parmentier 2007). During this dissolution process, two stars may escape the cluster in the same direction, with similar velocities. Although initially unbound, these two stars may form a new binary system after having left the gravitational potential of the cluster. In this process, the separation between the two stars is typically on the order of the size of their natal cluster at the time of dissolution, i.e., of order a parsec.

2. *N*-body simulations and results

It is theoretically possible to form a wide binary in the star cluster dissolution phase and it is important to find out whether this process happens frequently enough to account for the observed wide binary population. To this end, we carry out *N*-body simulations of evolving star clusters using the STARLAB package (Portegies Zwart *et al.* 2001). We randomly draw N single stars from the Kroupa (2002) initial mass function in the mass range $0.1 - 50$ M$_\odot$. The stars are given positions and velocities according to two models, (i) the spherical and homogeneous Plummer (1911) model and (ii) a fractal stellar distribution with fractal dimension 1.5, following the prescriptions of Goodwin & Whitworth (2004). The simulations are run until the star cluster is completely dissolved. We only consider the wide binary population in the semi-major-axis range 10^3 AU $< a <$ 0.1 pc, where the upper limit roughly corresponds to the stability limit in the Galactic field. We find the dependence on initial conditions by varying the cluster mass, size and morphology. Our main results are the following.

1. The resulting wide binary fraction is $1 - 30\%$, depending on the initial properties of the star cluster. The wide binary fraction increases strongly with increasing initial substructure of the cluster, and decreases with increasing cluster mass.

2. The dissolution of star clusters results in a binary population with a bimodal semi-major-axis distribution. At small separations, we find a *dynamical peak*, consisting of binaries formed dynamically in the centre of the star cluster. At large separations, we find a *dissolution peak*, which contains the binary systems formed during the dissolution process of the star cluster. As an example, we show in Figure 1 the binary population resulting from a star cluster with $N = 10^3$ single stars and an initial radius of 0.1 pc. The resulting semi-major axis scales linearly with the initial size of the star cluster.

3. Wide-binary formation is a process where two random stars encounter each other. When two stars are nearby in phase space, their mutual gravitational attraction (i.e., their combined mass) determines whether they form a new binary. The resulting mass-ratio distribution can therefore be characterised by gravitationally focused random pairing (e.g., Gaburov *et al.* 2008). The resulting eccentricity distribution is thermal (e.g., Heggie 1975) and there is no correlation between the orbital orientation and the stellar rotation vectors.

3. Implications for higher-order multiplicity

In the previous section we described simulations of star clusters that initially consist of single stars only. In reality, however, most stars (possibly even all stars) are formed as part of binary or multiple systems (e.g., Duquennoy & Mayor 1991; Fischer & Marcy 1992; Kroupa 1995; Kouwenhoven *et al.* 2005, 2007; Zinnecker & Yorke 2007). To first order, primordial binaries with separations much smaller than those of wide binaries can be dynamically approximated as single stars. Therefore, for a star cluster with primordial binaries we expect to find approximately the same wide-binary population as described above. But in this case, the wide 'binaries' are, in fact, wide triple or quadruple systems, where one or both of the subsystems consists of a primordial binary. Our N-body simulations of clusters with primordial binaries show that this is indeed the case. The ratio between the newly formed wide binaries, triples and quadruples, is a function of the primordial binary fraction. For example, for a primordial binary fraction of 50%, we expect a binary:triple:quadruple ratio of 1:2:1. By accurately measuring the multiplicity of wide-binary systems, we can therefore constrain the primordial binary fraction.

4. Summary

A significant fraction of the stars in the Galactic field and halo are part of wide ($a > 10^3$ AU) binary systems. The origin of these binary systems cannot be explained by clustered star formation, nor by dynamical capture in the Galactic field and halo. We propose that these binaries are formed during the dissolution process of young star clusters, where two initially unbound stars escape in a similar direction and form a bound wide binary (Kouwenhoven *et al.*, in prep). We test this hypothesis using N-body simulations and find that we are able to reproduce wide-binary fractions of $1 - 30\%$ (in the semi-major-axis range 10^3 AU $< a <$ 0.1 pc), depending on the initial conditions. Moreover, as most stars form as part of a binary system, we predict that a large fraction of the known wide 'binaries' are, in fact, triple or quadruple systems, and the ratio of wide binary, triple and quadruple systems therefore provides an indication of the primordial binary fraction.

Acknowledgements

MBNK was supported by the Peter and Patricia Gruber Foundation through a PPGF fellowship, the Kavli Foundation, the Peking University One Hundred Talent Fund (985) and by STFC (grant PP/D002036/1). RJP acknowledges support from STFC.

References

Bahcall, J. N. & Soneira, R. M. 1981, *ApJ*, 246, 122
Bastian, N., Gieles, M., Lamers, H. J. G. L. M., Scheepmaker, R. A., & de Grijs, R. 2005, *A&A*, 431, 905
Chanamé, J. & Gould, A. 2004, *ApJ*, 601, 289
Close, L. M., Richer, H. B., & Crabtree, D. R. 1990, *AJ*, 100, 1968
Fischer, D. A. & Marcy, G. W. 1992, *ApJ*, 396, 178
de Grijs, R. & Parmentier, G. 2007, *ChJA&A*, 7, 155
Duquennoy, A. & Mayor, M. 1991, *A&A*, 248, 485
Gaburov, E., Gualandris, A., & Portegies Zwart, S. 2008, *MNRAS*, 384, 376
Garnavich, P. M. 1988, *ApJ* (Letters), 335, L47
Goodman, J. & Hut, P. 1993, *ApJ*, 403, 271
Goodwin, S. P. & Whitworth, A. P. 2004, *A&A*, 413, 929
Heggie, D. C. 1975, *MNRAS*, 173, 729
Kroupa, P. 2001, *MNRAS*, 322, 231
Kouwenhoven, M. B. N., Brown, A. G. A., Zinnecker, H., Kaper, L., & Portegies Zwart, S. F, 2005, *A&A*, 430, 137
Kouwenhoven, M. B. N., Brown, A. G. A., & Kaper, L. 2007, *A&A*, 464, 581
Kouwenhoven, M. B. N., Brown, A. G. A., Goodwin, S. P., Portegies Zwart, S. F., & Kaper, L. 2009, *A&A*, 493, 979
Kroupa, P. 1995, *MNRAS*, 277, 1491
Lada, C. J. & Lada, E. A. 2003, *ARA&A*, 41, 57
Lépine, S. & Bongiorno, B. 2007, *AJ*, 133, 889
Mengel, S., Lehnert, M. D., Thatte, N., & Genzel, R. 2005, *A&A*, 443, 41
Poveda, A., Allen, C., & Hernández–Alcántara, A. 2007, in: W. I. Hartkopf, E. F. Guinan and P. Harmanec (eds.), *Binary Stars as Critical Tools and Tests in Contemporary Astrophysics*, Proc. IAU Symp. 240, p. 417
Plummer, H. C. 1911, *MNRAS*, 71, 460
Portegies Zwart, S. F., McMillan, S. L. W., Hut, P., & Makino, J. 2001, *MNRAS*, 321, 199
Wasserman, I. & Weinberg, M. D. 1991, *ApJ*, 382, 149
Zinnecker, H & Yorke, H. W. 2007, *ARA&A*, 45, 481

Star clusters: basic galactic building blocks
Proceedings IAU Symposium No. 266, 2009
R. de Grijs & J. R. D. Lépine, eds.

doi:10.1017/S1743921309991645

Kinematic parameters and membership lists of open clusters in the Bordeaux Carte du Ciel zone

Alberto Krone–Martins,[1,2] Caroline Soubiran,[2] Ramachrisna Teixeira[1] and Christine Ducourant[2]

[1]Instituto de Astronomia, Geofísica e Ciências Atmosféricas, Universidade de São Paulo, Rua do Matão, 1226, 05508-900, Cidade Universitária, São Paulo, Brazil

[2]Observatoire de Sciences d'Aquitaine, Université de Bordeaux 1, 2 rue de l'Observatoire, Floirac, France
email: algol@astro.iag.usp.br

Abstract. In the study of open clusters, the distinction between cluster and field members and the assignment of membership probabilities for each individual star in the relevant region are central issues. Here, we present an analysis based on the use of a fully automated method that relies on a hybrid heuristics, based on genetic algorithms and hill-climbing optimisation, as well as different probability distribution functions for the observables. We analysed the use of some variations for the parameterisation of those distribution functions. Finally, a catalogue comprising kinematic parameters and associated membership probability lists for all the open clusters in the Bordeaux PM2000 region was derived.

Keywords. open clusters and associations: general, methods: statistical, methods: data analysis

1. Introduction

Open clusters are known to be a most valuable tool for studies of our Galaxy and also of stellar astrophysics once the problem of selecting their physical members is solved. In the astronomical literature, open clusters have been used for many purposes, from determination of the spiral structure to investigations of star-formation mechanisms. Particularly, they play an important role as tracers of the dynamics (e.g., Frinchaboy & Majewski 2008) and chemical evolution of our Galaxy's disk (e.g., Friel 1995). Also, at the dawn of precise astronomical surveys such as delivered by *Gaia* (Perryman *et al.* 2001), their contribution to astrophysical studies tends to be of increasing importance.

The problem of membership selection is traditionally solved using kinematic information alone, but in principle membership determination could be done in multidimensional space, using (for example) spatial and kinematic information, as in Zhao *et al.* (1998), or even colour–magnitude diagram (CMD) isochrone information, as in Kharchenko *et al.* (2005). Nonetheless, we argue that when analysing these objects one should rely more on kinematics and as little as possible on a single-age, CMD-based analysis. This is so because it has been reported in the literature that star formation in some open clusters could be noncoeval: see, e.g., Eisenhauer *et al.* (1998) for NGC 3603 and Strobel (1992) for 14 other open clusters.

Keeping this scenario in mind, we set out to obtain open-cluster membership-probability lists and kinematic parameters using a fully automated method, a set of modified parameterisations for the probability distribution functions of Zhao *et al.* (1990, 2006) and high-quality proper-motion data from the Bordeaux PM2000 catalogue.

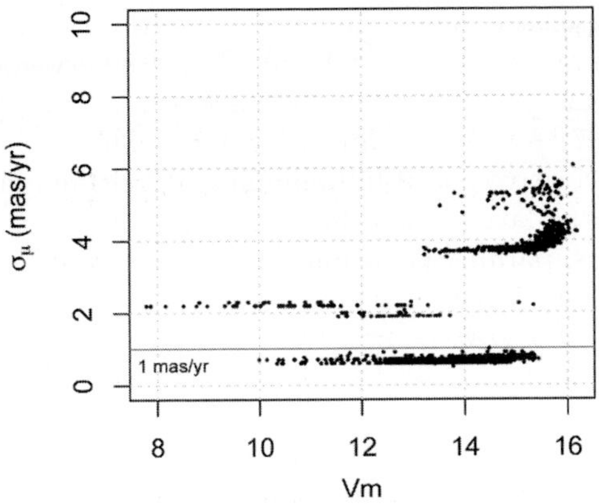

Figure 1. Example of the error distribution for the total proper motion as a function of the meridian V_M magnitude. The data was extracted from the PM2000 catalogue around the NGC 2682 cluster.

2. Data

The source for the proper-motion data is the PM2000 catalogue of Ducourant *et al.* (2006). It comprises $\sim$ 2.6 million stars in the declination range $+11^\circ \leqslant \delta_{J2000} \leqslant +18^\circ$, up to $V_M \sim 16.2$ mag. It contains positions and proper motions on the ICRS, as well as CCD meridian-circle V_M magnitudes. The PM2000 catalogue was derived from a compilation of systematic drift-scan observations in the Bordeaux Carte du Ciel zone with an automated meridian circle, the reduction of 512 Carte du Ciel plates (epoch $t \approx$ 1900) of the Bordeaux zone, scanned with the APM in Cambridge, and three catalogues: AC2000.2 ($t \approx 1907$), USNO-A2.0 ($t \approx 1950$) and the unpublished USNO Yellow Sky catalogue ($t \approx 1978$). The positional precision ranges from 50 to 70 mas, while the proper-motion precisions vary from 1.5 to 6 mas yr^{-1}, depending on magnitude. All data were analysed using a global iterative astrometric reduction process, as described in Teixeira *et al.* (1992) and Benevides–Soares & Teixeira (1992). An example of the errors on the total proper motion can be seen in Figure 1. Open cluster selection was based on the Dias *et al.* (2002) list: for all objects listed therein that were also found in the PM2000 declination range, we extracted PM2000 data.

3. Method

We adopted variations of the general form of the probability distribution function (PDF) for the proper motion of the stars in the region of the clusters. These were basically modified versions of the PDFs proposed by Zhao & He (1990), consisting of bivariate Gaussian mixtures. We considered circular and elliptical cluster PDFs and also took into account that the observational errors could be much greater than the internal dispersion, setting the latter to zero.

We also adopted another PDF that takes the (α, δ) positions of the individual stars into account, as was done by Zhao *et al.* (1998). However, here we directly used the stellar coordinates, leaving the cluster centre as a free parameter. For all parameterisations we constructed the logarithmic likelihood function to be optimised.

To find better global optima for this function, we chose to use heuristics based on evolutionary computing, genetic algorithm (GA). Nonetheless, to find the solution with pure GAs is a very expensive task from a computational point of view, as they are generally slow to converge. On the other hand, they are very good at finding the region where the optimum point lies. Therefore, we decided to adopt the code described in Mebane & Sekhon (2007) for the optimisation task. This code is based on a mixture between a GA and a derivative-based hill-climbing algorithm: while the former finds the region of the optimum, the latter finds the optimal point.

All adopted algorithms, parameterisations and data were validated through a comparison with the literature and Monte Carlo simulations.

4. Results

We applied the GA and all parameterisations described above to the selected PM2000 data. The kinematic parameters and membership probability lists for all known open cluster regions in our cluster input catalogue were obtained.

Next, we visually inspected the results using the proper-motion distributions (VPD), probability histograms and member/nonmember star charts. An example of the results for a poorly populated object can be seen in Figure 2.†

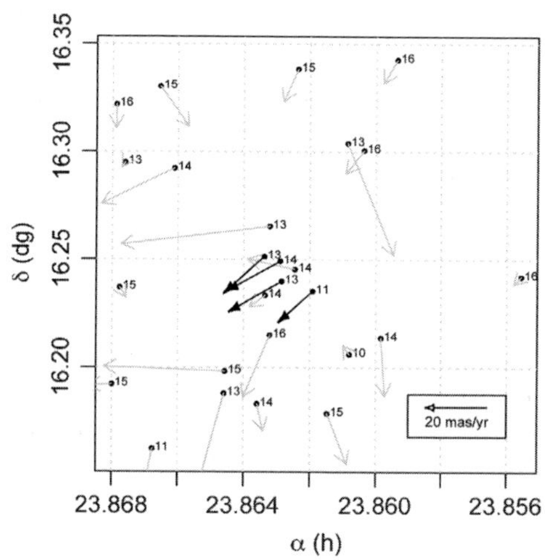

Figure 2. NGC 7772 star chart. The stars with dark arrows are considered members. The numbers indicate the meridian V_M magnitudes.

We realised that, in most cases, the parameterisation that considers the observed internal dispersion of the cluster as zero, allowed us to obtain the best diagrams, even if in several cases the solution obtained using other parameterisations had higher log-likelihood values. This probably means that for most objects we still do not have enough precision for the proper-motion data to disentangle the intrinsic internal dispersion from the observational errors, and that taking into account this parameter could lead one to possibly misleading interpretations of the observations.

We also note that using hybrid optimisation heuristics, besides making the analysis completely automated, allowed us to obtain kinematic parameters even for some very poorly populated objects, such as the open cluster remnant NGC 7772 (see Figure 2).

† We note that only the three stars sharing almost the same proper-motion vector have high-probability membership values using the adopted parameterisation.

5. Conclusions and perspectives

We obtained kinematic parameters and probability membership lists for all open clusters in the PM2000 catalogue. For some clusters, this was the first measurement of their kinematic parameters and membership lists in the literature.

Also, a fully automated system for determining the parameters and membership lists was developed. Although we only used kinematic and spatial data, the tools we have developed could be easily adapted to successfully use CMD, radial velocity or any other data. We note, however, that great care should be taken when choosing which observables to add to the analysis, because for any observable there must be an accompanying model. The latter could bias the analysis results.

More details, as well as the resulting kinematic parameters, membership lists and comments on some interesting objects will be published in a forthcoming journal article.

Acknowledgements

Part of this work was supported by the Brazilian agencies FAPESP (2006/04251-4) and CAPES (BEX0812/07-2) and the Brazilian–French cooperation agreement CAPES–COFECUB (587/07). This research has made use of ALADIN (Bonnarel *et al.* 2000), R (R Development Core Team 2008), *Vizier* (Ochsenbein *et al.* 2000) and the WEBDA database (Paunzen & Mermilliod 2008).

References

Benevides–Soares, P. & Teixeira, R. 1992 *A&A*, 253, 307

Bonnarel, F., Fernique, P., Bienaymé, O., Egret, D., Genova, F., Louys, M., Ochsenbein, F., Wenger, M., & Bartlett, J. G. 2000, *A&AS*, 143, 33

Dias, W. S., Alessi, B. S., Moitinho, A., & Lépine, J. R. D 2002 *A&A*, 389, 871 (data accessed through VIZIER, version 2002–2007)

Ducourant, C., *et al.* 2006 *A&A*, 448, 1235

Eisenhauer, F., Quirrenbach, A., Zinnecker, H., & Genzel, R. 1998, *ApJ*, 498, 278

Friel, E. D. 1995, *ARA&A*, 33, 381

Frinchaboy, P. M. & Majewski, S. R. 2008, *AJ*, 136, 118

Kharchenko, N. V., Piskunov, A. E., Röser, S., Schilbach, E., & Scholz, R.-D. 2005, *A&A*, 438, 1163

Mebane Jr., W. R. & Sekhon, J. S. 2007, *J. Stat. Softw.*, in press (`http://sekhon.berkeley.edu/`)

Ochsenbein, F., Bauer, P., & Marcout, J. 2000, *A&AS*, 143, 221

Paunzen, E. & Mermilliod, J.-C. 2008, `http://www.univie.ac.at/webda/`

Perryman, M., de Boer, K., Gilmore, G., Høg, E., Lattanzi, M., Lindegren, L., Luri, X., Mignard, F., Pace, O., & de Zeeuw, P. T. 2001, *A&A*, 369, 339

R Development Core Team, 2008, ISBN 3-900051-07-0, `http://www.R-project.org`

Strobel, A. 1992, *A&A*, 253, 374

Teixeira, R., Réquième, Y., Benevides–Soares, P., & Rapaport, M. 1992, *A&A*, 264, 307

Zhao, J. L. & He, Y. P. 1990, *A&A*, 237, 54

Zhao, J. L., Chen, L., & Wen, W. 2006, *ChJA&A*, 6, 435

Star clusters: basic galactic building blocks
Proceedings IAU Symposium No. 266, 2009
R. de Grijs & J. R. D. Lépine, eds.

doi:10.1017/S1743921309991657

Metallicities of intermediate-age and old star clusters in the Large Magellanic Cloud

Arūnas Kučinskas,[1,2] Vidas Dobrovolskas,[2] Algimantas Černiauskas,[3,1] Roma Lazauskaitė[3] and Toshihiko Tanabé[4]

[1]Vilnius University, Institute of Theoretical Physics and Astronomy, Vilnius, Lithuania
[2]Astronomical Observatory of Vilnius University, Lithuania
[3]Department of Physics, Vilnius Pedagogical University, Lithuania
[4]Institute of Astronomy, University of Tokyo, Japan

Abstract. We derive photometric metallicities for 56 intermediate-age and old star clusters in the Large Magellanic Cloud from the slopes of their red-giant branches in near-infrared color–magnitude diagrams (CMDs). The cluster sample covers the LMC bar and disk to a distance of ~ 2 kpc from the LMC center. The derived cluster metallicity distribution spans a range of [Fe/H] = +0.0...1.4 and is similar to the metallicity distribution of field stars. We find that the average cluster metallicities in certain fields of the LMC bar and disk may be different, but there seems to be no compelling evidence for a global cluster metallicity gradient in the LMC.

Keywords. Magellanic Clouds, galaxies: star clusters, galaxies: abundances

---

The full poster (in pdf format) is available at
`http://www.astro.iag.usp.br/~iaus266/Posters/pKucinskas.pdf`.

Star clusters: basic galactic building blocks
Proceedings IAU Symposium No. 266, 2009
R. de Grijs & J. R. D. Lépine, eds.

doi:10.1017/S1743921309991669

Super star clusters in HII galaxies

Patricio Lagos,[1] Eduardo Telles[2] and E. R. Carrasco[3]

[1]Instituto de Astrofísica de Canarias, Via Láctea s/n, 38200 La Laguna, Spain
email: plagos@iac.es

[2]Observatório Nacional, Rua José Cristino, 77, Rio de Janeiro, 20921-400, Brazil

[3]Gemini Observatory/AURA, Southern Operations Center, Casilla 603, La Serena, Chile

Abstract. We summarize our results based on observations with the NIRI camera on the *Gemini North* telescope of three HII galaxies (Mrk 36, UM 408 and UM 461), obtained to identify and determine the ages and masses of the elementary components (the star cluster population) of the starburst regions in compact HII galaxies. Our preliminary results indicate that the masses of the stellar clusters in these galaxies range from $\sim 10^4$ to $\sim 10^6$ $M_\odot$, with associated ages of a few Myr. The most massive star clusters fall in the so-called super star cluster category. The identification of these clusters suggests that the formation and evolution of massive star clusters is the dominant mode of star formation in these galaxies. Their spatial distribution and ages seem to indicate that star formation is simultaneous over these timescales in some of our objects. We also review our recent description of the spatial distribution of physical conditions in the HII galaxy UM 408 using the GMOS integral-field unit on *Gemini South*. The spatial distribution of the oxygen abundance does not show any significant variation or gradient across the galaxy on scales of hundreds of parsecs, within our observational uncertainties, confirming that this compact HII galaxy, like other previously studied dwarf irregular galaxies, is chemically homogeneous.

Keywords. galaxies: dwarf, galaxies: star clusters, galaxies: ISM, ISM: abundances

1. Introduction

HII galaxies are metal-poor galaxies characterized by their blue optical colors, small sizes ($\gtrsim$ 1 kpc), and low luminosities ($M_B > -18$ mag), while undergoing vigorous star formation. Their observed colors and (HII-like) spectra indicate that they are young starbursts (Sargent & Searle 1970). The hypothesis of these objects being young galaxies forming their first generation of stars in the local Universe has been ruled out, since an underlying red stellar population has been detected in most observed galaxies in this class (e.g., Telles & Terlevich 1997; Cairós *et al.* 2003).

HII galaxies can be classified into two classes of objects (Telles *et al.* 1997), luminous galaxies with an irregular outer shape and evident signs of disturbance (Type I), and compact, regular, low-luminosity objects (Type II). The present star-formation activity in most Type I galaxies, can be explained as the consequence of a merger or fusion of objects, but this scenario does not explain the star-formation activity in compact and regular galaxies. In fact, HII galaxies are isolated objects, without an excess of companions when their clustering properties are compared with more massive galaxies (Telles & Madox 2000). The lack of external perturbers in the most compact and isolated galaxies indicates that an additional mechanism other than tidal interactions must be considered to explain their present star-formation activity. The internal mechanism which may trigger star formation in these galaxies without an obvious external agent is as yet unknown, but may be related to the overall physical conditions of the interstellar medium (ISM), particularly the gas surface densities, in conjunction with stochastic effects, allowing star formation to start.

Once star formation has been initiated, the collective effect of stellar winds and supernova ejecta from young massive stellar clusters or super star clusters (SSCs) sweep up the ISM, producing a dispersion of metals (Tenorio–Tagle 1996) on scales of ~ 1 kpc and timescales of less than $\sim 10^7$ yr. This scenario has been invoked to explain the observed distribution of chemical abundances in HII galaxies (Lagos *et al.* 2009; and references therein).

The aim of this contribution is to show how high-resolution images and spectroscopy, obtained using 8m-class telescopes, can be used to study a sample of compact HII galaxies. We used near-infrared images (NIRI; Near-Infrared Imager) to resolve the star cluster population in the starburst regions and to study the clustered mode of star formation in compact galaxies. In addition, we took advantage of GMOS–IFU facilities to study the degree of chemical homogeneity in the ISM of a sample of compact HII galaxies. The spatial variations of the abundances should reflect the physical mechanisms involved in their recycling properties.

2. Stellar populations in HII galaxies: near-infrared photometry

To study the stellar populations in HII galaxies, we obtained broad-band J (1.25μm), H (1.65μm), $K_{\rm p}$ (2.12μm) and narrow-band Brγ images of the galaxies Mrk 36, UM 408 and UM 461 using NIRI on *Gemini North*. In these high-resolution images we detected a myriad of compact regions that were associated with star clusters. We compared the near-infrared broad-band images with the Brγ emission images to identify the young cluster population. Brγ emission was found in our sample of galaxies, but the majority of the clusters do not show evidence of strong emission. The properties (ages and masses) of these clusters were calculated by comparing the observed colors with STARBURST99 (Leitherer *et al.* 1999) models. In Figure 1a we show the $K_{\rm p}$-band image of the compact galaxy UM 408. In this image, we identify with circles the detected stellar clusters. Our preliminary results indicate that the masses of these clusters range from $\sim 10^4$ to $\sim 10^6$ $M_\odot$, with associated ages of a few Myr. We note that the clusters are practically coeval, at least in UM 408, with ages of ~ 5 Myr, suggesting that star formation was triggered simultaneously.

3. Spatial distribution of oxygen abundance: IFU spectroscopy

We started a program to study the spatial distribution of the physical conditions in compact HII galaxies. Thus far, we have analysed in detail the object UM 408 (Lagos *et al.* 2009). The observations were performed using GMOS–IFU on *Gemini South*, using the B600 and R600 gratings in one-slit mode, covering a total spectral range from ~ 3021 to 7225 Å. We calculated oxygen abundances in the regions where [OIII]λ4363 was detected. In Figure 1b we show the smoothed spatial distribution of 12+log(O/H) in the GMOS–IFU field of view (see the area in the $K_{\rm p}$-band image of the galaxy). Each aperture or pixel in Figure 1b corresponds to 0.2 arcsec. The two giant regions formed by clusters 4+5 and 2+3, labelled A and B in Lagos *et al.* (2009), show a difference in oxygen abundance of only Δ(O/H) = 0.02 dex, indicating that these regions have identical chemical properties, within the errors. The bulk of the observed data points (black dots in Figure 1c) are lying in a region of $\pm 2\sigma$ ($\sigma = 0.1$ dex) with respect to the average value. Therefore, the new metals formed in the current star-formation episodes are not observed and reside probably in the hot gas phase ($T \sim 10^7$ K), while the metals from previous star-formation events

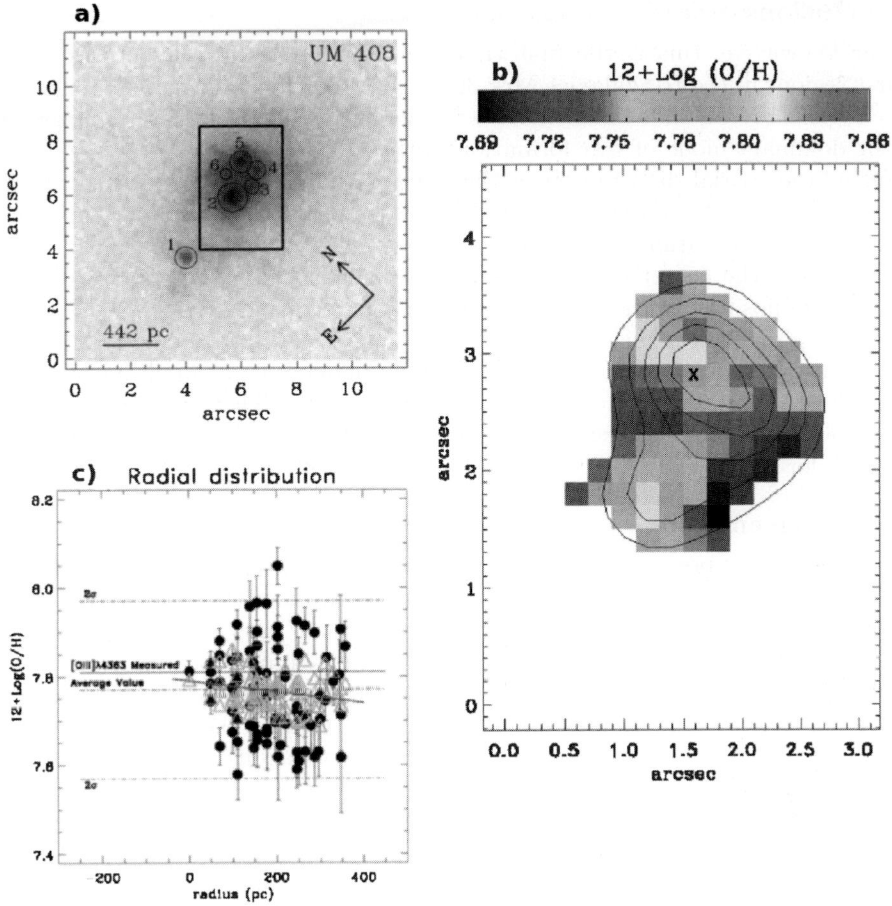

Figure 1. Correlation between the spatial distribution of oxygen abundances [$12+\log(\mathrm{O/H})$] and the position of the stellar clusters in the galaxy UM 408. *(a)* K_{p}-band image of the galaxy UM 408. The rectangle indicates the field of view of $3'' \times 4.4''$ used for the GMOS–IFU observation. *(b)* Smoothed spatial distribution. The isocontours display the Hα emission. The maximum Hα emission is found associated with region A and is indicated in the map by a cross. Hα contours are overplotted on the map. *(c)* Radial distribution. The central dotted line represents the $12+\log(\mathrm{O/H})$ average value of 7.77, and the solid line the integrated value of $12+\log(\mathrm{O/H}) = 7.81$ in the region where the [OIII]λ4363 emission lines were measured. The 2σ dispersion is represented by the dotted lines. The difference between the maximum and minimum values in the original data set (black dots) is 0.47 dex. Light blue triangles represents the smoothed $12+\log(\mathrm{O/H})$ radial distribution. A least-squares fit using the original data set is indicated by the red line. For more details, see Lagos *et al.* (2009).

are well mixed and homogeneously distributed throughout the whole extent of the galaxy. The absence of chemical overabundances in the ISM of UM 408 and in the dwarf galaxies studied in the literature leads to the conclusion that the population of young clusters are not producing localized overabundances. This agrees with theoretical estimates (e.g., Wofford 2009). On the other hand, we observed that there is a very marginal gradient of decreasing abundance from the center outwards, indicating that the highest abundance values are found near the peak of Hα emission.

4. Conclusions

To our knowledge, this is the first time that the elementary structures within the starburst regions in the HII galaxies Mrk 36 and UM 408 are resolved. The identification of these regions reveals that the formation and evolution of massive star clusters appears to be the dominant mode of star formation in these HII galaxies.

As far as the spatial distribution of the oxygen abundance is concerned, we did not detect localized overabundances associated with the positions of the star clusters in UM 408, within our observational uncertainties. However, a marginal gradient of decreasing abundance from the center outwards indicates that the highest abundance values are found near the position of the detected star clusters in this galaxy.

All results presented in this contribution are suggestive that compact HII galaxies are unevolved, low-metallicity dwarf galaxies possibly undergoing a simultaneous episode of star formation over their entire optically observed extent. They are a genuine example of the simplest starbursts occurring on galactic scales, possibly mimicking the properties one expects for young galaxies at high redshift.

Acknowledgements

PL acknowledges a postdoctoral grant from the Spanish MCINN within the ***Estallidos*** collaboration, partially funded by the Spanish MCINN under the Plan Nacional de I + D 2007 grant AYA2007-67965-C03-01: Estallidos (`http://www.iac.es/project/GEFE/estallidos`). Based on observations obtained at the Gemini Observatory, which is operated by the Association of Universities for Research in Astronomy, Inc., under a co-operative agreement with the NSF on behalf of the Gemini partnership: the National Science Foundation (USA), the Science and Technology Facilities Council (UK), the National Research Council (Canada), CONICYT (Chile), the Australian Research Council (Australia), Ministério da Ciência e Tecnologia (Brazil) and Ministerio de Ciencia, Tecnología e Innovación Productiva (Argentina).

References

Cairós, L. M., Caon, N., Papaderos, P., Noeske, K., Vílchez, J. M., Lorenzo, B. G., & Muñoz–Tuñón, C. 2003, *ApJ*, 593, 312

Lagos, P., Telles, E., Muñoz–Tuñón, C., Carrasco, E. R., Cuisinier, F., & Tenorio–Tagle, G. 2009, *AJ*, 137, 5068

Leitherer, C., *et al.* 1999, *ApJS*, 123, 3

Sargent, W. L. W. & Searle, L. 1970, *ApJ*, 162, 155

Telles, E. & Terlevich R. 1997, *MNRAS*, 286, 183

Telles, E., Melnick, J., & Terlevich, R. 1997, *MNRAS*, 288, 78

Telles, E. & Maddox, S. 2000, *MNRAS*, 311, 307

Tenorio–Tagle, G. 1996, *AJ*, 111, 1641

Wofford, A. 2009, *MNRAS*, 395, 1043

Star clusters: basic galactic building blocks
Proceedings IAU Symposium No. 266, 2009
R. de Grijs & J. R. D. Lépine, eds.

doi:10.1017/S1743921309991670

W UMa-type systems in globular clusters

Lifang Li, Fenghui Zhang and Zhanwen Han

National Astronomical Observatories/Yunnan Observatory, Chinese Academy of Sciences,
P.O. Box 110, Kunming, Yunnan Province, 650011, P. R. China
email: `llf@ynao.ac.cn`

Abstract. W UMa systems can be found everywhere in the Galaxy. They can be used as a distance tracer. Therefore, W UMa systems are very important to investigate the structure of the Galaxy. The distance to W UMa systems in globular clusters (GCs) is determined using a period–color–luminosity relation. It is found that the mean distance (r_a) of W UMa systems is consistent with their host cluster distances ($r_{\rm GC}$) deduced from their intrinsic distance moduli if $r_{\rm GC} \leqslant 10$ kpc. There is a significant difference between r_a and $r_{\rm GC}$ for $r_{\rm GC} \geqslant 10$ kpc. We discuss the reasons causing this deviation.

Keywords. binaries: eclipsing, globular clusters: general

1. Introduction

W UMa contact binaries (also known as EW systems; Rucinski 2000) are unique objects in which the luminosity is redistributed through their common envelope. W UMa systems are very easy to detect and identify owing to (i) the rather large amplitudes of their light changes and (ii) their short periods, so that limited-duration monitoring is sufficient (Rucinski 2000). W UMa systems can be found everywhere in our Galaxy. In fact, many EW systems have been detected in Galactic cluster searches (see Rucinski 1998). Meanwhile, the absolute magnitude of EW systems is related to their orbital periods and colors, implying that EW systems can be used as distance indicators and would play an important role in studying the structure of our Galaxy (Rucinski 2004).

2. EW systems as a distance indicator to globular clusters

Many EW systems have been found in GCs. We have collected data for the EW systems in 17 GCs: see Table 1. The absolute magnitude of EW systems is determined through a period–color–luminosity (PCL) relation (Rucinski 2000),

$$M_{V,\rm Ruc} = -4.44 \log P + 3.02(B-V)_0 + 0.12, \tag{2.1}$$

or

$$M_{V,\rm Ruc} = -4.43 \log P + 3.63(V-I)_0 - 0.31. \tag{2.2}$$

If the effect of metallicity must be taken into account, Rucinski (2000) provided metallicity corrections, $\delta M_V^{BV} = -0.3$[Fe/H] and $\delta M_V^{VI} = -0.12$[Fe/H], respectively. The distances to EW systems can then be calculated using

$$5 \log r = (m-M)_{V,\rm Ruc} + 5 - A_V. \tag{2.3}$$

This leads to a mean distance (r_a) of EW systems in a given GC. However, the color excess and interstellar extinction of each EW system are usually taken to be a mean value relavent to their host GC, except for EW systems in NGC 3201 and M12, in which these quantities have been obtained for individual systems (von Braun & Mateo 2002;

Table 1. Some parameters for 17 GCs.

Clusters	l (deg)	b (deg)	r_{GC} (kpc)	$E(B-V)$ (mag)	$(m-M)_V$ (mag)	[Fe/H] (dex)	N_{EW}	r_a (kpc)	$r_{a,met}$ (kpc)	Refs
NGC 104	305.90	−44.89	4.4	0.04	13.37	−0.76	7	4.1	3.68	1
NGC 288	152.28	−89.38	8.4	0.03	14.69	−1.24	2	12.81	10.8	1
NGC 4372	300.99	−9.88	7.1	0.39	15.01	−2.09	8	6.4	4.8	1
Rup 106	300.89	+11.67	18.5	0.20	17.25	−1.67	2	5.7	4.6	1
NGC 5139	309.10	+14.97	6.4	0.12	13.97	−1.62	18	5.44	4.97	1
NGC 5272	42.21	+78.71	12.2	0.01	15.12	−1.57	1	6.41	5.16	1
NGC 5466	42.15	+73.59	17.2	0.00	16.15	−2.22	2	16.21	11.91	1
NGC 6121	350.97	+15.97	5.9	0.36	12.83	−1.20	16	3.65	3.09	1
NGC 6362	325.55	−17.57	5.3	0.08	14.79	−0.95	4	3.91	3.43	1
NGC 6397	338.17	−11.96	2.96	0.18	12.36	−1.95	8	3.72	3.34	1,2,3
NGC 6441	353.53	−5.01	3.5	0.44	16.62	−0.53	13	5.46	5.09	1,4
NGC 6752	336.50	−25.63	5.2	0.04	13.13	−1.56	7	4.91	3.96	1
NGC 6838	56.74	−4.56	6.7	0.25	13.75	−0.73	8	5.09	4.60	1,5
NGC 6934	52.10	−18.90	16.5	0.09	16.09	−1.53	2	6.84	5.53	6
NGC 3201	277.23	+8.46	4.65	0.24	14.08	−1.54	9	4.81	4.42	7,8,9
M12	15.7	+26.3	4.9	0.26	13.46	−1.54	1	5.40	5.00	10
NGC 6388	345.6	−6.7	6.16	0.40	15.19	−1.20	5	5.26	4.84	11

Columns: l, Galactic latitude; b, Galactic longitude; $E(B-V)$, colour excess; r_{GC}, distance to the Sun; $(m-M)_V$, distance modulus; [Fe/H], metalicity; N_{EW}, total number of EW systems in the cluster; r_a, average distance of EW systems; $r_{a,met}$, average distance including the effect of metallicity.
References in Table 1: (1) Rucinski (2000), (2) Kaluzny *et al.* (2003), (3) Kaluzny *et al.* (2006), (4) Pritzl *et al.* (2001), (5) Park & Nemec (2000), (6) Kaluzny *et al.* (2001), (7) Layden *et al.* (2003), (8) von Braun & Mateo (2002), (9) Mazur *et al.* (2003), (10) von Braun *et al.* (2002), (11) Pritzl *et al.* (2002).

von Braun *et al.* 2002). In addition, the distance, r_{GC}, to the GCs can also be derived from their distance modulus, $(m-M)_V$, through Equation (2.3). These are also included in Table 1.

The relation between r_a and r_{GC} is shown in Figure 1. It can be seen that the values of r_a are consistent with r_{GC} for GCs with $r_{GC} \leqslant 10$ kpc, and there is a significant difference between the two distances for some GCs at greater distances: r_a is usually less than r_{GC} for these relatively distant GCs.

3. Discussion and conclusions

Although the average distance to the EW systems is very different from their parent cluster distance for some distant GCs, EW systems might be reliable distance indicators. We will now explore which effects might cause such differences. (i) Some EW systems found in the field of these GCs might not be cluster members. It is necessary to find a more accurate way to determine whether or not a star is indeed a cluster member. (ii) The stars in GCs are very crowded while the dust distribution in GCs is not necessarily uniform. There could be a significant difference in the reddening and extinction values for EW systems in GCs (von Braun & Mateo 2002; von Braun *et al.* 2002). (iii) Some GCs at large distances may suffer from significant effects related to their presence close to the center of the Galaxy, such as the high extinction associated with the Galactic center or reflection of its luminosity.

EW systems can be used as standard candles tracing small-scale structure, especially in our Galaxy (Rucinski 2004). Therefore, they would play an important role in studies of the structure of our Galaxy. However, color excesses and extinctions should be accurately

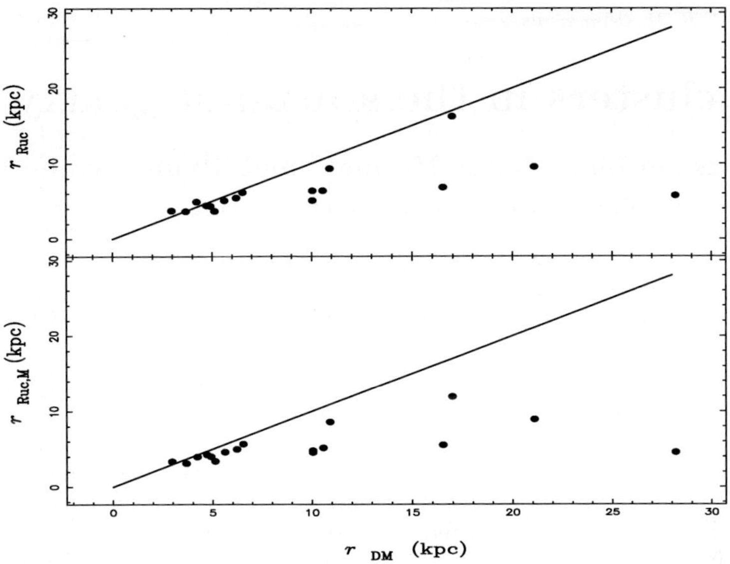

Figure 1. Relation between r_a and $r_{\rm DM} = r_{\rm GC}$. The top panel does not take into account the effects of metallicity, which are included in the bottom panel. The solid line represents $r_a = r_{\rm GC}$.

determined for each EW system separately, as done for NGC 3201 (von Braun & Mateo 2002), because the stars in GCs are very crowded and the dust distribution is usually not uniform. In fact, differential $E(V - I)$ variations of up to ~ 0.2 mag on a scale of arcminutes across NGC 3201 have been presented in the form of an extinction map (von Braun & Mateo 2001). To find accurate distances to EW systems in GCs through the PCL relation of Rucinski 2000, it is also necessary to find a way in which we can distinguish GC members from field stars.

Acknowledgements

This work was partially supported by the Chinese Natural Science Foundation (grants 10673029, 10773026, 10821061, 10521001 and 2007CB815406), and also by the Yunnan Natural Science Foundation (2007A113M).

References

Kaluzny, J., Olech, A., & Stanek, K. Z. 2001, *AJ*, 121, 1533

Kaluzny, J. & Thompson, I. B. 2003, *AJ*, 125, 2534

Kaluzny, J., Thompson, I. B., Krzeminski, W., & Schwarzenberg–Czerny, A. 2006, *MNRAS*, 365, 548

Layden, A. C. & Sarajedini, A. 2003, *AJ*, 125, 208

Mazur, B., Krzeminski, W., & Thompson, I. B. 2003, *MNRAS*, 340, 1205

Park, N. K. & Nemec, J. M. 2000, *AJ*, 119, 1803

Pritzl, B. J., Smith, H. A., Catelan, M., & Sweigart, A. V. 2001, *AJ*, 122, 2600

Pritzl, B. J., Smith, H.A., Catelan, M., & Sweigart, A. V. 2002, *AJ*, 124, 949

Rucinski, S. M. 1998, *AJ*, 116, 2998

Rucinski, S. M. 2000, *AJ*, 120, 319

Rucinski, S. M. 2004, *NewAR*, 48, 703

von Braun, K. & Mateo, M. 2001, *AJ*, 121, 1522

von Braun, K. & Mateo, M. 2002, *AJ*, 123, 279

von Braun, K., Mateo, M., Chiboucas, K., Athey, A., & Hurley–Keller, D. 2002, *AJ*, 124, 2067

Star clusters: basic galactic building blocks
Proceedings IAU Symposium No. 266, 2009
R. de Grijs & J. R. D. Lépine, eds.

doi:10.1017/S1743921309991682

Star clusters in the starburst galaxy M82

Sungsoon Lim,[1] Narae Hwang[1,2] and Myung Gyoon Lee[1]

[1]Department of Physics and Astronomy, Seoul National University, 599 Gwanak-ro, Gwanak-gu, Seoul, Republic of Korea
email: [slim, mglee]@astro.snu.ac.kr

[2]National Astronomical Observatory of Japan, 2-21-1 Osawa, Mitaka, Tokyo, Japan
email: narae.hwang@nao.ac.jp

Abstract. We present a study of star clusters in the starburst galaxy M82 using the *BVI* mosaic images taken with the *Hubble Space Telescope*'s Advanced Camera for Surveys. We have selected about 850 clusters with $V < 23$ mag based on their morphological information. The brightest cluster is as bright as $V \sim 16.5$ mag ($M_V \sim -11.2$ mag) and most clusters are fainter than $V = 18$ mag ($M_V = -9.7$ mag). The V-band luminosity function is represented by a power law with a slope $\alpha \sim -2.0$ in the range of $-9.5 < M_V < -7$ mag. The star clusters in M82 are mainly distributed in the galaxy's disk. However, about a dozen clusters are found far from the disk and are considered to belong to the halo of M82. The color–magnitude diagrams of star clusters show that most star clusters are highly reddened. We have derived the age of the star clusters using a spectral-energy-distribution fitting method with Bruzual & Charlot simple stellar population models. We discuss the age distribution as well as the photometric properties of the star clusters in regard to the formation history of the M82 cluster population.

Keywords. galaxies: individual (M 82), galaxies: star clusters

1. Introduction

M82 is the nearest starburst galaxy (except for IC10) and HI observation revealed that it is interacting with M81 and NGC 3077 (Yun *et al.* 1994). There are numerous studies of the star clusters in this galaxy, but most covered only a small part of the galaxy (e.g., de Grijs *et al.* 2001; Melo *et al.* 2005, etc.). Recently, Mayya *et al.* (2008) presented a study of 653 star clusters in the disk and central region of M82, based on the *Hubble Space Telescope (HST)*/Advanced Camera for Surveys (ACS) data. They concluded that disruption of star clusters in M82 is affected by the mass and size of the clusters, using the mass estimates derived assuming that the mean ages of the nuclear and disk clusters are, respectively, 8 and 100 Myr. In this study, we present a study of the star clusters in a wide field covering the entire disk of M82.

2. Data and method

We used mosaic images obtained by the Hubble Heritage Team using *HST*'s ACS Wide Field Channel (WFC) through the F435W (B), F555W (V) and F814W (I) filters in March 2006 (Mutchler *et al.* 2007). The field of view is about $12' \times 8'$ centered on M82 and covers its entire optical disk.

We used SExtractor (Bertin & Arnouts 1996) for star cluster detection and photometry. Since star clusters in M82 ($D \approx 3.5$ Mpc; Lee *et al.* 2009) appear slightly extended in the *HST* images, we selected star cluster candidates automatically using the structural parameters given by SExtractor (e.g., $2.5 <$ FWHM < 40 pixels, ellipticity < 0.8, area > 40 pixels). To select a clean cluster sample, however, we visually inspected

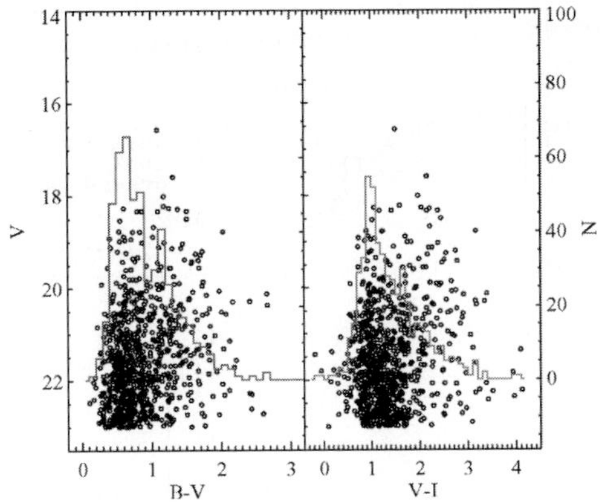

Figure 1. Color–magnitude diagrams (CMDs) of star clusters in M82. Solid lines show $(B-V)$ and $(V-I)$ color distributions for $V < 22$ mag, respectively.

the shapes of the radial profiles and 2D contours of star cluster candidates of $V < 23$ mag. We compared the photometric data of clusters with the theoretical spectral-energy distribution (SED) to derive the age of the star clusters. SED models were produced using the population synthesis model of Bruzual & Charlot (2003).

3. Results

We selected ~ 850 star clusters with $V < 23$ mag based on visual classification. The brightest star cluster is as bright as $V \sim 16.5$ mag ($M_V \sim -11.2$ mag). However, most are fainter than $V = 18$ mag ($M_V = -9.7$ mag). The color distribution of the star clusters shows two peaks at $(B-V) \sim 0.65$ and $(B-V) \sim 1.2$ mag, respectively, and a long tail extending to the red (Figure 1).

The color–magnitude diagram (CMD) of M82 star clusters displays somewhat different features from that of M51 star clusters (Hwang & Lee 2008). The range of $(B-V)$ and $(V-I)$ colors of star clusters in M82 is much broader than that of star clusters in M51. It is suspected that this may be due to different extinctions and different inclinations between M82 and M51.

We have investigated the V-band luminosity function (LF) of the star clusters (Figure 2). The faint cut at $M_V = -4.7$ mag ($V = 23$ mag) is the magnitude limit adopted for this study. The LF for $M_V > -7$ mag becomes flat, which is probably due to incompleteness effects. The logarithmic slope is $\sim -2.01 \pm 0.05$ in the range from -9.5 to -6.5 mag, and is in a good agreement with the result of Mayya *et al.* (2008). It is found that the LF slope of M82 star clusters is similar to that of Small and Large Magellanic Cloud (SMC and LMC) star cluster populations. However, it is flatter than the LF slopes of star cluster systems in other spiral galaxies, suggesting that a larger number of very bright star clusters exist in M82 compared with other late-type galaxies.

Most of the star clusters are located in the galaxy disk, but some are found outside the disk (Figure 3). The red star clusters, with $(B-V) > 0.9$ mag, are mostly distributed in the central region of the disk, suggesting that they are highly reddened due to the internal extinction in the galaxy. There are also many bright star clusters in the central region of the disk. They are good candidates for super star clusters (SSC). However, some may be unresolved star clusters that are located close to one another. Since M82 is an edge-on galaxy, a distinction between disk and halo populations can be made rather easily. The red clusters in the halo are probably globular clusters.

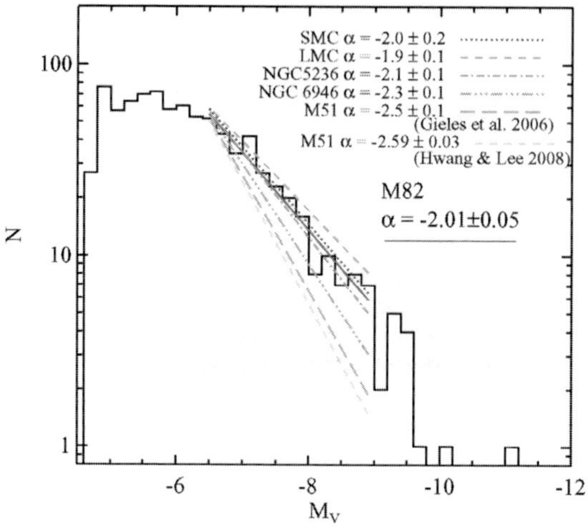

Figure 2. Luminosity function of star clusters in M82. The red solid line shows the power-law fit to the range $-9.5 < M_V < -6.5$ mag. The other lines show the luminosity function of star clusters in five other galaxies compiled from Gieles *et al.* (2006) and Hwang & Lee (2008). Each luminosity function was shifted to match that of the M82 star clusters for easy comparison.

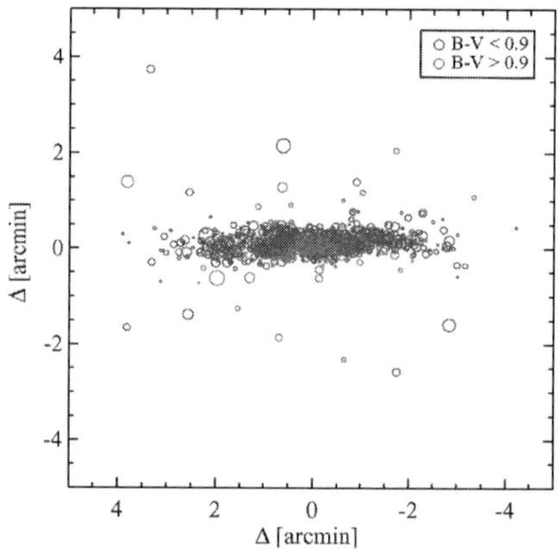

Figure 3. Spatial distribution of the star clusters. Larger symbols indicate brighter clusters. Blue and red open circles represent star clusters with $(B-V) < 0.9$ and > 0.9 mag, respectively.

The age distribution of star clusters shows three peaks, at $\log(t/\mathrm{yr}) = 6.5, 8.8$ and > 10 (3 Myr, 630 Myr and > 10 Gyr), respectively (Figure 4). The peak at 630 Myr is consistent with the epoch of the last encounter with M81 expected from theoretical simulations (Yun 1999). The existence of young clusters with $\log(t/\mathrm{yr}) = 1$ (~ 10 Myr) suggests that the recent star-formation episode has continued since about 10 Myr ago. The oldest population, $\log(t/\mathrm{yr}) > 10$, may consist of globular clusters.

4. Summary

• We have found about 850 star clusters in M82. Most are located in the disk, but some ($\sim 3\%$) are found outside the disk.

• The logarithmic slope of the star cluster luminosity function is about -2.0, similar to those for the SMC or LMC star cluster populations, but flatter than the luminosity function of M51 star clusters.

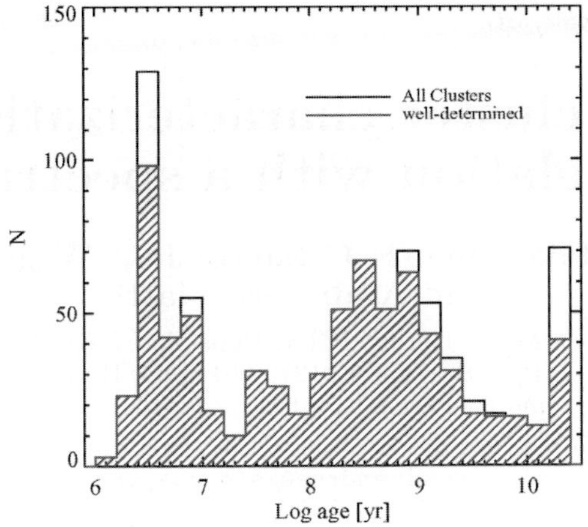

Figure 4. Age distribution of star clusters in M82. The blue shaded histogram represents star clusters with reliably fitted ages (small χ^2). Note that there are three dominant populations of clusters, with $\log(t/\mathrm{yr}) = 6.5, 8.8$ and > 10.

• The peak for the intermediate-age population corresponds to the theoretically expected epoch of the last encounter with M81. It is also consistent with the result derived from star clusters in M81 by Chandar *et al.* (2001). This suggests that M81 and M82 may have undergone a strong interaction about 100 Myr ago, triggering the formation of clusters in both galaxies.

Acknowledgements

This work is supported in part by a grant (R01-2007-000-20336-0) from the Basic Research Program of the Korea Science and Engineering Foundation. NH acknowledges support by a Grant-in-Aid for JSPS Fellows, No. 20-08325.

References

Bertin, E. & Arnouts, S. 1996, *A&AS*, 117, 393

Bruzual, G. & Charlot, S. 2003, *MNRAS*, 344, 1000

Chandar, R., Tsvetanov, Z., & Ford, H. C. 2001, *AJ*, 122, 1342

de Grijs, R., O'Connell, R. W., & Gallagher III, J. S. 2001, *AJ*, 121, 768

Gieles, M., Larsen, S. S., Bastian, N., & Stein, I. T. 2006, *A&A*, 450, 129

Hwang, N. & Lee, M. G. 2008, *AJ*, 135, 1567

Lee, M. G., *et al.* 2009, in prep.

Lim, S. 2008, *MSc Thesis*, Seoul National University

Mayya, D., Romano, R., Rodríguez–Merino, L. H., Luna, A., Carrasco, L., & Rosa–González, D. 2008, *ApJ*, 679, 404

Melo, V. P., Muñoz–Tuñón, C., Maíz–Apellániz, J., & Tenorio–Tagle, G. 2005, *ApJ*, 619, 270

Mutchler, M., *et al.* 2007, *PASP*, 119, 1

Yun, M. S., Ho, P. T. P., & Lo, K. Y. 1994, *Nature*, 372, 530

Yun, M. 1999, in: J. E. Barnes & D. B. Sanders (eds.), *Galaxy Interactions at Low and High Redshift*, Proc. IAU Symp. No. 186, p. 81

Star clusters: basic galactic building blocks
Proceedings IAU Symposium No. 266, 2009
R. de Grijs & J. R. D. Lépine, eds.

doi:10.1017/S1743921309991694

Open cluster characterization via cross-correlation with a spectral library†

Francisco F. S. Maia,[1] João F. C. Santos Jr.,[1] Wagner J. B. Corradi[1] and Andrés E. Piatti[2]

[1]Universidade Federal de Minas Gerais, ICEx, Dept. de Física, Av. Antônio Carlos, 6627, Belo Horizonte 31270-901, MG, Brazil
email: [ffsmaia, jfcsj, wag]@ufmg.br

[2]Instituto de Astronomía y Física del Espacio, CC67, Suc. 28, 1428 Buenos Aires, Argentina
email: andres@iafe.uba.ar

Abstract. We present a characterization method based on spectral cross-correlation to obtain the physical parameters of the controversial stellar aggregate ESO442–SC04. The data used was obtained with GMOS at the *Gemini South* telescope, and includes spectra of 17 stars in the central region of the object and 6 standard stars. FXCOR was used iteratively to obtain self-consistent radial velocities for the standard stars and average radial velocities for the science spectra. Spectral types, effective temperatures, surface gravities and metallicities were determined using FXCOR to correlate cluster spectra with the ELODIE spectral library and select the best correlation matches using the Tonry & Davis ratio. Analysis of the results suggests that the stars in ESO442–SC04 are not bound and, therefore, they do not constitute a physical system.

Keywords. open clusters and associations: general, open clusters and associations: individual (ESO 442–SC04)

1. Introduction

Open clusters are destroyed over time by the action of external forces (Galactic tidal field, collisions with molecular clouds) and as a consequence of their own dynamical and stellar evolution. Studies on the subject indicate that most open clusters dissolve in 0.5–2.5 Gyr (Portegies Zwart *et al.* 2001) and that finding clusters in a state of dissolution should be common. However, distinguishing such objects from a random overdensity of field stars is a difficult task, and many authors have been developing techniques to study and properly classify them (e.g., Pavani & Bica 2007).

ESO442–SC04 was identified by Bica *et al.* (2001) as a possible open cluster remnant, but was later classified by Carraro *et al.* (2005) as an asterism. We have applied an iterative method to characterize this object and derive the physical parameters of its stars.

2. Data

The Gemini Multi-Object Spectrograph (GMOS, *Gemini South*) was used to collect spectra of 36 selected stars in a $5' \times 5'$ region centered on ESO442–SC04, and 6 standard stars from Nordström *et al.* (2004). The collected spectra cover the spectral range 3875 Å − 5300 Å at a resolution of $R \approx 4000$ and signal-to-noise ratios (S/N) in the range $5 - 50$. Table 1 lists the observed standard stars and their properties (from the literature).

† The full poster (in pdf format) is available at
http://www.astro.iag.usp.br/~iaus266/Posters/pMaia.pdf.

Table 1. Literature data for the standard stars.

Object	Type	v_r (km s−1)	[Fe/H](±0.12)	T_{eff} (± 110) (K)
HD 104471	G0V	−7.2 ± 0.1	0.00	5984
HD 104982	G2V	10.5 ± 0.1	−0.40	5610
HD 105004	F8VI	121.6± 0.3	−0.79	5821
HD 107122	F1V	16.2 ± 3.3	−0.42	6576
HD 111433	F3IV	4.0 ± 0.6	+0.25	6471
CD-289374	...	30.4 ± 0.2	−1.18	4830

We adopted the ELODIE.3.1 stellar spectral library (Prugniel & Soubiran 2001) as template spectra for the cross-correlation technique to determine the physical parameters of the targets. The library includes 1962 spectra of 1388 stars, providing a broad coverage of the stellar atmospheric parameters $T_{\rm eff}$, log g and [Fe/H].

3. Radial-velocity determination

Initially, the FXCOR task was applied to the standard-star spectra, using the values given in Table 1 as a first guess of their radial velocities. A self-consistent set of solutions was then obtained by correcting the velocities and redoing the correlation to find consistent values for the full set. Table 2 shows the results of the correlations and the average radial velocities obtained.

Table 2. Calculated radial velocities for the standard stars. Bold-faced values represent the velocities adopted for each template in the correlation.

star/template	HD 104471	HD 104982	HD 105004	HD 107122	HD 111433	CD −289374
HD 104471	**−126.3±11.5**	−7.5±5.2	118.0±15.1	16.7±25.9	−24.6±11.4	30.4±5.7
HD 104982	−127.5±5.2	**−8.7±5.7**	114.0±13.4	9.4 ±24.7	−28.8±13.0	32.9±11.2
HD 105004	−122.7±15.8	−1.1±13.4	**121.6±0.3**	19.6±27.9	−20.6±18.0	37.5±16.6
HD 107122	−126.8±25.9	−1.9±24.7	118.2±27.9	**16.2±3.3**	−24.6±18.4	36.0±28.9
HD 111433	−126.3±11.5	−4.5±13.0	117.6±18.0	16.2±18.4	**−24.6±11.4**	33.7±17.8
CD −289374	−128.8±11.2	−8.7±5.2	114.5±16.6	10.6±3.3	−27.9±17.8	**30.4±0.2**
$\overline{V_r}$	−126.4±6.1	−5.4±5.4	117.3±7.0	14.8±9.5	−25.2±6.3	33.5±6.6

Next, FXCOR was run on the science spectra using the six standard-star spectra as templates. The radial velocities found for each template were averaged and the dispersion calculated for all science spectra. Table 4 shows the spectra used, their corresponding S/N and the radial velocities obtained.

4. Cross-correlation with spectral library

We used ELODIE spectral library templates for the standard stars and FXCOR to determine, through the sharpness of the correlation-function peak (Tonry & Davis ratio: TDR; Tonry & Davis 1979), the most similar templates for each star. Figure 1 shows the two best-matching spectra (found from correlation) for both the standard star HD 111443 and one of our science spectra. Their spectral types and TDR correlation parameters are also shown.

Spectral types were determined by collecting the ten most similar templates and summing over the TDR values of those with the same spectral type. The template with the highest TDR sum was adopted as the closest spectral type. Table 3 shows the resulting spectral types for the standard stars, compared with the relevant values from SIMBAD.

Effective temperatures, surface gravities and metallicities were calculated as the average values listed for the chosen templates, weighted by the TDR correlation value of each template. The adopted uncertainties correspond to the weighted standard deviation of

Table 3. Comparison of spectral types for standard stars.

	HD 104471	HD 104982	HD 105004	HD 107122	HD 111433	CD −289374
Determined	G0	G2V	F5V	F3V	F3V	G0
SIMBAD	G0V	G2V	F8IV	F1V	F3IV	—

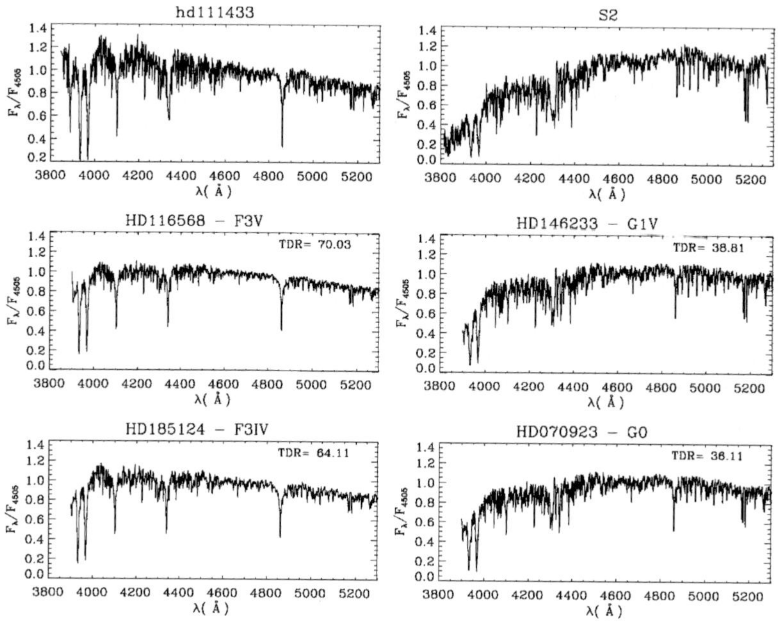

Figure 1. Best correlation templates for standard star HD 111443 *(left)* and science spectrum S2 *(right)*. The name of the template and TDR value are shown.

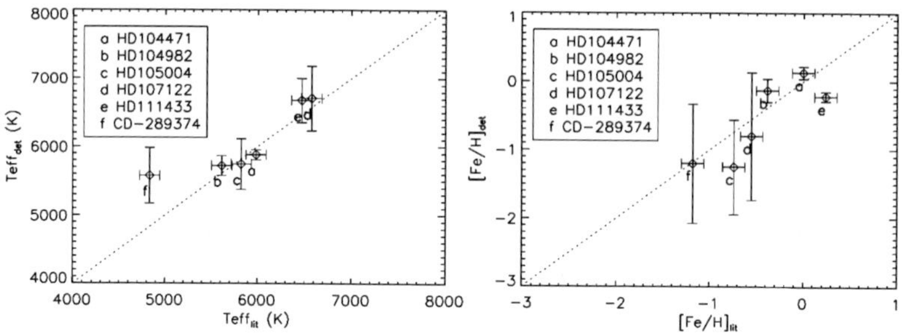

Figure 2. Comparison between our and literature parameters for the standard stars. Effective temperatures *(left)* and metallicities *(right)*.

the derived average. Figure 2 shows the relation between the effective temperatures and metallicities thus determined and the literature parameters.

We used the same method to determine the spectral types, effective temperatures, surface gravities and metallicities of our science spectra, as shown in Table 4.

5. Results

By employing cross-correlation techniques and using the ELODIE spectral library, we were able to determine heliocentric radial velocities, spectral types, effective temperatures, surface gravities and metallicities for 17 stars in the inner region of ESO442–SC04. These results are summarized in Table 4.

Table 4. Radial velocities, spectral types, effective temperatures, surface gravities and metallicities determined for the science spectra.

Spectrum	S/N	$\overline{v_r}$ (σ_{v_r}) (km s^{-1})	Sp. Type	$T_{\rm eff}$ ($\sigma_{T_{\rm eff}}$) (K)	log(g) ($\sigma_{\log(g)}$)	[Fe/H] ($\sigma_{[Fe/H]}$)
S2	15	75.8 (7.2)	G5V	5676 (117)	4.20 (0.21)	−0.01 (0.09)
S6	40	9.5 (5.2)	F5V	6347 (348)	4.24 (0.12)	−0.59 (0.27)
S9	20	25.1 (4.7)	F5V	6309 (416)	4.24 (0.08)	−0.52 (0.29)
S10	10	93.6 (10.3)	K2V	5213 (61)	4.48 (0.10)	−0.03 (0.14)
S12	20	227.2 (12.7)	G0	5380 (424)	3.14 (0.88)	−1.60 (0.66)
S14	25	110.0 (10.9)	G0	5673 (324)	3.75 (0.72)	−1.79 (0.57)
S15	30	78.5 (7.5)	G5V	5606 (115)	4.33 (0.11)	+0.02 (0.08)
S19	50	68.0 (8.2)	F3V	6834 (261)	4.28 (0.08)	−0.41 (0.29)
S20	25	74.4 (6.9)	G2V	5740 (63)	4.17 (0.18)	+0.03 (0.13)
S21	30	31.3 (5.6)	F7IV	5910 (460)	3.89 (0.75)	−0.99 (0.81)
S23	35	95.5 (8.5)	G0V	5765 (261)	3.99 (0.56)	−0.92 (0.67)
S25	6	145.9 (17.1)	F2IV	6480 (439)	4.14 (0.19)	−0.19 (0.16)
S28	30	15.0 (5.4)	F8V	6005 (306)	4.19 (0.19)	−0.57 (0.26)
S29	22	10.7 (7.1)	G8III	4877 (87)	2.76 (0.16)	−0.03 (0.15)
S30	12	40.0 (6.6)	G0III	5721 (69)	4.07 (0.18)	+0.16 (0.07)
S32	10	47.0 (6.7)	G8III	5011 (212)	2.76 (0.42)	−0.03 (0.13)
S33	12	142.4 (12.9)	G8V	5633 (117)	4.31 (0.18)	−0.01 (0.06)

2MASS data was used to construct a colour–magnitude and a spatial diagram showing the relative positions of the stars on the sky, along with the radial velocities, surface gravities and effective temperatures obtained for the targets. The objects diagrams are labelled according to their internal identification and coded with colors representing groups including a range of radial velocities: see Figure 3.

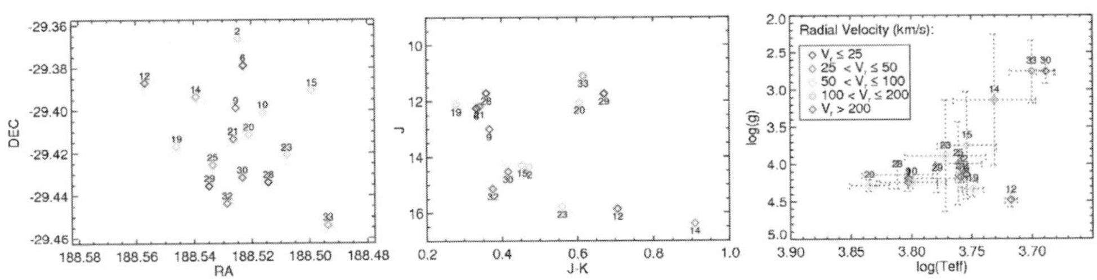

Figure 3. Positional chart *(left)*, color–magnitude diagram *(middle)* and surface gravity–temperature diagram *(right)* for the spectra. Stars are labeled by their internal identification shown in Table 4 and color-coded according to their radial velocities *(right legend)*.

6. Conclusion

We devised a spectral cross-correlation method to determine kinematic and astrophysical parameters for stars in the central region of ESO442–SC04. The radial velocities and metallicities exhibit dispersions greater than 50 km s^{-1} and 0.6 dex, respectively, suggesting that few, if any, of the stars are physically bound. Although their distribution in the CMD (Figure 3) suggests that they follow evolutionary sequences typical of old clusters, our results suggest that ESO442–SC04 does not constitute a physical system.

References

Bica, E., Santiago, B. X., Dutra, C. M., Dottori, H., de Oliveira, M. R., & Pavani, D. 2001, *A&A*, 366, 827

Carraro, G., Dinescu, D. I., Girard, T. M., & van Altena, W. F. 2005, *A&A*, 433, 143

Nordström, B., Mayor, M., Andersen, J., Holmberg, J., Pont, F., Jørgensen, B. R., Olsen, E. H., Udry, S., & Mowlavi, N. 2004, *A&A*, 418, 989

Pavani, D. B. & Bica, E. 2007, *A&A*, 468, 139

Portegies Zwart, S. F., McMillan, S. L. W., Hut, P., & Makino, J. 2001, *MNRAS*, 321, 199

Prugniel, P. & Soubiran, C. 2001, *A&A*, 369, 1048

Tonry, J. & Davis, M. 1979, *AJ*, 84, 1511

Star clusters: basic galactic building blocks
Proceedings IAU Symposium No. 266, 2009
R. de Grijs & J. R. D. Lépine, eds.

doi:10.1017/S1743921309991700

Analysis of the B and Be-star populations of the double cluster h and χ Persei

Amber N. Marsh,[1] M. Virginia McSwain[1] and Thayne Currie[2]

[1]Department of Physics, Lehigh University, Bethlehem, PA, USA
email: [anm506, mcswain]@lehigh.edu

[2]Harvard Smithsonian Center for Astrophysics, Cambridge, MA, USA
email: tcurrie@cfa.harvard.edu

Abstract. We present blue optical spectra of 92 members of h and χ Per obtained with the *WIYN* telescope at Kitt Peak National Observatory. From these spectra, several stellar parameters were measured for the B-type stars, including $v \sin i, T_{\rm eff}, \log g_{\rm polar}, M_\star$, and $R_\star$. Strömgren photometry was used to measure $T_{\rm eff}$ and $\log g_{\rm polar}$ for the Be stars. We also analyze photometric data of cluster members and discuss the near- to mid-infrared excesses of Be stars.

Keywords. open clusters and associations: individual (NGC 869, NGC 884), stars: emission-line, stars: Be

1. Introduction

NGC 869 and NGC 884 (h and χ Persei, respectively) comprise a well-known double cluster rich in massive B-type stars, and have been the focus of many studies over the years. Recent studies show that NGC 869 and NGC 884 have nearly identical ages of $\sim 13 - 14$ Myr, common distance moduli of d$M \sim 11.85$ mag, and common reddenings of $E(B-V) \sim 0.5 - 0.55$ (Currie *et al.* 2009; Slesnick *et al.* 2002; Bragg & Kenyon 2005).

Currie *et al.* (2008; hereafter C08) identified two populations of NGC 869 and NGC 884 stars with detected *Spitzer* MIPS 24 μm excess emission: 20 A and F-type stars with luminous debris-disk emission and 57 brighter, earlier stars with weaker excess emission. They identify most of the latter group as candidate Be stars. However, only 21 were previously listed as Be stars (e.g., Bragg & Kenyon 2002; Slesnick *et al.* 2002).

In this study, we analyze blue optical spectra of 92 early-type cluster members, including 16 candidate Be stars from C08, and investigate their near- to mid-infrared (IR) excesses. With continued monitoring of these stars in both the optical and IR regimes, we hope to explore these excesses as a reasonable means for identifying potential Be stars within clusters, as well as to investigate the transient natures of the disks surrounding the known Be stars in NGC 869 and NGC 884.

2. Overview

Blue optical spectra of 92 members of NGC 869 and NGC 884 were obtained on 2005 November 14–15 using the *WIYN 3.5m* telescope with the Hydra spectrograph. The observed spectra cover a wavelength range of 4250–4900 Å.

Shown in Figure 1 are samples of the model spectral fits used to measure values for $v \sin i, T_{\rm eff}$, and $\log g$ for B-type stars. $v \sin i$ was determined by comparing the He I $\lambda\lambda$4387, 4471,4713, and Mg II λ4481 lines with the Kurucz ATLAS9 models (Kurucz 1994) and taking a weighted average of these four values. For stars having $T_{\rm eff} \geqslant 15\,000$ K, the TLUSTY BSTAR2006 models (Lanz & Hubeny 2007) were used to find $T_{\rm eff}$ and $\log g$

using the Hγ line. For stars having $T_{\rm eff} \leqslant 15\,000$ K, the Kurucz ATLAS9 models were used (Kurucz 1994). The method of Huang & Gies (2006; hereafter HG06) was used to determine $\log g_{\rm polar}$. For Be stars, Strömgren photometry available from the WEBDA database was used to derive $T_{\rm eff}$ and $\log g_{\rm polar}$ based on the methods of McSwain *et al.* (2008). The masses and radii for all stars were determined from the Schaller *et al.* (1992) evolutionary tracks, which are shown plotted with $T_{\rm eff}$ and $\log g_{\rm polar}$ in Figure 2.

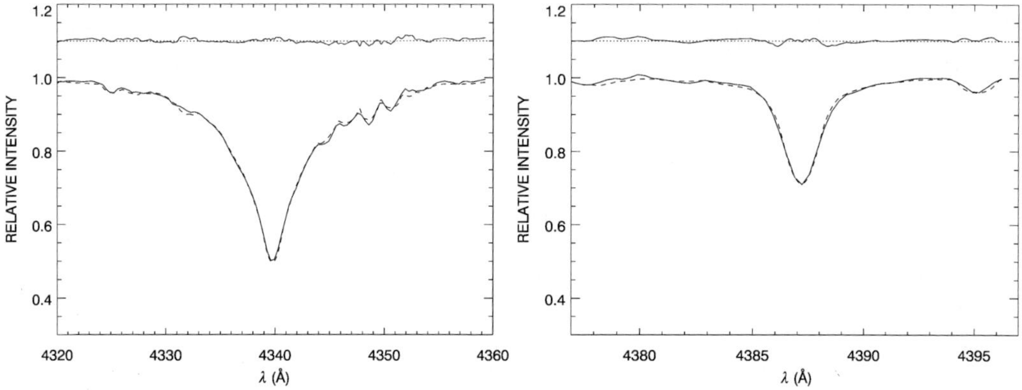

Figure 1. Sample spectral-line fits for NGC 869–90. Shown on the left is Hγ and on the right is HeIλ4378. The solid line is our observed spectrum while the dashed line displays our model fit to the line, with the computed residual shown above, shifted for clarity.

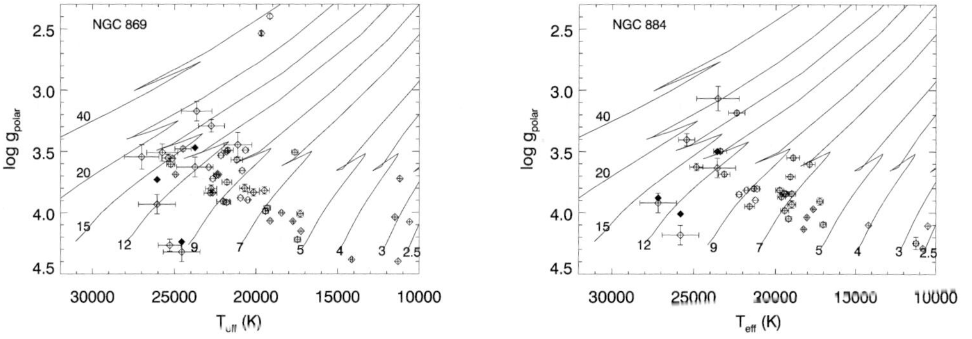

Figure 2. For both NGC 869 *(left)* and NGC 884 *(right)*, $T_{\rm eff}$ and $\log g_{\rm polar}$ are plotted with the evolutionary tracks of Schaller *et al.* (1992). The zero-age main-sequence mass of each evolutionary track is labeled along the bottom. Normal B-type stars are shown as open diamonds while Be stars are filled diamonds.

3. Results

Sixteen Be candidates from C08 are present in our sample or that of HG06. Three of these 16 stars show no evidence of circumstellar emission in our spectra, although all have been observed to be Be stars in the past (Keller *et al.* 2001). Ten of the C08 Be candidates in our spectra show emission. Stellar parameters for the remaining three candidates are found in HG06, thus we cannot comment on the presence of emission at the time of observation. In addition, we find Be emission in one star (No. 1772) that was not observed by C08, and we present results for one additional star (No. 1268) identified as a Be star by Keller *et al.* (2001). These results are summarized in Table 1.

Table 1. Measured physical parameters for Be stars

Cluster Star[1]	$v \sin i$ (km s^{-1})	$T_{\rm eff}$ (K)	$\log g_{\rm polar}$ (dex)	$M_\star$ ($M_\odot$)	$R_\star$ ($R_\odot$)	Be Cand.[2]	Comment	Ref.
NGC 869–49	172	23757	3.63	12.3	8.9	Y	Emission present	This work
NGC 869–517	178	...	...	...	...	Y	Emission present	This work
NGC 869–566	306	21183	3.57	10.4	8.7	Y	No emission present[3]	This work
NGC 869–846	205	22747	3.29	14.7	14.3	Y	Weak emission present	This work
NGC 869–847	87	27000	3.54	17.8	11.8	Y	Weak emission present	This work
NGC 869–1162	66	19175	2.40	33.6	60.8	Y	No emission present[3]	This work
NGC 869–1261	285	26065	3.93	12.0	6.2	Y	Strong emission present	This work
NGC 869–1268	151	24491	3.48	14.8	11.6	N	No emission present[3,4]	This work
NGC 869–1278	197	24562	4.32	9.0	3.4	Y	Emission present	This work
NGC 884–1772	379	...	...	...	...	...	Emission present	This work
NGC 884–1926	106	27190	3.92	13.5	6.7	Y	Strong emission present	This work
NGC 884–2091	236	...	...	...	...	Y	Emission present	This work
NGC 884–2138	153	23579	3.63	12.0	8.7	Y	Emission present	This work
NGC 884–2165	79	26571	4.03	11.9	5.5	Y	—	HG06
NGC 884–2402	141	28238	3.81	15.6	8.1	Y	—	HG06
NGC 884–2468	134	10500	4.11	2.7	2.4	Y	No emission present[3]	This work
NGC 884–2563	308	25820	4.18	10.7	4.4	Y	Strong emission present	This work
NGC 884–2949	168	18240	3.96	6.4	4.4	Y	—	HG06

Notes:
[1] Identification numbers from the WEBDA database. [2] Be candidate in C08. [3] Stars not showing emission in our observations are likely transient Be stars. [4] Identified as Be star by Keller *et al.* (2001).

Spectral-energy distributions (SEDs) for three stars in NGC 869 and NGC 884 are displayed in Figure 3. UBV magnitudes are from the WEBDA database, $JHK_{\rm s}$ are from the 2MASS survey, and *Spitzer* [8] and [24] μm are from C08. These magnitudes were converted to fluxes using the methods detailed in Bessell *et al.* (1998), Cohen *et al.* (2003), Colina *et al.* (1996), Reach *et al.* (2005), and Rieke *et al.* (2008). Assuming a constant $E(B-V) = 0.52$ mag for NGC 869 and NGC 884 (Bragg & Kenyon 2005; Slesnick *et al.* 2002), reddened blackbody curves have been overlaid on these plots to investigate their near- to mid-IR excesses. All three stars shown in Figure 3 are proposed Be candidates (C08), with NGC 884–2138 and NGC 869–49 having emission present in our optical spectra and observed near- to mid-IR excess. NGC 884–2165 is not included in our spectroscopic sample but has previously been identified as a Be star and has observed IR excess (Keller *et al.* 2001).

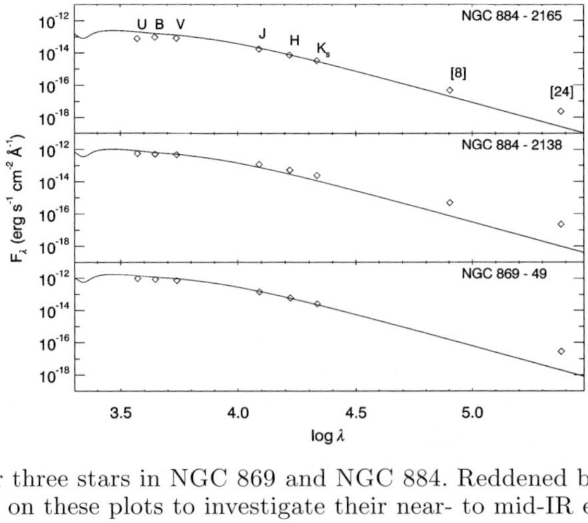

Figure 3. SEDs for three stars in NGC 869 and NGC 884. Reddened blackbody curves are overlaid on these plots to investigate their near- to mid-IR excesses.

4. Conclusions and further work

We have measured the physical parameters of 77 B-type stars and 15 Be stars in NGC 896 and NGC 884. Sixteen Be candidates from C08 are present in our sample or that of HG06. Of these 16 Be candidates, three stars show no evidence of emission in our optical data and are likely transient Be stars. Ten of these Be candidates show emission in our spectra. Those Be candidates without emission in our spectra should be monitored in the future to further investigate their transient nature.

In the future, IRAC 3.6–5.8 μm data will be combined with the optical and IR fluxes used here to investigate the observed SEDs. We will fit the new SEDs using modern flux models rather than blackbody curves. Modifications accounting for variable reddening throughout the clusters will also be made. These new SED fits can then be used to model the Be disk sizes and temperatures.

Acknowledgements

We are grateful for travel support provided by the American Astronomical Society and the International Astronomical Union. We would like to thank Yale University for providing access to the *WIYN* telescope at KPNO. Institutional support was provided by Lehigh University. This work was also supported by NASA DPR No. NNX08AV70G.

References

Bessell, M. S., Castelli, F., & Plez, B. 1998, *A&A*, 333, 231
Bragg, A. & Kenyon, S. 2002, *AJ*, 124, 3289
Bragg, A. & Kenyon, S. 2005, *AJ*, 130, 134
Cohen, M., Wheaton, W. A., & Megeath, S. T. 2003, *AJ*, 126, 1090
Colina, L., Bohlin, R. C., & Castelli, F. 1996, *AJ*, 112, 307
Currie, T., Evans, N. R., Spitzbart, B. D., Irwin, J., Wolk, S. J., Hernandes, J., Kenyon, S. J., & Pasachoff, J. M. 2009, *AJ*, 137, 3210
Currie, T., Kenyon, S. J., Balog, Z., Rieke, G., Bragg, A., & Bromley, B. 2008, *ApJ*, 672, 558
Huang, W. & Gies, D. R. 2006, *ApJ*, 648, 591
Keller, S. C., Grebel, E. K., Miller, G. J., & Yoss, K. M. 2001, *AJ*, 122, 248
Kurucz, R. L. 1994, *Kurucz CD-ROM 19, Solar Abundance Model Atmospheres for 0, 1, 2, 4, 8 km s*$^{-1}$, Cambridge: Smithsonian Astrophysical Observatory
Lanz, T. & Hubeny, I. 2007, *ApJS*, 169, 83
McSwain, M. V., Huang, W., Gies, D. R., Grundstrom, E. D., & Townsend, R. H. D. 2008, *ApJ*, 672, 590
Reach, W. T., Megeath, S. T., Cohen, M., Hora, J., Carey, S., Surace, J., Willner, S. P., Barmby, P., Wilson, G., Glaccum, W., Lowrance, P., Marengo, M., & Fazio, G. G. 2005, *PASP*, 117, 978
Rieke, G. H., Blaylock, M., Decin, L., Engelbracht, C., Ogle, P., Avrett, E., Carpenter, J., Cutri, R. M., Armus, L., Gordon, K., Gray, R. O., Hinz, J., Su, K., & Willmer, C. N. A. 2008, *AJ*, 135, 2245
Schaller, G., Schaerer, D., Meynet, G., & Maeder, A. 1992, *A&AS*, 96, 269
Slesnick, C., Hillenbrand, L. A., & Massey, P. 2002, *ApJ*, 576, 880

Star clusters: basic galactic building blocks
Proceedings IAU Symposium No. 266, 2009
R. de Grijs & J. R. D. Lépine, eds.

doi:10.1017/S1743921309991712

Young open clusters in the Milky Way and Small Magellanic Cloud

C. Martayan

[1]European Organisation for Astronomical Research in the Southern Hemisphere, Alonso de Cordova 3107, Vitacura, Casilla 19001, Santiago 19, Chile
email: cmartaya@eso.org

[2]GEPI, Observatoire de Paris, CNRS, Université Paris Diderot, 5 place Jules Janssen, 92195 Meudon Cedex, France

Abstract. NGC 6611, Trumpler 14, Trumpler 15, Trumpler 16, and Collinder 232 are very young open clusters located in star-formation regions in the Eagle Nebula and Carina in the Milky Way, and NGC 346 in the Small Magellanic Cloud. With different instrumentation and techniques, it has been possible to detect and classify new Herbig Ae/Be and classical Be stars and to provide new tests/comparisons of the Be stars' appearance models. Special (He-strong) stars in these star-formation regions are also discussed.

Keywords. open clusters and associations: individual (Trumpler 14, Trumpler 15, Trumpler 16, Collinder 232), Magellanic Clouds, galaxies: star clusters, stars: early-type, stars: pre–main-sequence, stars: emission-line, Be, stars: evolution, dust, extinction

1. Observations

We used the ESO–WFI (Baade *et al.* 1999) in slitless mode (focusing on Hα, $R \sim 1000$) to observe the Galactic open cluster NGC 6611 in the Eagle Nebula. This setup allows us to disentangle the stars, based on emission lines from their circumstellar disks, from the nebular lines due to diffuse emission in the surrounding nebula. We observed 100 stars in NGC 6611, pre-selected from our WFI catalogue, using the *VLT*–FLAMES/GIRAFFE facilities in MEDUSA mode (Pasquini *et al.* 2002). We also pre-selected B-type stars in the open clusters Trumpler 14, 15, 16, and Collinder 232 in the Carina region from our astrophotometric catalogues based on the EIS pre-FLAMES survey images (Momany *et al.* 2001). We then observed $\sim$ 200 OB-type stars with *VLT*–FLAMES/GIRAFFE. We used the LR2, HR4, HR5, HR6, and HR15 settings with resolutions from $R = 6500$ to 18 000. These setups were chosen for the Balmer lines, Helium I and II, Mg II, and Si III. The open cluster NGC 346 in the Small Magellanic Cloud was also observed with the ESO–WFI in slitless mode (Hα). More than 50 emission-line stars were found, half of them were detected for the first time: see Martayan *et al.* (2009).

2. Determination of the interstellar reddening and correction of the magnitudes

To correct the magnitudes of the stars for interstellar reddening, we measured the equivalent widths of the interstellar lines at 443.0, 450.2, and 661.3 nm. We then used the calibration from Herbig (1975) to obtain the values of the interstellar reddening, $E(B-V)$ for each star observed with GIRAFFE. A reddening map for Trumpler 14, 15, and 16 is shown in Figure 1. Based on this estimate of $E(B-V)$, we corrected the 2MASS JHK magnitudes as well as the *Spitzer* magnitudes. However, *Spitzer* data

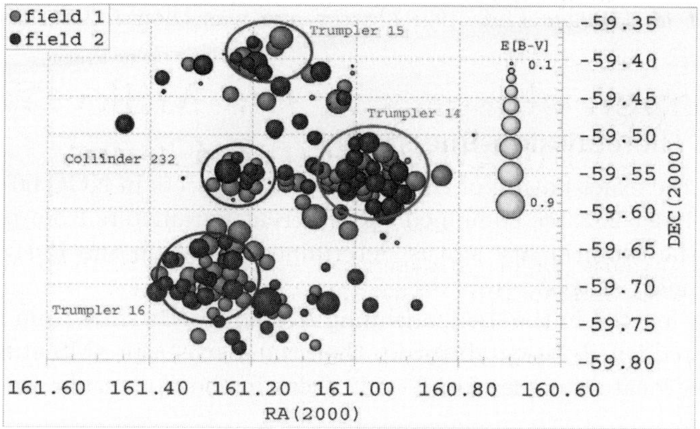

Figure 1. Map of the interstellar reddening in the Carina region for stars observed in Trumpler 14, 15, 16, and Collinder 232.

are only available for the stars in NGC 6611 but not for those in Trumpler 14, 15, and 16. Two kinds of dereddened colour–magnitude diagrams were obtained for NGC 6611 (see Martayan *et al.* 2008) For the Trumpler 14, 15, and 16 clusters, the 2MASS diagram is shown in Figure 2. The emission-line stars are indicated by red triangles. The status of several of these is obvious: they are clearly pre-main-sequence (pre-MS) or MS objects. However, for the stars located in the intermediate regions, complementary analysis must be done, using the evolutionary status determined from the fundamental stellar parameters and compared to the age of the open clusters.

3. Determination of fundamental parameters

The fundamental parameters were determined using the GIRFIT code of Frémat *et al.* (2006). This code fits the observed spectrum with theoretical templates obtained from the TLUSTY and SYNSPEC codes of Hubeny & Lanz (1995). For each open cluster member star, we obtained the most appropriate age using theoretical evolutionary tracks

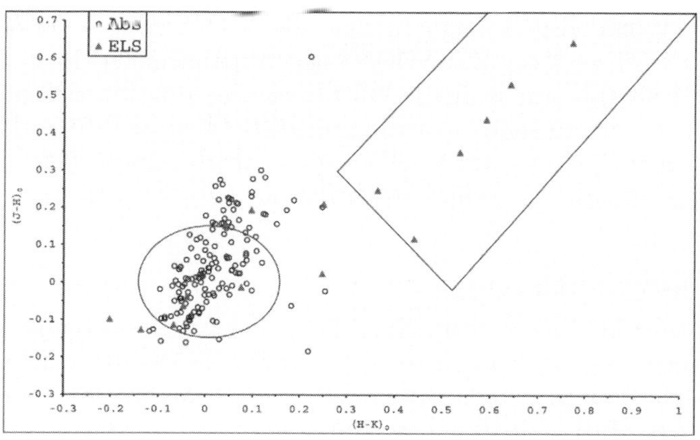

Figure 2. Dereddened 2MASS colour–colour diagram for stars observed in the Trumpler 14, 15, 16, and Colinder 232 open clusters. The large rectangular area from Hernández *et al.* (2005) corresponds to the area containing pre-main-sequence objects like Herbig Ae/Be stars.

from Schaller *et al.* (1992). The open cluster age was then redetermined using these stars.

4. Nature of the emission-line stars

To determine the exact nature of the emission-line objects in NGC 6611, Trumpler 14, 15, 16, and Collinder 232, we combined the different parameters from the colour–colour diagrams with the evolutionary status determined based on the Hertzsprung–Russell-diagram evolutionary tracks:

- if the star is located in the 'box' outlining Herbig Ae/Be stars from Hernández *et al.* (2005) in the dereddened 2MASS diagram, then the star is a pre-MS star.
- if the star is located in the class I or II areas in the *Spitzer* diagram, then the star is a pre-MS star.
- if the star is located in an intermediate region, the evolutionary status obtained from the fundamental parameters combined with their membership of a given open cluster must be taken into account:
    - ○ open cluster member with an age equivalent to that of the cluster: MS star;
    - ○ open cluster member with an apparently older age than that of the cluster: pre-MS star;
    - ○ not a member of the open cluster: uncertain.
- if the star is not present in the boxes of pre-MS stars and if its evolutionary status is incompatible with the age of its host open cluster, it is probably a classical Be star (with a decretion disk instead of an accretion disk for pre-MS stars like Herbig Ae/Be stars).

All details regarding NGC 6611 are available in Martayan *et al.* (2008).

5. Special stars

Among the emission-line stars of NGC 6611, two are of special interest. The first is W601, which was found to host a magnetic field by Alecian *et al.* (2008). They also found that this star is a He-strong star. Thanks to *VLT*–GIRAFFE observations of W601 and W080, it was possible to compare the spectral features of these two stars. We found that W080 is a spectral twin of W601 and is probably also a He-strong star. Moreover, its broad, deep spectral lines cannot be reproduced by the models (Martayan *et al.* 2008). To discover whether W080 hosts a magnetic field like W601, two sets of observations were obtained at the *CFHT* with the ESPADONS spectropolarimeter. However, the faintness of W080 implies that this star is at the limit of the capabilities of this instrument. The preliminary results seem to show no strong magnetic field in W080. However, the data are currently being analysed to find a potentially weaker magnetic field and/or to provide upper limits to its intensity as well as confirmation of its He-strong status.

6. Conclusions

We found 11 emission-line stars in NGC 6611 and its vicinity, nine of whih are new Herbig Ae/Be stars. In the open clusters Trumpler 14, 15, 16, and Collinder 232 and their vicinity in the Carina nebula, six of the 16 emission-line stars seem to be Herbig Ae/Be stars. For 10 of the remaining emission-line stars, the status is uncertain for two and eight are probably MS and hence classical Be stars. In NGC 346, complementary observations at higher spectral resolution are needed to determine the exact nature of the 50 emission-line stars found.

References

Alecian, E., *et al.* 2008, *A&A* (Letters), 481, L99

Baade, D., *et al.* 1999, *ESO Messenger*, 95, 15

Frémat, Y., Neiner, C., Hubert, A.-M., Floquet, M., Zorec, J., Janot–Pacheco, E., & Renan de Medeiros, J. 2006, *A&A*, 451, 1053

Herbig, G. H. 1975, *ApJ*, 196, 129

Hernández, J., Calvet, N., Hartmann, L., Briceño, C., Sicilia–Aguilar, A., & Berlind, P. 2005, *AJ*, 129, 856

Hubeny, I. & Lanz, T. 1995, *ApJ*, 439, 875

Martayan, C., Floquet, M., Hubert, A. M., Neiner, C., Frémat, Y., Baade, D., & Fabregat, J. 2008, *A&A*, 489, 459

Martayan, C., Baade, D., & Fabregat, J. 2009, *A&A*, in press (arXiv:0909.2303)

Momany, Y., *et al.* 2001, *A&A*, 379, 436

Pasquini, L., *et al.* 2002, *ESO Messenger*, 110, 1

Schaller, G., Schaerer, D., Meynet, G., & Maeder, A. 1992, *A&AS*, 96, 269

Star clusters: basic galactic building blocks
Proceedings IAU Symposium No. 266, 2009
R. de Grijs & J. R. D. Lépine, eds.

doi:10.1017/S1743921309991724

Evolution and appearance of Be stars in SMC clusters

C. Martayan,[1,2] D. Baade,[3] Y. Frémat[4] and J. Zorec[5]

[1]European Organisation for Astronomical Research in the Southern Hemisphere, Alonso de Cordova 3107, Vitacura, Casilla 19001, Santiago 19, Chile
email: cmartaya@eso.org

[2]GEPI, Observatoire de Paris, CNRS, Université Paris Diderot, 5 place Jules Janssen, 92195 Meudon Cedex, France

[3]European Organisation for Astronomical Research in the Southern Hemisphere, Karl–Schwarzschild–Str. 2, 85748 Garching b. München, Germany

[4]Royal Observatory of Belgium, 3 avenue circulaire, 1180 Brussels, Belgium

[5]Institut d'Astrophysique de Paris, UMR7095, CNRS, Université Marie & Pierre Curie, 98bis Boulevard Arago, 75014 Paris, France

Abstract. Star clusters are privileged laboratories for studying the evolution of massive stars (OB stars). One particularly interesting question concerns the phases during which the classical Be stars occur, which—unlike HAe/Be stars—are not pre-main-sequence objects, nor supergiants. Rather, they are extremely rapidly rotating B-type stars with a circumstellar decretion disk formed by episodic ejections of matter from the central star. To study the impact of mass, metallicity, and age on the Be phase, we observed Small Magellanic Cloud (SMC) open clusters with two different techniques: (i) with the ESO–WFI in slitless mode, which allowed us to find the brighter Be and other emission-line stars in 84 SMC open clusters, and (ii) with the *VLT*–FLAMES multifiber spectrograph to determine accurately the evolutionary phases of Be stars in the Be-star-rich SMC open cluster NGC 330. Based on a comparison to the Milky Way, a model of Be stellar evolution, appearance as a function of metallicity and mass, and spectral type is developed, involving the fractional critical rotation rate as a key parameter.

Keywords. stars: emission-line, Be, stars: fundamental parameters, stars: statistics, stars: early-type, stars: evolution, surveys, galaxies: star clusters, Magellanic Clouds

1. Observations and spectral analysis

We performed an Hα survey with the ESO–WFI (Baade *et al.* 1999) in slitless mode in the central parts of the Small Magellanic Cloud (SMC). Three million low-resolution spectra were obtained. They were extracted from the images with SExtractor (Bertin & Arnouts 1996) and we developed Album (Martayan *et al.* 2009) to identify spectra exhibiting Hα emission. After astrometric calibration, the extracted sources were cross-matched with the ogle-ii photometric catalogue (Udalski *et al.* 1998). Within the area observed, results for 84 SMC open clusters are presented. In particular, the results are compared with those from McSwain & Gies (2005) in Galactic open clusters. The ratios of clusters with and without classical Be stars are compared with respect to metallicity (Z), spectral type, and age. Observations were also obtained with *VLT*–FLAMES (Pasquini *et al.* 2002) of the Be-star-rich SMC open cluster NGC 330 and its vicinity. LR2 spectra (395–455 nm, $R = 6400$, Hϵ, Hδ, Hγ, HeI447.1nm, MgII448.1nm) and LR6 spectra (650–770 nm, $R = 8600$, Hα) were aquired between October 2003 and September 2004. For each star, the fundamental parameters were determined with girfit (Frémat *et al.* 2006), taking account of fast-rotation effects with FASTROT (Frémat *et al.* 2005). In particular, rotational velocities were determined as well as the statistical, fractional angular breakup velocity for Be stars of different ages (see Martayan *et al.* 2007).

2. Observed metallicity effect on the relative frequency of Be stars

The metallicity of the SMC is significantly lower ($0.001 < Z < 0.009$; Cioni *et al.* 2006) than that of the Galaxy ($Z = 0.020$). Figure 1 shows the fraction of Be to all B-type stars by spectral type, separately for the SMC and the Milky Way. It clearly indicates that early-type Be stars are 3–5 times as frequent in the SMC as in the Milky Way (beyond B2–B3, the SMC sample is incomplete). Note that the age ranges are comparable. The large difference in frequency is obviously most easily attributed to the difference in metallicity.

3. Appearance of Be stars in open clusters as a function of age

Figure 2 displays the fraction of open clusters with Be stars in the SMC and the Milky Way versus age. It appears that Be stars are preferentially hosted by young open clusters. A first decrease in the fraction of open clusters with Be stars is seen around 30–40 Myr (corresponding to early-type Be stars reaching the terminal-age main sequence; TAMS) is followed by an increase of open clusters with Be stars. The terminal decrease in older open clusters is also due to the late-type Be stars arriving on the TAMS.

4. Metallicity-dependent evolution of rotational velocities

To explain the overall shape of Figure 2, the temporal evolution of the fractional angular velocity ($\frac{\Omega}{\Omega_c}$) of Be stars can be reconstructed from observed rotational velocities (here we use *VLT*-FLAMES spectra) and theoretical evolutionary tracks from Maeder & Meynet (2001). From our SMC data and results obtained by Zorec *et al.* (2005) in the Milky Way, Be stars have a minimal $\frac{\Omega}{\Omega_c}$ of 0.7. If/when B-type stars rotate more slowly, Be symptoms do not seem to develop.

The evolution of the $\frac{\Omega}{\Omega_c}$ depends on stellar mass and metallicity (see figure 5 in

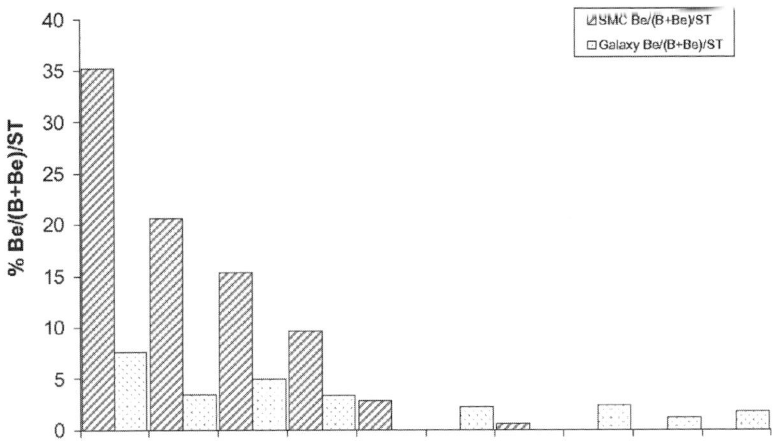

Figure 1. Relative abundances of Be stars in the SMC (hatched histogram) and the Galaxy (dot-filled histogram) as a function of spectral type.

Martayan *et al.* 2007), which holds the key to our understanding of the distribution seen in Figure 2:

• In massive Milky Way Be stars, $\frac{\Omega}{\Omega_c}$ is sufficient to develop Be-star symptoms at the beginning of the main sequence. Due to wind-driven mass loss, the angular momentum decreases until $\frac{\Omega}{\Omega_c}$ drops to below the threshold of 0.7 and the initial Be-star appearance will be lost after $\sim 5-10$ Myr. This explains the small decrease observed in the Milky Way for open clusters with Be stars (cf. Mathew *et al.* 2008).

• Intermediate-mass Be stars seem to retain a sufficently high level of $\frac{\Omega}{\Omega_c}$ during their entire main-sequence evolution to preserve their appearance as Be stars without interruption. The Be-star attributes will disappear once the TAMS is reached at ~40 Myr, which explains the lower fraction of Galactic open clusters with Be stars of this age.

• Rapidly rotating low-mass B stars can appear as Be stars at the beginning of the main sequence but will lose this status quickly through a reorganization of the internal angular momentum. Subsequently, the standard evolution of $\frac{\Omega}{\Omega_c}$ with time (increasing radius with small angular-momentum loss), will let them reach an $\frac{\Omega}{\Omega_c}$ above 0.7, so that at an age of 40–50 Myr they are again recognizeable as Be stars. This can explain the second increase of the number of open clusters with Be stars at an age of 40 Myr. The low-mass Be stars will then begin to reach the TAMS and the number of open clusters with Be stars decreases with age.

• In the SMC, the evolution of $\frac{\Omega}{\Omega_c}$ of intermediate- and low-mass Be stars is similar to that in the Milky Way because in this mass range mass loss is too small to significantly alter the angular momentum or its internal distribution. This explains the evolution of open clusters with Be stars and ages between 30 and 100 Myr. Still older open clusters that host Be stars probably had multiple star-formation episodes or the Be stars may be blue stragglers. Note that, for the Milky Way, McSwain & Gies (2005) suspect that 75% of the Be stars are binaries.

• The evolution of more massive Be stars is different in the SMC than in the Milky Way. Because of the lower SMC metallicity, mass and angular-momentum loss are also lower, and $\frac{\Omega}{\Omega_c}$ increases with time as in low-mass Be stars. They will reach the 70% limit

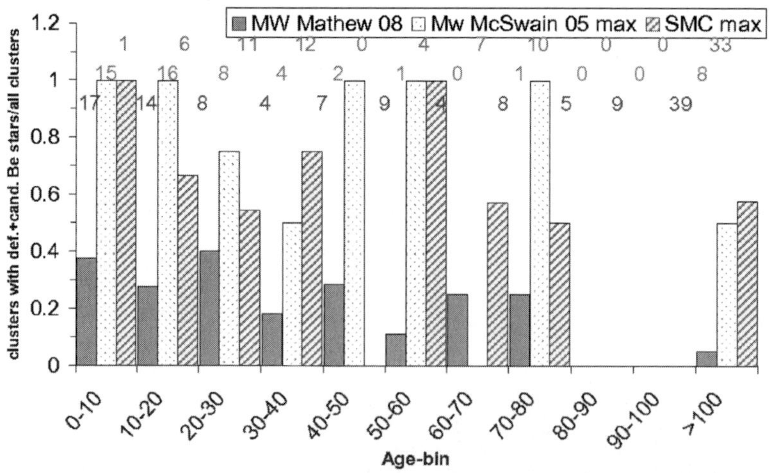

Figure 2. Fraction of open clusters with Be-star members in the SMC (hatched histograms) and the Milky Way (dot-filled histograms: data from McSwain & Gies 2005; filled histograms: data from Mathew *et al.* 2008) as a function of cluster age. The number of clusters studied is indicated.

at $\sim$4 – 5 Myr and appear again as Be stars, in contrast to massive Be stars in the Milky Way. Therefore, in the SMC very young open clusters are expected to contain Be stars.

However, for observational verification of this difference in the fraction of Be stars between very young open clusters in the Milky Way and SMC, a larger sample of open clusters younger than 10 Myr is needed. For such a comparison, one must carefully discriminate between classical Be stars and Herbig Ae/Be stars, which are still on the pre-main sequence and owe their emission lines to their parental accretion disk. A method to achieve this separation is described in Martayan *et al.* (2008).

The inferred long-term transitions between the Be and B phases will probably not be abrupt, and one may speculate whether the much faster and repetitive Be to B, and B to Be transformations observed in many Be stars are part of long-term transitions. If so, one would expect that Be stars with stable and those with more volatile circumstellar disks should not show significant differences otherwise. In fact, this is what McSwain *et al.* (2009) reported recently.

5. Conclusions

We conducted a study of open clusters with Be stars using two data sets,

• a low-resolution Hα survey of 84 SMC open clusters. It indicates that Be stars are more abundant in the SMC than in the Milky Way. The fraction of open clusters with Be stars reaches a local minimum at ages around 30–40 Myr and eventually declines for clusters older than 70–80 Myr.

• Fundamental stellar parameters determined from medium-resolution spectra of B and Be stars in the SMC open cluster NGC 330 and its vicinity allowed us to study the evolution of the fractional critical angular velocity, $\frac{\Omega}{\Omega_c}$, of Be stars. Its dependence on stellar mass and metallicity can explain the distribution with age of the fraction of open clusters with Be stars and the differences between the Galaxy and the SMC.

All data and a more comprehensive discussion are presented in Martayan *et al.* (2009).

References

Baade, D., *et al.* 1999, *ESO Messenger*, 95, 15

Bertin, E. & Arnouts, S. 1996, *A&AS*, 117, 393

Cioni, M.-R. L., Girardi, L., Marigo, P., & Habing, H. J. 2006, *A&A*, 452, 195

Frémat, Y., Zorec, J., Hubert, A.-M., & Floquet, M. 2005, *A&A*, 440, 305

Frémat, Y., Neiner, C., Hubert, A.-M., Floquet, M., Zorec, J., Janot–Pacheco, E., & Renan de Medeiros, J. 2006, *A&A*, 451, 1053

Maeder, A. & Meynet, G. 2001, *A&A*, 373, 555

Martayan, C., Frémat, Y., Hubert, A.-M., Floquet, M., Zorec, J., & Neiner, C. 2007, *A&A*, 462, 683

Martayan, C., Floquet, M., Hubert, A. M., Neiner, C., Frémat, Y., Baade, D., & Fabregat, J. 2008, *A&A*, 489, 459

Martayan, C., Baade, D., & Fabregat, J. 2009, *A&A*, in press (arXiv:0909.2303)

Mathew, B., Subramaniam, A., & Bhatt, B. C. 2008, *MNRAS*, 388, 1879

McSwain, M. V. & Gies, D. R. 2005 *ApJS*, 161, 118

McSwain, M. V., Huang, W., & Gies, D. R. 2009, *ApJ*, 700, 1216

Pasquini, L., *et al.* 2002, *ESO Messenger*, 110, 1

Udalski, A., Szymański, M., Kubiak, M., Pietrzyński, G., Woźniak, P., & Żebruń, K. 1998, *AcA*, 48, 147

Zorec, J., Frémat, Y., & Cidale, L. 2005, *A&A*, 441, 235

Star clusters: basic galactic building blocks
Proceedings IAU Symposium No. 266, 2009
R. de Grijs & J. R. D. Lépine, eds.

doi:10.1017/S1743921309991736

Chemical abundances of the LMC cluster Hodge 11 and its surrounding field

Renee Mateluna, Douglas Geisler and Sandro Villanova

Departamento de Astronomía, Universidad de Concepción, Esteban Iturra s/n Barrio Universitario, Concepción, Chile
email: renee@astro-udec.cl

Abstract. We have obtained high-resolution *VLT*/FLAMES spectra of a number of giants in the LMC cluster Hodge 11 and the surrounding field. We present results of a detailed model-atmosphere abundance analysis for a variety of elements, including Fe, Ca, Ti, Si, Sc, Ni and Cr.

Keywords. galaxies: abundances, galaxies: star clusters

1. Introduction

The Large Magellanic Cloud (LMC) exhibits an array of star-formation processes and episodes. This galaxy provides an excellent opportunity to study the formation and chemical-enrichment history of similar galaxies in great detail. Additionally, the ΛCMD paradigm of galaxy formation suggests that galaxies like the Milky Way form their halos through the accretion/merger of numerous dwarf satellites. Therefore, understanding the formation history and chemical evolution of the LMC is an important step in understanding these merger events and their role in galaxy formation.

One of the most direct ways to learn about a galaxy's chemical evolution and star-formation history is through the study of its clusters. Coupled with the fact that their ages are relatively easily determined, clusters are excellent tools for deriving the age/abundance relation of the host galaxy. Grocholski *et al.* (2006) present a extensive study using the FORS2 facility and obtained Ca-triplet-based metallicities for a sample of 29 clusters, spanning the full range of ages and abundances (see Figure 1). However, for the LMC a huge gap in the cluster age–metallicity relation is present (see Fig. 1; DaCosta 1991, 2002; Geisler *et al.* 1997) with only one known cluster in it. If we want to investigate the chemical evolution during this period, we must use field stars.

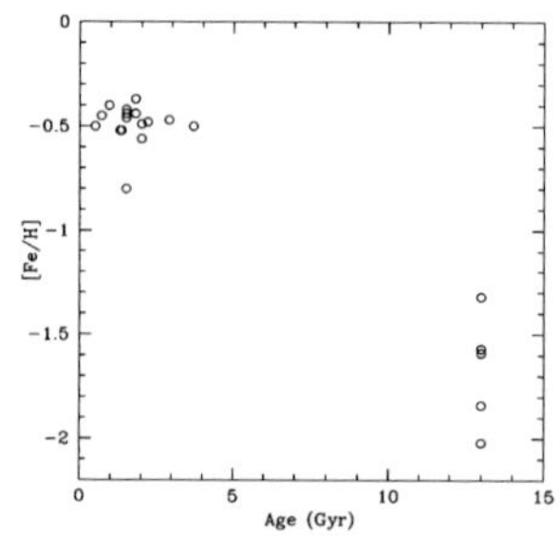

Figure 1. Age/metallicity relation obtained from FORS2 Ca-triplet observations (Grocholski *et al.* 2006) of a sample of LMC clusters.

2. Data and results

The data was obtained with the high-resolution spectrograph FLAMES (GIRAFFE: $R \sim 25\,000$ and UVES: $R \sim 47\,000$) on the *VLT*. The abundances for Fe, Ca and Ti were obtained with an LTE (local thermodynamic equilibrium) model-atmosphere analysis. Ca and Ti (α elements) abundances for Hodge 11 ([Fe/H] $= -1.9$ dex) and field stars are presented in comparison with Pompeia *et al.* (2008) data for the general LMC field (see Figure 2).

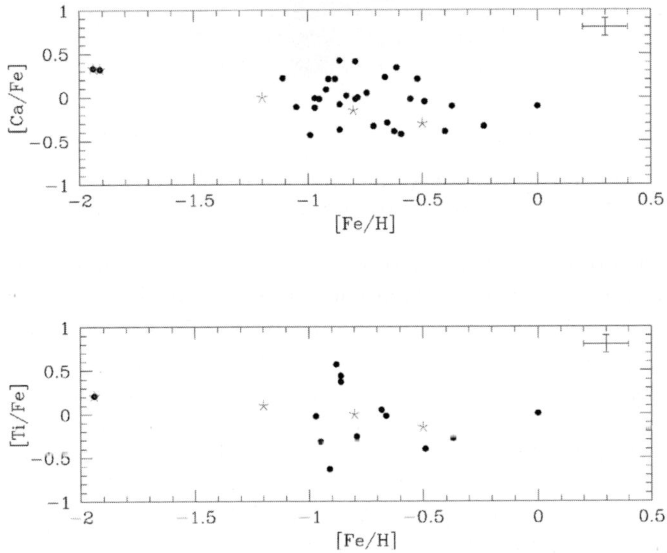

Figure 2. Abundances of Ca and Ti compared to Fe for Hodge 11 and the surrounding field stars. Black dots: our data. The crosses are the mean values for the LMC from Pompeia *et al.* (2008).

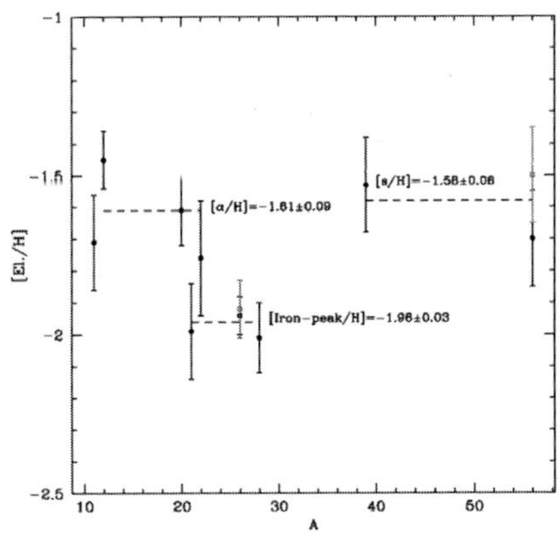

Figure 3. Abundances for Na, α, iron-peak and s-process elements for the cluster Hodge 11. Black and red points are from UVES and GIRAFFE data, respectively.

3. Conclusions and future work

The derived metallicity is in excellent agreement with the Ca-triplet value of the FORS2 sample (Figure 3) and the obtained abundances so far are in good agreement

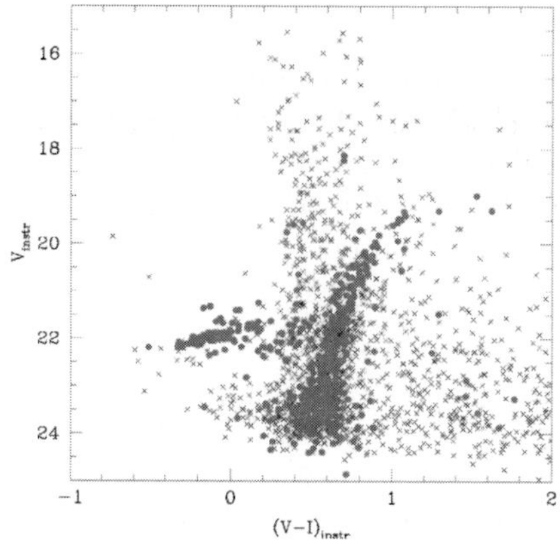

Figure 4. *LCO 1m* telescope wide-field photometry in the NGC 2257 field. Stars within 2′ from the cluster center are indicated by red points.

with the results of Pompeia *et al.* (2008) for the LMC field stars (Figure 2). The next step is to obtain high-resolution spectroscopy of a subsample of clusters. We have data for four (NGC 2257, Hodge 11, NGC 1841, and NGC 1718) with GIRAFFE and UVES (*VLT*/FLAMES). Spectroscopic data will be coupled with deep wide-field photometry obtained with MOSAIC at the *CTIO 4m* to have a homogeneous and complete database and derive atmospheric parameters (an example of wide-field photometry in the field of NGC 2257 is shown in Figure 4, based on data from the *LCO 1m* telescope). These clusters were selected because they span the full range of age (2–13 Gyr) and metallicities (−0.4 to −2.0 dex) of the FORS2 sample. Our goal is to derive abundances for many elements, including Na, α (Si, Ca, Ti), iron-peak (V, Cr, Fe, Ni, Cu), and s-process (Y and Ba). The detailed age/abundance relations for these elements, along with information about the relative contribution from asymptotic giant branch stars and Type I and II supernovae, will be used to help constrain the chemical evolution of the LMC.

Acknowledgements

RM thanks Conicyt for a PhD scholarship, SOCHIAS, the ALMA–Conicyt fund (project no. 31070007), the Universidad de Concepción, the MECESUP fund and CATA–BASAL for their financial support to attend this meeting.

References

Da Costa, G. S. 1991, in: R. Haynes & D. Milne, *The Magellanic Clouds*, Proc. IAU Symp. No. 148, p. 183

Da Costa, G. S. 2002, in: D. Geisler, E. K. Grebel & D. Minniti, *Extragalactic Star Clusters*, Proc. IAU Symp. No. 207, p. 83

Geisler, D., Bica, E., Dottori, H., Clairá, J. J., Piatti, A. E., & Santos Jr., J. F. C. 1997, *AJ*, 114, 1920

Grocholski, A. J., Cole, A. A., Sarajedini, A., Geisler, D., & Smith, V. V. 2006, *AJ*, 132, 1630

Pompeia, L., Hill, V., Spite, M., Cole, A., Primas, F., Romanicello, M., Pasquini, L., Cioni, M.-R., & Smecker Hane, T. 2008, *A&A*, 480, 379

Star clusters: basic galactic building blocks
Proceedings IAU Symposium No. 266, 2009
R. de Grijs & J. R. D. Lépine, eds.

doi:10.1017/S1743921309991748

Fundamental stellar properties from eclipsing binary analyses

E. F. Milone,[1] R. Stagg,[1] P. Guhathakurta,[2] D. A. VandenBerg[3] and J. Kallrath[4,5]

[1]Physics & Astronomy Department, University of Calgary, AB, Canada
[2]Lick Observatory/University of California at Santa Cruz, CA, USA
[3]Department of Physics and Astronomy, University of Victoria, Victoria, BC, Canada
[4]Astronomy Department, University of Florida, Gainesville, FL, USA
[5]BASF A.G., GVC/S-B009, Ludwigshafen, Germany

Abstract. The properties of the binary components of the eclipsing variables found in the 47 Tuc core region as part of a search for extrasolar planets have been re-examined following reanalysis of the image photometry and calibration. This is part of an ongoing study to find fundamental properties of stars of different metallicity in a variety of environments through eclipsing-binary analyses.

Keywords. binaries: eclipsing, stars: fundamental parameters, Hertzsprung–Russell diagram, globular clusters: individual (47 Tucanae)

The full poster (in pdf format) is available at
`http://www.astro.iag.usp.br/~iaus266/Posters/pMilone.pdf`.

Star clusters: basic galactic building blocks
Proceedings IAU Symposium No. 266, 2009
R. de Grijs & J. R. D. Lépine, eds.

doi:10.1017/S174392130999175X

A global optimization technique for the fitting of color–magnitude diagrams

Hektor Monteiro, Wilton S. Dias and Thiago C. Caetano
Universidade Federal de Itajubá, Itajubá, MS, Brazil
email: hektor.monteiro@gmail.com

Abstract. We present a new technique to analyze color–magnitude diagrams, such as those of stellar clusters, based on the cross-entropy global-optimization procedure. The method uses theoretical isochrones available in the literature and minimizes an objective function that describes the quality of a fit to the color–color and color–magnitude diagrams. The objective function is based on the two-dimensional distances of the points in the color–magnitude and color–color diagrams to the theoretical curves and is modified by weights that take into account the stellar distance to the observed cluster center, observed magnitude uncertainties and the King profile of the cluster, among others. The parameters determined simultaneously are distance, reddening, age and metalicity. The method uses a Monte Carlo approach to obtain uncertainties on the determined parameters for the cluster. We present results for 10 well-studied open clusters and show that the results are consistent with previous studies.

Keywords. open clusters and associations: general methods: numerical, methods: statistical

1. Introduction

Galactic open clusters are a key class of objects used for a wide range of investigations owing to their broad age and distance span. However, main-sequence 'fitting' to determine those parameters in most studies to date has been done with subjective 'by eye' fitting. This is mainly due to the fact that the isochrones have no simple parametric form and are usually in the form of tabulated points for a set of fundamental parameters such as age and metallicity. The traditional 'fitting' method used is to first determine the reddening by adjusting the zero-age main sequence (ZAMS) to the observed color–color diagram—usually $(B-V)$ versus $(U-B)$—of the cluster and subsequently, keeping this value fixed, adjusting the distance and age using the observed color–magnitude diagram (CMD) and tabulated isochrones. The subjectivity in the determination of the reddening is particularly problematic since it affects the subsequent determination of the distance and age of the cluster. It is also important to point out that in some cases the reddening is not determined through the use of U-band photometry, which can further compromise the results obtained.

In this paper, we present a new technique to fit models to open cluster photometric data using a geometric criterion to define the goodness of fit and a global optimization algorithm known as cross entropy (CE) to find the best-fitting isochrone.

2. The cross-entropy optimization method

The CE procedure provides a simple, adaptive way of estimating the optimal reference parameters. Basically, the CE method involves an iterative procedure where in each iteration the following is done:

1. Random generation of the initial parameter sample, respecting pre-defined criteria;
2. Selection of the best candidates based on some mathematical criterion;

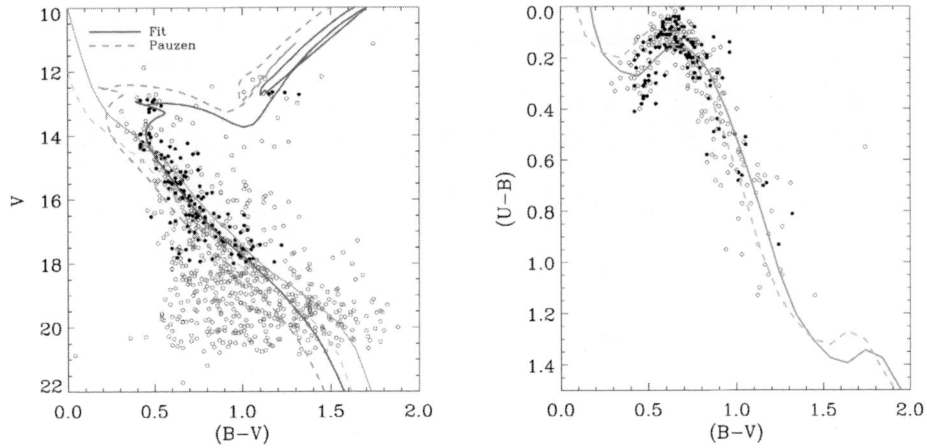

Figure 1. Example of the final fit results for NGC 2355.

3. Random generation of updated parameter samples from the previous best candidates to be evaluated in the next iteration;

4. Repeat steps (*b*) and (*c*) until a pre-specified stopping criterion is met.

For isochrone fitting, we require that a good solution be one that minimizes the sum of the minimum distances of the data points in the $(V, B-V)$ plane to the model. Ideally, any point would sit on the model curve and the variances as well as the minimal distances would be zero for a perfect solution. The sum of the variances is also modified by weights for each data point. Thus, our objective function is given by:

$$S = \sum_i \min \left[\frac{(V_i - I_V)^2}{\sigma^2_{V_i}} + \frac{((B-V)_i - I_{(B-V)})^2}{\sigma^2_{(B-V)_i}} \right] \times W_i, \tag{2.1}$$

where I is the tabulated isochrone function, V the observed V-band magnitude, $(B-V)$ the color index, σ_{V_i} the error on the V magnitude of point i, $\sigma_{(B-V)_i}$ the error on the $(B-V)$ color of point i and W_i is the weight of point i based on its probability of belonging to the cluster.

The tabulated isochrones are defined by four parameters, i.e., age distance, extinction and metallicity. The parameter ranges are pre-defined by the user and should be representative of the problem being optimized. In our isochrone fitting we defined the parameter space as follows:

1. **Age**: from log(age/yr) = 6.6 to 10.15;
2. **Distance**: from 1 to 10 000 pc;
3. $E(B-V)$: from 0.0 to 3.0 mag.

To determine the weights of each observed star, we must first remove the contamination from stars that do not belong to the cluster. The contamination is usually from field stars that can be at different distances and have different reddening values than the cluster. Therefore, we introduce schemes to filter out contaminating stars and determine which stars are more likely to belong to the cluster in a way that can be easily reproduced given simple and clear parameters.

The weight estimation for each star is performed by comparing the characteristics of the stars in an area around the ith star in the CMD defined by a box with dimensions $3\sigma_{V_i} \times 3\sigma_{(B-V)_i}$. We then calculate the average and standard deviation of V and $(B-V)$ for the stars that fall within this box and that are at a radius within the cluster region defined through the stellar density profile. We then calculate the probability of this star

Table 1. Basic parameters for the investigated clusters newly determined here. The numbers in the final column are the WEBDA reference codes, $E(B-V)$ is the extinction, d the distance to the cluster, log(Age) the logarithmic age (in years), Z the adopted metallicity and $\chi^2_{\rm red}$ the reduced χ^2. The literature values are from Paunzen & Netopil (2006).

	Fit					Literature				
Cluster	$E(B-V)$ (mag)	Distance (pc)	log(Age) (yr)	Z	$\chi^2_{\rm red}$	$E(B-V)$ (mag)	Distance (pc)	log(Age) (yr)	$\chi^2_{\rm red}$	Ref.
NGC 2477	0.26 ± 0.01	1349 ± 20	8.95 ± 0.06	0.019	2.9	0.26 ± 0.08	1227 ± 166	8.94 ± 0.11	3.3	152
NGC 7044	0.55 ± 0.05	3383 ± 215	9.35 ± 0.11	0.019	1.9	0.63 ± 0.06	3097 ± 145	9.26 ± 0.08	3.1	62
NGC 2266	0.16 ± 0.01	3194 ± 90	8.95 ± 0.06	0.008	2.2	0.10 ± 0.01	3490 ± 180	8.87 ± 0.04	3.9	41
Berkeley 32	0.14 ± 0.02	3364 ± 79	9.65 ± 0.14	0.008	3.1	0.15 ± 0.03	3491 ± 401	9.54 ± 0.08	4.7	40
NGC 2682	0.02 ± 0.02	831 ± 25	9.50 ± 0.06	0.019	2.8	0.05 ± 0.02	820 ± 47	9.61 ± 0.09	3.8	335
	0.05 ± 0.02	826 ± 64	9.55 ± 0.09	0.019	2.6	0.05 ± 0.02	820 ± 47	9.61 ± 0.09	4.7	31
	0.06 ± 0.01	910 ± 16	9.40 ± 0.05	0.019	3.5	0.05 ± 0.02	820 ± 47	9.61 ± 0.09	4.8	54
NGC 6791	0.07 ± 0.01	3597 ± 67	10.10 ± 0.06	0.03	1.5	0.17 ± 0.04	4418 ± 743	9.89 ± 0.11	2.9	100
	0.06 ± 0.01	3467 ± 117	10.15 ± 0.05	0.03	2.1	0.17 ± 0.04	4418 ± 743	9.89 ± 0.11	4.2	75
NGC 2506	0.05 ± 0.01	2844 ± 237	9.20 ± 0.08	0.008	1.2	0.06 ± 0.04	3315 ± 219	9.22 ± 0.11	1.5	284
	0.05 ± 0.01	3031 ± 86	9.30 ± 0.05	0.008	3.1	0.06 ± 0.04	3315 ± 219	9.22 ± 0.11	2.6	163
NGC 2355	0.24 ± 0.01	1681 ± 44	9.10 ± 0.06	0.008	2.6	0.14 ± 0.06	2086 ± 163	8.92 ± 0.07	6.3	217
	0.18 ± 0.01	1780 ± 55	9.00 ± 0.07	0.008	2.4	0.14 ± 0.06	2086 ± 163	8.92 ± 0.07	4.1	44
Melotte 105	0.54 ± 0.14	2062 ± 334	8.45 ± 0.24	0.019	1.4	0.48 ± 0.05	2094 ± 159	8.35 ± 0.09	1.4	289
	0.56 ± 0.05	2196 ± 209	8.35 ± 0.08	0.019	1.3	0.48 ± 0.05	2094 ± 159	8.35 ± 0.09	1.3	32
Tr01	0.61 ± 0.12	2598 ± 360	7.90 ± 0.34	0.019	1.1	0.57 ± 0.04	2356 ± 511	7.48 ± 0.08	1.3	320
	0.59 ± 0.04	2464 ± 153	6.85 ± 0.56	0.019	2.2	0.57 ± 0.04	2356 ± 511	7.48 ± 0.08	2.2	86

being a cluster member,

$$W_i = e^{\frac{-(V_i - \langle V_c \rangle)^2}{2\sigma^2_{V_c}}} \times e^{\frac{-((B-V)_i - \langle (B-V)_c \rangle)^2}{2\sigma^2_{(B-V)_c}}}, \qquad (2.2)$$

where V_i is the V magnitude of the $i^{\rm th}$ star, $\langle V_c \rangle$ the average V magnitude of the cluster stars within the 3σ error box, $(B-V)_i$ the color index of the $i^{\rm th}$ star and $\langle (B-V)_c \rangle$ the average color index of the cluster stars within the 3σ error box.

3. Application to observed clusters

The observational data was obtained from the WEBDA† catalog for the clusters selected from the standard list of Paunzen & Netopil (2006). We only selected clusters with U-band photometry, so that we could determine the reddening using the color–color diagram of open clusters, as done traditionally. We did not use data sets that contained mixed observations from different authors, but used only those that originated from a single source. For the standard list of clusters, only 13 satisfied these conditions. For some clusters we also obtained multiple data sets that satisfied our criteria and those were also used. From the 13 final clusters, we selected a sample of 10 to which we applied our fitting technique. Figure 1 shows an example of the final fit for NGC 2355 and Table 1 includes the final results for all clusters studied.

4. Conclusions

We developed a new method to fit isochrones to open cluster data, limiting ourselves to well-studied objects with tabulated Padova isochrones and keeping metallicities fixed to values determined in the literature. The results using the observed data available in the literature show that the parameters determined through our technique are consistent with results obtained by other authors. The Monte Carlo error estimate obtained by our method also gives us a better understanding of the reliability of the results, given the filtering and fitting criteria together with the $\chi^2_{\rm red}$.

We have shown that our method is reliable and robust and although the results are consistent with literature values for the clusters studied, we believe that there is room for improvement in the accepted parameter values for the set of clusters adopted as standards. The possibility of re-evaluating previous results with more quantifiable means is important, as it removes the subjectivity inherent to most open cluster studies to date.

Acknowledgements

The authors would like to thank CNPq (grant 302762/2007-8), CAPES (project CAPES–GRICES 040/2008) and FAPEMIG (APQ-00090-08).

References

Paunzen, E. & Netopil, M. 2006, *MNRAS*, 371, 1641

† available from `http://obswww.unige.ch/webda`

Star clusters: basic galactic building blocks
Proceedings IAU Symposium No. 266, 2009
R. de Grijs & J. R. D. Lépine, eds.

doi:10.1017/S1743921309991761

Probing the structure of the Galactic disk through open clusters

Xiaoying Pang and Chenggang Shu

Key Laboratory for Astrophysics, Shanghai Normal University, Shanghai, P. R. China

Abstract. The WEBDA database of open clusters (hereafter OCs) in the Galaxy contains 970 OCs, of which 911 have age determinations, 920 have distance measurements, and 911 have color-excess data. Base on the statistical analysis of global properties of open clusters, we investigate disk properties such as the height above the Galactic plane. We find that old open clusters (age $\geqslant 1$ Gyr) are preferentially located far from the Galactic plane with $\langle |z| \rangle \sim 394.5$ pc. They lie in the outer part of the Galactic disk. The young open clusters are distributed in the Galactic plane almost symmetrically with respect to the Sun, with a scale height perpendicular to the Galactic plane of 50.5 pc. The age distribution of open clusters can be fit approximately with a two-component exponential decay function: one component has an age scale factor of 225.2 Myr, and the other consists of longer-lived clusters with an age scale of 1.8 Gyr, which are smaller than those derived by Janes & Phelps (1994) of 200 Myr and 4 Gyr for the young and old OCs, respectively. As a consequence of completeness effects, the observed radial distribution of OCs with respect to Galactocentric distance does not follow the expected exponential profile. Instead, it falls off both for regions external to the solar circle and more sharply towards the Galactic Center, which is probably due to giant molecular cloud disruption in the center. We simulate the effects of completeness, assuming that the observed distribution of the number of OCs with a given number of stars above the background is representative of the intrinsic distribution of OCs throughout the Galaxy. Two simulation models are considered, in which the intrinsic number of the observable stars are distributed (i) assuming the actual positions of the OCs in the sample, and (ii) random selection of OC positions. As a result, we derive completeness-corrected radial distributions which agree with an exponential disk throughout the observed Galactocentric distance in the range of 5–15 kpc, with scale lengths in the range of 1.6–2.8 kpc.

Keywords. Galaxy: disk, open clusters and associations: general

---

The full poster (in pdf format) is available at
`http://www.astro.iag.usp.br/~iaus266/Posters/pPang.pdf`.

Star clusters: basic galactic building blocks
Proceedings IAU Symposium No. 266, 2009
R. de Grijs & J. R. D. Lépine, eds.

doi:10.1017/S1743921309991773

Metallicity analysis of an SMC cluster sample using CaII-triplet spectroscopy

M. C. Parisi,[1] A. Grocholski,[2] D. Geisler,[3] A. Sarajedini[4] and J. J. Clariá[1]

[1]Observatorio Astronómico de Córdoba, Argentina
email: [celeste, claria]@oac.uncor.edu

[2]Space Telescope Science Institute, Baltimore, MD, USA
email: aarong@stsci.edu

[3]Universidad de Concepción, Chile
email: dgeisler@astro-udec.cl

[4]University of Florida, Gainesville, FL, USA
email: ata@astro.ufl.edu

Abstract. The metallicities of 15 Small Magellanic Cloud (SMC) clusters are determined from the Calcium II-triplet lines in spectra obtained using the FORS2 multi-object spectrograph on the *VLT* (Paranal, Chile). The metallicity distribution, metallicity gradient and age–metallicity relationship in the SMC are determined and analyzed using new and literature data for these and other, previously studied SMC objects. Only the main results regarding these topics are presented and discussed.

Keywords. galaxies: star clusters, techniques: spectroscopic

1. Introduction

We have obtained near-infrared spectra covering the CaII-triplet (CaT) lines for a large number of stars associated with 15 Small Magellanic Cloud (SMC) clusters using *VLT*/FORS2. We have targeted clusters that cover as large a range in ages as possible so as to sample the age–metallicity relation over a wide baseline. We also attempted to select clusters that were in relatively uncrowded fields and that were spread around the galaxy to cover as wide an area and radial range as possible to search for any global effects such as gradients. Spectrocopic targets were chosen based on the instrumental color–magnitude diagram (CMD), assigning highest priority to stars lying along the apparent cluster giant branch. We used slits that were $1''$ wide and $8''$ long. In all fields, we obtained a single 900 s exposure with a typical seeing of less than $1''$. Pixels were binned 2×2, yielding a plate scale of $0.25''$ pixel^{-1}. The spectra have a dispersion of ~ 0.85 Å pixel^{-1} with a characteristic rms scatter of ~ 0.06 Å, and cover a range of ~ 1600 Å in the CaT region. Signal-to-noise ratios ranged from ~ 10 to ~ 70 pixel^{-1}.

2. Radial velocities, equivalent widths and metallicities

To measure the radial velocities (RVs) of our program stars, we performed cross correlations between their spectra and the spectra of 32 bright Milky Way open and globular cluster template giants using IRAF tasks. We used the template stars of Cole *et al.* (2004), who observed these stars with a setup very similar to ours. Typical RV errors are 7.5 km s^{-1}.

Table 1. Derived SMC cluster properties.

Cluster	N	[Fe/H] (dex)	$\sigma_{[Fe/H]}$ (dex)	a (°)
BS 121	5	−0.66	0.07	1.496
HW 47	4	−0.92	0.04	3.502
HW 84	4	−0.91	0.05	5.513
HW 86	4	−0.61	0.06	7.345
L 4	9	−1.08	0.04	3.265
L 5	5	−1.25	0.05	3.092
L 6	7	−1.24	0.03	3.124
L 7	7	−0.76	0.06	2.888
L 17	8	−0.84	0.03	1.718
L 19	7	−0.87	0.03	1.564
L 27	7	−1.14	0.06	1.392
L 106	7	−0.88	0.06	7.877
L 108	6	−1.05	0.05	4.460
L 110	9	−1.03	0.05	5.323
L 111	8	−0.82	0.03	7.830

To measure equivalent widths, we used a previously written FORTRAN program (see Cole *et al.* 2004 for details), following the procedure of Armandroff & Zinn (1988). To derive the metallicities of our entire cluster sample, we adopted the Cole *et al.* (2004) relationship (see Parisi *et al.* 2009 for more details). Our estimate of the total metallicity error per star ranges from 0.09 to 0.35 dex, with a mean of 0.17 dex.

Cluster members were selected using a combination of their positions relative to the cluster center and their location in the CMD, abundances and RVs (see Parisi *et al.* 2009). We used these cluster members to calculate the mean metallicity of our cluster sample. We determine mean cluster metallicities to 0.05 dex (random error), from a mean of 6.4 members per cluster. The results are given in Table 1, together with the semi-major axis, a (see Section 3). Metallicity errors correspond to the standard error of the mean (s.e.m.).

3. Metallicity results

The metallicity distribution (MD), metallicity gradient and age–metallicity relation (AMR) are investigated, combining our clusters with those observed by Da Costa & Hatzidimitriou (1998) and Glatt *et al.* (2008) (also using the CaT) and one cluster with a detailed, high-resolution metallicity to compile a sample of 25 clusters on a homogeneous metallicity scale.

We derive a mean metallicity for our CaT sample of −0.94 dex, with a standard deviation $\sigma = 0.19$ dex, while for the full sample these values become −0.96 and 0.19 dex, respectively. The mean values are in very good agreement with each other and the global mean value of −1 dex found by Carrera *et al.* (2008) from CaT spectra of a large number of field giants. The MD is shown in Figure 1 (left top panel), in which we also show the MD for Large Magellanic Cloud (LMC) clusters derived by Grocholski *et al.* (2006) in our CaT study (left bottom panel). The SMC clusters fall in a rather small metallicity range of < 0.8 dex, from −0.6 to −1.4 dex, and are concentrated in the 0.5 dex range from −0.75 to −1.25 dex. This is unlike the LMC clusters, which cover ~ 2 dex in metallicity (Grocholski *et al.* 2006), with higher and lower metallicities than found in their SMC counterparts. The broad characteristics of the MD of the Magellanic

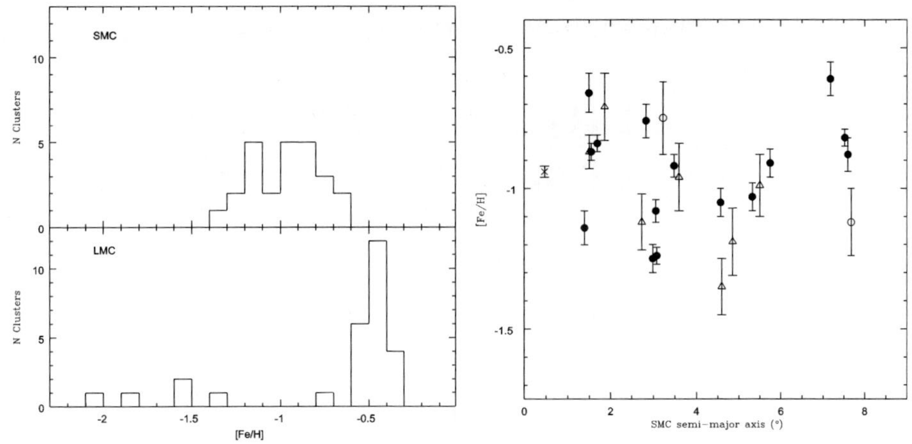

Figure 1. *(left)* Metallicity distribution (MD) of SMC clusters *(top panel)*: 15 from the present work, six from Da Costa & Hatzidimitriou (1998), three from Glatt *et al.* (2008) and NGC 330 Gonzalez & Wallerstein (1999). The histogram in the bottom panel corresponds to the MD for LMC clusters derived from our CaT investigation (Grocholski *et al.* 2006). *(right)* Metallicity versus projected radius (semi-major axis a) for the SMC clusters. Open circles represent clusters from Da Costa & Hatzidimitriou (1998) and triangles represent clusters from Glatt *et al.* (2008). Clusters from our CaT sample are represented by filled circles. NGC 330 is shown by a cross. No clear trend is evident.

Cloud cluster systems have been known for some time. Da Costa (1991) noted that (sic) "The LMC managed to make metals but no clusters during the age gap while the SMC managed to make clusters but no metals". These curious facts are now even more evident and corroborated in much greater detail than known at that time, but their explanation remains as mysterious. There is a suggestion for bimodality in the MD, with peaks at [Fe/H] ~ -0.9 and -1.15 dex, but more clusters are needed to corroborate this result.

In the right panel of Figure 1 we plot metallicity versus the semi-major axis a value for our cluster sample and for the full sample. The parameter a is the semi-major axis that an ellipse would have if it were centered on the SMC center, aligned with the bar, had a b/a ratio of 1/2 and one point of its trajectory coincided with the cluster position. No clear trend is evident. Dividing our sample at 4°, as did Piatti *et al.* (2007a,b), we find for the 15 inner clusters a mean metallicity and standard deviation of -0.94 and 0.19, respectively, while the 10 outer clusters have -1.00 and 0.21. The difference is not significant. To check to see if this could possibly be due to an inverse age-gradient effect, we also checked the mean ages of the two divisions. The inner clusters are 3.1 (1.9) Gyr and the outer clusters 4.4 (3.4) Gyr old. Thus, this cannot be the cause of a lack of an observed metallicity gradient. We conclude that any true metallicity gradient in the SMC cluster system must be relatively weak. Additional data are required to ascertain the existence and strength of any gradient. If the gradient is indeed minimal, Zaritsky *et al.* (1994)'s suggestion that a strong bar weakens any disk gradient is a viable explanation.

The AMR (Figure 2) shows evidence for three phases: a very early ($>$ 11 Gyr) phase in which the metallicity reached ~ -1.2 dex, a long intermediate phase from $\sim$ 10 to 3 Gyr ago, in which the metallicity only slightly increased although a number of clusters formed, and a final phase from 3 to 1 Gyr ago, in which the rate of enrichment was substantially faster. We find good overall agreement with the model of Pagel & Tautvaišienė (1998; solid line), which assumes a burst of star formation at 4 Gyr. A hybrid infall + outflow model of Carrera (2005; long-dashed line) also fits the data reasonably well. The simple

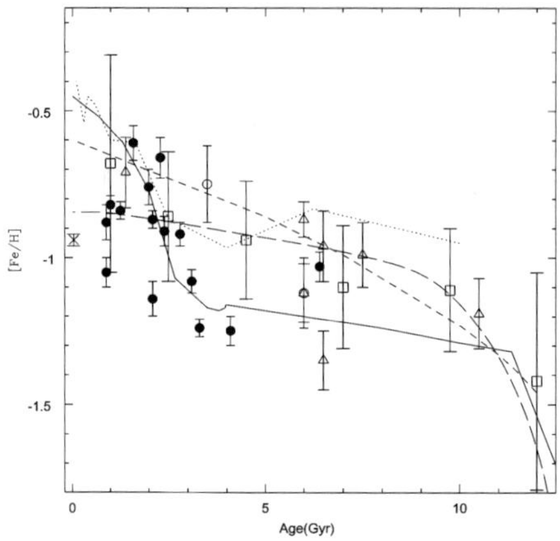

Figure 2. Age–metallicity relation. Symbols are the same as in Figure 1. The mean metallicity in six age bins calculated by Carrera *et al.* (2008) is also shown (squares). Lines are explained in the text (Section 3).

closed-box model of Da Costa & Hatzidimitriou (1998; short-dashed line) yields a much poorer fit and the AMR derived by Harris & Zaritsky (2004; dotted line) is significantly offset to higher metallicities for intermediate-age clusters. A number of different lines of evidence point to the likelihood of a burst in the SMC star- and cluster-formation intensity about 3 Gyr ago. The cause of such a burst is currently a source of much speculation. The suggestion by Bekki *et al.* (2004) that it is due to a close passage of the SMC and LMC is intriguing but requires better knowledge of their orbits, especially proper motions, to be definitively tested.

References

Armandroff, T. E. & Zinn, R. 1988, *AJ*, 96, 92

Bekki, K., Couch, W. J., Beasley, M. A., Forbes, D. A., Chiba, M., & Da Costa, G. S. 2004, *ApJ* (Letters), 610, L93

Carrera, R., Gallart, C., Aparicio, A., Costa, E., Méndez, R. A., & Noël, N. E. D. 2008, *AJ*, 136, 1039

Carrera, R. 2005, PhD Thesis, Departamento de Astrofísica, Universidad de La Laguna, Spain

Cole, A. A., Smecker–Hane, T. A., Tolstoy, E., Bosler, T. L., & Gallagher, J. S. 2004, *MNRAS*, 347, 367

Da Costa, G. S. 1991, in: R. Haynes & D. Milne (eds.), *The Magellanic Clouds*, Proc. IAU Symp. No. 148, p. 183

Da Costa, G. S. & Hatzidimitriou, D. 1998, *AJ*, 115, 1934

Glatt, K., *et al.* 2008, *AJ*, 136, 1703

Gonzalez, G. & Wallerstein, G. 1999, *AJ*, 117, 2286

Grocholski, A. J., Cole, A. A, Sarajedini, A., Geisler, D., & Smith, V. 2006, *AJ* 132, 1630

Harris, J. & Zaritsky, D. 2004, *AJ*, 127, 1531

Pagel, B. E. J. & Tautvaišienė, G. 1998, *MNRAS*, 299, 535

Parisi, M. C., Grocholski, A. J., Geisler, D., Sarajedini, A., & Clariá, J. J. 2009, *AJ*, 138, 517

Piatti, A. E., Sarajedini, A., Geisler, D., Clark, D., & Seguel, J. 2007a, *MNRAS*, 377, 300

Piatti, A. E., Sarajedini, A., Geisler, D., Gallart, C., & Wischnjewsky, M. 2007b, *MNRAS* (Letters), 381, L84

Zaritsky, D., Kennicutt Jr., R. C., & Huchra, J. P. 1994, *ApJ*, 420, 87

Star clusters: basic galactic building blocks
Proceedings IAU Symposium No. 266, 2009
R. de Grijs & J. R. D. Lépine, eds.

doi:10.1017/S1743921309991785

Open cluster remnants in low-density Galactic fields

D. B. Pavani,[1] L. O. Kerber,[2] E. Bica[3] and W. J. Maciel[4]

[1]Universidade Federal de Pelotas (UFPel), Campus Universitário, s/n, Caixa Postal 354, CEP 96010-900 Pelotas, RS, Brazil
email: dpavani@if.ufrgs.br

[2]Universidade Estadual de Santa Cruz (UESC), Campus Soane Nazaré de Andrade km 16 Rodovia Ilhéus–Itabuna CEP 45662-000, Ilhéus, Bahia, Brazil
e-mail: [kerber, maciel]@astro.iag.usp.br

[3]Universidade Federal do Rio Grande do Sul (UFRGS), Av. Bento Goncalves 9500, Caixa Postal 15051, CEP 91501-970 Porto Alegre, RS, Brazil
email: bica@if.ufrgs.br

[4]Instituto de Astronomia, Geofísica e Ciências Atmosféricas da Universidade de São Paulo (IAG/USP), Rua do Matão, 1226, Cidade Universitária, CEP 05508-090 São Paulo/SP, Brazil

Abstract. Open cluster remnants (OCRs) are fundamental objects to investigate open cluster dissolution processes (e.g., Bica *et al.* 2001; Carraro 2002; Pavani *et al.* 2003; Carraro *et al.* 2007; Pavani & Bica 2007). They are defined as poorly populated concentrations of stars, with enough members to show evolutionary sequences in colour–magnitude diagrams (CMDs) as a result of the dynamical evolution of an initially more massive physical system. An OCR is intrinsically poorly populated, which makes its differentiation from field-star fluctuations difficult. Among the possible approaches to establish the nature of OCRs, we adopted CMD analysis combined with a robust statistical tool applied to 2MASS data. In addition, photometry is the main information source available for possible OCRs (POCRs). We developed a statistical diagnostic tool to analyse the CMDs of POCRs and verify them as physical systems, explore membership probabilityies taking into account field contamination and derive age, distance and reddening values in a self-consistent way. We present the results of our analysis of 88 POCRs that are part of a larger sample that is widely distributed across the sky, with a significant density contrast of bright stars compared to the Galactic field. The 88 objects are projected onto low-density Galactic fields, at relatively high latitudes ($|b| > 15°$). Studies of larger POCR samples will provide a better understanding of OCR properties and constraints for theoretical models, including new insights into the evolution of open clusters and their dissolution rates. The results of this ongoing survey will provide a general picture of these fossil stellar systems and their connection to Galactic-disk evolution.

Keywords. open clusters and associations: general, infrared: general, methods: statistical

1. The sample

The objects were selected from open cluster catalogues and visual inspection of Digitised Sky Survey maps. All objects are underpopulated compared with the 'usual' open clusters, so that availability of proper-motion, radial-velocity or spectral studies is rare. However, photometry is the main information source available for POCRs and the 2MASS database provides objects and fields in a homogeneous manner, together with photometric errors and completeness estimates. Figure 1 shows an overall view of the complete sample. The sample contains 36 objects selected from the NGC catalogue, 32 from the ESO–LV, 7 from Ruprecht, 5 from IC, 5 from Collinder and 3 from elsewhere.

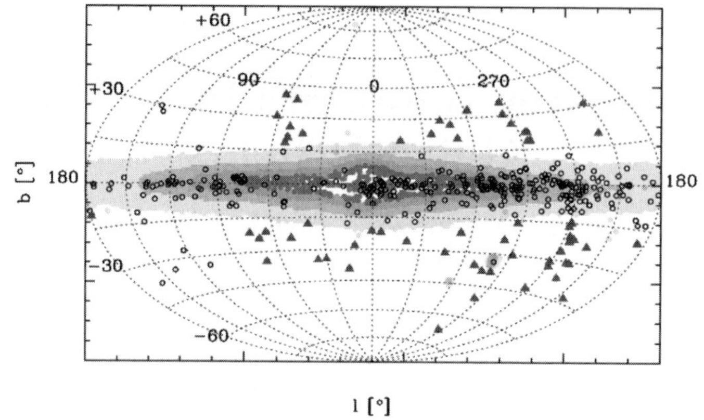

Figure 1. Whole-sky Aitoff projection showing, in Galactic coordinates, the positions of the present POCR sample (triangles) and the larger sample of POCRS (open circles). 2MASS background stellar densities are shown at four levels, corresponding to the 20th, 40th, 60th and 80th percentiles of the $\log(\rho_{\rm field})$ distribution.

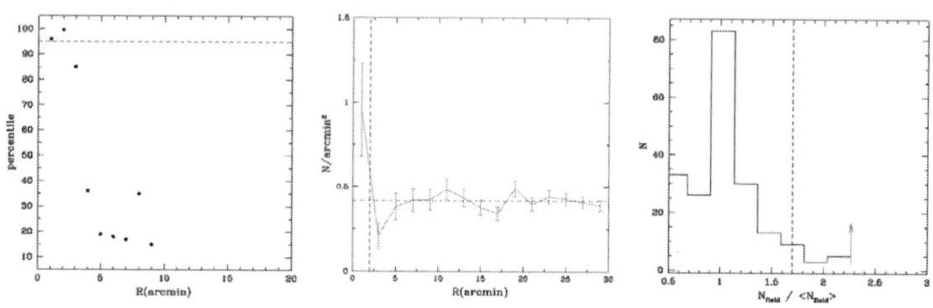

Figure 2. *(left)* Example of the search for the largest radius that maximizes the number of stars in the POCR compared with the counts in the field-star distribution (percentiles). The horizontal dashed line indicates the 95th percentile. *(middle)* Example of radial stellar density profiles compared with the stellar-field density averages (horizontal dashed line). $R_{\rm cut}$ is also shown (vertical dashed line). *(right)* Example of star-count distributions for the field-star samples with the same radius as the object, compared with the POCR ($N_{\rm POCR}$; arrow).

2. Method

The method assigns to each star a membership probability and finds a set of best isochrone fits, maximizing the number of fitted stars. Applying the same method to the randomly selected neighbouring fields, it objectively tests whether or not POCRs can be understood as field fluctuations.

(*a*) **Determining the POCR cutoff radius.** This corresponds to star-count analysis of the POCR sample ($N_{\rm POCR}$) compared with the field ($N_{\rm field}$). We use neighbouring-field samples randomly selected from concentric background regions as control fields. The number of comparison fields depends on the ratio of the field-to-POCR areas. The process to determine the cutoff radius ($R_{\rm cut}$) is shown in Figure 2.

(*b*) **Taking into account field-star contamination.** This method was inspired by the field-star decontamination process of Mighell *et al.* (1996), Kerber *et al.* (2002) and Kerber & Santiago (2005). We adapted their central ideas and applied them to poorly populated objects. Therefore, we assume that the CMD of a control field is representative of the field. By a comparison with the POCR CMD, we compute the probability for any POCR star to be a member. We developed a semi-automated method to test a suitable number of possible isochrone solutions (for solar metallicity). First, we estimate initial

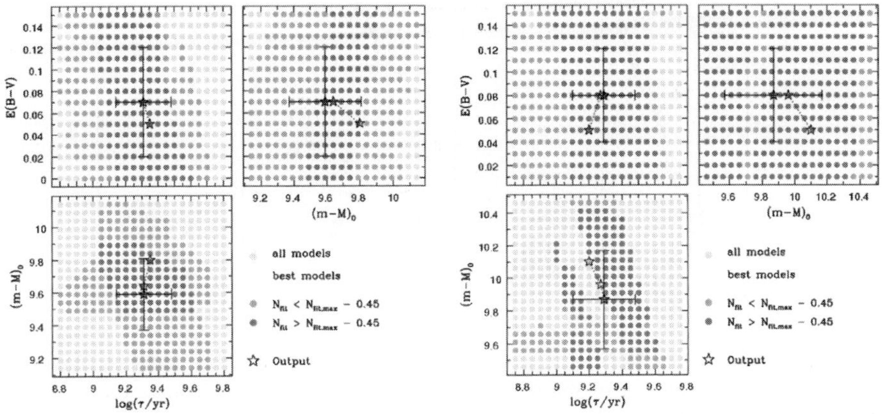

Figure 3. Example of the parameter space explored for ESO324–SC15, which was identified as an OCR *(left)* and NGC 272 *(right)*, classified as a field fluctuation. The evolution of the output solution (open stars) is shown.

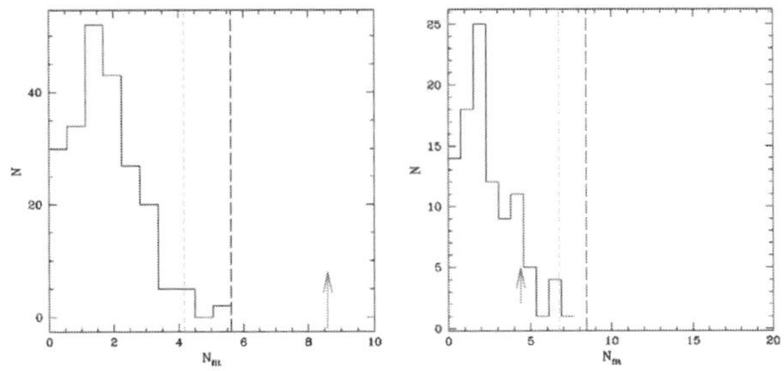

Figure 4. Examples of the distributions of the isochrone-fitted stars for the field-star samples compared with the object (arrow).

guesses for each physical parameter by eye and we subsequently explore a large parameter space in equal steps. Therefore, for each POCR we typically test 8000 isochrone solutions. The parameter space explored is shown in Figure 3. These figures show the best solutions and the evolution of the average (output) solution after each iteration.

(c) **Statistical comparison with field stars.** The statistical comparison with the field and the final characterization are illustrated in Figure 4, which presents the number of isochrone-fitted stars, taking into account the probability distribution for the field stars ($n_{\rm fit}$) compared with the value for each object ($n_{\rm fit,POCR}$). The number of fitted stars for ESO324–SC15 are not reproduced by their neighbouring field samples, thus being classified as OCR by the criterion. On the other hand, the same figure also reveals that NGC 272 can be understood as a field stellar fluctuation (FF). In Figure 5 we present CMDs for two OCRs: ESO435–SC48—log(age/yr) $= 9.3 \pm 0.17, E(B-V) = 0.07 \pm 0.05$ mag, $(m-M)_0 = 9.59 \pm 0.22$ mag—and ESO324–SC15—log(age/yr) $= 9.1 \pm 0.1$, $E(B-V) = 0.08 \pm 0.05$ mag, $(m-M)_0 = 9.60 \pm 0.23$ mag.

3. Results and concluding remarks

Figure 6 provides an overview of the properties of the OCRs. The method assigns to each star a membership probability and finds a set of best isochrone fits, maximizing the number of fitted stars. The method objectively tests whether or not the possible open

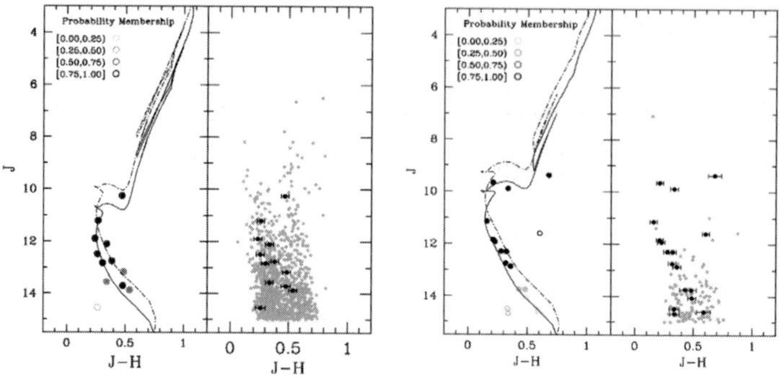

Figure 5. J versus $(J-H)$ CMD of photometric OCRs: ESO435–SC48 *(left)* and ESO324–SC15 *(right)* according to their stellar probability membership. Padova isochrones (Girardi *et al.* 2002) of solar metallicity are shown as well as a shifted version including binaries of equal masses (dashed line). *(right)* Cluster stars overplotted on the stars in the control field ($R = 30'$).

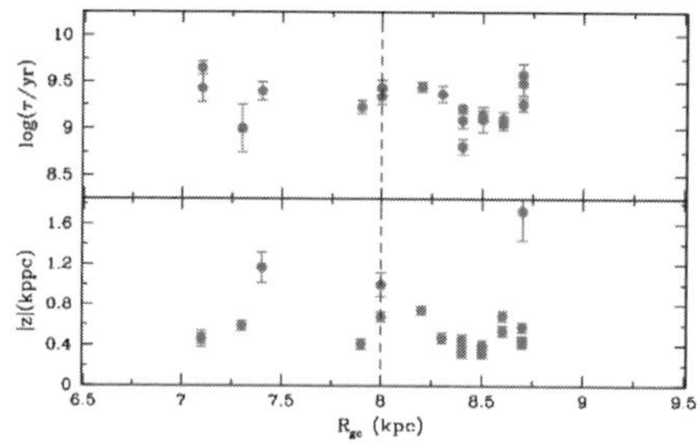

Figure 6. The objects are shown as a function of height from the plane $|z|$ and age, as a function of Galactocentric distance R_{gc} *(right)*. The Sun is assumed to be at $R_{gc} = 8$ kpc (Reid 1993).

cluster remnant can be understood as a field fluctuation. The objective criteria suggested that 23 objects were OCRs, 45 FFs, and 20 were classified as borderline objects. These results are consistent with their behaviour in parameter space (Figure 3).

References

Bica, E., Santiago, B. X., Dutra, C. M., Dottori, H., de Oliveira, M. R., & Pavani, D. 2001, *A&A*, 366, 827

Carraro, G. 2002, *A&A*, 385, 471

Carraro, G., de La Fuente Marcos, R., Villanova, S., Moni Bidin, C., de La Fuente Marcos, C., Solivella, G., & Baumgardt, H. 2007, *A&A*, 466, 931

Girardi, L., Bertelli, G., Bressan, A., Chiosi, C., Groenewegen, M. A. T. Marigo, P., Salasnich, B., & Weiss, A. 2002, *A&A*, 391, 185

Kerber, L. O, Santiago, B. X., Castro, R., & Valls–Gabaud, D. 2002, *A&A*, 390, 121

Kerber, L. O. & Santiago, B. X. 2005, *A&A*, 435, 77

Miguell, K. J., Rich, R. M., Shara, M., & Fall, S. M. 1996, *A&A*, 111, 2314

Pavani, D. B., Bica, E., Ahumada, A. V., & Clariá, J. J. 2003, *A&A*, 399, 113

Pavani, D. B. & Bica, E. 2007, *A&A*, 468, 139

Reid, M. J. 1993, *ARA&A*, 31 345

Star clusters: basic galactic building blocks
Proceedings IAU Symposium No. 266, 2009
R. de Grijs & J. R. D. Lépine, eds.

doi:10.1017/S1743921309991797

Intermediate-mass black holes in Galactic globular clusters

Carolina Pepe and Leonardo J. Pellizza

Instituto de Astronomía y Física del Espacio, UBA–CONICET, C.C. 67, Suc. 28 (C1428ZAA), Buenos Aires, Argentina
email: [carolinap, pellizza]@iafe.uba.ar

Abstract. Over the last few years, different observations have suggested the existence of intermediate-mass ($\sim 10^3$ M$_\odot$) black holes in the centers of globular clusters. However, the issue is still a matter of debate, as current observations have alternative explanations. We previously developed a hydrodynamical model for the interstellar medium in these systems to explain the luminosity of the central X-ray source found in NGC 6388, assuming a black hole accreting from the insterstellar medium. Here, we explore the predictions of our model regarding the flow of the interstellar matter in the inner cluster regions and find that the density and velocity profiles could help to determine the presence of a central black hole as well as its mass.

Keywords. ISM: kinematics and dynamics, accretion, accretion disks

1. Introduction

Over the last few years, different observations have suggested the existence of black holes with masses of $10^2 - 10^4$ M$_\odot$, intermediate between the masses of black holes formed by the collapse of high-mass stars and those of the supermassive black holes found in the centers of galaxies (Miocchi 2007). For this reason, they have been called intermediate-mass black holes (IMBHs). First, ultraluminous X-ray sources (ULXs) have been detected in nearby galaxies (Fabbiano 2006). The luminosities of these sources ($L > 10^{40}$ erg s^{-1}) are well in excess of the Eddington luminosity for stellar-mass objects. As this limit is proportional to the accreting object's mass, these luminosities could be interpreted as sub-Eddington emission of more massive objects, i.e., IMBHs. On the other hand, stellar density profiles and stellar dynamics in the centers of globular clusters suggest the presence of central objects with masses in the range of interest (van den Bosch *et al.* 2006; McLaughlin *et al.* 2006; Miocchi 2007; Noyola *et al.* 2008). However, the evidence is not conclusive and there exist alternative models to explain the observations (Fabbiano 2006).

One way to detect black holes is through the emission of matter they accrete, usually observed in X-rays, or that of matter ejected in the form of jets, observed as radio emission. Maccarone *et al.* (2004) investigated this emission, assuming that the IMBHs in globular clusters accrete from the interstellar medium (ISM), but found that neither radio nor X-ray emission could be detected at present. Their model assumes that the ISM density is constant and the accretion proceeds at 0.1% of the Bondi–Hoyle rate. However, an extended X-ray source has been detected in the center of NGC 6388 by Nucita *et al.* (2008), with properties that are consistent with those expected for accretion onto an IMBH. This observation motivated us to develop a hydrodynamical model for the ISM, to explore a more realistic scenario for the accretion onto any possible IMBH (Pepe & Pellizza 2008). We have shown that the accretion regime differs from that of Bondi–Hoyle accretion because of the constant injection of matter by the cluster red giants. We have also shown that the luminosity of the X-ray source in the center of NGC 6388 can be explained by this flow if an efficiency of 1.5×10^{-4}, similar to that found by other authors, is assumed.

The existence of ISM in globular clusters has been demonstrated by several authors (Freire *et al.* 2001; Boyer *et al.* 2006), hence its flow could become a valuable tool to explore the existence of IMBHs. In the present sudy, we explore the predictions of our model regarding the behaviour of observable quantities such as the velocity and density of the ISM in the inner region of the cluster, where the influence of the IMBH is most important. We aim at determining the presence of an IMBH and its mass.

2. The model

To determine the presence of IMBHs in the centers of globular clusters, we studied the dynamics of the ISM using our hydrodynamical model (Pepe & Pellizza 2008). The ISM is generated by mass loss from the red giants in the cluster. The cluster's gravitational field can be described by a King model (King 1966). The values of the model parameters for NGC 6388 were taken from Harris (1996). Following Scott & Rose (1975), we assumed that the ISM is an ideal gas, and that its flow can be considered as steady, spherically symmetric and isothermal. Under these hypotheses, the flow is governed by continuity and Euler's equations. The former is

$$\frac{1}{r^2}\frac{\mathrm{d}}{\mathrm{d}r}(\rho r^2 u) = \alpha\rho^*, \tag{2.1}$$

where r is the radial coordinate, u and ρ the velocity and density of the flow, respectively, and ρ^* is the stellar density. The right-hand side describes the gas injection by the stars at a fractional rate α. Euler's equation is

$$\rho u\frac{\mathrm{d}u}{\mathrm{d}r} = -\frac{k_{\mathrm{B}}T}{\mu}\frac{\mathrm{d}\rho}{\mathrm{d}r} - \frac{G\rho}{r^2}(M(r) + M_{\mathrm{BH}}) - \alpha u\rho^*, \tag{2.2}$$

where G is the gravitational constant, k_{B} is Boltzmann's constant, μ the mean molecular mass, $M(r)$ the stellar mass inside a radius r, and M_{BH} the central IMBH mass. It is assumed here that the material is injected with zero velocity into the flow. As shown by Pepe & Pellizza (2008), Equations (2.1) and (2.2) can be combined into

$$\frac{\mathrm{d}\psi}{\mathrm{d}\xi} = \frac{\psi}{\psi^2 - \psi_{\mathrm{s}}^2}\left(\frac{2\psi_{\mathrm{s}}^2}{\xi} - \frac{\Omega'(\xi)(\psi_{\mathrm{s}}^2 + \psi^2)}{\Omega(\xi) - \Omega(\xi_{\mathrm{st}})} - \frac{9(\Omega(\xi) + \Omega_{\mathrm{BH}})}{\xi^2}\right), \tag{2.3}$$

where the dimensionless variables $\xi = r/r_0$, $\psi = u/\sigma$, $\Omega(\xi) = M(r)/4\pi\rho_0 r_0^3$ and $\Omega_{\mathrm{bh}} = M_{\mathrm{bh}}/4\pi\rho_0 r_0^3$ have been introduced. Here, r_0 is the cluster's King radius, ρ_0^* its central density, $\sigma^2 = 4\pi G\rho_0 r_0^2/9$ the velocity dispersion and ψ_{s} the nondimensional sound speed in the medium. Here, ξ_{st} is the nondimensional stagnation radius, at which $\psi = 0$. Equation (2.3) was integrated numerically using the algorithm described in Pepe & Pellizza (2008), but extending the integration inwards of the inner sound radius, and for different values of the IMBH mass in the range $10^3 - 10^4$ $\mathrm{M}_\odot$. A model with $M_{\mathrm{BH}} = 0$ was also computed to investigate the properties of the flow in the no-IMBH case.

3. Results

Figure 1 shows the results of integrating Equation (2.3) for different IMBH masses, $M_{\mathrm{BH}} = 1,\ 3,\ 5,\ 8,\ 10 \times 10^3$ $\mathrm{M}_\odot$ and $T = 10^4$ K, for the inner region of the cluster. In Pepe & Pellizza (2008) we showed that when an IMBH is present in the center of a

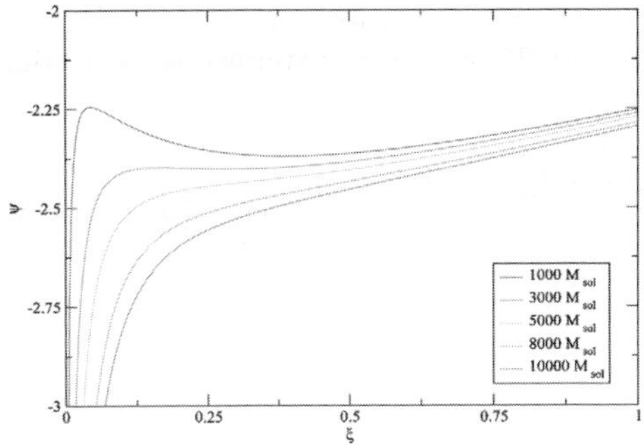

Figure 1. Velocity profile for the ISM in NGC 6388 in the inner region. As the IMBH mass grows, the matter falls faster onto the black hole.

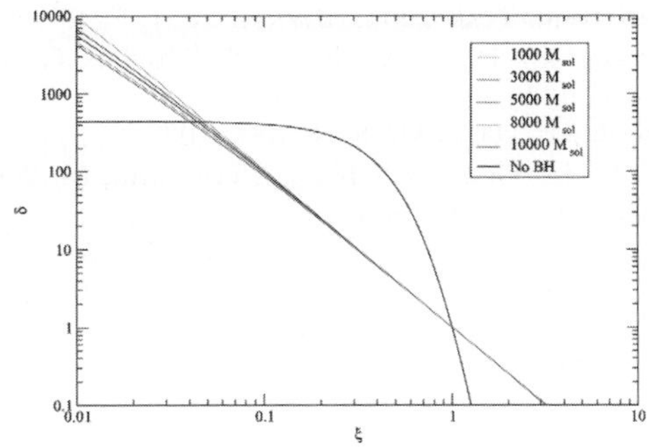

Figure 2. Density profile for the ISM in NGC 6388. The steep profile corresponds to the IMBH case and it differs from the flat profile of the no-IMBH case.

globular cluster, the flow develops a stagnation radius; outside it, the matter is ejected as a wind while inside it is accreted by the black hole. However, no difference could be seen on the cluster scale for different IMBH masses. The behaviour of the flux in the outer regions is mostly determined by the cluster's gravitational pull. Figure 1 shows that in the central region of the cluster the effect of the IMBH becomes observable, as a variation in the velocity profiles with the IMBH mass is observed. The higher the mass, the faster the matter falls onto the IMBH, as one might expect. Figure 2 shows the density profiles for different IMBH masses ($M_{\rm BH} = 0,\ 1,\ 3,\ 5,\ 8,\ 10 \times 10^3\ {\rm M}_\odot$). The density profile changes from flat to steep for $M_{\rm BH} \neq 0$. This could be an observational indicator of the presence of an IMBH in globular clusters with such a density profile. However, varying the IMBH mass produces little change in the density profile.

4. Conclusions

The flow of the interstellar matter ejected by red giants in a globular cluster with a central IMBH was calculated for the cluster inner regions by integration of the hydrodynamic equations in the presence of the cluster and IMBH gravitational fields. We found that the density and velocity profiles of the ISM show differences between the scenarios with and without a IMBH. These could provide a feasible way to observationally determine the presence of such objects in globular clusters. Moreover, if the gas velocity could be measured with the resolution needed, the mass of the IMBH could be estimated. A remarkable fact is that the ISM properties are affected by the presence of the IMBH up to radii comparable to the King radius, while the dynamics of the stars are influenced in a much smaller region.

References

Boyer, M. L., Woodward, C. E., van Loon, J. T., Gordon, K. D., Evans, A., Gehrz, R. D., Helton, L. A., & Polomski, E. F. 2006, *AJ*, 132, 1415

Fabbiano, G. 2006, *ARA&A*, 44, 323

Freire, P. C., Camilo, F., Lorimer, D. R., Lyne, A. G., Manchester, R. N., & D'Amico, N. 2006, *MNRAS*, 326, 901

Harris, W. E. 1996, *AJ*, 112, 1487

King, I. R. 1966, *AJ*, 71, 276

Lanzoni, B., Dalessandro, E., Ferraro, F. R., Miocchi, P., Valenti, E., & Rood, R. T. 2007, *ApJ*, 668, 139

Maccarone, T. J. 2004, *MNRAS*, 351, 1049
McLaughlin, D. E., Anderson, J., Meylan, G., Gebhardt, K., Pryor, C., Minniti, D., & Phinney, S. 2006, *ApJS*, 166, 249
Miocchi, P. 2007, *MNRAS*, 381, 103
Noyola, E., Gebhardt, K., & Bergmann, M. 2008, *ApJ*, 676, 1008
Nucita, A. A., De Paolis, F., Ingrosso, G., Carpano, S., & Guainazzi, M. 2008, *A&A*, 478, 763
Pepe, C. & Pellizza, L. J. 2008, *BAAA*, 51, in press
Scott, E. H. & Rose, W. K. 1975, *ApJ*, 197, 147
van den Bosch, R., de Zeeuw, T., Gebhardt, K., Noyola, E., & van de Ven, G. 2006, *ApJ*, 641, 852

Star clusters: basic galactic building blocks
Proceedings IAU Symposium No. 266, 2009
R. de Grijs & J. R. D. Lépine, eds.

doi:10.1017/S1743921309991803

Abundances of seven red giants in the open cluster NGC 3114†

C. B. Pereira and C. Quireza

Observatório Nacional, Rua General José Cristino 77, 20921-400 São Cristóvão, Rio de Janeiro, RJ Brazil
email: [claudio, quireza]@on.br

Abstract. We present a chemical analysis of seven red giants in the open cluster NGC 3114. Our main goal is to investigate the chemical composition of this cluster, which is not yet available in the literature. We employed the FEROS spectrograph on the ESO *2.2m telescope*. Atmospheric parameters and metallicity were derived from the measured equivalent widths of several iron lines using the spectral code MOOG and Kurucz model atmospheres. We obtained the abundances of O, Al, Ca, Mg, Si, Ti, Ni, Cr, Sc, Y, Zr, La, Ce and Nd by measuring the equivalent widths of the absorption lines of these elements. A mean metallicity of $[Fe/H] = 0.05 \pm 0.13$ relative to the Sun was determined from the data of the red-giant members. This result is in good agreement with the Galactic-disk radial distribution of iron traced by open clusters. We did not find any intrinsic star-to-star scatter in the [element/Fe] ratios for the stars in this cluster. We compare our results with investigations of other open clusters. An age of 8.2 Gyr is derived from isochrone fits.

Keywords. open clusters and associations: individual (NGC 3114), Galaxy: abundances

1. Introduction

Open clusters probe the structure and evolution of the Galactic disk. They are tools for the investigation of properties like age, interstellar dust and metallicity distribution, the age–metallicity relation, as well as the dynamics of the disk. They are also good targets for studies of stellar evolution. Relative to field stars, star clusters should allow more accurate evaluations of their age, metallicity, distance and velocity, since such properties should be common for all of their members.

In this paper, we present spectroscopic observations of seven red giants in the open cluster NGC 3114. This cluster is located in the Carina spiral feature, a region of heavy contamination by field stars (Carraro & Patat 2001). There are some photometric studies for NGC 3114 (Jankowitz & McCosh 1963; Schneider & Weiss 1988; Clariá *et al.* 1989; Sagar & Sharpless 1991; Carraro & Patat 2001; Paunzen *et al.* 2003), but high-resolution data dedicated to abundance studies are not available in the literature. Thanks to the radial-velocity surveys of González & Lapasset (2001) and, more recently, Mermilliod *et al.* (2008), several stars have been identified as members of NGC 3114, including the seven red giants analyzed here. Our aim is to provide accurate abundances for stars in this cluster.

Spectral-line observations are summarized in Section 2. In Section 3, we describe our analysis of the data and the main results. In Section 4, we discuss the astrophysical implications of our results. A brief conclusion follows in Section 5.

† Based on observations with the *2.2m telescope* at the European Southern Observatory (La Silla, Chile).

Table 1. Log of the observations and relevant information of the target stars.

ID[a]	HD	V^a (mag)	$(B-V)^a$	RV[a] km s^{-1}	RV[b] km s^{-1}	Date obs	Exp (s)
6	87109	7.6	1.29	-1.43 ± 0.23	-1.31 ± 0.49	2008 Apr 9	600
150	87479	7.9	1.17	-2.19 ± 0.51	-1.71 ± 0.61	2008 Apr 9	420
170	87526	7.3	0.89	-1.95 ± 0.40	-2.38 ± 0.17	2008 Apr 9	600
181	87566	8.3	1.28	-2.18 ± 0.10	-2.27 ± 0.23	2008 Apr 10	420
238	304859	8.5	1.27	-1.72 ± 0.18	-1.40 ± 0.34	2008 Dec 22	600
262	87833	8.6	1.16	-1.20 ± 0.23	-1.22 ± 0.17	2008 Apr 10	600
283	304864	7.7	1.25	-1.73 ± 0.31	-1.41 ± 0.36	2008 Dec 22	600

[a] Mermilliod *et al.* (2008); [b] this work.

2. Observational data

2.1. *Properties of NGC 3114*

NGC 3114 (C 1001–598, Cl VDBH 86; $\alpha = 10^h02^m.0$, $\delta = -60°06'$ (J2000.0); $l = 283°$, $b = -04°$), is a young open cluster projected onto the outskirts of the Carina complex (Feinstein 1995). Carraro & Patat (2001) presented a CCD *UBVRI* photometric study of the cluster and obtained magnitudes down to $V \simeq 22.0$ mag. Their results for the age, heliocentric distance and reddening were, respectively, 3.0×10^8 years, 920 ± 50 pc and $E(B-V) = 0.07 \pm 0.03$ mag. Similar values were also found by González & Lapasset (2001), who derived an age of 1.6×10^8 years, a distance modulus $(V - M_V) = 9.8 \pm 0.2$ mag and a reddening of $E(B-V) = 0.07 \pm 0.01$ mag.

Our sample includes seven red giants originally selected from the radial-velocity catalog of Mermilliod *et al.* (2008). For the seven stars in our sample, González & Lapasset (2001) computed membership probabilities greater than the 95% level.

2.2. *Observations*

We obtained high-resolution spectra for these seven stellar members of NGC 3114 using the ESO *2.2m telescope* (La Silla, Chile) with the FEROS echelle spectrograph (Kaufer *et al.* 1999). The wavelength coverage is from 4000 to 9200 Å. Table 1 lists the observation log and additional information about the observed stars. The stars in our sample closely share the same radial velocity (RV). Our values are close to the results of Mermillod *et al.* (2008), who provided $\langle \mathrm{RV} \rangle = -1.72$ km s^{-1}. We note, however, that González & Lapasset (2001) estimated values that are systematically lower than ours, by a factor of about two ($\langle \mathrm{RV} \rangle = -3.8 \pm 0.3$ km s^{-1}).

3. Analysis and results

The spectra of the red giants in our sample show many atomic absorption lines of Fe I and Fe II as well as transitions due to Ca I, Na I, Mg I, Al I, Sc II, Ti I, V I, Cr I, Co I, Ni I, Zn I, Cu I, Y II, Ba II, La II, Ce II, Nd II and Eu II. We have chosen a set of lines sufficiently unblended to yield reliable abundances.

The determination of atmospheric parameters, effective temperatures (T_{eff}), surface gravities ($\log g$), microturbulent velocities (ξ) and metallicity ([Fe/H]; throughout, we use the notation $[\mathrm{X/H}] = \log(N(\mathrm{X})/N(\mathrm{H}))_\star - \log(N(\mathrm{X})/N(\mathrm{H})_\odot)$ is prerequisite for determination of the final abundance. These parameters were derived from the measured equivalent widths of the iron lines, using the spectral code MOOG and Kurucz model atmospheres. To derive these parameters, we followed the procedures described in previous studies investigating the spectra of barium and symbiotic stars (Pereira 2005; Pereira & Roig 2009). The final adopted atmospheric parameters of our sample stars are given in Table 2.

The internal errors associated with our adopted effective temperatures and microturbulent velocities can be determined from the uncertainty in the slope of the Fe I abundance

Table 2. Adopted atmospheric parameters, Fe I and Fe II metallicities with the corresponding standard deviations and number of lines employed.

Star	T_{eff} (K)	$\log g$ (dex)	ξ km s^{-1}	[Fe I/H]$\pm\sigma$(#)	[Fe II/H]$\pm\sigma$(#)
HD 87109	4 700	1.1	2.0	−0.05±0.12 (47)	−0.04±0.12 (11)
HD 87479	5 000	1.3	1.9	+0.33±0.15 (40)	+0.31±0.15 (7)
HD 87526	5 300	1.5	2.9	−0.74±0.10 (71)	−0.74±0.08 (13)
HD 87566	4 500	1.6	1.6	+0.01±0.15 (54)	+0.01±0.14 (11)
HD 87833	4 900	2.0	1.9	−0.05±0.16 (45)	−0.03±0.15 (10)
HD 304859	4 500	1.6	1.6	+0.07±0.18 (51)	+0.08±0.09 (10)
HD 304864	4 700	1.1	2.2	+0.01±0.15 (34)	+0.01±0.26 (6)

Table 3. Mean elemental abundances ([X/Fe]), dispersion and the number of lines used for the abundance determination.

[X/Fe]	HD 87109	HD 87479	HD 87526	HD 87566	HD 87833	HD 304859	HD 304864
O	−0.45	—	+0.30	−0.21	−0.15	−0.37	< −0.31
Mg	+0.18(2)	+0.17(1)	+0.60(2)	+0.17(2)	+0.20±0.17(3)	—	—
Al	−0.03±0.19(4)	−0.04±0.20(4)	+0.17±0.16(3)	−0.16±0.18(4)	+0.07±0.17(5)	+0.16±0.32(4)	—
Si	+0.03±0.22(9)	−0.08±0.19(6)	+0.36±0.17(8)	0.00±0.21(9)	+0.07±0.18(6)	+0.31±0.21(8)	+0.16±0.21(7)
Ca	+0.04±0.18(3)	−0.09±0.17(3)	+0.13±0.12(8)	0.00±0.21(3)	−0.10±0.15(5)	+0.10±0.25(3)	−0.11(2)
Sc	+0.05±0.19(4)	−0.10±0.20(5)	−0.01±0.15(6)	0.00±0.25(6)	−0.15±0.12(4)	+0.1±0.23(6)	+0.20±0.23(3)
Ti	−0.08±0.20(4)	−0.10±0.07(3)	−0.15±0.13(4)	−0.27±0.19(3)	+0.05±0.19(5)	−0.13(2)	−0.22(2)
Cr	−0.20(2)	+0.06(1)	+0.18±0.13(4)	−0.05(2)	−0.13(2)	−0.11(2)	−0.20(2)
Ni	+0.03±0.28(7)	+0.01±0.19(5)	−0.10±0.20(9)	−0.07±0.22(6)	−0.13±0.17(7)	−0.08±0.22 (6)	−0.15±0.24 (6)
Y	+0.04±0.26(3)	—	+0.02±0.14(4)	−0.02±0.31(4)	−0.04±0.08(3)	−0.04±0.25(4)	—
Zr	+0.09±0.17(4)	−0.21±0.28(3)	—	−0.16±0.18(4)	+0.20±0.09(4)	−0.06±0.20(4)	−0.06±0.08(2)
La	+0.29±0.22(6)	−0.17±0.17(4)	+0.02±0.22(4)	+0.26±0.23(2)	+0.25±0.19(5)	+0.25±0.23(4)	+0.29(2)
Ce	+0.02±0.26(5)	—	+0.13±0.27(11)	+0.17±0.26(5)	+0.17±0.19(5)	+0.23±0.26(7)	+0.12(2)
Nd	+0.22±0.28(7)	—	+0.26±0.24(10)	+0.34(2)	+0.26±0.14(10)	+0.18±0.20(5)	+0.31±0.17(4)

versus excitation potential and Fe I versus reduced equivalent width (W_λ/λ) relations. The standard deviation in $\log g$ was set by changing this parameter around the adopted solution until the difference between Fe I and Fe II mean abundance differed by exactly one standard deviation of the [Fe I/H] mean value. We estimated typical uncertainties in the atmospheric parameters on the order of ±120 K, ±0.2 dex and ±0.3 km s^{-1}, respectively, for T_{eff}, $\log g$ and ξ.

3.1. *Abundance analysis*

Chemical element abundances of the target stars were determined with the local-thermodynamic-equilibrium (LTE) model-atmosphere techniques. In brief, equivalent widths are calculated by integration through a model atmosphere and compared with the observed equivalent widths. The calculation is repeated, changing the abundance of the element, until a match is achieved. The line-synthesis code MOOG was used to carry out the calculations. The results are given in Table 3, where we list the abundance ratio of element to iron ([X/Fe]), its standard deviation and the number of lines employed for each species (#). These ratios were based on a solar iron abundance of $\log \epsilon_{\odot}(\mathrm{Fe}) = 7.52$.

4. Discussion

4.1. *HD 87526: a metal-poor star in NGC 3114*

Compared to the metallicity of the other six red giants, HD 87526 (G3) is a metal-poor star. According to González & Lapasset (2001), HD 87526 is located in the Hertsprung gap, where many binaries (red giant + main sequence star) are encountered in other clusters, although they could not detect any variation based on four radial-velocity measurements.

4.2. *Metallicity and metal distribution*

From the mean of [Fe I/H] values in Table 2, we obtained a metallicity of [Fe/H]= $+0.05 \pm 0.13$ dex, which corresponds to solar metallicity ($Z = 0.02$). We have excluded

HD 87526 from this average, since the metallicity we found for this star is very low ([Fe/H]= −0.74 dex) compared to the other cluster members and we believe it may not be representative of cluster's mean metallicity. Clariá *et al.* (1989) derived a mean photometric metallicity of [Fe/H]= −0.04, in good agreement with ours.

Our metallicity is also in accordance with Sestito *et al.*'s (2008) radial gradient ([Fe/H] versus galactocentric distance) for open clusters.

4.3. *Masses and ages*

Using our results for $T_{\rm eff}$, $\log g$ and [Fe/H], and adopting the values of distance modulus, reddening and selective absorption from González & Lapasset (2001) and the bolometric correction of Alonso *et al.* (1999; their equation 18), we estimated masses for each of our sample objects. The values that we found vary from 0.84 to 2.71 $M_\odot$. The errors, however, are quite large (about 50%).

Ages were estimated using stellar isochrones from Bertelli *et al.* (1994) for solar chemical composition ($Z = 0.020$, $Y = 0.280$). We found good agreement with the curve for $\log \tau = 8.2$ ($\tau = 1.6 \times 10^8$ yr), in accordance with González & Lapasset (2001).

5. Conclusions

We derived the chemical composition of seven red giants members in the open cluster NGC 3114. We obtained atmospheric parameters as well as the abundances of O, Mg, Al, Si, Ca, Ti, Ni, Cr, Sc, Y, Zr, La, Ce and Nd. We did not find any intrinsic star-to-star scatter in the [element/Fe] ratios for the stars in this cluster.

We derived a mean metallicity of [Fe/H]= 0.05 ± 0.13 dex, consistent with the solar value. This is also in accordance with the Galactic-disk metallicity distribution for open clusters. Finally, we estimated an age of $\tau = 1.6 \times 10^8$ yr ($\log \tau = 8.2$) for NGC 3114.

References

Alonso, A., Arribas, S., & Martínez–Roger, C. 1999, *A&AS*, 140, 261
Bertelli, G., Bressan, A., Chiosi, C., Fagotto, F., & Nasi, E. 1994, *A&AS*, 106, 275
Carraro, G. & Patat, F. 2001, *A&A*, 379, 136
Clariá, J. J., Lapasset, E., & Minniti, D. 1989, *A&AS*, 78, 363
Feinstein, A. 1995, *Rev. Mex. A&A*, 2, 57
González, J. F. & Lapasset, E. 2001, *AJ*, 121, 2657
Jankowitz, N. E. & McCosh, C. J. 1963, *MNSSA*, 22, 18
Junqueira, S. & Pereira, C. B., 2001, *AJ*, 122, 360
Kaufer, A., Stahl, O., Tubbesing, S., Nørregaard, P., Avila, G., Francois, P., Pasquini, L., & Pizzella, A. 1999, *ESO Messenger*, 95, 8
Mermilliod, J. C., Mayor, M., & Udry, S. 2008, *A&A*, 485, 303
Paunzen, E., Pintado, O. I., & Maitzen, H. M. 2003, *A&A*, 412, 721
Pereira, C. B. 2005, *AJ*, 129, 2469
Pereira, C. B. & Roig, F. 2009, *AJ*, 137, 118
Sagar, R. & Sharples, R. M. 1991, *A&AS*, 88, 47
Schneider, H. & Weiss, W. W. 1988, *A&AS*, 75, 353
Sestito, P., Bragaglia, A., Randich, S., Pallavicini, R., Andrievsky, S. M., & Korotin, S. A. 2008, *A&A*, 488, 943

Star clusters: basic galactic building blocks
Proceedings IAU Symposium No. 266, 2009
R. de Grijs & J. R. D. Lépine, eds.

doi:10.1017/S1743921309991815

Local starbursts seen when the Universe was 2–4 Gyr old

S. M. Petty,[1,2] D. F. de Mello,[1,2,3] J. P. Gardner[2] and J. S. Gallagher[4]

[1]The Catholic University of America, Washington, DC, USA
[2]NASA/Goddard Space Flight Center, Greenbelt, MD, USA
[3]The Johns Hopkins University, Baltimore, MD, USA
[4]University of Wisconsin, Madison, WI, USA

Abstract. We explore the multiwavelength properties of three nearby starburst galaxies: NGC 3079, NGC 7673, and Mrk 08. We established that each of these galaxies has similar rest-frame far-ultraviolet (FUV) morphologies as Lyman-break galaxies (LBGs) at $z \sim 1.5$ and 4, when the age of the Universe was ~ 4.3 and ~ 1.6 Gyr, respectively. LBGs are at an important stage in galaxy evolution when the Universe had a peak in the star-formation-rate density. Many LBGs are primarily composed of star-forming clumps, i.e., stellar clusters, with a significant lack of older stellar populations. Here, we present the comparison of the spectral-energy distributions (SEDs) of three nearby starburst galaxies with those of typical LBGs. From our nearby sample, each object has been artificially redshifted to observe what the galaxies would look like at $z \sim 1$ to 4 in the rest-frame FUV. NGC 3079 is an edge-on Seyfert 2 galaxy. It has a bright bulge and is interacting with two other galaxies, with extended H I only along NGC 3079. The redshifting process changes its appearance, so that at high z it looks like a chain galaxy with multiple knots of star formation and no bulge. NGC 7673 has extended H I and the star formation is mostly within the inner optical region in the multiple star-forming clumps defining the galaxy morphology. In the FUV, the galaxy looks highly compact with little detail resolved. As it is artificially redshifted, the galaxy continues to look more spherical. Mrk 8 is a merging pair, with the two galaxies observable in the visible spectrum. It is classified as a Wolf–Rayet galaxy, which suggests a very young burst, and is composed of several large star-forming regions. The FUV image does not resolve the separate galaxies, and the appearance remains similar for each redshift. We use the Gini coefficient, M_{20}, and the Sérsic index to quantify the morphologies. The SEDs of the objects have similarities with LBG stellar population models. Because these local galaxies can be studied in more detail, they act as a bridge between nearby observations of starburst galaxies and high-z starburst galaxies such as LBGs.

Keywords. galaxies: individual (Mrk 8, NGC 3079, NGC 7673), galaxies: starburst, galaxies: evolution

---

The full poster (in pdf format) is available at
`http://www.astro.iag.usp.br/~iaus266/Posters/pPetty.pdf`.

Star clusters: basic galactic building blocks
Proceedings IAU Symposium No. 266, 2009
R. de Grijs & J. R. D. Lépine, eds.

doi:10.1017/S1743921309991827

Washington photometry of five star clusters in the Large Magellanic Cloud

Andrés E. Piatti,[1] Doug Geisler,[2] Ata Sarajedini[3] and Carme Gallart[4]

[1]Instituto de Astronomía y Física del Espacio, CC 67, Suc. 28, 1428, Ciudad de Buenos Aires, Argentina
email: andres@iafe.uba.ar

[2]Grupo de Astronomía, Departamento de Astronomía, Universidad de Concepción, Casilla 160-C, Concepción, Chile
email: dgeisler@astro-udec.cl

[3]Department of Astronomy, University of Florida, PO Box 112055, Gainesville, FL 32611, USA
email: ata@astro.ufl.edu

[4]Instituto de Astrofísica de Canarias, Calle Vía Láctea, E-38200, La Laguna, Tenerife, Spain
email: carme@iac.es

Abstract. We present CCD photometry in the Washington-system C and T_1 passbands down to $T_1 \sim 22.5$ mag in the fields of NGC 1697, SL 133, NGC 1997, SL 663, and OHSC 28, five mostly unstudied star clusters in the LMC. Cluster radii were estimated from star counts in appropriately sized boxes distributed throughout the entire observed fields. We perform a detailed analysis of field-star contamination and derive cluster colour–magnitude diagrams (CMDs). Based on the best fits of isochrones computed by the Padova group to the $(C - T_1, T_1)$ CMDs, the $\delta(T_1)$ index and the 'standard giant-branch' procedure, we derive metallicities and ages for the five clusters. With the exception of NGC 1697 (age = 0.7 Gyr, [Fe/H] = 0.0 dex), the remaining four clusters are of intermediate age (from 2.2 to 3.0 Gyr) and relatively metal poor ([Fe/H] = −0.7 dex). We combine our sample with clusters with ages and metallicities on a similar scale and examine relationships between position in the LMC, age and metallicity. We confirm previous results that clusters younger than $\sim$ 1 Gyr were formed during an outside-in process; this occurred after a burst of cluster formation that took place mainly in the outer disk and peaked $\sim$ 2 Gyr ago. Finally, the cluster and field age–metallicity relations (AMRs) show evidence for a metallicity offset but do overlap, particularly on the upper-envelope side of the cluster AMR.

Keywords. galaxies: individual (Large Magellanic Cloud), galaxies: star clusters, techniques: photometric

1. Estimates of the fundamental parameters of the LMC clusters

Figure 1a shows the observed colour–magnitude diagram (CMD) of NGC 1697. To construct the cluster radial profiles, we computed the number of stars per unit area at a given radius, r, as is shown in Figure 1b. We then estimated background levels, the radii of the FWHM ($r_{\rm FWHM}$), and the cluster radii ($r_{\rm cl}$). Cluster dimensions are relatively small: four of the clusters have FWHMs between 50 and 110 pixels, while that of OHSC 28 is only 25 pixels, and they fade into the stellar field populations at a distance of twice or three times $r_{\rm FWHM}$.

Figure 2a shows the $r_{\rm cl}$-extracted CMD for SL 663 (upper right). We also show a close-up schematic finding chart of the cluster. We statistically cleaned the cluster CMDs of stars that can potentially belong to the foreground field or have relatively large $\sigma(C-T_1)$. The resulting SL 663 field CMD is shown in the bottom left panel of Figure 2a. We then

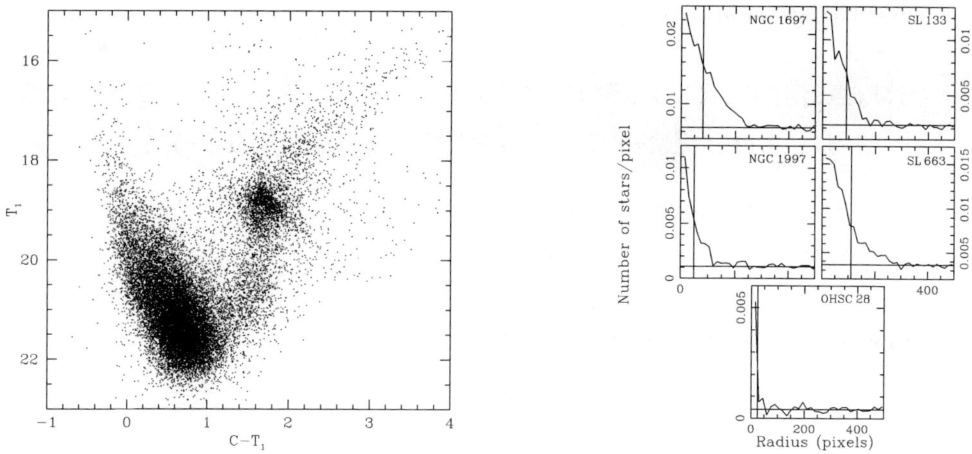

Figure 1. *(a)* $(C - T_1,\ T_1)$ colour–magnitude diagram of all measured stars in the field of NGC 1697. *(b)* Stellar density profiles of the selected clusters. The horizontal lines correspond to the background levels far from the clusters, while the vertical lines indicate $r_{\rm FWHM}$.

subtracted from each cluster CMD, for different bins with sizes $[C - T_1, T_1] = (0.5,\ 0.2)$ mag, the corresponding number of stars counted in the field CMDs. In the bottom right panel of Figure 2a, we show the CMD of the remaining cluster stars. In the subsequent analysis, we use these CMDs to estimate the fundamental parameters of each cluster.

Cluster reddening values were estimated by interpolating the extinction maps of Burstein & Heiles (1982). As for the cluster distance moduli, we adopt for all the clusters a value for the LMC distance modulus of $(m - M)_0 = 18.50 \pm 0.10$ mag. We then selected a set of isochrones and superimposed them onto the cluster CMDs, once they were properly shifted. Finally, we chose the isochrone which best reproduces the cluster's main features. Figure 2b shows the results of the isochrone fits. We also derived the ages of the four clusters older than 1 Gyr from the $\delta(T_1)$ index—calculated by determining the difference in T_1 magnitude between the red-giant clump and main-sequence turnoff in the cluster CMD—and equation (4) of Geisler *et al.* (1997). We then averaged the ages obtained from the isochrone fits and those derived from the $\delta(T_1)$ index since they are in very good agreement.

We followed the 'standard giant-branch' (SGB) procedure of inserting absolute M_{T_1} magnitudes and intrinsic $C - T_1$ colours for the clusters into figure 2 of Geisler & Sarajedini (1999) to obtain (by interpolation) the clusters' metal abundances ([Fe/H]). These derived metallicities were corrected for age effects using the prescription given in Geisler *et al.* (2003). The corrected metallicities for the cluster sample show excellent agreement with the Z values of the isochrones which best resemble the cluster features. Olszewski *et al.* (1991) and Grocholski *et al.* (2006) found [Fe/H] $= 0.43 \pm 0.20$ dex and 0.54 ± 0.05 dex for NGC 1997 and SL 663, respectively, from the spectra of three and eight stars taken at the infrared CaII triplet. Bearing in mind our metallicity error and that of Grocholski *et al.* (see also their figure 12), we find good agreement between both estimates of SL 663's metal abundance.

2. Analysis of cluster chemical evolution in the LMC

To investigate the chemical evolution of the LMC, we added the clusters studied here to a list of 54 selected LMC clusters. We plot in Figure 3a the cluster ages and metallicities as a function of their deprojected angular distance for four metallicity intervals, [Fe/H]

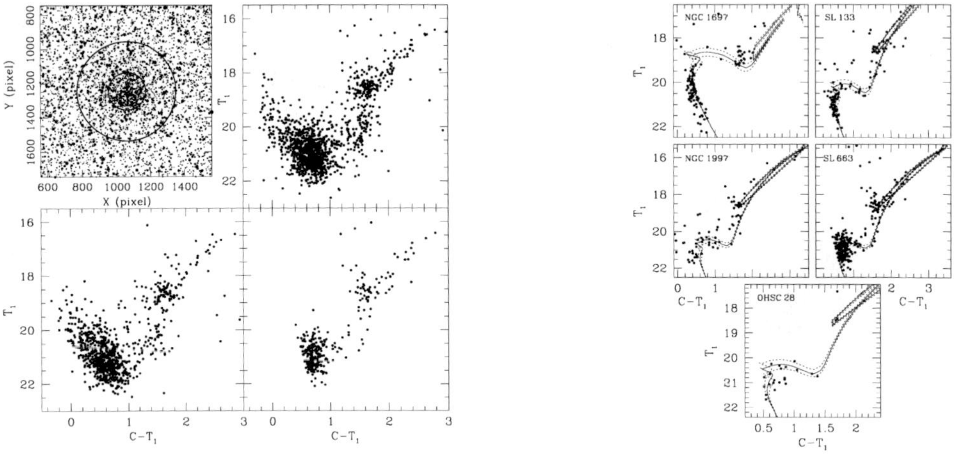

Figure 2. *(a)* Schematic finding chart of SL 663 *(upper left)*, with three extracted CMDs (see text for details). *(b)* ($C - T_1$, T_1) CMDs for the selected star clusters. Isochrones from Girardi *et al.* (2002) are overplotted. The solid and dashed lines correspond to the derived cluster ages and to the ages taking into account their associated errors.

< -1.20 (pentagons), $-1.20 \leqslant$ [Fe/H] $\leqslant -0.80$ (□), $-0.80 <$ [Fe/H] < -0.4 (○) and [Fe/H] $\geqslant -0.4$ dex (△), respectively. We included, in the upper panel, an enlargement for young clusters with deprojected distances $< 6°$. The bottom panel of Figure 3a reveals that the most metal-rich clusters are preferentially located in the inner disk, where they were probably formed, since most are younger than 1 Gyr (see top panel). In contrast, clusters with [Fe/H] < -0.5 dex exist across the entire disk out to 10°. Thus, the distance from the galaxy centre alone does not appear to be the main variable describing the cluster spatial–metallicity relationship. Our result agrees with Grocholski *et al.* (2006). However, in contrast to our Figure 3, they found that clusters with [Fe/H] > -1.0 dex show a tight distribution ([Fe/H] $= -0.48 \pm 0.09$ dex), with no tail towards solar metallicities. According to the metallicity distribution of our larger cluster sample, we find a noticeable peak at [Fe/H] ~ -0.7 dex and a notable concentration of more metal-rich clusters.

The top panel of Figure 3a shows that the most metal-poor clusters are also the oldest ones in the LMC, as expected. There also exists a quiescent period of cluster formation between 11 and 3 Gyr ago, before a large number of intermediate-age clusters (IACs) were mainly formed in the outer disk, starting $\sim$ 3 Gyr ago. The enlargement in the top panel shows that subsequent cluster formation has been concentrated to the inner disk. Indeed, here a possible gradient exists, in which the younger clusters are formed closer to the galaxy centre. We find in this last result a hint for an outside-in cluster-formation scenario. Note that this putative outside-in formation took place in the inner disk after the burst formation process was triggered mainly in the outer disk beginning $\sim$ 3 Gyr ago.

Figure 3b shows the resulting age–metallicity relation (AMR), where open circles represent individual clusters, while filled triangles and circles represent average metallicities, for a sample of stars in the bar (Cole *et al.* 2005) and the disk field (Carrera *et al.* 2008), respectively. From the metal-poor end of the AMR, we find that several old clusters have [Fe/H] $\leqslant -2.2$ dex, while the mean metallicity of the poorest field stars reaches [Fe/H] $= -1.4$ dex. Even so, there are also old clusters within the metallicity range of the oldest field-star populations. Towards the region of the IACs, we find a similar behaviour of

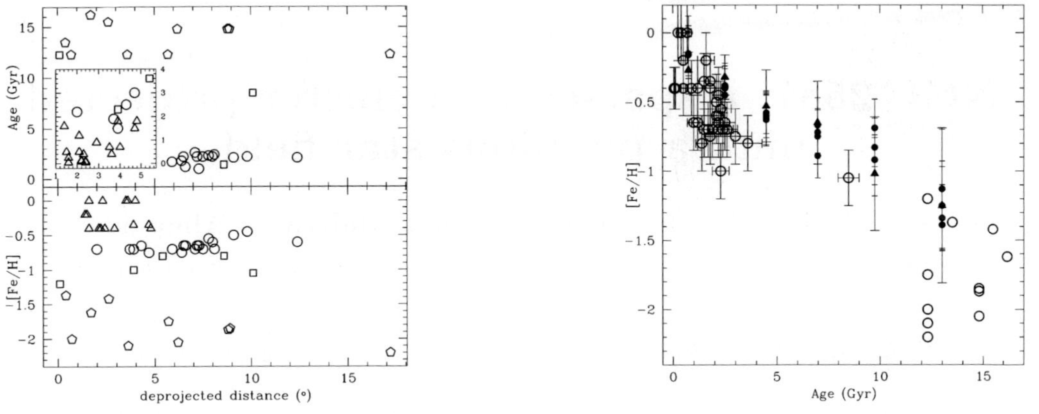

Figure 3. *(a)* Relationships between LMC cluster ages and metallicities and the deprojected distance for [Fe/H] < −1.20 (pentagons), −1.20 < [Fe/H] < −0.80 (□), 0.8 < [Fe/H] < −0.4 (◯) and [Fe/H] > −0.4 dex (△). *(b)* Age–metallicity relation for LMC clusters (open circles) and that derived by Carrera *et al.* (2008) for the LMC bar (filled triangles) and disk (filled circles) systems.

cluster and field-star abundances in the sense that the field-star metallicities tend to delimit the upper envelope of the cluster metallicities. The offset between the clusters and the field population may reflect a metallicity-scale difference. For ages younger than 1 Gyr, it seems that clusters and field stars share, on average, similar metallicity ranges. Note that in the cluster age gap, the only cluster present is at the metal-poor end of the field-star distribution. This result agrees, in turn, with theoretical models that suggest the bar and IACs formed as a result of a close encounter with the Smal Magellanic Cloud 4 Gyr ago (Bekki & Chiba 2005). In this scenario, that encounter peaked ∼ 2 Gyr ago and triggered the formation of a large number of clusters in both Magellanic Clouds (MCs). Recently, Bekki (2008) proposed that the MC system has a common halo, produced either by dynamical coupling that started ∼ 4 Gyr ago or by a remnant of a small group of galaxies destroyed through tidal stripping by the Galaxy. Unfortunately, we cannot favour either of these suggested scenarios based on the available cluster data. However, since the AMR for the open clusters in the Galaxy does not show enhanced cluster formation like that which occurred in the MCs 2 Gyr ago, perhaps both MCs have been interacting more like members of a binary system than isolated entities of a group of galaxies which includes the Milky Way.

References

Bekki, K. 2008, *ApJ* (Letters), 684, L87

Bekki, K. & Chiba, M. 2005, *MNRAS*, 356, 680

Burstein, D. & Heiles, C. 1982, *AJ*, 87, 1165

Carrera, R., Gallart, C., Hardy, E., Aparicio, A., & Zinn, R. 2008, *AJ*, 135, 836

Cole, A. A., Tolstoy, E., Gallagher III, J. S., & Smecker–Hane, T. 2005, *AJ*, 129, 1465

Geisler, D., Bica, E., Dottori, H., Clariá, J. J., Piatti, A. E., & Santos Jr., J. F. C. 1997, *AJ*, 114, 1920

Geisler, D. & Sarajedini, A. 1999, *AJ*, 117, 308

Girardi, L., Bertelli, G., Bressan, A., Chiosi, C., Groenewegen, M. A. T., Marigo, P., Salasnich, B., & Weiss, A. 2002, *A&A*, 391, 195

Grocholski, A. J., Cole, A. A., Sarajedini, A., Geisler, D., & Smith, V. V. 2006, *AJ*, 132, 1630

Olszewski, E., Schommer, R., Suntzeff, N., & Harris, H. C. 1991, *AJ*, 101, 515

Star clusters: basic galactic building blocks
Proceedings IAU Symposium No. 266, 2009
R. de Grijs & J. R. D. Lépine, eds.

doi:10.1017/S1743921309991839

NGC 2587: a sparse open cluster projected onto a populous star field

Andrés E. Piatti,[1] Juan J. Clariá[2] and Andrea V. Ahumada[2,3]

[1]Instituto de Astronomía y Física del Espacio, Casilla 67, Suc. 28, 1428 Buenos Aires, Argentina
email: andres@iafe.uba.ar

[2]Observatorio Astronómico de Córdoba, Laprida 854, 5000 Córdoba, Argentina
email: claria@oac.uncor.edu

[3]European Southern Observatory, Alonso de Córdova 3107, Santiago, Chile
email: aahumada@eso.org

Abstract. We present CCD $BVI_{\rm KC}$ photometry in the field of the open cluster NGC 2587. We developed a new method to clean statistically the colour–magnitude diagrams. NGC 2587 is found to be slightly younger than the Hyades and probably of solar metallicity. From 18 probable members with measured proper motions, we derived the following mean cluster values: $\mu_\alpha = -4.3 \pm 3.6$ mas yr^{-1} and $\mu_\delta = -2.5 \pm 3.4$ mas yr^{-1}. Colour excesses, $E(B-V) = 0.10$ and $E(V-I) = 0.15$ mag, and a heliocentric distance of 3.7 ± 0.7 kpc are obtained. The interstellar extinction in the cluster direction is found to follow the normal reddening law.

Keywords. open clusters and associations: individual (NGC 2587), techniques: photometric

1. Introduction

No photometric study has yet been published of NGC 2587, a relatively bright open cluster in Puppis. This is one of the reasons that justifies the current work. Although Bustos Fierro (2007) (hereafter BF07) measured proper motions within a 2° × 2° field of a *Carte du Ciel* plate containing NGC 2587, he neither determined the basic cluster parameters nor verified its physical reality. The present study attempts at shedding light on the nature of this object by using high-quality photometric data obtained with the 0.9m telescope at Cerro Tololo Inter-American Observatory (Chile).

2. Cluster centre and angular radius

The coordinates of the cluster centre were determined by fitting Gaussian distributions to the stellar-density profiles in the X and Y directions, using tasks in IRAF and applying the statistical method described by Piatti *et al.* (2006). The coordinates of the cluster centre were $(X_c, Y_c) = (1300 \pm 10, 1190 \pm 10)$ pixels, equivalent to $\alpha(2000) = 8^{\rm h}23^{\rm m}27^{\rm s}, \delta(2000) = -29°30'34"$. Although these values agree well with those estimated by eye, BF07 and Palma et al (2008) centred the cluster 5.9′ to the southeast and 1.2′ to the northeast from our value, respectively. The radial stellar-density profile of NGC 2587 is not very well approximated by a Gaussian. However, 75% of the cluster members (see Section 3) are found to be located within 5′ from the cluster centre. Since the most distant member lies at about 8′ from this centre, we adopted the latter value as a rough cluster radius estimate.

3. A new statistical method to clean colour–magnitude diagrams

Figure 1 shows the colour–magnitude diagrams (CMDs) of all observed stars in the cluster field. It is difficult to accurately identify the cluster main sequence (MS) from the CMDs: a crowded MS rises up from the magnitude limit at $V \sim 16$. A few apparent clumps of red stars are also identified at $(V, V-I) \sim (12.5, 1.2), (13.2, 1.5)$ mag as well as their counterparts in the $(V, B-V)$ CMD. From the various MSs possible, that of the cluster appears to be the one which reaches the brightest magnitude and bluest colour. Thus, if we want to estimate the cluster's fundamental parameters from the CMDs, we must first disentangle the cluster MS stars from those belonging to the surrounding field. Note that both cluster and field stars are affected by nearly the same reddening, which precisely causes the overlap of their MSs. This becomes an additional difficulty when attempting to clean the cluster CMDs. NGC 2587 does not appear to be a concentrated or rich stellar aggregate. On the contrary, bearing in mind the cluster appearance on the sky and the results obtained by BF07 and Palma *et al.* (2008), the apparently small number of cluster stars would be spread over the field. Precisely due to the low density and sparse distribution of cluster stars and to the populous star field onto which the cluster is projected, any cluster CMD constructed from circular extractions around its centre would be highly contaminated by field stars. Since NGC 2587 seems to cover most of the observed $13.6' \times 13.6'$ area, we developed an alternative method to clean statistically the CMDs, separately. The method relies on the following assumptions: (i) the cluster is a low-stellar-density object, (ii) cluster stars are spread throughout the entire observed field and (iii) the cluster is projected onto a populous star field. Considering these three assumptions on the whole implies that any extracted CMD of stars located within a unit area at any place of the observed field should be mainly represented by the stellar field CMD. Thus, we divided the observed region covering [0:2000,0:2000] pixels into 64 boxes of 250 pixels on a side and constructed for each the corresponding CMDs. At first glance, there are some differences among the various box-extracted CMDs, which reveals a certain inhomogeneity in the spatial distribution of the stellar brightness and colours.

The proposed method consists of adopting one of the 64 box-extracted CMDs as a 'star-field CMD' to statistically clean the remaining 63 box-extracted CMDs. We repeated the cleaning process using as star-field CMD each of the 64 CMDs. At the end of this process, each box-extracted CMD was individually cleaned using a different star-field CMD each

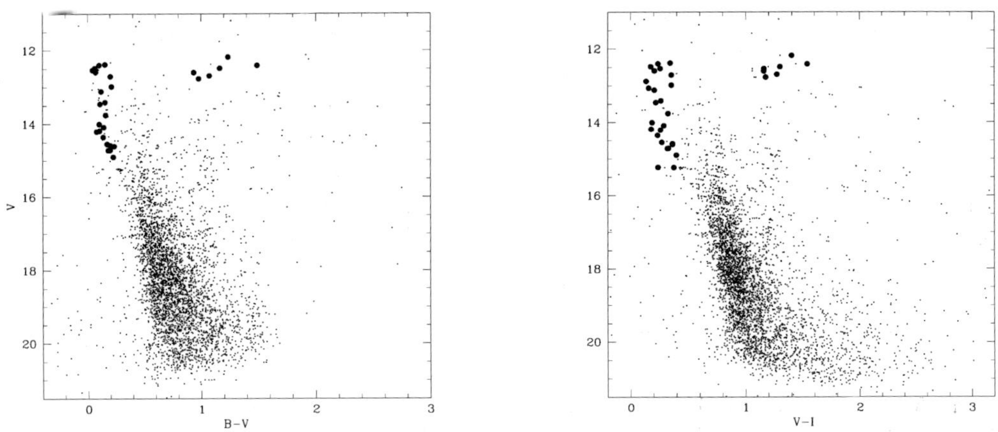

Figure 1. $(V, B-V)$ and $(V, V-I)$ diagrams for the stars measured in the field of NGC 2587. Filled circles represent stars adopted as probable cluster members.

time. The cleaning was performed using bins of $[\Delta V, \Delta(B-V)] = [\Delta V, \Delta(V-I)] = (1.0, 0.2)$ mag. We carried out the cleaning using bins of (0.5, 0.2) mag and (0.5, 0.1) mag, as well as boxes of 500 × 500 pixels to compare residuals. When comparing the 63 cleaned CMDs for a given box with the observations, we find the residuals from box-to-box variations and the fiducial CMD features of that box. In other words, a star that appears nearly 63 times in the cleaned CMDs of a given box has many chances (statistically speaking) of being part of the intrinsic characteristics of the CMD of that box. Moreover, a star that is eliminated nearly 63 times has magnitude and colours within the typical values found in the field-star CMDs. Thus, the fewer times a star is removed in a given box, the higher its probability of being an inherent characteristic of that box. Therefore, we decided to adopt as a probable cluster member of NGC 2587, any star that has at least 2 measures of $(B-V)$ and $(V-I)$, and has also been removed fewer than 20% of the times. From the 4406 measured stars, a total of 32 fulfill such requirements. Figure 2 shows the resulting cleaned CMDs (filled circles) superimposed onto the observed data (dots). Among the observed stars brighter than $V = 12$ mag, only HD 70927 has a known MK spectral type. According to Houk & Cowley (1975), this is an F7/8 II-type star. Therefore, its visual absolute magnitude is $M_V = -2.3$ (Straizys 1992). Assuming HD 70927 were affected by the same reddening as NGC 2587, it would clearly be a foreground object located $\sim$ 1.4 kpc from the Sun. As for the remaining observed bright stars, their positions in the CMDs indicate that they are very unlikely to be members.

4. Cluster fundamental parameters

We fitted theoretical isochrones of Lejeune & Schaerer (2001) to the cleaned CMDs. First, we independently fitted the zero-age main sequence (ZAMS) to the $(V, B-V)$ and $(V, V-I)$ CMDs for each selected metallicity and derived $E(B-V)$, $E(V-I)$ and $V-M_V$. The fit for $Z = 0.020$ best matches the nonevolved stellar sequence. Next, we selected isochrones of log t (yr) greater than 8.0 and used the derived pairs of $[V-M_V, E(B-V)]$ and $[V-M_V, E(V-I)]$ values to estimate the cluster's age. The isochrone of log t (yr) = 8.70 (t = 500 Myr) turned out to most accurately reproduce the cluster features. To match

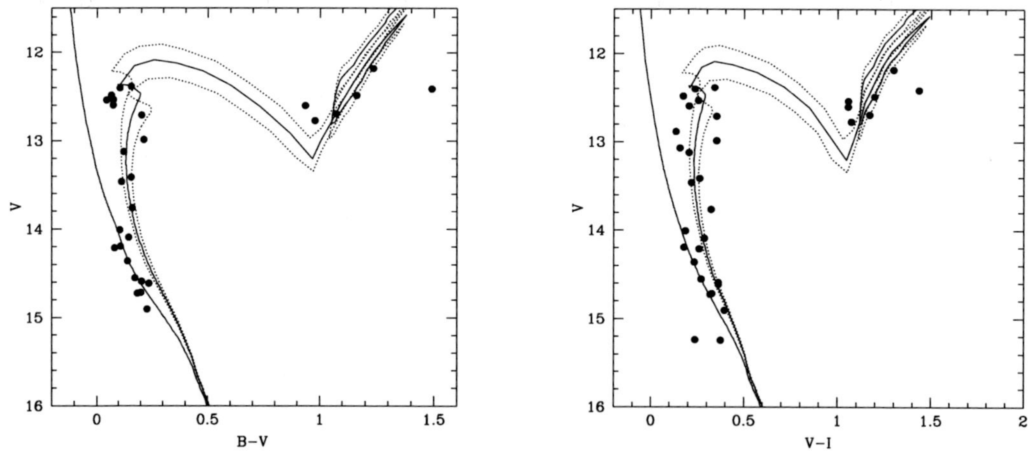

Figure 2. $(V, B-V)$ and $(V, V-I)$ diagrams for the probable members of NGC 2587. The zero-age main sequence and the adopted isochrones from Lejeune & Schaerer (2001), computed taking into account overshooting, are overplotted as solid lines. For comparison purposes, we included as dotted lines the isochrones associated with the cluster's age uncertainties.

this isochrone, we used $E(B-V)$, $E(V-I)$ and $V-M_V$ values of 0.10, 0.15 and 12.50 mag, respectively, which were derived from the ZAMS fit. In Figure 2 we overplot the ZAMS and the isochrone of log t (yr) = 8.70 (solid lines) for $Z = 0.020$ on the cluster CMDs and two additional isochrones of log t (yr) = 8.65 and 8.75 (dotted lines), for comparison purposes. We compared the observed $E(V-I)/E(B-V)$ ratio (= 1.50 ± 0.40) with that from the normal extinction law (Dean *et al.* 1978) and found very good agreement. We used $R = A_V/E(B-V) = 3.2$ (Straizys 1992) to derive a cluster heliocentric distance of 2.7 ± 0.7 kpc and a distance from the Galactic plane of 210 pc. Assuming the Sun's distance from the Galactic centre to be 8.5 kpc, we derived a Galactocentric distance to the cluster of ~ 9.8 kpc.

5. Analysis of proper motions

BF07 measured proper motions (PMs) of 4172 stars in the cluster region. We found that 20 of our 32 probable members are in his list. Except for two stars with remarkably different μ_δ values, the remaining 18 stars have rather similar PMs, with average values of $\mu_\alpha = -4.3 \pm 3.6$ and $\mu_\delta = -2.5 \pm 3.4$ mas yr^{-1}. This result is in excellent agreement with the membership status adopted for these stars. For this reason, we assumed these mean values as the cluster mean PMs. The bright star HD 70927, considered to be a nonmember owing to its M_V, has a motion which differs by a little more than 1σ from the motion accepted for the cluster members. BF07 derived membership probabilities between 50 and 70% for 23 stars which he considered cluster members. However, only two of these stars appear to be located along the isochrone which best reproduces the cluster features. The remaining 21 stars apparently belong to the foreground/background field. Since BF07 constructed the cluster PM distributions from stars spread across a region which includes part of the cluster area and part of the field region, the resulting mean PMs do not represent those of the cluster, but are a mixture of cluster and field PMs.

References

Bustos Fierro, I. H. 2007, *PhD Thesis*, Universidad Nacional de Córdoba, Argentina (BF07)
Dean, F. J., Warren, P. R., & Cousins, A. W. J. 1978, *MNRAS*, 183, 569
Houk, N. & Cowley, A. P. 1975, *University of Michigan Catalogue of Two-Dimensional Spectral Types for the HD stars*
Lejeune, T. & Schaerer, D. 2001, *A&A*, 366, 538
Palma, T., Ahumada, A. V., Clariá, J. J., & Bica, E. 2008 *AN*, 329, 392
Piatti, A. E., Clariá, J. J., & Ahumada, A. V. 2006 *MNRAS*, 367, 599
Straizys, V. 1992 *Multicolor Stellar Photometry*, Pachart Publishing House: Tucson, Arizona

Star clusters: basic galactic building blocks
Proceedings IAU Symposium No. 266, 2009
R. de Grijs & J. R. D. Lépine, eds.

doi:10.1017/S1743921309991840

Hα imaging of X-ray sources in selected globular clusters with *SOAR*

Pawel Pietrukowicz

Pontifícia Universidad Católica de Chile, Av. Vicuna MacKenna 4860, Casilla 306, Santiago, Chile

Nicolaus Copernicus Astronomical Center, ul. Bartycka 18, 00716 Warsaw, Poland

Abstract. I present results of a search for cataclysmic variables (CVs) and chromospherically active binaries (ABs) as counterparts to X-ray sources detected with *Chandra* in six Galactic globular clusters (GCs): M 4, M 28, M 30, M 71, M 80, and NGC 6752. Binary systems play a critical role in the evolution of GCs, serving as an internal energy source countering the tendency of GC cores to collapse. Theoretical studies predict dozens of CVs in the cores of some GCs (e.g., 130 for M 28, 40 for M 30). A number of such binaries are also expected outside the core radius. However, few CVs are known so far in GCs. Using the 4.1m *SOAR* telescope, I have found 25 stars with Hα excess in the observed clusters. Six are candidate CVs, five are candidate active binaries. The other 14 objects with Hα excess are probably foreground/background stars or extragalactic sources.

Keywords. binaries: close, stars: dwarf novae, novae, cataclysmic variables, Hertzsprung–Russell diagram, globular clusters: individual (M 4, M 28, M 30, M 71, M 80, NGC 6752)

The full poster (in pdf format) is available at `http://www.astro.iag.usp.br/~iaus266/Posters/pPietrukowicz.pdf`.

Star clusters: basic galactic building blocks
Proceedings IAU Symposium No. 266, 2009
R. de Grijs & J. R. D. Lépine, eds.

doi:10.1017/S1743921309991852

The population of young stars in Orion A: X-rays and IR properties

Ignazio Pillitteri,[1] S. J. Wolk,[1] L. Allen,[2] S. T. Megeath,[3] R. A. Gutermuth[1] and the SOXS collaboration

[1]Smithsonian Astrophysical Observatory/Harvard University, 60 Garden St. Cambridge, MA 02138, USA

[2]National Optical Astronomy Observatories, USA

[3]Department of Physics & Astronomy, University of Toledo, OH, USA

Abstract. Stars in the very early stages of their formation are characterized by strong infrared excess and X-ray emission. We present the results of the survey of Orion A in both the infrared and X-rays obtained with the *Spitzer* and *XMM/Newton* observatories. We study the spectral-energy distribution class of the young stellar object (YSO) population using infrared colors from 2MASS and *Spitzer* (IRAC and MIPS) and by means of X-ray fluxes, luminosities and plasma temperatures. We discuss clustering properties and spatial segregation among different infrared YSO classes to trace their formation history.

Keywords. stars: activity, stars: formation, open clusters and associations: individual (Orion A, L1641), X-rays: stars, infrared: stars

---

The full poster (in pdf format) is available at
`http://www.astro.iag.usp.br/~iaus266/Posters/pPilletteri.pdf`.

Star clusters: basic galactic building blocks
Proceedings IAU Symposium No. 266, 2009
R. de Grijs & J. R. D. Lépine, eds.

doi:10.1017/S1743921309991864

Clusters and stellar angular-momentum evolution

M. H. Pinsonneault, D. M. Terndrup and P. Denisenkov

The Ohio State University, Columbus, OH, USA

Abstract. The origin and evolution of stellar rotation has proven to be both important and challenging. Data obtained in star clusters has already provided key constraints on the role of protostellar disks and the timescales for angular-momentum loss and internal transport. Recent data sets also provide empirical support for the idea that a wide range of rotation rates converge on the main sequence. We evaluate the prospects for rotation–mass–age relationships and the role of open clusters in calibrating them.

Keywords. stars: rotation, stars: spots, stars: statistics

The full poster (in pdf format) is available at
`http://www.astro.iag.usp.br/~iaus266/Posters/pPinsonneault.pdf`.

Star clusters: basic galactic building blocks
Proceedings IAU Symposium No. 266, 2009
R. de Grijs & J. R. D. Lépine, eds.

doi:10.1017/S1743921309991876

MASSCLEAN (MASSive CLuster Evolution and ANalysis Package): description, tests, and results

Bogdan Popescu and M. M. Hanson

Department of Physics, University of Cincinnati, Cincinnati, OH, USA
email: popescb@mail.uc.edu, margaret.hanson@uc.edu

Abstract. MASSCLEAN is a new, sophisticated and robust stellar cluster image and photometry simulation package. This package can create color–magnitude diagrams and standard FITS images in any of the traditional optical and near-infrared bands based on cluster characteristics input by the user, including but not limited to distance, age, mass, radius and extinction. At the limit of very distant, unresolved clusters, we have checked the integrated colors created in MASSCLEAN against those from other simple stellar population models, with consistent results. Because the algorithm populates the cluster with a discrete number of tenable stars, it can be used as part of a Monte Carlo method to derive the probabilistic range of characteristics (integrated colors, for example) consistent with a given cluster mass and age. We present the first ever mass-dependent integrated colors as a function of age, derived from over 100 000 Monte Carlo runs, which can be used to improve the current age-determination methods for stellar clusters.

Keywords. Galaxy: stellar content, open clusters and associations: general

---

1. Introduction

Star clusters are the building blocks of galaxies. To analyze the known population of star clusters, to search for new clusters, and to derive the selection effects of current near-infrared survey searches for Milky Way clusters, we created a simulation package.

The MASSCLEAN package† (Popescu & Hanson 2009; Hanson & Popescu 2007, 2008, 2009) is built using numerous well-established theoretical and empirical models for stars and stellar clusters, starting with the Kroupa–Salpeter initial mass function (IMF) for the stellar mass distribution (Salpeter 1955; Kroupa 2002). The user has the option to choose between two stellar evolutionary models, the Geneva (Lejeune & Schaerer 2001) or the Padova models (Marigo *et al.* 2008). The extinction model is based on Cardelli *et al.* (1989), and a King model is used for the spatial distribution (King 1962). An optional stellar field can be generated using the 'SKY' model (Bahcal & Soneira 1984). MASSCLEAN is built in a modular way and can combine observational data and simulations using an unprecedented level of accuracy.

2. Images: Galactic and extragalactic simulations

MASSCLEAN generates FITS images, using SKYMAKER (Bertin 2009). The simulated J-band image for NGC 3603 is presented in Figure 1b, and the 2MASS‡ image is presented in Figure 1a.

† The package is freely available under GNU General Public License at: `http://www.physics.uc.edu/~bogdan/massclean/`

‡ This publication makes use of data products from the Two Micron All Sky Survey, which is a joint project of the University of Massachusetts and the Infrared Processing and

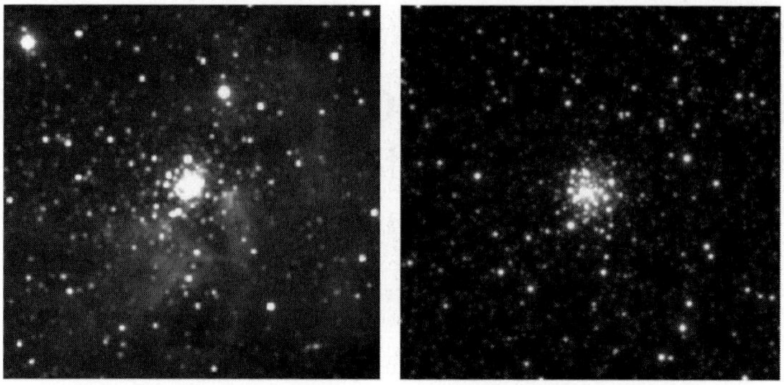

(a) Real image (2MASS). (b) Simulated image.

Figure 1. NGC 3603 J-band images (4.4×4.4 arcmin^2).

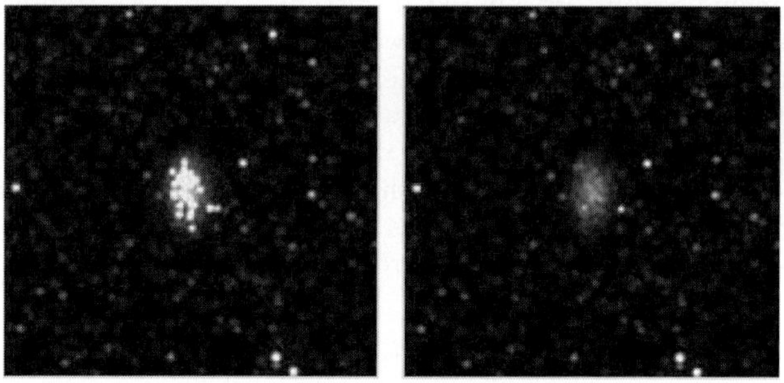

(a) log(age/yr) = 6.0. (b) log(age/yr) = 8.0.

Figure 2. 10^5 $M_\odot$ simulated cluster at the distance of M31 and the resolution of *HST*/ACS.

In Figure 2 we present a simulated 10^5 $M_\odot$ cluster at the distance of M31 and the resolution of *HST*/ACS. The images show the cluster at ages of 1 and 100 million years.

We are using MASSCLEAN to help us derive the selection effects of current near-infrared survey searches for Milky Way clusters (Hanson *et al.* 2009, in prep.). As a proof of concept, we present an example image simulation of 45 randomly generated clusters of various ages, masses, concentrations, and extinction values. The V and K-band images are shown in Figures 3a and b, respectively (note that no stellar field is included). The positions of the clusters along with the scaled angular size of their cores is shown in Figure 3c.

3. Integrated colors and ages

MASSCLEAN follows the evolution of every star in the cluster, providing full photometry at all ages. We have tested that at the infinite-mass limit the integrated colors are consistent with standard simple stellar population (SSP) models. A log(age/yr) step of 0.05 has been used to allow comparison with all other models. The integrated colors generated by MASSCLEAN using Padova 2008 evolutionary models for solar metalicity are presented in Figure 4a. The results from other SSP codes (Marigo *et al.* 2008; Bruzual & Charlot 2003; Maraston 2005; Kotulla *et al.* 2009) are also displayed, along with the

Analysis Center/California Institute of Technology, funded by the National Aeronautics and Space Administration and the National Science Foundation.

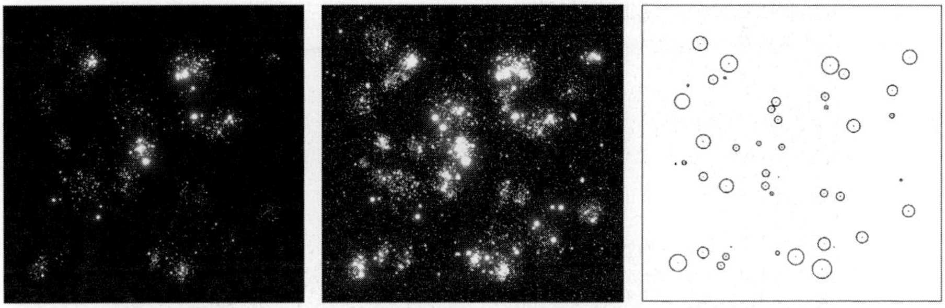

(a) V-band image. (b) K-band image. (c) Positions.

Figure 3. 45 simulated clusters.

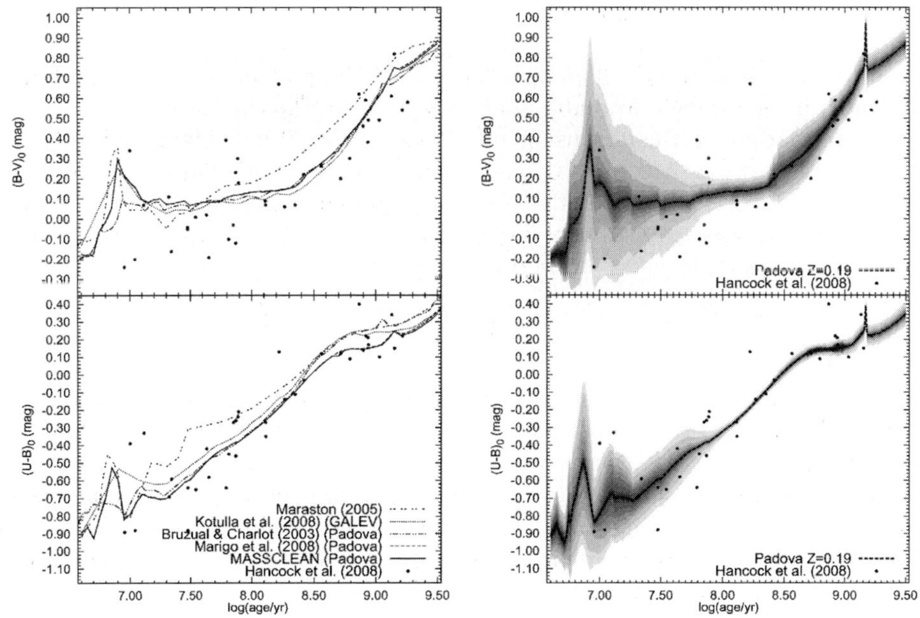

(a) Comparison with modern SSP codes. (b) Mass-dependent integrated colors.

Figure 4. MASSCLEAN integrated colors.

Milky Way clusters from Hancock *et al.* 2008. The MASSCLEAN integrated colors follow Marigo *et al.* (2008), since they both use the same evolutionary models for their input.

Of course, real star clusters have a finite mass. They do not have fully populated isochrones at all ages and their integrated colors may not always correspond to the ones computed in the infinite-mass limit (Bruzual 2002; Cerviño & Luridiana 2004, 2006; Fagiolini *et al.* 2007; Lançon & Mouhcine 2000; Lançon & Fouesneau 2009). The dispersion of the integrated colors (1σ) due to different values of mass and fluctuations in the IMF is presented in Figure 4b. $(B-V)_0$ versus log(age/yr) and $(U-B)_0$ versus log(age/yr) are presented for a few values of the cluster's mass. The step in log(age/yr) is 0.01. The color dispersion for clusters with masses in the $2 \times 10^3 - 2.5 \times 10^5$ $M_\odot$ range is based on about 100 000 Monte Carlo runs (which corresponds to about 15 million cluster models). We present, for the first time, mass-dependent integrated colors for clusters. The entire set

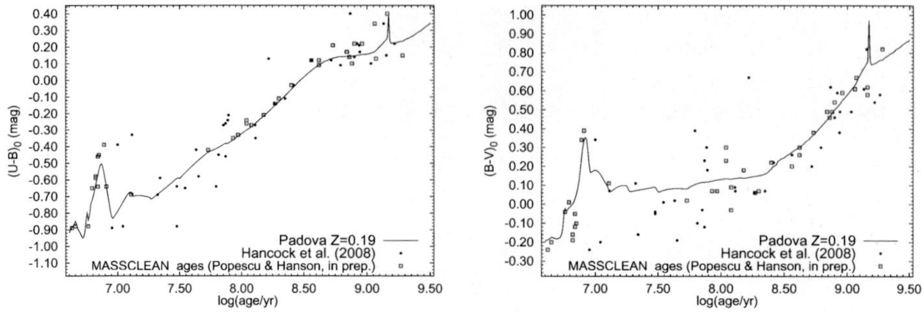

(a) $U-B$ colors as a function of age. (b) $B-V$ colors as a function of age.

Figure 5. Real clusters, with masses of just a few thousand solar masses, are expected to be redder than SSP models predict using the 'infinite-mass assumption,' particularly in the age range 10^6-10^8 yr. We are testing this result using integrated colors of Milky Way clusters with well-constrained ages (Popescu & Hanson, in prep.).

of mass-dependent integrated colors using both Padova and Geneva models and different metallicities will be publicly available soon (Popescu & Hanson, in prep.).

Age determination of clusters using integrated colors is usually based only on the infinite-mass limit. Using our mass-dependent integrated colors and magnitudes, we expect to provide a more realistic model. As an initial demonstration, we present new age determinations based on MASSCLEAN integrated colors for numerous Milky Way clusters in Figure 5 (Popescu & Hanson, in prep.).

4. Summary

We have developed a sophisticated cluster image and photometry simulation package which we are using to test against current search algorithms. We predict that clusters over certain age ranges (particularly when the red-supergiant phase is just starting) and clusters with low stellar densities (even though possibly of very high total mass) are being missed by the current searches, particularly when combined with high extinction. Our mass-dependent integrated colors for clusters should provide a better description for known clusters and a more accurate age determination.

References

Bahcall, J. N. & Soneira, R. M. 1984, *ApJS*, 55, 67
Bertin, E. 2009, *MmSAI*, 80, 422
Bruzual, G. 2002, in: D. Geisler, E. K. Grebel & D. Minniti (eds.), *Extragalactic Star Clusters*, Proc. IAU Symp. No. 207, p. 616
Bruzual, G. & Charlot, S. 2003, *MNRAS*, 344, 1000
Cardelli, J. A., Clayton, G. C., & Mathis, J. S. 1989, *ApJ*, 345, 245
Cerviño, M. & Luridiana, V. 2004, *A&A*, 413, 145
Cerviño, M. & Luridiana, V. 2006, *A&A*, 451, 475
Fagiolini, M., Raimondo, G., & Degl'Innocenti, S. 2007, *A&A*, 462, 107
Hancock, M., Smith, B. J., Giroux, M. L., & Struck, C. 2008, *MNRAS*, 389, 1470
Hanson, M. M. & Popescu, B. 2007, *BAAS*, 39, 836
Hanson, M. M. & Popescu, B. 2008, in: F. Bresolin, P. A. Crowther & J. Puls, *Massive Stars as Cosmic Engines*, Proc. IAU Symp. No. 250, p. 307
Hanson, M. M. & Popescu, B. 2009, *BAAS*, 41, 690
Hanson, M. M., Popescu, B., Larsen, S. S., & Ivanov, V. 2009, in: I. F. Corbett (ed.), *Highlights of Astronomy*, Vol. 14, in press

King, I. 1962, *AJ*, 67, 471

Kotulla, R., Fritze, U., Weilbacher, P., & Anders, P. 2009, *MNRAS*, 396, 462

Kroupa, P. 2002, *Science*, 295, 82

Lançon, A. & Mouhcine, M. 2000, in: A. Lançon & C. Boily, *Massive Stellar Clusters*, ASP Conf. Ser. 211, p. 34, San Francisco: ASP

Lançon, A. & Fouesneau, M. 2009, in: C. Leitherer, P. D. Bennett, P. W. Morris & J. T. van Loon (eds.), *Hot and Cool: Bridging Gaps in Massive Star Evolution*, ASP Conf. Ser., in press (arXiv:0903.4557), San Francisco: ASP

Lejeune, T. & Schaerer, D. 2001, *A&A*, 366, 538

Maraston, C. 2005, *MNRAS*, 362, 799

Marigo, P., Girardi, L., Bressan, A., Groenewegen, M. A. T., Silva, L., & Granato, G. L. 2008, *A&A*, 482, 833

Popescu, B. & Hanson, M. M. 2009, *AJ*, submitted (arXiv:0811.4210)

Salpeter, E. E. 1955, *ApJ*, 121, 161S

Skrutskie, M. F., *et al.* 2006, *AJ*, 131, 1163

Star clusters: basic galactic building blocks
Proceedings IAU Symposium No. 266, 2009
R. de Grijs & J. R. D. Lépine, eds.

doi:10.1017/S1743921309991888

The molecular cloud and embedded young stellar population associated with IRAS 18236–1205

Ricardo Retes, Abraham Luna, Divakara Mayya and Luis Carrasco

Instituto Nacional de Astrofísica, Óptica y Electrónica (INAOE), Mexico

Abstract. We test a membership method to select embedded young stellar objects (YSOs) from a Galactic molecular cloud with ongoing massive star formation using multiband analysis. We select and discuss the embedded stellar population in the molecular cloud associated with IRAS 18235−1205, a small, geometrically well-defined Galactic molecular cloud. The IRAS source has infrared fluxes characteristic of an UCHII region, CS($J = 2 - 1$) emission, and methanol and water maser emission, suggesting that this region is a good candidate for studies of young, massive star formation. The selection method of embedded stellar populations is based on the spatial distribution of ^{13}CO($J = 1 - 0$) and *Spitzer*/MIPS 24 μm point sources. Photometric analysis using near/mid-infrared images are used to test our selection criteria. Three objects are associated with the IRAS source; two have a characteristic spectral-energy distribution (SED) of a Class I/0 object (protostar) and the third has an SED of Class II.

Keywords. ISM: molecular clouds, stars: formation, stars: pre–main-sequence, infrared: stars

---

The full poster (in pdf format) is available at
`http://www.astro.iag.usp.br/~iaus266/Posters/pRetes.pdf`.

Star clusters: basic galactic building blocks
Proceedings IAU Symposium No. 266, 2009
R. de Grijs & J. R. D. Lépine, eds.

doi:10.1017/S174392130999189X

Internal dynamics of the NGC 3603 young cluster

Boyke Rochau,[1] Wolfgang Brandner,[1] Andrea Stolte,[2] Mario Gennaro[1] and Thomas Henning[1]

[1]Max-Planck-Institut für Astronomie, Königstuhl 17, 69117 Heidelberg, Germany

[2]I. Physikalisches Institut, University of Cologne, Zülpicher Str. 77, 50937 Köln, Germany

Abstract. Starburst clusters are spectacular young and dense stellar systems containing copious numbers of massive O-type and Wolf–Rayet stars. These objects depict a stellar population comprising stars over the entire mass range, making them ideal objects to study cluster formation and evolution as well as to test and calibrate stellar evolutionary models. Among the Galactic spiral-arm starburst clusters, NGC 3603 Young Cluster (YC), which is located in the Carina spiral arm, is the youngest and most compact one. Galactic starburst clusters also serve as templates for extragalactic super star clusters as observed, e.g., in the Antennae galaxies. Based on two epochs (obtained 10 yr apart) of high-quality astrometric *HST*/WFPC2 observations, we derive individual proper motions for several hundred stars. The proper motions enable us to distinguish cluster members from field stars, and to derive a first estimate of the internal cluster dynamics. Photometry of the clean sample of cluster stars allows us to create color–magnitude diagrams (CMDs) that are not contaminated by field stars, and to compare the sequence of cluster members with different sets of theoretical isochrones and evolutionary tracks. We clearly identify a lower-mass pre-main-sequence and a higher-mass main-sequence stellar population in NGC 3603. A particular focus of our comparison with theoretical isochrones is the exact shape and extent of the pre-main-sequence/main-sequence transition region in NGC 3603 YC. The proper motions also contain dynamical information. From this, we derive a velocity dispersion, which leads to the first dynamical mass estimate of the NGC 3603 cluster. We also investigate whether or not NGC 3603 YC is virialized, and study evidence for primordial versus dynamical mass segregation. Past and future evolution of NGC 3603 YC can be studied and will tell us about the clusters 'fate' and its long-term survival chances.

Keywords. astrometry, Hertzsprung–Russell diagram, stars: kinematics, stars: pre–main-sequence, open clusters and associations: individual (NGC 3603 YC)

The full poster (in pdf format) is available at
`http://www.astro.iag.usp.br/~iaus266/Posters/pRochau.pdf`.

Star clusters: basic galactic building blocks
Proceedings IAU Symposium No. 266, 2009
R. de Grijs & J. R. D. Lépine, eds.

doi:10.1017/S1743921309991906

Nonradial pulsations in the open cluster NGC 3766

Rachael M. Roettenbacher, Ernest C. Amouzou and M. Virginia McSwain

Department of Physics, Lehigh University, 16 Memorial Drive E, Bethlehem, PA, USA
email: [rmr207, eca2, mcswain]@lehigh.edu

Abstract. Nonradial pulsations (NRPs) are a proposed mechanism for the formation of decretion disks around Be stars. They are important tools to study the internal structure of stars. NGC 3766 has an unusually large fraction of transient Be stars, so it is an excellent location to study the formation mechanism of Be-star disks. High-resolution spectroscopy can reveal line-profile variations from NRPs, allowing measurements of both the degree, l, and azimuthal order, m. However, spectroscopic studies require large amounts of time with large telescopes to achieve the necessary high signal-to-noise ratio and time-domain coverage. On the other hand, multicolor photometry can be performed more easily with small telescopes to measure l only. Here, we present representative light curves of Be stars and nonemitting B stars in NGC 3766 from the *CTIO 0.9m* telescope in an effort to study NRPs in this cluster.

Keywords. open clusters and associations: individual (NGC 3766), stars: emission-line, Be, stars: oscillations (including pulsations)

1. Introduction

Be stars are a class of nonsupergiant B-type stars with Balmer and other line-emission features due to an equatorial decretion disk. The disk is likely the result of a combination of the star's rapid rotation (near the critical limit) and nonradial pulsations (NRPs; Porter & Rivinius 2003).

NRPs are spherical harmonic waves traversing the surface of a star. These pulsations can be found in multiple frequencies on the surface simultaneously (Rivinius *et al.* 2003). There are two primary classes of NRP modes, g and p modes. g modes are described by a low-frequency pulsation that has gravity as its restoring force. The dominant oscillation in this mode is transverse across the surface. p modes are dominated by high-frequency, radial oscillations with a pressure-restoring force (De Ridder 2001). These modes in main-sequence, pulsating B stars are driven by the κ mechanism (Gutiérrez–Soto *et al.* 2007).

Temperature and flux gradients are established between the dimmer, cooler material on the peaks of the pulsations and the brighter, warmer material in the troughs. The flux variations over the stellar surface are then observed as either ripples within photospheric absorption-line profiles or as periodic variations in magnitude. A large, high-resolution spectroscopic study would reveal both the degree, l, and the azimuthal order, m, but such studies are challenging due to the need for large amounts of time on a large telescope. Photometry, which is easily performed with data gathered by small telescopes, only measures l (Rivinius *et al.* 2003).

McSwain *et al.* (2008) previously showed that NGC 3766, an open cluster in Centaurus, is rich in transient Be stars. In an effort to detect NRPs and study the formation of these

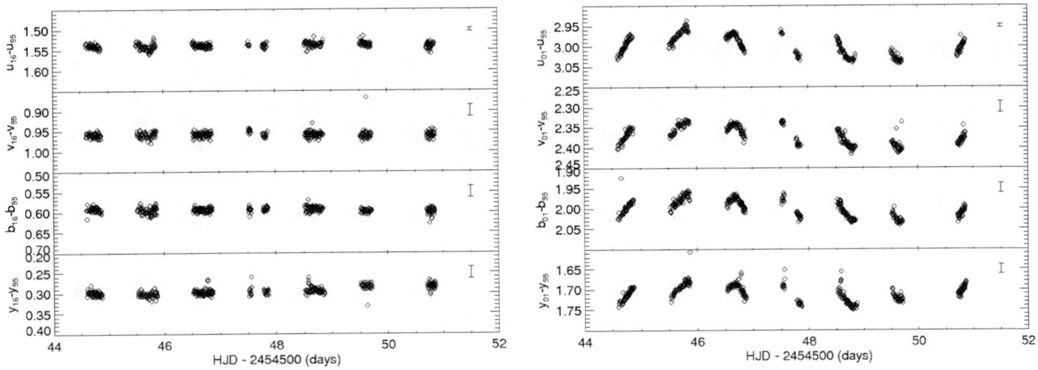

Figure 1. On the *left*, Strömgren *uvby* differential magnitudes are plotted for two reference stars, Nos. 16 and 95, in NGC 3766. On the *right* are differential magnitudes for No. 1 in the same format. Representative error bars are shown in the upper right of each plot.

transient disks, we are currently performing a long-term photometric study of the cluster. Here, we present preliminary differential light curves that reveal magnitude variations of several Be stars that are consistent with NRPs.

2. Observations and data analysis

We observed the cluster NGC 3766 using the *CTIO 0.9m* telescope and SITe 2048 CCD from 2008 March 19 to 24. The CCD was used in quad readout mode without binning. We used Strömgren *uvby* filters and exposure times of 120, 20, 10, and 10 s, respectively. No additional standard stars were measured. Sky flats were used to calibrate the *u* and *v* filters, while dome flats were used for the *b* and *y* filters.

The Strömgren *uvby* photometric data were zero-corrected and flat-field-corrected using standard routines in IRAF using the QUADRED package. The DAOFIND and PHOT procedures were used to automatically identify the stars and perform aperture photometry. We used the numbering scheme established by Ahmed (1962) and found in the WEBDA database to identify the stars. Differential magnitudes were determined for the variable Be stars identified in Balona & Engelbrecht (1986) and McSwain *et al.* (2008) and for four reference stars, Nos. 16, 95, 111, and 147. The reference stars have constant differential magnitudes within the measured errors, shown in Figure 1 (left). Figures 1 (right) and 2 show differential light curves for Be stars Nos. 1, 20, and 63 with variations in magnitude consistent with NRPs.

3. Results and further work

Nineteen of the 25 Be stars measured at this preliminary stage have magnitude variations consistent with NRPs with periods on the order of several hours. For example, the *u*-band light curve of No. 20 was folded using a period of about 6.98 hours, shown in Figure 3.

Additional data collected at the *CTIO 0.9m* telescope in 2008 June, 2009 February, and 2009 May on NGC 3766 will be analyzed in conjunction with these data to perform a period search and identify the various periods that are present. In the future, we will construct a model for the nonradial pulsations that will allow theoretical light curves

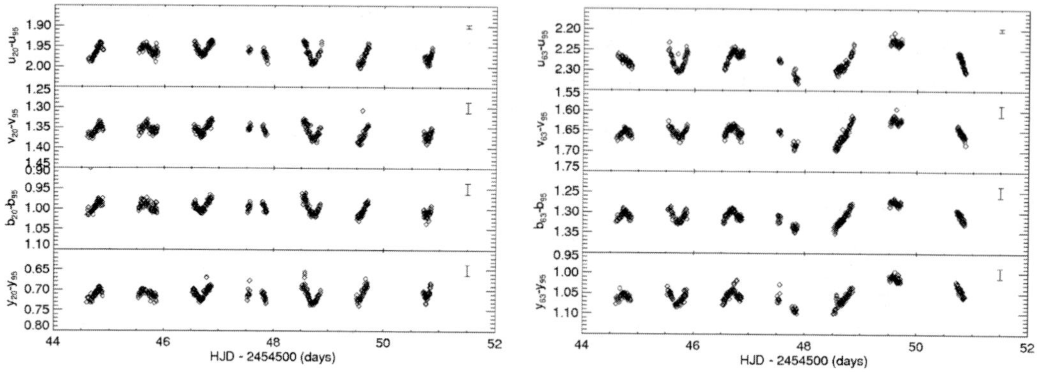

Figure 2. Differential photometry for Nos. 20 and 63 in the same format as in Figure 1.

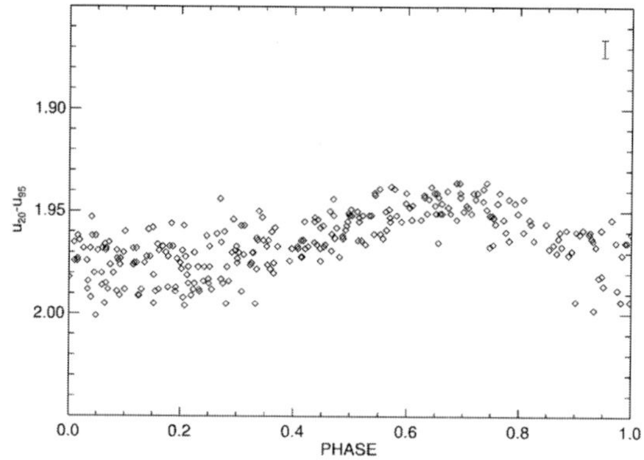

Figure 3. The folded u-band light curve for No. 20 using a period of 6.98 hours.

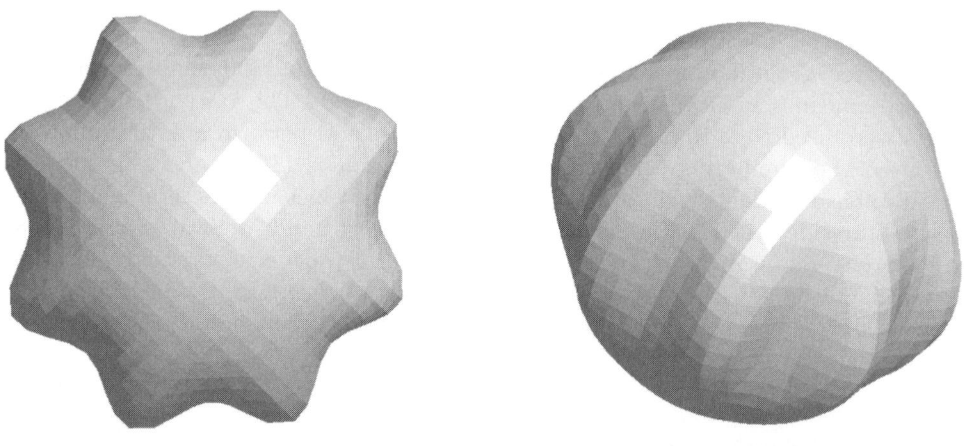

Figure 4. A pole-on and an inclined model of a sphere distorted by nonradial pulsations, with $l = m = 8$. This is a highly exaggerated model of both the amplitude and high order of pulsations for display purposes.

to be generated and compared to the observed light curves. Assuming slowly rotating stars, the nonradial pulsations can be modeled by spherical harmonic perturbations of a spherical stellar surface, given by Buta & Smith (1979) as

$$\frac{\delta R}{R} \sim A_\ell^m Y_\ell^m(\theta, \phi) e^{\imath \sigma_\ell^m t},$$

where $\delta R/R$ is the fractional change in radius, A_ℓ^m the amplitude associated with spherical harmonic $Y_\ell^m(\theta, \phi)$, and σ_ℓ^m is the frequency of the mode. Pole-on and inclined examples of spherical harmonics (with greatly exaggerated amplitudes) are presented in Figure 4. The most commonly observed mode of pulsation in Be stars is $l = |m| = 2$ (Rivinius *et al.* 2003), but for illustrative purposes, $l = m = 8$ is presented.

Acknowledgements

We gratefully acknowledge travel support from the International Astronomical Union, the American Astronomical Society, and NASA DPR number NNX08AV70G. We also thank Charles Bailyn and the SMARTS Consortium for their help in scheduling these observations. We are also grateful for an institutional grant from Lehigh University and a US Department of Education GAANN Fellowship.

References

Ahmed, F. 1962, *Publ. Roy. Obs. Edinburgh*, 3, 57

Balona, L. A. & Engelbrecht, C. A. 1986, *MNRAS*, 219, 131

Buta, R. J. & Smith, M. A. 1979, *ApJ*, 232, 213

De Ridder, J. 2001, *PhD Thesis*, Katholieke Univ. Leuven

Gutiérrez–Soto, J., Fabregat, J., Suso, J., Lanzara, M., Garrido, R., Hubert, A.-M., & Floquet, M. 2007, *A&A*, 476, 927

McSwain, M. V., Huang, W., Gies, D. R., Grundstrom, E. D., & Townsend, R. H. D. 2008, *ApJ*, 672, 590

Porter, J. M. & Rivinius, T. 2003, *PASP*, 115, 1153

Rivinius, T., Baade, D., & Štefl, S. 2003, *A&A*, 411, 229

Star clusters: basic galactic building blocks
Proceedings IAU Symposium No. 266, 2009
R. de Grijs & J. R. D. Lépine, eds.

doi:10.1017/S1743921309991918

Extragalactic star-forming regions and open clusters in the Milky Way

F. Sakhibov,[1,2] A. S. Gusev,[3] N. V. Kharchenko[4] and A. E. Piskunov[5]

[1]Giessen–Friedberg University for Applied Sciences, Friedberg, Germany
[2]Institute of Astrophysics, Academy of Sciences of Tajikistan, Dushanbe, Tajikistan
[3]Sternberg Astronomical Institute, Moscow, Russia
[4]Main Astronomical Observatory National Academy of Sciences of Ukraine, Kiev, Ukraine
[5]Institute of Astronomy of the Russian Academy of Sciences, Moscow, Russia

Abstract. We have carried out a comparative analysis of the evolution of integrated photometric parameters of young star-forming complexes in spiral and irregular galaxies and Galactic open clusters. We find that when the interstellar extinction is properly taken into account, the extragalactic complexes observed as giant HII regions and open clusters in the Milky Way form a single evolutionary sequence of objects evolving at different stages.

Keywords. open clusters and associations: general, galaxies: star clusters

---

The full poster (in pdf format) is available at
`http://www.astro.iag.usp.br/~iaus266/Posters/pSakhibov.pdf`.

Star clusters: basic galactic building blocks
Proceedings IAU Symposium No. 266, 2009
R. de Grijs & J. R. D. Lépine, eds.

doi:10.1017/S174392130999192X

Extreme-mass-ratio solar-type binaries and their implications for cluster environs

R. G. Samec,[1] Christa M. Labadorf,[1] E. R. Figg,[1] Trevor S. Loflin,[1] Gregory R. Behn,[1] Nathan A. Hawkins,[1] Heather A. Chamberlain,[1] D. R. Faulkner[2] and W. Van Hamme[3]

[1]Astronomy Group, Physics and Engineering Department, Bob Jones University, 1700 Wade Hampton Boulevard, Greenville, SC 29614, USA

[2]University of South Carolina, Lancaster, 476 Hubbard Dr., Lancaster, SC 29720, USA

[3]Physics Department, Florida International University, 11200 SW 8th St., Miami, FL 33199, USA

Abstract. During the past decade, the number known shallow-amplitude, yet totally eclipsing, extreme-mass-ratio binaries has increased from one (AW UMa, which is now in doubt) to about a dozen. Statistics are accumulating that will tell us the nature of these once rare systems. These individual interacting field binaries are believed to be the progenitors of FK Comae-type fast-rotating subgiants, which are similar to the cluster stars called 'variable blue stragglers.' We review our recent observations and new analyses of the individual systems, V409 Hya, GSC 1283 0053 Ori, GSC 2537 520 CVn, using the 2007 Wilson Code and summarize results from our previous observations (V802 Aql, V902 Sgr) and those from the literature (e.g., CK Boo, GR Vir, HV Aqr, UY UMa, EM Pis, CU Tau, TV Mus) to disclose a preliminary picture of the character of this interesting class of binary stars.

Keywords. magnetic fields, binaries: eclipsing, blue stragglers

---

The full poster (in pdf format) is available at
`http://www.astro.iag.usp.br/~iaus266/Posters/pSamec.pdf`.

Acknowledgements

Our recent CCD observations were taken with the Lowell 0.81 m telescope through our association with NURO (the National Undergraduate Research Observatory) and the SARA (Southeastern Association for Research in Astronomy) 0.9 m telescope. We gratefully acknowledge travel support received through the small research grant program of the American Astronomical Society.

Star clusters: basic galactic building blocks
Proceedings IAU Symposium No. 266, 2009
R. de Grijs & J. R. D. Lépine, eds.

doi:10.1017/S1743921309991931

The spatial distribution of stars in open clusters

Néstor Sánchez and Emilio J. Alfaro

Instituto de Astrofísica de Adalucía–CSIC, Apdo. 3004, E-18080, Granada, Spain
email: `[nestor, emilio]@iaa.es`

Abstract. We study the internal spatial structure of 16 open clusters in the Milky Way spanning a wide range of ages. For this, we use the minimum-spanning-tree method (the Q parameter, which enables one to classify the stellar distribution as either radially or fractally clustered), King-profile fitting, and the correlation dimension (D_c) for those clusters with fractal patterns. On average, clusters with fractal-like structure are younger than those exhibiting radial stellar-density profiles. There is a significant correlation between Q and the cluster age measured in crossing-time units. For fractal clusters, there is a significant correlation between fractal dimension and age. These results support the idea that stars in newly born clusters likely follow the fractal patterns of their parent molecular clouds, and eventually evolve towards more centrally concentrated structures. However, stellar clusters as old as ~ 100 Myr can exist that have not totally destroyed their fractal structure. Finally, we have found the intriguing result that the lowest fractal dimensions obtained for the open clusters seem to be considerably smaller than the average value measured in Galactic molecular cloud complexes.

Keywords. ISM: structure, methods: statistical, open clusters and associations: general, stars: formation

---

1. Introduction

The hierarchical structure observed in some open clusters is presumably a consequence of their formation in a turbulent medium with an underlying fractal structure (Elmegreen & Scalo 2004). Otherwise, open clusters having central stellar concentrations with radial stellar-density profiles likely reflect the dominant role of gravity, either on the primordial gas structure or as a result of rapid evolution from a more structured state (Lada & Lada 2003). Therefore, an analysis of the stellar distribution may yield information on the formation process and early evolution of open clusters. It is necessary, however, that this kind of analysis is done by measuring the cluster structure in an objective, quantitative, and systematic way. Here, we study the internal spatial structure in a sample of 16 open clusters spanning a wide range of ages.

2. Procedure

1. We first used VizieR (Ochsenbein *et al.* 2000) to search for catalogs containing both positions and proper motions of stars in open cluster regions.

2. We applied a robust nonparametric method to assign cluster membership (Cabrera–Caño & Alfaro 1990). This method makes no a priori assumptions about cluster and field-star distributions.

3. We fitted King (1962) profiles to the radial density distribution of cluster members. From these fits, we obtained both the core (R_c) and tidal radius (R_t).

4. Next, we used the minimum-spanning-tree technique (see Figure 1) to calculate the dimensionless parameter Q (see details in Cartwright & Whitworth 2004 and Schmeja &

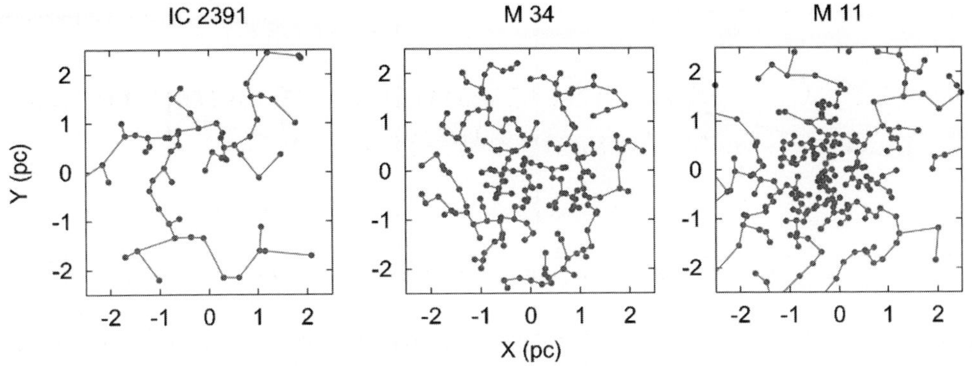

Figure 1. The minimum spanning tree is the set of straight lines connecting the points such that the sum of their lengths is at minimum. Here, we show minimum spanning trees for three open clusters, from which we can calculate the structure parameter Q. Star positions are indicated by blue circles and red lines represent the tree. The value of Q quantifies the stellar spatial distribution. In IC 2391 the stars are distributed following an irregular fractal pattern ($Q = 0.77 < 0.8$), in M 34 they are distributed roughly homogeneously ($Q = 0.8$), and in M 11 the stars follow a radial density profile ($Q = 1.02 > 0.8$).

Table 1. Properties of the sample clusters.

Name	log T	D	N_s	R_c	R_t	Q	D_c
IC 2391	7.661	175	62	1.46	2.65	0.77	1.74 ± 0.20
M 11	8.302	1877	289	1.98	4.49	1.02	...
M 34	8.249	499	181	0.11	1.73	0.80	2.04 ± 0.05
M 67	9.409	908	354	2.21	5.92	0.98	...
NGC 188	9.632	2047	1459	2.90	10.57	0.91	...
NGC 581	7.336	2194	526	1.38	11.86	0.76	1.79 ± 0.06
NGC 1513	8.110	1320	156	1.55	7.73	0.72	1.82 ± 0.09
NGC 1647	8.158	540	683	1.23	8.86	0.70	1.94 ± 0.02
NGC 1817	8.612	1972	277	3.39	11.97	0.79	1.94 ± 0.04
NGC 1960	7.468	1318	311	2.96	8.77	0.87	...
NGC 2194	8.515	3781	228	3.17	10.31	0.85	...
NGC 2548	8.557	769	168	2.61	9.16	0.90	...
NGC 4103	7.393	1632	799	0.72	10.74	0.78	1.85 ± 0.04
NGC 4755	7.216	1976	196	1.11	3.50	0.94	...
NGC 5281	7.146	1108	80	0.62	2.44	0.84	...
NGC 6530	6.867	1330	145	1.43	7.47	0.67	1.74 ± 0.09

T: cluster age (Myr), D: distance (pc), N_s: number of members, R_c: core radius (pc), R_t: tidal radius (pc), Q: structure parameter, D_c = correlation dimension.

Klessen 2006). The value $Q \simeq 0.8$ separates radial ($Q > 0.8$) from fractal-type clustering ($Q < 0.8$).

5. Finally, we calculated the correlation dimension (D_c) and its associated uncertainty by employing an algorithm which gives reliable results (Sánchez *et al.* 2007a; Sánchez & Alfaro 2008).

3. Main results

Table 1 summarizes the relevant data (ages and distances were taken from the WEBDA database).

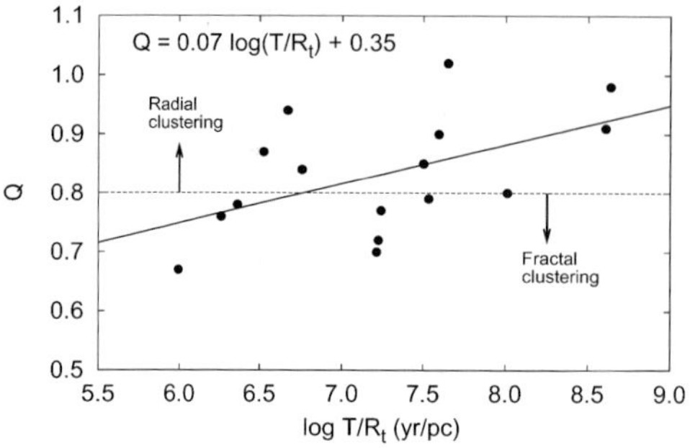

Figure 2. Structure parameter Q as a function of logarithmic age divided by the tidal radius, which is nearly proportional to age in crossing-times units. The dashed line at $Q = 0.8$ roughly separates radial from fractal clustering. The best linear fit (equation at the top) is represented by a solid line.

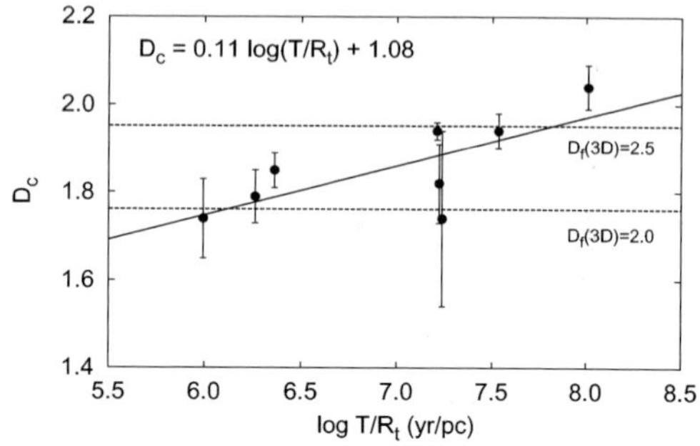

Figure 3. Calculated correlation dimension as a function of age (in crossing-time units). The best linear fit (equation at the top) is represented by a solid line. For reference, horizontal dashed lines indicate the values corresponding to three-dimensional distributions with fractal dimensions of $D_{\rm f} = 2.0$ and 2.5.

On average, stars in young clusters tend to be distributed following clustered, fractal-like patterns ($Q < 0.8$), while older clusters tend to exhibit radial stellar-density profiles ($Q > 0.8$). However, the statistical analysis indicates that there is no significant correlation between Q and $\log(T)$. If, instead, we consider the variable $T/R_{\rm t}$, which is proportional to the cluster age measured in crossing-time units (assuming nearly the same typical velocity dispersion for the open clusters), a significant correlation is observed (Figure 2). We also observe significant correlations (confidence levels > 96%) between $D_{\rm c}$ and T (cluster age) and also between $D_{\rm c}$ and $T/R_{\rm t}$ (age in crossing-time units) for those clusters with internal substructure (Figure 3).

4. Discussion

Our results support the idea that stars in newly born clusters likely follow the fractal patterns of their parent molecular clouds, and that they eventually evolve towards more centrally concentrated structures (see Schmeja & Klessen 2006; Schmeja *et al.* 2008, 2009; Sánchez *et al.* 2007a, 2009). However, this seems to be only an overall trend. The very young cluster σ Orionis (age ~ 3 Myr) exhibits a radial density gradient with $Q \simeq 0.88$ (Caballero 2008). On the other hand, Table 1 shows open clusters as old as ~ 100 Myr that have not totally destroyed their clumpy structure (for example, both NGC 1513 and NGC 1647 have $Q \sim 0.7$). Goodwin & Whitworth (2004) simulated the dynamical evolution of young clusters and showed that survival of the initial substructure depends strongly on the initial velocity dispersion. Fractal clusters with a low velocity dispersion tend to erase their substructure rather quickly. However, if the velocity dispersion is high, such that the cluster remains supported against its own gravity or even expands, significant levels of substructure can survive for several crossing times. Thus, our results give some observational support to Goodwin & Whitworth's (2004) simulations.

From Figure 3, we can see that clusters with the smallest correlation dimensions ($D_{\rm c} = 1.74$) would have three-dimensional fractal dimensions around $D_{\rm f} \sim 2.0$ (estimated from previous papers; see, e.g., figure 1 in Sánchez & Alfaro 2008). This is a very interesting result, because this value is considerably smaller than the average value estimated for Galactic molecular clouds in recent studies, which is $D_{\rm f} \simeq 2.6 - 2.7$ (Sánchez *et al.* 2005, 2007b). Young, newly born stars will probably reflect the conditions of the interstellar medium from which they formed. Therefore, a group of stars born from the same cloud, i.e., born at almost the same place and time, should have a fractal dimension similar to that of the parent cloud. If the fractal dimension of the interstellar medium has a nearly universal value around 2.6–2.7, then how can some clusters exhibit such small fractal dimensions? Perhaps some clusters develop some kind of substructure starting from an initially more homogeneous state. This possibility has been confirmed in numerical simulations (Goodwin & Whitworth 2004), although some coherence in the initial velocity dispersion is required. Another explanation is that this difference is a consequence of a more clustered distribution of the densest gas from which stars form on the smallest spatial scales in the molecular cloud complexes, according to a multifractal scenario (Chappell & Scalo 2001). Perhaps the star-formation process itself modifies in some (unknown) way the underlying geometry generating distributions of stars that can be very different from the distribution of gas in the parental clouds. Finally, one possibility is that the fractal dimension of the interstellar medium in the Galaxy does not have a universal value and therefore some regions form stars distributed following more clustered patterns. There is no a priori reason to assume that $D_{\rm f}$ has nearly the same value everywhere in the Galaxy, independent of either the dominant physical processes or environmental variables. Recent simulations of supersonic isothermal turbulence by Federrath *et al.* (2009) have shown that compressive forcing yields fractal dimensions for the interstellar medium significantly smaller ($D_{\rm f} \sim 2.3$) compared to solenoidal forcing ($D_{\rm f} \sim 2.6$). Thus, $D_{\rm f}$ could be very different from region to region in the Galaxy depending on the main physical processes driving the turbulence. At least on galactic scales, it has been shown that there are significant differences in the fractal dimension of the distribution of star-forming sites among galaxies, contrary to the universal picture previously claimed in the literature (see Sánchez & Alfaro 2008), so that the possibility of a nonuniversal fractal dimension for the interstellar medium in the Galaxy cannot, in principle, be ruled out.

References

Caballero, J. A. 2008, *MNRAS*, 383, 375
Cabrera–Caño, J. & Alfaro, E. J. 1990, *A&A*, 235, 94
Cartwright, A. & Whitworth, A. P. 2004, *MNRAS*, 348, 589
Chappell, D. & Scalo, J. 2001, *ApJ*, 551, 712
Elmegreen, B. G. & Scalo, J. 2004, *ARA&A*, 42, 211
Federrath, C., Klessen, R. S., & Schmidt, W. 2009, *ApJ*, 692, 364
Goodwin, S. P. & Whitworth, A. P. 2004, *A&A*, 413, 929
King, I. 1962, *AJ*, 67, 471
Lada, C. J. & Lada, E. A. 2003, *ARA&A*, 41, 57
Ochsenbein, F., Bauer, P., & Marcout, J. 2000, *A&AS*, 143, 221
Sánchez, N., Alfaro, E. J., & Pérez, E. 2005, *ApJ*, 625, 849
Sánchez, N., Alfaro, E. J., Elias, F., Delgado, A. J., & Cabrera–Caño, J. 2007a, *ApJ*, 667, 213
Sánchez, N., Alfaro, E. J., & Pérez, E. 2007b, *ApJ*, 656, 222
Sánchez, N. & Alfaro, E. J. 2008, *ApJS*, 178, 1
Sánchez, N. & Alfaro, E. J. 2009, *ApJ*, 696, 2086
Schmeja, S. & Klessen, R. S. 2006, *A&A*, 449, 151
Schmeja, S., Kumar, M. S. N., & Ferreira, B. 2008, *MNRAS*, 389, 1209
Schmeja, S., Gouliermis, D. A., & Klessen, R. S. 2009, *ApJ*, 694, 367

Star clusters: basic galactic building blocks
Proceedings IAU Symposium No. 266, 2009
R. de Grijs & J. R. D. Lépine, eds.

doi:10.1017/S1743921309991943

Astrometry and photometry of stars and star clusters in M33

Izaskun San Roman,[1] Ata Sarajedini,[1] Carme Gallart[2] and Antonio Aparicio[2]

[1]Department of Astronomy, University of Florida, 211 Bryant Space Science Center, Gainesville, FL 32611-2055, USA
email: [izaskun, ata]@astro.ufl.edu

[2]Instituto de Astrofísica de Canarias, Vía Láctea, s/n, E38200 La Laguna, Tenerife, Spain

Abstract. We present positions and magnitudes of stars and star clusters in a $1^\circ \times 1^\circ$ area centered on M33. The survey is based on deep archival ground-based images using the MegaPrime camera on the 3.6m *Canada–France–Hawaii telescope.* We provide u', g', r', i', z' magnitudes by performing standard profile-fitting photometry of these images and apply image classification algorithms such as SExtractor. We also present a catalog of extended sources by applying visual-inspection classification. This complete catalog provides promising targets for deep photometry with *HST* and for high-resolution spectroscopy to study the structure and star-formation history of the disk and halo of M33.

Keywords. galaxies: individual (M 33), galaxies: spiral, galaxies: star clusters, galaxies: stellar content

1. Introduction

The diverse set of issues required to understand the formation history of a galaxy involves the need for a comprehensive study of its stars and star clusters. In particular, the ages, metallicities and kinematics of star clusters are representative of the galaxy-formation process. These correlations underscore the importance of compiling complete catalogs of stars and star clusters in nearby galaxies. At a distance of $\sim$ 870 kpc, M33 is the only nearby late-type spiral galaxy providing a connection between the cluster populations of earlier-type spirals and the numerous, nearby later-type dwarf galaxies. An up-to-date catalog of star clusters in M33 can be found in Sarajedini & Mancone (2007; SM hereafter) who merged all modern catalogs of star clusters and cluster candidates. However, as discussed by the authors, the sample of clusters suffer from severe incompleteness. A more recent catalog produced by Zloczewski *et al.* 2008 (hereafter ZKH) was derived from the same data used in this study. They present a catalog of 4780 extended sources, including 3554 new candidate stellar clusters. While employing a comparable method, we achieve somewhat different results.

2. Observations and reduction

Processing of the CCD frames was performed as a part of the *CFHT* Queue Service Observing Mode using the ELIXIR pipeline. To facilitate the search of objects, only the best images available were analyzed, consisting of 15 g', 14 r', 15 u', 28 i' and 3 z' with median seeing values of $< 0.8''$ in g', r' and i' and $\sim 0.6''$ in u' and z'. We divided each final, combined image into two subfields, including an overlapping area, to deal with the spatial variability of the point-spread function (PSF).

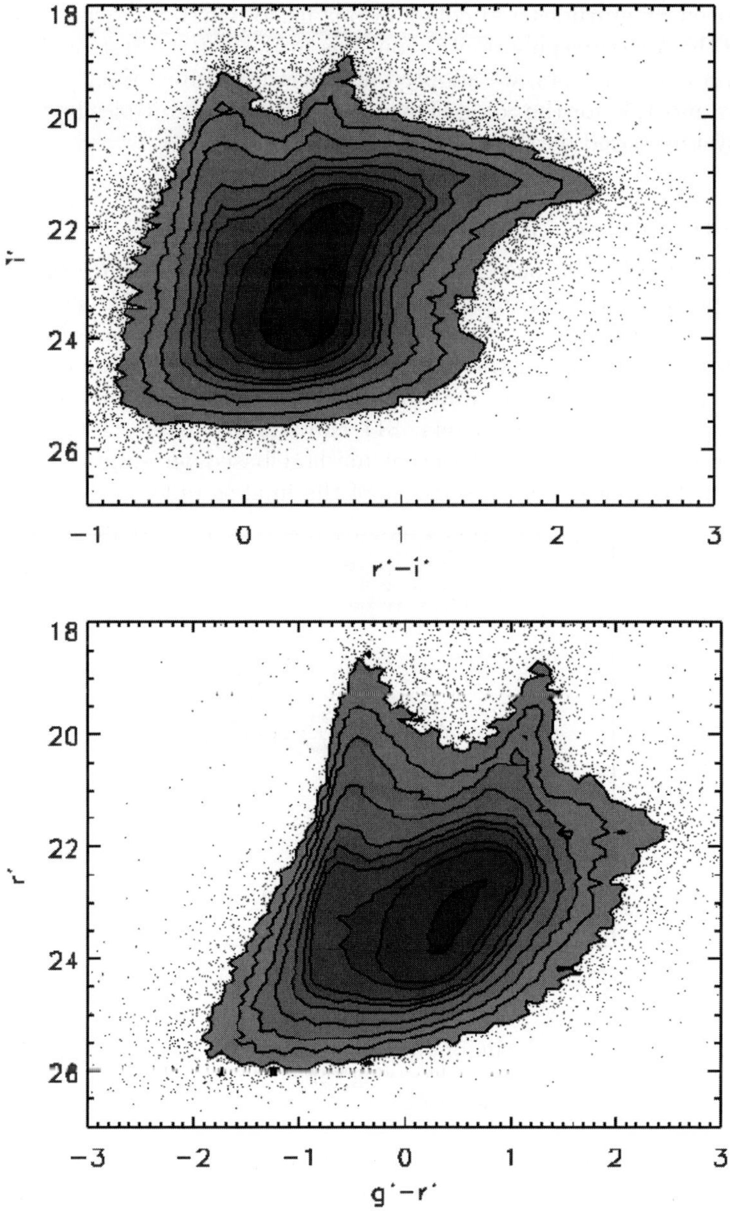

Figure 1. Isodensity contour plot (Hess diagram) of the upper half of M33. The outermost contour corresponds to a density of 50 stars per 0.1 mag box, with a step between each of 200, while for the bottom panel the box sampling has been reduced in the innermost contours to try to see the red clump. The color–magnitude diagrams display a broad red-giant branch with the tip near $i' \sim 21$ mag and a young main sequence at $(g' - r') \sim -0.5$ mag, reaching $r' \sim 20$ mag.

We perform standard profile-fitting photometry on these images using the DAOPHOT/ALLSTAR routines (Stetson 1994). A detailed description of how the PSFs were constructed has been presented in Sarajedini *et al.* (2006). Appropriate aperture corrections were added to obtain the instrumental magnitudes. These magnitudes have been calibrated to the SDSS system by applying the transformations provided by the ELIXIR pipeline.

The integrated magnitudes and colors of our list of candidate clusters have been calculated using the aperture-photometry routines in DAOPHOT (Stetson 1987). To be consistent with previous authors, we adopted an aperture radius of 2.2″ for the magnitude measurements and 1.5″ for the colors. Once again, these magnitudes have been calibrated to the SDSS system based on the ELIXIR transformation.

3. Candidate clusters: testing the catalog

Our detection method is based on the fact that at the distance of M33, nonstellar objects are expected to be more extended than the PSF. After subtracting the stellar PSF from all sources in our frames, extended objects leave a doughnut-shaped appearance, as they are undersubtracted in the wings and oversubtracted in the center. After visual inspection of the residual images and analysis of the FWHM, this technique leaves us with a total of 2998 extended objects: 811 candidate clusters, 1969 galaxies and 218 unknown objects. An additional rejection method based on ellipticity will be applied. Figure 2 shows the photometric properties of the final sample.

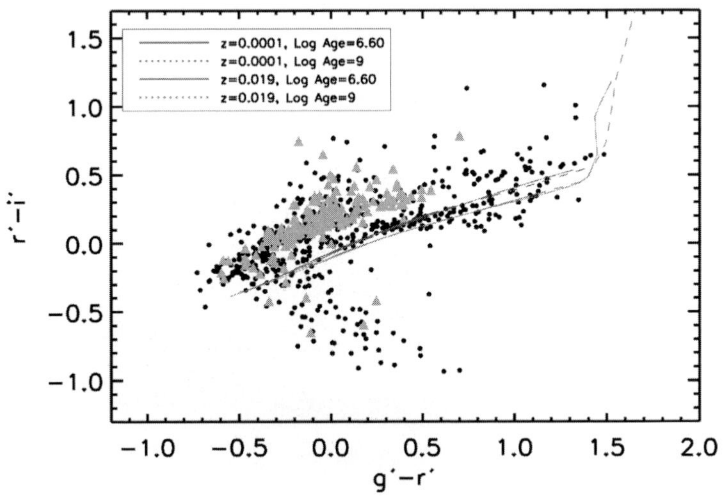

Figure 2. Color–magnitude diagram of our stellar candidate clusters (black circles) compared with the known star clusters in the SM catalog (green circles). Overplotted are single stellar population models from Girardi *et al.* (2002).

The 12 *HST*/ACS fields examined in San Roman *et al.* (2009) included 72 of the candidate clusters in the present catalog, of which 51 turned out to be genuine star clusters. This suggests that around ∼ 70% of our 811 proposed candidates will be actual stellar clusters. However, from the 349 guaranteed clusters listed in the updated SM, our catalog only recovers 210 objects, implying missing objects mostly in the center of the galaxy. This is not surprising since the method is less effective in extremely crowded regions.

Comparison with the similar study of ZKH reveals unexpected discrepancies. From the total 1752 common objects between both catalogs, only 124 sources were classified as candidate clusters by both authors. As San Roman *et al.* (2009) argue, the total 'true' cluster candidates in the ZKH catalog would not be greater than ∼ 40%. This suggests a systematic misidentification in the candidate-object pattern or a defective PSF sustraction. Figure 3 shows the photometric differences between the two studies, which are not statistically significant.

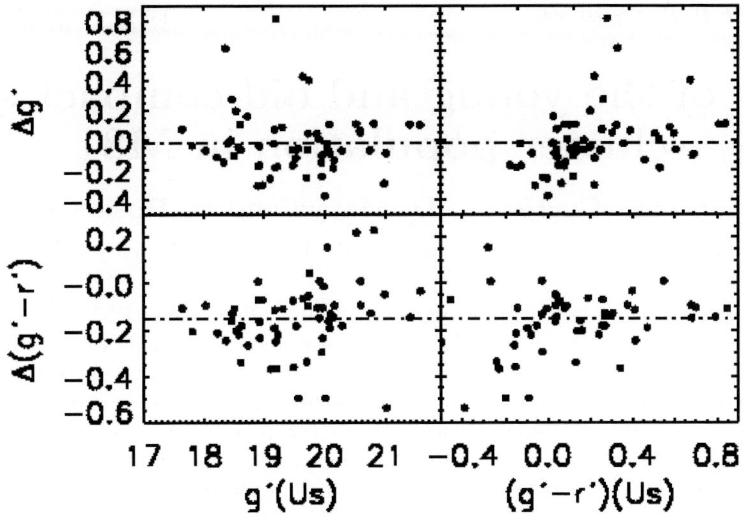

Figure 3. Comparison of the integrated photometry of common candidate clusters between the present study and ZKH. The dashed lines represent the mean differences of $\langle \Delta g' \rangle = -0.035 \pm 0.032$ and $\langle \Delta(g' - r') \rangle = -0.171 \pm 0.019$ mag, where the uncertainties are standard errors of the mean.

Acknowledgements

We gratefully acknowledge support from the US National Science Foundation through grant AST-0707277.

References

Sarajedini, A., Barker, M. K., Geisler, D., Harding, P., & Schommer, R. 2006, *AJ*, 132, 1361
Sarajedini, A. & Mancone, C. L. 2007, *AJ*, 134, 447 (SM)
San Roman, I., Sarajedini, A., Garnett, D. R., & Holtzman, J. A. 2009, *ApJ*, 699, 839
Zloczewski, K., Kaluzny, J., & Hartman, J. 2008 *AcA*, 58, 23. (ZKH)

Star clusters: basic galactic building blocks
Proceedings IAU Symposium No. 266, 2009
R. de Grijs & J. R. D. Lépine, eds.

doi:10.1017/S1743921309991955

Study of the young and old compact stellar cluster population in M81

M. Santiago–Cortés, Y. D. Mayya and D. Rosa–González

Coordinación de Astrofísica, Instituto Nacional de Astrofísica, Óptica y Electrónica, Mexico

Abstract. We study the population of compact stellar clusters in M 81 using the *HST*/ACS images in the filters F435W, F606W and F814W that cover a total field of view of approximately 340 square arcmin, the largest area covered to date for this galaxy. We present details about the selection criteria, which were based on both morphological and photometrical features. The extracted sample of compact stellar clusters shows the presence of two cluster populations, a blue cluster group (young) with more than 300 objects, and a red cluster group (old) containing 138 objects. Surprisingly, the young group lacks clusters more massive than 10 000 $M_{\odot}$ which are present in large numbers in its neighbor M 82. The luminosity function of the young group follows a power-law distribution with an index of 2.0. The luminosity function of the red group closely resembles that of the globular clusters in the Milky Way. Assuming an age of 5 Gyr, these red clusters have masses between 0.1 and 10×10^6 $M_{\odot}$.

Keywords. catalogs, galaxies: individual (M 81), galaxies: spiral, galaxies: star clusters

---

The full poster (in pdf format) is available at
`http://www.astro.iag.usp.br/~iaus266/Posters/pSantiago-Cortes.pdf`.

Star clusters: basic galactic building blocks
Proceedings IAU Symposium No. 266, 2009
R. de Grijs & J. R. D. Lépine, eds.

doi:10.1017/S1743921309991967

Astrophysical properties of binary star clusters in the Small Magellanic Cloud†

João F. C. Santos Jr.,[1] Alex A. Schmidt[2] and Eduardo Bica[3]

[1]Departamento de Física, ICEx, Universidade Federal de Minas Gerais, Av. Antônio Carlos 6627, Belo Horizonte 31270-901, MG, Brazil
email: jsantos@fisica.ufmg.br

[2]Departamento de Matemática, Universidade Federal de Santa Maria, Av. Roraima 1000, Santa Maria 97105-900, RS, Brazil

[3]Departamento de Astronomia, Universidade Federal do Rio Grande do Sul, Av. Bento Gonçalves 9500, Porto Alegre 91501-970, RS, Brazil

Abstract. To study the evolution of binary star clusters, we have imaged seven systems in the Small Magellanic Cloud with the *SOAR 4m* telescope using B and V filters. The sample contains pairs with well-separated components ($d < 30$ pc) as well as systems that apparently have merged, as evidenced by their unusual structures. By employing isochrone fitting to their color–magnitude diagrams, we have determined reddening values, ages and metallicities, and by fitting King models to their radial stellar-density profiles we estimated core radii. Disturbances of the density profiles are interpreted as evidence of interactions. Properties such as the distances between their components and their age differences are addressed in terms of the timescales involved, to assess the physical connection of the system. In two cases, the age difference is more than 50 Myr, which suggests a chance alignment, capture or sequential star formation.

Keywords. Magellanic Clouds, galaxies: star clusters

1. Introduction

Differently from the Galaxy, the low-density environment of the Magellanic Clouds favors the gravitational interaction of single clusters, which form binary (or multiple) groupings and will eventually merge or be disrupted. Pieces of evidence for these scenarios include: (i) many binary clusters in the Magellanic Clouds have been identified and recognized as true physically connected systems (Bhatia & Hatzidimitriou 1988; Bhatia *et al.* 1991) and (ii) deep and precise colour–magnitude diagrams (CMDs) for isolated clusters in the Large Magellanic Cloud have revealed multiple population sequences (Mackey *et al.* 2008). If these facts are combined, their results indicate that we are witnessing different stages of cluster evolution in a context where cluster groupings and mergers are not uncommon. Evidence of cluster binarity and mergers has been reported in both Clouds, which are very rich in cluster pairs and multiples (Dieball *et al.* 2002; Bica *et al.* 2008; Carvalho *et al.* 2008). We determine astrophysical and structural parameters of SMC binary clusters by fitting isochrones to the CMDs and King functions to the radial-density profiles with the purpose of providing constraints on the evolution of such systems.

† The full poster (in pdf format) is available at
http://www.astro.iag.usp.br/~iaus266/Posters/pSantos.pdf.

Table 1. Cluster sample and derived parameters.

Cluster	α_{2000} (h:m:s)	δ_{2000} (° : ′ : ″)	$E(B-V)$ (mag)	age (Myr)	Z	R_c (pc)	Δage (Myr)	separation (pc)	age (lit.) (Myr)
NGC 220	0:40:31	−73:24:10	0.15 ± 0.03	70 ± 10	0.004	2.5 ± 1.0	–	–	100 ± 23[a] 65 ± 13[b] 126[c] 70 − 100[d]
NGC 222	0:40:44	−73:23:00	0.15 ± 0.03	70 ± 10	0.004	1.5 ± 0.5	0	28	100 ± 23[a] 70 ± 7[b] 100[c] 70 − 100[d]
NGC 241	0:43:33	−73:26:25	0.05 ± 0.01	80 ± 10	0.004	2.0 ± 1.0	–	–	79 ± 18[a] 200[c]
NGC 242	0:43:38	−73:26:37	0.05 ± 0.01	80 ± 10	0.004	–	0	9	79 ± 18[a] 63[c]
B 78	0:54:45	−72:07:46	0.10 ± 0.02	45 ± 5	0.004	3.3 ± 2.2	–	–	
L 51	0:54:54	−72:06:46	0.10 ± 0.02	40 ± 5	0.004	1.5 ± 0.3	5	17	
IC 1611	0:59:48	−72:20:02	0.07 ± 0.01	140 ± 30	0.002	2.2 ± 1.1	–	–	158 ± 37[a] 100 ± 20[b] 126[c] 100 ± 20[e]
IC 1612W=H 86-186	0:59:57	−72:22:24	0.07 ± 0.01	120 ± 30	0.002	2.5 ± 0.5	–	–	180 ± 20[b] 126[c]
IC 1612E=IC 1612	1:00:01	−72:22:08	0.07 ± 0.01	60 ± 10	0.002	2.0 ± 1.0	60	8	50 ± 24[a] 100 ± 50[b] 100[c]
NGC 376	1:03:53	−72:49:34	0.09 ± 0.02	50 ± 10	0.002	3.0 ± 1.3	–	–	32 ± 7[a] 20 ± 2 16[c] 25 ± 10[e]
K 50	1:04:36	−72:09:38	0.03 ± 0.01	50 ± 10	0.004	2.5 ± 1.2	–	–	20 ± 5[a] 8[c]
NGC 422	1:09:25	−71:46:00	0.06 ± 0.01	110 ± 30	0.004	1.4 ± 0.5	–	–	
IC 1641	1:09:39	−71:46:07	0.06 ± 0.01	500 ± 30	0.004	2.9 ± 1.0	390	20	

[a]Pietrzynski & Udalski (1999), [b]de Oliveira *et al.* (2000), [c]Chiosi *et al.* (2006), [d]Matteucci *et al.* (2002), [e]Piatti *et al.* (2007).

2. Observations

The *SOAR* optical imager (SOI) mounted in a bent-Cassegrain configuration on the 4.1m *SOAR* telescope (Cerro Pachón, Chile) was employed to observe a sample of seven cluster merger candidates. Images in Bessel *BV* filters were obtained on photometric nights (Oct/Nov 2007), using the SOI mini-mosaic of two E2V 2×4k CCDs (1 pixel = 15μm) to cover a $5.2 \times 5.2\,\mathrm{arcmin}^2$ field of view. The CCDs were binned 2×2, yielding a scale of 0.154 arcsec pixel^{-1}. The average seeing was $\sim$ 0.95 arcsec in B and $\sim$ 0.8 arcsec in V. Two images in each filter were obtained, with single exposure times of 480 s in B and 195 s in V. The spatial scale is 1 arcsec = 0.3 pc. The CCD frames were reduced with IRAF software. STARFINDER (Diolaiti *et al.* 2000) was used to perform PSF photometry. Table 1 presents the cluster sample and includes the parameters derived here and from the literature. The V-band image of the interesting noncoeval pair IC 1641/NGC 422 is shown in Figure 1.

3. Radial density profiles and structural parameters

We built the radial density profile (RDP) for each merger candidate component by employing star counts in circular rings around the cluster center. For the sake of uniformity, all RDPs are based on rings 5″ wide. To obtain an optimized RDP, we limit the data to a threshold magnitude for which the difference between the cluster central density and the adjacent field density is maximum. The cluster center is tweaked in the process. All stars fainter than this threshold are disregarded in the analysis. We employed two-parameter (central density σ_0, core radius R_c) King profiles (King 1966) to derive

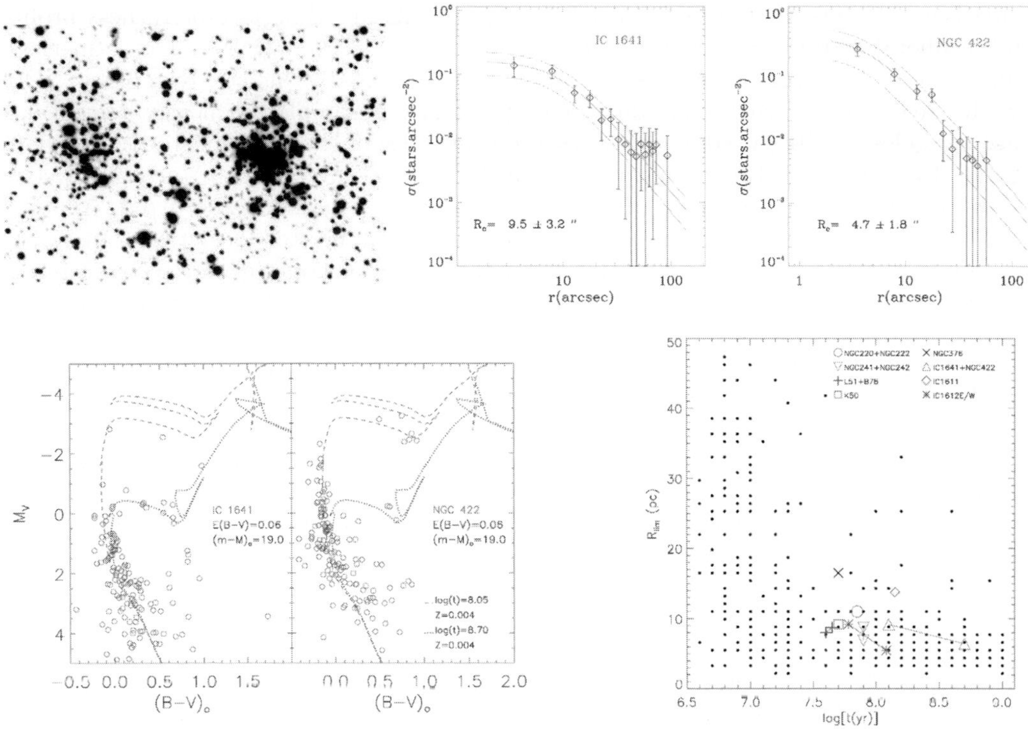

Figure 1. From top left to bottom right: (i) V image extraction ($2.6' \times 1.7'$) of pair IC 1641 *(left)* / NGC 422 *(right)*; north is up and east to the left; (ii) The RDP on a logarithmic scale with a two-parameter King-profile fit superimposed, for which R_c is indicated; 1σ dispersions of the fits are plotted as dotted lines; (iii) Isochrone fits to the CMDs of IC 1641 and NGC 422; for comparison, the best-matching isochrone for IC 1641 is also plotted in the CMD of NGC 422 and vice versa; (iv) Limiting radius (Bica *et al.* 2008) as a function of age; the dots represent SMC clusters from Chiosi *et al.* (2006) and other symbols identify the present sample of binary/merger clusters.

structural properties. The background was determined by fitting a constant to the outermost four rings, which was then subtracted from the overall stellar density before the King profile fitting was performed. Figure 1 shows the RDP for clusters IC 1641/NGC 422.

4. CMDs and astrophysical parameters

We obtained astrophysical parameters by fitting Padova isochrones (Girardi *et al.* 2002) to the V versus $(B-V)$ CMDs. To minimize contamination of background stellar fields, a circular region defined by a limiting radius ($R_{\rm lim}$) covering the cluster central area was selected. $R_{\rm lim}$ is defined as the radius where the cluster stellar density begins to stand out from the background. For all clusters in the sample, good fits were obtained for $(m-M)_0 = 19.0 \pm 0.1$ mag. Reddening, age and metallicity corresponding to the best-matching isochrones are presented in Table 1. Figure 1 shows isochrone fits to the observed CMDs of IC 1641 and NGC 422.

5. Discussion

In general, our age estimates agree with CMD-based age determinations in the literature (see Table 1). The ages of L51/B78 and NGC422/IC1641 have been determined using isochrone fits for the first time. Because of a disturbed surface-brightness profile and isophotal features, Carvalho *et al.* (2008) classified K50 and NGC 376 as mergers and the pairs IC 1612E/W and NGC 241/242 as true binary clusters. Except for NGC 422/IC 1641 and IC 1612E/W, all cluster pairs in our sample are nearly coeval and have ages close to 10^7 yr, consistent with the timescales of the dynamical models. NGC 422/IC 1641 and IC 1612E/W have ages that differ by 390 Myr and 60 Myr respectively, implying that, if they are interacting, a capture process may have occurred. Another possibility is that the oldest cluster in the pair triggered the star formation that led to the youngest through supernova explosions and shell expansion. Such a mechanism has also been proposed as a plausible explanation by Vallenari *et al.* (1998) for the LMC pair NGC 1903/SL 357, although their age difference is 550 Myr.

As regards the timescales involved, the expected effect of the gravitational interaction for clusters with a mass of $10^5\,M_\odot$ and a half-mass radius of 10 pc, is that they will merge in 10^7 yr if their separation is ~ 25 pc (Sugimoto & Makino 1989). Models for less massive clusters in the LMC tidal field predict a few $\times 10^7$ yr for the pair lifetime, with separations varying from ~ 6 pc for cluster masses of $10^3\,M_\odot$ and ~ 14 pc for $10^4\,M_\odot$ (Bhatia 1990). According to these models, after this timescale either the cluster pair merges due to mutual gravitational interaction or is disrupted by the galactic tidal field and, less probably, by passing giant molecular clouds.

Clusters resulting from pairs that have merged are expected to show structural properties which differ from those of single clusters. The plot of $R_{\rm lim}$ versus age indicates that K50, NGC 376 and IC 1611 have a larger $R_{\rm lim}$ than the mean locus of single clusters, which may be additional evidence of their merger status (Mackey & Gilmore 2003).

Acknowledgements

We thank the Brazilian agencies CNPq and FAPEMIG for financial support.

References

Bhatia, R. K. 1990, *PASJ*, 42, 757
Bhatia, R. K. & Hatzidimitriou, D. 1988, *A&A*, 230, 215
Bhatia, R. H., Read, M. A., Tritton, S., & Hatzidimitriou, D. 1991, *A&AS*, 87, 335
Bica, E., Bonatto, C., Dutra, C. M., & Santos Jr., J. F. C. 2008, *MNRAS*, 389, 678
Carvalho, L., Saurin, T. A., Bica, E., Bonatto, C., & Schmidt, A. A. 2008, *A&A*, 485, 71
Chiosi, E., Vallenari, A., Held, E. V., Rizzi, L., & Moretti, A. 2006, *A&A*, 452, 179
de Oliveira, M. R., Dutra, C. M., Bica, E., & Dottori, H. 2000, *A&AS*, 146, 57
Dieball, A., Müller, H., & Grebel, E. K. 2002, *A&A*, 391, 547
Diolaiti, E., Bendinelli, O., Bonaccini, D., Close, L., Currie, D., & Parmeggiani, G. 2000, *A&AS*, 147, 335
Girardi, L., Bertelli, G., Bressan, A., Chiosi, C., Groenewegen, M. A. T., Marigo, P., Salasnich, B., & Weiss, A. 2002, *A&A*, 391, 195
King, I. 1966, *AJ*, 71, 64
Mackey, A. D., Broby Nielsen, P., Ferguson, A. M. N., & Richardson, J. C. 2008, *ApJ* (Letters), 681, L17
Mackey, A. D. & Gilmore, G. F. 2003, *MNRAS*, 338, 120
Matteucci, A., Ripepi, V., Brocato, E., & Castellani, V. 2002, *A&A*, 387, 861
Piatti, A. E., Sarajedini, A., Geisler, D., Clark, D., & Seguel, J. 2007, *MNRAS*, 377, 300
Pietrzyński, G. & Udalski, A. 1999, *AcA*, 49, 157
Sugimoto, D. & Makino, J. 1989, *PASJ*, 41, 1117
Vallenari, A., Bettoni, D., & Chiosi, C. 1998, *A&A*, 331, 506

Star clusters: basic galactic building blocks
Proceedings IAU Symposium No. 266, 2009
R. de Grijs & J. R. D. Lépine, eds.

doi:10.1017/S1743921309991979

Sulfur in the globular clusters 47 Tuc and NGC 6752

L. Sbordone,[1,2] A. Chieffi,[3] M. Limongi,[4] E. Caffau,[2] H.-G. Ludwig[1,2] and P. Bonifacio[1,2,5]

[1]CIFIST Marie Curie Excellence Team, France

[2]GEPI, Observatoire de Paris, CNRS, Université Paris Diderot, Place Jules Janssen, 92190 Meudon, France

[3]INAF – IASF, via del Fosso del Cavaliere 100, 00133 Rome, Italy

[4]INAF – Osservatorio Astronomico di Roma, via di Frascati 33, 00040 Monte Porzio Catone, Italy

[5]INAF – Osservatorio Astronomico di Trieste, via G. B. Tiepolo 11, 34131 Trieste, Italy

Abstract. The light elements Li, O, Na, Al, and Mg are known to show star-to-star variations in the globular clusters 47 Tuc and NGC 6752. We have investigated the behavior of the α element sulfur, for which no previous measurements exist in any Galactic globular cluster. We used high-resolution UVES spectra of S I multiplet 1 around 923 nm, and determined S abundances by means of ATLAS static plano-parallel models. NLTE corrections were applied and 3D corrections were also computed from CO5BOLD 3D hydrodynamical models. Sulfur has been measured in four subgiant stars in NGC 6752, leading to an average value of [S/Fe] $= +0.49 \pm 0.15$ dex, consistent with what is observed in field stars of similar metallicity. In 47 Tuc, we measured S in four turnoff (TO) and five subgiant (SG) stars, for an average value of [S/Fe] $= 0.18 \pm 0.14$ dex. While the measurement errors are consistent with a constant value among all cluster stars analyzed, we detected a highly significant correlation with sodium abundance, as well as a tentative one with silicon. The sulfur–sodium correlation is difficult to explain in terms of nucleosynthesis. Given its high statistical significance, it is also difficult to dismiss it as fortuitous. Until better data for more stars are available, the question as to its origin remains open.

Keywords. stars: abundances, globular clusters: individual (47 Tuc, NGC 6752)

The full poster (in pdf format) is available at
`http://www.astro.iag.usp.br/~iaus266/Posters/pSbordone.pdf`.

Star clusters: basic galactic building blocks
Proceedings IAU Symposium No. 266, 2009
R. de Grijs & J. R. D. Lépine, eds.

doi:10.1017/S1743921309991980

Hierarchical clustering in the Local Group dwarf galaxy NGC 6822

S. Schmeja,[1] D. A. Gouliermis,[2] R. S. Klessen,[1] W. J. G. de Blok[3] and F. Walter[2]

[1]Zentrum für Astronomie der Universität Heidelberg, Heidelberg, Germany

[2]Max-Planck-Institut für Astronomie, Heidelberg, Heidelberg, Germany

[3]Department of Astronomy, University of Cape Town, Cape Town, South Africa

Abstract. Star formation appears to be clumped into a hierarchy of structures, from giant molecular clouds down to individual cores and clusters, which are often hierarchical themselves, showing significant substructure. This has been demonstrated for our Galaxy through the application of sophisticated statistical methods, in particular the nearest-neighbour density and the minimum spanning tree (MST), to different star-forming regions. Here we present our analysis of clustered star formation as demonstrated through the detection of structures of young stellar populations in the dwarf star-forming galaxy NGC 6822.

Keywords. galaxies: Local Group, galaxies: star clusters, galaxies: stellar content, galaxies: structure

---

The full poster (in pdf format) is available at
`http://www.astro.iag.usp.br/~iaus266/Posters/pSchmeja.pdf`.

Star clusters: basic galactic building blocks
Proceedings IAU Symposium No. 266, 2009
R. de Grijs & J. R. D. Lépine, eds.

doi:10.1017/S1743921309991992

Refining the true parameters of the open cluster NGC 4852

Gladys Solivella,[1] Edgard Giorg,[1] Rubén Vázquez[1] and Giovanni Carraro[2]

[1]Facultad de Ciencias Astronómicas y Geofísicas, UNLP–IALP–CONICET, Argentina
[2]ESO, Santiago, Chile

Abstract. NGC 4852 is a moderately compact cluster centered at $\alpha_{2000} = 13:00:09; \delta = -59:36:48$, located near the center of an Hα superring. This cluster forms part of an extended region including young stellar aggregates inside a circle with a radius of 3 degrees, where many show an abundance of emission line stars. In the field of this cluster, two stars of known type exist: Wray 15−1039 (emission-line object) and CD −58:4845 (emission-line star). We do not yet know whether the Be phase is transient or whether it is just what randomly happens in some hot stars. It appears that Be star may be found even in clusters as old as 70 Myr with a high occurrence rate in clusters of 25–27 Myr old. A recent photometric survey in NGC 4852 down to $V = 22 - 23$ mag established that NGC 4852 is about 200 − 250 Myr old, located at 1.1 kpc from the Sun and with a mean $E(B - V) = 0.45$ mag. Since the presence of potential Be-type stars in the cluster area suggests it may be a very young object instead of moderately old, we decided to carry out spectroscopy for 33 selected stars and CCD *UBVI* photometry for the bright objects in the cluster area. This way, we attempt to clarify their evolutionary state and include them in the framework of emission-line stars and open clusters. From our analysis, we agree with the cluster distance and reddening determined by earlier studies, but we derive that the age of NGC 4852 is younger than 40 Myr.

Keywords. stars: fundamental parameters, open clusters and associations: individual (NGC 4852)

The full poster (in pdf format) is available at
`http://www.astro.iag.usp.br/~iaus266/Posters/pSolivella.pdf`.

Star clusters: basic galactic building blocks
Proceedings IAU Symposium No. 266, 2009
R. de Grijs & J. R. D. Lépine, eds.

doi:10.1017/S1743921309992006

Membership status of TWA22AB

Ramachrisna Teixeira,[1] Christine Ducourant,[2] Gael Chauvin,[3] Alberto G. O. Krone–Martins,[1] Mickael Bonnefoy[3] and Inseok Song[4]

[1]Instituto de Astronomia, Geofísica e Ciências Atmosféricas, Universidade de São Paulo, Rua do Matão 1226, Cidade Universitária, São Paulo, Brazil
email: teixeira@astro.iag.usp.br

[2]Laboratoire d'Astrophysique de Bordeaux, Université de Bordeaux I, France

[3]Laboratoire d'Astrophysique, Observatoire de Grenoble, France

[4]Department of Physics & Astronomy, University of Georgia, GA USA

Abstract. Using the *ESO NTT*/SUSI2 telescope, we observed TWA22AB during five different observing runs over 1.2 years to measure its trigonometric parallax and proper motion. HARPS at the *ESO 3.6m* telescope was also used to measure the system's radial velocity over 2 years. Based on trigonometric-parallax, proper-motion and radial-velocity measurements, we re-analyzed the membership of TWA22AB of the young, nearby associations TW Hydrae, β Pictoris and Tucana–Horologium.

Keywords. stars: distances, open clusters and associations: general, stars: individual (TWA22AB), binaries: general

1. Introduction

TWA22AB was initially regarded as a member (Song *et al.* 2003) of the TW Hydrae association (TWA; Kastner *et al.* 1997). In addition to being one of the youngest ($\approx$ 8 Myr) and nearest ($\approx$ 20 pc) stars to Earth, TWA22AB has proved to be very interesting once it was resolved as a tight, very-low-mass binary (Bonnefoy *et al.* 2009). This binary can serve as a very useful dynamical calibrator for pre-main-sequence evolutionary models (Bonnefoy *et al.* 2009). However, its membership of the TWA has been questioned recently (Mamajek 2005) due to a lack of accurate kinematic measurements.

Based on trigonometric-parallax, proper-motion and radial-velocity measurements, we aim here to re-analyze the membership of TWA22 of the young, nearby associations TWA, β Pictoris (β Pic; Barrado y Navascués *et al.* 1999) and Tucana–Horologium (Tuc–Hor; Zuckerman & Webb 2000; Torres *et al.* 2000).

We report our analysis based on several complementary criteria, resulting in a greater chance of TWA22AB being a member of β Pic without excluding the possibility to be a member of the TWA because of large uncertainties in the data of its members. In any case, the probability that it belongs to one of the young and nearby known associations is low (Teixeira *et al.* 2009).

2. Astrometric parameters for TWA22AB and radial velocity

Using the *ESO NTT*/SUSI2 telescope, we observed TWA22AB during five different observing runs over 1.2 years to measure its trigonometric parallax and proper motion. This is part of a larger project to measure trigonometric parallaxes and proper motions of most known TWA members at submilliarcsec level (Ducourant *et al.* 2008). HARPS at the *ESO 3.6m* telescope was also used to measure the system's radial velocity over 2 years.

Table 1. Astrometric parameters and radial velocity for TWA22AB.

RA h m s	Dec ° ′ ″	π mas	d pc	$\mu_\alpha \cos(\delta)$ mas yr^{-1}	μ_δ mas yr^{-1}	$V_{\rm rad}$ km s^{-1}
10 17 26.79	−53 54 26.5	57.0±0.7	17.5±0.2	−175.8±0.8	−21.3±0.8	14.8±2.1

3. Membership analysis

To test the membership of TWA22AB, our first approach is to statistically compare its Galactic space motion with the mean *UVW* values for TWA, β Pic, and Tuc–Hor members. Only stars with known trigonometric parallaxes were considered. For TWA, our sample set includes five members. For β Pic, we used 26 and for Tuc–Hor 18 members (Teixeira *et al.* 2009).

The calculated mean space motions of TWA, β Pic and Tuc–Hor are reported in the table below, together with the spatial heliocentric coordinates and velocities for TWA22AB. The mean values derived here for the associations are in good agreement with published values.

Table 2. Heliocentric coordinates (X, Y, Z) and Galactic velocities (U, V, W) of TWA22 AB, together with mean values for TWA, β Pic and Tuc–Hor. Positive $X(U)$ points to the Galactic center, $Y(V)$ is positive in the direction of Galactic rotation and $Z(W)$ is positive towards the north Galactic pole.

Name	X	Y	Z	U	V	W	Age	d
		pc			km s^{-1}		Myr	pc
TWA22	3.5 ±0.2	−17.2±0.2	0.7±0.2	−8.0±0.4	−17.1±2.1	−9.0±0.2	⩽10(1)	17.5±0.2
TWA	14.4±8.7	−46.9±5.3	22.5±2.5	−10.0±2.0	−17.6±1.4	−4.8±1.1	8(2)	55±7
β Pic	9.2±28.9	−7.5±13.0	−13.2±6.5	−10.9±1.9	−16.1±1.0	−9.0±1.2	12 (3)	34±14
Tuc–Hor	14.1±17.7	−19.4±8.0	−34.9±3.5	−9.5±1.7	−20.6±1.7	−0.6±2.6	27 (4)	46±5

To test the membership of TWA22AB of TWA, β Pic and Tuc–Hor based on its space motion, we applied a χ^2 test with three degrees of freedom, using their space-motion measurements. We find that the probabilities that TWA22AB's space motion is compatible with the mean space motion of β Pic, TWA and Tuc–Hor are 50, 1 and 0.5%, respectively. An alternative approach is to perform a k–NN analysis in *UVW* space, where we have computed the distance of TWA22AB to all members of these associations. Among the k nearest neighbours to TWA22AB, the fraction of members for a given association gives the membership probability to that group. This k–NN analysis corroborates that TWA22AB is more likely a member of β Pic than of TWA or Tuc–Hor. Both calculations tend to reject TWA and Tuc–Hor as a host association for TWA22AB.

Another approach is to count the number of TWA, β Pic and Tuc–Hor members (or 'neighbours') within a sphere of fixed radius in *UVW* space, centered on TWA22AB's value and on each individual member of these associations. Radii are selected proportional to the velocity dispersions of the associations. We found that the *UVW* spatial densities of TWA and Tuc–Hor members around TWA22AB are significantly smaller than the average for both associations. On the other hand, the *UVW* spatial density of β Pic members around TWA22AB is similar to the average density in β Pic, as shown in Figure 1.

Finally, another way to test the membership of TWA22AB of TWA, β Pic and Tuc–Hor is to use the trace-back technique to compare the Galactic space position of TWA22 AB

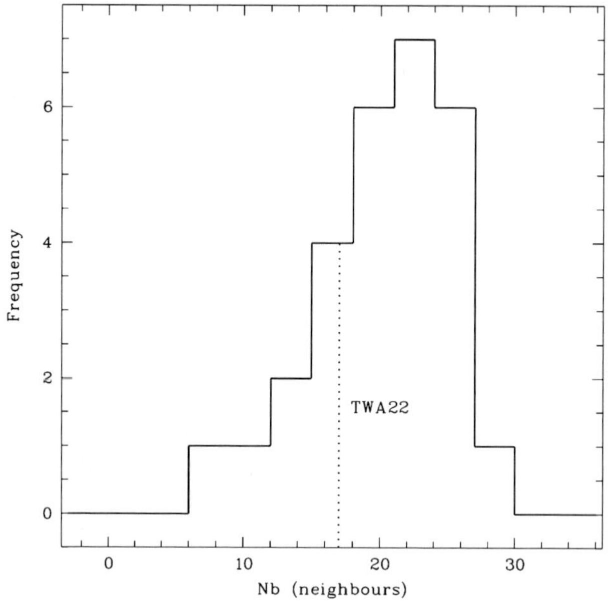

Figure 1. Number of β Pic members in a sphere of 2σ radius in the UVW space around each member. (σ: velocity dispersion for this association.)

and TWA, β Pic and Tuc–Hor backwards in time. To avoid using any uncertain values for the solar peculiar motion by transforming heliocentric velocities into local standard-of-rest velocities, we decided to work in a reference system centered, along time, on TWA22AB instead. We present, in Figure 2, the distance between TWA22AB and the center of each association in time. We note that TWA22AB is always closer to β Pic than to TWA or Tuc–Hor.

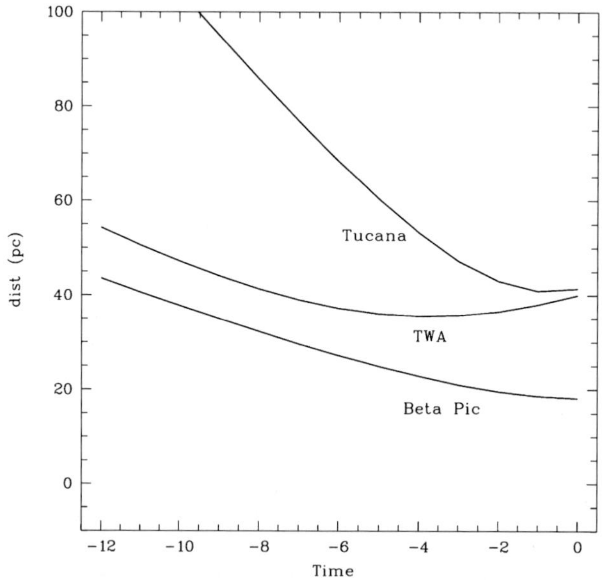

Figure 2. Distances between TWA22AB and three moving groups (TWA, β Pic and Tuc–Hor) for the past 12 Myr.

4. Conclusions

Motivated by the importance of the young, very-low-mass astrometric binary TWA 22AB as an important calibration point for stellar theoretical calculations, we measured its precise trigonometric parallax (57.0±0.7 mas), proper motions (-175.8 ± 0.8 mas yr^{-1}, -21.3 ± 0.8 mas yr^{-1}), and radial velocity ($V_{\rm rad} = 14.8\pm2.1$ km s^{-1}). These parameters are fundamental to determine the physical properties of the tight binary system.

Our high-quality astrometric measurements along with HARPS radial-velocity measurements allow us to discuss the membership of TWA22 AB of nearby associations. Our kinematic study shows that membership of TWA22AB of known young, nearby associations can be excluded, except for the β Pictoris and TW Hydrae associations. Membership probabilities based on the system's space motion or the use of the trace-back technique also support a possible membership of TWA22AB of the β Pictoris association. Membership of the TW Hydrae association cannot be fully excluded due to the current lack of precise parallax measurements for most of its members.

Our results are, to some extent, inconclusive as regards the membership of TWA22AB of TWA or β Pic, but they are consistent with that from age analysis. The location of TWA22AB in a color–magnitude diagram supports an age of about 10 Myr, but cannot be determined precisely enough to distinguish between 8 (TWA age) and 12 Myr (β Pic moving group age). Precisely known trigonometric distances of many more TWA members, an aim of our larger astrometric program to observe all known TWA members, should soon improve this situation.

References

Barrado y Nevascués, D., Stauffer, J. R., Song, I., & Caillault, J. P. 1999, *ApJ* (Letters), 520, L123

Bonnefoy, M., Chauvin, G., Dumas, C., Lagrange, A.-M., Beust, H., Desort, M., Texeira, R., Ducourant, C., Beuzit, J.-L., & Song, I. 2009, *A&A*, 506, 799

Ducourant, C., Teixeira, R., Krone–Martins A., Le Campion, J. F., & Chauvin, G. 2008, *RMxAC*, 34, 29

Kastner, J. H., Zuckerman, B., Weintraub, D. A., & Forveille, T. 1997, *Science*, 277, 67

Mamajek, E. 2005,*ApJ*, 634, 1385

Song, I., Zuckerman, B., & Bessel, M. S. 2003, *AJ*, 599, 342

Song, I., Schneider, G., Zuckerman, B., Farihi, J., Becklin, E. E., Bessell, M. S., Lowrance, P., & Macintosh, B. A. 2006, *ApJ*, 652, 724

Teixeira, R., Ducourant, C., Chauvin, G., Krone–Martins, A. G. O., Bonnefoy, M., & Song, I. 2009, *A&A*, 503, 281

Torres, C. A. O., da Silva, L., Quast, G. R., de la Reza, R., & Jilinski, E. 2000, *AJ*, 120, 1410

Zuckerman, B., & Webb, R. A. 2000, *ApJ*, 535, 959

Star clusters: basic galactic building blocks
Proceedings IAU Symposium No. 266, 2009
R. de Grijs & J. R. D. Lépine, eds.

doi:10.1017/S1743921309992018

SACY: open clusters connected to young nearby associations

C. A. O. Torres,[1] G. R. Quast,[1] C. H. F. Melo,[2] M. Sterzik,[2] R. de la Reza[3] and Licio da Silva[3]

[1]Laboratório Nacional de Astrofísica/MCT, Brazil
[2]ESO, Santiago, Chile
[3]Observatório Nacional/MCT, Brazil

Abstract. In recent years. we have spectroscopically observed thousands of counterparts of selected *ROSAT* bright sources in the SACY project (Search for Associations Containing Young stars). We demonstrated that hundreds of the young stars found this way belong to nearby young, loose associations, some detected only in SACY, with ages in the range from 5 to 100 Myr. For their ages, these associations show no trace of their primitive nurseries, which poses some difficulties as to their origins. Nevertheless, there are some associations clearly associated with young open clusters, such as the ε Cha association (6 Myr) with the η Cha cluster, and the Argus association with IC 2391 (40 Myr). There seem to exist also more subtle connections, such as the AB Dor association (75 Myr) with the Pleiades. Here, we present evidence of the connections of young associations to open clusters, to understand their origin and their possible insertion in the cluster infant-mortality and Gould Belt scenarios.

Keywords. open clusters and associations: individual (η Cha, ε Cha, IC 2391, Argus association), solar neighborhood, stars: pre–main-sequence, stars: kinematics, surveys

---

The full poster (in pdf format) is available at
`http://www.astro.iag.usp.br/~iaus266/Posters/pTorres.pdf`.

Star clusters: basic galactic building blocks
Proceedings IAU Symposium No. 266, 2009
R. de Grijs & J. R. D. Lépine, eds.

doi:10.1017/S174392130999202X

Pre-main-sequence stars in the stellar association N 11 in the Large Magellanic Cloud: clustering properties

Antonella Vallenari, Rosanna Sordo and Emanuela Chiosi

INAF, Padova Observatory, Vicolo Osservatorio 5, 35122 Padova, Italy
email: antonella.vallenari@oapd.inaf.it

Abstract. Magellanic Clouds are of extreme importance to study the star-formation process in low-metallicity environments. Here, we discuss the clustering properties of the pre-main-sequence candidates and young embedded stellar objects in N 11, located in the Large Magellanic Cloud. Deep archival *HST*/ACS photometry is used to derive color–magnitude diagrams of the associations in N 11 and of the foreground field population. These data are complemented by archival infrared *Spitzer* data which allow detection of young embedded stellar objects. The spatial distribution of the pre-main-sequence candidates and young embedded stellar objects is discussed. The degree of clustering is derived using the minimal-spanning-tree method. No significant difference is found in clustering degree of young blue main-sequence stars and faint pre-main-sequence candidates, suggesting that they might be part of the same formation process.

Keywords. Magellanic Clouds, stars: pre–main-sequence, stars: formation

---

1. Introduction

The Magellanic Clouds are ideal laboratories to study the process of star formation in detail owing to their proximity. In the Large Magellanic Cloud (LMC), several regions of active star formation are found. N 11 is the second largest nebula in the LMC after the 30 Doradus nebula. It is located in the north-western corner of the LMC. Bica *et al.*'s (1999) catalog lists 49 clusters and associations in the entire region of N 11. In the area studied, about 13 objects are found (see Figure 1). This region is often presented as one of the best examples of triggered star formation (Mokiem *et al.* 2007). The basic idea is that the association LH 9 has triggered star formation in LH 10 and LH 13. Here, we make use of *HST* ACS/WFC archival data complemented by *Spitzer* archival data (SAGE project; Meixner *et al.* 2006). While *HST* observations can provide information about faint, exposed pre-main-sequence candidates (PMSs), infrared data allow to detect embedded young stellar objects (YSOs). This paper is part of a project aiming to cast light on the process of field and cluster star formation in the Magellanic Clouds (see, among others, Chiosi & Vallenari 2006, 2007).

2. PMS detection in N 11

A large population of pre-main-sequence candidates is found in N 11. Their masses are in the range 1.3–2 $M_\odot$ and they are compatible with having ages from 2 to 10 Myr. YSOs with ages between 0.1 and 1 Myr are found intermixed with PMSs. More details can be found in Vallenari (2008) and Vallenari *et al.* (2009). Here, we summarize the most relevant results.

2.1. *Age determination*

Our analysis of the *HST* data implies that PMSs in LH 9 are slightly older than elsewhere in N 11. Prolonged star formation is detected in LH 9 South, with ages from 2 to 10 Myr, while LH 9 North, LH 10 (center) and LH 13 seem to be younger (with mean ages of 5, 2 and 2 Myr, respectively). A large age spread from 1 to 6 Myr is found based on spectroscopy of blue stars in these regions. The presence of YSOs in LH 9 South, LH 13 (type I, with ages < 0.1 Myr) and LH 9 (North), LH 10 (Type II, ages $< 1-2$ Myr) suggests that star formation in the region is still ongoing. Mokiem *et al.* (2007) estimate that the time needed for a supernova shock to cross the distance between LH 9 and LH 10 is on the order of 0.1 Myr. Since the lifetime of massive stars is on the order of 3 Myr, the age difference is in agreement with the idea that LH 9 has triggered the formation of a first stellar generation in LH 10. This cannot be excluded on the basis of our discussion, even if more generations of stars are found in the region.

2.2. *Spatial distribution*

We compare the location of PMS candidates selected on the basis of ACS photometry with the Hα-emission map of Mac Low *et al.* (1998), with the CO cloud distribution (Israel *et al.* 2003) and with Herbig Ae/Be-star concentrations (Hatano *et al.* 2006). The most conspicuous concentrations of PMSs correspond to the locations of OB stars. As for Herbig Ae/Be stars, they are found in all these locations, confirming the presence of a population with ages in the range of 1–3 Myr (see Figure 1). YSOs are found inside or close to the CO clouds.

3. Clustering properties of the PMSs and the field population

Interstellar turbulence is found to be very efficient in sweeping up molecular gas forming massive structures, which in turn can undergo a large-scale gravitational collapse. As a consequence, clusters are formed in a hierarchical fashion with subclusters which eventually merge to build up the final, condensed object (Schmeja *et al.* 2009). For this reason, the clustering properties of the stars in a region can provide information about the formation and evolution of the stellar structures. A powerful method used in the literature for statistical analysis of the spatial distribution of the stars in a cluster is the minimal spanning tree (MST). The MST is the unique network of straight lines joining a set of points such that it minimizes the total length of all lines without creating closed loops. Schmeja & Klessen (2006) demonstrate that the structure of a cluster can be derived using the parameter $Q = m/s$, comparing the mean length, m, of its MST to the normalized correlation length, s, defined as the mean separation between the stars. Table 1 shows the Q values for the region studied. In all N 11 fields, the values of $Q < 0.8$ are typical of the products of hierarchical star formation. Different clusters/associations have slightly different values of Q, suggesting that they might be at slightly different evolutionary stages. Q is found to be independent of the age of the clusters. This result is not in agreement with the behavior found in Galactic star-forming regions (Schmeja *et al.* 2009). These results are quite difficult to interpret. They can be explained as the footprints of inhomogeneity in the original turbulent cloud. However, it should be kept in mind that this comparison is complicated by the presence of a large age dispersion detected in the fields.

Inside N 11, no significant difference is found in the clustering degree of brighter blue main-sequence stars ($m_{\rm F435W} < 20$ mag) and fainter red stars ($m_{\rm F435W} > 20$ mag), suggesting that they might be part of the same formation process.

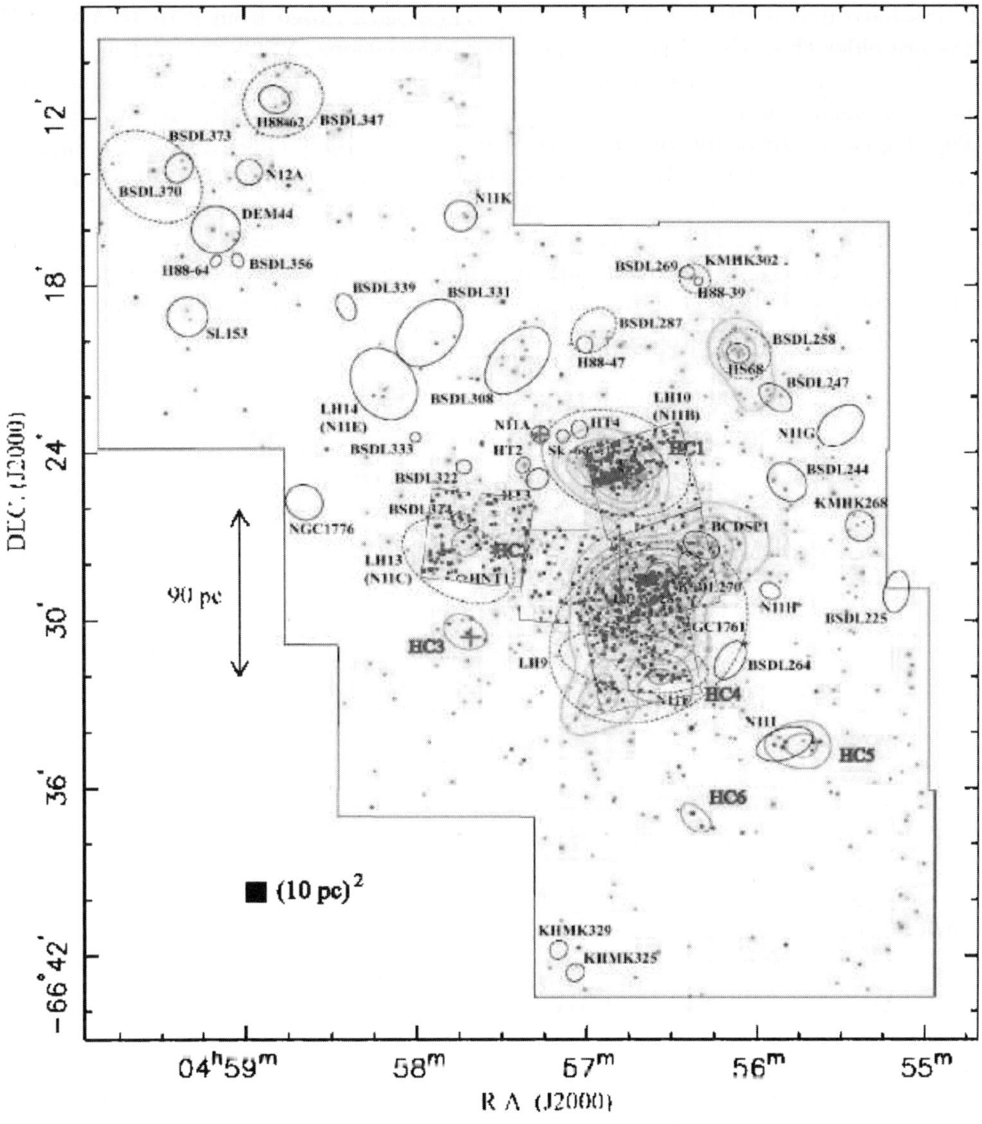

Figure 1. Map of N 11, where the clusters/associations from Bica et al.'s (1999) catalog are indicated. Red circles (marked HC1 to HC6) and cyan contours show the Herbig Ae/Be and OB candidates, respectively (from Hatano *et al.* 2006). Large boxes show the ACS/WFC fields studied in this paper. Small red dots show the PMS candidates selected on the basis of ACS photometry.

4. Conclusions

In this paper, we discuss the formation process of clusters and stars in N 11, and analyze the age and the clustering properties of the clusters/associations found in the region.

PMS and YSO candidates associated with N 11 are discovered based on *HST* ACS/WFC photometry and archival *Spitzer* data. While *HST* observations can provide information about faint, exposed pre-main-sequence candidates, infrared data allow us to detect embedded young stellar objects.

PMSs have masses from 1.3 to 2.0 $M_{\odot}$. Their ages range from 2 to 10 Myr. LH 9 PMSs are older than the objects in the other associations. YSOs (types I and II) with ages < 1 Myr and Herbig Ae/Be stars with ages in the range 1–3 Myr are found in the same locations as the PMSs.

The degree of clustering of the stars is discussed to get clues about the formation process. PMSs exhibit a higher degree of clustering in comparison with the full stellar sample. For all clusters/associations in N 11, the values of $Q \leqslant 0.8$ indicate hierarchical star formation. In turbulent regimes, it is expected that the clusters evolve from hierarchical structures towards more concentrated configurations. This behavior is not found in the region studied, where Q shows no age dependence. This might reflect inhomogeneity in the original turbulent cloud.

Table 1. Q values for all stars in the cluster areas given by Bica *et al.*

Object	$\bar{m}$	$\bar{s}$	Q
NGC 1769	0.86	1.33	0.65
NGC 1763	0.73	0.98	0.75
NGC 1761	0.77	1.18	0.66
NGC 1760	0.73	1.20	0.56
HD 32228	0.73	0.93	0.79
BSDL 324	0.80	1.11	0.73
BSDL 270	2.45	3.64	0.67
BCDSP 1	0.78	1.19	0.66

References

Bica, E. L. D., Schmitt, H. R., Dutra, C. M., & Oliveira, H. L. 1999, *AJ*, 117, 238
Chiosi, E. & Vallenari, A. 2006, *A&A*, 452, 179
Chiosi, E. & Vallenari, A. 2007, *A&A*, 466, 165
Hatano, H., *et al.* 2006, *AJ*, 132, 2653
Israel, F. P., de Graauw, T., Johansson, L. E. B., Booth, R. S., Boulanger, F., Garay, G., Kutner, M. L., Lequeux, J., Nyman, L.-A., & Rubio, M. 2003, *A&A*, 401, 99
Mac Low, M. M., Chang, T. H., Chu, Y.-H., Points, S. D., Smith, R. C., & Wakker, B. P, 1998, *ApJ*, 493, 260
Mokiem M. R., *et al.* 2007, *A&A*, 465, 1003
Meixner, M., *et al.* 2006, *AJ*, 132, 2268
Schmeja, S. & Klessen, R. S. 2006, *A&A*, 449, 151
Schmeja, S., Gouliermis, D. A., & Klessen, R. S. 2009, *ApJ*, 694, 367
Vallenari, A., 2008 *MmSAI*, 79, 448
Vallenari, A., Sordo, S., & Chiosi, E. 2009, *A&A*, submitted

Star clusters: basic galactic building blocks
Proceedings IAU Symposium No. 266, 2009
R. de Grijs & J. R. D. Lépine, eds.

doi:10.1017/S1743921309992031

Bridging galactic star formation from intermediate to local epochs

Elysse N. Voyer,[1,2] Duilia F. de Mello,[1,2] Jonathan P. Gardner,[2] Brian Siana[3] and Harry Teplitz[4]

[1]The Catholic University of America, Washington, DC, USA

[2]NASA's Goddard Space Flight Center, Greenbelt, MD, USA

[3]California Institute of Technology, Pasadena, CA, USA

[3]NASA/IPAC Infrared Science Archive, Pasadena, CA, USA

Abstract. Starburst galaxies trace the star-formation history of the Universe throughout cosmological time. Several studies have shown that the star-formation-rate density of the Universe begins to steadily decline after the Universe had aged $\sim$ 5.8 Gyr (redshifts $z \sim 1$). However, we do not yet fully understand the mechanism behind this shift in star formation at intermediate z. One possibility is that during this epoch galaxies underwent 'downsizing,' a shift in star formation being dominated by high- to lower-mass galaxies. Rest-frame ultraviolet (UV) observations of starburst galaxies reveal regions of young stellar clusters where massive O and B stars dominate the luminosity. Observations in the FUV ($\sim$ 1500 Å) can be used to detect starburst galaxies at $z < 1$ because the bright end of the rest-frame FUV spectrum is not redshifted much, and is observable in the FUV filter. Alternatively, the rest-frame FUV light from starburst galaxies at higher redshifts is shifted to longer wavelengths and must be observed in redder filters. We present a study of the starburst population at intermediate z from FUV data taken with *Hubble*'s Solar Blind Channel (SBC) of the Advanced Camera for Surveys (ACS). The number counts of FUV galaxies as a function of magnitude provide a direct statistical measure of the density and evolution of starbursts, and subsequently of the stellar clusters formed within these galaxies' environments. We present a comparison between the FUV starburst-galaxy counts at this epoch, and the local FUV counts of starbursts observed with *GALEX*.

Keywords. galaxies: evolution, galaxies: starburst, galaxies: statistics

---

The full poster (in pdf format) is available at
`http://www.astro.iag.usp.br/~iaus266/Posters/pVoyer1.pdf`

Star clusters: basic galactic building blocks
Proceedings IAU Symposium No. 266, 2009
R. de Grijs & J. R. D. Lépine, eds.

doi:10.1017/S1743921309992043

Probing star formation at intermediate z

E. N. Voyer,[1,2] D. F. de Mello,[1,2] C. Quirk,[1] B. Siana,[3] J. P. Gardner[2] and H. Teplitz[4]

[1]The Catholic University of America, Washington, DC, USA
[2]NASA Goddard Space Flight Center, Greenbelt, MD, USA
[3]California Institute of Technology, Pasadena, CA, USA
[3]NASA/IPAC Infrared Science Archive, Pasadena, CA, USA

Abstract. We summarize the latest results of an ongoing project aimed at connecting starburst galaxies in the local Universe and Lyman-break galaxies (LBGs) in the distant Universe using rest-frame ultraviolet (UV) images. We are quantifying star formation in the local Universe using *GALEX* data of interacting galaxies and we are using *Hubble Space Telescope U*-band and optical images of the Ultra Deep Field to quantify star formation at intermediate ($z \sim 1$) and high ($z > 2$) redshifts, respectively. We are measuring sizes of star-forming clumps in all redshift ranges and searching for evolutionary effects. We chose the rest-frame UV to do this work because young and massive stars radiate most their energy in the UV and also because LBGs are selected to be UV-bright. We discuss whether all galaxies go through an LBG phase and whether disks show any sign of size of evolution at $z < 1$.

Keywords. galaxies: evolution, galaxies: starburst

The full poster (in pdf format) is available at
`http://www.astro.iag.usp.br/~iaus266/Posters/pVoyer2.pdf`.

Star clusters: basic galactic building blocks
Proceedings IAU Symposium No. 266, 2009
R. de Grijs & J. R. D. Lépine, eds.

doi:10.1017/S1743921309992055

The massive star-forming region Cygnus OB2

N. J. Wright and J. J. Drake

Harvard–Smithsonian Center for Astrophysics, 60 Garden Street, Cambridge, MA 02138, USA

Abstract. We present results from a catalogue of 1696 X-ray point sources detected in the massive star-forming region Cygnus OB2, the majority of which have optical or near-infrared associations. We derive ages of 3.5 and 5.25 Myr for the stellar populations in our two fields, in agreement with recent studies that suggest that the central 1–3 Myr-old OB association is surrounded and contaminated by an older population with an age of 5–10 Myr. The fraction of sources with protoplanetary disks, as traced by K-band excesses, is unusually low. Although this has previously been interpreted as due to the influence of the large number of OB stars in Cyg OB2, contamination from an older population of stars in the region could also be responsible. An initial mass function is derived and found to have a slope of $\Gamma = -1.27$, in agreement with the canonical value. Finally, we introduce the recently approved *Chandra* Cygnus OB2 Legacy Survey that will image a 1 square degree area of the Cygnus OB2 association to a depth of 120 ks, likely detecting $\sim 10\,000$ stellar X-ray sources.

Keywords. open clusters and associations: individual (Cygnus OB2), stars: pre–main-sequence, X-rays: stars

1. Introduction

Massive star-forming regions (SFRs) are thought to be the major star factories in the Universe (Elmegreen 1985), each containing hundreds to thousands of OB stars and millions of low-mass stars. Understanding how these regions form and evolve is vital to our comprehension of the factors controlling the stellar initial mass function and the structure and evolution of our Galaxy. Incisive studies of massive SFRs are hindered by their scarcity and great distances, resulting in both an inability to probe the low-mass end of the stellar spectrum and insufficient spatial resolution to diagnose the physical processes at work. Therefore, it is important to make full use of the very few nearby opportunities we have to pursue these issues. Cygnus OB2 is one such region, at a distance of only 1.45 kpc (Hanson 2003), and is one of the most massive OB associations in our Galaxy with a rich and diverse stellar population (e.g., Massey & Thompson 1991). X-ray observations are highly effective in separating young association members from older field stars because pre-main-sequence (pre-MS) stars are thousands of times more luminous in X-rays than MS stars (Preibisch & Feigelson 2005). To penetrate the high extinction in the direction of Cygnus OB2 as well as removing the contamination from Galactic field stars, we have used *Chandra* observations of Cyg OB2 to study this massive star-forming region on our doorstep.

2. Observations

We have combined X-ray observations of two *Chandra* fields in the center of Cygnus OB2 (Figure 1) with optical and near-infrared (near-IR) data from the *INT* Photometric

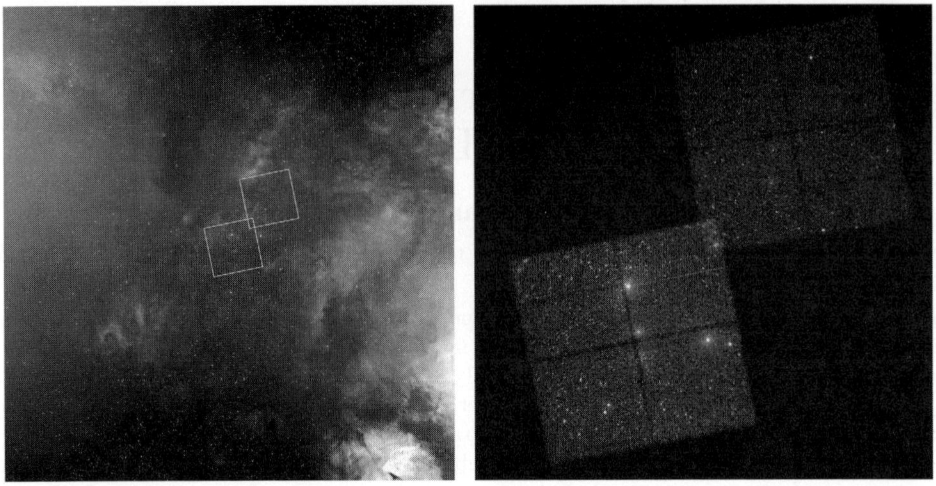

Figure 1. *(left)* Hα image of the Cygnus OB2 region mosaicked from IPHAS observations and displayed using a logarithmic intensity scale. The image size is 2.67×2.85 square degrees and centered on (RA, Dec) = (20:33:00, +41:15:00) with north up and east to the left. The $17' \times 17'$ fields of view of the two *Chandra*/ACIS observations discussed in this paper are shown as white squares. *(right)* Grey-scale image (approximately $30' \times 30'$) of the two *Chandra*/ACIS-I observations of Cygnus OB2. Image intensity is proportional to the logarithm of the X-ray count density.

Hα Survey (IPHAS; Drew *et al.* 2005), the 2 Micron All Sky Survey (Cutri *et al.* 2003), and the United Kingdom Infrared Deep Sky Survey (Lawrence *et al.* 2007) surveys. Our catalogue (Wright & Drake 2009) contains 1696 X-ray sources with broad-band fluxes, spectral fits, and variability information, of which 1501 (89%) have optical or near-IR counterparts. From an analysis of the Poissonian false-source probabilities, we estimate that this catalogue contains $< 1\%$ of false sources, and an even lower fraction when only sources with optical or near-IR associations are considered. A Monte Carlo simulation of the Bayesian cross-matching scheme also allowed the various sources of error in this process to be quantified. IPHAS photometry was employed to remove contamination from foreground stars, $\sim 5\%$ of the catalogue. Simulations of the extragalactic X-ray population were used to estimate its impact on the catalogue. The majority of extragalactic sources are not expected to have optical or near-IR counterparts due to the high Galactic extinction in Cygnus.

3. Near-IR stellar properties

The near-IR color–color and color–magnitude (Figure 2) diagrams were used to study the stellar populations in the two *Chandra* fields. Pre-MS stellar isochrones (Siess *et al.* 2000) were used to estimate mean ages of 3.5 and 5.25 Myr for the two fields. Considerable spread was observed around the isochrones, despite the small photometric errors, potentially suggesting a large age spread in the region. These results are in agreement with recent studies (e.g., Comeron *et al.* 2008; Drew *et al.* 2008) that suggest that the central association of OB stars with an age of 1–3 Myr is surrounded and contaminated by an older population with an age of $\sim 5 - 10$ Myr. Protoplanetary disks are traced using the near-IR K-band excess, although only 5.8% (central field) and 7.6% (north-western field) of sources are found to have inner disks. These low disk fractions could be attributed to the disrupting effects of the strong stellar winds and ultraviolet radiation

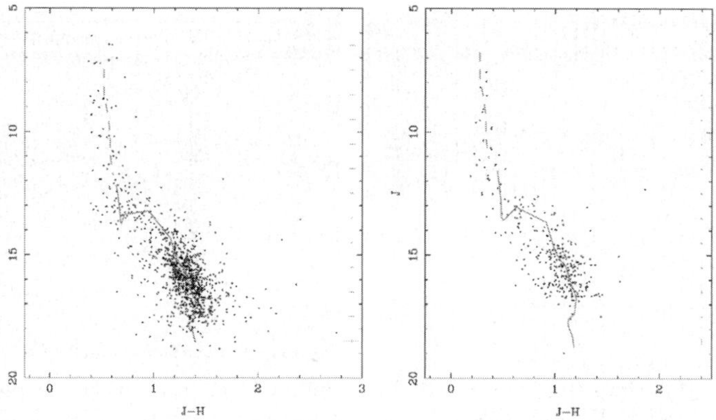

Figure 2. Near-IR color–magnitude diagrams for sources in the central *(left)* and north-western *(right)* fields. Zero-age MS (dashed red line) and pre-MS (full red line) tracks are shown. Mean photometric errors per magnitude bin are shown on the right of each figure.

from the OB stars in the vicinity, or they may be due to the influence of older stellar populations in the region.

4. X-ray stellar properties

The X-ray luminosity function (XLF) of *Chandra* sources was compared to that of the well-studied Orion Nebula Cluster (Getman *et al.* 2005). We find good agreement between the two, suggesting we are not missing any significant component. Within the limits of X-ray observations (i.e., that late-B and early A-type stars do not all emit X-rays), we find that we are complete to 1 $M_\odot$ in both fields. The distribution of median X-ray-source energies provides no evidence for a population of hard X-ray sources that would originate from highly embedded sources. This lack of embedded sources suggests that star formation in these two fields has ceased.

5. The initial mass function

Using masses derived from existing spectroscopy and fits to the near-IR CMD, we derived stellar masses for all our sources. The resulting mass function (Figure 3), complete to 1 $M_\odot$, has a slope of $\Gamma = -1.27 \pm 0.16$, in good agreement with the canonical value (Kroupa 2002). There is no evidence for a shallow mass function or a high-mass turnover in this mass range, as has been predicted for some massive star-forming regions. There is only a small difference in slope between the two fields (Figure 3), suggesting that significant mass segregation has yet to occur despite the age of the region. The mass function is steeper at higher masses with $\Gamma = -2.19 \pm 0.07$, which could indicate a steepening of the mass function or could be due to contamination by a population of older stars in the region that have lost their highest-mass members.

6. The *Chandra* Cygnus OB2 Legacy Survey

These observations cover only a very small region in the centre of Cygnus OB2, which Knödlseder (2000) suggest may be as large as 2 degrees wide. The recently approved *Chandra* Legacy Survey of Cygnus OB2 (PI J. J. Drake) will extend these observations to a 1 degree-wide region (Figure 5) with a depth of 120ks, achieving a completeness down to 0.5 $M_\odot$ and detecting $\sim 10\,000$ cluster members. These observations, combined

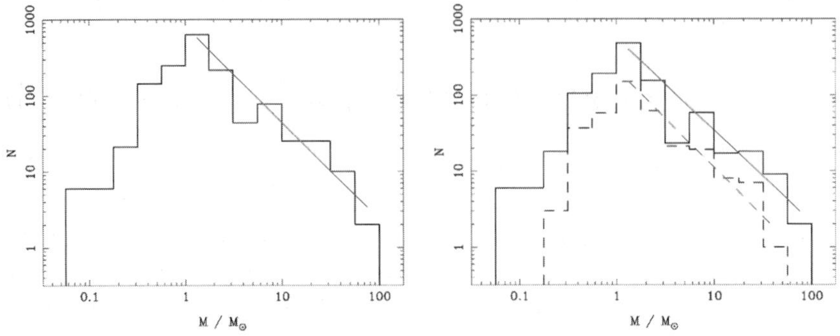

Figure 3. Mass functions for sources in both fields *(left)* and in the separate fields (*right*; central field: full line, NW field: dashed line). Power-law fits are shown in red, excluding the bins containing A- and B-type stars. The power-law indices are $\Gamma = -1.27 \pm 0.16$ (combined fields), $\Gamma = -1.22 \pm 0.16$ (central field) and $\Gamma = -1.29 \pm 0.16$ (NW field).

with existing (e.g., *VLA*, IPHAS, UKIDSS, *Spitzer*) and approved (*e-Merlin*, *VERITAS*, *Herschel*) surveys of the region, will provide a huge dataset for deep studies of star and planet formation in a massive star-forming region.

7. Conclusions

First results are presented from the recently published catalogue of X-ray sources in the massive star-forming region Cygnus OB2 (Wright & Drake 2009). The catalogue contains 1696 X-ray point sources, of which 1501 (89%) have optical or near-IR counterparts. The XLF of sources indicates that, within the limits of an X-ray-selected sample, we are complete to $\sim$1 $M_\odot$. The near-IR color–magnitude diagram suggests a potential age spread across the region, and best-fitting ages of 3.5 and 5.25 Myr support recent suggestions of multiple epochs of star formation in Cygnus OB2. The mass function is measured and found to have a slope of $\Gamma = -1.27$, in agreement with the canonical value and providing no evidence for a shallower slope, as predicted for some massive SFRs.

References

Comerón, F., Pasquali, A., Figueras, F., & Torra, J. 2008, *A&A*, 486, 453
Cutri, R. M., *et al.*, 2003, *The IRSA* 2MASS *Point Source Catalog*
Drew, J. E., *et al.*, 2005, *MNRAS*, 362, 753
Drew, J. E., Greimel, R., Irwin, M. J., & Sale, S. E. 2008, *MNRAS*, 386, 1761
Elmegreen, B. G., 1985, in: D. C. Black & M. S. Matthews, *Protostars and Planets II*, p. 33, Tucson, Arizona: University of Arizona Press
Getman, K. V., *et al.* 2005, *ApJS*, 160, 319
Knödlseder, J., 2000, *A&A*, 360, 539
Kroupa, P., 2002, *Science*, 295, 82
Lawrence, A., *et al.*, 2007, *MNRAS*, 379, 1599
Hanson, M. M., 2003, *ApJ*, 597, 957
Massey, P. & Thompson, A. B., 1991, *AJ*, 101 1408
Preibisch, T. & Feigelson, E. D. 2005, *ApJS*, 160, 390
Siess, L., Dufour, E., & Forestini, M., 2000, *A&A*, 358, 593
Wright, N. J. & Drake, J. J., 2009, *ApJS*, 184, 84

Star clusters: basic galactic building blocks
Proceedings IAU Symposium No. 266, 2009
R. de Grijs & J. R. D. Lépine, eds.

doi:10.1017/S1743921309992067

Super star clusters and their emission lines

Richard Wünsch,[1] Jan Palouš,[1] Guillermo Tenorio–Tagle,[2] Sergiy Silich[2] and Casiana Muñoz–Tuñón[3]

[1]Astronomical Institute, Academy of Sciences of the Czech Republic, Prague, Czech Republic

[2]Instituto Nacional de Astrofísica Óptica y Electrónica, AP 51, 72000 Puebla, Mexico

[3]Instituto de Astrofísica de Canarias, E 38200 La Laguna, Tenerife, Spain

Abstract. Super star clusters (SSCs) are young massive objects typically observed in starburst galaxies. They consist of millions of stars occupying volumes only several parsecs across. Recent infrared observations of SSCs reveal strong emission lines with moderately supersonic widths (FWHM $\sim 50-100$ km s^{-1}). An additional, much broader and weaker line component is present in some cases. Using 2D and 3D hydrodynamic simulations, we study models of SSCs in the so-called *bimodal regime*, which occurs if the stellar density inside the cluster exceeds a certain limit. We confirm the existence of the bimodal solution predicted analytically and suggest an explanation for double-component line profiles: the narrow component is formed close to the cluster center by *repressurizing shocks* compressing the rarefied gas cooled down due to the thermal instability into dense cold clumps, while the broad component is created by the wind, which cools down at a certain distance from the cluster.

Keywords. galaxies: star clusters, HII regions, ISM: bubbles, ISM: lines and bands, hydrodynamics

---

The full poster (in pdf format) is available at
`http://www.astro.iag.usp.br/~iaus266/Posters/pWunsch.pdf`.

Star clusters: basic galactic building blocks
Proceedings IAU Symposium No. 266, 2009
R. de Grijs & J. R. D. Lépine, eds.

doi:10.1017/S1743921309992079

Blue stragglers in star clusters and the conventional SSP models

Yu Xin,[1] Richard de Grijs,[2,3] Licai Deng[3] and Pavel Kroupa[1]

[1]Argelander Institute for Astronomy, University of Bonn, Bonn, Germany

[2]Department of Physics and Astronomy, University of Sheffield, Sheffield, UK

[3]National Astronomical Observatories, Chinese Academy of Sciences, Beijing, China

Abstract. The presence of blue straggler stars (BSs) as secure members of star clusters poses a major challenge to the conventional picture of simple stellar population (SSP) models. The models are based on the stellar evolution theory of single stars, while the major formation mechanisms of BSs are all correlated with stellar interactions. Based on a sufficient working sample including 100 Galactic open clusters, one Galactic globular cluster, and seven Magellanic Cloud star clusters, we discuss the modifications of the properties of broad-band colors and Lick indices of the standard SSP models due to BS populations.

Keywords. blue stragglers, galaxies: star clusters

The full poster (in pdf format) is available at
`http://www.astro.iag.usp.br/~iaus266/Posters/pXin.pdf`.

Star clusters: basic galactic building blocks
Proceedings IAU Symposium No. 266, 2009
R. de Grijs & J. R. D. Lépine, eds.

doi:10.1017/S1743921309992080

IC 361, a distant intermediate-age cluster in Camelopardalis†

Justas Zdanavičius,[1] Richard P. Boyle,[2] Frederick J. Vrba,[3] Kazimieras Zdanavičius[4] and Stanislava Bartašiūtė[1]

[1]Astronomical Observatory of Vilnius University, Čiurlionio 29, Vilnius, LT-03100, Lithuania
email: `[justas, kz]@itpa.lt`

[2]Vatican Observatory Research Group, Steward Observatory, Tucson, AZ 85721, USA
email: `boyle@ricci.as.arizona.edu`

[3]US Naval Observatory Flagstaff Station, P.O. Box 1149, Flagstaff, AZ 86002, USA
email: `fjv@nofs.navy.mil`

[4]Institute of Theoretical Physics and Astronomy, Vilnius University, Goštauto 12, Vilnius, LT-01108, Lithuania

Abstract. We have undertaken CCD photometry in the eight-color *Vilnius* + I system for the open cluster IC 361 ($l = 147.5°$, $b = 5.7°$) located in the constellation of Camelopardalis. Based on multicolor data, estimates of distance moduli, foreground reddening and metallicity have been obtained for individual stars measured in the field of the cluster. This allowed us to eliminate most of the field stars and to identify the probable cluster members, from which we derived the true distance modulus $(m - M)_0 = 12.7$ mag and metallicity [Fe/H] = −0.3 dex. The interstellar extinction is found to be nonuniform across the field, with values of A_V ranging from 1.9 to 2.6 mag. A comparison of the color–magnitude diagrams with Padova isochrones yields $(m - M)_0 = 12.6$ mag, or a cluster distance of 3.3 kpc, and an age of 1.0 Gyr. Therefore, IC 361 appears to be a mildly metal-deficient cluster of intermediate age, located as far as, or just beyond, the Perseus spiral arm.

Keywords. techniques: photometric, open clusters and associations: individual (IC 361)

1. Introduction

The open cluster IC 361 is well detached from the field, but because of faintness it has been very poorly studied, if not completely forgotten. The only photometric information published to date is that by Piccirillo & Stein (1978), who obtained photoelectric UBV and photographic BV data for 19 and 32 stars in the field of the cluster, respectively, and identified the contours of the main sequence and a few red-giant stars. Based on these preliminary data they adopted a distance of 2.5 kpc, a high reddening value, $E(B - V) = 0.55$ mag, and suggested the age in the range from 0.5 to 1 Gyr. IC 361 lies in the second Galactic quadrant ($\ell = 147.5°$, $b = 5.7°$), in the immediate vicinity of the Camelopardalis dark clouds. Therefore, it is not surprising that the cluster could be subject to considerable interstellar reddening. If so, multicolor intermediate-band photometry can provide improved determinations of the fundamental cluster parameters.

† The full poster (in pdf format) is available at `http://www.astro.iag.usp.br/~iaus266/Posters/pZdanavicius.pdf`.

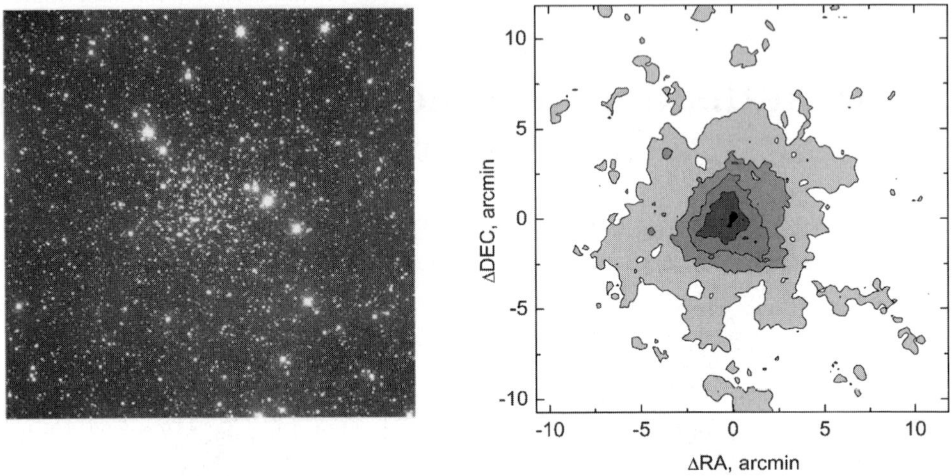

Figure 1. *(left)* Image of the field of IC 361 (20′ × 20′) in the I filter (Flagstaff 2K CCD). *(right)* Isodensity contours of stars measured in the I filter.

2. Observations and data reductions

CCD observations in seven filters ($UPXYZVS$) of the intermediate-band *Vilnius* system and in the I band were carried out in 1995 with a 2K×2K CCD camera on the 1 m telescope of the USNO Flagstaff Station, giving an unvignetted field of 20′ × 20′ (Figure 1, left panel). Additional observations of the central part of the field (12′ × 12′) in the *Vilnius* filters UYV were obtained in 2008 with a 4K CCD camera at the 1.8m *Vatican Advanced Technology Telescope (VATT)* on Mt. Graham (Arizona). Repeated exposures in the $UPXYZVS$ filters were also taken in March 2009 using the same 1 m telescope of the USNO Flagstaff Station, but these new observations have not yet been included in the present analysis.

The CCD data frames were reduced using the IRAF packages, by combining both the aperture and point-spread-function method. The calibration equations were obtained by observing the standard field of M67 (Laugalys *et al.* 2004). In the Flagstaff field, a total of 1450 stars brighter than $V = 18.5$ mag were measured in seven *Vilnius* filters, while in the I filter the number of stars measured reaches 7250. With the *VATT* telescope, a limiting magnitude of $V = 19.5$ mag was obtained with a good signal-to-noise ratio. The color–magnitude diagrams (CMDs), $V, V-Y$ and $V, V-I$, for stars common to the *VATT* and Flagstaff data sets are shown in Figure 2.

3. Results

The coordinates of the cluster center, corresponding to the density peaks in our photometry, are

$$\alpha = 4^{\rm h}18^{\rm m}55.0^{\rm s},\ \delta = +58^\circ 14' 53'' \ (2000).$$

According to the surface-density distribution of stars measured in the I filter, the cluster has a radius of 7′ and a fairly strong central concentration (Figure 1, right panel).

The classification of stars in terms of spectral class, absolute magnitude and interstellar reddening, $E(Y-V)$, was performed using the reddening-free Q parameters. In calculating the Qs, we used the color-excess ratios corresponding to the normal interstellar extinction law. Since we applied several classification methods, such as SIGMAQ, COMPAR, XQKLAS, and XDX, the values derived from an average of these classifications

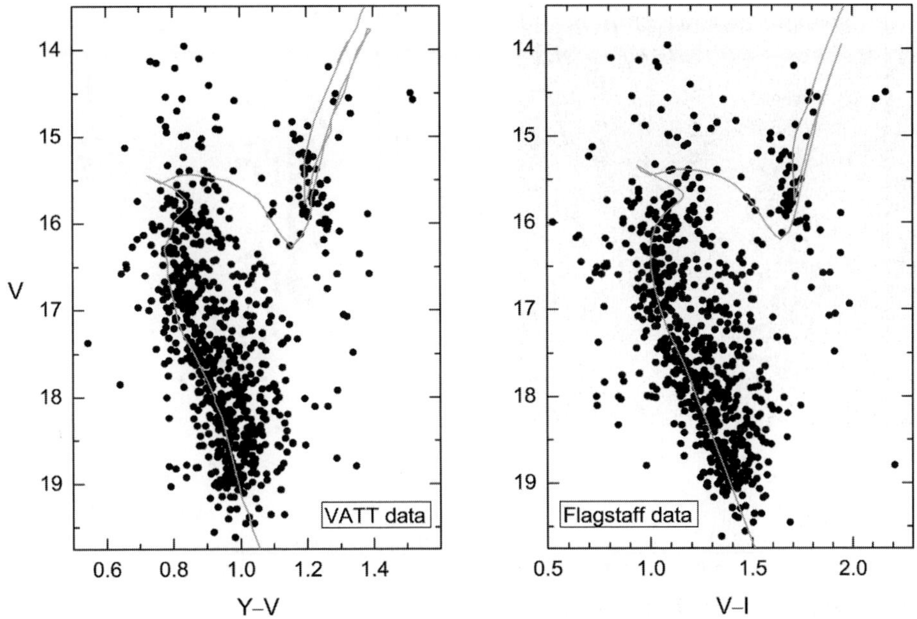

Figure 2. Uncleaned CMDs for stars common to the *VATT* and Flagstaff data sets. The Padova isochrones shown are for $Z = 0.010$, $\log(t/\text{yr}) = 9.0$, and $(m - M)_V = 14.7$ mag (see Figure 3); taken for $V, Y - V$ from Bressan & Tautvaišienė (1996) and for $V, V - I$ from Marigo *et al.* (2008).

were taken as final results. The methods applied give an accuracy for the spectral class on the order of ± 1 decimal spectral subclass. The accuracy of M_V depends on spectral type and varies from ± 0.2 to ± 0.5 mag. To calculate A_V from the derived color excesses $E(Y - V)$, a ratio of $R = 4.16$ was used.

Metallicities were determined only for the red-giant stars, using the dereddened colors $P - X$, $X - Y$, and $P - Y$, calibrated in terms of [Fe/H]. Most main-sequence stars are of spectral types earlier than F0, i.e., they fall outside the limits of the *Vilnius*-system calibration in [Fe/H].

The most probable photometric members of the cluster were selected by combining the following precepts:

(1) statistical removal of field stars using the V_0, Q_{XYZ} diagram;
(2) concentration of stars in the A_V versus $(m - M)_V$ diagram;
(3) consistency of the classification results obtained from different methods.

Within 7′ from the cluster center, the number of statistically identified members down to $I = 19.6$ mag amounts to ~ 800. However, combining all adopted criteria reduces the number of the most probable members to only 200. These stars were used to derive the main cluster parameters. By averaging the individual values we find

- the mean distance modulus $(m - M)_0 = 12.7 \pm 0.2$ mag,
- the mean metallicity [Fe/H]$= 0.3 \pm 0.1$ dex (only the red giants were used),
- the mean extinction $A_V = 2.1$ mag, with a clear indication of variable extinction ranging from 1.9 to 2.6 mag.

The CMDs of the probable members are shown in Figure 3, which give further evidence of the presence of nonuniform interstellar extinction in the direction of the cluster. Displayed in the $V_0, (Y - V)_0$ diagram are the same stars as in the left panel, but corrected for individual reddening and extinction estimates. The scatter in the $V_0, (Y - V)_0$

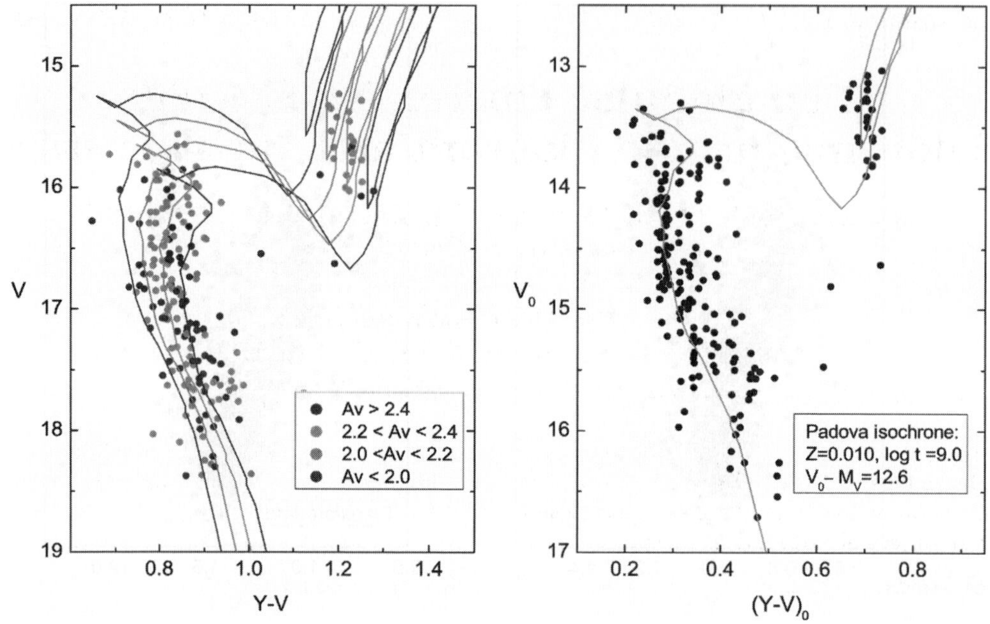

Figure 3. CMDs for probable photometric members, with the best-fitting isochrone shown: $Z = 0.010$, $\log(t/\mathrm{yr}) = 9.0$, and $(m - M)_0 = 12.6$ mag. *(left)* $V, Y - V$ diagram with the same isochrone adjusted for different values of A_V. *(right)* The same stars in the $V_0, (Y - V)_0$ diagram.

diagram is, however, similar to that seen in the left panel, but the stars are more concentrated around the best-fitting isochrone. This $Z = 0.010$ isochrone gives the following cluster parameters:

- a distance modulus $(m - M)_0 = 12.6$ mag, or a distance of 3.3 kpc,
- an age of 1.0 ± 0.15 Gyr.

4. Conclusion

We conclude that IC 361 is a distant intermediate-age cluster located in, or just beyond, the Perseus spiral arm. Its mild metal deficiency is in agreement with the Galactic metallicity gradient. However, the presence of high and variable extinction, also confirmed by *IRAS* 100μm data, complicates a more detailed study of this beautiful open cluster.

Acknowledgements

JZ acknowledges support through the Lithuanian State Science and Studies Foundation grant S-17/2008.

References

Bressan, A. & Tautvaišienė, G. 1996, *BaltA*, 5, 239

Laugalys, V., Kazlauskas, A., Boyle, R. P., Vrba, F. J., Philip, A. G. D., & Straižys, V. 2004, *BaltA*, 13, 1

Marigo, P., Girardi, L., Bressan, A., Groenewegen, M. A. T., Silva, L., & Granato, G. L. 2008, *A&A*, 482, 883

Piccirillo, J. & Stein W. L. 1978, *AJ*, 83, 971

Star clusters: basic galactic building blocks
Proceedings IAU Symposium No. 266, 2009
R. de Grijs & J. R. D. Lépine, eds.

doi:10.1017/S1743921309992092

The globular cluster NGC 5286: color–magnitude diagram and variable stars

M. Zorotovic,[1,2] M. Catelan,[2] M. Zoccali,[2] H. A. Smith[3] and B. J. Pritzl[4]

[1]ESO, Santiago, Chile
[2]Pontifícia Universidad Católica, Santiago, Chile
[3]Michigan State University, East Lansing, MI, USA
[4]University of Wisconsin, Oshkosh, WI, USA

Abstract. We present *BV* photometry and the results of a search for stellar variability in the globular cluster NGC 5286, which has tentatively been associated with the Canis Major dwarf spheroidal galaxy. Our results indicate an Oosterhoff type II for the cluster, which is unusual for an object of extragalactic origin.

Keywords. globular clusters: individual (NGC 5286), stars: variables: other

---

The full poster (in pdf format) is available at
`http://www.astro.iag.usp.br/~iaus266/Posters/pZorotovic.pdf`.

Author Index

Object Index

Subject Index